An Introduction to Financial Markets

A Quantitative Approach

Paolo Brandimarte

This edition first published 2018
© 2018 John Wiley & Sons, Inc.

All rights reserved. No part of this publication may be reproduced, stored in a retrieval system, or transmitted, in any form or by any means, electronic, mechanical, photocopying, recording or otherwise, except as permitted by law. Advice on how to obtain permission to reuse material from this title is available at http://www.wiley.com/go/permissions.

The right of Paolo Brandimarte to be identified as the author of this work has been asserted in accordance with law.

Registered Office
John Wiley & Sons, Inc., 111 River Street, Hoboken, NJ 07030, USA

Editorial Office
111 River Street, Hoboken, NJ 07030, USA

For details of our global editorial offices, customer services, and more information about Wiley products visit us at www.wiley.com.

Wiley also publishes its books in a variety of electronic formats and by print-on-demand. Some content that appears in standard print versions of this book may not be available in other formats.

Limit of Liability/Disclaimer of Warranty: In view of ongoing research, equipment modifications, changes in governmental regulations, and the constant flow of information relating to the use of experimental reagents, equipment, and devices, the reader is urged to review and evaluate the information provided in the package insert or instructions for each chemical, piece of equipment, reagent, or device for, among other things, any changes in the instructions or indication of usage and for added warnings and precautions. While the publisher and authors have used their best efforts in preparing this work, they make no representations or warranties with respect to the accuracy or completeness of the contents of this work and specifically disclaim all warranties, including without limitation any implied warranties of merchantability or fitness for a particular purpose. No warranty may be created or extended by sales representatives, written sales materials or promotional statements for this work. The fact that an organization, website, or product is referred to in this work as a citation and/or potential source of further information does not mean that the publisher and authors endorse the information or services the organization, website, or product may provide or recommendations it may make. This work is sold with the understanding that the publisher is not engaged in rendering professional services. The advice and strategies contained herein may not be suitable for your situation. You should consult with a specialist where appropriate. Further, readers should be aware that websites listed in this work may have changed or disappeared between when this work was written and when it is read. Neither the publisher nor authors shall be liable for any loss of profit or any other commercial damages, including but not limited to special, incidental, consequential, or other damages.

Library of Congress Cataloging-in-Publication Data is available

ISBN: 978-1-118-01477-6

Printed in the United States of America

10 9 8 7 6 5 4 3 2 1

Contents

Preface *xv*
About the Companion Website *xix*

Part I
Overview

1 Financial Markets: Functions, Institutions, and Traded Assets *1*
    1.1 What is the purpose of finance? *2*
    1.2 Traded assets *12*
        1.2.1 The balance sheet *15*
        1.2.2 Assets vs. securities *20*
        1.2.3 Equity *22*
        1.2.4 Fixed income *24*
        1.2.5 FOREX markets *27*
        1.2.6 Derivatives *29*
    1.3 Market participants and their roles *46*
        1.3.1 Commercial vs. investment banks *48*
        1.3.2 Investment funds and insurance companies *49*
        1.3.3 Dealers and brokers *51*
        1.3.4 Hedgers, speculators, and arbitrageurs *51*
    1.4 Market structure and trading strategies *53*
        1.4.1 Primary and secondary markets *53*
        1.4.2 Over-the-counter vs. exchange-traded derivatives *53*
        1.4.3 Auction mechanisms and the limit order book *53*
        1.4.4 Buying on margin and leverage *55*
        1.4.5 Short-selling *58*
    1.5 Market indexes *60*
    Problems *63*
    Further reading *65*
    Bibliography *65*

2 Basic Problems in Quantitative Finance *67*
    2.1 Portfolio optimization *68*
        2.1.1 Static portfolio optimization: Mean–variance efficiency *70*
        2.1.2 Dynamic decision-making under uncertainty: A stylized consumption–saving model *75*
    2.2 Risk measurement and management *80*

             2.2.1   Sensitivity of asset prices to underlying risk factors   *81*
             2.2.2   Risk measures in a non-normal world: Value-at-risk   *84*
             2.2.3   Risk management: Introductory hedging examples   *93*
             2.2.4   Financial vs. nonfinancial risk factors   *100*
    2.3   The no-arbitrage principle in asset pricing   *102*
             2.3.1   Why do we need asset pricing models?   *103*
             2.3.2   Arbitrage strategies   *104*
             2.3.3   Pricing by no-arbitrage   *108*
             2.3.4   Option pricing in a binomial model   *112*
             2.3.5   The limitations of the no-arbitrage principle   *116*
    2.4   The mathematics of arbitrage   *117*
             2.4.1   Linearity of the pricing functional and law of one price   *119*
             2.4.2   Dominant strategies   *120*
             2.4.3   No-arbitrage principle and risk-neutral measures   *125*
    S2.1   Multiobjective optimization   *129*
    S2.2   Summary of LP duality   *133*
    Problems   *137*
    Further reading   *139*
    Bibliography   *139*

Part II
Fixed-income assets

3   Elementary Theory of Interest Rates   *143*
    3.1   The time value of money: Shifting money forward in time   *146*
             3.1.1   Simple vs. compounded rates   *147*
             3.1.2   Quoted vs. effective rates: Compounding frequencies   *150*
    3.2   The time value of money: Shifting money backward in time   *153*
             3.2.1   Discount factors and pricing a zero-coupon bond   *154*
             3.2.2   Discount factors vs. interest rates   *158*
    3.3   Nominal vs. real interest rates   *161*
    3.4   The term structure of interest rates   *163*
    3.5   Elementary bond pricing   *165*
             3.5.1   Pricing coupon-bearing bonds   *165*
             3.5.2   From bond prices to term structures, and vice versa   *168*
             3.5.3   What is a risk-free rate, anyway?   *171*
             3.5.4   Yield-to-maturity   *174*
             3.5.5   Interest rate risk   *180*
             3.5.6   Pricing floating rate bonds   *188*
    3.6   A digression: Elementary investment analysis   *190*
             3.6.1   Net present value   *191*
             3.6.2   Internal rate of return   *192*

CONTENTS   vii

          3.6.3   Real options   *193*
    3.7 Spot vs. forward interest rates   *193*
          3.7.1   The forward and the spot rate curves   *197*
          3.7.2   Discretely compounded forward rates   *197*
          3.7.3   Forward discount factors   *198*
          3.7.4   The expectation hypothesis   *199*
          3.7.5   A word of caution: Model risk and hidden assumptions   *202*
    S3.1 Proof of Equation (3.42)   *203*
    Problems   *203*
    Further reading   *205*
    Bibliography   *205*

4 Forward Rate Agreements, Interest Rate Futures, and Vanilla Swaps   *207*
    4.1 LIBOR and EURIBOR rates   *208*
    4.2 Forward rate agreements   *209*
          4.2.1   A hedging view of forward rates   *210*
          4.2.2   FRAs as bond trades   *214*
          4.2.3   A numerical example   *215*
    4.3 Eurodollar futures   *216*
    4.4 Vanilla interest rate swaps   *220*
          4.4.1   Swap valuation: Approach 1   *221*
          4.4.2   Swap valuation: Approach 2   *223*
          4.4.3   The swap curve and the term structure   *225*
    Problems   *226*
    Further reading   *226*
    Bibliography   *226*

5 Fixed-Income Markets   *229*
    5.1 Day count conventions   *230*
    5.2 Bond markets   *231*
          5.2.1   Bond credit ratings   *233*
          5.2.2   Quoting bond prices   *233*
          5.2.3   Bonds with embedded options   *235*
    5.3 Interest rate derivatives   *237*
          5.3.1   Swap markets   *237*
          5.3.2   Bond futures and options   *238*
    5.4 The repo market and other money market instruments   *239*
    5.5 Securitization   *240*
    Problems   *244*
    Further reading   *244*
    Bibliography   *244*

6 Interest Rate Risk Management   *247*
    6.1 Duration as a first-order sensitivity measure   *248*
          6.1.1   Duration of fixed-coupon bonds   *250*

       6.1.2 Duration of a floater *254*
       6.1.3 Dollar duration and interest rate swaps *255*
6.2 Further interpretations of duration *257*
       6.2.1 Duration and investment horizons *258*
       6.2.2 Duration and yield volatility *260*
       6.2.3 Duration and quantile-based risk measures *260*
6.3 Classical duration-based immunization *261*
       6.3.1 Cash flow matching *262*
       6.3.2 Duration matching *263*
6.4 Immunization by interest rate derivatives *265*
       6.4.1 Using interest rate swaps in asset–liability management *266*
6.5 A second-order refinement: Convexity *266*
6.6 Multifactor models in interest rate risk management *269*
Problems *271*
Further reading *272*
Bibliography *273*

Part III
Equity portfolios

7 Decision-Making under Uncertainty: The Static Case *277*
  7.1 Introductory examples *278*
  7.2 Should we just consider expected values of returns and monetary outcomes? *282*
       7.2.1 Formalizing static decision-making under uncertainty *283*
       7.2.2 The flaw of averages *284*
  7.3 A conceptual tool: The utility function *288*
       7.3.1 A few standard utility functions *293*
       7.3.2 Limitations of utility functions *297*
  7.4 Mean–risk models *299*
       7.4.1 Coherent risk measures *300*
       7.4.2 Standard deviation and variance as risk measures *302*
       7.4.3 Quantile-based risk measures: V@R and CV@R *303*
       7.4.4 Formulation of mean–risk models *309*
  7.5 Stochastic dominance *310*
  S7.1 Theorem proofs *314*
       S7.1.1 Proof of Theorem 7.2 *314*
       S7.1.2 Proof of Theorem 7.4 *315*
Problems *315*
Further reading *317*
Bibliography *317*

8 Mean–Variance Efficient Portfolios *319*
  8.1 Risk aversion and capital allocation to risky assets *320*

            8.1.1 The role of risk aversion *324*
      8.2 The mean–variance efficient frontier with risky assets *325*
            8.2.1 Diversification and portfolio risk *325*
            8.2.2 The efficient frontier in the case of two risky assets *326*
            8.2.3 The efficient frontier in the case of n risky assets *329*
      8.3 Mean–variance efficiency with a risk-free asset: The separation property *332*
      8.4 Maximizing the Sharpe ratio *337*
            8.4.1 Technical issues in Sharpe ratio maximization *340*
      8.5 Mean–variance efficiency vs. expected utility *341*
      8.6 Instability in mean–variance portfolio optimization *343*
      S8.1 The attainable set for two risky assets is a hyperbola *345*
      S8.2 Explicit solution of mean–variance optimization in matrix form *346*
      Problems *348*
      Further reading *349*
      Bibliography *349*

9 Factor Models *351*
      9.1 Statistical issues in mean–variance portfolio optimization *352*
      9.2 The single-index model *353*
            9.2.1 Estimating a factor model *354*
            9.2.2 Portfolio optimization within the single-index model *356*
      9.3 The Treynor–Black model *358*
            9.3.1 A top-down/bottom-up optimization procedure *362*
      9.4 Multifactor models *365*
      9.5 Factor models in practice *367*
      S9.1 Proof of Equation (9.17) *368*
      Problems *369*
      Further reading *371*
      Bibliography *371*

10 Equilibrium Models: CAPM and APT *373*
      10.1 What is an equilibrium model? *374*
      10.2 The capital asset pricing model *375*
            10.2.1 Proof of the CAPM formula *377*
            10.2.2 Interpreting CAPM *378*
            10.2.3 CAPM as a pricing formula and its practical relevance *380*
      10.3 The Black–Litterman portfolio optimization model *381*
            10.3.1 Black–Litterman model: The role of CAPM and Bayesian Statistics *382*
            10.3.2 Black-Litterman model: A numerical example *386*
      10.4 Arbitrage pricing theory *388*
            10.4.1 The intuition *389*
            10.4.2 A not-so-rigorous proof of APT *391*
            10.4.3 APT for Well-Diversified Portfolios *392*
            10.4.4 APT for Individual Assets *393*

           10.4.5 Interpreting and using APT   *394*
    10.5 The behavioral critique   *398*
           10.5.1 The efficient market hypothesis   *400*
           10.5.2 The psychology of choice by agents with limited rationality   *400*
           10.5.3 Prospect theory: The aversion to sure loss   *401*
    S10.1 Bayesian statistics   *404*
           S10.1.1 Bayesian estimation   *405*
           S10.1.2 Bayesian learning in coin flipping   *407*
           S10.1.3 The expected value of a normal distribution   *408*
    Problems   *411*
    Further reading   *413*
    Bibliography   *413*

Part IV
Derivatives

11 Modeling Dynamic Uncertainty   *417*
    11.1 Stochastic processes   *420*
           11.1.1 Introductory examples   *422*
           11.1.2 Marginals do not tell the whole story   *428*
           11.1.3 Modeling information: Filtration generated by a stochastic process   *430*
           11.1.4 Markov processes   *433*
           11.1.5 Martingales   *436*
    11.2 Stochastic processes in continuous time   *438*
           11.2.1 A fundamental building block: Standard Wiener process   *438*
           11.2.2 A generalization: Lévy processes   *440*
    11.3 Stochastic differential equations   *441*
           11.3.1 A deterministic differential equation: The bank account process   *442*
           11.3.2 The generalized Wiener process   *443*
           11.3.3 Geometric Brownian motion and Itô processes   *445*
    11.4 Stochastic integration and Itô's lemma   *447*
           11.4.1 A digression: Riemann and Riemann–Stieltjes integrals   *447*
           11.4.2 Stochastic integral in the sense of Itô   *448*
           11.4.3 Itô's lemma   *453*
    11.5 Stochastic processes in financial modeling   *457*
           11.5.1 Geometric Brownian motion   *457*
           11.5.2 Generalizations   *460*
    11.6 Sample path generation   *462*
           11.6.1 Monte Carlo sampling   *463*
           11.6.2 Scenario trees   *465*
    S11.1 Probability spaces, measurability, and information   *468*

Problems   *476*
Further reading   *478*
Bibliography   *478*

12 Forward and Futures Contracts   *481*
12.1 Pricing forward contracts on equity and foreign currencies   *482*
    12.1.1 The spot–forward parity theorem   *482*
    12.1.2 The spot–forward parity theorem with dividend income   *485*
    12.1.3 Forward contracts on currencies   *487*
    12.1.4 Forward contracts on commodities or energy: Contango and backwardation   *489*
12.2 Forward vs. futures contracts   *490*
12.3 Hedging with linear contracts   *493*
    12.3.1 Quantity-based hedging   *493*
    12.3.2 Basis risk and minimum variance hedging   *494*
    12.3.3 Hedging with index futures   *496*
    12.3.4 Tailing the hedge   *499*
Problems   *501*
Further reading   *502*
Bibliography   *502*

13 Option Pricing: Complete Markets   *505*
13.1 Option terminology   *506*
    13.1.1 Vanilla options   *507*
    13.1.2 Exotic options   *508*
13.2 Model-free price restrictions   *510*
    13.2.1 Bounds on call option prices   *511*
    13.2.2 Bounds on put option prices: Early exercise and continuation regions   *514*
    13.2.3 Parity relationships   *517*
13.3 Binomial option pricing   *519*
    13.3.1 A hedging argument   *520*
    13.3.2 Lattice calibration   *523*
    13.3.3 Generalization to multiple steps   *524*
    13.3.4 Binomial pricing of American-style options   *527*
13.4 A continuous-time model: The Black–Scholes–Merton pricing formula   *530*
    13.4.1 The delta-hedging view   *532*
    13.4.2 The risk-neutral view: Feynman–Kač representation theorem   *539*
    13.4.3 Interpreting the factors in the BSM formula   *543*
13.5 Option price sensitivities: The Greeks   *545*
    13.5.1 Delta and gamma   *546*
    13.5.2 Theta   *550*
    13.5.3 Relationship between delta, gamma, and theta   *551*

        13.5.4 Vega   *552*
13.6 The role of volatility   *553*
        13.6.1 The implied volatility surface   *553*
        13.6.2 The impact of volatility on barrier options   *555*
13.7 Options on assets providing income   *556*
        13.7.1 Index options   *557*
        13.7.2 Currency options   *558*
        13.7.3 Futures options   *559*
        13.7.4 The mechanics of futures options   *559*
        13.7.5 A binomial view of futures options   *560*
        13.7.6 A risk-neutral view of futures options   *562*
13.8 Portfolio strategies based on options   *562*
        13.8.1 Portfolio insurance and the Black Monday of 1987   *563*
        13.8.2 Volatility trading   *564*
        13.8.3 Dynamic vs. Static hedging   *566*
13.9 Option pricing by numerical methods   *569*
Problems   *570*
Further reading   *575*
Bibliography   *576*

14 Option Pricing: Incomplete Markets   *579*
  14.1 A PDE approach to incomplete markets   *581*
        14.1.1 Pricing a zero-coupon bond in a driftless world   *584*
  14.2 Pricing by short-rate models   *588*
        14.2.1 The Vasicek short-rate model   *589*
        14.2.2 The Cox–Ingersoll–Ross short-rate model   *594*
  14.3 A martingale approach to incomplete markets   *595*
        14.3.1 An informal approach to martingale equivalent measures   *598*
        14.3.2 Choice of numeraire: The bank account   *600*
        14.3.3 Choice of numeraire: The zero-coupon bond   *601*
        14.3.4 Pricing options with stochastic interest rates: Black's model   *602*
        14.3.5 Extensions   *603*
  14.4 Issues in model calibration   *603*
        14.4.1 Bias–variance tradeoff and regularized least-squares   *604*
        14.4.2 Financial model calibration   *609*
Further reading   *612*
Bibliography   *612*

Part V
Advanced optimization models

15 Optimization Model Building   *617*
  15.1 Classification of optimization models   *618*

15.2	Linear programming *625*	
	15.2.1 Cash flow matching *627*	
15.3	Quadratic programming *628*	
	15.3.1 Maximizing the Sharpe ratio *629*	
	15.3.2 Quadratically constrained quadratic programming *631*	
15.4	Integer programming *632*	
	15.4.1 A MIQP model to minimize TEV under a cardinality constraint *634*	
	15.4.2 Good MILP model building: The role of tight model formulations *636*	
15.5	Conic optimization *642*	
	15.5.1 Convex cones *644*	
	15.5.2 Second-order cone programming *650*	
	15.5.3 Semidefinite programming *653*	
15.6	Stochastic optimization *655*	
	15.6.1 Chance-constrained LP models *656*	
	15.6.2 Two-stage stochastic linear programming with recourse *657*	
	15.6.3 Multistage stochastic linear programming with recourse *663*	
	15.6.4 Scenario generation and stability in stochastic programming *670*	
15.7	Stochastic dynamic programming *675*	
	15.7.1 The dynamic programming principle *676*	
	15.7.2 Solving Bellman's equation: The three curses of dimensionality *679*	
	15.7.3 Application to pricing options with early exercise features *680*	
15.8	Decision rules for multistage SLPs *682*	
15.9	Worst-case robust models *686*	
	15.9.1 Uncertain LPs: Polyhedral uncertainty *689*	
	15.9.2 Uncertain LPs: Ellipsoidal uncertainty *690*	
15.10	Nonlinear programming models in finance *691*	
	15.10.1 Fixed-mix asset allocation *692*	
Problems *693*		
Further reading *695*		
Bibliography *696*		

16 Optimization Model Solving *699*

    16.1 Local methods for nonlinear programming *700*
        16.1.1 Unconstrained nonlinear programming *700*
        16.1.2 Penalty function methods *703*
        16.1.3 Lagrange multipliers and constraint qualification conditions *707*
        16.1.4 Duality theory *713*
    16.2 Global methods for nonlinear programming *715*

- **16.2.1** Genetic algorithms *716*
- **16.2.2** Particle swarm optimization *717*
- **16.3** Linear programming *719*
  - **16.3.1** The simplex method *720*
  - **16.3.2** Duality in linear programming *723*
  - **16.3.3** Interior-point methods: Primal-dual barrier method for LP *726*
- **16.4** Conic duality and interior-point methods *728*
  - **16.4.1** Conic duality *728*
  - **16.4.2** Interior-point methods for SOCP and SDP *731*
- **16.5** Branch-and-bound methods for integer programming *732*
  - **16.5.1** A matheuristic approach: Fix-and-relax *735*
- **16.6** Optimization software *736*
  - **16.6.1** Solvers *737*
  - **16.6.2** Interfacing through imperative programming languages *738*
  - **16.6.3** Interfacing through non-imperative algebraic languages *738*
  - **16.6.4** Additional interfaces *739*

Problems *739*

Further reading *740*

Bibliography *741*

Index *743*

Preface

This book arises from slides and lecture notes that I have used over the years in my courses *Financial Markets and Instruments* and *Financial Engineering*, which were offered at Politecnico di Torino to graduate students in Mathematical Engineering. Given the audience, the treatment is naturally geared toward a mathematically inclined reader. Nevertheless, the required prerequisites are relatively modest, and any student in engineering, mathematics, and statistics should be well-equipped to tackle the contents of this introductory book.[1] The book should also be of interest to students in economics, as well as junior practitioners with a suitable quantitative background.

We begin with quite elementary concepts, and material is introduced progressively, always paying due attention to the practical side of things. Mathematical modeling is an art of selective simplification, which must be supported by intuition building, as well as by a healthy dose of skepticism. This is the aim of remarks, counterexamples, and financial horror stories that the book is interspersed with. Occasionally, we also touch upon current research topics.

Book structure

The book is organized into five parts.

1. Part One, **Overview**, consists of two chapters. Chapter 1 aims at getting unfamiliar readers acquainted with the role and structure of financial markets, the main classes of traded assets (equity, fixed income, and derivatives), and the main types of market participants, both in terms of institutions (e.g., investments banks and pension funds) and roles (e.g., speculators, hedgers, and arbitrageurs). We try to give a practical flavor that is essential to students of quantitative disciplines, setting the stage for the application of quantitative models. Chapter 2 overviews the basic problems in finance, like asset allocation, pricing, and risk management, which may be tackled by quantitative models. We also introduce the fundamental concepts related to arbitrage theory, including market completeness and risk-neutral measures, in a simple static and discrete setting.

2. Part Two, **Fixed-income assets**, consists of four chapters and introduces the simplest assets depending on interest rates, starting with plain bonds. The fundamental concepts of interest rate modeling, including the term

---

[1] In case of need, the mathematical prerequisites are covered in my other book: *Quantitative Methods: An Introduction for Business Management*. Wiley, 2011.

structure and forward rates, as well as bond pricing, are covered in Chapter 3. The simplest interest rate derivatives (forward rate agreements and vanilla swaps) are covered in Chapter 4, whereas Chapter 5 aims at providing the reader with a flavor of real-life markets, where details like day count and quoting conventions are relevant. Chapter 6 concludes this part by showing how quantitative models may be used to manage interest rate risk. In this part, we do not consider interest rate options, which require a stronger mathematical background and are discussed later.

3. Part Three, **Equity portfolios**, consists of four chapters, where we discuss equity markets and portfolios of stock shares. Actually, this is not the largest financial market, but it is arguably the kind of market that the layman is more familiar with. Chapter 7 is a bit more theoretical and lays down the foundations of static decision-making under uncertainty. By static, we mean that we make one decision and then we wait for its consequences, finger crossed. Multistage decision models are discussed later. In this chapter, we also introduce the basics of risk aversion and risk measurement. Chapter 8 is quite classical and covers traditional mean–variance portfolio optimization. The impact of statistical estimation issues on portfolio management motivates the introduction of factor models, which are the subject of Chapter 9. Finally, in Chapter 10, we discuss equilibrium models in their simplest forms, the capital asset pricing model (CAPM), which is related to a single-index factor model, and arbitrage pricing theory (APT), which is related to a multifactor model. We do not discuss further developments in equilibrium models, but we hint at some criticism based on behavioral finance.

4. Part Four, **Derivatives**, includes four chapters. We discuss dynamic uncertainty models in Chapter 11, which is more challenging than previous chapters, as we have to introduce the necessary foundations of option pricing models, namely, stochastic differential equations and stochastic integrals. Chapter 12 describes simple forward and futures contracts, extending concepts that were introduced in Chapter 4, when dealing with forward and futures interest rates. Chapter 13 covers option pricing in the case of complete markets, including the celebrated and controversial Black–Scholes–Merton formula, whereas Chapter 14 extends the basic concepts to the more realistic setting of incomplete markets.

5. Part Five, **Advanced optimization models**, is probably the less standard part of this book, when compared to typical textbooks on financial markets. We deal with optimization model building, in Chapter 15, and optimization model solving, in Chapter 16. Actually, it is difficult to draw a sharp line between model building and model solving, but it is a fact of life that advanced software is available for solving quite sophisticated models, and the average user does not need a very deep knowledge of the involved algorithms, whereas she must be able to build a model. This is the motivation for separating the two chapters.

Needless to say, the choice of which topics should be included or omitted is debatable and based on authors' personal bias, not to mention the need to keep a book size within a sensible limit. With respect to introductory textbooks on financial markets, there is a deeper treatment of derivative models. On the other hand, more challenging financial engineering textbooks do not cover, e.g., equilibrium models and portfolio optimization. We aim at an intermediate treatment, whose main limitations include the following:

- We only hint at criticism put forward by behavioral finance and do not cover market microstructure and algorithmic trading strategies.
- From a mathematical viewpoint, we pursue an intuitive treatment of financial engineering models, as well as a simplified coverage of the related tools of stochastic calculus. We do not rely on rigorous arguments involving self-financing strategies, martingale representation theorems, or change of probability measures.
- From a financial viewpoint, by far, the most significant omission concerns credit risk and credit derivatives. Counterparty and liquidity risk play a prominent role in post-Lehman Brothers financial markets and, as a consequence of the credit crunch started in 2007, new concepts like CVA, DVA, and FVA have been introduced. This is still a field in flux, and the matter is arguably not quite assessed yet.
- Another major omission is econometric time series models.

Adequate references on these topics are provided for the benefit of the interested readers.

My choices are also influenced by the kind of students this book is mainly aimed at. The coverage of optimization models and methods is deeper than usual, and I try to open readers' critical eye by carefully crafted examples and counterexamples. I try to strike a satisfactory balance between the need to illustrate mathematics in action and the need to understand the real-life context, without which quantitative methods boil down to a solution in search of a problem (or a hammer looking for nails, if you prefer). I also do not disdain just a bit of repetition and redundancy, when it may be convenient to readers who wish to jump from chapter to chapter. More advanced sections, which may be safely skipped by readers, are referred to as *supplements* and their number is marked by an initial "S."

In my Financial Engineering course, I also give some more information on numerical methods. The interested reader might refer to my other books:

- P. Brandimarte, *Numerical Methods in Finance and Economics: A MATLAB-Based Introduction* (2nd ed.), Wiley, 2006
- P. Brandimarte, *Handbook in Monte Carlo Simulation: Applications in Financial Engineering, Risk Management, and Economics*, Wiley, 2014

Acknowledgements

In the past years, I have adopted the following textbooks (or earlier editions) in my courses. I have learned a lot from them, and they have definitely influenced the writing of this book:

- Z. Bodie, A. Kane, and A. Marcus, *Investments* (9th ed.), McGraw-Hill, 2010
- J.C. Hull, *Options, Futures, and Other Derivatives* (8th ed.), Prentice Hall, 2011
- P. Veronesi, *Fixed Income Securities: Valuation, Risk, and Risk Management*, Wiley, 2010

Other specific acknowledgements are given in the text. I apologize in advance for any unintentional omission.

Additional material

Some end-of-chapter problems are included and fully worked solutions will be posted on a web page. My current URL is

- http://staff.polito.it/paolo.brandimarte/

A hopefully short list of errata will be posted there as well. One of the many corollaries of Murphy's law states that my URL is going to change shortly after publication of the book. An up-to-date link will be maintained on the Wiley web page:

- http://www.wiley.com/

For comments, suggestions, and criticisms, all of which are quite welcome, my e-mail address is

- paolo.brandimarte@polito.it

PAOLO BRANDIMARTE

Turin, September 2017

About the Companion Website

This book is accompanied by a companion website:

www.wiley.com/go/brandimarte/financialmarkets

The website includes:
- Solutions manual for end-of-chapter problems

Part One

Overview

Chapter One

Financial Markets: Functions, Institutions, and Traded Assets

Providing a simple, yet exhaustive definition of finance is no quite easy task, but a possible attempt, at least from a conceptual viewpoint, is the following:[1]

Finance is the study of how people and organizations allocate scarce resources over time, subject to uncertainty.

This definition might sound somewhat generic, but it does involve the two essential ingredients that we shall deal with in practically every single page of this book: *Time* and *uncertainty*. Appreciating their role is essential in understanding why finance was born in the past and is so pervasive now. The time value of money is reflected in the interest rates that define how much money we have to pay over the time span of our mortgage, or the increase in wealth that we obtain by locking up our capital in a certificate of deposit issued by a bank. It is common wisdom that the value of $1 now is larger than the value of $1 in one year. This is not only a consequence of the potential loss of value due to inflation.[2] A dollar now, rather than in the future, paves the way to earlier investment opportunities, and it may also serve as a precautionary cushion against unforeseen needs. Uncertainty is related, e.g., to the impossibility of forecasting the return that we obtain from investing in stock shares, but also to the risk of adverse movements in currency exchange rates for an import/export firm, or longevity risk for a worker approaching retirement. As we show in Chapter 2, we may model issues related to time and uncertainty within a mathematical framework, applying principles from financial economics and tools from probability, statistics, and optimization theory. Before doing that, we need a more concrete view

---

[1] This definition is taken from [2].

[2] This holds under common economic conditions; the exception to the rule is deflation, which is (at the time of writing) a possibility in Euroland. In this book, we will assume that the standard economic conditions prevail.

in order to understand how financial markets work, which kinds of assets are exchanged, and which actors play a role in them and what their incentives are. We pursue this "institutional" approach to get acquainted with finance in this chapter. Some of the more mathematically inclined students tend to consider this side of the coin modestly exciting, but a firm understanding of it is necessary to put models in the right perspective and to appreciate their pitfalls and limitations.

In Section 1.1, we discuss the role of time and uncertainty in a rather abstract way that, nevertheless, lays down some essential concepts. A more concrete view is taken in Section 1.2, where we describe the fundamental classes of assets that are traded on financial markets, namely, stock shares, bonds, currencies, and the basic classes of derivatives, like forward/futures contracts and options. In order to provide a proper framework, we also hint at the essential shape of a balance sheet, in terms of assets, liabilities, and equity, and we emphasize the difference between standardized assets traded on regulated exchanges and less liquid assets, possibly engineered to meet specific client requirements, which are traded over-the-counter. In Section 1.3, we describe the classes of players involved in financial markets, such as investment/commercial banks, common/hedge/pension funds, insurance companies, brokers, and dealers. We insist on the separation between the institutional form and the role of those players: A single player may be of one given kind, in institutional terms, but it may play different roles. For instance, an investment bank can, among many other things, play the role of a prime broker for a hedge fund. Furthermore, depending on circumstances, players may act as hedgers, speculators, or arbitrageurs. The exact organization of financial markets is far from trivial, especially in the light of extensive use of information technology, and a full description is beyond the scope of this book. Nevertheless, some essential concepts are needed, such as the difference between primary and secondary markets, which is explained in Section 1.4. There, we also introduce some trading strategies, like buying on margin and short-selling, which are essential to interpret what happens on financial markets in practice, as well as to understand some mathematical arguments that we will use over and over in this book. Finally, in Section 1.5 we consider market indexes and describe some basic features explaining, for instance, the difference between an index like the Dow Jones Industrial Average and the Standard & Poor 500.

1.1 What is the purpose of finance?

If you are reading this book, chances are that it is because you would like to land a rewarding job in finance. Even if this is not the case, one of the reasons why we aim at finding a good job is because we need to earn some income in order to purchase goods and services, for ourselves and possibly other people we care about. Every month (hopefully) we receive some income, and we must plan its use. The old grasshopper and ant fable teaches that we should actually

1.1 What is the purpose of finance?

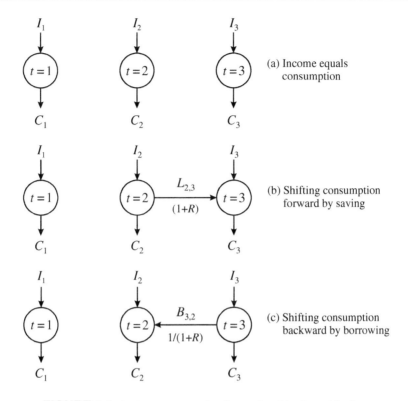

FIGURE 1.1 Shifting consumption forward and backward in time.

plan ahead with care. Part of that income should be saved to allow consumption at some later time. Sometimes, we might need to use more income than we are earning at present, e.g., in order to finance the purchase of our home sweet home.

Now, imagine a world in which we cannot "store" money, and we have to consume whatever our income is *immediately*, no more, no less, just like we would do with perishable food, if no one had invented refrigerators and other conservation techniques. This unpleasing situation is depicted in Fig. 1.1(a). There, time is discretized in $T = 3$ time periods, indexed by $t = 1, \ldots, T$.[3] The income during time period t is denoted by I_t, and it is equal to the consumption C_t during the same period:

$$I_t = C_t, \qquad t = 1, \ldots, T.$$

---

[3] Sometimes, time discretization requires careful thinking about events. Do we earn income at the beginning or at the end of a time period? In other words, is income earned during time period t immediately available for consumption during the same time period? We may argue that income during time period t is available for consumption only during time period $t+1$. We shall discuss more precise notation and concepts in Section 2.1.2. Here, for the sake of simplicity, we assume that every event during a time *period* is concentrated at some time *instant*. We sometimes use the rather awkward term *epoch* to refer to a specific point in time. We also often use the term time *bucket* to refer to a time period delimited by two time instants.

This state of the matter is not quite satisfactory, if we have excess income in some period and would like to delay consumption to a later time period. In Fig. 1.1(b), part of income I_2, denoted by $L_{2,3}$, is shifted forward from time period 2 to time period 3. This results in an increase of C_3 and a decrease of C_2. The amount of income saved can be regarded as money invested or lent to someone else. By a similar token, we might wish to anticipate consumption to an earlier time period. In Fig. 1.1(c), consumption C_2 is increased by shifting income backward in time from time period $t = 3$, which means borrowing an amount of money $B_{3,2}$, to be used in time period $t = 2$ and repaid in time period $t = 3$. Savers and borrowers may be individuals or institutions, and we may play both of these roles at different stages of our working life. Clearly, all of this may happen if there is a way to match savers and borrowers, so that all of them may improve their consumption timing. This is one of the many roles of financial markets; more specifically, we use the term **money markets** when the time span of the loan is short. In other cases, the investment may stretch over a considerable time span, especially if savers/borrowers are not just households, but corporations, innovative startups, or public administrations that have to finance the development of a new product, the building of a new hospital, or an essential infrastructure. In this case, we talk about **capital markets**.

Needless to say, if we accept to delay consumption, it is because we expect to be compensated in some way. Informally, we exchange an egg for a chicken; formally, we earn some interest rate R along the time period involved in the shift.[4] We may interpret the shift as a flow of money over a network in time but, unlike other network flows involved in transportation over space, we do not have exact conservation of flows. With reference to Fig. 1.1(b), we have the following flow balance equations at nodes 2 and 3:

$$C_2 = I_2 - L_{2,3},$$
$$C_3 = I_3 + L_{2,3}(1 + R),$$

stating that we give up an amount $L_{2,3}$ of consumption at time 2 in exchange for an increase $(1 + R)L_{2,3}$ in later consumption. The factor $1 + R$ is a gain associated with the flow of money along the arc connecting node $t = 2$ to node $t = 3$. This is what the **time value** of money is all about. The exact value of the interest rate R, as we shall see in Chapter 3, may be related to the possibility of default (i.e., the borrower may not repay the full amount of his debt) and to inflation risk, among other things.

Clearly, there must be another side of the coin: The increase in later consumption must be paid by a counterparty in an exchange. We delay consumption while someone else anticipates it. With reference to Fig. 1.1(c), we have

---

[4] In financial practice, whenever an interest rate is quoted, it is always an *annual* rate. For now, let us associate the rate with an arbitrary time period.

1.1 What is the purpose of finance?

the following flow balance equations at nodes 2 and 3:

$$C_2 = I_2 + \frac{B_{3,2}}{1+R},$$
$$C_3 = I_3 - B_{3,2}.$$

Note that we are expressing the borrowed amount $B_{3,2}$ in terms of the money at time $t = 3$, when the debt is repaid; in other words $B_{3,2}$ is a flow out of node 3. This is not essential at all: If we use money at time $t = 2$, i.e., we consider the flow $B_{3,2}^*$ *into* node 2, the flow balance would simply read

$$C_2 = I_2 + B_{3,2}^*,$$
$$C_3 = I_3 - B_{3,2}^*(1+R).$$

The two sides of the coin must be somehow matched by a market mechanism. In practice, funds are channeled by financial intermediaries, which must be compensated for their job. In fact, there is a difference between lending and borrowing rates, called **bid–ask** (or bid–offer) **spread**, which applies to other kinds of financial assets as well. Lending and borrowing money through a bank is what we are familiar with as individuals, whereas a large corporation and a sovereign government have the alternative of raising funds by issuing securities like bonds, typically promising the payment of periodic interest, as well as the refund of the capital at some prespecified point in time, the maturity of the bond. Corporations may also raise funds by issuing stock shares. Buying a stock share does not mean that we lend money to a firm; hence, we are not entitled to the payment of any interest. Rather, we own a share of the firm and may receive a corresponding share of earnings that may be distributed in the form of dividends to stockholders. However, the amount that we will receive is random and no promise is made about dividends, as they depend on how well the business is doing, as well as the decision of reinvesting part of the earning in new business ventures, rather than distributing the whole of it.

After being first issued, securities like bonds and stock shares may be exchanged among market participants, at prices that may depend on several underlying risk factors. Since the values of these factors are not known with certainty, the future prices of bonds and stock shares are random. In fact, time is intertwined with another fundamental dimension in finance, namely, **uncertainty**. When we lend or borrow money at a given interest rate, the future cash flows are known with certainty, if we do not consider the possibility of a default on debt. However, when we buy a stock share at time $t = 0$ and plan to sell it at time $t = T$, randomness comes into play. Let us denote the initial price by $S(0)$.[5] The future price $S(T)$ is a **random variable**, which we may denote as $S(T, \omega)$ to emphasize its dependence on the random outcome (scenario) ω. We recall that, in probability theory, a random variable is a function mapping underlying random outcomes, corresponding to future scenarios or states of nature, to numeric values. Let ω_i, $i = 1, \ldots, m$, denote the i-th outcome, which

---

[5] Depending on notational convenience, we shall write $S(t)$ or S_t, as no ambiguity should arise.

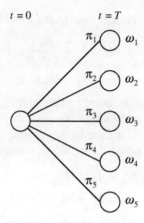

FIGURE 1.2 Representing uncertain states of the world by a scenario fan.

occurs with probability π_i. For the sake of simplicity, we are considering a discrete and finite set of possible outcomes, whereas later we will deal extensively with continuous random variables. A simple way to depict this kind of discrete uncertainty is by a scenario fan like the one depicted in Fig. 1.2. Therefore, $S(T,\omega)$ is a random variable, and we associate a future price $S(T,\omega_i)$ with each future state of the world. The corresponding holding period return is defined as follows.

DEFINITION 1.1 (Holding period return) *Let us consider a holding period $[0,T]$, where the initial asset price is $S(0)$ and the terminal random asset price is $S(T,\omega)$. We define the* **holding period** *return as*

$$R(\omega) \doteq \frac{S(T,\omega) - S(0)}{S(0)} \qquad (1.1)$$

and the **holding period** *gain as*

$$G(\omega) \doteq \frac{S(T,\omega)}{S(0)} = 1 + R(\omega). \qquad (1.2)$$

The gain and the holding period return (return for short) are clearly related. A return of 10% means that the stock price was multiplied by a gain factor of 1.10.

Remark. The term *gain* is not so common in finance textbooks. Usually, terms like *total return* or *gross return* are used, rather than gain. On the contrary, terms like *rate of return* and *net return* are used to refer to (holding period) return. The problem is that these terms may ring different bells, especially to practitioners. We may use the qualifier "total" when we want to emphasize a return including dividend income, besides the capital gain related to price changes. Terms like "gross" and "net" may be related with taxation issues, which we shall always disregard. This is why we prefer using "gain," even though this usage is less common. We shall not confuse gain, which is a multiplicative factor, with profit/loss, which is an additive factor and is expressed in monetary terms. Furthermore, we shall reserve the term "rate of return" to the case of *annual* returns. For instance, interest rates are always quoted annually, even though they may be applied to different time periods by a proper scaling. We will use the term "return," when the holding period may be arbitrary. The following example illustrates a further element of potential confusion when talking about return.

Example 1.1 Different shades of return

Consider a holding period consisting of two consecutive years. In year one, the return from investing in a given stock share is $+10\%$; in year two the return is -10%. What was the "average" return?

As it turns out, the question is stated in a very imprecise way. It might be tempting to say that, trivially, the average return was 0%, the familiar arithmetic mean of $+10\%$ and -10%. However, we cannot really add returns like this. Over the two years, the gain was

$$G = (1 + 0.10) \times (1 - 0.10) = 0.99,$$

i.e., we have lost money, as the holding period return was -1% [we may recall the rule $(1+x)(1-x) = 1 - x^2$]. Indeed, the problem is that the very term "average" is ambiguous. If what we actually mean is the *expected value* of the annual return, which we may estimate by a *sample mean*, then we may say that the arithmetic average is, in fact

$$\overline{R}_a = \frac{0.10 - 0.10}{2} = 0.$$

But if we mean an average *over time*, we should deal with a sort of geometric average over two years:

$$(1 + 0.10) \times (1 - 0.10) = (1 + \overline{R}_g)^2 \quad \Rightarrow \quad \overline{R}_g = -0.5013\%.$$

We may also notice that, in this case, an average should refer to a standard time interval, usually one year. Indeed, we should not confuse the holding period return with an *annual* (rate of) return. We will need a way to annualize a generic holding period return.

Returns and gains are random variables. Hence, a natural question is: How should we model uncertain returns? There is a huge amount of work carried out on this subject, including plenty of empirical investigation. The next example shows that there cannot be any single convenient answer.

Example 1.2 One distribution does not fit all

The most familiar probability distribution is, no doubt, the normal. Can we say that the distribution of return from a stock share is normal? Empirical investigation tends to support a different view, as the normal distribution is symmetric and features thin tails, i.e., it tends to underestimate the probability of extreme events. In any case, the worst stock return we may experience is -100%, or (-1), i.e., we lose all of our investment (this is related to the limited liability property of stock shares, discussed later). In other words, the worst gain is 0, and a stock share price can never be negative. Since the support of a normal random variable is unbounded, $(-\infty, +\infty)$, according to this uncertainty model there is always a nonzero probability of observing an impossible price.

However, let us discuss the matter from a very limited viewpoint, namely, convenience. One nice feature of a normal distribution is that if we add normal variables, we get another normal (to be precise, we should be considering *jointly normal* variables). This is nice when we add returns from different stock shares over the same time period. If we have invested 30% of our wealth in stock share a and 70% in stock share b, the holding period return for the portfolio is

$$R_p = 0.3 R_a + 0.7 R_b, \qquad (1.3)$$

where we denote the return of stock shares a and b by R_a and R_b, respectively, and R_p is the portfolio return. To justify Eq. (1.3), let us consider:

- Initial stock prices $S_a(0)$ and $S_b(0)$
- Initial wealth $W(0)$
- Stock prices $S_a(T)$ and $S_b(T)$ at the end of the holding period
- Wealth $W(T)$ at the end of the holding period

Then, if initial wealth is split as we have assumed, we may write

$$W(0) = \frac{0.3 \times W(0)}{S_a(0)} \cdot S_a(0) + \frac{0.7 \times W(0)}{S_b(0)} \cdot S_b(0)$$
$$= N_a \cdot S_a(0) + N_b \cdot S_b(0),$$

where N_a and N_b are the number of stock shares a and b, respectively, that we buy. At the end of the holding period, we have

$$W(T) = N_a \cdot S_a(T) + N_b \cdot S_b(T)$$
$$= N_a \cdot (1 + R_a) \cdot S_a(0) + N_b \cdot (1 + R_b) \cdot S_b(0)$$

$$= W(0) \cdot \left[1 + \frac{N_a \cdot R_a \cdot S_a(0) + N_b \cdot R_b \cdot S_b(0)}{W(0)}\right]$$
$$= W(0) \cdot (1 + 0.3 \cdot R_a + 0.7 \cdot R_b)$$
$$= W(0) \cdot (1 + R_p),$$

which gives Eq. (1.3).

If R_a and R_b are jointly normal, then R_p is a nice normal, too, which is quite convenient. Furthermore, if the holding period return R is normal, so is the corresponding stock price $S(T) = S(0) \cdot (1 + R)$. However, imagine that we take a different perspective. Rather than considering two stock shares over one time period, let us consider one stock share over two consecutive time periods. In other words, we take a *longitudinal* view (a single variable over multiple time periods) rather than a *cross-sectional* view (multiple variables over a single time period). Let us denote by $R(1)$ and $R(2)$ the two holding period returns of that single stock share, over the two consecutive time periods. As we have mentioned, in this case we should not add returns, but rather multiply gains $G(1)$ and $G(2)$ to find the holding period gain

$$\begin{aligned} G &= G(1) \cdot G(2) \\ &= [1 + R(1)] \cdot [1 + R(2)] \\ &= 1 + R(1) + R(2) + R(1) \cdot R(2). \end{aligned}$$

The last expression involves a product of returns. Unfortunately, if $R(1)$ and $R(2)$ are normal, their product is not. Hence, the holding period gain G is not normal, and the same applies to the holding period return $R = G - 1$. We may only say that the holding period return is approximately normal if the single-period returns are small enough to warrant neglecting their product.

One way out is to consider the **logarithmic return**, or **log-return** for short,
$$r \doteq \log(1 + R) \equiv \log G,$$
where we use log rather than ln to denote natural logarithm. It is interesting to note that, given the well-known Taylor expansion (Maclaurin series, if you prefer)

$$\log(1 + x) \approx x - \frac{x^2}{2} + \frac{x^3}{3} - \frac{x^4}{4} + \cdots,$$

for a small x, the log-return can be approximated by the return. Since

$$\log\left[\left(1+R(1)\right)\cdot\left(1+R(2)\right)\right] = \log\left(1+R(1)\right) + \log\left(1+R(2)\right)$$
$$= r(1) + r(2),$$

we see that log-returns are additive, and if we assume that they are normal, we preserve normality over time.

Since
$$S(T) = S(0) \cdot G = S(0) \cdot e^r,$$

the normality of the log-return r implies that the gain and the stock prices are lognormally distributed, i.e., they may be expressed as the exponential of a normal random variable. On the one hand, this is nice, as it is consistent with the fact that we cannot observe negative stock prices. Furthermore, the product of lognormals is lognormal, which is nice in the longitudinal sense. Unfortunately, this is not nice in the cross-sectional sense, since the sum of lognormals is not lognormal, and we get in trouble when we consider the return of a portfolio of different stock shares.

To summarize, whatever modeling choice we make, some complication will arise. On the one hand, normal returns/gains (and stock prices) simplify the analysis of a portfolio over a single holding period, but they are empirically questionable and complicate the analysis over multiple time periods. On the other hand, lognormal gains (and stock prices) are fine for dynamic modeling of a single stock share, but they complicate the analysis of a portfolio. We may conclude that, whatever we choose, we have to accept some degree of approximation somewhere. The alternative is to tackle complicated distributions by numerical methods.

Beside risky assets, we shall also consider a risk-free (or riskless) asset. This is a peculiar asset for which $S(T,\omega)$ is actually a constant across states of the world. A concrete example is a safe bank account, whereby

$$B(T,\omega) = B(0) \cdot (1 + R_f),$$

for every state of the world (or scenario) $\omega \in \Omega$. The rate R_f will be referred to as **risk-free return**. If the holding period T is one year, we may refer to the annual risk-free return as the **risk-free rate**. The above framework to depict uncertainty does not only apply to stock shares, but to other financial and nonfinancial assets as well, like bonds, commodities, foreign currencies, etc. As we shall see in Chapter 2, uncertainty motivates some basic problems in finance, like portfolio optimization and risk management.

1.1 What is the purpose of finance?

Now, if we may invest wealth at some prespecified and risk-free interest rate, why should we bother with risky investments? The answer is that risk is associated with the hope of a larger return, i.e., risky assets come with a **risk premium**. Some investors are willing to assume a limited amount of risk in exchange for the possibility of an increase in future consumption. Other investors, however, would just like to get rid of risks they bear:

- Imagine a nonfinancial firm subject to research, development, and production costs incurred in some currency, for a range of products that are exported and sold in another currency. For instance, the firm might sign a contract for the design and construction of a production plant, where the overall price offered to the client is in US dollars, but actual costs are incurred in euro. Adverse fluctuations in currency exchange rates may well wipe out profit margins. As we shall see, firms may hedge this risk away using certain derivative assets, such as forward and futures contracts.
- With reference to Fig. 1.2, let us assume that the state corresponding to outcome ω_5 is a "bad state," i.e., a state in which we will be able only to afford a very low consumption level, possibly because an adverse event occurs (like illness, accident, or loss of job). Then, we might consider purchasing shares of an *insurance* contract, i.e., an asset whose value is strictly positive when ω_5 occurs, 0 otherwise. We assume that the insurance payoff is 1 in the bad state, but any other value will do, if assets are perfectly divisible and we may scale investments up and down at will.

More generally, an investor may shape the probability distribution of her wealth according to her taste and appetite for risk. A market participant with a given risk exposure may change it, and this is the essential function of risk management. Clearly, for any player hedging a risk exposure away, there must be another market participant willing to assume that risk or part of it. With respect to this uncertainty dimension, financial markets play the role of a **risk transfer** mechanism. For instance, insurance companies do that in exchange for a premium, and rely on *risk pooling* and *reinsurance* contracts to manage the resulting risk exposure.[6] One of the main problems in this context is the definition of a fair insurance premium, which is a standard task in actuarial mathematics.

Note that an insurance contract is an asset from the viewpoint of the policy owner, but not a tradable one, as we cannot sell our life insurance policy. However, an insurance company, for which insurance contracts are a liability, may pool and sell them to interested investors, using a process called **securitization**, which is the creation of liquid securities from illiquid assets.[7] By doing this, the

---

[6] Risk pooling may be considered as a corollary of the law of large numbers. If we aggregate a large number of small and independent risks, the overall risk should be reduced. This happens, e.g., with car insurance policies. Risk pooling may fail miserably with strongly correlated risks. Reinsurance, in a sense, is an opposite mechanism, by which a large risk is fractioned and sold to third parties.

[7] See Section 1.2.2.

risk may be fractioned and sold to investors who are willing to bear part of the risk for a given price. Securitization is a good example to illustrate the pros and cons of financial innovation. On the one hand, it allows to create securities that may offer enhanced return to holders, which is good in a regime of low interest rates. Furthermore, it may allow to insure against catastrophic risks that could not be insured otherwise. On the other one, just consider the damage inflicted to financial markets by the creation of illiquid and opaque mortgage-backed securities, bundling subprime mortgages and leading to the 2008 financial crisis. The same considerations apply to derivatives, which may be used in quite different ways by players having different views about the future or different attitudes toward risk (like hedgers and speculators, discussed later in this chapter). In all of these cases, the fundamental recurring themes are asset pricing and risk management, which we start considering in Chapter 2.

In later chapters, we will also see that making decisions under uncertainty is no trivial task, and that in real life, things are complicated by the fact that the two dimensions that we have considered, time and uncertainty, are actually intertwined. The resulting picture, illustrated in Fig. 1.3, is a **scenario tree**, where uncertainty unfolds progressively over time. The tree consists of a set of nodes n_k, $k = 0, 1, \ldots, 14$. Node n_0 is the root of the tree and represents the current state of the world. Then, over three time instants, $t = 1, 2, 3$, we observe a sequence of realizations of random variables representing financial risk factors. The outcomes ω_i of the sample space are associated with **scenarios**, i.e., sequences of nodes in the tree. For instance, scenario ω_3 corresponds to the sequence of nodes

$$(n_0, n_1, n_4, n_9).$$

More formally, each scenario is a **sample path** of a stochastic process. We also see that the probability of a scenario depends on *conditional* probabilities of events. For instance, the conditional probability of node n_4 at time $t = 2$, given that we are at node n_1 at time $t = 1$, is $\pi_{4|1}$. Hence, the *unconditional* probability of scenario ω_3 is

$$P(\omega_3) = \pi_{1|0} \cdot \pi_{4|1} \cdot \pi_{9|4}.$$

Since we are at state n_0, we may write π_1 rather than $\pi_{1|0}$, but we must be careful in distinguishing conditional and unconditional probabilities. Stochastic processes and the generation of scenario trees are discussed in Chapter 11. Dynamic policies in such a context must allow for a way to adapt a strategy to contingencies, and this leads to challenging multistage optimization models discussed in Chapter 15.

1.2 Traded assets

Finance revolves around buying and selling assets, pricing them, and assessing the involved risk. But what are assets, exactly? Open any page of a financial

1.2 Traded assets

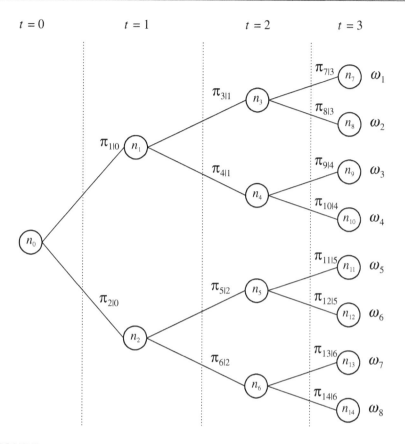

FIGURE 1.3 A scenario tree generalizes the scenario fan of Fig. 1.2 by unfolding uncertainty progressively over time.

journal and you will read about assets such as stock shares, bonds, or derivatives. Indeed, these are the assets that we will mostly deal with in this book; yet, it is essential to get a broader picture. Generally speaking, an **asset** is *anything that can be transformed into money by its owner*:

- A financial institution, like a pension fund, may rely on a portfolio of bonds as an asset: The stream of coupon payments is used to pay pensions to retired workers.
- A nonfinancial firm uses machines and other equipments to produce items for sale. These items may be innovative products protected by a patent; the patent is another asset that may be sold.
- An individual may use her human capital, possibly a Ph.D. title, to land a good, rewarding, and hopefully well-paid job. Unlike other assets, a Ph.D. title is not marketable.

- An insurance policy is an asset that can be transformed into money, but only when a prespecified event occurs, since it cannot be freely traded.

We see that assets may be both tangible or intangible objects that can be transformed into a sequence of cash flows, and they come in plenty of different forms. To start putting some order, let us introduce a few basic features to help us in classifying assets:

Real vs. financial. The bonds owned by a pension fund are financial in nature, whereas manufacturing equipments are real.

Risky vs. risk-free. Stock shares are considered as risky assets, as their future price and their dividend income are not known with certainty. A certificate of deposit issued by a very solid bank is perceived as a risk-free asset, since we know exactly how much money we are going to collect at maturity, even though the risk of default cannot be ruled out with absolute certainty.

Liquid vs. illiquid. Liquidity refers to the possibility of selling an asset quickly *and* at a fair price; both sides of the coin are relevant. If we need a lot of money immediately, we may sell our home; however, if we really want to do it quickly, we may be forced to accept a price that is possibly much lower than its fair value. A similar consideration applies to manufacturing equipments, which may be very specific and difficult to sell for a fair price. On the contrary, most stock shares are very liquid and actively traded on regulated exchanges. Shares of common funds are liquid and can be redeemed on short notice, whereas shares of hedge funds may require several weeks to be liquidated.

Tradable vs. nontradable. Most financial assets are easily traded on markets, but we cannot sell our own insurance policy. The fact that an asset is nontradable does not diminish its importance. For instance, when we age, we lose a fraction of human capital, as the sheer number of future cash flows that we obtain from our job gets less and less. If our human capital is a rather safe asset, then we may initially consider tilting our strategic asset allocation toward reasonably risky stock shares. When we age, it is a common advice that we should rebalance the portfolio toward safer assets.

Exchange-traded vs. over-the-counter. Stock shares are traded on regulated exchanges, just like some simple and standardized classes of derivatives (vanilla options and futures contracts). Sometimes, we need a more specific kind of asset for risk management purposes, which may be tailored by an investment bank according to our requirements. When an investment bank engineers a very specific asset, this is sold over-the-counter (**OTC**), rather than on regulated exchanges. Plain vanilla options are examples of exchange-traded derivatives, whereas exotic options are OTC assets. Unfortunately, a tailored OTC asset will be harder to sell. Typically, we may only sell it back to the original issuer by closing the contract, and its price is less easy to quantify as it is not related to a transparent demand–offer mechanism.

1.2 Traded assets

In the rest of this section, we outline the most common forms of financial assets, namely, stock shares (equity), bonds, and derivatives, as well as foreign currencies and hybrid assets. Before doing so, it is useful to lay down a (very) simplified view of a balance sheet, showing the connection between assets, liabilities, and equity. We also discuss briefly the difference between assets and securities.

1.2.1 THE BALANCE SHEET

A cornerstone of corporate finance is the **balance sheet**, one of the fundamental documents periodically issued by firms, which is used by investors and stakeholders to assess the health state of the firm. The essence of a balance sheet may be schematically represented in the following tabular form:

Assets	Liabilities
	Equity

which involves three sections:

1. **Assets**, as we have seen, can be transformed to positive cash flows, i.e., future payments that the firm will receive. Hence, the asset side of the balance sheet lists what the firm "owns."
2. **Liabilities**, on the other hand, are related to negative cash flows, i.e., future payments that will have to be covered. Hence, the liability side of the balance sheet lists what the firm "owes."
3. **Equity** is defined as the difference between the total value of the assets and the total value of the liabilities:

$$\text{Assets} - \text{Liabilities} = \text{Equity}.$$

Equity must be positive. When equity is negative, it means that the assets will not be able to generate sufficient cash flows in order to pay the liabilities, and bankruptcy occurs.

Example 1.3 The balance sheet and financial ratios

> Let us consider the extremely simplified and fictional balance sheet of a firm, reported in Table 1.1. On the asset side, we have current assets, which are liquid assets, like cash, or assets that can be converted to cash in the short term, like accounts receivable (money that will be received from customers). Fixed assets are less liquid and can be converted to cash, but not so quickly. In the case of equipment, the value may be questionable, and affected by depreciation and amortization standards, which may be chosen according to tax management

Table 1.1 Fictional balance sheet (in $ millions) for Example 1.3.

Assets		Liabilities	
Current assets		Current liabilities	
Cash	$80	Accounts payable	$300
Accounts receivable	$120	Long-term debt	$1800
Fixed assets			
Equipment	$2500		
Total assets	$2700	Total liabilities	$2100
		Total equity:	$600

policies. The liability side can also be partitioned into short-term liabilities, like accounts payable (money that must be paid to suppliers), and long-term debt (possibly bonds). We may check that the two sides of the balance sheet, total assets and total liabilities plus equity, are matched. If ten million shares are outstanding, the **book value** of each stock share should be

$$\frac{\$600}{10} = \$60.$$

This is the book value of the firm, which need not correspond to the market value. If the market value of each share is $40, then we say that the **book-to-market** ratio is

$$\frac{\$60}{\$40} = 1.5.$$

A ratio larger than 1 should suggest that the stock share is underpriced.

Based on the balance sheet, different ratios may be computed in order to measure the financial well-being and the solvency of a firm. A natural ratio is

$$\text{Total debt ratio} = \frac{\text{Total liabilities}}{\text{Total assets}} = \frac{\$2100}{\$2700} \approx 0.78.$$

More specific ratios consider only short-term items. In general, we aim at measuring the degree of **leverage** of a firm (or bank), i.e., the ratio of debt to equity.

Another fundamental accounting document, which we shall not discuss in detail, is the **income statement**, which links sales to net income, taking costs and taxes into account. Let us assume that net income is $200 (million) for our fictional firm. Net income is used to define important ratios:

- **Return on assets (ROA)**, the ratio of net income to total assets:

$$\frac{\$200}{\$2700} \approx 0.074$$

- **Return on equity (ROE)**, the ratio of net income to total equity:

$$\frac{\$200}{\$600} \approx 0.33$$

- **Earnings per share (EPS)**, the ratio of net income to shares outstanding:

$$\frac{\$200}{10} = \$20$$

- **Price-to-earnings (PE)**, the ratio of price per share to earning per share:

$$\frac{\$40}{\$20} = 2$$

These ratios are also used to classify stock shares as follows:

- **Value stocks** are stocks that look undervalued, but could deliver long-term profits to shareholders. They may feature low PE and price-to-book ratios.
- **Growth stocks**, on the contrary, look overvalued with respect to current market, but they may promise further growth opportunities due to expanding markets, new products, etc. They are generally rather volatile.

Furthermore, some of these ratios may be used in the multifactor models of Chapter 9.

We should always keep in mind that the ratios we have just defined may vary considerably across different industry sectors. Hence, rather than considering their absolute values, we should compare them against those of similar firms. By the same token, depending on the nature of the firm we are considering, the exact kind of items listed in a balance sheet may be very different, financial or nonfinancial, tangible or intangible, fairly easy or very difficult to evaluate, liquid or illiquid, as well as short or long term. It is also important to realize that the cash flows associated with assets and liabilities may be deterministic or stochastic, as the following examples illustrate.

Manufacturing firms. Specialized equipment for production is an asset, but a rather illiquid one, just as account receivables (payments to be received from clients). However, account receivables are usually short-term assets, and in this sense they contribute to firm's liquidity, even though they are not marketable. Assets may also be somewhat intangible, like customer goodwill or the portfolio of knowledge embedded in human resources. On the other hand, money that the firm owes to suppliers (accounts payable) contributes to short-term liabilities. Liabilities also include money that the firm has borrowed from a bank to finance its short-term operations or its long-term research and development programs. Alternatively, a large corporation may issue bonds to finance itself. Note that such corporate bonds are liabilities for the issuer, but they are assets from the viewpoint of bondholders, which may be financial intermediaries or individual investors.

Banks. Assets and liabilities for a bank tend to have a financial nature, but they need not be marketable. One such example is mortgages, unless they are pooled by a securitization process and sold as mortgage-backed securities. It is important to understand how the uncertain balance between assets and liabilities may be a source of risk for banks. Traditional mortgages that the bank has contracted with its clients and kept in its balance sheet are long-term assets, whereas the deposits are short-term liabilities, since the client may withdraw money whenever she feels like it. This maturity mismatch may result in considerable exposure to interest rate risk, since short- and long-term assets or liabilities react in different ways to changes in interest rates.[8] Banks with a proprietary trading desk may hold any kind of financial asset, including bonds and stock shares. A bank may finance its operations using deposits, but since they result in short-term and uncertain liabilities, they may issue certificates of deposits or bonds, which appear in the liability side of its balance sheet.

Insurance companies. A life insurance company receives periodic payments that may be invested in financial assets, whose cash flows will be used to pay, e.g., pensions and annuities, which appear on the liability side. The financial assets may be more or less risky, just like the liabilities. A life insurer faces longevity risk and, possibly, inflation risk if pension benefits are inflation-indexed. By a similar token, a non-life insurer collects premia from policyholders and is subject to stochastic liabilities related to, e.g., loss of property and car accidents.

These examples just give a vague idea of the variety of assets and liabilities that may appear on balance sheet. The picture is complicated by the fact that the exact way in which items are listed is far from trivial, and it is affected by accounting standards and regulations, having an impact on tax payments. Moreover, there may be little agreement on how assets and liabilities are exactly valued. This results in a possibly remarkable discrepancy between the

---

[8] We consider interest rate risk management in Chapter 6.

book value, i.e., the value reported in the balance sheet, and the actual market value of an item. This is inevitable, when cash flows are stochastic, requiring a suitable valuation model. Whatever model we choose, model and correlation risk come into play. To understand correlation risk, let us consider an insurance company. If there is a large number of uncorrelated and relatively small risk exposures, like car accidents, the overall value of the liabilities may be fairly predictable. However, considerable risk is faced when insuring properties with large values, or when an unexpected increase in correlation among risks introduces a remarkable amount of volatility.[9] By the same token, if a balance sheet includes derivatives, which one among the many conflicting valuation models should be used? And how should we estimate their parameters?[10]

Given this complexity, the accounting profession has (reasonably) given priority to standardization and consistency, rather than financial accuracy and mathematical sophistication, issuing a set of debatable guidelines and rules. As the reader can imagine, this is beyond the scope of this book, and a thorough discussion of accounting documents like balance sheet and income statement can be found in corporate finance books. Nevertheless, a bit of understanding of the balance sheet is also necessary for anyone interested in quantitative models of financial markets. The two primary assets that we describe in the following, stock shares and bonds, are clearly related to the balance sheet. Whatever assets and liabilities are listed and how exactly, a fundamental principle applies: If a firm is liquidated and closed down, assets are sold, generating funds that are used to pay the outstanding liabilities. If any equity remains, this money is distributed to stockholders (also called shareholders). This is why stock shares are referred to as equity, and stock markets as equity markets: Stock shares represent residual claims on equity.[11] Note that creditors, possibly bondholders, have priority over shareholders, and there is a pecking order for creditors as well. Bond indentures describe bond features like collateralization, i.e., if firm's assets are locked as a guarantee against default, and the level of seniority (priority in the pecking order) associated with the bond. Clearly, these features

---

[9] An example of unexpected increase in correlation is the increase of defaults on mortgage payments, when a generalized economic crisis leads to an increase in unemployment. In this case, default is not due to strictly individual issues, like illness or delinquency. The same may apply when an increase in the interest rates makes floating-rate mortgages more expensive, making default the only possible choice for some homeowners.

[10] Another issue with derivatives is their exact purpose. In fact, derivatives may be used to manage risk exposures and improve the balance sheet. However, they may also be used for quite risky speculation and, sometimes, drawing the line between the two uses is difficult. A pension fund might use derivatives in a defensive manner, but in order to prevent reckless behavior by fund managers, their use may be prohibited altogether. By the same token, at the time of writing, there is considerable controversy, here in Italy, about how public authorities have used interest rate derivatives in order to manage public debt. Many risk management strategies have backfired, which is always a possibility and, per se, is no evidence of reckless management. The problems are: (a) the appropriateness and size of the exposure that was assumed and (b) the suspiciously high prices that were paid to investment banks.

[11] By the way, it should be clear why sovereign governments may issue bonds, but not stock shares.

have an impact on the riskiness and value of bonds. Furthermore, the stylized structure of the balance sheet provides us with a useful representation of trading strategies[12] and an essential link between financial markets and corporate finance.

1.2.2 ASSETS VS. SECURITIES

Sometimes, it may be useful to draw a line separating assets from securities. **Securities** are assets that can be readily purchased or sold on financial markets, like stock shares and bonds; we may also say that securities are tradable or marketable financial assets. On the contrary, a health insurance policy is an asset, but it cannot be sold by its owner and cannot be considered as a security. Note that this does not mean that nontradable assets have no value. We should also note that the line between assets and securities is often not so clear. For instance a mortgage is an asset for a bank, but an illiquid one, as we said. However, pools of mortgages may be transformed into liquid securities by securitization, whereby tradable mortgage-backed securities are created. Other kinds of asset-backed securities (ABS) have been created and traded. By a similar token, a commodity like oil is not, per se, a security, even though it can be traded. The point is that an individual investor cannot really buy and store oil. However, she can take a position related to oil price by using derivatives written on oil and other nonfinancial commodities. Real estate funds have also been created to enable retail investors to take a stake in this family of alternative assets, like residential or commercial real estate. Therefore, in this book, we will not insist too much on the difference, but we will use the term "security" when the liquidity feature of an asset needs to be emphasized.

We will investigate liquid securities in some detail, but we should always keep in mind that liquidity is not only related to the specific kind of assets per se, but to market conditions as well. In conditions of stress, market liquidity may be severely reduced, putting a lot of pressure on market players in need for cash.

Example 1.4 The liquidity trap in thin markets

> In a deep and liquid market, a trade has little impact on prices, but markets may get thin and, needless to say, they have a nasty habit of doing so at the least favorable moment. Consider a hedge fund financing the purchase of assets by borrowing money. We will see later that this strategy is called margin trading. In the balance sheet of the hedge fund, the borrowed money contributes to the liability side, whereas the purchased assets are on the asset side. Equity, which is

---

[12] See Sections 1.4.4 and 1.4.5.

the difference of the two sides of the balance sheet, will float with the value of assets, whereas the liabilities are what they are. Quite often, hedge funds purchase rather illiquid and risky assets, either to earn some additional return, or as a part of a complex trading strategy. In fact, this is why we can redeem shares of a common fund at short notice, but doing so with a hedge fund requires much more time, as complex trading strategies involving illiquid assets are not so easy to unwind.

Under market stress, a flight to quality may occur, whereby market participants sell risky assets in order to rebalance their portfolios toward safer assets, like sovereign bonds of a quite solid country. As a result, asset values may be considerably reduced, eroding equity of hedge funds. The thinner the market, the larger this effect.

Well-intended regulations specify that a minimum safety cash margin must be maintained in order to preserve equity. Hence, when equity is eroded, the fund may be forced to liquidate assets to raise additional cash. But when this happens in bad times, a vicious feedback cycle may arise. We need to sell illiquid assets to raise cash, which in turn leads to further a reduction in the market price of the assets, forcing additional sales. It may even be the case that potential buyers are aware of the state of the matter and have a strong incentive to wait for a further reduction of the price asked by a fund in desperate need of liquidity.

This liquidity trap was a key factor in the famous near-collapse of Long Term Capital Management (LTCM) in 1998. As a consequence of Russian default of bonds, market nervousness ensued, leading to a drop in the market prices of risky securities, with a huge impact on the highly leveraged portfolio of the fund. Similar issues arose in the more recent subprime mortgage crisis: Illiquid assets could not be liquidated because of a market crunch. Thus, investors in need of cash were forced to sell liquid securities, like stock shares, leading to a collapse in equity markets as well.

Example 1.5 Are you on-the-run?

Sometimes, there are slight differences in the liquidity of otherwise equivalent securities. Treasury bonds, i.e., bonds issued by sovereign governments, are issued and sold on markets at regular time intervals in order to finance public spending and debt. The most recently issued bonds are called *on-the-run*, whereas their older relatives are

> called *off-the-run*. On-the-run bonds are more actively traded, and liquid, and this has an impact on their price. Some traders may try to take advantage of this price differential by suitable trading strategies, buying the cheaper bonds and short-selling the more expensive ones.

1.2.3 EQUITY

As we have pointed out, stock shares represent residual claims on the equity of a firm, i.e., what remains after liquidating assets and paying liabilities. This is why we use terms like "equity markets" and, as we shall see later, "equity derivatives." Stock shares are risky assets, as suggested by the randomness in the holding period return of Eq. (1.1). The holding period return as defined there involves only a **capital gain**, i.e., a return related to a price change. However, there is also a possible source of income in the form of dividends distributed to shareholders. If we denote by D the dividend paid during the holding period $(0, T)$, the corresponding holding period return is

$$R(\omega) = \frac{S(T,\omega) + D - S(0)}{S(0)}. \tag{1.4}$$

Dividends may be random or not, depending on the length of the holding period. Dividends are announced with some advance with respect to the ex-dividend date,[13] but they are uncertain for the not-so-close future. Actually, if the holding period is long enough, the exact timing with which dividends are paid is also relevant, as they may be reinvested in the stock itself or other assets. Thus, to be more precise, we should consider D in Eq. (1.4) as the value projected forward to time T. For instance, if a dividend of €0.60 will be paid in two months and the holding period is six months,

$$D = 0.60 \times e^{r \times 4/12},$$

where r is the (continuously compounded) annual interest rate, which we use to shift the cash flow four months forward.[14] Care must be taken with respect to taxation, as dividend income and capital gains might be taxed in a different way. We should also mention that an important topic in corporate finance is the

---

[13] To be precise, the ex-dividend date does not necessarily coincide with the date on which a dividend is paid. Since stock shares change hand continuously, a rule must be established to specify who is going to receive the dividend. If we buy the stock share after the ex-dividend date, when the stock share is said to go "ex-dividend," we are not entitled to receive the dividend, but the previous shareholder is, even if the dividend will be paid later.

[14] This operation is the reverse of cash flow discounting, and we will discuss such issues in Chapter 3.

dividend policy, i.e., the strategy by which a firm decides whether earning will be reinvested or distributed in the form of dividends.

When we are stockholders, we are actually the owners of shares of a firm, which is not the case for bondholders. This raises an important issue: Are we responsible for illegal behavior by the board of directors or damage caused by defective products? The answer is no, since stock shares are **limited liability** assets. This is essential, especially for large corporations, in order to enable separation between management and ownership. Apart from legal implications, this feature implies that stock prices cannot be negative and that the worst-case return from holding stock shares is -100%. From a mathematical viewpoint, as we have already mentioned, this also implies that a widely used distribution like the normal, which features an unbounded support, the whole real line $\mathbb{R} = (-\infty, +\infty)$, cannot be a model for stock returns, but an approximation at best.[15]

Although this will not play a major role in this book, we must keep in mind that stock shares have not only an economic nature, but a legal one as well. In practice, there may be different kinds of stock shares associated with the same firm, like **common** and **preferred** stock shares. The difference may be in voting rights, which may not be associated with preferred stock shares. On the other hand, preferred stock shares come with the "promise" of a given dividend, whereas common stocks do not have any such guarantee. The holder of a preferred share has priority over holders of common stocks in terms of dividend payments; however, if no dividend is paid, this does not involve any default on the part of the firm. On the contrary, if interest on debt is not paid, a default occurs, with the possibility of the firm being declared bankrupt. This is a relevant consideration when a firm has to decide on the best way to raise capital, by issuing either stock shares or debt. The cost of servicing debt is tax-deductible, which may yield some advantage in terms of taxation. Issuing new stock shares may dilute property, and it may not be taken well by markets, resulting in a sudden drop in the stock price. On the other hand, issuing debt increases the possibility of bankruptcy. This choice of the capital structure is a fundamental topic in corporate finance.

There are other important features of stock shares that are worth mentioning:

- Not all stock shares are publicly traded. Some may be kept under the control of original owners of a firm in order to have the final say in matters of management. Furthermore, not all firms are listed on financial markets, since this requires an expensive process, as some standard requirements must be met in order to be quoted. Private equity funds may be used to invest in privately held firms.
- Unlike other assets, like bonds or options, stock shares do not have a maturity. However, unlike energy in physics, stock shares may be created and destroyed. Sometimes, new equity is floated in order to raise

---

[15] As we shall see later, the normal distribution may also be unsatisfactory for other reasons, as it is symmetric and thin-tailed.

additional capital. Sometimes, equity disappears when shares are repurchased by the firm itself.[16] It may also be the case that a firm is delisted or acquired by another firm (not to mention the unpleasing event of bankruptcy).

- Other exceptional events that may have a relevant impact on the price of a stock share are:
  - Stock splits: Two or more stock shares of the same firm are created out of a single one. Stock splits may occur when the stock price is quite large. Increasing the number of outstanding stock shares and reducing their price may improve liquidity and lower the bid–ask spread. Sometimes, reverse splits occur, which may require some adjustments, e.g., to deal with owners of an odd number of shares when the reverse split is 1-for-2. A reverse split may occur when the stock price is very low. For instance, a low price may even preclude the listing of a share on a stock market, and a reverse split may be a corrective action to avoid delisting.
  - Spinoffs: A firm is separated in two firms, and two different stock shares are created out of each stock share of the original firm.
  - Mergers and acquisitions: Two firms are merged into a single one, with a corresponding merging of pre-existing stock shares.

Once again, all of these operations have rationales and features that are discussed in detail by books about corporate finance. We observe that their impact on stock prices must be properly accounted for. If a stock share is currently traded at a price of $100 and a 2-for-1 split occurs, the new resulting price will be something like $50, which clearly does not imply a return of -50%. Stock market indexes, discussed later, should take all of this into due account. By the same token, derivative contracts must clearly specify how these events are dealt with.[17] A stock split has no effect on the *market capitalization* of a firm, which is given by the total number of shares outstanding, times their market price.

1.2.4 FIXED INCOME

Floating stock shares is one way a firm can raise the capital it needs. An alternative is to borrow money, which does not necessarily mean literally borrowing money from a bank. A common way to raise capital in mature financial markets is issuing a bond. Bonds are also issued by sovereign governments, as well as by local authorities: Examples are US treasury bonds and municipal bonds. A

---

[16] Stock repurchase may have different motivations, as it may be a way to compensate shareholders without issuing dividends, or a way to reduce the number of outstanding shares, when they are deemed to trade at a too low price.

[17] For instance, we shall see that a typical call option suffers from a drop in the underlying asset price. Usually, call options are not protected against payment of dividends, but they are against stock splits.

bond is a security that, in its simplest form, may be described by the following main features:

- The **face value** F, also called nominal or par value, which is the amount that the issuer promises to pay back to the bondholder.
- The **maturity** T, i.e., the time at which the face value will be paid back.
- The **coupon rate** c, which is the interest rate applied to the face value to define periodic interest payments that are paid to the bondholder. These payments are called "coupons" for historical reasons, as bonds were physical pieces of paper with coupons that were detached to request payment of periodic interest.

If $c = 0$, i.e., no coupon is paid along the bond life, we have a **zero-coupon** bond, often referred to as a "zero." If $c > 0$, we have a **coupon-bearing bond**. Usually, coupons are paid twice a year, but different frequencies may be arranged.

Example 1.6 A plain coupon-bearing bond

> Let us assume that $F = \$10{,}000$, $T = 5$, measured in years, and semiannual coupons are paid, with rate $c = 4\%$. Note that coupon rates, like all interest rates, are always quoted *annually*, but should be adjusted to the actual period they refer to. In this case, since frequency is semiannual, the actual coupon rate is 2% for six months. This means that along the bond life there will be ten cash flows to the bondholder. At times $t = k \times 0.5$, $k = 1, 2, \ldots 9$, measured in years, the cash flow will be
> $$\frac{c}{2} \times F = \$200,$$
> whereas the final cash flow at $T = 5$ includes both the last coupon and the face value, amounting to $\$10{,}200$.

The choice between funding alternatives depends on the circumstances. Most firms would not issue a bond for a short-term cash need,[18] whereas for a long-term project, issuing bonds may be a better alternative, at least for a suitably sized firm. Debt securities are liabilities from the viewpoint of the issuing firm, which has an impact on both taxes and the risk of bankruptcy. On the one hand, the cost of servicing debt is tax-deductible; on the other one, however, this increases the risk of default. As we have hinted at before, the choice between issuing debt or equity is affected by a tradeoff related to these and other issues, such as the dilution of property, etc.

---

[18] A possible alternative is issuing commercial paper.

Usually, zeros are short-term bonds, whereas coupons are paid for longer-term maturities. For instance, US treasury bonds may be classified as:

- T-bills, zeros, with maturities up to one year
- T-notes, coupon-bearing, with maturities up to ten years
- T-bonds, coupon-bearing, with longer maturities

Some long-term zeros are in fact traded, but they are often created synthetically by stripping coupons of long-term coupon-bearing bonds. This is an example of a financial engineering practice known as **unbundling cash flows**.[19]

Usually, when a bond is issued, the coupon rate more or less reflects the current level of interest rates. If we compare the uncertainty in dividends of a stock share against a fixed-rate coupon bond, i.e., a bond where c is declared and fixed, we may understand why bond markets are referred to as fixed-income markets. Bonds are the basic fixed-income securities, but, as we shall see, this name also refers to quite different securities whose cash flows depend on the level of interest rates. Indeed, the term "fixed-income" is quite a bit misleading. To begin with, we may have bonds whose coupon rate is not fixed, but depends on the time-varying level of interest rates. We refer to these bonds as **floating-rate bonds**, or **floaters**. Other bonds pay coupons affected by other variables, like inflation or even a stock market index (we talk of **linkers**, in such a case). In fact, we use the term "fixed-income markets" to refer to a wide array of securities related to interest rates. They include interest rate derivatives, such as swaps and options, as well as hybrid securities, like convertible and callable bonds, discussed later.

While the cash flows of a fixed-rate bond are supposed to be known with certainty, the bond price itself is affected by the following risk factors:

- **Default risk**. The bond issuer may default on the coupon payments or even on the reimbursement of the face value, totally or partially. In fact, not all bonds are created equal: Collateral guarantees and bond indentures, which may also specify the order in which bondholders are refunded in case of bankruptcy, are relevant. In the event of default, part or all of the face value or coupons may be lost. Debt restructuring may even result in a change of maturity.
- **Inflation risk**. This is relevant for long-term bonds. Some bonds pay real-interest[20] coupons, i.e., the coupon rate (or the face value) is adjusted according to inflation.
- **Foreign-exchange risk**. This is obviously relevant if we invest in foreign bonds, which may be denominated in a foreign currency.
- **Interest rate risk**. We will explore the inverse relationship between bond prices and interest rates: When interest rates increase, bond prices

---

[19] See Section 1.2.6.4.
[20] In Section 3.3, we shall see that a *nominal* interest rate may be eroded by a high inflation rate. The real interest rate is adjusted for inflation and should reflect actual purchasing power.

1.2 Traded assets

go down and vice versa. A rather paradoxical result is that a floating-rate bond, with uncertain cash flows, may be less risky than a fixed-rate bond.[21]

Interest rate risk is relevant when we do not plan to hold a bond until maturity. If we sell the bond, there may be considerable uncertainty about its future price. In fact, defining a holding period return for bonds is actually complicated, since we should also specify how coupons are used exactly. In an asset–liability management problem, they might just be used to pay a stream of liabilities. If they are reinvested, there is uncertainty about the future interest rates at which this will be done, resulting in **reinvestment risk**.

1.2.5 FOREX MARKETS

Another huge market is the foreign exchange market (FOREX market for short), where currencies are exchanged. The involved risk factor is the exchange rate between pairs of currencies, which is relevant also for international equity and fixed-income portfolios. Nonfinancial firms are also subject to currency exchange variability, which explains the number of FOREX derivatives available. FOREX markets are also the terrain of plenty of speculative short-term trading.

The institutional arrangements behind FOREX markets are not trivial but, given their limited role in this book, we leave the related issues to the references. There is one, somewhat annoying, detail that we have to mention. If we read a stock market quote and the price of a stock share is, say, $12, we interpret this as the price of one share. Dimensionally, the quote is dollars per share. Hence, if we buy 3 shares at that price, from a dimensional viewpoint we spend

$$12 \, \frac{\text{dollar}}{\text{share}} \times 3 \text{ shares} = 36 \text{ dollars}.$$

We would probably never think of a quote in terms how many shares we can buy with 1 dollar, although sometimes, given an available budget, we must find out how many shares we may afford to buy; hence, we would not consider share per dollar as a sensible unit. When we buy commodities, the specific measurement unit plays a more explicit role. We might buy a certain kind of vegetables for, say, €3.2 per kilogram. Considering dimensions (i.e., units of measurement), if we buy 2 kg, we pay

$$3.2 \, \frac{\text{euro}}{\text{kg}} \times 2 \text{ kg} = 6.4 \text{ euro}.$$

In this case, too, measurement units have a straightforward interpretation when figuring out prices and cash flows.

Now, if the exchange rate between USD and EUR is quoted as

$$\text{EUR/USD} = 1.1166,$$

---
[21] See Section 3.5.6.

what does it mean? At the time of writing, the value of €1 is larger than the value of $1, and indeed the above ratio tells that we may buy 1.12 dollars with 1 euro, allowing for some rounding and neglecting transaction costs. Alternatively, we may say that the price of 1 euro is 1.12 dollars. This depends on the perspective we take, and since FOREX quotes always involve two monetary units, some clarification is in order.

Quoting exchange rates is about stating an equivalence between two amounts denominated in different countries. We might write something like

$$\text{EUR } 1 = \text{USD } 1.1166, \tag{1.5}$$

or, equivalently

$$\text{USD } 1 = \text{EUR } 1/1.1166 = \text{EUR } 0.8956. \tag{1.6}$$

Let us focus on the first case, Eq. (1.5). We say that

- EUR is the **base currency**, and we consider EUR 1 as a *fixed* number
- USD is the **quoted currency**, and we consider USD 1.1166 as a *variable* number

In a currency pair, written as EUR/USD, the currency to the left of the slash is the base currency, and the currency to the right is the quoted currency.

Depending on which currency is considered as domestic, there are two types of quotes:

- In **direct quotation** the domestic currency is the quoted currency, i.e., a variable amount of the domestic currency is quoted against a fixed amount of foreign currency. This kind of quotation is also called *normal* or *uncertain for certain*.
- In **indirect quotation** the domestic currency is the base currency, i.e., a fixed amount of domestic currency is quoted against a variable amount of foreign currency. This quotation is also called *reciprocal* or *certain for uncertain*.

For instance, a Eurozone bank quoting as in Eq. (1.5) would use an *indirect* quote. A difficulty with FOREX markets is that different quotations are used on different markets. The choice may depend on the following:

- A matter of perspective, i.e., what our domestic currency is
- A matter of convenience: for instance, a quote like EUR/JPY 115.261, stating that 1 euro corresponds to 115.261 Japanese yen is convenient, whereas the reciprocal would be less convenient
- A matter of priority, as the choice is influenced by the fact that there are some "major" currencies which are more widely traded than other ones
- A matter of local conventions, since, for instance, the conventions in the UK are different from the conventions in the USA

1.2 Traded assets

A further complication is introduced by bid–ask spreads. The quote in Eq. (1.5) would more likely read as

$$\text{EUR/USD } 1.1165/67,$$

stating that the bank is bidding 1.1165 dollars to buy 1 euro, and asking 1.1167 dollars to sell 1 euro. Sometimes, the "three Bs rule" is invoked:

The market maker Buys the Base currency at the Bid (low) price.

Indeed, the Eurozone bank would buy 1 euro for 1.1165 dollars. Needless to say, a Eurozone bank will also quote an exchange rate like USD/CHF, which does *not* involve any domestic currency, to add to the confusion. In this case, we have to come up with a *cross-rate*, starting with a mix of direct or indirect quotes.

An indirect quote may also be considered as a "quantity quotation," in the sense that it gives the quantity of foreign currency needed to buy one unit of the domestic currency. The direct quote may be considered as a "price quotation," i.e., the price of one unit of foreign currency in terms of the domestic currency. In this book, we will deal extensively with derivative pricing, including forward/futures contracts on currencies. For the sake of uniformity, we will always interpret ratios as *prices*, just as we do in commodity prices, rather than currency pairs. Hence, assuming that we are US investors, we would consider a price like

$$1.1166 \text{ dollars per euro,}$$

stating that the price of €1 is $1.1166, so that if we want to buy €200.00, we have to pay

$$\frac{\$}{\text{€}} 1.1166 \times \text{€}200.00 = \$223.32.$$

Note that this is the contrary with respect to a base/quoted currency pair. A Eurozone investor would consider that as a price at which a euro is sold. We will neglect bid–ask spreads, and no ambiguity should arise.

1.2.6 DERIVATIVES

Stock shares and bonds are, in a sense, primary assets. They need not be primary *risk factors*, as we may build a model relating their prices (or returns) to underlying risk factors like inflation, oil price, and interest rates.[22] However, the relationship between risk factors and stock share prices/returns is represented by a mathematical model, possibly estimated by statistical methods, on which there may be no general agreement.

An incredibly large class of assets has been created on top of primary assets, collectively known as **derivatives**. A derivative security is a financial asset

---

[22] See Chapter 9 on factor models.

deriving its value from some other variable, *by an explicit formula that is written in a contract*. For instance, if S_t is the random price of an asset at time t,[23] a typical derivative features a **payoff** $f(S_T)$ at a well-defined time T, the **maturity** of the derivative, for some well-defined function $f(\cdot)$ of the underlying asset price at maturity. We insist again on the fact that the function $f(\cdot)$ is explicitly written in the contract. More complex derivatives feature a payoff depending on the whole price path until maturity.

If the derivative is written on a stock share or a bond with price S_t, we say that the latter is the **underlying asset**. However, S_t can be something else, not necessarily a primary asset. For instance, we may consider:

- The price of a nonfinancial asset, e.g., a commodity like gold or oil, provided that a well-defined price is quoted on exchanges
- A risk factor that is not the price of a traded asset, but a financially relevant variable nevertheless, like an interest rate or a market index,[24] or even an elusive variable like volatility
- A risk factor that is not related to financial assets or prices, as in weather derivatives
- The price of another derivative, as in compound options or swaptions

We observe that there is room for a considerable variety of derivatives, as they may depend on a combination of underlying variables, and they also differ in terms of the function defining the payoff.

Derivatives may be used for opposite purposes, namely, risk hedging and speculation. Originally, they were meant to be risk transfer mechanisms and have quite a long history, definitely predating the development of quantitative finance. However, they have become quite controversial assets, as the volume of derivatives outstanding is so huge that it often larger than the market of the underlying primary assets.[25]

There are different issues related to derivatives, which may be tackled by quantitative finance models:

- Pricing[26]: What is the fair value of a derivative, and how is it related to underlying risk factors?

---

[23] Depending on convenience, we will write S_t or $S(t)$; we will not stick to a single notation, as no ambiguity actually arises.

[24] See Section 1.5.

[25] Statistics published by the Bank for International Settlements in 2014 estimated a total notional amount of OTC derivatives of about $630 trillion. Interest rate swaps accounted for $381 trillion. These numbers are impressive but misleading, since the notional amount of an interest rate swap, as we shall see, overestimates the value of the derivative and the actual cash flows that will occur. Nevertheless, there is no doubt that this is a huge market.

[26] As is common in the literature, we will use the term **pricing**, even though **valuation** would be more correct. The fair value of the derivative is only a component of the actual price asked by a bank issuing derivatives, since this will include a profit margin and some buffer against residual risk that cannot be hedged away in real life.

- Hedging: If a market player like an investment bank writes, i.e., creates a derivative, how can it manage the ensuing risk?
- Portfolio management: How can a derivative be used to change the characteristics of a portfolio, increasing or reducing its exposure to selected risk factors?

A wide array of derivatives is traded, but here we just want to introduce the three basic families: Forward contracts, futures contracts, and vanilla options.

We should also notice that, beside quantitative issues, there is a host of regulatory and legal issues related to derivatives, not to mention how should they be accounted for in financial statements of banks and firms. These are, however, outside the scope of this book.

1.2.6.1 Forward contracts

A **forward contract** is an arrangement between two counterparties, which at time t_0 agree to buy and sell, respectively, an asset at a prespecified forward price $F(t_0, T)$ at a later date T, the maturity of the contract. The part agreeing to buy the asset is said to hold the **long position** in the contract, whereas the part agreeing to sell is said to hold the **short position** in the contract. Note that the contract is symmetric, in the sense that *both* parties are forced to comply with what they have agreed.

The current **spot price** of the underlying asset when the contract is written, denoted by $S(t_0)$, is a known number, whereas the spot price $S(T)$ at maturity is uncertain. At time t_0 the **forward price** $F(t_0, T)$ is established once for all. During the time interval (t_0, T) the spot price $S(t)$ will change randomly. By the same token, the forward price $F(t, T)$, observed at time t, for delivery at time T, $t \leq T$, will change as well. This is the forward price for new forward contracts written at a later time $t > t_0$, but the forward price in previously arranged contracts will *not* change. As we shall see, the value of a contract will depend on the difference between the fixed $F(t_0, T)$ and the uncertain $F(t, T)$ along the life of the contract.

Both the spot price $S(t)$ and the forward price $F(t, T)$ are stochastic processes, which are arguably correlated in some way. We should find a way to model the relationship between spot and forward price, which may be a nontrivial task. However, we can immediately see that the following spot–forward convergence condition must hold at maturity:

$$S(T) = F(T, T). \tag{1.7}$$

In fact, $F(T, T)$ is the forward price for an *immediate* delivery at time $t = T$, and it must be the same as the spot price, otherwise two prices would be quoted for the same item.[27]

---

[27] Formally, this is an example of the law of one price, which is a consequence of the no-arbitrage principle that we investigate in Section 2.3.

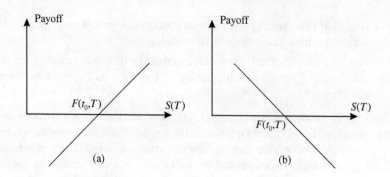

FIGURE 1.4 Payoff from forward contracts: (a) long position, (b) short position.

The value of the contract arises from the payoff that will result at maturity. The payoff for the long position is

$$S(T) - F(t_0, T).$$

To see why, observe that, if $S(T) > F(t_0, T)$, the long position may buy the underlying asset at the delivery price $F(t_0, T)$ and sell it at the current spot price $S(T)$ at maturity, earning a profit. Note, however, that the payoff may well be negative, since the long position has to buy at the delivery price, even when this is larger than the prevailing spot price. Going the other way around, the payoff for the short position is

$$F(t_0, T) - S(T).$$

Clearly, the sum of the two payoffs is zero: The profit for the long position is just the loss for the short position, and vice versa (this is a zero-sum bet, in some sense). The payoffs are illustrated in the two diagrams of Fig. 1.4, for the long and short positions, respectively. The long position benefits from an increase in the spot price, whereas the short position benefits from a decrease in the spot price. As we shall see in later chapters, it is common jargon to say that an investor is "long a variable" if she gains from an increase in the variable, and is "short a variable" if she gains from a decrease in the variable. The variable may be the price of an asset, an interest rate, and whatnot. When the underlying variable is not really the price of a deliverable asset, the contract is settled in cash, i.e., an amount corresponding to the payoff is exchanged (when our payoff is negative, it means that we owe money to our counterparty). In Chapter 2, we will see that, using no-arbitrage pricing principles, the forward delivery price is selected in such a way that the value of the contract is initially zero for both parties. Thus, payoff and profit coincide, as nothing is paid when entering the contract. After inception of the contract, the spot price $S(t)$ and the forward price $F(t, T)$ will change, and this will affect the value of the contract, which may drift away from zero.

The fact that the value of a forward contract is initially zero explains why it may be so attractive for a speculator, at least in principle. In the case of speculation, the profit from a successful trade in the underlying asset is limited

1.2 Traded assets

by the fact that we have to actually *buy* (or short-sell) it on the spot market, which may be limited by the available budget. In principle, nothing is needed to enter into a forward contract, as the forward price is determined in such a way that the value of contract at inception is zero. Thus, the return is not really defined, as the denominator is zero! In practice, it may be the case that some collateral has to be posted, and in any case there are transaction costs. Nevertheless, we will see how a considerable leverage may be obtained with derivatives in general, magnifying both profit and loss opportunities.

The other side of the coin is that a forward contract may be used to eliminate or reduce risk, i.e., for hedging purposes. Assume that we will have to buy the underlying asset at time T in the future. Since $S(T)$ is uncertain, we face some risk, but by entering into a *long* position, we will be able to buy at $F(t_0, T)$ no matter what, eliminating uncertainty altogether. If the contract is settled in cash rather than by buying the underlying asset, the net cash flow at time T will be

$$\underbrace{S(T) - F(t_0, T)}_{\text{payoff}} \underbrace{- S(T)}_{\text{purchase cost}} = -F(t_0, T), \tag{1.8}$$

which is negative, since we are buying the asset, and is equivalent to a contract for physical delivery of the asset. A typical case in which derivatives are settled in cash is when the underlying is a nontradable asset like a stock market index. In other cases, physical delivery would be possible in principle, but it might be avoided because of transportation costs and the like.

Example 1.7 A long hedge

Suppose that in six months we will need 500 ounces of gold, and that the current (time $t = 0$) forward price for delivery in 0.5 years (six months) is

$$F(0, 0.5) = 1250 \ \$/\text{ounce}.$$

Then, we may enter into a long position for 500 ounces to lock that price. As a practical remark, we shall see that real-life contracts may be given for standardized sizes, such as, e.g., 100 ounces. If the contract is settled by physical delivery, we shall buy gold at 1250 dollars per ounce, no matter what. The corresponding (negative) cash flow is

$$-1250 \ \$/\text{ounce} \times 500 \ \text{ounces} = -\$625{,}000.$$

If the contract is settled in cash, and the spot price at maturity turns out to be 1150 \$/ounce, our cash flow will be

$$[(1150 - 1250) - 1150] \ \$/\text{ounce} \times 500 \ \text{ounces} = -\$625{,}000,$$

the same as before. Note that, in this case, we buy at a cheaper spot price, but this is compensated by a loss on the long forward position.

If we have to sell the underlying asset, we should enter into a *short* position, which just implies a change in sign in Eq. (1.8).

A perfect hedge results if a forward contract matching both the desired maturity and the underlying asset, as well as the contract size, can be agreed. By the way, note that "perfect hedge" means that risk is completely eliminated, not that the outcome is necessarily a pleasing one. If we take the long position, as in Example 1.7, we will regret our decision if the spot price at maturity turns out to be lower than the delivery price. By the same token, a short position is not a nice place to be if the underlying spot price increases. Still, risk management should be assessed *a priori*, i.e., *ex-ante*, not *ex-post*. The main feature of forward contracts is that they are actually a private arrangement between the two counterparties, typically a firm and an investment bank. Forward contracts are not securities freely traded on regulated exchanges, but rather an OTC agreement. This implies both advantages and disadvantages. On the positive side, the details of an OTC contract may be tailored according to quite specific needs. On the negative side:

- Since there is no quoted price, which is driven by demand and offer, pricing a specific contract may be troublesome. Hence, a firm in need for a hedge might adopt a strategy of competitive pricing, which means asking around for multiple quotes to compare them. A possibly better alternative is to establish long-term relationships with a single, trustworthy bank.

- The contract is not standardized, hence it is not liquid. Unwinding the position may be difficult if the hedging needs change. This typically implies assessing the value of the contract and closing it before maturity by a cash settlement. Note that this is the result of a negotiation process, possibly implying the valuation of an illiquid contract, and not the immediate sale of a security on regulated and liquid markets.

- A further issue with forward contracts is counterparty risk. There is only one cash flow, at maturity, possibly a huge one. Imagine that we hold a short position in a forward contract written on an asset whose price is dropping dramatically. We are about to collect a remarkable payoff, but what if the long position walks away? In fact, only creditworthy firms are accepted as partners in a forward agreement, but counterparty risk is not completely eliminated.

The solution to liquidity and counterparty risk issues is represented by futures contracts, which are the exchange-traded equivalent of forward contracts.

1.2.6.2 Futures contracts

Futures contracts are quite similar to forward contracts, in the sense that the delivery of an underlying asset or commodity is arranged for a future date, at

1.2 Traded assets

a prespecified **futures price**[28] $F(t,T)$ that is continuously quoted on regulated exchanges. Futures contracts have specific features aimed at easing the difficulties with forward contracts, namely, liquidity and counterparty risk:

- Standardization, to improve liquidity
- Daily marking-to-market through a clearinghouse, to ease counterparty risk

In order to improve liquidity, futures contracts are standardized. This means that the range of available underlying assets and delivery dates is limited and cannot be arranged to perfectly suit very specific needs. For instance, certain contracts are only available for quarterly delivery, i.e., maturing at four months per year. This makes the use of futures contracts in hedging more difficult, but it results in a deeper market, where it is easy to buy and sell futures contracts. Furthermore, a liquid market is less subject to manipulation and cornering.

Example 1.8 Cornering in futures markets

> Cornering is an illegal practice, whereby speculators accumulate a significant amount of the underlying asset. When maturity is approached, the short positions will be forced to buy the asset at large prices to honor their contracts, if the supply is limited. To circumvent this difficulty, contracts should be arranged only for underlying assets with a sufficiently deep market, or alternatively a range of underlying assets may be eligible for delivery, rather than a single one. For instance, in futures contracts on bonds, a whole range of bonds may be delivered, not only a specific one. Clear rules define the equivalence among similar, but not identical, bonds and the coefficients by which the delivery price is modified if necessary. For instance, bonds with comparable maturities, but different coupon rates may be included in the range for acceptable delivery.

The two essential features of futures contracts aimed at easing counterparty risk are:

1. The existence of a **clearinghouse**. The clearinghouse consists of a group of solid financial institutions, and it steps between the long and the short positions. The institutional arrangement is depicted in Fig. 1.5. Actually, if we hold the long position, we do not really "see" any corresponding short position in the contract. We only deal with the clearinghouse, which assumes the counterparty risk.

---

[28] Please note the essential difference between the *future* spot price, which is uncertain, and the *futures price* associated with a derivative contract.

FIGURE 1.5 The institutional arrangement of a futures contract. The clearinghouse manages the margin accounts of the long and short positions.

2. The contracts are **marked to market** daily. This means that, rather than settling the contract at maturity, daily cash flows are exchanged at the end of every trading day. Indeed, both long and short positions are required to post some margin, in the form of cash or some collateral, on an account managed by the clearinghouse. If the futures price moves unfavorably, we will lose some amount of money immediately, rather than at maturity. The loss is sustained by the margin account, where daily profits are also collected in the case of a favorable movement. There is a minimum amount that must be maintained on the margin account, the maintenance margin. If the account level falls below the **maintenance margin**, a **margin call** is issued. Failure to comply with the margin call by posting more cash or collateral on the margin account has the consequence that our contract is immediately closed out and assumed by the clearinghouse.

We should note that the actual exposure of the clearinghouse is related to *net* position, balancing long and short positions. One proof that the mechanism does work occurred on October 19th, 1987, a day remembered as the Black Monday of 1987, when a loss in excess of 20% in the Dow Jones index occurred. The S&P500 index sustained a similar drop, with a corresponding shock on index futures. Indeed, some brokers who were members of the clearinghouse went bankrupt on that day, but the clearinghouse survived and all contracts were honored.

We will analyze later the full details of futures contracts, as well as their use for hedging and speculation. For now, we just clarify the mechanics of daily marking-to-market.[29] Imagine that, at time t_0, an arbitrary moment within a trading day, we enter into a long position in a futures contract at price $F(t_0, T)$. Say that, at the end of the day, corresponding to time t_1, when the futures prices are settled and marking-to-market takes place, the settlement price is $F(t_1, T)$. The cash flow for the long position at the end of the first day is, for each contract,

$$F(t_1, T) - F(t_0, T),$$

which is positive if there is an increase in the futures price. The corresponding cash flow for the short position is $F(t_0, T) - F(t_1, T)$. In general, if the settlement price at the end of day t_k is larger than the corresponding price of the

---

[29]The picture is a bit simplified here.

1.2 Traded assets

previous day, i.e., if
$$F(t_k, T) > F(t_{k-1}, T),$$
money is drawn from the margin account of the short position and deposited into the margin account of the long position, and vice versa if there is a decrease in the futures price. The marking-to-market mechanism generates a series of daily cash flows for the long position:

$$F(t_1, T) - F(t_0, T),$$
$$F(t_2, T) - F(t_1, T),$$
$$F(t_3, T) - F(t_2, T),$$
$$\vdots$$
$$F(t_m, T) - F(t_{m-1}, T),$$

where $t_m \equiv T$. The last cash flow may also be expressed as

$$S(T) - F(t_{m-1}, T),$$

since the futures price at maturity, $F(T, T)$, converges to the spot price. If we sum these cash flows, we obtain a telescoping sum:

$$\sum_{i=1}^{m} \Big[F(t_1, T) - F(t_{i-1}, T)\Big] = F(t_m, T) - F(t_0, T)$$
$$= S(T) - F(t_0, T). \tag{1.9}$$

Thus, the net sum of cash flows looks just like the payoff from a forward contract. A similar expression, with a change in sign, applies to the short position. Now, in the light of this result, one could wonder whether there is a significant difference between forward and futures contracts. Indeed, there is a subtle but important difference between the two: The daily cash flows may be reinvested immediately at some interest rate, when positive. Negative cash flows, i.e., losses, may also be financed at some interest rate. We will prove in Section 12.2 that, if interest rates are deterministic, the forward and the futures price are the same. However, if the interest rate moves randomly, this will have an effect, especially if there is a definite correlation between futures prices and interest rates. This is especially the case with interest rate futures. Thus, forward and futures prices need not be identical.

Liquidity has another, possibly surprising, effect. As a general rule, futures contracts do *not* result in the actual delivery of the underlying asset, and most futures contracts are closed before maturity. To close a futures contract, all we have to do is entering into an offsetting position: A long position is closed by entering into an equivalent short position, and vice versa. This feature is essential both for hedgers and speculators, who do not really want to buy the underlying asset, especially if the price of the underlying asset is only a proxy for the actual risk factor that they are exposed to. For instance, a firm that is

Table 1.2 An illustration of the mechanics of futures markets. All data are in $.

Day	Trade price	Settlement price	Daily gain	Cumulative gain	Account balance	Margin call
1	1350				16,000	
1		1346	−800	−800	15,200	
2		1330	−3200	−4000	12,000	
3		1334	800	−3200	12,800	
4		1315	−3800	−7000	9000	1000
5		1304	−2200	−9200	7800	2200
6		1320	3200	−6000	13,200	
7		1330	2000	−4000	15,200	
8		1328	−400	−4400	14,800	
9	1338		2000	−2400	16,800	

exposed to risk factors related to energy or transportation costs may consider using oil futures as a suitable hedging instrument, but they would certainly not be interested in the actual trade of oil.

Example 1.9 Mechanics of futures markets

Table 1.2 illustrates a possible scenario in a trade on gold futures. On day 1, when the gold futures price is $1350 per ounce, we enter a long position for two contracts, whose unit size is 100 ounces (hence, each contract specifies the purchase of 100 ounces of gold at a total price of $135,000). The initial margin required by the broker is $8000 per contract, hence, we have to deposit $16,000 on the margin account immediately. The maintenance margin is $5000 per contract. At end of day 1, the futures is settled at $1346. Hence, we have a cash flow

$$\$(1346 - 1350) \times 200 = -\$800$$

which is actually a loss, as the futures price declined and we hold a long position. In Table 1.2, we list the settlement price for a sequence of days, resulting in daily gains, which are cumulated. The margin account falls below the maintenance margin at the end of day 4. After marking-to-market, the margin account balance is only $9000, and $1000 have to be posted in order to restore the maintenance margin. We get another margin call after the settlement of the next day. At some time during day 9, when the futures price is $1338, we close the contract, with a total loss of $2400.

1.2.6.3 Vanilla options

Options, like forward and futures contracts, concern buying or selling an asset in the future at a predetermined price. However, options are more complicated contracts, since they are asymmetric. In forward and futures contracts, the long and the short position have symmetric obligations, in the sense that both of them are forced to buy and sell, respectively, the underlying asset at the agreed price, whether they like it or not. This results in linear payoff functions, and there is a unique given price such that the value of the contract is zero at its inception. On the contrary, options feature nonlinear, possibly complicated payoffs. Furthermore, options involve two quite different roles, the option **writer** and the option **holder**, making the contract asymmetric. The option writer is the counterparty originally creating the option, which is sold to the holder. To get the point, let us focus on the simplest family of options, namely, **vanilla** options. Two kinds of vanilla options are traded, call and put options.

- In a **call option**, the option holder has the right, *but not the obligation*, to buy the underlying asset from the option writer, in the future, at a fixed price K called the **strike price**.
- In a **put option**, the option holder has the right, *but not the obligation*, to sell the underlying asset to the option writer, in the future, at a fixed strike price K.

We immediately notice the asymmetric nature of options: The holder has the right to a choice, and the option writer will have to comply, no matter what. The writer of a call option will be forced to sell the asset if the holder exercises the call option, and the writer of a put option will be forced to buy the asset if the holder exercises the put option. This immediately suggests that: (a) the payoff will be nonlinear, (b) the option writer should be compensated for this obligation, and (c) the option will have a positive value at its inception, unlike linear contracts. In the case of options, the jargon is misleadingly different from the case of futures: The option writer is said to hold the **short position** in the contract, whereas the option holder holds the **long position**. Since the option can be a call or a put, in this case the terminology does not refer to who buys or sells the underlying asset. The long position should be understood as the side of the contract that profits from an increase in the value of some asset. The long position in a futures profits from an increase in the futures price, and the long position in an option profits from an increase in the option value, for both call and put options. The short position, on the contrary, profits from a drop in the futures price or in the option value. Clearly, the option writer earns a profit if the option expire worthless, without being exercised by the holder, who paid the option premium. As with forwards/futures, the contract can be settled in cash, rather than by actual delivery of the underlying asset, if this is not tradable, or it is not convenient to do so.

If the option can be exercised only at a prespecified time T, the option **maturity**, the option is said to be **European-style**. If the option can be exercised at any time before and including a time T, which in this case is an expiration date,

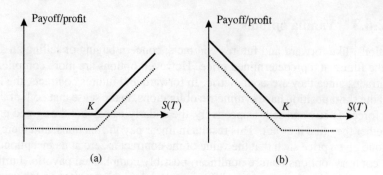

FIGURE 1.6 Payoff (continuous line) and profit (dotted line) from long positions in options: (a) call option, (b) put option.

rather than a maturity, the option is said to be **American-style**. The payoff of a European-style call option, from the holder viewpoint, is

$$\max\{S(T) - K, 0\}.$$

To see why, consider that the holder will exercise only if it is convenient to do so, which is the case if the spot price $S(T)$ at maturity is larger than the strike price K. In such a case, the holder may buy the asset at K from the option writer and sell it immediately at $S(T)$ on the spot market. By the same token, the payoff of the put option, from the holder viewpoint, is

$$\max\{K - S(T), 0\}.$$

If $K > S(T)$, the option holder may buy the asset on the spot market at $S(T)$, and force the option writer to take delivery at the strike price K. The payoff and profit to the long position for a call and a put option, respectively, are depicted in the diagrams of Fig. 1.6. We immediately observe that the payoffs are nonlinear (piecewise linear, to be precise). Furthermore, payoff and profit are not the same thing, unlike the case of linear contracts with initial zero value. Since the payoff cannot be negative, it must be the case that an option has some positive value at time $t = t_0$, when the option is written, which is the fair price that the writer should ask.[30] Thus, the profit to holders is the payoff shifted down by the option price. While there is only one "right" forward/futures price, such that the initial value of the contract is zero, options with different strike prices are traded. We should expect that the price of a call option, all other factors being equal, is a decreasing function of the strike price, whereas the price of a put option is an increasing function of the strike price.

The diagrams for the short position are just the diagrams of Fig. 1.6 turned upside down, as shown in Fig. 1.7. The option writer is compensated by earning

---

[30]We stress again that we confuse "value" and "price." Option pricing models, as we shall see, yield a *fair value*. The actual price will account for profit and some additional fudge against the risk born by the writer.

1.2 Traded assets

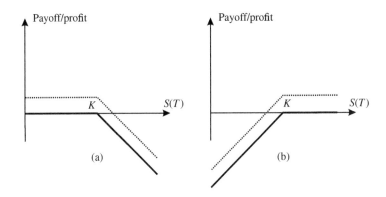

FIGURE 1.7 Payoff (continuous line) and profit (dotted line) from short positions in options: (a) call option, (b) put option.

the option price (or premium), but it is important to realize a key difference between the two roles. If the option expires worthless, the holder will lose the whole option premium, but this is the worst that can happen. Figure 1.7(a) shows that there is no bound on the potential loss for a call writer. Thus, two essential tasks in quantitative finance are finding the fair value of options and devising ways to hedge the risk of writing options. A significant portion of this book is devoted to these two problems, and we will find out that they are tightly linked. We will also see that pricing American-style derivatives is, as a general rule, much more complicated. To see why, consider the case of an American-style put option if $S(t) < K$ at $t < T$, before maturity. The option payoff, if the option is exercised early at time t, is the same as the European-style option, with $S(T)$ replaced by $S(t)$. Hence, the option holder could earn a positive payoff, $K - S(t)$, by exercising early, but is this really an optimal choice? Should the option holder exercise immediately, or wait for a better opportunity? The answer is not really trivial, as it implies the solution of specific kind of dynamic stochastic optimization problem, an optimal stopping problem.

We observe that the option payoffs for call and put options only depend on the value of the underlying asset at maturity (or the early exercise date for American-style options). The payoff is a simple piecewise linear function that does not depend on the whole history of the underlying asset price. This is why these simple options are called *vanilla*.[31] Vanilla options are commonly traded on regulated exchanges, but several OTC variants, involving multiple assets and more complicated payoff functions are commonly engineered. These options are often called **exotic** options.

Just like futures, options may be used for both hedging and speculation purposes. Let us illustrate these uses by two simple examples.

---

[31] Vanilla is the most basic ice cream flavor.

Example 1.10 A protective put

Let us consider a protective put strategy. We hold an asset, with value $S_0 = S(t_0)$, but we are concerned with a possible loss over the holding period $[t_0, T]$. One way to hedge risk is buying a put option with strike K. Then, the overall portfolio value at maturity is the sum of the asset value and the option value,

$$S_T + \max\{K - S_T, 0\} = \max\{K, S_T\}.$$

If we look at the total payoff, it seems that the larger the strike, the better. Clearly, this is too good to be true. Indeed, we should not forget that the protection from the put option does not come for free, and it is a safe guess that a put option with a larger strike price will be more expensive, too. On the contrary, hedging with forward or futures contracts can be achieved at no initial cost. However, we give up the whole upside potential (if S_T grows), whereas this is partially retained by hedging with options.

Example 1.11 A bullish speculation

The current price of an asset is $S_0 = \$100$, and we have a strong belief that it will rise in the near future. One possible strategy is simply to buy the asset. If we are right and, say, the asset price at some later time T turns out to be $S_T = \$120$, the holding period return is

$$\frac{120 - 100}{100} = 20\%.$$

Now let us assume that a call option with strike price $K = \$100$ costs $\$5$. If we buy the call option, the return in the above scenario is a stellar

$$\frac{\max\{120 - 100, 0\} - 5}{5} = \frac{15}{5} = 300\%.$$

Clearly, there must be some other side of the coin. To get a feeling, let us assume that we are wrong and the underlying asset price goes down by 1%. The percentage loss, if we invest in the asset itself, will be a not too painful 1%: We may be fairly disappointed, but this is a loss we may well live with. However, the call option return is

$$\frac{\max\{99 - 100, 0\} - 5}{5} = \frac{-5}{5} = -100\%,$$

since the option expires worthless and we lose the whole premium.

1.2.6.4 Hybrid securities, bundling/unbundling, and securitization

So far, we have considered simple assets like stock shares, plain bonds, and vanilla options. We have hinted at the possibility of creating more complex assets, like exotic options featuring different payoff structures. One such example is an **Asian option**, whose payoff depends on some form of average. The most natural Asian option involves an arithmetic average over time of the price of a single underlying asset, like

$$\max\left\{\frac{1}{N}\sum_{k=1}^{N} S(t_k) - K, 0\right\},$$

where $t_k, k = 1, \ldots, N$, is a sequence of sampling instants. Such options are usually not traded on regulated exchanges, but sold OTC. It may sound surprising, but we may find both European- and American-style Asian options. The point is that *Asian* refers to the form of the payoff, whereas the other labels refer to the possibility of early exercise.

By assembling or disassembling assets and cash flows, a whole world of possibly quite complex assets can be created by financial engineering. The building blocks are often stock shares, bonds, and options, and the basic procedures include:

- Cash flow bundling and unbundling
- Addition of option-like features to traditional assets
- Securitization

Let us illustrate the idea with a few concrete examples.

Convertible bonds. A convertible bond is a corporate bond with an optional component: The holder has the right to exercise an option to transform it into a prespecified number of stock shares of the same firm. We may regard this kind of asset as a security bundling a bond and a sort of call option on a stock share. To be precise, the bundled derivative is often not really an option, but rather a **warrant**. The difference is that when a warrant is exercised, a brand new set of shares is created, diluting equity. Convertible bonds may be appealing to issuers as a way to raise capital when the stock share price is perceived by the management as unjustifiably low. Given the embedded option, the price of the bond will be higher than otherwise. If the stock share price rises, new stock shares will be issued and the company will stop servicing debt. Otherwise, the firm will be able to deduct the cost of debt servicing from profit, with a tax advantage. Convertible bonds may be appealing to investors as well, when assessing the company risk is difficult, and they offer upside potential if the firm grows.[32]

---

[32]Convertibles may also be interesting for sophisticated investors looking for arbitrage opportunities, which we introduce later. See, e.g., [4].

Callable bonds. A callable bond is a bond that may be bought back by the issuer, at a given price, subject to certain limitations. For instance, a bond may be declared noncallable for a given number of years after its issuance. When an investor buys a callable bond, she is essentially selling a call option back to the bond issuer. Hence, all other factors being equal, a callable bond is cheaper than a noncallable one. The bond issuer will find the call opportunity convenient if there is a drop in interest rates, since it may refinance debt by issuing brand new bonds with a reduced coupon rate. This is bad news for the bondholder, as she will be subject to reinvestment risk: She is forced to get her capital back just when the bond value is increasing because interest rates are dropping,[33] and she will have to reinvest in new bonds with lower coupons (or old bonds with higher price and corresponding lower yield).

Structured bonds. A structured bond typically offers a coupon that is not linked to interest rates, but rather to another index, like a stock market index. Even if the index return turns out to be negative, the repayment of face value of the bond is guaranteed. This shows that a structured bond includes an option element. Indeed, structured bonds were also used to circumvent regulations forbidding mutual/pension funds from investing in derivatives.

Example 1.12 A structured bond

A rather fancy, but real-life example of a structured bond is the following:

- Bond maturity is four years.
- At maturity, the payment of the face value is guaranteed, plus a single coupon; the coupon, too, will be paid at maturity, and no periodic coupon will be paid.
- The coupon is linked to the monthly average value of a basket of ten stock shares in the telecommunication industry; since maturity is 4 years, 48 monthly observations of ten stock prices are involved in the average.
- The average return of the portfolio might well be negative, but in this case the coupon will just be zero, and no loss will be sustained.
- It will be possible to ask for the anticipated payment of the coupon every six months, starting from the end of year 2.
- It is also possible to ask for the anticipated repayment of the face value, but this implies a reduction with respect to the face value.

---

[33] We will explore the inverse relationship between bond prices and interest rates in Chapters 3 and 6.

1.2 Traded assets

> This looks like a very complicated security, but it may be assembled by **bundling** a zero-coupon bond and an exotic option. The zero ensures the payment of the face value, which is reduced if early repayment is requested. The option is a complicated version of a call. Let $S_j(t_i)$ be the price of each underlying stock share, indexed by $j = 1, \ldots, 10$, at time $t_i = i/12$, where $i = 0, 1, 2, \ldots 48$. Note that we are considering one year as the time unit, as customary in finance, and for the sake of simplicity we are assuming that one year consists of 12 *identical* months, which is not really the case. Finally, let us consider the following payoff:
>
> $$\max\left\{0, \frac{1}{48}\sum_{i=1}^{48}\sum_{j=1}^{10} S_j(t_i) - K\right\},$$
>
> where K, the strike, is just the initial value of the portfolio,
>
> $$K = \sum_{j=1}^{10} S_j(t_0).$$
>
> This option has three features:
>
> - It is a **rainbow** option, as it is written on multiple underlying assets.
> - It is an **Asian** option, since its payoff is related to the average price, rather than to a single price at maturity (or early exercise).
> - It is a **Bermudan-style** option, since it features early exercise opportunities, but only at a limited set of epochs, corresponding to $t = 2, 2.5, 3, 3.5$ years; thus, it is halfway between American- and European-style options.

Long-maturity zeros. As we have seen, zero-coupon bonds are typically associated with short maturities. However, we may find zeros maturing in 30 years. These zeros are often created by investment banks that hold long-term, coupon-bearing sovereign bonds, and strip the coupons creating zeros. This **coupon stripping** procedure is an example of the more general idea of **cash flow unbundling**. As we shall see, the availability of a rich array of zeros is useful in asset–liability management. Furthermore, they are quite sensitive to changes in the interest rates, and can be used for speculation and hedging purposes.

Mortgage-backed securities. When a bank issues a mortgage to a homeowner, it creates an asset in its balance sheet. This asset, however, is not liquid. In order to create a marketable security, the cash flows from a pool of mortgages can be bundled together by a securitization procedure, in order to create a mortgage-backed security that can be traded. A mortgage-backed security can be risky, as homeowners may default on payments or, on the contrary, they might repay debt early, if interest rates move in their favor. The former issue exposes the investor to default risk, whereas the latter issue creates reinvestment risk. Despite these risks, these securities promised a larger yield than other bonds, which made them quite popular when they were introduced. Default risk should be somehow mitigated by risk pooling. It seem sensible to say that a limited amount of defaults in a pool of mortgages should be not too much of a trouble. The idea was pushed to the limit when subprime mortgages, i.e., mortgages offered to homeowners with a high chance of default, were securitized. Unfortunately, risk pooling works when risks are not quite correlated. When the subprime mortgage crisis erupted in 2007, correlations increased sharply, proving that, indeed, some of the underlying risks were not fully understood. As we discuss in Section 5.5, the matter was further complicated by tranching procedures, whereby different layers of securities with different risk levels are assembled, possibly by a second round of securitization. The ensuing crisis lead to the demise of Lehman Brothers and to a revision of financial engineering practices that, at the time of writing, is not quite settled yet.

1.3 Market participants and their roles

After discussing securities that are actually traded on financial markets, let us take a more concrete look at who market participants are and their roles. In Section 1.1, we have described the role of financial markets in terms of consumption timing and risk transfer, which underlines the following functions performed by financial markets:

- To channel available funds from lenders to borrowers.
- To transfer risk, both for individuals and corporations.

Actually there are many other important functions, which includes the ones listed below:

- To provide a payment mechanism (e.g., by bank drafts and credit cards). We will not consider this side of the financial system, but we have to bear in mind that this is one of the main historical reasons behind the creation of finance during the Renaissance in Italy, when the needs of traders facing travel risk had to be met.
- To provide financial services, including the creation and sale of securities like bonds by both public and private issuers, as well as offering advice to firms in matters of financial management.
- To create market liquidity, i.e., the possibility of buying and selling assets quickly *and* at a fair price, as well as to offer portfolio adjustment facilities. The actual complexity of the information technology infrastructure,

1.3 Market participants and their roles

needed to actually perform trading on markets and to take care of asset custody, should not be underestimated.

- To enable the separation between ownership and management in large corporations, which cannot always be effectively managed by family owners. Corporate growth would otherwise be impossible much beyond the size of a firm owned by the original founders. This is a controversial matter, as it may create problems related to bad incentives and agency issues. Moreover, the excess of financialization of the economy is under scrutiny, and with good reason.
- It has been claimed that the financial system also plays an information role, since the wide availability of financial data may be used to gather valuable knowledge. It may be argued that this is a bit debatable, in the light of speculation excess and some market anomalies studied by behavioral finance.

In concrete terms, all of these functions (and others) are carried out by an interconnected network of actors including

- Households and private investors
- Large corporations and smaller firms
- Governments and other public agencies, including local authorities like municipalities
- Financial intermediaries like banks, brokers, dealers, market makers, etc.
- Financial service providers like financial advisory firms, common funds, hedge funds, pension funds, insurance companies, etc.
- Regulatory and supervisory agencies, like the SEC (Security Exchange Commission) in the USA, the equivalent CONSOB[34] in Italy, the Basel Committee, central banks, etc.

All of these actors are connected by markets, which we may think as a platform on which transactions can be executed, either over-the-counter or on a computer network. We will discuss a bit of market structures later, but it is fundamental to immediately understand the two basic structures: **Primary markets** and **secondary markets**. When securities are created, they are first sold on primary markets. For instance, a corporation may float equity, possibly by an **initial public offering** (IPO), which needs some support from investment banks, under the scrutiny of regulatory bodies. By a similar token, a government may sell bonds to institutional investors using an auction mechanism. Households do not have direct access to primary markets, but they operate on secondary markets, where securities may be freely traded after they are issued.

---

[34]Commissione Nazionale per le Società e la Borsa.

Example 1.13 Selling vs. writing options

> A common source of confusion is the nature of the obligations related to selling an option, like a vanilla call or put. If we sell a call option, are we obligated to sell the underlying asset to the holder if the option is exercised? The source of confusion is the kind of market on which the option is sold. The *writer* is the one selling the option first on *primary* markets. Then the option, assuming it is an exchange-traded one, may change hands on secondary markets, but the obligation is only assumed by the original writer. If we buy and then sell an option, we are just selling the rights to a new holder, incurring a profit or a loss. To avoid any ambiguity, we shall always use the term "option writing" when we mean "selling on primary markets," collecting the option premium and assuming the obligations stated in the contract. When we talk about "selling" an option, it will always refer to secondary markets, as part of a trading strategy.

Some specific market players, like brokers and dealers, make sure that there is sufficient liquidity on markets. Before discussing some of these actors in more detail, let us underline that they play different, but not mutually exclusive, roles. For instance, governments may be net savers or net borrowers, just like households. However, we know well that our role can change over time, since we may borrow money at the beginning of our working life (e.g., under the form of a mortgage) and, hopefully, we become savers as our career progresses. Of particular interest are some key roles that may be played by investors, non-financial firms, and financial intermediaries: Hedgers, speculators, and arbitrageurs. These will be discussed later.

1.3.1 COMMERCIAL VS. INVESTMENT BANKS

Banks come in many forms, including retail banks mostly dealing with households, commercial banks offering services to small-medium firms, and large investment banks. Investment banks are often involved in mergers and acquisitions, and they also act as underwriters to help corporations in raising capital by floating equity or issuing bonds. Usually these securities are bought by investment banks on primary markets, and then sold on secondary markets. Furthermore, there are different legal entities, like banks floating their own equity and credit unions. Here, we just want to draw the line between **deposit-** and **non-deposit taking** banks.

A deposit-taking institution, like a retail or commercial bank, may also collect funds from households, who deposit money on accounts that may be more or less protected against bankruptcy. Bankruptcy may result from careless credit distribution decisions by the bank, from the difficulty to collect loans back due to economic stagnation, or, in extreme cases, from risky proprietary trading, i.e., trading that the bank carries out itself, rather than on behalf of its clients. The recent trend has been to reduce this kind of government-backed protection.

Large investment banks are non-deposit taking, and must raise capital by other means, like floating equity, issuing bonds, or borrowing money from other banks. As we shall see in Section 1.4.4, using debt rather than equity has the results of boosting ROE by a leverage mechanism. The leverage ratio measures the amount of debt used with respect to own equity. Roughly speaking, if ROA is 5%, a leverage ratio of 2 will double ROE. Unfortunately, the same happens in the case of loss, and when excessive debt is used, bankruptcy risk is considerably increased.[35]

Since risky proprietary trading activity may hurt clients of deposit-taking banks, a line between the two types of banks was drawn in the USA after the 1929 crash (Glass–Steagall Act), which is why investment banks could not take deposits. This line has been blurred in the last decades. Furthermore, the increasing interconnection among market players has increased systemic risk, i.e., the possibility that the collapse of a large institution affects many others by a domino effect.[36]

1.3.2 INVESTMENT FUNDS AND INSURANCE COMPANIES

Individual investors may feel that they lack the information required to make sound investment decisions. Furthermore, it may be difficult to properly diversify the risk exposure with a limited budget, as transaction costs preclude the possibility of many small investments in a broad set of securities. Hence, they may purchase shares of **mutual funds**, that are supposedly managed by skilled professionals who, in exchange for a fee, should provide good return opportunities to their clients. Shares are continuously created in the case of an open-end fund. Shares are destroyed when a client redeems her shares of a mutual fund. On the contrary, closed-end funds have a given number of shares that may be traded.

There are two basic kinds of fund manager. **Active** managers try to earn extra return by skill and by pursuing, for instance, stock-picking and market-timing strategies. The actual performance of active managers is the subject of a good amount of controversy. As an alternative, we may consider a **passive** manager, who will not try to do any better than the market as a whole, but will just provide a diversified portfolio tracking a broad index. We shall see that the passive view has some theoretical support by equilibrium models like the capital asset pricing model. Clearly, the fee required by a passive manager should be definitely small with respect to the cost of an active fund.

---

[35] Apparently, the leverage ratio of Lehman Brothers before their collapse was something like 20. LTCM, too, had reduced equity by forcing investors out before their near collapse.

[36] The collapse of Lehman Brothers affected hedge funds, among other things, as they acted as prime brokers for these funds.

The ultimate active fund is a **hedge fund**. Despite the misleading name,[37] hedge fund managers pursue possibly very risky and nonstandard investment strategies in order to earn extra return. When we buy a share of a mutual fund, we are client of the fund. On the contrary, when we buy a share of a hedge fund, we are *partners* of the hedge fund, which has a different legal nature. This is related to the high risk involved, and in fact only wealthy individuals are allowed to expose themselves to this level of risk. Furthermore, due to the complex trading strategies and the use of possibly illiquid assets, it may take a considerable amount of time to redeem shares of a hedge fund. These barriers are somewhat circumvented by funds of hedge funds.

The ultimate passive fund is an **exchange-traded fund**, **ETF** for short, which is a just passive fund tracking an index. In order to reduce costs, ETF shares are not distributed through a commercial network, unlike passive mutual funds, but they traded on exchanges, just like stock shares. This opens a thorny issue, since the ETF is supposed to track an index, but an uncontrolled demand–offer mechanism might cause its value to drift away from the fair one.[38] Market-makers guarantee the necessary liquidity and make sure that the ETF value is kept in line. Furthermore, a deviation from the fair value would create an arbitrage opportunity, which will be exploited by skilled investors, assuming liquid and well-functioning markets.

There are other non-deposit taking financial intermediaries that are engaged in fund management, namely, pension funds and insurance companies. These intermediaries face difficult asset–liability management problems, as they collect pension contributions and insurance premia that must be properly invested in assets, in order to generate cash flows and meet an *uncertain* stream of liabilities. A **non-life insurance** company may deal with, e.g., a stream of car accidents or other property damage. A **life insurance** company faces a similar task as a pension fund. Liabilities are uncertain because of longevity risk and, possibly, inflation-indexation. A **defined-benefit pension fund** must guarantee to retired workers an income that depends on the received wages, according to prespecified rules. The contribution level may be increased over time, depending on contingencies. Recently, because of decreasing interest rates and increasing life expectancy, there has been a shift toward **defined-contribution pension funds**, in which there is no defined income, and considerable risk is borne by the retired worker. It may seem that a defined-benefit fund is much preferable from the worker's viewpoint. However, we should also consider that with a company defined-benefit fund it is more difficult to transfer vested benefits, if a worker changes the employer. Furthermore, a firm with a well-funded pension fund might become the target of hostile takeovers.

---

[37]Risk hedging means *reducing* risk.

[38]This cannot happen with a mutual fund, whose **net asset value** (**NAV**) is evaluated and reported daily by the fund management.

1.3.3 DEALERS AND BROKERS

A fundamental requirement of financial markets is liquidity. Hence, there is a need for institutional players that are continuously available to buy and sell an asset. These market-makers are also referred to as **specialists**. This role may be played by **dealers** and **brokers**. It may be the case that the same institution is both a dealer and a broker, but the two functions are different. To understand the difference, think of a real estate agent. Her role is to connect two counterparties, but she does not really own an inventory of houses and apartments. This is the role of a broker. The broker has no inventory of assets, and as such she does not suffer from inventory risk. A commission on the trade is paid to the broker to compensate her. There are also primary brokers associated with hedge funds, which may need large trades. On the contrary, when we travel around the world and exchange currencies at an airport, we do business with a dealer. The dealer *does* keep an inventory of the assets she trades. Clearly, this inventory entails some risk. In fact, the dealer is compensated by enforcing a bid–ask spread:[39]

- The **bid price** is the price at which the dealer is willing to buy the asset from us.
- The **ask price** is the price at which the dealer is willing to sell the asset to us.

Needless to say, the ask price is larger than the bid price, and their difference is a measure of market liquidity. We have seen an example of bid–ask spread in Section 1.2.5 on foreign exchange, and we will see similar examples in the case of stock shares when we discuss market mechanisms. The same applies to interest rates, as the rates at which we may lend or borrow money, when dealing with a bank, are quite different.

Bid–ask spreads are a form of **market friction**. Other market frictions are represented by taxes and by transaction costs associated with trades. These fees may have a fixed and/or a variable component. In general, thanks to the use of information technology, transaction costs have been reduced over the years.[40] For the sake of simplicity, we will usually ignore such frictions, which may be a sensible approximation for large institutional investors.

1.3.4 HEDGERS, SPECULATORS, AND ARBITRAGEURS

Market participants are often engaged in risk transfer, which is the traditional purpose of insurance contracts. More recently, a huge market of derivative assets has been developed, connecting **hedgers** and **speculators**. Hedgers are exposed to risk factors, like interest rates and currency exchange rates, and would

---

[39] You may also hear the term bid–offer, but I personally prefer bid–ask, since the difference between bid and ask sounds much clearer to me than the difference between bid and offer.

[40] Some argue that the reduction of market frictions and the related increase of transaction frequency is far from being a blessing, as it may lead to market instability. High-frequency algorithmic trading strategies are often blamed for this.

like to reduce or eliminate that exposure. One possibility is to lock a rate for the future by forward contracts, for instance. Speculators, on the contrary, have a definite view about the direction that markets will take, and they are willing to take a bet on it. Thus, speculators may be willing to "buy volatility" from hedgers. We should realize that these two roles are not mutually exclusive. Indeed, there are multiple sources of risk that affect the value of a portfolio of assets, and a market participant might want to shape her portfolio in such a way that it is made less sensitive to risk factors on which she does not feel like betting, and more sensitive to other factors on whose direction she is more confident. Hence, she will increase the exposure to some risk factors, behaving as a speculator, and at the same time she will reduce the exposure to other risk factors, behaving as a hedger. As a concrete example, an investor may feel that she is good at picking stock shares that will perform better than the market as a whole, but she is unsure about the market direction. In the case of a market crash, being good at stock-picking and lose less than the market may only be a partial consolation. As we shall see, she will be interested in taking risks that are specific to some firms, while getting rid of systematic market risk. Hedgers and speculators need models to quantify uncertainty in risk factors and to understand how different sources of risk affect asset prices. On the one hand, we need tools to measure risk. On the other hand, we also need risk management approaches and decision models to find the best hedging strategy.[41]

Pricing models are also needed to check the consistency of the prices of assets that depend on common risk factors. For instance, derivatives written on the same underlying asset should be somehow related. If prices are inconsistent, trading strategies may be devised in order to take advantage of price misalignment. In technical terms, we talk of arbitrage opportunities, which are exploited by **arbitrageurs**. In liquid and well-functioning markets, it may be argued that arbitrage opportunities should not last long, as arbitrageurs will be quick in detecting and exploiting them, bringing prices back in line. We will investigate the mathematics of arbitrage in Section 2.3. There, we shall take a simplistic view of markets, ignoring market frictions, modeling errors, and liquidity issues. Nevertheless, we will be able to develop powerful pricing models based on the idea that there should be no arbitrage opportunity in market equilibrium. As usual, market reality is definitely more complex, and the actual arbitrage strategies may be not so sharp and may fail to work for an array of reasons. It may also be argued that arbitrageurs are sort of parasites taking advantage of what other market participants do, without really contributing to any growth in the real economy. Nevertheless, arbitrageurs play a vital role to the correct market functioning by keeping prices in line. One example that we have already hinted at is the need to ensure consistency between prices of an ETF share and the index that the fund is supposed to track.

---

[41] See Section 2.2.

1.4 Market structure and trading strategies

Quantitative finance relies on mathematical models that are, by necessity, an abstraction of market reality. Finding the right level of abstraction and detail simplification is definitely an art rather than a science, and we are engaged in a quest for those models that are inevitably wrong, but hopefully useful. While we may not be interested in an overly detailed view of market structures and the institutional mechanisms by which a trade is executed, we must be aware of some fundamental features that we outline here.

1.4.1 PRIMARY AND SECONDARY MARKETS

We have already hinted at the difference between primary and secondary markets. A primary market is where a security is first traded. The exact mechanism depends on the kind of security. For instance, an auction mechanism is used to introduce new government bonds on the market, but the auction is restricted to institutional investors. In the case of a stock, we should distinguish an **IPO**, i.e., the initial public offering of shares of a firm that is first quoted on the market, from a **seasoned offering**, where further equity is floated by an already quoted firm. An IPO may be a costly business, as several requirements are typically set by regulators and must be met by a firm floating equity on exchanges. A pool of investment banks is involved in the process, which may also include so-called "road shows" to present the offering to investors. Anyway, we must keep in mind that shares need not be traded on an exchange. Some firms are kept private and possibly owned by private equity funds. A significant part of equity can also be kept by the original owners to maintain control over management, and only the rest are floated and are outstanding on secondary markets.

1.4.2 OVER-THE-COUNTER VS. EXCHANGE-TRADED DERIVATIVES

Not all assets are traded on regulated exchanges, as some are traded OTC. For instance, forward agreements are negotiated directly between the two counterparts, unlike futures. Another example of OTC derivatives are exotic options with possibly quite complicated payoffs. The advantage of an OTC agreement is that it may be tailored to meet specific risk hedging requirements. The disadvantage is that the lack of a quoted price may put a firm or a public administration at disadvantage. Furthermore, nonstandardized assets are rather illiquid, which means that unwinding the position may be expensive, if not impossible.

1.4.3 AUCTION MECHANISMS AND THE LIMIT ORDER BOOK

As we have mentioned, auctions are used, for instance, when selling sovereign bonds on primary markets. Here, we consider secondary markets for stock shares and describe an auction mechanism based on the **limit order book**.

Table 1.3 A five-level limit order book for a liquid stock share.

Bid		Ask	
Quantity	Price	Price	Quantity
130	77.26	77.28	8881
137	77.25	77.31	273
5855	77.23	77.33	115
300	77.22	77.34	272
13,080	77.16	77.35	738

When a market participant issues an order, she may specify a limit price, i.e., the maximum (minimum) price at which she is willing to buy (sell) an asset. The limit order book is structured on two columns:

- On the left, we observe the buy orders, associated with limit prices sorted in decreasing order. The top level line reports the highest bid price, as well as the related quantity (possibly related to different orders).

- On the right, we observe the sell orders, associated with limit prices sorted in increasing order. The top level orders are associated with the smallest ask price.

A five-level limit order book is reported in Table 1.3. The two top quotes are called the **inside quotes**. When limit prices cross each other, a trade takes place. Otherwise, no trade is executed. Orders need not specify a limit price, as an investor may just issue an order to be executed at the best available price. It may happen that a large order is executed at different prices, when its size exceeds the quantity available in an inside quote. The spread between the inside quotes reflects liquidity. We may notice that the bid–ask spread in Table 1.3 is quite small. Table 1.4 tells a rather different story, as there is a much larger spread, especially in percentage terms, between the inside quotes. A large spread typically comes with less trades during a day and lower volumes.[42]

Quite often, price-contingent orders are used. There are two features: (1) the kind of order, which may be buy or sell, and (2) the activation condition, which is related to the price going above or below a threshold level. Therefore, we have four basic types of price-contingent orders.

- The **stop-loss** order is a selling order to be activated when the price goes below a limit. The rationale behind the order is clear: We hold an asset, and in case of a drop in price we want to cut losses and get rid of it.

---

[42]The data reported here are not quite recent but real. They refer to the Paris stock exchange in 2010, and the first share is a large and well-known French cosmetics producer, whereas the second one is a less traded producer of containers for overseas shipping.

1.4 Market structure and trading strategies

Table 1.4 A five-level limit order book for a rather illiquid stock share.

Bid		Ask	
Quantity	Price	Price	Quantity
108	21.91	22.20	100
55	21.90	22.30	206
110	21.85	22.35	232
260	21.88	22.40	100
54	21.77	22.50	100

- The **limit-sell** order is a sell order activated by a price going above the threshold. In this case, the idea is that we hold an asset, and we sell it when a target profit has been achieved.
- The **limit-buy** order is a buy order activated when the price goes below a limit. This means that the asset is cheap enough to be bought. This is related to a contrarian strategy, a strategy trying to buy undervalued stocks.
- The **stop-buy** order is a buy order activated when the price goes above the threshold. The rationale is that the price is high enough to signal an increasing trend. This is related to a momentum strategy, i.e., a strategy trying to chase increasing trends.

High-frequency analysts build models at the limit order level, considering both prices and volume, with the aim of developing algorithmic trading strategies. Other models at this microstructure level concern the optimal execution of a large trade in order to minimize market impact. We shall not consider this operational level in this book.

1.4.4 BUYING ON MARGIN AND LEVERAGE

Buying on margin is a leveraged strategy aimed at boosting returns using debt. In corporate finance, **leverage** refers to the ratio of debt over equity. Here we have a similar use, as leverage means buying an asset by only partially using our own capital, and borrowing the rest from a broker or a bank.

Imagine that we have a strong view about a specific stock share, a bullish one in particular. As we have already mentioned, we may use derivatives, rather than just going long the asset to take advantage of our view, but it may very well be the case that derivatives written on that specific stock share are not available. Hence, we may resort to leverage, more specifically, to **buying on margin**. We have already met the term *margin* when dealing with futures contracts and margin accounts. Here we refer to posting the asset itself as a collateral of our debt, plus some cash acting as a buffer and guaranteeing the broker that we will repay the debt even if the asset price drops.

The mechanism revolves around the concept of **margin ratio**. Margin requirements specify an initial margin ratio, as well as a **maintenance margin**. To grasp the idea, it is useful to refer to the asset–liability–equity triad. In this case:

- The asset is the amount of stock shares that we have purchased.
- The liability is the money we owe to the broker.
- Equity, as usual, is their difference.

In this case, the margin ratio is defined as the ratio of the value of equity to the value of assets. If the margin ratio falls below the maintenance margin, we get a margin call, which means that we have to post additional cash (or other collateral). Failure to do so will result in our position being liquidated by the broker. This is best illustrated by an example.

Example 1.14 Margin trading

Say that the current price of a stock share of Boom Corp is $100, and we buy 100 shares, for a total amount of $10,000. To finance the trade, we borrow $4,000 from the broker. The initial situation is as follows:

Assets	Liabilities
Stock $10,000	Loan from broker $4000
	Equity
	$6000

The initial margin ratio is

$$\frac{\text{Equity}}{\text{Assets}} = \frac{\$6000}{\$10,000} = 60\%,$$

and let us assume that the maintenance margin ratio is 30%. Note that, for the sake of simplicity, we are not considering the interest payment to the broker. If things turn sour and the stock price falls to $70 per share, the new balance sheet will be

Assets	Liabilities
Stock $7000	Loan from broker $4000
	Equity
	$3000

and the margin ratio now is just

$$\frac{\$3000}{\$7000} = 43\%.$$

1.4 Market structure and trading strategies

> A natural question is: How far can a stock price fall before getting a margin call? If we let P be price of the stock, the margin ratio is
>
> $$\frac{100P - \$4000}{100P}.$$
>
> The limit price is obtained by setting this ratio to 30% and solving for P, which yields $P_{\text{lim}} = \$57.14$.

The effect of leverage is to boost both profit and loss. To see this, imagine that our view in the previous example is correct and that the Boom Corp price rises by 30%. The relevant return is return on equity (ROE), rather than return on assets (ROA). ROA is the usual rate of return that we consider when dealing with portfolio management (30% in this case), but ROE is boosted by the fact that we need to invest only a fraction of the value of the assets. To be a bit more realistic, let us assume that the broker requires an interest rate of 3% (over the holding period of the trade). ROE is

$$\frac{\$10,000 \times 0.30 - \$4000 \times 0.03}{\$6000} = 48\%.$$

The more leverage we apply, the better, in the rosy scenario. For instance, if we increase initial leverage to 50%, ROE is

$$\frac{\$10,000 \times 0.30 - \$5000 \times 0.03}{\$5000} = 57\%,$$

to be compared with the 30% of a normal trade. In practice, with a 50% leverage we double return, which is eroded by the 3% interest. Clearly, there must be another side of the coin. Imagine that we are wrong and price plummets to $70. With a 50% leverage, ROE is

$$\frac{-\$10,000 \times 0.30 - \$5000 \times 0.03}{\$5000} = -63\%,$$

i.e., we double loss and on top of it we have to pay interest on debt.

The example we have considered is rather stylized. The understanding of margin trading arrangements is essential to interpret what happens in real life, since it is one of the factors of relevant events like the LTCM demise. The required cash can also be obtained by posting securities, typically through a repo agreement,[43] i.e., a repurchase agreement. This is a sort of collateralized loan, as relatively safe securities are sold to a counterparty, with the agreement to repurchase them later at a given price. It is easy to see that this boils down

---

[43] See Section 5.4.

to borrowing money for a given interest rate. Things may really go badly as feedback effects may arise for large trades on illiquid assets. Imagine that the asset involved in a margin trade loses value because of a market crash, and we start getting margin calls. This may be associated with a reduction in market liquidity, which implies that we have to sell illiquid securities in order to raise cash. Unfortunately, selling assets in an illiquid market may have a large impact, leading to a further reduction in asset values, so that we are caught in a feedback cycle, potentially leading to bankruptcy. The larger the leverage, the more difficult it is to get out of such a situation.

1.4.5 SHORT-SELLING

Short-selling, like buying on margin, is a strategy that can be used for speculative purposes. In this case, however, the bet is a bearish one, as short-selling profits from a drop in the asset value. The mechanics of a short sale is as follows:

- At time $t = 0$ we borrow the asset through a dealer/broker, then we sell it and deposit the proceeds plus required margin into an account.
- At time $t = T$ we close out the position by buying the stock and returning it to the party from which is was borrowed.

If the asset is a stock share that pays a dividend in the interval $(0, T)$, a corresponding cash amount must be paid as well. A similar consideration applies to bonds and coupons. Therefore, when considering profit from a short sale, there is a change in sign with respect to the usual case:

$$\text{Profit} = \text{initial price} - (\text{ending price} + \text{dividends}).$$

Of course, the trade will result in a loss if the asset price increases. Furthermore, borrowing the asset may be expensive, as we must compensate the broker and/or the asset holder, and possibly limited to a short time interval. Sometimes, a **short-squeeze** occurs, i.e., the short position is forced to close the trade at a very unfavorable time, just when the asset price is rocketing.

In margin trading, we borrow cash to buy an asset, whereas in short-selling we borrow the asset itself and raise cash, which must be kept into an account with the broker until the trade is closed. As usual, the lender of the asset protects herself by requiring the deposit of a margin, in addition to the proceeds of the short sale. In this case, the margin ratio is defined as equity divided by the value of the assets owed, which is a liability in this case. This definition of margin ratio differs from the one we used in the case of buying on margin. As a mnemonic help, the margin ratio is always defined by dividing equity by the side that is sensitive to the current value of the traded stock shares, i.e., the asset side when buying on margin and the liability side when short-selling.

1.4 Market structure and trading strategies

▣ Example 1.15 A short trade

We are strongly bearish about stock shares of DotBomb. Hence, we sell 1000 shares at the current price of $100. The proceeds, $100,000, are deposited in our margin account, together with some collateral. If the initial margin required is 50%, we must deposit a corresponding amount in cash or rather safe securities, e.g., T-bills. The initial situation is as follows:

Assets	Liabilities
Cash + T-bills $150,000	Short position in stock $100,000
	Equity
	$50,000

If we are right, and DotBomb falls to $70, we can close our position out and earn $30,000 (neglecting commissions and interest). If, however, DotBomb rises to $110, the new situation is

Assets	Liabilities
Cash + T-bills $150,000	Short position in stock $110,000
	Equity
	$40,000

and the margin ratio drops to

$$\frac{\$40,000}{\$110,000} = 36\%.$$

How much can the stock price increase, before we get a margin call? If the maintenance ratio is 30%, we must find a stock price P such that

$$\frac{\$150,000 - 1000P}{1000P} = 30\%.$$

By solving for P, we obtain $P_{\text{lim}} = \$115.38$.

This example, too, is somewhat stylized. We are not considering the cost of the trade, and the fact that it may be expensive to keep the short position open for a long time. Just like with buying on margin, a realistic assessment of the return of a trade should be based on ROE. In this case, the actual investment is the additional margin that has to be posted, which may be rewarded at a given interest rate. When short-selling an asset is difficult or expensive, a short position may also be created by taking a position in a derivative. For instance, we may sell futures contracts written on the asset, as we shall see in more detail in Chapter 12.

Short-selling is a highly controversial strategy, as it is believed by some to be a way to manipulate and depress the markets. A famous case in point is George Soros' bet against the GB pound in 1992, when the UK was forced to leave the European Exchange Rate Mechanism. When markets crash, short sales are often prohibited, as they are considered by some as a nasty way to generate a vicious feedback cycle.[44] A different view maintains that short-selling is essential to preserve liquidity, as well as to keep prices in check when markets roar. We should also consider that short-selling may play a role in hedging and is not necessarily a speculative strategy. As it happens with many matters in finance, the jury is out. We have to stress the fact that in later chapters, especially when discussing pricing by no-arbitrage, we will assume that unlimited short sales are possible. Clearly, this is a rather idealized view of real markets.

1.5 Market indexes

We are all familiar with stock market indexes like Dow Jones Industrial Average (DJIA), NASDAQ, Dax, and Nikkei, which are often mentioned on newspapers and TV news. Some of them have a quite long history (the DJIA has been computed since 1896), and they are also the underlying factor of several traded derivatives, like index futures and options. Indexes are tracked and replicated by index funds and ETFs.[45] Most widely known indexes are related to a geographic area, such as a national stock market, but some indexes, like MSCI (Morgan Stanley Capital International), refer to world markets. On the other hand, we may also use more specific indexes, related to a given industry sector, or even nonfinancial markets, as is the case of indexes for the real estate market.

Indeed, there is a wide variety of indexes, beyond the familiar ones for stock markets. For instance, the EURIBOR and LIBOR rates are actually indexes, as they are an average of a set of interbank offered rates. We may also use bond market indexes, which are a bit more problematic, since bonds, unlike stock shares, have a maturity. Thus, the pool of bonds in an index must be continuously updated. Furthermore, we shall see that the volatility of a bond price gets smaller and smaller as the maturity is approached. An increasingly important index, VIX, tracks stock market volatility. Intuition would suggest that such an index should be built by estimating the standard deviation of stock market return by familiar descriptive statistics. Unfortunately, the usefulness of such a backward-looking index would be questionable. A forward-looking index, may rely on the implied volatility of a set of traded options.[46]

---

[44] Short-selling strategies have always been controversial, as illustrated by significant historical cases reported in [12]. In the USA, in the midst of political discussions about possible prohibition of short-selling, the practice was even deemed "unAmerican."

[45] ETFs may also be short or leveraged. A short ETF allows to take a short position in the index, profiting from a market drop. A leveraged ETF multiplies profits and losses by a given factor.

[46] We will discuss implied volatility in Section 13.6. Here, it suffices to say that it is a volatility such that option prices predicted by a mathematical model match the actual market prices.

1.5 Market indexes

As we can imagine, the definition of a suitable index is far from trivial and involves managing extraordinary events as well, such as stock splits, mergers/acquisitions, delistings, etc. In order to get a feeling for the involved issues, it is interesting to compare two well-known indexes:

- The Dow Jones Industrial Average, which includes 30 blue chips, and is a **price-based** index.
- The Standard & Poor's S&P500 index, which is a broadly based index involving 500 stock shares, and is a **market-value-weighted** index.

Generally speaking, we may consider a set of m stock share prices S_k, $k = 1, \ldots, m$, and define an index

$$I = \frac{1}{D} \sum_{k=1}^{m} w_k S_k,$$

for a given set of **weights** w_k, and a **divisor** D. Usually, when defining an average, we assume that weights add up to 1, which in this case would be obtained by choosing D as the sum of weights. Actually, the divisor is initially chosen in such a way that the resulting index assumes a "nice" value, say, 100 or 1000. More importantly, the divisor is changed when the index composition is changed to reflect new market conditions,[47] or when events such as spinoffs or mergers/acquisitions take place. The defining features of the aforementioned indexes are:

- Weights w_k are all set to 1 for the DJIA, i.e., the index essentially tracks a portfolio consisting of one stock share for each name in the index.
- Weights in the S&P500 index correspond to the number of outstanding stock shares (free-float only); hence, the portfolio reflects the actual market capitalization of each firm.

Example 1.16 Price-based vs. market-value-weighted indexes

Consider the following scenario:

- Stock share A has an initial price $S_A(0) = \$25$, at time $t = 0$, which is increased by 20% to $S_A(T) = \$30$ at time $t = T$. The total market capitalization is $500 million (hence, 20 million shares are outstanding).

---

[47] An interesting market anomaly is the plunge in the price of stock shares that are dropped from an index. Rationally, this should not imply anything in terms of intrinsic firm value, but the consequent reduction in trading activity on that stock share may have a significant effect.

- Stock share B has an initial price $S_B(0) = \$100$, which drops by 10% to $S_B(T) = \$90$ at time $t = T$. The total market capitalization is $100 million (hence, one million shares are outstanding).

A price-based index would initially be

$$\frac{25 + 100}{2} = 62.5,$$

where we assume a divisor $D = 2$, which is really inconsequential when considering percentage changes in the index. At the end of the time horizon, the new index value would be

$$\frac{30 + 90}{2} = 60,$$

with a drop of 4%. Note that the price drop of the more expensive stock share dominates here, but this does not reflect the true market weights. Let us consider a market-value-weighted index, with initial value

$$\frac{25 \times 20 \cdot 10^6 + 100 \times 1 \cdot 10^6}{10^6} = 600,$$

where we set $D = 10^6$. The new index value would be

$$\frac{30 \times 20 \cdot 10^6 + 90 \times 1 \cdot 10^6}{10^6} = 690,$$

with an increase of 15%.

We notice a relevant difference in the behavior of the two indexes. The difference may also be reflected in the way the index is adjusted when something new happens. Consider, for instance, a 2-for-1 stock split. Clearly, a market-value-weighted index would not be affected, but an adjustment would be needed for the price-based index in order to preserve the continuity in its value.

Example 1.17 Index adjustments

Let us consider how to manage an index for a stock market on which two stocks are traded. Company A has 50 shares outstanding, with current price $2, and company B has 10 shares outstanding, with current price $10. The current value of a price-based index is 6, whereas the value of a market-value-weighted index is 100. Let us consider the following scenario: The price of Company A's stock increases to

1.5 Market indexes

$4 per share, and Company B's stock splits 2 for 1 and is priced at $5. How will the values of the price-based and market-value-weighted indexes change?

To begin with, we have to find the divisors. The current divisor for the price-based index is clearly $D = 2$, since

$$\frac{2 + 10}{2} = 6.$$

Then, it is important to notice that, actually, the second stock price did not change. The drop from $10 to $5 merely reflects the split. After the change in price of the first share, without considering the stock split, the new index would be

$$\frac{4 + 10}{2} = 7.$$

The new divisor is changed in order to reflect the split without introducing a discontinuity in the index:

$$\frac{4 + 5}{D'} = 7 \quad \Rightarrow \quad D' = \frac{9}{7}.$$

The divisor for the market-value-weighted index is found as follows:

$$\frac{2 \times 50 + 10 \times 10}{D} = 100 \quad \Rightarrow \quad D = 2.$$

However, the stock split is inconsequential for this index and does not require any adjustment in the divisor. Hence, the new index value is

$$\frac{50 \times 4 + 20 \times 5}{2} = 150.$$

An important observation is that indexes are not adjusted when dividends are paid. This is relevant, as a stock share price experiences a corresponding drop when a dividend is paid, and this will also affect the index, as well as derivatives written on the index. In option pricing models, the collective dividend behavior of the stock shares in the index may be approximated by a continuous-time dividend yield. Also note that the index is nondimensional and should be regarded as a number, rather than as a price. When defining derivative payoffs, the index must be multiplied by a given number in order to define a monetary payoff. For instance, the S&P500 index is multiplied by 250 to be converted into a monetary value.

Problems

1.1 Consider assets A_1 and A_2, whose holding period returns R_1 and R_2, in five possible scenarios, are given in the following table:

Scenario	Probability	R_1	R_2
ω_1	0.2	0.03	0.09
ω_2	0.2	0.17	0.16
ω_3	0.3	0.28	0.10
ω_4	0.2	0.05	0.02
ω_5	0.1	−0.04	0.16

Note that the probabilities are not equal, and that returns are not given as a percentage (if you prefer, you might also write, e.g, $R_1(\omega_1) = 3\%$). Find the expected value and the standard deviation of the returns of the two assets, as well as their (Pearson) coefficient of correlation.

1.2 We are pursuing a short-selling strategy, where we have shorted 300 shares of XYZ, at price €40. The initial margin required by the broker is 50% of the overall value, and the maintenance margin is 25%. What is the limit price of the stock before we are slapped with a margin call?

1.3 Consider a European-style call option maturing in five months, with strike price $K = $ €40, written on a stock share with current price $S(0) = $ €35. We (very unrealistically) assume that the uncertainty about the stock price at maturity $T = 5/12$ may be represented by eight equally likely scenarios: $S(T) \in \{20, 25, 30, 35, 40, 45, 50, 55\}$. Find the expected value of the option payoff.

1.4 Let us consider a market index for a tiny market, on which just 3 stocks are traded. In this market, 50,000 shares of the first firm are outstanding, 100,000 of the second one, and 80,000 of the third one. The index is a weighted-average of the three stock prices, reflecting the capitalization of the three firms. The current stock prices are €50, €30, and €45, respectively. To make the index easy to read and nondimensional, it is divided by a divisor (established once for all and kept constant in time; we rule out exceptional events like those described in Example 1.17); assume that with that choice of divisor, the index now is 118. We also assume that the stock shares do not pay any dividend.

The following table lists the stock prices (in EUR) for a three-day scenario (a single sample path):

Day	1	2	3
Price of stock 1	52	48	45
Price of stock 2	28	25	30
Price of stock 3	43	40	39

Find the corresponding scenario for the index value.

Further reading

- Most general books on financial asset management, like [1] and [5], have one or more sections on financial institutions and market mechanisms. More detail is provided in specific texts like [8]; see also [14].
- Many useful pieces of information on financial institutions are also given in [11], with a nice twist toward risk management.
- Market microstructure is dealt with in [7].
- The first chapters of [13] specifically cover the market structure for bond and debt markets. Bond markets, including bonds with embedded options, are also treated in [6].
- An adequate discussion of FOREX markets is provided by [15].
- A full understanding of how financial markets and institutions work cannot be achieved without some knowledge of real stories. The case of Long Term Capital Management is described, among others, in [9]. Another very useful reading is [3].
- While this book is concerned with financial markets, it is also essential to acquire some background knowledge on corporate finance, which is provided, among many others, by [10].

Bibliography

1 Z. Bodie, A. Kane, and A. Marcus. *Investments* (9th ed.). McGraw-Hill, New York, 2010.

2 Z. Bodie and R. Merton. *Finance*. Prentice Hall, Upper Saddle River, NJ, 1999.

3 R. Bookstaber. *A Demon of Our Own Design: Markets, Hedge Funds, and the Perils of Financial Innovation*. Wiley, New York, 2008.

4 N.P. Calamos. *Convertible Arbitrage: Insights and Techniques for Successful Hedging*. Wiley, Hoboken, NJ, 2003.

5 K. Cuthbertson and D. Nitzsche. *Investments*. Wiley, Chichester, 2008.

6 F.J. Fabozzi. *Bond Markets: Analysis and Strategies* (9th ed.). Pearson, Upper Saddle River, NJ, 2015.

7 L. Harris. *Trading and Exchanges: Market Microstructure for Practitioners*. Oxford University Press, New York, 2003.

8 P. Howells and K. Bain. *Financial Markets and Institutions* (5th ed.). Pearson Education, Harlow, 2007.

9 R. Lowenstein. *When Genius Failed: The Rise and Fall of Long-Term Capital Management*. Random House Trade Paperbacks, New York, 2001.

10 S.A. Ross, R.W. Westerfield, J.F. Jaffe, and B.D. Jordan. *Corporate Finance: Core Principles and Applications* (4th ed.). McGraw-Hill, New York, 2014.

11 A. Saunders and M.M. Cornett. *Financial Markets and Institutions* (6th ed.). McGraw-Hill, New York, 2014.

12 R. Sloan. *Don't Blame the Shorts: Why Short Sellers Are Always Blamed for Market Crashes and How History Is Repeating Itself.* McGraw-Hill, New York, 2009.

13 S.M. Sundaresan. *Fixed Income Markets and Their Derivatives* (2nd ed.). South Western College Publishing, Cincinnati, OH, 2002.

14 S.R. Veale, editor. *Stocks, Bonds, Options, Futures: Investments and their Markets* (2nd ed.). Prentice Hall, Paramus, NJ, 2001.

15 J. Walmsey. *The Foreign Exchange and Money Markets Guide* (2nd ed.). Wiley, New York, 2000.

Chapter Two

Basic Problems in Quantitative Finance

In the following chapters, we will discuss at length how quantitative finance methods may be used to solve practically relevant problems. Before embarking on a detailed investigation, it is important to get a broad overview of the most relevant themes and their mutual relationships. Bearing this in mind, in this chapter we introduce simplified versions of the following problems:

Portfolio optimization. Given a set of risky financial assets, whose return is uncertain, we must decide the fraction of wealth allocated to each of them, in order to find a satisfactory risk–reward tradeoff, while complying with a set of constraints.

Risk measurement and management. *Measuring* risk is not only essential for financial firms when selecting a portfolio of assets. Nonfinancial firms are subject to financial risk as well, in terms of exposure to adverse movements in interest or currency exchange rates. The next logical step is *managing* risk, which typically involves hedging some risk factors away by suitable policies, and possibly taking a position on those risk factors on which we believe we can place a reasonable bet.

Asset pricing. Finding the fair price of a financial asset is useful when we want to determine whether it is under- or overpriced, in order to drive portfolio decisions and, possibly, to detect arbitrage opportunities (a concept that we will formalize in this chapter). Another typical application is dealing with over-the-counter (OTC) derivatives, for which quoted prices on regulated exchanges are not available; prices of OTC derivatives are quoted on request by investment banks, and we might wonder whether the asked price is fair or not. Finally, we need an asset pricing model for risk management, too, since we need to assess the relationships between a set of underlying risk factors and the price of assets ranging from fairly simple bonds to quite complex derivatives.

To introduce the first of these themes, in Section 2.1 we describe the classical Markowitz approach to static portfolio optimization in terms of mean–variance efficiency. We also outline possible generalizations, in terms of alternative ways of tackling the risk–reward tradeoff, as well as dynamic models. In Section 2.2, we stress the role of models quantifying the sensitivity of portfolio value or return with respect to underlying risk factors. These models play a key role in risk management policies by immunization, among other things. From a theoretical viewpoint, Sections 2.3 and 2.4 are perhaps the key portion of this chapter, as we introduce a powerful pricing approach based on absence of arbitrage opportunities. First, by a set of simple but quite relevant examples discussed in Section 2.3, we will be immediately able to appreciate both its appeal and its limitations. Then, we delve a bit into the mathematics of arbitrage in Section 2.4.

In this chapter we also include two supplements. Section S2.1 lays down the fundamental concepts of multiobjective optimization, which is the foundation of mean–risk portfolio optimization models. Then, in Section S2.2 we summarize duality for linear programming (LP), which is used in Section 2.4. LP duality[1] is essential in many computational approaches, but here we use it to relate the feasibility of an LP model to the boundedness of another one. The existence of arbitrage opportunities would lead to an unbounded profit in a certain LP, but the fact that its dual is feasible precludes the existence of such a money-making machine.

2.1 Portfolio optimization

In Chapter 1, we have introduced the essential families of securities (equity, fixed income, and derivatives) that are traded on financial markets. A natural problem, then, is how to allocate wealth among those different assets, shaping our portfolio. This gives rise to a wide array of decision problems, differing with respect to a few essential features:

The role of time. We may tackle static or dynamic problems. In a static problem, we only consider two time instants, now and the end of the holding horizon, whereas in a dynamic problem we have to make a sequence of decisions, possibly taking the unfolding of random events into account. Furthermore, we may formulate models in discrete or continuous time. This has an obvious impact on the complexity of the problem and the way the inherent uncertainty is represented.

The hierarchical level. Problems may have a more strategic or operational flavor. At a strategic level, we may want to allocate wealth to broad families of assets, like domestic/foreign stock shares vs. domestic/foreign bonds. At a lower hierarchical level, say, a tactical one, individual stock picking may be considered. Going down the decision hierarchy, the time horizon is normally shorter and shorter. At the operational level, we might even

---

[1] We deal extensively with duality in Sections 16.1.4 and 16.3.2.

2.1 Portfolio optimization

consider the optimal trade execution in real time, which involves market micro-structure issues, in order to minimize the price impact of a trade. Algorithmic trading strategies are also concerned with real-time execution issues.

Asset only vs. asset–liability management problems. A private investor may consider a portfolio problem in terms of pure asset allocation, but an insurance company or a bank must take liabilities into consideration as well. Therefore, we need a joint characterization of the risk factors affecting both assets and liabilities. Liabilities might also be considered at the individual level, too, as we may wish to plan our personal investments accounting for future consumption decisions that we have already planned. This should not be confused with consumption–saving problems, where future consumption plans are an output of the decision model. This last category of models is more commonly addressed in the economic literature.

Complexity of the market model. In the simplest static model, we may just represent the uncertainty in asset returns over the holding period as a multivariate distribution. This is not as trivial as we might think, as some features, like time-varying correlations, fat tails, and asymmetric (skewed) distributions may be difficult to model. In a dynamic model, we might also wish to deal with path dependencies, jumps, and stochastic volatilities. Additional issues adding realism to a model are related to transaction costs, bid–ask spreads, as well as the market impact of a trade on an illiquid market.

Representation of uncertainty. Classical uncertainty models rely on the traditional tools of probability and statistics. The more complicated the model, the more statistical estimation issues arise. However, statistical estimation is based on past data, and the past is not always the major concern in finance. More sophisticated models may try to incorporate subjective views about the future, as well as uncertainty about the uncertainty itself, commonly called model ambiguity. Essentially, we have a single and well-defined probability distribution in a classical probabilistic decision problem, but we have several ones in a problem with distributional uncertainty, and we have none in a robust optimization problem, which relies on a different, nonstochastic framework to represent uncertainty.[2]

Risk–reward tradeoff. In finance, the quest for increased return has to be tempered with due attention to the corresponding increase in risk. On the one hand, we might wish to model the degree of risk aversion at an individual level. In the traditional economic literature, this is addressed by the introduction of utility functions.[3] On the other hand, if we think of professionally managed portfolios for a set of clients, objective risk measures may be more relevant. Thus, we have to select first a way to measure risk, which

---
[2] See Section 15.9.
[3] See Section 7.3.

is nontrivial per se, especially in dynamic models, and then a way to trade risk against expected reward. This tradeoff is particularly difficult when long-term wealth management objectives are traded off against short-term risk.

Depending on how we address all of these issues, we may end up with optimization models that can be easily solved by a commercial spreadsheet on a laptop, or with models that are essentially intractable. We start here by considering a very basic, but quite relevant, static model.

2.1.1 STATIC PORTFOLIO OPTIMIZATION: MEAN–VARIANCE EFFICIENCY

In this book we will consider the use of optimization models for portfolio decisions, as well as their pitfalls and limitations. When building an optimization model we have to specify three ingredients:[4]

1. **Decision variables**, expressed as real numbers collected into a vector $\mathbf{x} \in \mathbb{R}^n$
2. **Constraints**, like equalities, inequalities, and other restrictions defining the feasible set $S \subseteq \mathbb{R}^n$, to which the vector of decision variables should belong
3. **Objective function** $f(\cdot)$, to be minimized or maximized

By assembling these building blocks, we obtain an optimization problem in a rather abstract form,

$$\min (\text{or max}) \quad f(\mathbf{x})$$
$$\text{s.t.} \quad \mathbf{x} \in S,$$

where s.t. stands for "subject to." The process of optimization model building is usually nonlinear, in the sense that there are successive iterations in which these elements are added and refined, in no specific order, especially when the model builder works with a client who is trying to rationalize her problem.

There are different sets of decision variables that may be used in portfolio optimization. In simple models, we may consider the asset weights, i.e., the fraction of wealth that we allocate to each asset. However, in an asset–liability model we must consider the need to generate cash flows matching liabilities, which leads to a different set of decision variables, such as how many units of each asset we hold, i.e., the actual number of stock shares or bonds. We may also consider the amount of money allocated to each asset. Furthermore, when modeling transaction costs within a dynamic setting, we may also need decision variables representing the amount of assets that we buy or sell at each trading

---

[4] We deal with optimization model building for finance in Chapter 15. For a general introduction to deterministic and stochastic optimization models, you may also see [2, Chapters 12–13].

2.1 Portfolio optimization

date. Naturally, in dynamic models, decision variables are usually indexed by time.

For now, let us stick to a simple model based on a static set of portfolio weights, which we denote by w_i, $i = 1, \ldots, n$, where n is the number of assets that we are considering for inclusion in the portfolio. Portfolio weights may be collected into vector $\mathbf{w} \in \mathbb{R}^n$. With this choice of decision variables, a natural constraint is

$$\sum_{i=1}^{n} w_i \equiv \mathbf{1}^\mathsf{T}\mathbf{w} = 1,$$

where $\mathbf{1} \in \mathbb{R}^n$ is a column vector with all elements set to 1, and we use T to denote vector and matrix transposition.[5] This constraint looks pretty innocent, as it states that weights add up to 1, but it actually defines a *fully invested* portfolio. If we do not require full investment, we may rewrite the constraint as $\mathbf{1}^\mathsf{T}\mathbf{w} \leq 1$. If we rule out short-selling, we also require a non-negativity condition,

$$w_i \geq 0, \quad i = 1, \ldots, n.$$

We may also represent this condition in the compact form $\mathbf{w} \geq \mathbf{0}$, where $\mathbf{0}$ is a vector collecting n zeros, and the inequality is interpreted componentwise as usual. These two constraints imply $w_i \leq 1$. We relax the condition if we allow short-selling.[6] If we include a risk-free asset, it is customary to denote its weight as w_0. If we allow borrowing money to buy stock shares, we may relax the non-negativity constraint on this weight; as a result, in principle we might allow asset weights to be larger than 1. As the reader can imagine, this corresponds to very risky portfolios.[7] In practice, additional constraints may be enforced in order to limit exposure to individual assets or sets of assets, defined on the basis of geography or industrial sectors. These simple constraints may be expressed as linear inequalities. For instance, let us consider a small universe of $n = 50$ assets and imagine that the subset

$$\mathcal{I} = \{4, 8, 15, 16, 23, 42\}$$

corresponds to a specific industry (e.g., consumer electronics) or a geographic region. We might enforce a lower or an upper bound on the sector exposure, say 5% and 25%, respectively, by requiring

$$0.05 \leq \sum_{i \in \mathcal{I}} w_i \leq 0.25,$$

---

[5] In this book, we will always assume that vectors are *column* vectors. A row vector will be denoted as the transposed vector $\mathbf{x}^\mathsf{T}$.

[6] See Section 1.4.4 for trading strategies based on short-selling and buying on margin.

[7] The excessive use of leverage by investment banks is well known and led, e.g., to the collapse of Lehman Brothers. However, individual investors may be prone to the same mistakes. For instance, in August 2015, Chinese stock markets faced considerable downturn, which hit Chinese small investors very hard, since many borrowed money to invest in a euphoric market that had experienced an astonishing growth over the recent past. The consequent reduction in the capacity of consumption of many Chinese small investors was deemed potentially responsible for severe economic consequences for everyone outside China as well.

which is just a pair of *linear* inequalities.

Defining the objective function may be quite tricky in financial problems. On the one hand, we would like to achieve a large return. On the other hand, we would like to keep risk under control. Whatever our aim is, we cannot do anything without a model of uncertainty. In our case, we need to represent uncertainty in the return of each asset. Let us denote the random return of asset i, over the holding period, by R_i.

> **A note on notation.** In most books on probability theory, random variables are denoted by uppercase letters, like X, whereas a lowercase letter x refers to a realization of the random variable. Treatments emphasizing measure-theoretic approaches to probability use $X(\omega)$ to insist on the fact that a random variable is actually a function mapping outcomes ω in a sample space Ω to numerical values. In the economic literature, an alternative notation is sometimes adopted, where one uses a tilde to denote the random variable: $\tilde{\epsilon}$ is the random variable, and ϵ is its realization. This is certainly convenient when using Greek letters. In this book, we will use letters like R, r, $\tilde{R}$, and $\tilde{r}$ quite liberally, choosing notation for the sake of convenience. The reason is that we shall refer to concepts like holding period returns, annualized returns, excess returns, nominal interest rates, and real interest rates, in terms of both random variables and realizations. Since it is easy to run out of suitable letters, we will not stick to a single notation throughout the book. Nevertheless, the context will make what we mean quite clear, and no ambiguity shall arise.

We assume, for the moment, that we are able to build a suitable characterization of the joint distribution of the returns of all assets. Then, the portfolio return, denoted by R_p, can be easily[8] related to our decision variables w_i,

$$R_p = \sum_{i=1}^{n} w_i R_i.$$

The larger the return, the better. However, maximizing the return makes no sense, as R_p is a random variable. When we choose portfolio weights, we shape the probability distribution of the portfolio return, but we do not define its exact value. Then, to make a decision, we must define a way to rank the probability distributions. An obviously relevant quantity is the expected value of portfolio return. Let us denote the expected return of individual assets as[9]

$$\mu_i \doteq \mathrm{E}[R_i], \quad i = 1, \ldots, n.$$

---

[8] See the discussion in Example 1.2 for details.
[9] We use $\doteq$ rather than $=$ when defining something, and $\equiv$ to refer to an identity.

2.1 Portfolio optimization

The expected value of a linear combination of random variables is just the corresponding linear combination of expected values. Hence, we find

$$\mu_p \doteq \mathrm{E}[R_p] = \sum_{i=1}^{n} w_i \mu_i = \boldsymbol{\mu}^\mathsf{T} \mathbf{w},$$

where we collect expected returns into vector $\boldsymbol{\mu} \in \mathbb{R}^n$.

It seems that a sensible objective would be to maximize the expected portfolio return, which leads to the following model:

$$\begin{aligned} \max \quad & \boldsymbol{\mu}^\mathsf{T} \mathbf{w} \\ \text{s.t.} \quad & \mathbf{1}^\mathsf{T} \mathbf{w} = 1 \\ & \mathbf{w} \geq \mathbf{0}. \end{aligned} \qquad (2.1)$$

This is actually a linear programming (LP) problem, and a powerful and reliable technology is available to solve even large-scale LPs, based on the simplex algorithm or interior point methods. However, a little thought reveals that the solution of the above problem is quite trivial: Allocate the whole wealth to the asset with the largest expected return. Since common wisdom suggests that portfolios should be suitably diversified, it is clear that there is something missing: The degree of risk aversion that is likely to impact any investment choice. As we shall discuss in Chapter 7, when we only consider the expected value of a random outcome, we are said to be risk-neutral decision makers. However, most investors are not risk-neutral, but more or less risk-averse. One way to take risk aversion into account is to introduce a risk measure, i.e., a function mapping random outcomes into a single number capturing risk, which is traded off against expected return. This will lead us to the formulation of mean–risk models.

As we learn in basic statistics, one way to characterize the dispersion of a random variable is by its standard deviation or, equivalently, its variance. Recalling again basic properties of linear combinations of random variables, the variance of portfolio return as a function of portfolio weights is

$$\sigma_p^2 \doteq \mathrm{Var}(R_p) = \sum_{i=1}^{n} \sum_{j=1}^{n} w_i \sigma_{ij} w_j = \mathbf{w}^\mathsf{T} \boldsymbol{\Sigma} \mathbf{w},$$

where $\sigma_{ij} \doteq \mathrm{Cov}(R_i, R_j)$ is the covariance between returns of assets i and j, and covariances are collected into the square covariance matrix $\boldsymbol{\Sigma} \in \mathbb{R}^{n \times n}$. Note that the diagonal of this matrix collects the return variances, $\mathrm{Var}(R_i) \doteq \sigma_i^2 \equiv \sigma_{ii}$. There is a good reason to consider standard deviation as the risk measure, rather than variance: Standard deviation is measured in the same units as return. However, variance may be mathematically more convenient, as it leads to quadratic programming (QP) problems, which are easy to solve, just like LPs.[10] Actually, the kind of optimization problem we end up with depends

---
[10] See Section 15.1 for a classification of optimization models.

on how we represent the tradeoff between risk and reward. We discuss details in Supplement S2.1, where we outline basic approaches to multiobjective optimization, and there are essentially three choices:

1. Form a linear combination of the two objectives, and solve the problem

$$\begin{aligned} \max \quad & \boldsymbol{\mu}^T \mathbf{w} - \lambda \cdot \mathbf{w}^T \boldsymbol{\Sigma} \mathbf{w} \\ \text{s.t.} \quad & \mathbf{1}^T \mathbf{w} = 1 \\ & \mathbf{w} \geq \mathbf{0}, \end{aligned}$$

where λ is a coefficient related to risk aversion. The maximization of this risk-adjusted expected return leads to a quadratic programming problem, if all constraints are linear.

2. Maximize expected return, subject to an upper bound on risk. The risk budget may be expressed in terms of variance,

$$\mathbf{w}^T \boldsymbol{\Sigma} \mathbf{w} \leq \beta,$$

or standard deviation,

$$\sqrt{\mathbf{w}^T \boldsymbol{\Sigma} \mathbf{w}} \leq \gamma.$$

The first choice leads to a quadratically constrained quadratic program (QCQP; to be precise, the resulting model is a subcase of a QCQP, as the objective function is linear), while the second one leads to a second-order cone programming (SOCP) problem. The unfamiliar reader should not worry about these definitions,[11] but we remark that QPs are easier to solve than QCQPs, and that SOCPs have become efficiently solvable only after recent algorithmic breakthroughs.

3. The last possibility is to minimize risk subject to a lower bound $\mu_{\min}$ on expected return:

$$\begin{aligned} \min \quad & \mathbf{w}^T \boldsymbol{\Sigma} \mathbf{w} \\ \text{s.t.} \quad & \mathbf{1}^T \mathbf{w} = 1 \\ & \boldsymbol{\mu}^T \mathbf{w} \geq \mu_{\min} \\ & \mathbf{w} \geq \mathbf{0}. \end{aligned}$$

This last possibility also leads to a QP.

Whatever choice we make, a clear difficulty is choosing a sensible value for the involved parameters, λ, β, γ, or $\mu_{\min}$, in such a way to find a good tradeoff. As we shall see, the most common choice is the last one: By changing the lower bound on expected return, we generate a set of portfolios, which is called the mean–variance efficient frontier.[12] From a computational viewpoint, this

---

[11] More information of these families of convex optimization models is given in Chapter 15.
[12] See Chapter 8.

2.1 Portfolio optimization

approach is quite convenient. Furthermore, we do not need to specify the whole joint distribution of returns, as we just need first- and second-order moments. This simplicity should not mask the limitations and the difficulties of such a model:

- Variance and standard deviation are symmetric risk measures, since they penalize both under- and above-average returns in the same way, even though the former may correspond to a huge loss and the latter to a welcome extra-profit. Hence, they may not be quite adequate to deal with possibly asymmetric distributions of returns.
- Estimating the covariance matrix Σ is no easy task.
- What about *forecasting* expected returns μ for the future, rather than collecting past averages that might not be relevant anymore?

We will address all of these issues later on.

Despite all of its pitfalls, the mean–variance model plays a pivotal role in financial theory, leading to a body of knowledge known as Modern Portfolio Theory, and it is associated with the name of its inventor, Harry Markowitz.[13] From a theoretical viewpoint, mean–variance efficiency is also related to an important, albeit controversial, equilibrium model, the capital asset pricing model (CAPM), which we shall discuss in Chapter 10.

2.1.2 DYNAMIC DECISION-MAKING UNDER UNCERTAINTY: A STYLIZED CONSUMPTION–SAVING MODEL

The Markowitz mean–variance portfolio model is static, since we have to make a single decision at time $t = 0$, and then we just cross our fingers and wait. In real life, a static model is repeatedly solved, possibly updating the relevant parameters when new information is acquired, but this is not made explicit in the model. On the contrary, in a dynamic model we explicitly account for the dynamics of uncertain factors, as well as for the possibility of adapting decisions along the way. When uncertain factors are realized, we gather additional information and may revise the plan. If this possibility is explicitly accounted for in the model, we end up with a challenging multistage decision problem under uncertainty. We shall discuss approaches to deal with such problems in Chapter 15. Here, we introduce a simple example in order to get acquainted with the related issues and to discuss important concepts for discrete-time dynamic modeling.

As we shall see in Chapter 11, we may build dynamic models based on a continuous- or a discrete-time framework. Time discretization may be a computational necessity, but it may also be a choice dictated by how decisions are

---

[13] For his work on portfolio theory, Harry Markowitz was awarded the Nobel Prize in Economics in 1990. What is less known is that he was also involved in the development of SIMSCRIPT, an early programming language for discrete-event simulation, and was the recipient of the John von Neumann Theory Prize for his work in operations research.

made in the actual problem. Here, we consider a time horizon $[0, T]$, where we make decisions at the beginning of time intervals (or time *buckets*) of length δt, like months, years, or days. With a slight abuse of notation, we deal with time instants indexed by $t = 0, 1, \ldots, T$, where the time horizon T is assumed deterministic.[14] The building elements of a dynamic decision model under uncertainty are:

- A set of **control/decision variables** (e.g., the amounts of each asset that we buy or sell), denoted by $\mathbf{x}_t$, $t = 0, \ldots, T - 1$. Note that we do not decide anything at the last time instant, $t = T$, but we just observe the final outcome.

- A set of **state variables** (e.g., the current level of wealth or the holdings of a set of assets after portfolio rebalancing or before). We denote state variables by $\mathbf{S}_t$, $t = 0, \ldots, T$; $\mathbf{S}_T$ is the terminal state. We use a capital $\mathbf{S}$ to emphasize the random nature of states.[15] The initial state $\mathbf{s}_0$ is given, but the state evolution is random and is modeled as a stochastic process. Some states may be affected by our decisions, whereas the evolution is purely exogenous for other states.

- A set of exogenous risk factors, playing the role of **disturbances** and affecting the transitions among states. We denote the vector of random variables corresponding to risk factors by $\boldsymbol{\xi}_t(\omega)$, $t = 1, \ldots, T$. Note that here time indexing starts from $t = 1$, as the first realization of the risk factors occurs *after* we make the initial decision $\mathbf{x}_0$. The last realization of the risk factor, $\boldsymbol{\xi}_T$, occurs after we have made the last decision $\mathbf{x}_{T-1}$, and it leads to the terminal state $\mathbf{S}_T$.

- A set of **cost/reward functions**, used in defining the objective function, depending on control and state variables. We may have a set of functions f_t, $t = 0, \ldots, T - 1$ to define the performance along the state trajectory, and a function F_T to assign a value (cost or reward) to the terminal state $\mathbf{S}_T$.

It is important to understand the difference between a multiperiod decision model and a truly dynamic model under uncertainty. In a *multiperiod* but *static* problem, at time $t = 0$ we assign a value to all of the decision variables $\mathbf{x}_t$. Thus, decisions are deterministic functions of time. In a dynamic problem under uncertainty, we do not plan everything in advance. We make decisions along the way, and we adapt them as uncertainty unfolds: Decisions are a sequence of random variables, as they depend on the uncertain states that we will observe

---

[14] To be precise, we should choose a time unit, one year as a rule, and consider time instants of form $t_i = t \cdot \delta t$, $i = 0, 1, \ldots, I$, where $T = I \cdot \delta t$. Here, for the sake of simplicity, we are confusing the integer index with the time instant.

[15] Arguably, we should do the same for the control variables $\mathbf{x}_t$, since they will be random variables, too, if decisions are adapted to the random occurrence of states.

2.1 Portfolio optimization

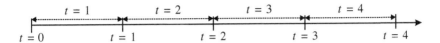

FIGURE 2.1 Illustrating time conventions.

in the future.[16] We emphasize that time indexing reflects the fact that at each time instant t, $t = 0, \ldots, T - 1$, we first observe the state $\mathbf{S}_t$, and then we apply a control action $\mathbf{x}_t$, and then we will observe a realization $\boldsymbol{\xi}_{t+1}(\omega)$ of the risk factors, which will lead to the next state $\mathbf{S}_{t+1}$. A rather generic and abstract formulation of the resulting optimization problem is the following:

$$\min \quad \mathrm{E}\left[\sum_{t=0}^{T-1} f_t(\mathbf{S}_t, \mathbf{x}_t) + F_T(\mathbf{S}_T)\right] \quad (2.2)$$
$$\text{s.t.} \quad \mathbf{x}_t \in A_t(\mathbf{S}_t)$$
$$\mathbf{S}_{t+1} = \Phi_t\big(\mathbf{S}_t, \mathbf{x}_t, \boldsymbol{\xi}_{t+1}(\omega)\big),$$

where we denote by $A_t(\mathbf{S}_t)$ the set of feasible control actions at time t, possibly depending on the current state, and by Φ_t the state transition function at time t. In this framework, the sequence of decisions $\mathbf{x}_t$ is a stochastic process, as it depends on the state trajectory. Note, however, that we are allowed to adapt decisions to the observed state now, but *not* to foresee the future state trajectory. For now, we avoid formalizing this requirement explicitly, but it is useful to define and visualize a suitable convention to define time instants and time periods, in order to avoid common confusion and ambiguity when dealing with discrete-time models:

- We consider *time instants* indexed by $t = 0, 1, 2, \ldots$ At these time instants, we observe the system state and make a decision.
- By a *time interval* t, we mean the time interval between time instants $t - 1$ and t. After the decision at time instant $t - 1$, the system evolves and a new state is reached at time t. During the time interval, the random disturbance will be realized, influencing the transition to the new state.

These definitions are illustrated in Fig. 2.1. Note once more that, with this timing convention, we emphasize the fact that noise is realized *during* time interval t, *after* making the decision at time instant $t - 1$.

To illustrate the framework, as well as the timing conventions, let us consider a stylized consumption–saving problem, which may be stated as follows:

---

[16]We are implicitly assuming that we may define a suitable set of state variables, collecting all of the necessary information to analyze possible future evolution. Formally, we deal with Markov processes, discussed in Chapter 11. In more complicated cases, we may have to keep track of the whole observed history of the risk factors.

- At time $t = 0$, the decision maker (or agent) is endowed with an initial wealth W_0 and has to decide the consumed amount C_0. It is not possible to borrow money, so $0 \le C_0 \le W_0$. Note that C_0 is a decision to be made, and not a given liability. What is not consumed is the saved amount, $S_0 = W_0 - C_0$.

- A second decision to be made at time $t = 0$ is the allocation of the saved amount S_0 between a risky and a risk-free asset; let $\alpha_0 \in [0, 1]$ be the fraction allocated to the risky asset, R_1 the rate of return of the risky asset over the first time interval spanning from $t = 0$ to $t = 1$, and r_f the return of the risk-free asset. We use R_1, rather than R_0, to emphasize that this piece of information is not known when the decision is made. For the sake of simplicity, the risk-free return and the distribution of risky returns are both assumed constant over time. Furthermore, we do not consider any intertemporal dependence between risky returns, which are assumed independent over time (they are a sequence of i.i.d. random variables).

- At time instant $t = 1$, the available wealth W_1 is the sum of capital and labor (noncapital) income. Capital income depends on the realized random return R_1 of the risky asset and the allocation decision α_0. The portfolio gain is
$$\alpha_0(1 + R_1) + (1 - \alpha_0)(1 + r_f).$$
Labor income, denoted by L_1, may be random, too, as it may depend on an uncertain employment state. Again, our notational choice emphasizes that this piece of information is not known when the first consumption–saving decision is made at $t = 0$. We multiply the portfolio gain by the saved amount S_0, add labor income, and express the available wealth at the end of the first time period as follows:
$$\begin{aligned} W_1 &= S_0 \left[\alpha_0 (1 + R_1) + (1 - \alpha_0)(1 + r_f) \right] + L_1 \\ &= (W_0 - C_0) \left[1 + r_f + \alpha_0 (R_1 - r_f) \right] + L_1. \end{aligned}$$
Then, again, W_1 is split into consumption C_1 and saving S_1, an allocation α_1 is chosen, and the process is repeated.

- Wealth is a state variable, and the underlying state of employment may be another one. We assume that the state of employment at time t, denoted by λ_t, may take one among three values in the set $\mathcal{L} = \{\alpha, \beta, \eta\}$, and $L_\alpha > L_\beta > L_\eta$. We may interpret η as "unemployed," α as "fully employed," and β as an intermediate situation. The dynamics of this state is modeled by a matrix of time-independent transition probabilities, with elements
$$\pi_{ij} = \mathrm{P}\{\lambda_{t+1} = j \mid \lambda_t = i\}, \qquad i, j \in \mathcal{L}. \tag{2.3}$$
This is the *conditional* probability that the employment state at time $t + 1$ will be j, given that it was i at time t. The initial employment state λ_0 is given, and we assume that the corresponding income is already included in the initial wealth W_0. We note that the next employment state transition depends only on the current state, and not on the whole past history. Thus,

2.1 Portfolio optimization

we only need to specify conditional probabilities and, technically, we are using a model based on a **discrete-time Markov chain**, to be discussed in Chapter 11. Furthermore, the employment state is not influenced by consumption–saving decisions. The sample path of the employment state is purely exogenous in this model, whereas wealth is partially endogenous, as it depends on consumption and portfolio allocation decisions.

- All of the above holds, with the natural adjustment of time subscripts, for all of the time instants up to $t = T - 1$.

Now that we have defined control and state variables and their dynamics, we need to define the objective function, which is a bit trickier, as it requires to address a risk–reward tradeoff. The consumption C_t is a random variable, so that maximizing consumption does not make mathematical sense. We could maximize its expected value $\mathrm{E}[C_t]$ (actually, the sum of these expected values over time), but this may lead to poor and quite risky decisions. Here, we follow standard approaches in economics and assume that the risk–reward tradeoff is dealt with by introducing a **utility function**. We shall consider decision-making under uncertainty and utility functions in more detail later, in Chapter 7. For now, we may just rely on intuition, and assume an increasing and concave utility function depending on consumption, $u(C_t)$. The function should be increasing, as the more we consume, the better. Furthermore, we shall see that by choosing a concave utility function, we may model risk aversion. In the limit, if $u(\cdot)$ is the identity function, we revert back to the expected value of consumption, which characterizes risk-neutral decision-makers. Hence, we might consider the maximization of the sum of the expected utilities $\mathrm{E}[u(C_t)]$ over time. Furthermore, we may also consider a term $q(W_T)$, accounting for utility from bequest, i.e., the terminal wealth W_T that we may leave to the beloved ones. This additional term represents a value of the terminal state. Actually, there is another tradeoff that we should address, involving immediate vs. future consumption. To account for this, we also introduce a **subjective discount factor** $\beta \in (0, 1)$ and consider the following maximization problem:

$$\max \quad \mathrm{E}\left[\sum_{t=0}^{T-1} \beta^t u(C_t) + \beta^T q(W_T)\right].$$

When β is large, the impact of later consumption is relevant; when β is small, i.e., we discount future consumption more heavily, the consequence is that we are somewhat greedy and emphasize immediate satisfaction. If we assume that, at time $t = T$, the terminal wealth W_T is fully consumed, i.e., we add a terminal consumption decision $C_T = W_T$, the problem boils down to

$$\max \quad \mathrm{E}\left[\sum_{t=0}^{T} \beta^t u(C_t)\right].$$

Consumption–saving models are common in economics.[17] We will see that asset–liability management problems, where liabilities are stochastic but exogenous, rather than endogenous decisions, are more relevant in quantitative finance.[18]

2.2 Risk measurement and management

Risk measurement and management is another major topic in quantitative finance. In Section 2.1.1, we have considered the standard deviation of portfolio return as a possible risk measure. However, there are different issues that have to be considered:

Defining a risk measure. Standard deviation is a symmetric risk measure, as it is based on squared deviations from the expected value. However, while we certainly do not like deviations toward huge losses, we are not likely to complain about windfalls and unexpected profits. Thus, we should look for alternative definitions leading to asymmetric measures. We shall learn that some sensible properties of risk measures may not be satisfied by seemingly reasonable risk measures. The issue becomes even thornier when considering *dynamic* risk measures.

Defining a risk model. The mean–variance model is deceptively simple: We just define a risk measure depending on asset returns. However, this requires an estimate of a possibly large covariance matrix, which is by no means easy to obtain. It might be preferable to rely on a statistical model of returns, like a **factor model**, leading to more reliable estimates and fundamental insights into the structure of risk, which may be decomposed in **common** and **specific risk factors**. This is also relevant when we consider the impact of common risk factors on the prices of several assets depending on them, as is the case with bond prices depending on interest rates and with complex portfolios of derivatives written on the same underlying assets. It is also important to realize that, although we will focus primarily on financial risk factors related to volatile equity markets, interest rates, and currency exchange rates, some nonfinancial sources of risk are also relevant. A list of examples includes volume, regulatory, and operational risk.

Risk management. Risk measurement is useful in monitoring the consequence of decisions, but we need a proper way to *make* those decisions. Depending on the specific problem that we are tackling, this may require the choice of suitable assets to build a hedge and the choice of their mix, or the definition

---

[17] For an extensive discussion of such models see, e.g., [3].
[18] This kind of models may be tackled by stochastic programming with recourse or by stochastic dynamic programming. See Chapter 15.

of a portfolio optimization model that is likely to be more complex than the static mean–variance model.

Statistical and computational issues. The precise way in which we address the above issues has a significant impact on how, in concrete terms, we deal with computational challenges in risk measurement and management. For instance, measuring the risk of a derivative portfolio may call for repeated repricing of derivatives subject to uncertain risk factors, which in turn may require extensive Monte Carlo simulation runs. The computational effort is related to the need for reliable estimates of the selected risk measure. Furthermore, there is little point in formulating a sophisticated decision model, if there is no robust and efficient way of solving it.

In the following sections, we make the above points more concrete by simple examples, paving the way for a deeper treatment in later chapters.

2.2.1 SENSITIVITY OF ASSET PRICES TO UNDERLYING RISK FACTORS

We often need to evaluate how asset prices depend on a set of risk factors. There is a huge variety of models that are used in finance, but a fundamental line must be drawn between linear and nonlinear models.

2.2.1.1 Linear risk factor models

Asset prices themselves, or their rates of return, may be considered as the primary risk factors we have to deal with. They are related by the simple relationship

$$S(T) = S(0) \cdot (1 + R),$$

where R is the holding period return over the time span $[0, T]$. However, we must also consider how the returns of different assets are related one to another, e.g., by estimating their covariance matrix. Given a joint sample of two random variables, observed at times $t = 1, \ldots, T$, say, R_{it} and R_{jt}, which are the returns of two stock shares over consecutive holding periods, the estimation of the covariance σ_{ij} may be accomplished by calculating the sample covariance,

$$S_{ij} = \frac{1}{T-1} \sum_{t=1}^{T} (R_{it} - \overline{R}_i)(R_{jt} - \overline{R}_j),$$

where $\overline{R}_i$ and $\overline{R}_j$ are the two sample means for each asset. This seems pretty trivial, but a little thought shows that this is not the case at all. What if we are considering a universe of $n = 500$ assets? The covariance matrix is a symmetric matrix consisting of 250,000 entries. A rough cut and imprecise calculation suggests that the sample size T must be large enough to estimate about 125,000 parameters with sufficient accuracy. To be precise, we need $n(n-1)/2$ covariances and n variances; hence, we need an estimate of

$$n + \frac{n \cdot (n-1)}{2} = \frac{n \cdot (n+1)}{2} = 125{,}250$$

parameters. Unfortunately, even if a suitably large data set were available, many data would be so old as to be irrelevant. An alternative strategy is to introduce a statistical model trying to capture the source of covariance explicitly. One such model is a simple linear factor model,

$$R_i = \alpha_i + \sum_{k=1}^{m} \beta_{ik} F_k + \epsilon_i, \quad i = 1, \ldots, n, \qquad (2.4)$$

relating the random return of asset i to a set of m **common risk factors** F_k, $k = 1, \ldots, m$, with $m \ll n$, and one specific risk factor ϵ_i for each stock share. The reader with a minimal statistical background will recognize this as a linear regression model, where α_i is a constant contributing to expected return, and β_{ik} is the sensitivity of stock share i to risk factor k. For instance, the oil price should be a relevant risk factor for stock shares in the automotive industry, whereas this common risk factor might play a less relevant role for stocks in the telecommunication industry. We will see how stock betas may be used to shape the exposure of a portfolio to common risk factors. The **specific risk factor** ϵ_i is related to the peculiarities of a stock share. The contribution of specific risk factors to overall portfolio risk may be arguably reduced by proper diversification, whereas more subtlety is required when dealing with common, systematic risk factors. A linear factor model may ease statistical issues. It is easy to see that, if all of the factors are mutually uncorrelated,[19] the covariance can be expressed as

$$\sigma_{ij} = \sum_{k=1}^{m} \beta_{ik} \beta_{jk} \sigma_k^2, \quad i \neq j,$$

where σ_k^2 is the variance of factor F_k. When considering variance, we find

$$\sigma_i^2 \equiv \sigma_{ii} = \sum_{k=1}^{m} \beta_{ik}^2 \sigma_k^2 + \sigma_{\epsilon i}^2,$$

where $\sigma_{\epsilon i}^2$ is the variance of the specific factor. We observe that with this structure, the estimation of the covariance matrix requires the estimation of $m \times n$ betas, m common factor variances, and n specific variances. In the case we were considering, if $n = 500$ and $m = 3$, this amounts to 2003 parameters, with a reduction of two orders of magnitude with respect to the naive approach. Apart from statistical issues, a factor model provides us with very useful insights about the structure of a risk exposure, and it may be used to change the

---

[19]We shall consider the matter in full detail in Chapter 9.

2.2 Risk measurement and management

exposure itself, possibly making a portfolio insensitive to some risk factors or, on the contrary, to expose the portfolio to some risk factors on which we are confident to make a good bet.

2.2.1.2 Nonlinear risk factor models

Linear models may be used when dealing with an equity portfolio, but they may be not so adequate when dealing with a portfolio of derivatives. Consider the price at time t of a vanilla call option written on an underlying stock share, whose current price is S_t. The option price will depend, among other factors, on S_t, but in a nonlinear way. At present, we are not yet equipped to consider the kind of dependence exactly, but a quick look at the nonlinear payoff depicted in Fig. 1.6 provides us with some intuition about the kind of relationship between the option price and S_t.

An easier case in which we may grasp the nonlinear relationship between an asset price and the underlying risk factors is the case of a zero-coupon bond and the level of interest rates. We will consider bond pricing and the impact of interest rate risk at length in Chapter 3, but the nature of the relationship is easy to grasp in terms of discounted cash flows. If a zero with face value F matures in exactly T years, and the relevant annual interest rate is $R_{0,T}$, the fair bond price is

$$P_z(0; R_{0,T}, T) = \frac{F}{(1 + R_{0,T})^T}. \tag{2.5}$$

The notation $P_z(0; R_{0,T}, T)$ suggests that this is the price at time $t = 0$ of a zero, maturing at time $t = T$, depending on the interest rate $R_{0,T}$.[20] By "relevant" interest rate we mean an annual rate which applies to an investment horizon of T years, for an investment with a default risk comparable to that of the zero-coupon bond. It is important to realize that $R_{0,T}$, like any interest rate, refers to a *single* year; on the contrary, when referring to stock markets, we often use a holding period return, which need not be annualized. In this case, the interest rate $R_{0,T}$ may also be interpreted as an annual **yield**, i.e., the annual growth factor of the initial investment in the zero. Later, this pricing equation will be better justified by no-arbitrage arguments, but for now it is enough to understand how future cash flows should be discounted in order to find their present value.

Equation (2.5) immediately shows that the bond price is a function of the interest rate, and that an increase in $R_{0,T}$ will imply a drop in $P_z(0; R_{0,T}, T)$. To get a feeling for this kind of risk, let us consider three zeros, all with face value $F = 100$, maturing in 3, 10, and 30 years, respectively, and let us assume, for the sake of simplicity, that there is a single relevant interest rate applying to all maturities. For instance, if $R_{0,T} = 4\%$, the price of the first zero is

$$P_z(0; 0.04, 3) = \frac{100}{(1 + 0.04)^3} = 88.90.$$

---

[20] We shall introduce and motivate a more complete notation in Chapter 3. In fact, the price of a coupon-bearing bond depends on a whole range of interest rates with different maturities.

Table 2.1 Interest rate risk and the effect of zero-coupon bond maturity.

T (years)	3	10	30
$P_z(0; 0.04, T)$	88.90	67.56	30.83
$P_z(0; 0.05, T)$	86.38	61.39	23.14
% loss	−2.83	−9.13	−24.96

If the rate increases to 5%, the new price is

$$P_z(0; 0.05, 3) = \frac{100}{(1 + 0.05)^3} = 86.38,$$

with a corresponding loss of

$$\frac{86.38 - 88.90}{88.90} = -2.83\%.$$

In Table 2.1, we observe the effect of this increase in the interest rate for the three maturities. We clearly see that maturity plays a key role, and that a long-maturity zero is a quite risky asset. Maturity also plays a role in the risk of a coupon-bearing bond, but we will see that the coupon rate is also relevant. Thus, we need some specific measures of risk for bonds. More generally, when there is a nonlinear dependence with respect to a risk factor, we may approximate it with a Taylor expansion to the first or second order, which requires first- and second-order derivatives. We shall see that, for a bond, these sensitivities with resect to interest rates are captured by duration and convexity. By a similar token, in the case of options, we will consider sensitivity measures such as delta, gamma, and vega, which relate the option price to the current price of the underlying asset and the current level of its volatility. Sensitivities to risk factors provide us with useful information and play a relevant role in strategies to immunize a portfolio with respect to selected risk factors, possibly in an approximate way.

2.2.2 RISK MEASURES IN A NON-NORMAL WORLD: VALUE-AT-RISK

Variance and standard deviation are the two dispersion measures that we learn in basic statistics. Standard deviation of return is referred to as **volatility** in finance, and it seems a simple and relevant risk measure. While volatility is certainly relevant, it might not tell the whole story. The key issue is that standard deviation is a *symmetric* risk measure, taking into account extra potential for profit in the same way as extra potential for loss. As such, standard deviation may be a suitable risk measure in the case of symmetric distributions, most notably the normal. Indeed, two parameters, expected value and standard deviation, tell everything we need to know about a normal distribution. However,

2.2 Risk measurement and management

empirical analysis does not quite support the view that returns are normally distributed.

A first issue is **skewness**, i.e., lack of symmetry. Probability distributions for profit–loss are not symmetric in general, as the following examples suggest:

- Consider the payoff of a call option, as shown in Fig. 1.6. Even if we assume that the distribution of the underlying asset price at maturity is normal, the nonlinearity of the payoff changes the nature of its distribution, which is skewed.
- The option payoff cannot be negative, unlike a normal variable. Another relevant risk factor that cannot take negative values is the volatility itself, which is a fundamental risk factor for derivatives.
- Interest rates cannot be negative as well, and the nonlinearity of the relationship between the price of a zero-coupon bond and the interest rate contributes to the non-normality of bond returns.

But how can we *measure* skewness? Let us recall that the expected value is the first-order moment of a random variable, whereas variance is the second-order central moment.[21] A formal definition of a skewness coefficient relies on a third-order central moment:

$$\gamma \doteq \frac{\mathrm{E}\bigl[(X-\mu)^3\bigr]}{\sigma^3}. \tag{2.6}$$

Essentially, this is the third-order moment of the standardized variable. Standardization is necessary, as we want to capture skewness irrespective of location and scale.[22] The odd exponent, unlike the case of variance, preserves the sign of deviations, and the information provided by skewness is illustrated in Fig. 2.2. The density on the left has positive skewness, and it is skewed to the right. The other density has negative skewness, and it is skewed to the left. As we can imagine, skewness is zero for a symmetric distribution, and a nonzero skewness suggests lack of normality.

A less obvious, but quite relevant feature of a normal distribution is its thin tails. This means that extreme events are not quite likely. We learn in basic statistics that, if $X \sim \mathsf{N}(\mu, \sigma^2)$,

$$\mathrm{P}\{\mu - 3\sigma \leq X \leq \mu + 3\sigma\} \approx 0.9973,$$

which means that most realizations are within three deviations from the expected value. We also learn that Student's t distributions are heavier tailed, i.e., they feature more significant uncertainty, especially with few degrees of freedom. Heavy (or fat) tails are relevant in finance, as they are related to extreme events like stock market crashes (and rallies). Fat tails are measured by the

---

[21] The moment of order k of a random variable X is defined as $m_k \doteq \mathrm{E}[X^k]$. The central moment of order k of X is defined as $M_k \doteq \mathrm{E}[(X-\mu)^k]$, where $\mu = \mathrm{E}[X]$.
[22] It is easy to see that skewness of a random variable X (as well as kurtosis, to be defined shortly) is insensitive to affine transformation $a + bX$, subject to the restriction $b > 0$.

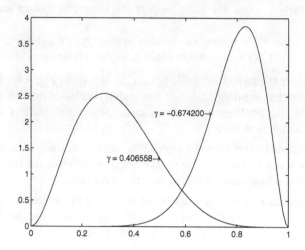

FIGURE 2.2 An illustration of positive and negative skewness.

kurtosis coefficient, which relies on a fourth-order central moment:

$$\kappa \doteq \frac{\mathrm{E}\left[(X-\mu)^4\right]}{\sigma^4}. \tag{2.7}$$

We observe that the fourth power disregards the sign of deviations with respect to the expected value, unlike skewness, but it emphasizes *large* deviations much more than the second power of variance. It can be shown (using, e.g., the moment generating function) that kurtosis is always 3 for a normal distribution, whatever σ is. On the contrary, other distributions show fatter tails, a property captured by a kurtosis in excess of 3. The information provided by kurtosis is illustrated in Fig. 2.3. The distribution with kurtosis $\kappa = 9$ is a Student's t, and it features fatter tails than the standard normal, which has $\kappa = 3$.

When a profit–loss distribution features significant skewness and kurtosis, standard deviation is less appropriate as a risk measure, and alternatives may be defined. Furthermore, from a different viewpoint, practitioners might appreciate risk measures expressed in a more straightforward way, i.e., in terms of a potential *monetary* loss. This sets any kind of risk on a common ground and is more easily perceived by management. One way to cope with these issues is to define asymmetric, quantile-based risk measures in monetary terms. A well-known example is value-at-risk.

2.2.2.1 A quantile-based risk measure: Value-at-risk

One way to define asymmetric risk measures, overcoming some of the difficulties with standard deviation, is to resort to quantile-based risk measures. The idea is to focus on the bad tail of the distribution, where losses are incurred. Let us consider the portfolio loss L_H, over a holding period $[0, H]$, and evaluate its

2.2 Risk measurement and management

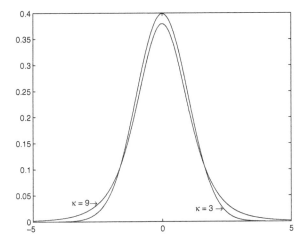

FIGURE 2.3 An illustration of kurtosis.

quantiles. Note that loss is measured in monetary terms, which sets all kinds of risk on a common ground. For the sake of simplicity, let us assume that L_H is a continuous random variable. Then, we define the **value-at-risk**, with time horizon H, at confidence level $1 - \alpha$, as a number $\text{V@R}_{1-\alpha,H}$ such that

$$P\{L_H \leq \text{V@R}_{1-\alpha,H}\} = 1 - \alpha. \tag{2.8}$$

For instance, if we set $\alpha = 0.05$, we obtain V@R at 95%, for time horizon H. Quite often, short horizons are considered like, e.g., one day. The probability that the loss exceeds V@R is α.[23] It is common to denote value-at-risk as VaR, where the last capital letter should avoid confusion with variance.[24] However, we prefer the less ambiguous notation V@R. Informally, V@R aims at measuring the maximum portfolio loss one could suffer, over a given time horizon, within a given confidence level.

When considering a discrete set of scenarios, we should consider L_H as a discrete random variable and modify the definition of value-at-risk as

$$\text{V@R}_{1-\alpha,H} = \inf\left\{V \in \mathbb{R} \mid P\{L_T \leq V\} \geq 1 - \alpha\right\}. \tag{2.9}$$

This is just the definition of the **generalized inverse** of the cumulative distribution function (CDF) of loss, which boils down to Eq. (2.8) in the case of a continuous random variable with a continuous and strictly increasing (hence, invertible) CDF. Actually, one could also define another measure by using strict

---

[23] We stick to the statistical convention that α is the small area associated with the tail of a probability density function (PDF), but sometimes the opposite notation is adopted.
[24] Arguably, the lowercase letter in the middle should also avoid confusion with VAR, which usually refers to vector autoregressive models in econometrics and time series analysis.

inequality in Eq. (2.9),

$$\inf\{V \in \mathbb{R} \mid P\{L_T < V\} \geq 1 - \alpha\}.$$

Since such technicalities are not relevant, for a continuous random variable with an invertible CDF, and are not needed to grasp the essentials of value-at-risk, we defer a more detailed treatment to Section 7.4.3.

Actually, there are two possible definitions of V@R, depending on the reference wealth that we use in defining loss. Let W_0 be the initial portfolio wealth. If R_H is the random return over the holding period, the future wealth is

$$W_H = W_0(1 + R_H).$$

Its expected value is

$$\mathrm{E}[W_H] = W_0(1 + \mu),$$

where μ is the expected holding period return. The absolute loss over the holding period is related to the initial wealth:

$$L_H^a = W_0 - W_H = -W_0 R_H. \tag{2.10}$$

The quantile of absolute loss at level $1 - \alpha$ is the **absolute** V@R at that confidence level. We define a **relative** V@R if we take the *expected* future wealth as a reference in evaluating the relative loss

$$L_H^r = \mathrm{E}[W_H] - W_H = W_0(\mu - R_H). \tag{2.11}$$

Quite often, we are interested in small holding periods, like one day. The reason is that V@R may be used by banks to decide how much cash they should set aside as a fudge against short-term losses. If the bank is in need for cash in a very short time period, it may be difficult to liquidate assets to generate liquidity, especially when markets crash. If the holding period return is short, it turns out that *drift is dominated by volatility*. Technically, this means that the expected return is small compared to standard deviation. This may be shown by resorting to continuous-time models for stock share prices, like geometric Brownian motion. Since such models will be treated later, in Chapter 11, we provide here an intuitive justification.

Example 2.1 The square-root rule

> Let us consider a sequence of i.i.d. (independent and identically distributed) variables over time, X_t, $t = 1, \ldots, H$. Let μ and σ be the expected value and standard deviation, respectively, for each variable X_t. If we sum variables over the H time periods, we define a new variable,
>
> $$Y = \sum_{t=1}^{H} X_t,$$

2.2 Risk measurement and management

and find
$$E[Y] = \sum_{t=1}^{H} E[X_t] = \mu H, \qquad (2.12)$$

and
$$\sigma_Y = \sqrt{\sum_{t=1}^{H} \text{Var}(X_t)} = \sigma\sqrt{H}. \qquad (2.13)$$

Note that we are writing the variance of a sum as a sum of variances, which is wrong in general, but it is fine under the assumption of independence (actually, lack of correlation is enough). Thus, we find that the expected value scales linearly with time, whereas the standard deviation scales with the square root of time. The two functions are compared in Fig. 2.4. We may observe that, for a large H, the expected value dominates, but roles are reversed for a small H going to zero. We shall refer to this fact as the **square-root rule**.

This seems to suggest that, for a small time period, volatility does dominate drift, but the conclusion is not really warranted. To begin with, random returns should be multiplied, rather than summed. However, if we consider summing log-returns, or an approximation valid for small daily returns, we may claim that the reasoning makes sense. The assumption of independence of returns over time is bit more critical, but there are models based on this idea. Probably, the most delicate point is that we should consider "slicing" a random variable, i.e., expressing a given Y as a sum of terms $X_{\delta t}$, for a small time interval δt. If we add independent (uncorrelated) normals, we do get a normal. But can we slice a normal into smaller normals? We should resort to a more sophisticated concept related to *self-similar* stochastic processes. Indeed, we will use such a process, the Wiener process, based on a normal distribution. We cannot come up with a rigorous analysis here, but empirical data do suggest the idea that, on a short time interval of length δt, drift goes to zero more rapidly than volatility does. Put another way, we cannot reject the null hypothesis that return over δt is zero.

Assuming that the square-root rule holds, for a short time period H we have
$$\mu = E[R_H] \approx 0.$$

Hence, Eq. (2.11) boils down to Eq. (2.10), and there is no difference between absolute and relative value-at-risk. In this book, we will always consider absolute V@R. Nevertheless, we should keep in mind that relative V@R may be more relevant to longer-term risks, as those faced by a pension fund.

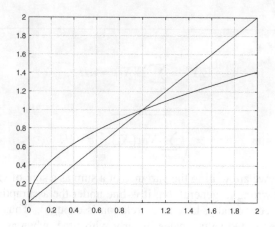

FIGURE 2.4 An illustration of the square-root rule.

The loss L_H is a random variable depending on a set of risk factors, possibly through nonlinear transformations linked to bond or option pricing models. In fact, estimating V@R for a real-life portfolio is a nontrivial exercise requiring:

- The definition of a set of risk factors and the characterization of their *joint* distribution. Note that correlations may not be enough. Correlations and volatilities do characterize a multivariate normal distribution, but we cannot take normality for granted. Incidentally, as we show in the examples below, under a normality assumption there would be little point in using V@R, as this would provide us with little additional information with respect to standard deviation.
- The definition of pricing models linking asset prices to the underlying risk factors.
- The estimation of a quantile, most likely by numerical methods, like Monte Carlo sampling.

These tasks may be daunting in practice, but let us just illustrate V@R with a couple of small examples.

Example 2.2 Elementary V@R calculation

We have invested $100,000 in Quacko Corporation stock shares, whose daily volatility is 2%. This means that the volatility of asset return is

$$\sigma_H = \sqrt{\text{Var}(R_H)} = 0.02,$$

2.2 Risk measurement and management

> where the holding period H is one day and R_H is the daily return. For the sake of simplicity, let us assume that the return is normally distributed. Given the square-root rule, we have
>
> $$R_H \sim \mathsf{N}(0, \sigma_H^2) = \mathsf{N}(0, 0.02^2).$$
>
> Loss over the holding period is
>
> $$L_H = -W_0 \cdot R_H,$$
>
> where the initial wealth W_0 is \$100,000. The expected value of loss is $\mathrm{E}[L_H] = 0$, and its volatility is
>
> $$\sigma_L = \sqrt{\mathrm{Var}(L_H)} = \sqrt{(-W_0)^2 \cdot \mathrm{Var}(R_H)} = W_0 \cdot \sigma_H.$$
>
> We know that quantiles of a normal distribution are related to quantiles $z_{1-\alpha}$ of a standard normal variable. If we want daily V@R with 99% confidence level, we find
>
> $$\begin{aligned} \mathrm{V@R}_{0.99,1} &= z_{0.99} \times \$100{,}000 \times 0.02 \\ &= 2.3263 \times \$100{,}000 \times 0.02 \\ &= \$4652.70. \end{aligned}$$
>
> Therefore, we are "99% sure" that we will not lose more than \$4652.70 in one day.

The assumption of normality of returns can be dangerous, as the normal distribution has a relatively low kurtosis; alternative distributions have been proposed, featuring fatter tails, in order to better account for tail risk, which is what we are concerned about in risk management. Nevertheless, the calculation based on the normal distribution is so simple and appealing that it is tempting to use it even when we should rely on more realistic models. In practice, we are not interested in V@R for a single asset, but in V@R for a whole portfolio. Again, the normality assumption streamlines our task considerably.

Example 2.3 V@R in multiple dimensions

> Suppose that we hold a portfolio of two assets: We have invested $W_A = \$10{,}000$ in stock share A and $W_B = \$20{,}000$ in stock share B. We assume that daily returns are jointly normal with the following

parameters:

$$\sigma_A = 2\%, \quad \sigma_B = 3\%, \quad \rho = 0.4,$$

where ρ is the correlation coefficient. We are interested in the daily V@R, so we disregard drift again, with confidence level 99%.

We first compute the volatility of the holding period return of the whole portfolio:

$$\begin{aligned}\sigma_P &= \sqrt{W_A^2 \sigma_A^2 + 2 W_A W_B \rho \sigma_A \sigma_B + W_B^2 \sigma_B^2} \\ &= 10000 \times \sqrt{1^2 \cdot 0.02^2 + 2 \cdot 1 \cdot 2 \cdot 0.4 \cdot 0.02 \cdot 0.03 + 2^2 \cdot 0.03^2} \\ &= \$704.2727.\end{aligned}$$

Therefore

$$\text{V@R}_{0.99,1}^P = z_{0.99} \cdot \sigma_P = 2.326348 \times 704.2727 \approx \$1638.38.$$

It is interesting to compare the risk of the overall portfolio with the risks of the two individual positions:

$$\begin{aligned}\text{V@R}_{0.99,1}^A &= z_{0.99} \cdot W_A \cdot \sigma_A = 2.326348 \times 10000 \times 0.02 \\ &\approx \$465.27, \\ \text{V@R}_{0.99,1}^B &= z_{0.99} \cdot W_B \cdot \sigma_B = 2.326348 \times 20000 \times 0.03 \\ &\approx \$1395.81, \\ \text{V@R}_{0.99,1}^P &= 1638.38 \\ &\leq 1861.08 = \text{V@R}_{0.99,1}^A + \text{V@R}_{0.99,1}^B.\end{aligned}$$

We observe that the risk of the portfolio is less than the sum of the two risks. The amount of this reduction depends on the correlation. The two quantities would be the same for perfect positive correlation, $\rho = 1$, whereas overall risk would be minimized with perfect negative correlation, $\rho = -1$.

In Example 2.3, we take advantage of the *stability* property of the normal distribution, i.e., the sum of jointly normal random variables is still a normal random variable. In general this is not true, and even if we assume that the underlying risk factors are normally distributed (which does not make sense for volatility risk and correlation risk), the portfolio loss would not be normally distributed because of nonlinearities. Monte Carlo methods may be used in more realistic cases.

2.2 Risk measurement and management

What we have just illustrated is the simplest parametric approach to V@R estimation. It is simple because of the assumption of normality, because we are able to capture the joint distribution of the risk factors, and because the risk model is trivially linear (indeed, the asset prices themselves are the risk factors). One of the most difficult issues is to capture the joint dependence of the risk factors. Correlations tell the whole story in the normal case, but not in general. Furthermore, the correlation themselves may change over time, especially when markets crash. Hence, we should also mention that a completely different route may be taken, based on historical V@R. Rather than assuming a specific joint distribution, we may rely on a *nonparametric* approach based on resampling historical data. The advantage of historical data is that they should naturally capture dependence. Hence, we may combine them, according to *bootstrapping* procedures, to generate future scenarios and estimate V@R by historical simulation.

A further effect of the normality assumption is that we observe a reduction in risk through portfolio diversification. This makes sense, and it is formally referred to as **subadditivity** property. If we consider two random variables X and Y, representing two losses, the subadditivity property of a risk measure $\mathcal{R}$ mapping random variables to real numbers is

$$\mathcal{R}(X+Y) \leq \mathcal{R}(X) + \mathcal{R}(Y),$$

i.e., the risk of the sum (the aggregate portfolio) is not larger than the sum of risks (of the individual portfolios). In the normal case, V@R is essentially related to standard deviations, since these are used to find quantiles. It is easy to see that the standard deviation of a sum of two random variables X and Y is not larger than sum of standard deviations:

$$\begin{aligned} \sigma_{X+Y} &= \sqrt{\sigma_X^2 + 2\rho\sigma_X\sigma_Y + \sigma_Y^2} \\ &\leq \sqrt{\sigma_X^2 + 2\sigma_X\sigma_Y + \sigma_Y^2} \\ &= \sqrt{(\sigma_X + \sigma_Y)^2} = \sigma_X + \sigma_Y, \end{aligned} \quad (2.14)$$

where the inequality depends on the upper bound on the correlation between X and Y, $\rho \leq 1$. Hence, we see that V@R is subadditive under a normality assumption. Unfortunately, this depends on the fact that V@R does not really tell us anything more than standard deviation in the normal case, but the normality assumption itself is not quite realistic. In fact, as we shall see in Section 7.4.3, value-at-risk is *not* subadditive in general and lacks a fundamental feature that characterizes **coherent** risk measures, which will be introduced in Section 7.4.1.

2.2.3 RISK MANAGEMENT: INTRODUCTORY HEDGING EXAMPLES

The ability to measure risk is certainly necessary in finance, but it may of little use without a way to *manage* risk. We must devise strategies to reduce risk exposures or eliminate them altogether, if we can and wish doing so. This can be done by shaping the portfolio in such a way as to obtain a specific exposure to underlying risk factors, or by including additional assets in the portfolio, like derivative contracts. We can introduce some essential concepts by simple examples involving linear contracts like forward and futures contracts. More sophisticated models are needed when considering nonlinear instruments like options.

2.2.3.1 Perfect hedging and forward contracts

Perfect hedging is achieved when risk is eliminated completely, i.e., when future cash flows, or returns, are made deterministic. Imagine that we hold N units of an asset with current price S_0, which we plan to hold for a time period of length T. The future asset price S_T is uncertain, but if we sell N forward contracts (i.e., take a short position), with price $F_0 \equiv F(0, T)$ and maturing at T, at the end of the holding period we will be able to sell the asset at the arranged forward price. Alternatively, if the contract is settled in cash, the total net cash flow at time T is

$$\underbrace{N \cdot S_T}_{\text{Sell asset}} + \underbrace{N \cdot (F_0 - S_T)}_{\text{Short position payoff}} = N \cdot F_0, \tag{2.15}$$

which is equivalent to selling the asset directly to the counterparty of the forward contract, i.e., the long position. If risk is measured by standard deviation, we clearly see that risk has completely eliminated, i.e., we have a perfect hedge. This does not imply that perfect hedging is the best choice, since we have also eliminated any potential for additional profit. If the price S_T turns out to be large, the loss on the short position eliminates all of the profit from the spot trade.

If we wish to retain a portion of the upside potential, we may take an alternative policy and rely on options. If we buy N put options with price P_0 and strike K, the equivalent cash flow at time T is

$$N \cdot S_T + N \cdot \max\{K - S_T, 0\} - N \cdot P_0 \cdot e^{rT} = N \cdot [\max\{K, S_T\} - P_0 \cdot e^{rT}].$$

Note that, in order to take the time value of money into proper account, we have to project the cash flow $-N \cdot P_0$ forward in time, which requires multiplying it by a factor involving a risk-free rate r.[25] If the maturity T of the hedge is short, this may be neglected. Depending on the realization of the random variable S_T, the cash flow may be positive or negative; therefore, we have not eliminated the possibility of a loss related to paying the premium for an option that we might not use, but we retain some upside potential. Apparently, it is a clever choice to choose a put option with a large strike price K; needless to say, a large strike

---

[25] See Section 3.1 for an extensive discussion. In this case, the "growth" factor is a positive exponential function, as we assume a continuously compounded interest rate. Alternatively, we could project cash flows backward in time, multiplying future cash flows by a negative exponential, playing the role of a discount factor back to time $t = 0$.

price will be reflected in a large price P_0, making the cost of this insurance excessive. The choice must balance the potential for down- and upside, and it may depend on individual risk preferences.

2.2.3.2 Minimum variance hedging and futures contracts

Equation (2.15) shows that a perfect hedge may be achieved, but this assumes that a forward contract for the desired underlying asset and maturity can be arranged. However, forward contracts are traded OTC, and not on regulated exchanges. Thus, the resulting hedge may be expensive[26] and not quite liquid, which means that unwinding positions when necessary may be awkward. As we have seen, we may resort to futures contracts, but since these are standardized, a perfect hedge is not quite feasible, since we cannot match the risk exposure and the maturity perfectly. Nevertheless, we may consider a futures contract with a maturity close to our time horizon, written on an asset whose price process is correlated to the price process of the asset we are interested in, resulting in a cross-hedging strategy.[27]

Example 2.4 Cross-hedging

> Consider a firm that needs to purchase plastic for packaging the goods it produces. Arranging a forward contract on plastic may not be quite reasonable, but since plastic is made out of oil, the oil price is a correlated risk factor. We do not really want to take a long position in a forward contract on oil and actually buy it. However, we may take a long position in oil futures and then close our position, before the maturity of the hedging instruments, without having to take the physical delivery of oil.

If the hedge consists of a futures contract, the overall cash flow will not be zero because of asset and maturity mismatches; nevertheless, we might be satisfied by minimizing its variance. To be precise, when using futures contracts, we should also account for another relevant difference with respect to forward contracts: Due to daily marking-to-market, there is a sequence of daily cash flows. This difference may be relevant or not, depending on the time horizon, as well as the volatility of interest rates and their correlation with the price processes S_t and F_t. In principle, the hedge should be adjusted dynamically, by a strategy called *tailing the hedge*. We will see how to deal with this in Chapter 12. For

---

[26] It may seem that a hedge based on a forward contract cannot be "expensive," since no cash flow occurs at the beginning. However, the forward price F_0 might be not quite fair, and there could be a bid–ask spread between the prices offered to short and long positions.

[27] We will have a more detailed look at hedging with linear contracts, like forward and futures, in Section 12.3.

the sake of simplicity, let us ignore the issue and assume that a futures contract essentially behaves like a forward contract.

We may establish a cross-hedge with the aim of minimizing the resulting cash flow variance. In such a case, we must also determine a suitable hedging ratio h, i.e., the number of the futures contracts that we buy or sell for each unit of the asset. Unlike the case of perfect hedging, h need not be 1. So, let us consider a hedge with maturity T_1, based on a futures contract maturing at time T_2, where $T_1 < T_2$, written on some underlying standardized asset. By maturity of the hedge, we mean the time instant at which we close the futures contracts, by taking an opposite position, and sell our assets (or, alternatively, we assess the value of our portfolio). We assume again that we hold N units of an asset that we wish to sell at T_1. Using the same reasoning that leads to Eq. (2.15), the cash flow at time T_1 is

$$C_1 = N \cdot [S_{T_1} + h \cdot (F_0 - F_{T_1})]. \tag{2.16}$$

Since the future contract matures at T_2, after the maturity T_1 of the hedge, and it is typically written on a different underlying asset, we do not have spot–futures price convergence; hence, $F_{T_1} \neq S_{T_1}$ in general. Note that we may easily close a futures contract before maturity and that the resulting cash flow for a short position, if we neglect the time value of money, is the difference between the initial and terminal futures prices, when the hedge is closed. We may easily find the hedge ratio h that minimizes the variance of the total cash flow,[28]

$$\min_h \quad \text{Var}[S_{T_1} + h \cdot (F_0 - F_{T_1})]$$
$$= \text{Var}(S_{T_1}) + h^2 \text{Var}(F_{T_1}) - 2h\text{Cov}(S_{T_1}, F_{T_1}),$$

where we get rid of the irrelevant N. This is clearly a convex parabola, as a function of h, and the first-order optimality condition,

$$2h\text{Var}(F_{T_1}) - 2\text{Cov}(S_{T_1}, F_{T_1}) = 0,$$

yields the optimal hedge ratio,

$$h^* = \frac{\text{Cov}(S_{T_1}, F_{T_1})}{\text{Var}(F_{T_1})}. \tag{2.17}$$

The hedging ratio h^* will be negative, if the correlation between S_{T_1} and F_{T_1} is negative, in which case we should take a long position in the futures, even though we want to sell our assets. If we denote the standard deviation of the futures price at T_1 by σ_F, the standard deviation of the spot price at T_1 by σ_S, and their (Pearson) correlation coefficient by ρ, we may write

$$h^* = \rho \cdot \frac{\sigma_S}{\sigma_F}. \tag{2.18}$$

---

[28] If we do not really want to sell the asset, we may consider the variance of $\delta S + h \cdot \delta F$, where $\delta S \doteq S_{T_1} - S_0$ and $\delta F \doteq F_{T_1} - F_0$. Thus, we are interested in the variation of wealth, rather than in cash flows. Since prices at time $t = 0$ are known, they do not contribute to variance, and the two expressions are equivalent.

2.2 Risk measurement and management

This looks very similar to what we are familiar with from the theory of simple linear regression models. There, the optimal slope of the regression line is given by the ratio of the covariance between the target and the explanatory variable to the variance of the explanatory variable. This slope, obtained by least-squares minimization of residuals, minimizes the variance unexplained by the model. Here, we have a similar expression for the hedge ratio, which minimizes the residual variance of the hedged portfolio. Essentially, we regress the spot price on the price of the hedging instrument. By the way, we observe that, if there is no hedge mismatch in terms of underlying asset price and maturity of the futures contract, then we have perfect correlation and identical variances, due to the convergence of spot and futures prices, and we revert back to perfect hedging, where $h^* = 1$.

Clearly, we should find hedging instruments that are as much correlated as possible with our risk exposure (positively or negatively, it does not really matter, as this is only reflected in a change of sign in h^*). Nothing forbids, in principle, using more than one hedging instrument. The mathematics involved is not quite different, and it basically requires solving a system of linear equations. However, as we shall show in Example 2.5, we should not forget that there might be nonfinancial risk factors at play. Indeed, in practice, hedging is not such a simple problem as Eq. (2.18) might suggest:

- Real-life contracts have a standardized volume, so the hedging ratio may have to be somehow rounded, resulting in an over- or under-hedging error.
- We need an estimate of parameters like standard deviations and correlations. These may be obtained from past data, but there is no guarantee that these parameters are constant in time, especially in stressed market conditions. Estimation and modeling errors might affect the performance of the hedge.
- Sometimes, we cannot disregard the impact of daily marking-to-market of futures contracts, which calls for dynamic adjustments. Since this may be costly to implement, further hedging errors might result.
- In the minimum-variance formulation, we end up with a least-squares problem, a very simple convex optimization problem. We observe that the degree of risk aversion and the possible market views do not play any role in determining the solution. However, more sophisticated optimization models may be necessary to better match the hedge with investors' views and appetite for risk.

2.2.3.3 First-order immunization

So far, we have mostly considered linear models. The factor model of Eq. (2.4) is linear, and the hedging equations involving forward/futures contracts are also linear, since their payoff diagrams are linear.[29] Due to linearity, the required

---

[29] See Fig. 1.4.

mathematical machinery is rather simple. However, the payoff from vanilla call and put options is not linear at all. Actually, it is *piecewise* linear at maturity, but if we close the hedge before maturity by selling the options, we should expect a more complicated dependency on the underlying risk factors. The matter is even more involved when considering exotic derivatives. Also bonds may be considered as interest rate derivatives, with a price depending in a nonlinear way on the relevant risk factor, the interest rate. Actually, as we shall see, the whole term structure of interest rates is involved, as different interest rates apply to different maturities. Thus, we are lead to consider *nonlinear* hedging problems.

One possibility to tackle such problems is to approximate the nonlinear problem by linearization strategies; another one is to resort to possibly sophisticated optimization models. Let us consider the first idea, linearization, which is pursued by taking advantage of Taylor expansions to the first order. Let us illustrate the concept in a generalized and abstract framework, whereby the value V of a portfolio depends on several risk factors, which we denote by R_i, $i = 1, \ldots, m$. When the underlying risk factors change by an amount δR_i, there will be a corresponding change δV in the portfolio value. The change in the value of V may be approximated to the first order as follows:

$$\delta V \approx \sum_{i=1}^{m} \frac{\partial V}{\partial R_i} \cdot \delta R_i.$$

Now let us consider m hedging instruments (assets) with unit prices H_j, $j = 1, \ldots, m$, which are sensitive to the same risk factors as V. Observe that the number of hedging instruments matches the number of risk factors. We may approximate the change δH_j in H_j in the same way as δV,

$$\delta H_j \approx \sum_{i=1}^{m} \frac{\partial H_j}{\partial R_i} \cdot \delta R_i, \qquad j = 1, \ldots, m.$$

If we include ϕ_j units of each hedging instrument in the overall portfolio, the value of the hedged portfolio is

$$V^H = V + \sum_{j=1}^{m} \phi_j H_j.$$

The coefficients ϕ_j play the same role as the hedging ratio h when using futures (see Eq. 2.16). Note that some additional budget may be required to set up the hedge, as we may need to pay for the insurance, e.g., in the case of call or put options. Actually, this need not be the case, as some derivatives (like interest rate swaps), have zero initial value, just like forward and futures contracts. For the moment, let us neglect this issue.

2.2 Risk measurement and management

The approximated change in value of the hedged portfolio is

$$\delta V^H = \delta V + \sum_{j=1}^{m} \phi_j \cdot \delta H_j$$

$$= \sum_{i=1}^{m} \frac{\partial V}{\partial R_i} \cdot \delta R_i + \sum_{j=1}^{m} \left(\phi_j \sum_{i=1}^{m} \frac{\partial H_j}{\partial R_i} \cdot \delta R_i \right)$$

$$= \sum_{i=1}^{m} \left(\frac{\partial V}{\partial R_i} + \sum_{j=1}^{m} \phi_j \frac{\partial H_j}{\partial R_i} \right) \cdot \delta R_i. \qquad (2.19)$$

One clear issue is: How can we obtain the required partial derivatives, i.e., first-order sensitivities to risk factors? In simple cases, like plain bonds and vanilla options under a suitable market model, they may be obtained by differentiating an explicit pricing formula. Alternatively, they may be obtained by numerical and/or statistical methods, possibly by fitting a linear regression model by least-squares. In fact, the coefficients of a linear regression model may be considered as first-order sensitivities of an approximate linearized model.

The portfolio is approximately immunized, to the first order, if the condition $\delta V^H = 0$ is met for whatever perturbation δR_i may occur. Hence, all of the coefficients multiplying the factors δR_i in Eq. (2.19) must be set to zero, which requires the solution of a system of m linear equations in the m unknown variables ϕ_j:

$$\frac{\partial V}{\partial R_i} + \sum_{j=1}^{m} \phi_j \frac{\partial H_j}{\partial R_i} = 0, \qquad i = 1, \ldots, m.$$

Depending on the specific context, first-order immunization may translate to approaches known as duration matching for fixed-income portfolios and delta-hedging for option trading, as we shall see. As we may expect, the actual performance of such policies may not be completely satisfactory:

- We are immunizing to the first order, but when the underlying factors change, first-order sensitivities change, for nonlinear hedging instruments. Hence, hedge adjustments may be needed, and we may incur in significant transaction costs.

- We may improve performance by using second-order sensitivities, too. In the case of bonds, second-order sensitivities are related to bond convexity, and in the case of options they are related to option gamma. Again, we may use explicit formulas or numerical methods to obtain these sensitivities, but the fact remains that we are perfectly hedged for *small* perturbations. Practitioners might argue that an imperfect, but more robust hedge might be preferred.

- We have matched the number of hedging instruments and the number of risk factors. However, there may be factors that are simply not hedgeable. In such a case, we may still try to minimize the variance of the hedge, or another suitable risk measure. Furthermore, the number of risk factors can be too large to be practical. In interest rate risk management, there is an infinite number of risk factors in principle, i.e., an interest rate for each possible maturity. However, we may take advantage of the relationships among risk factors to reduce the complexity from both a computational and financial perspective.[30]

- Last but not least, whenever we use a model, like a factor model or a pricing model, we are subject to model risk. The model itself may be wrong, or our estimates of critical parameters may turn out to be inadequate to cope with changed market conditions.

2.2.4 FINANCIAL VS. NONFINANCIAL RISK FACTORS

So far, we have only considered *financial* risk related to, e.g., uncertain stock share returns and the impact of interest rate on bond prices. We have also mainly assumed the viewpoint of a financial institution or an individual investor. From a mathematical perspective, this is enough motivation to develop quantitative tools to tackle the related problems. However, we must bear in mind that financial risk factors have an impact on nonfinancial players and that nonfinancial risk factors have an impact on financial players, too. In this section, we give a broader view of risk categories, with some examples showing the role of nonfinancial factors.

Beside **market risk** and **interest rate risk**, we should consider **currency risk** (foreign-exchange risk, if you prefer). This is relevant to financial institutions investing in foreign assets, as well as to nonfinancial firms with international operations. Signing a contract that prescribes a future payment in a foreign currency implies exposure to adverse movement in the exchange rate. Hence, nonfinancial firms may also be interested in using derivatives to mitigate this exposure. **Inflation risk** is another important category. We should realize that risk factors need not be uncorrelated. There is an interplay between inflation risk and interest rate and currency risk.

We should also be concerned with **counterparty** and **credit risk**. Counterparty risk is relevant, e.g., in OTC markets: The counterparty in a forward contract may fail to comply with his obligations. Credit risk is related to the possibility of default on a loan, and has a severe impact on both corporate and noncorporate bonds. Both categories of risk may have to do with the financial health of a nonfinancial player, as well as with general economic conditions.

There are also specific risk categories that are relevant to specific assets. For instance, **volatility risk** refers to the impact of volatility on derivative prices. Vanilla call and put options are both sensitive to a change in volatility, in the same way: The larger the volatility, the larger the price. This should not be

---

[30] A well-known data reduction technique that is frequently proposed is principal component analysis. See, e.g., [2, Chapter 17].

confused with market risk, as volatility risk is nondirectional,[31] i.e., it is not related to prices going either up or down.

When building sophisticated pricing models, we are introducing **model risk**. Model risk may be considered as nonfinancial in the sense that it is not related with market variables, but it is clearly related to financial market modeling. Other nonfinancial risk categories are **regulatory** and **political risk**. Changes in bank regulations and government policies may have an impact on financial markets.

On the corporate side, a form of risk which is important to financial institutions is **volume risk**. In the hedging examples we have considered, we have taken for granted the knowledge of N, i.e., the units of the asset we have to buy or sell. However, this amount may be related to business contingencies, as the following example illustrates.

Example 2.5 Volume risk in forward contracts

In Section 2.2.3.1, we have considered perfect hedging, and we have taken for granted that we know the *size* of the required hedge. Consider the case of a US firm that will need an amount of N euro in six months. The firm could buy the euros now, but this would be bad for liquidity. An alternative is to take a long position in a forward contract for an amount N, locking the price.

So far, so good. But we should ask why the firm needs that amount of currency. Imagine that the firm has anticipated the need to buy a set of components from a supplier in the eurozone, to assemble a given number of equipments for a client in the USA. The amount N depends on how many items the client plans to order, which determines the number of components needed. What happens if, maybe because of economic recession, the client cuts the order by a significant amount? Now the US manufacturer has to buy an excessive amount of euros, at a locked forward price, and the hedge is not perfect anymore. If the dollar drops with respect to euro, no harm done: The firm will end up with a windfall payoff, but speculation on currencies is not its core activity. If, on the contrary, the euro drops, the firm will have to buy euros that it does not need, at a large price, just when the business is turning sour. Chances are that the risk manager will find himself in an uncomfortable spot when trying to explain this to his boss. Overlooking volume risk may be dangerous, indeed.

In practice, quantifying a risk exposure is far from trivial. In a complex organization, it may be difficult to assess how much hedging is really needed.

---

[31] See Section 13.6.

This is especially true when considering a multinational firm with many different currencies involved. Sometimes, opposite cash flows in the same currencies may provide a sort of natural hedge, but this may be nontrivial to assess by checking accounting statements. It is also difficult to associate volume risk with reliable probabilities. For instance, imagine that the volume risk is related to business expansion opportunities, depending on whether some contracts will be signed or not by potential customers. Historical data may be of very little help, and subjective probabilities may be difficult to assess. Robust models might be needed, and an approximate hedge which works more or less well in any scenario might be preferable to a perfect hedge working perfectly only in one scenario.

Last, but not least, we should also consider **operational risk**. Loosely speaking, this is a catch-all risk category accounting for risk due to errors, system malfunctions, and wrong data/information creeping in the system. A few examples may illustrate the relevance of operational risk:

- Order execution critically relies on information technology (IT) infrastructure. Imagine the effect of a computer network crashing at a most critical moment. Such an event may be due to a system malfunction or to a catastrophic event, like a flood or an earthquake. Proper countermeasures should be taken in order to ensure business continuity.
- Algorithmic trading relies on sophisticated algorithms, executed at maximum speed. War histories abound on the dire consequences of obscure software bugs or improper installation of wrong software versions.
- In other cases, strange market behavior may be attributed to human error. A well-known example is the *fat finger* mistake, whereby a wrong number is entered. Imagine adding a trailing '0' to the price of a trading order.
- In a more and more interconnected world of big data, the consequences of a mistake or wrong information may be remarkable. In 2008, stock shares of United Continental Holdings Inc. (UAL, the parent company of United Airlines) plunged by 75%, allegedly because an investors' newsletter reported news of UAL filing for bankruptcy. Actually, that was news six years old, popping up from a Google search, and it had nothing to do with relevant market information. Imagine the effect of these search engine issues on trading strategies based on sentiment analysis.

2.3 The no-arbitrage principle in asset pricing

We have argued that we need mathematical models to find the fair value of assets, especially derivatives. The output of valuation models is just one ingredient in determining the actual price, even though we use the term *asset pricing* rather than *asset valuation*. The actual price will include a profit margin, as well as some fudge to allow for model errors and risk factors that cannot be hedged. The cornerstone of such models is the **no-arbitrage principle**. Informally, an arbitrage opportunity is a trading strategy that requires no initial commitment of money and will result in a riskless profit. The basic idea is that such a nice money-making machine (or free lunch, if you prefer) should not exist, since market participants will immediately take advantage of any opportunity like that, making it disappear quite fast. The idea is somewhat related to the **efficient market hypothesis** (**EMH**) in financial economics.[32] Despite its simplicity and appeal, the principle is not free from controversy. In particular, it is criticized by behavioral financial economists, who argue against a completely rational approach to finance.[33]

Example 2.6 A behavioral joke

> An illuminating joke on the controversy about the EMH goes more or less like this. A well-known professor of the rational school of thought is walking and chatting with a student. The student notices a banknote lying on the sidewalk and points it out to the professor, who replies: "There is no banknote on the sidewalk, as if there were one, someone would have already picked it up."

2.3.1 WHY DO WE NEED ASSET PRICING MODELS?

Several assets, like stocks, bonds, and exchange-traded derivatives, are quoted on regulated markets, where prices are driven by the interplay between demand and offer. So, one may well wonder why we should bother developing mathematical models to find prices that we should better read on a computer screen. Indeed, pricing assets seems like a daunting task, which should take into account several issues:

- Uncertain risk factors
- Liquidity issues
- Wealth and risk aversion of market participants
- Information asymmetries and different market views

At best, we may hope to come up with a reasonably simplified model that does not produce blatantly absurd prices.

Actually, there are several reasons for the development of asset pricing models:

---

[32] See Chapter 10 for a related discussion of equilibrium models. Roughly speaking, the EMH states that asset prices immediately incorporate all relevant information, and there is no bias due to irrational behavior or information asymmetry.

[33] See Section 10.5.

- Pricing quoted assets is useful to calibrate a model in terms of underlying factors. For instance, a bond price depends on interest rates. Hence, if we are provided with a pricing model, we may estimate the underlying interest rates by observing bond prices. This is also quite relevant when we want to infer something about unobservable parameters like volatility, on the basis of observed option prices. Historical price volatility may be estimated on the basis of past asset prices, but what we may really need is an estimate of the *current* volatility (even better, a forecast of future volatility). Given a model relating option prices with volatility, we may come up with an *implied volatility*, which is a market consensus view, rather than a historical volatility.
- If we have a well-defined view about the underlying factors of the price of a security, we might observe a discrepancy between the quoted and fair prices and take advantage of it by a suitably designed trading strategy. For instance, if we believe that the current level of implied volatility is too low and that it is going to increase, we could buy derivatives that are long volatility (i.e., their price is increasing with respect to volatility). If markets move as we predict, a considerable profit might result.
- Some assets, like OTC derivatives, are not really quoted. When a firm asks an investment bank to engineer a derivative for a specific hedging requirement, it may wish to check whether the asked price is reasonably fair. Pricing models may provide the firm with an estimate of the fair value.
- In risk management applications, we need a model telling us how asset prices are expected to react to moves in underlying risk factors. For instance, we do not want to consider the price of each bond or derivative as a risk factor by itself, as it would be quite difficult to come up with a sensible model accounting for their correlations. Rather, we should focus on a limited number of common factors, affecting the whole set of securities, and assess price sensitivities by a parsimonious model.

In this book, we will always price securities under the assumption that no arbitrage opportunity is available. As we shall see, this does not necessarily guarantee the uniqueness of prices.

2.3.2 ARBITRAGE STRATEGIES

There are different ways of defining an **arbitrage strategy** in theory, and different ways of carrying it out in practice. All of them, however, lead to a safe way of making money, without the need of an initial capital. This should not be confused with investing in the risk-free asset, where we do make a sure profit, but we need some initial capital. Such a nice money-making machine, as suggested by economic common sense, should not exist. Or, at the very minimum, it should exist for a very short amount of time, since someone will take immedi-

2.3 The no-arbitrage principle in asset pricing

ately advantage of it, pushing prices back in line. We may distinguish different types of arbitrage:

- **Instantaneous arbitrage**, where we trade at a given time instant, without committing any resource, and we gain an immediate riskless profit.
- **Static arbitrage**, which involves trading at two time instants, say, $t = 0$ and $t = T$, and results in the creation of a riskless profit without committing any resource.
- **Dynamic arbitrage**, which involves trading at multiple time instants and will not be considered in this chapter.

Example 2.7 An instantaneous arbitrage

Suppose that a stock share is traded on two markets, one with prices denominated in euros and the other one in dollars. Now imagine that we observe the following prices:

- The current exchange rate is 1.34\$/€ (by which we mean that the price of one € is \$1.34)
- The stock share price is €50 on the first market and \$68 on the second market

It is easy to see that these prices are not in line. If we take the first two as correct, the price of that stock share should be

$$€50 \times 1.34 \frac{\$}{€} = \$67.$$

We may buy the stock share on market 1 and sell it immediately on market 2, with a risk-free profit of \$1. If many players pursue the same strategy, the ensuing pressure on prices will push them back in line and eliminate the misalignment.

As one can imagine, instantaneous arbitrage also implies that three currency exchange rates involving three currencies should be consistent. Let us denote by $S_{x/y}$ the price of currency y in units of x. Then, given currencies a, b, and c, we must have:

$$S_{a/b} = S_{a/c} \times S_{c/b}.$$

A violation of this relationship allows a *triangular arbitrage*. Note that we are not considering transaction costs, as well as execution uncertainty due to delays, which may affect the actual execution of an arbitrage strategy.

Example 2.8 A static arbitrage with an immediate profit

Consider the following situation:

- The price of a stock share is currently $55. The stock share will not pay any dividend in the next six months.
- Two options, a call and a put, are traded on that stock share, maturing in six months and with the same strike price, $60. The price of the call option is $2, and the price of the put is $7.
- The risk-free rate, with semiannual compounding, is currently 4% (this means that by lending money for six months we gain 2% of the capital; see Section 3.1).

Now, consider the following trade at time $t = 0$:

- Write the put option to a counterparty (which will be the holder of the put option).
- Short the stock share.
- Buy the call option (we are the holders of the call option).
- Buy a riskless zero-coupon with face value, $60, which is equivalent to investing

$$\frac{60}{1 + 0.02} \approx \$58.83$$

for six months, at the risk-free rate.

Note that the overall cash flow at time $t = 0$ is

$$7 + 55 - 2 - 58.83 = 1.17.$$

We are actually making some money, which is equivalent to saying that we have bought a portfolio with negative value.

At maturity, at time $t = 0.5$ measured in years, one of the following three cases will occur:

1. If the stock price is exactly $60 no option is exercised, and we use the face value of the bond to buy the stock share and close the short position.

2. If the stock price is larger than $60, the put is not exercised by its holder, and we use the call to buy the stock share at $60 and close the short position, where the cash needed is provided by the zero-coupon bond.

3. If the stock price is smaller than $60, the put is exercised by its holder: We have to buy the stock share at $60, and we use it to close the short position; again, the cash needed is provided by the zero. The call is not exercised.

2.3 The no-arbitrage principle in asset pricing

> Whatever the case, we always break even and the net cash flow at maturity is zero.

The trade that we have used may look somewhat mysterious, but it is related to the **put–call parity** relationship that we will consider later, in Chapter 13. In this kind of arbitrage, there is a portfolio with negative value at time $t = 0$, and its value is always zero at $t = T$. Buying the portfolio results in an immediate profit now, with no possibility of losing money in the future. We may consider another kind of arbitrage in which the net cash flow is zero now, and we may only make a profit in the future.

Example 2.9 A static arbitrage with deferred profit

Let us consider two assets, a and b, with current price

$$S_a(0) = S_b(0) = 1,$$

at time $t = 0$. At a later time $t = T$, the asset prices, denoted by $S_a(T,\omega)$ and $S_b(T,\omega)$, are random variables whose value depends on the realized scenario ω. Let us assume that there are three possible scenarios ω_i, $i = 1, 2, 3$, with prices given in Table 2.2. Let us consider a portfolio in which we hold an amount h_a and h_b of the two assets, respectively. We notice that profit/loss is given as follows, depending on the scenario/state of nature:

State ω_1: $\quad h_a(2-1) + h_b(1-1) = h_a$
State ω_2: $\quad h_a(0-1) + h_b(0-1) = -h_a - h_b$
State ω_3: $\quad h_a(2-1) + h_b(2-1) = h_a + h_b.$

If we choose $h_a = 1$ and $h_b = -1$, the net cash flow at $t = 0$ is zero, we have a profit in state ω_1, and no profit/loss is incurred in the remaining states. More generally, in this example any portfolio with $h_a + h_b = 0$, $h_a > 0$, is an arbitrage strategy.

We may observe that the source of the anomaly is that asset a dominates asset b, state by state. We will consider conditions precluding this anomaly in Section 2.4.2.

In practice, the term "arbitrage" may refer to strategies that are actually somewhat risky, as they rely on price misalignments spotted by a specific asset pricing model. Such an arbitrage strategy is subject to model risk. Another source of potential trouble, in an arbitrage strategy unfolding over time, is the need for liquidity in the short term. A strategy might be potentially successful

Table 2.2 Asset prices at time $t = T$ in state ω, for Example 2.9.

State	ω_1	ω_2	ω_3
Price $S_a(T, \omega_i)$	2	0	2
Price $S_b(T, \omega_i)$	1	0	2

in the long term, but result in unsustainable short-term losses because of margin calls on leveraged or short positions. In this book, we consider *deterministic* and somewhat idealized arbitrage strategies, whereas *statistical* arbitrage strategies are often pursued in practice.

We will investigate the nature of arbitrage strategies, as well as their relationship with dominant strategies, in Section 2.3, for the simple case of a finite number of possible scenarios (states of nature), i.e., a discrete probability distribution. The concept may be extended to the continuous case, at the price of a more sophisticated mathematical machinery. Here, we want to consider a few examples illustrating practically relevant consequences of the assumption that arbitrage opportunities cannot exist for a long time in well-functioning markets.

2.3.3 PRICING BY NO-ARBITRAGE

Several pricing models assume that arbitrage opportunities do not exist. To be more precise, the pricing principle based on lack of arbitrage opportunities assumes that they may exist, but arbitrageurs are very quick to take advantage of them and their trading strategies make prices realign. Pricing by no-arbitrage is a relatively simple principle, leading to much simpler pricing approaches than full-fledged equilibrium models, as it does not require any information about wealth endowments, market views, and degrees of risk aversion of market participants.

In Section 2.4, we discuss in more detail the mathematics of (no-)arbitrage, but parts of its consequences are rather intuitive:

- We cannot have two different risk-free interest rates in the same economy. Otherwise, one would borrow at the lower rate to invest at the higher one. Clearly, this assumes a market with no bid–ask spread on rates, i.e., rates to borrow or lend money are the same.

- Assets or portfolios that will have the same value in the future, whatever scenario occurs, must have the same value now. This is essentially a **law of one price**, which again assumes a frictionless market with no transaction costs.

- Assuming that we may bundle and unbundle assets and cash flows freely, i.e., no friction is involved and we can synthesize an asset as a linear combination of other assets, we will also see that pricing is a linear operator.

2.3 The no-arbitrage principle in asset pricing

We may get a clue about the power of no-arbitrage by looking at a few simple, but instructive examples.

Example 2.10 Relative bond pricing

Consider two bonds maturing in five years. The first one has coupon rate 9% (assume a single annual coupon) and its price is 104.36 (face value is 100). The second one has coupon rate 7% (annual coupon, again) and its price is 96.3. What is the price of a zero-coupon bond maturing in five years?

It seems that we have very little information, but we may easily find the only price of the zero that is in line with the other two bond prices. We have just to realize that the two coupon-bearing bonds have synchronized cash flows:

$$C_{1,t} = 7, \quad C_{2,t} = 9, \quad t = 1,2,3,4$$
$$C_{1,5} = 107, \quad C_{2,5} = 109,$$

whereas the zero has only one cash flow at maturity, $C_{z5} = 100$.

We may easily build a portfolio consisting of the two coupon-bearing bonds, replicating the cash flow of the zero, by solving the following set of linear equations:

$$9x_1 + 7x_2 = 0$$
$$109x_1 + 107x_2 = 100,$$

where x_1 and x_2 are the amount of the two bonds in the replicating portfolio. Solving the system yields

$$x_1 = -\frac{7}{2}, \quad x_2 = \frac{9}{2}.$$

Note that the first position takes a negative value, i.e., we should sell the first bond short. The value of the replicating portfolio is

$$-\frac{7}{2} \times 104.36 + \frac{9}{2} \times 96.3 = 68.09,$$

which must be the price of the zero, if we rule out arbitrage opportunities. Note that we are implicitly assuming that default risk is negligible for all of the bonds involved, that short-selling is possible, and the market is frictionless.

Example 2.10 is a straightforward application of the law of one price. Let us investigate another example in this vein, involving options.

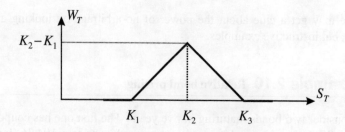

FIGURE 2.5 A butterfly spread.

Example 2.11 Butterfly spreads and the law of one price

Consider a weird derivative with the payoff W_T depicted in Fig. 2.5, depending on the price S_T of a certain underlying asset at time $t = T$. The breakpoint prices, $K_1 < K_2 < K_3$, are such that

$$K_2 = \frac{K_1 + K_3}{2}.$$

The payoff is piecewise linear, and the slopes on the range with strictly positive payoff are $+1$ and -1. This payoff corresponds to a common trading strategy, called **butterfly spread**. When K_2 is close to the current price S_0 of the underlying asset, the strategy is essentially a bet on low volatility, i.e., we make a profit if the asset price does not move away from S_0. Now imagine that, at time $t = 0$, three call options on the same asset are available, with strike prices K_1, K_2, and K_3, respectively. These three options mature at time $t = T$, and let $C_i(0)$, $i = 1, 2, 3$ denote their prices.

It is easy to see that the butterfly spread may be synthesized by taking a long position in one option with strike K_1, a long position in one option with strike K_3, and a short position in two options with strike K_2. To see this, observe that the value of the portfolio of call options matches the butterfly spread for any price S_T. This is summarized in Table 2.3. When $S_T < K_1$, all of the call options have zero payoff (we say that they are out-of-the-money) and are not exercised at maturity; hence, the total payoff from the option portfolio is zero, as in Fig. 2.5. For $S_T \in [K_1, K_2)$, the first option has a positive payoff (we say that it is in-the-money), and the total payoff is $S_T - K_1$. The other cases are treated similarly, as shown in Table 2.3. Summing the payoffs of the three options in each possible case yields the payoff of Fig. 2.5.

Therefore, we have a portfolio and an asset with the same value in each possible state in the future. Then, by the law of one price, the

2.3 The no-arbitrage principle in asset pricing

Table 2.3 Decomposing a butterfly spread.

Scenario	$S_T < K_1$	$K_1 \leq S_T < K_2$	$K_2 \leq S_T < K_3$	$K_3 \leq S_T$
Payoff option 1	0	$S_T - K_1$	$S_T - K_1$	$S_T - K_1$
Payoff option 2	0	0	$-2(S_T - K_2)$	$-2(S_T - K_2)$
Payoff option 3	0	0	0	$S_T - K_3$
Total payoff	0	$S_T - K_1$	$K_3 - S_T$	0

initial value of the asset must be the initial value of the portfolio,

$$C_1(0) - 2C_2(0) + C_3(0),$$

no more, no less. Otherwise, the law of one price would be violated and we could make an immediate profit by shorting the more expensive portfolio and buying the cheaper one, knowing that we will always break even at maturity. In this case, we would essentially buy a portfolio with a negative initial value, which means that we earn a profit by buying it, but with zero value (no commitment at all) for the future.

Example 2.12 Cash-and-carry arbitrage and forward prices

Suppose that the current spot price for an asset is $S_0 = \$50$, the current forward price for delivery in one year is $F_0 = \$53$, and the annual risk-free interest rate is 3%, with annual compounding. At time $t = 0$ we may:

1. Borrow $50 to buy the asset.
2. Enter into a long position to sell the asset in one year at the forward price.

In one year, we will have to repay

$$\$50 \times 1.03 = \$51.5.$$

Hence, we may sell the asset at $F_0 = \$53$, cashing in a risk-free difference of $1.5. In this case, we have a zero net cash flow at time $t = 0$, and a sure profit at maturity.

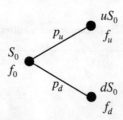

FIGURE 2.6 One-step binomial model for option pricing.

The last example is particularly puzzling. Common sense would suggest that the fair forward price should be related to the expected value of the spot price at maturity. However, under suitable market assumptions, the example suggests that no-arbitrage implies a different forward price. We will have more to say about pricing forward contracts in Chapter 12.

2.3.4 OPTION PRICING IN A BINOMIAL MODEL

Options are nonlinear instruments and, as we shall see in Chapter 13, their pricing involves a quite sophisticated mathematical machinery. However, we may get a clue on how we may price an option by adopting the simplest uncertainty model we may think of, the single-step binomial model. Consider a single time interval of length T. The underlying asset price at the beginning of the time step is S_0; the price S_T at the end of this period is a random variable, taking values uS_0 or dS_0, where $d < u$, with probabilities p_u and p_d, respectively. The single-step binomial model, which is essentially a coin-flipping model, is illustrated in Fig. 2.6.

An option is written on the asset, and its payoff can be f_u or f_d, depending on the outcome. For instance, for a call option with strike K, we have

$$f_u = \max\{0, uS_0 - K\},$$
$$f_d = \max\{0, dS_0 - K\}.$$

We would like to find the current option price f_0. Common sense would suggest that the fair option value should be related to the expected value of its payoff. Since, however, the payoff will be received at time $t = T$, we should discount it somehow. If the interest rate is continuously compounded, we shall see that this requires a discount factor consisting of a negative exponential. Hence, a seemingly sensible guess is (please note the question mark!)

$$f_0 \stackrel{?}{=} e^{-rT} \cdot \mathrm{E}[f_T] = e^{-rT} \cdot [p_u f_u + p_d f_d]. \qquad (2.20)$$

However, Example 2.12 suggests some caution with this intuition, since it may fail for forward contracts. Moreover, in Example 2.10, we have solved a pricing problem very easily by a replication strategy, and we can try with the same approach here.

2.3 The no-arbitrage principle in asset pricing

Assume that a riskless asset is traded, with initial price $B_0 = 1$ and future price $B_u = B_d = e^{rT}$, where r is the continuously compounded risk-free rate. This asset may essentially be regarded as a risk-free bank account or a zero-coupon bond maturing at T.[34] Using the two traded assets, we may set up a portfolio replicating the option payoff. In order to find this **replicating portfolio**, let us denote the number of stock shares in the portfolio by Δ and the amount of cash by Ψ. The initial value of the replicating portfolio is

$$\Pi_0 = \Delta S_0 + \Psi,$$

and its future value, depending on the realized state, will be either

$$\Pi_u = \Delta S_0 u + \Psi e^{rT} \quad \text{or} \quad \Pi_d = \Delta S_0 d + \Psi e^{rT}.$$

To find the composition (Δ, Ψ) of the replicating portfolio, we require two conditions, matching the option payoff state by state:

$$\Delta S_0 u + \Psi e^{rT} = f_u,$$
$$\Delta S_0 d + \Psi e^{rT} = f_d.$$

Solving this system of two linear equations in two unknown variables, we find

$$\Delta = \frac{f_u - f_d}{S_0(u - d)}, \tag{2.21}$$

$$\Psi = e^{-rT} \cdot \frac{u f_d - d f_u}{u - d}. \tag{2.22}$$

In order to avoid arbitrage, by the law of one price, the initial value of this portfolio must be exactly f_0:

$$\begin{aligned} f_0 &= \Delta S_0 + \Psi \\ &= \frac{f_u - f_d}{u - d} + e^{-rT} \cdot \frac{u f_d - d f_u}{u - d} \\ &= e^{-rT} \left\{ \frac{e^{rT} - d}{u - d} f_u + \frac{u - e^{rT}}{u - d} f_d \right\}. \end{aligned} \tag{2.23}$$

It is important to note that this relationship does *not* depend on the objective probabilities p_u and p_d. In particular, the option price is *not* the discounted expected value of the payoff, which could have been a seemingly reasonable guess, expressed in Eq. (2.20). However, Eqs. (2.20) and (2.23) do look quite similar. The latter can be interpreted as an expected value, provided that we introduce new "probabilities,"

$$\pi_u = \frac{e^{rT} - d}{u - d}, \qquad \pi_d = \frac{u - e^{rT}}{u - d} = 1 - \pi_u.$$

---

[34] If we assume a constant and continuously compounded risk-free rate r, there is no difference between depositing $1 in a bank account, which will grow to $\$e^{rT}$, and buying a bond with face value $\$e^{rT}$ at the current price of $1. However, things are quite different if we consider a bank account with a nonconstant interest rate, exposing us to reinvestment risk. Hence, it is better to interpret the risk-free asset as a bank account. We will clarify this matter in Section 14.3.3.

However, can we really say that these are probabilities? A first check is that, indeed, they add up to 1,

$$\pi_u + \pi_d = 1,$$

which is fine. However, probabilities should also be non-negative. It is easy to see π_u and π_d are positive if $d < e^{rT} < u$, which *must* be the case if there is no arbitrage strategy involving the riskless and the risky asset. If we had

$$e^{rT} < d < u,$$

then the risk-free asset would always be outperformed, and we could make an unbounded profit by borrowing cash and investing it in the risky stock share; the future value of the stock shares, in this case, will always be larger than the debt we have to repay. On the contrary, if

$$d < u < e^{rT},$$

then we should sell the stock share short and invest the proceeds in the risk-free asset, with the guarantee that we will have enough cash to buy back the stock share and close the short position. Thus, assuming that no arbitrage opportunity exists, we may interpret π_u and π_d as probabilities and write the option price as the discounted expected value of payoff, under a new probability measure defined by those probabilities:

$$f_0 = e^{-rT} \cdot \mathrm{E}_{\mathbb{Q}_n}[f_T] = e^{-rT} \cdot (\pi_u f_u + \pi_d f_d). \qquad (2.24)$$

The notation $\mathrm{E}_{\mathbb{Q}_n}[\cdot]$ points out that expectation is taken with respect to a different probability measure. The probability measure $\mathbb{Q}_n$ is also called **risk-neutral**. To understand why, let us write the expected value of S_T under probabilities π_u and π_d:

$$\begin{aligned}
\mathrm{E}_{\mathbb{Q}_n}[S_T] &= \pi_u S_0 u + \pi_d S_0 d \\
&= S_0 \cdot \left(\frac{e^{rT} - d}{u - d} u + \frac{u - e^{rT}}{u - d} d \right) \\
&= S_0 e^{rT}. \qquad (2.25)
\end{aligned}$$

We see that the return of the risky stock share is exactly the risk-free rate r. This would be true in a world of risk-neutral investors, who do not require any compensation for bearing risk and only care about expected values, as we shall see in Chapter 7. In such a risk-neutral world, the expected return of any asset would just be the risk-free rate.

Another interesting insight is obtained if we consider the ratio between prices S_t and B_t. Under $\mathbb{Q}_n$, a rearrangement of Eq. (2.25) yields, recalling that $B_0 = 1$,

$$\frac{S_0}{B_0} = \mathrm{E}_{\mathbb{Q}_n}\left[\frac{S_T}{B_T}\right],$$

i.e., the expected value of the ratio at time $t = T$ is just the current value of the ratio. As we shall see in Chapter 11, this is a property characterizing a

2.3 The no-arbitrage principle in asset pricing

family of stochastic processes called **martingales**. We leave it as an exercise (see Problem 2.6) to check that the ratio f_t/B_t is a martingale, too, under $\mathbb{Q}_n$. The condition

$$\frac{f_0}{B_0} = \mathrm{E}_{\mathbb{Q}_n}\left[\frac{f_T}{B_T}\right]$$

implies the pricing relationship of Eq. (2.24). This is why $\mathbb{Q}_n$ is also known as **equivalent martingale measure**. In our simple setting, dividing S_t by B_t essentially amounts to *discounting* the asset price. However, the idea is more general, and we may consider dividing by the price of an almost arbitrary asset, playing the role of a **numeraire**, under the condition that its price is strictly positive (to avoid trouble with division by zero). Alternative numeraires are associated with different probability measures, which leads to powerful pricing approaches.

We observe that the replication argument boils down to the trivial fact that we are expressing the vector $[f_u, f_d]^\mathsf{T}$ in the bidimensional space $\mathbb{R}^2$ as a linear combination of two linearly independent vectors $[S_u, S_d]^\mathsf{T}$ and $[e^{rT}, e^{rT}]^\mathsf{T}$ forming a basis. When an arbitrary payoff can be replicated by a set of spanning assets, we say that the market is **complete**. The trouble with this assumption is that market completeness implies that derivatives are redundant, so why should we bother with them? As one can imagine, real-life markets are **incomplete**, i.e., there are payoffs/derivatives that cannot be replicated by trading in elementary assets. We shall investigate the matter in more detail in Section 2.4, but the bottom line is the following:

- A market model does not allow arbitrage opportunities if and only if there exists an equivalent martingale measure.
- The equivalent martingale measure is unique if and only if the market model is complete.

Depending on the market model, proving all of this may be relatively simple or quite demanding but, as we have seen, the essential insights can be obtained by a simple binomial model. Clearly, a replication argument works under the assumption of market completeness. In this case, the equivalent martingale measure is unique and may be interpreted as risk-neutral. As we shall observe in Section 13.3, the argument may be recast in terms of hedging the risk of option writing. This second approach may have the advantage of applying to incomplete markets, too. Furthermore, arbitrage-free but incomplete markets allow the existence of multiple equivalent martingale measures. Hence, incomplete markets require a calibration procedure to find the "right" martingale measure. This amounts to finding a model matching the market prices of traded derivatives, and then applying the model to price OTC securities, as we shall see in Chapter 14.

2.3.5 THE LIMITATIONS OF THE NO-ARBITRAGE PRINCIPLE

Whenever we use the no-arbitrage principle, it is a good idea to remind ourselves of the assumptions, hidden or otherwise, that we are making. In fact, there are some limitations in its use, since we neglect features of real-life markets.

- We neglect market frictions, i.e., bid–ask spreads and transaction costs, as well as the effect of taxes and the difference in interest rates for borrowing and lending. Usually, the effect of transaction costs and other frictions is that there is a range of arbitrage-free prices, rather than a unique price.
- Some trades may work in the long term, but may be adversely affected by liquidity issues. This may happen if prices move in an unfavorable way in the short term, with the potential effect of receiving a sequence of nasty margin calls. We also have to cope with limits to short-selling, a usual ingredient of arbitrage strategies, as well as inventory costs (which are relevant when trading commodities).
- Another feature of real-life markets is the presence of market players with bounded rationality, as well as asymmetric information. Some role may be played by noise trader risk, i.e., the risk of price movements caused by uninformed traders, which may be irrational from the viewpoint of an informed trader, but may well affect prices in the short term.
- Another assumption we make is that we can observe prices and immediately operate on markets. Actually, an order must be issued and executed in an environment featuring faster and faster dynamics, due to the pervasive role of information technology. As a result, a trade may suffer from execution uncertainty.
- Last, but not least, we should be aware of model risk. Some assets may look relatively mispriced according to a pricing model, but we may not be sure that the model itself and the estimates of its parameters are quite correct.

A well-known real-life example of the above issues is represented by the debacle of the Long Term Capital Management (LTCM) hedge fund in 1998. The fund used convergence strategies, based on the detection of price misalignments, under the assumption that, sooner or later, security prices will be brought back in line (this is an extremely simplified view of the actual strategies employed). Under extreme market conditions, however, models may break down and prices may take unexpected routes. In the LTCM case, this was caused by a potential default on Russian bonds, which caused a flight to quality, i.e., massive sales of risky assets to invest in safe ones. Thus, prices did not converge at all, and the gaps widened, leading to massive losses. With highly leveraged positions, these losses cannot be sustained because of liquidity issues, even if convergence does take place in the long term.

2.4 The mathematics of arbitrage

The mathematics of arbitrage relies on stochastic models for asset prices, interest rates, and other risk factors. In Section 2.3.4, we have seen how, in a simple binomial setting, where we assume market completeness, we may replicate any payoff by a portfolio consisting of two spanning assets. Market completeness in the binomial case means that the payoffs of the two spanning assets are linearly independent and are a basis for the two-dimensional space $\mathbb{R}^2$. Clearly, two states do not make an excellent model of uncertainty. If we increase the number of scenarios to m, we may represent uncertainty more accurately, but if we pursue the same approach as before, we would need a set of m spanning assets forming a basis for $\mathbb{R}^m$. This does not seem practically sensible, if we are considering an option written on a single underlying asset. As we shall see in Chapter 13, the trick is to introduce *dynamic* replicating portfolios, based on cash and the underlying asset. If we allow for trading in *continuous time*, we may even be able to cope with a continuous random variable modeling uncertainty. We will deal with such models in Chapter 11, but developing the mathematics of arbitrage in that context requires tools from stochastic calculus and functional analysis, beyond the scope of this book. Here, we restrict the analysis to *static* trading strategies and *finite-dimensional* models, which allow us to use simpler tools from linear algebra and linear programming. This is sufficient to get the fundamental insights, without bothering too much with advanced mathematical machinery. Nevertheless, the uninterested reader may safely skip this section.

We consider a single-period market, where trading occurs at dates $t = 0$ and $t = T$. The sample space Ω consists of m possible states of the world (scenarios) $\Omega = \{\omega_1, \omega_2, \ldots, \omega_m\}$, with probability measure $p(\omega) > 0$, $\forall \omega \in \Omega$. Hence, we deal with quite simple stochastic processes with sample paths $Y(t, \omega)$, $t \in \{0, T\}$.[35] Since the initial state is known, when we refer to time $t = 0$ we may suppress dependence of the process on ω to improve readability. Thus, we will write $Y(0)$ and $Y(T, \omega)$ when referring to the initial and terminal states, respectively. On this market, $n + 1$ securities are traded:

- n "risky" securities, indexed by $i = 1, \ldots, n$, with price process $S_i(t, \omega)$, where the initial price is strictly positive, $S_i(0) > 0$, and the terminal price is non-negative, $S_i(T, \omega) \geq 0$. This non-negativity condition is satisfied by stock share prices, since equity shares are a limited liability asset, as well as by derivatives with a non-negative payoff, like call and put options.
- A "bank account," associated with a price process $B(t, \omega)$, such that $B(0) = 1$ and $B(T, \omega) > 0$.

---

[35] A stochastic process is a sequence of random variables over time. Here, we just have *one* random variable at time $t = T$, but it is a good idea to strive for generality.

Note that the bank account could correspond to a truly risk-free asset, where there is no randomness and $B(T,\omega) = B(T)$ for every ω, but this need not be the case. The strict positivity condition on B is important when we use it as a numeraire, i.e., when we consider the **discounted price processes**

$$S_i^*(t,\omega) \doteq \frac{S_i(t,\omega)}{B(t,\omega)}, \qquad i = 1,\ldots,n.$$

The discounted price process for the bank account is, trivially,

$$B_i^*(t,\omega) = 1.$$

For the sake of convenience, we will assume that $B(T,\omega) = 1 + r$, where r is a risk-free holding period return.

The questions that we want to address in this simple market model are:

- Is the market model sensible? A sensible market model should not allow for riskless money-making machines.

- How can we price **contingent claims**, i.e., securities offering a defined payoff $X(\omega)$ for each state ω at time $t = T$? The contingent claim is a contract C, associated with a function $X : \Omega \to \mathbb{R}$. In concrete terms, this will be a derivative with a well-defined payoff.

A static trading strategy may be described by the vector $\mathbf{h} = [h_0, h_1, \ldots, h_n]^T \in \mathbb{R}^{n+1}$, representing the holding of each security, where h_0 refers to the bank account. We fix $\mathbf{h}$ at time $t = 0$ and see the result at time $t = T$, at the end of the holding period. In this static setting, the trading strategy boils down to a portfolio; things are more complicated in a dynamic setting. In any case, there should not exist a trading strategy creating riskless money out of nothing.

The payoffs of all of the contingent claims form a **linear space** of random variables.[36] The numerical values taken by $X(\omega)$ can be collected into a payoff vector in $\mathbb{R}^m$, which we identify with the function itself. Thus, in our finite-dimensional setting, the space of random variables boils down to the linear space of vectors in $\mathbb{R}^m$, on which we may take arbitrary linear combinations. Note that the basic securities in the market model have a given initial price, but contingent claims have not. We would like to define a sensible **pricing functional** Π, mapping the random variable $X(\omega)$ (the payoff of the contingent claim) to a real number (its price).

---

[36]From a mathematical viewpoint, the term *linear space*, which is equivalent to *vector space* in our setting, refers to a set of objects that can be linearly combined: If $\mathbf{x}$ and $\mathbf{y}$ belong to a linear space L, so does the linear combination $a\mathbf{x} + b\mathbf{y}$, where a and b are real numbers. From a financial viewpoint, this means that we may bundle and unbundle assets forming other assets, without incurring in any transaction cost.

2.4 The mathematics of arbitrage

Let us define the vector of initial prices of the $n+1$ traded securities,

$$\mathbb{V} = \begin{bmatrix} 1 \\ S_1(0) \\ S_2(0) \\ \vdots \\ S_n(0), \end{bmatrix} \in \mathbb{R}^{n+1},$$

and the matrix of asset prices (payoffs) at $t = T$,

$$\mathbb{Z} = \begin{bmatrix} 1+r & S_1(T,\omega_1) & \cdots & S_n(T,\omega_1) \\ 1+r & S_1(T,\omega_2) & \cdots & S_n(T,\omega_2) \\ \vdots & \vdots & \ddots & \vdots \\ 1+r & S_1(T,\omega_m) & \cdots & S_n(T,\omega_m) \end{bmatrix} \in \mathbb{R}^{m \times (n+1)}.$$

Note that we choose to associate columns with assets and rows with states of the world. For a given trading strategy $\mathbf{h}$, the initial portfolio value is $\mathbb{V}^T\mathbf{h}$, and the vector of terminal values in each state of the world is $\mathbb{Z}\mathbf{h}$. Trading strategies generate payoffs that are linear combinations of the columns of $\mathbb{Z}$. These columns span a linear subspace of $\mathbb{R}^m$, but not necessarily the whole space.

A traded security is redundant if its payoff may be generated by a trading strategy based on the other securities. We say that the market is **complete** if any contingent claim may be replicated by a trading strategy. Otherwise, the market is **incomplete**. Clearly, the market is complete if the set of columns in $\mathbb{Z}$ (securities) spans not only a linear subspace, but the whole linear space $\mathbb{R}^m$. This is the case if the rank of $\mathbb{Z}$ is m, i.e., the matrix has full row-rank.

2.4.1 LINEARITY OF THE PRICING FUNCTIONAL AND LAW OF ONE PRICE

Let us consider contingent claims C, C_a, and C_b, whose respective payoffs are related by

$$X(\omega) = \alpha X_a(\omega) + \beta X_b(\omega).$$

The financial counterpart of this linear combination is related to the possibility of:

- **Bundling** securities, whereby we buy α units of C_a and β units of C_b, and sell one unit of C
- **Unbundling** securities, whereby we buy one unit of C and sell α units of C_a and β units of C_b.

In either case, the net payoff will be zero. If we compare the prices of the three contingent claims, we see that the pricing functional Π must be linear:

$$\Pi(X) \equiv \Pi(\alpha X_a + \beta X_b)$$
$$= \alpha \Pi(X_a) + \beta \Pi(X_b).$$

If there are no transaction costs, any difference would lead to a money-making machine (sell the expensive claim and buy the cheap one) ensuring an immediate profit with no net obligation in the future.

The linearity of the pricing functional also implies a common sense result: A contingent claim with zero payoff in every state must have zero price. All of this is essentially related to the **law of one price**: Two contingent claims with the same payoff in every state must have the same price. If the law of one price is violated, then we can easily build money-making machines. Note, however, that we may build money-making machines even if the law of one price is satisfied. As a simple example, consider a binomial model where

$$e^{rT} < d < u,$$

i.e., the risk-free gain is smaller than the gain of the risky security in every state. The law of one price is not violated, since the payoffs of the two secuties are different (in fact, linearly independent). Nevertheless, such a model would imply a clear arbitrage opportunity, as we have seen, based on borrowing cash and investing in the risky security. A similar consideration applies to the securities in Table 2.2. These two cases have one thing in common: The payoff of a security dominates, state by state, the payoff of another security. Therefore, the law of one price is a *necessary*, but not a *sufficient* condition for sensible market models and pricing mechanisms. We have to investigate in more detail trading strategies and define two concepts: Dominant strategies and arbitrage opportunities.

2.4.2 DOMINANT STRATEGIES

For a given trading strategy h, we define the **value process** (a stochastic process),

$$V(t,\omega) \doteq h_0 B(t,\omega) + \sum_{i=1}^n h_i S_i(t,\omega),$$

for $t \in \{0, T\}$, and the additive **gain**[37] (a random variable),

$$G(\omega) \doteq V(T,\omega) - V(0) = h_0 r + \sum_{i=1}^n h_i \cdot \delta S_i(\omega),$$

---

[37] In this book, we refer to gain as a *multiplicative* factor. In this section, gain is defined in additive terms of profit/loss, i.e., as a difference of values. In order to conform to existent literature, we avoid the introduction of another term, but no ambiguity should arise.

2.4 The mathematics of arbitrage

where $\delta S_i(\omega) \doteq S_i(T, \omega) - S_i(0)$. Since discounting is so relevant in pricing, we will use the bank account as a numeraire and define the discounted price and value processes as

$$S_i^*(t, \omega) \doteq \frac{S_i(t, \omega)}{B(t, \omega)}, \qquad V^*(t, \omega) \doteq h_0 + \sum_{i=1}^{n} h_i S_i^*(t, \omega),$$

respectively, as well as the **discounted gain**,

$$G^*(\omega) \doteq V^*(t, \omega) - V^*(0) = \sum_{i=1}^{n} h_i \cdot \delta S_i^*(\omega),$$

where $\delta S_i^*(\omega) \doteq S_i^*(T, \omega) - S_i^*(0)$. Note that the discounted price process for the bank account is by construction constant over time, $B^*(T, \omega) = B^*(0) = 1$, for every $\omega \in \Omega$. Hence, the discounted gain for the bank account is always zero. Furthermore, since $B(0) = 1$,

$$V^*(0) = V(0) = h_0 + \sum_{i=1}^{n} h_i S_i(0).$$

If the law of one price holds, we cannot find two trading strategies $\widehat{\mathbf{h}}$ and $\widetilde{\mathbf{h}}$, associated with value processes $\widehat{V}$ and $\widetilde{V}$, respectively, such that

$$\widehat{V}(T, \omega) = \widetilde{V}(T, \omega), \quad \forall \omega \in \Omega,$$

but $\widehat{V}(0) > \widetilde{V}(0)$. A market where the law of one price is violated cannot be in equilibrium, as we may generate money out of nothing by buying the cheaper security and short-selling the more expensive one. We would earn an immediate positive cash flow and break even in every future state of the world. However, even if the law of one price holds, we may still build a money-making machine by looking for a **dominant strategy**. A trading strategy $\widehat{\mathbf{h}}$ is dominant if there exists another strategy $\widetilde{\mathbf{h}}$, such that

$$\widehat{V}(0) = \widetilde{V}(0), \qquad \widehat{V}(T, \omega) > \widetilde{V}(T, \omega), \quad \forall \omega \in \Omega. \tag{2.26}$$

By buying the dominant strategy and selling the dominated one, we have a net zero cash flow now, and we make money for sure in the future. Again, this would be a money-making machine.

Since we may take linear combinations of trading strategies, it is easy to see that there exists a dominant strategy if and only if there exists a trading strategy such that

$$V(0) = 0, \qquad V(T, \omega) > 0, \quad \forall \omega \in \Omega. \tag{2.27}$$

To see this, just take the difference $(\widehat{V} - \widetilde{V})$ in Eq. (2.26). We may also build a different type of money-making machine. Let us assume that a bank account exists in our market model. Then, there exists a dominant strategy if and only if there exists a trading strategy such that

$$V(0) < 0, \qquad V(T,\omega) \geq 0, \quad \forall \omega \in \Omega. \qquad (2.28)$$

The two conditions in Eqs. (2.27) and (2.28) are not equivalent if we consider a market with no bank account. For instance, let us consider two contingent claims (allowing for negative payoffs) such that

$$\mathbb{V} = \begin{bmatrix} 1 \\ 1 \end{bmatrix} \qquad \mathbb{Z} = \begin{bmatrix} 3 & 2 \\ -3 & -2 \end{bmatrix}.$$

In this market model, we may easily find a dominant strategy according to Eq. (2.28). For instance, the portfolio $[2, -3]^\top$ has

$$V(0) = 2 \times 1 - 3 \times 1 < 0,$$

and payoffs

$$V(T, \omega_1) = 2 \times 3 - 3 \times 2 = 0,$$
$$V(T, \omega_2) = 2 \times (-3) - 3 \times (-2) = 0.$$

However, we cannot find a dominant strategy according to Eq. (2.27), since this requires

$$h_1 \times 1 + h_2 \times 1 = 0 \quad \Rightarrow \quad h_1 = -h_2,$$

which implies

$$V(T, \omega_1) = h_1 \times 3 - h_1 \times 2 = h_1,$$
$$V(T, \omega_2) = h_1 \times (-3) - h_1 \times (-2) = -h_1,$$

and these two payoffs cannot be both strictly positive.

However, the existence of a bank account security, which we take as a risk-free asset for the sake of simplicity, ensures that we may transform any dominant strategy of one type into a dominant strategy of the other type. To see this, let us assume that the condition (2.27) holds for a strategy $\mathbf{h}$, and let us rewrite it in terms of discounted value process,

$$V^*(0) = 0, \qquad V^*(T,\omega) > 0, \quad \forall \omega \in \Omega, \qquad (2.29)$$

which is equivalent to Eq. (2.27), since the price process of the bank account is strictly positive. Note the implication

$$V^*(0) = 0 \quad \Longrightarrow \quad h_0 = -\sum_{i=1}^{n} h_i S_i^*(0).$$

Furthermore, Eq. (2.29) implies that the discounted gain for $\mathbf{h}$ is also strictly positive,

$$G^*(\omega) > 0, \quad \forall \omega \in \Omega.$$

2.4 The mathematics of arbitrage

Now, let us define the minimum discounted gain for **h** as $\delta \doteq \min_{\omega \in \Omega} G^*(\omega)$, and consider a strategy $\widetilde{\mathbf{h}}$ where

$$\widetilde{h}_0 = h_0 - \delta = -\sum_{i=1}^{n} h_i S_i^*(0) - \delta,$$

$$\widetilde{h}_i = h_i, \qquad i = 1, \ldots, n.$$

As a consequence, the discounted gain $\widetilde{G}^*$ for the strategy $\widetilde{h}$ is

$$\widetilde{G}^*(\omega) = G^*(\omega) - \delta.$$

The intuitive idea is to subtract δ from the discounted gain of portfolio **h**, so that the condition $G^*(\omega) > 0$ becomes $\widetilde{G}^*(\omega) \geq 0$ for portfolio $\widetilde{\mathbf{h}}$. The same amount is subtracted from the initial value of portfolio **h**, which is zero, so that the initial value of portfolio $\widetilde{\mathbf{h}}$ is strictly negative. This amounts to borrowing an additional amount δ of cash, whose discounted value is still δ at time $t = T$. Then, we have

$$\widetilde{V}^*(0) = -\delta < 0,$$
$$\widetilde{V}^*(T, \omega) = \widetilde{V}^*(0) + \widetilde{G}^*(\omega) = -\delta + \widetilde{G}^*(\omega) \geq 0, \quad \forall \omega \in \Omega,$$

showing that strategy $\widetilde{\mathbf{h}}$ satisfies the conditions (2.28). We may also go the other way around, by reversing the argument and transforming a strategy satisfying Eq. (2.28) into a strategy satisfying Eq. (2.27). Thus, within our framework, we may associate dominant strategies with the conditions (2.28).

Why should we prefer the formulation of Eq. (2.28) to Eq. (2.27)? The point is that the inequalities $V(T, \omega) \geq 0$ on future payoffs are not strict, and they will define a *closed* set, when we use them to write an optimization model. On the other hand, a strict inequality like $V(T, \omega) > 0$ defines an *open* set, so that the existence of a minimum or a maximum cannot be guaranteed (but only an infimum or a supremum). In fact, in concrete, how can we look for a dominant strategy? We may solve the following linear program (LP):

$$\min \quad \mathbb{V}^\mathsf{T} \mathbf{h} \qquad (2.30)$$
$$\text{s.t.} \quad \mathbb{Z}\mathbf{h} \geq \mathbf{0},$$

whereby we minimize the initial value of the portfolio, subject to a non-negativity condition on the terminal payoff. Note that this LP is feasible, as $\mathbf{h} = \mathbf{0}$ is a feasible solution. However, it could be unbounded below, leading to an infinite profit. If this is the case, it means that the market model allows for arbitrage opportunities. How can we find a condition precluding this? One possibility is to take advantage of LP duality.[38] The LP (2.30) has a dual:

$$\max \quad \mathbf{0}^\mathsf{T} \boldsymbol{\pi} \qquad (2.31)$$
$$\text{s.t.} \quad \mathbb{Z}^\mathsf{T} \boldsymbol{\pi} = \mathbb{V}$$
$$\boldsymbol{\pi} \geq \mathbf{0},$$

where $\boldsymbol{\pi} \in \mathbb{R}^m$ is the vector of dual variables. If the dual is infeasible, the primal is unbounded below, and there is a dominant strategy. However, if the dual is feasible, its objective value is just zero, which must also be the value of the dual, ruling out a dominant strategy.

To summarize our findings, the absence of a dominant strategy is related to the existence of a **non-negative pricing functional**. In fact, the equality constraints in problem (2.31) may be read as

$$1 = (1+r) \sum_{j=1}^m \pi(\omega_j), \qquad (2.32)$$

for the bank account (the asset corresponding to $i = 0$), and as

$$S_i(0) = \sum_{j=1}^m \pi(\omega_j) S_i(T, \omega_j), \qquad i = 1, \ldots, n, \qquad (2.33)$$

for the risky securities. Thus, we observe that absence of dominant strategies leads us to a linear, non-negative pricing functional, giving the current value of an asset as a linear combination of the payoffs in the possible future states. Note that the law of one price requires that the pricing functional is linear, and an additional non-negativity condition is required to rule out dominant strategies. If we rescale $\boldsymbol{\pi}$, introducing $\mathbf{q} = \boldsymbol{\pi} \cdot (1+r)$, we obtain what we may interpret as a probability measure, since Eq. (2.32) implies

$$\sum_{j=1}^m q(\omega_j) = 1,$$
$$q(\omega) \geq 0, \qquad \forall \omega \in \Omega.$$

Furthermore, this probability measure allows to express the security prices in Eq. (2.33) as a discounted expected value:

$$S_i(0) = \sum_{j=1}^m q(\omega_j) \cdot \frac{S_i(T, \omega_j)}{1+r}, \qquad i = 1, \ldots, n. \qquad (2.34)$$

Let us explore the more general implications of the above reasoning in terms of pricing: The existence of dominant strategies would imply illogical

---

[38] See Supplement S2.2 for a quick overview, and Section 16.1.4 for a more thorough treatment.

2.4 The mathematics of arbitrage

asset prices, and this is precluded by the existence of a linear and non-negative pricing functional, associated with vector **q**. Since, intuitively, we have to discount future cash flows somehow in asset pricing, it turns out to be quite convenient to work with discounted values. Let us observe that we may ensure consistency in prices if there is a non-negative vector $\mathbf{q} = [q(\omega_1), q(\omega_2), \ldots, q(\omega_m)]^\mathsf{T} \in \mathbb{R}^m$, such that for every trading strategy, we have

$$V^*(0) = \sum_{j=1}^m q(\omega_j) V^*(T, \omega_j) = \sum_{i=j}^m q(\omega_j) \frac{V(T, \omega_j)}{B(T, \omega_j)}. \quad (2.35)$$

This is just obtained from Eq. (2.34), multiplying asset prices S_i by the respective portfolio holdings h_i, in order to find the portfolio value. To see the financial implication of Eq. (2.35), let us compare two trading strategies with payoffs $\widehat{V}(T, \omega)$ and $\widetilde{V}(T, \omega)$, respectively, such that $\widehat{V}(T, \omega) > \widetilde{V}(T, \omega)$ for every state ω. We may observe that if the pricing functional **q** in Eq. (2.35) is non-negative, a trading strategy featuring a larger payoff than another one, in every future state, cannot have a smaller initial value, and therefore it cannot be dominant.

If we denote by $\mathrm{E}_{\mathbb{Q}_n}[\,\cdot\,]$ the expectation under the probability measure defined by **q**, we observe that the expected value of the discounted value process under $\mathbb{Q}_n$ is constant:

$$V^*(0) = \mathrm{E}_{\mathbb{Q}_n}[V^*(T, \omega)].$$

We have already met this kind of condition in the binomial model. In the more general multiperiod case, this will be referred to as a **martingale** property.

2.4.3 NO-ARBITRAGE PRINCIPLE AND RISK-NEUTRAL MEASURES

The existence of a non-negative pricing functional precludes the existence of dominant trading strategies. But what if we weaken the condition in Eq. (2.27)? Let us formally define an **arbitrage opportunity** as a trading strategy such that:

$$\begin{aligned} &V(0) = 0, \\ &V(T, \omega) \geq 0, \quad \forall \omega \in \Omega, \\ &\mathrm{E}[V(T, \omega)] > 0. \end{aligned} \quad (2.36)$$

Here, we do not require that the payoff if strictly positive in every future state, but only that it is strictly positive in at least one state and non-negative in the other states. Since the requirements defining an arbitrage opportunity are weaker than those defining a dominant strategy, the conditions precluding the existence of a dominant strategy must be strengthened in order to preclude the existence of an arbitrage opportunity. As it turns out, the pricing measure/functional must be *strictly* positive.

Observe first that the conditions of Eq. (2.36) may be restated in terms of discounted values,

$$V^*(0) = 0,$$
$$V^*(T,\omega) \geq 0, \quad \forall \omega \in \Omega, \quad (2.37)$$
$$\mathrm{E}[V^*(T,\omega)] > 0,$$

or, equivalently, in terms of discounted gain,

$$G^*(\omega) \geq 0, \ \forall \omega \in \Omega; \quad \mathrm{E}[G^*(\omega)] > 0. \quad (2.38)$$

These conditions essentially state that the discounted gain of the portfolio is never negative, and it is strictly positive in at least one state, so that the expected value is strictly positive (we assume states ω with strictly positive probability under the real measure). Also recall that the discounted gain of the risk-free asset is always zero, so we may just focus on the n risky assets.

Now, let us write an LP model whose aim is to generate an arbitrage opportunity satisfying condition (2.38):

$$\min \ \sum_{i=1}^{n} 0 \cdot h_i \quad (2.39)$$

$$\text{s.t.} \ \sum_{j=1}^{m} y_j = 1 \quad (2.40)$$

$$y_j = \sum_{i=1}^{n} G_i^*(\omega_j) h_i, \quad j = 1, \ldots, m \quad (2.41)$$

$$y_j \geq 0, \quad j = 1, \ldots, m,$$

where $G_i^*(\omega_j) \doteq \delta S_i^*(\omega_j)$ is the discounted gain of asset i in state j. The objective function of Eq. (2.39) is identically zero, but this is not really essential, since we want to check if a trading strategy we would love really exists.[39] Equation (2.41) introduces auxiliary variables y_j, representing the discounted gain from the strategy **h** for each state of the world; these variables are required to be non-negative, as in Eq. (2.38), and are introduced for the sake of convenience. Equation (2.40) may look arbitrary, but since we may scale trading strategies at will, it is just a convenient way of requiring strict positivity of the expected discounted gain in at least one state.[40] Now, let us associate a dual variable π_0 with constraint (2.40), and dual variables π_j with each constraint (2.41), and write the dual of the LP problem (2.39). To this aim, it is useful to

---

[39]Quite often, we use powerful optimization methods to solve a *feasibility* problem, i.e., to find a solution satisfying a set of demanding constraints. In this case, it is a common practice to use a dummy objective identically zero.

[40]Once again, beware of strict inequalities in optimization, as they define *open* sets. Existence of an optimal solution can be guaranteed when the objective function is continuous and the feasible set is *closed* and bounded.

write the equality constraints (2.40) and (2.41) in matrix form, so that we may observe the shape of the technological matrix of the primal LP problem (2.39), which is transposed in the dual:

$$\begin{bmatrix} 0 & 0 & \cdots & 0 & 1 & 1 & \cdots & 1 \\ G_1^*(\omega_1) & G_2^*(\omega_1) & \cdots & G_n^*(\omega_1) & -1 & 0 & \cdots & 0 \\ G_1^*(\omega_2) & G_2^*(\omega_2) & \cdots & G_n^*(\omega_2) & & -1 & \cdots & 0 \\ \vdots & \vdots & \ddots & \vdots & \vdots & \vdots & \ddots & \vdots \\ G_1^*(\omega_m) & G_2^*(\omega_m) & \cdots & G_n^*(\omega_m) & 0 & 0 & \cdots & -1 \end{bmatrix} \begin{bmatrix} h_1 \\ \vdots \\ h_n \\ y_1 \\ \vdots \\ h_m \end{bmatrix} = \begin{bmatrix} 1 \\ 0 \\ \vdots \\ 0 \end{bmatrix}.$$

Note that the first n columns of the matrix correspond to the unrestricted variables h_i, and the last m columns correspond to the non-negative variables y_j (in case you got lost, you may wish to recall that subscript i refers to assets, and subscript j refers to states). Note that the right-hand side vector has a one in the first position, corresponding to constraint (2.40), and is zero otherwise, as in constraint (2.41). By transposing the matrix and taking into account the non-negativity restrictions on variables y_j, we find the following dual LP:

$$\begin{aligned} \max \quad & \pi_0 \\ \text{s.t.} \quad & \sum_{j=1}^m G_i^*(\omega_j)\pi_j = 0, \quad i = 1,\ldots,n & (2.42) \\ & \pi_0 - \pi_j \le 0, \quad j = 1,\ldots,m. & (2.43) \end{aligned}$$

Note that the dual has a trivial feasible solution, with value $\pi_0 = 0$, where all dual variables are set to zero. Thus, we are not in the pathological case in which both primal and dual LPs are infeasible. Hence, the primal will be infeasible (i.e., there is no arbitrage opportunity) if and only if the dual is unbounded above. If we can let $\pi_0 \to +\infty$, while satisfying constraints (2.42) and (2.43), it must be the case that we find strictly positive values π_j, one per state j, such that constraint (2.42) is satisfied. If they exist, the values $\pi_j > 0$ may be rescaled to values $q_j \equiv q(\omega_j) > 0$ in such a way that their sum is 1, yielding a strictly positive probability measure $q(\omega)$ such that

$$\mathrm{E}_{\mathbb{Q}_n}[G^*(\omega)] = \sum_{j=1}^m G^*(\omega_j)q(\omega_j) = 0, \quad (2.44)$$

where the notation $\mathrm{E}_{\mathbb{Q}_n}[\,\cdot\,]$ points out that the expectation is taken with respect to this measure, and not using the probabilities $p(\omega)$ of the market model.

The probability measure $\mathbb{Q}_n$ is called **risk-neutral**, since the expected return of risky assets under $\mathbb{Q}_n$ is just the risk-free return. To see why, observe that Eq. (2.44) states that the expected *discounted* gain is zero, which is the case if the expected holding period return is, in fact, the risk-free rate r. Also note that, in the more general case of a different numeraire with price process $B(t,\omega)$, if the expected discounted gain is zero, then the discounted price processes are martingales. Thus, we conclude that there is no arbitrage strategies if and only if there exists a strictly positive risk-neutral (martingale) probability measure. We remark again that non-negativity of the probability measure is needed to rule out the existence of dominant strategies. However, since arbitrage strategies are weaker than dominant strategies, we have to tighten the non-negativity condition to strict positivity.

Note that this probability measure need not be unique in general. When considering generic numeraire assets, the term **equivalent martingale measure** is preferred, since the discounted price process of securities is a martingale under such a measure. For any trading strategy, if $\mathbb{Q}_n$ is a martingale measure we have

$$\begin{aligned}
V(0) = V^*(0) &= \mathrm{E}_{\mathbb{Q}_n}[V^*(0)] = \mathrm{E}_{\mathbb{Q}_n}[V^*(T,\omega) - G^*(\omega)] \\
&= \mathrm{E}_{\mathbb{Q}_n}\left[\frac{V(T,\omega)}{B(T,\omega)}\right] - \underbrace{\mathrm{E}_{\mathbb{Q}_n}[G^*(\omega)]}_{=\,0\text{ by Eq. (2.44)}} \\
&= \mathrm{E}_{\mathbb{Q}_n}\left[\frac{V(T,\omega)}{B(T,\omega)}\right].
\end{aligned} \qquad (2.45)$$

Thus, under a martingale measure, the initial value of a trading strategy is just the discounted expected value of its payoff.

The last issue we have to consider is market **completeness**. If the market is complete, any contingent claim can be replicated by a trading strategy. However, if the market is not complete, there are contingent claims that cannot be replicated. Consider the subspace of contingent claims that can be replicated by a trading strategy. This is the subspace of the **attainable** payoffs. Absence of arbitrage opportunities implies the absence of dominant strategies, which in turn implies the law of one price. Hence, under a no-arbitrage assumption, the price of any attainable contingent claim is just the discounted expected value of its payoff, under any risk-neutral measure. If the number of states is equal to the number of linearly independent securities, then any contingent claim is attainable and can be replicated, i.e., the market is complete. The law of one price makes sure that the value of an attainable contingent claim must be the same under every risk-neutral measure. It is not difficult to prove that if the market is complete, then the risk-neutral measure is unique, so that we find one well-defined price for all contingent claims. If a contingent claim is not attainable, however, it can be shown that its discounted expected value is *not* the same for the whole set of risk-neutral measures. Hence, we do not find a unique arbitrage-free price, but a range of such prices. In practice, when the market is incomplete, we may pursue a calibration strategy, whereby a risk-neutral measure is selected, matching market prices of exchange-traded derivatives, and used to price OTC derivatives.[41]

---

[41] See Chapter 14.

The bottom line of our reasoning can be summarized as follows:

- There are no arbitrage opportunities in the market, if we can find a strictly positive equivalent martingale probability measure.
- If the market is complete, then this probability measure is unique and may be interpreted as risk-neutral, if we use a risk-free asset as a numeraire (which amounts to the usual discounting). If the market is not complete, then there are multiple martingale measures.

These findings generalize and are consistent with the insights we gathered from the binomial pricing model. Further generalization may be obtained by considering *dynamic* trading strategies, possibly in continuous time. The essential messages do not change, but continuous-time trading requires sophisticated mathematical machinery, since pathologies may occur that have to be ruled out. This requires a more careful and demanding treatment, which is beyond the scope of this book.

S2.1 Multiobjective optimization

When trading off expected reward and risk, it may be difficult to find a compromise solution between conflicting requirements. This is a common issue in multiobjective optimization, where we have to trade off conflicting requirements that cannot be reduced to a single performance measure. In this book, we will limit our treatment to two objectives, which may be visualized on a plane. A common approach, in multiobjective optimization, is to trace the frontier of efficient, or nondominated, solutions.

Let $\mathbf{x} \in S \subseteq \mathbb{R}^n$ be a vector representing our decision, which is constrained to be in the feasible set S, and let $\pi(\mathbf{x})$ denote the expected reward (e.g., wealth or return) and $\xi(\mathbf{x})$ denote the corresponding risk measure (like standard deviation). From a mathematical perspective, each feasible solution is characterized by a pair of objective values, which can be depicted as illustrated in Fig. 2.7, on a mean–risk plane. Note that good solutions are on the North–West corner, where expected reward is maximized and risk is minimized. Formally, we could consider a "vector" optimization problem:

$$\text{"max"} \begin{bmatrix} \pi(\mathbf{x}) \\ -\xi(\mathbf{x}) \end{bmatrix} \quad (2.46)$$
$$\text{s.t.} \quad \mathbf{x} \in S,$$

where we consider $-\xi(\mathbf{x})$, since risk should be minimized. However, stated as such, the problem has no meaning, and this is why we quote "max." The difficulty is that vectors on a plane are not a well-ordered set. If we consider the three solutions represented by black bullets in Fig. 2.7, there is no objective way to spot the best one, as the choice may depend on the degree of risk aversion

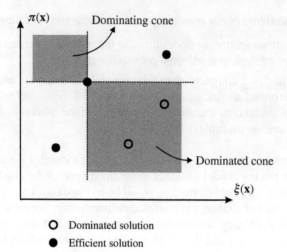

FIGURE 2.7 Schematic illustration of the concept of efficient solution.

of the decision maker. However, it stands to reason that the two solutions represented by hollow circles should not be considered, as there is an alternative which is better from both points of view.

DEFINITION 2.1 *Given the vector optimization problem (2.46), a feasible solution* $\mathbf{x}^*$ *is said to be an* **efficient** *or* **nondominated** *solution, if there is no other solution* $\tilde{\mathbf{x}} \in S$ *such that*

$$\pi(\tilde{\mathbf{x}}) \geq \pi(\mathbf{x}^*) \quad \text{and} \quad \xi(\tilde{\mathbf{x}}) \leq \xi(\mathbf{x}^*)$$

with a strict inequality for at least one of the two objectives. The set of nondominated solutions is called the **efficient frontier**.

In Fig. 2.7, two shaded regions are displayed. They are two cones,[42] rooted at a specific solution. One is the cone of dominated solutions, i.e., the solutions that are dominated by the vertex of the cone. The two dominated solutions, in fact, are located in a dominated cone and are not efficient. The dominating cone is the cone of solutions that dominate the vertex of the cone. For that efficient solution, the dominating cone is empty. The efficient frontier is the set of solutions with an empty dominating cone. When dealing with a continuous mathematical program, the efficient frontier might be a continuous curve, as illustrated in Fig. 2.8. In fact, this is the qualitative shape that is obtained by solving the mean–variance portfolio optimization problem described in Section 2.1.1.

In order to trace the efficient frontier, we need to find a way to recast the problem so that it can be tackled by standard optimization software. To this

---

[42] We define a cone formally in Section 15.5. Here we are dealing with a shifted cone, really.

S2.1 Multiobjective optimization

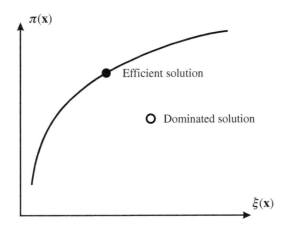

FIGURE 2.8 The efficient frontier in the continuous case.

aim, we can scalarize the problem according to some strategy, boiling the vector problem down to a family of single-objective optimization problems depending on one or more parameters. The first and perhaps more intuitive approach is to devise a weighted linear combination of the two objectives. One possibility, in our mean–risk framework, would be to introduce a parameter $\gamma \in [0, 1]$, which expresses the relative importance of the objectives. Then, we let γ span its range, and we solve a sequence of problems with objective

$$\max \gamma \cdot \pi(\mathbf{x}) - (1 - \gamma) \cdot \xi(\mathbf{x}).$$

One difficulty with this approach is the interpretation of γ. It is a bit easier to introduce a single parameter λ and solve a sequence of scalarized problems:

$$\max \quad \pi(\mathbf{x}) - \lambda \xi(\mathbf{x}) \qquad (2.47)$$
$$\text{s.t.} \quad \mathbf{x} \in S.$$

In this case, λ is related to risk aversion and, even though we obtain the pure risk minimization case, corresponding to $\gamma = 0$, only in the limit when $\lambda \to +\infty$, it may be easier to find guidelines in value selection. For instance, in the mean–variance case, values of λ in the interval $[2, 4]$ are considered sensible.[43]

This approach, based on weighted combinations or a risk aversion parameter, is clearly intuitive and guarantees that all of the solutions that we generate are efficient. However, there is no guarantee that *all* of the efficient solutions

---

[43] A word of caution is needed in this case. Even though we may consider standard deviation as a risk measure, variance is used in the model for the sake of computational convenience, as this leads to a convex quadratic programming problem, which can be solved very efficiently. Conceptually, this does not change the matter, as variance and standard deviation are closely related. However, we should consider that the efficient frontier is plotted on a plane which involves a transformed version of the underlying risk measure. The mentioned sensible range of values for λ applies when variance is used in the scalarized objective, not standard deviation.

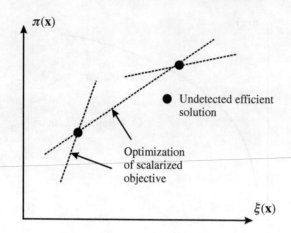

FIGURE 2.9 A simple scalarization may not be able to detect all of the efficient solutions.

will be generated in this way. The issue is illustrated in Fig. 2.9, where the dotted lines correspond to level curves of the scalarized objective, for different values of λ. In fact, the objective function (2.47) is constant along lines on the (ξ, π) plane, whose slope depends on the risk aversion coefficient λ; maximization of the scalarized objective requires moving to North–West. It is easy to see that two out of the three efficient solutions can be detected by suitably setting the risk aversion coefficient, but we cannot generate the third one. This does not occur in Fig. 2.8, where the plot of the efficient frontier looks essentially like the boundary of a convex set. A situation similar to Fig. 2.8 may occur when dealing with discrete optimization problems, or when dealing with arbitrary objective functions, possibly lacking suitable concavity/convexity properties. An interesting case, portfolio optimization with cardinality constraints, is described in Sections 8.2.3 and 15.4.1.

An alternative scalarization approach is based on the idea of transforming one objective into a constraint. In the mean–risk case, we may minimize risk, subject to a constraint on minimum expected reward,

$$\begin{aligned} \min \quad & \xi(\mathbf{x}) \\ \text{s.t.} \quad & \mathbf{x} \in S, \\ & \pi(\mathbf{x}) \geq \alpha, \end{aligned}$$

or maximize expected reward, subject to a risk budget,

$$\begin{aligned} \max \quad & \pi(\mathbf{x}) \\ \text{s.t.} \quad & \mathbf{x} \in S, \\ & \xi(\mathbf{x}) \leq \beta. \end{aligned}$$

We may trace the efficient frontier by solving a sequence of scalarized problems for varying values of α or β. It is worth noting that this second approach does

S2.2 Summary of LP duality

not suffer from the aforementioned difficulty in generating the efficient frontier. Actually, the choice of the scalarization approach usually depends on computational convenience, as well as the ease in choosing and interpreting the value of the involved parameter.

S2.2 Summary of LP duality

Optimization model building and solving are described in more detail in Chapters 15 and 16, which include sections on duality in mathematical programming and, more specifically, duality in linear programming (LP duality). In this Supplement, we just give the bare essentials of LP duality, as is relevant to this chapter.

Example 2.13 A trivial LP problem

> To begin with, what follows is an example of an LP problem:
>
> $$\begin{align} \max \quad & x_1 + 2x_2 \tag{2.48} \\ \text{s.t.} \quad & x_1 + x_2 \leq 4 \\ & x_1, x_2 \geq 0, \end{align}$$
>
> where s.t. stands for "subject to" the list of constraints. We observe that all decision variables occur in a linear fashion (no powers, no products, no fancy functions), and it is easy to see that the optimal solution is $x_1^* = 0$, $x_2^* = 4$, with optimal value of the objective function $f^* = 8$.

All LP models can be recast in the standard form:

$$\begin{align} \min \quad & \mathbf{c}^T \mathbf{x} \tag{2.49} \\ \text{s.t.} \quad & \mathbf{A}\mathbf{x} = \mathbf{b} \\ & \mathbf{x} \geq \mathbf{0}, \end{align}$$

where $\mathbf{x}, \mathbf{c} \in \mathbb{R}^n$, $\mathbf{A} \in \mathbb{R}^{m \times n}$, and $\mathbf{b} \in \mathbb{R}^m$. To leave room for optimization, the linear system of equations defining the constraints must be underdetermined, i.e., we must have $m < n$.

Example 2.14 Recasting an LP problem in standard form

> For instance, we may recast problem (2.48) in standard form by changing the sense of the objective and by introducing a non-negative slack

variable s to transform the inequality into an equality:

$$\begin{aligned}\min \quad & -x_1 - 2x_2 \\ \text{s.t.} \quad & x_1 + x_2 + s = 4 \\ & x_1, x_2, s \geq 0.\end{aligned} \qquad (2.50)$$

The standard form is algorithmically convenient, and it also allows to analyze LPs in full generality without bothering too much about specific cases.

Given any optimization problem, one of the following cases may occur:[44]

- The problem is feasible and there is a finite optimal solution. Note that the optimal solution need not be unique.[45]
- The problem is infeasible, i.e., the feasible set is empty. In this case, we conventionally say that the minimum value is $+\infty$, and the maximum value is $-\infty$ in the case of a maximization problem.
- The optimum is unbounded, i.e., we can reduce the cost (or maximize profit) without bound, while staying within the feasible set. In this case, we conventionally say that the minimum value is $-\infty$, and the maximum value is $+\infty$ in the case of a maximization problem.

Sometimes, we are only interested in finding a solution satisfying a set of constraints. In such a case, we resort to optimizing a fictional objective, which turns out to be a nice computational way to solve a difficult system of equalities and inequalities. Alternatively, we may wish to show that the feasible set is empty, i.e., we cannot find a solution with some desired features. This is what happens in the mathematics of arbitrage, when we want to show that a certain trading strategy cannot exist, under certain conditions. Duality theory may come in handy in this setting.

Any (primal) LP problem is associated with a dual problem. In the case of problem (2.49), its dual is

$$\begin{aligned}\max \quad & \mathbf{b}^\mathsf{T} \boldsymbol{\pi} \\ \text{s.t.} \quad & \mathbf{A}^\mathsf{T} \boldsymbol{\pi} \leq \mathbf{c},\end{aligned} \qquad (2.51)$$

where $\boldsymbol{\pi} \in \mathbb{R}^m$ is the vector of dual variables. We notice a simple pattern: The cost vector $\mathbf{c}$ and the right-hand side vector $\mathbf{b}$ are swapped, and the matrix $\mathbf{A}$ is transposed. Duality may be introduced as a special case of more general La-

---

[44]To be precise, it might happen that we find a finite inf but not a finite min, if the feasible set is open, as in the case $\min x$, s.t. $x > 2$. However, we disregard this case.

[45]In the LP case, in fact, we may have an infinite number of optimal solutions, since any convex combination of two optimal solutions has the same value, given the linearity of the objective function, and is feasible, given the convexity of the feasible set.

grangian duality for nonlinear programming,[46] or by using separation theorems for convex sets (rephrased in the form of Farkas' lemma). Under quite general conditions, weak duality applies, stating that the maximum of the dual problem cannot be larger than the minimum of the primal problem. In the LP case, a stronger duality condition applies and the following cases are possible:

- The primal and the dual are both feasible, in which case the two optimal objectives are the same.
- The primal and the dual are both infeasible (a rather uncommon and pathological case).
- The primal is infeasible and the dual is unbounded (above).
- The dual is infeasible and the primal is unbounded (below).

Furthermore, it turns out that the dual of the dual problem is just the original primal problem.

Example 2.15 Finding the dual of a standard LP

In the case of problem (2.50), we have

$$\mathbf{c} = \begin{bmatrix} -1 \\ -2 \\ 0 \end{bmatrix}, \quad \mathbf{A} = \begin{bmatrix} 1 & 1 & 1 \end{bmatrix}, \quad \mathbf{b} = \begin{bmatrix} 4 \end{bmatrix}.$$

Hence, we immediately find the dual:

$$\max \quad 4\pi$$
$$\text{s.t.} \quad \begin{bmatrix} 1 \\ 1 \\ 1 \end{bmatrix} \pi \leq \begin{bmatrix} -1 \\ -2 \\ 0 \end{bmatrix}.$$

Here, we have a single dual variable π, since our trivial LP has one constraint, and we should just maximize π, subject to simple upper bounds. By taking the most restrictive bound, we find

$$\pi^* = -2,$$

corresponding to an objective value of -8, which is the same as the primal value [recall the change in sign of the objective with respect to the original problem (2.48)].

---

[46] See Section 16.1.4.

Table 2.4 Rules to find the dual of an arbitrary LP.

Primal (minimize)	Dual (maximize)
ith constraint $\geq b_i$	ith variable ≥ 0
ith constraint $\leq b_i$	ith variable ≤ 0
ith constraint $= b_i$	ith variable unrestricted
jth variable ≥ 0	jth constraint $\leq c_j$
jth variable ≤ 0	jth constraint $\geq c_j$
jth variable unrestricted	jth constraint $= c_j$

Finding the dual of an LP by first recasting it in standard form may be quite inconvenient. In Table 2.4, we summarize the rules to transform a primal problem in general form into the corresponding dual. In this table, we assume that the primal problem is in minimization form, so that the corresponding dual is in maximization form; however, the two columns may be swapped (if we build the dual of the dual problem, as we said, we just find the original primal).

Example 2.16 Infeasible dual of an unbounded primal

Let us change the sense of the inequality in problem (2.48):

$$\begin{aligned} \max \quad & x_1 + 2x_2 \\ \text{s.t.} \quad & x_1 + x_2 \geq 4 \\ & x_1, x_2 \geq 0. \end{aligned} \qquad (2.52)$$

Clearly, the problem is unbounded. Applying the rules of Table 2.4, swapping the columns corresponding to primal and dual problems, we immediately find its dual:

$$\begin{aligned} \min \quad & 4\pi \\ \text{s.t.} \quad & \begin{bmatrix} 1 \\ 1 \end{bmatrix} \pi \geq \begin{bmatrix} 1 \\ 2 \end{bmatrix} \\ & \pi \leq 0, \end{aligned}$$

which is clearly infeasible, since the first two conditions on the dual variable π are not compatible with the third one.

> As a further check, we may work with the equivalent standard form of problem (2.52),
>
> $$\min \quad -x_1 - 2x_2$$
> $$\text{s.t.} \quad x_1 + x_2 - s = 4$$
> $$x_1, x_2, s \geq 0,$$
>
> whose dual is
>
> $$\max \quad 4q$$
> $$\text{s.t.} \quad \begin{bmatrix} 1 \\ 1 \\ -1 \end{bmatrix} q \leq \begin{bmatrix} -1 \\ -2 \\ 0 \end{bmatrix},$$
>
> with dual variable q. The variable substitution $\pi = -q$ yields the same infeasible dual as above.

Problems

2.1 You are the manager of a pension fund, and your fee depends on the achieved annual return. You can play it safe, and allocate wealth to a risk-free portfolio earning 4% per year (with annual compounding). Alternatively, you can pursue an active portfolio management strategy, whose return is modeled by a normal random variable with expected value 8% and standard deviation 10%. Your fee depends on the realized performance, according to the following table:

Annual return R	Fee
$R \leq 0\%$	\$0
$0\% \leq R < 3\%$	\$50,000
$3\% \leq R < 9\%$	\$100,000
$12\% \leq R$	\$200,000

- Assume that you do not care about your own risk, so that you just consider expected values. Which one is the better strategy for you?
- What is the standard deviation of your fee, if you take the active strategy?

2.2 The annual returns of two stock shares are represented by the following linear factor model:

$$R_1 = 0.03 + 1.2 R_m + \epsilon_1,$$
$$R_2 = 0.04 + 0.8 R_m + \epsilon_2,$$

where the common factor R_m represents a systematic market risk (e.g., the return on a stock market index), and ϵ_1 and ϵ_2 are specific risk factors. We assume that all of the risk factors are uncorrelated and normally distributed with the following parameters (expected value and standard deviation):

Factor	μ	σ
R_m	0.04	0.25
ϵ_1	0	0.30
ϵ_2	0	0.40

You have invested 40% of your wealth in the first asset and 60% in the second one. Find the probability that the realized annual return is negative, i.e., you lose money.

2.3 You have bought on margin 100 zero-coupon bonds, with face value €1000, maturing in three years. At present, the annual yield (annual return with annual compounding) of the bonds is 4.3%. The initial margin ratio is 50%, and the maintenance margin is 20%. Assuming that we neglect the passage of time, for which yield will you get a margin call?

2.4 You hold a long position in an asset, whose price is correlated with two commodity prices. The two commodities are the underlying assets of two futures contracts maturing at time T_F. There is no futures contract available on your asset, which you are going to sell at time $T_H < T_F$. Thus, you want to build a minimum variance hedge based on the two futures. Assuming that you know all of the relevant statistical information, and that we disregard issues related to marking-to-market, margin calls, etc., find the optimal hedge ratios.

2.5 The annual return of a risky financial portfolio, denoted by R_p, can be described by the following linear regression model:

$$R_p = 0.057 + 3.4 F_1 - 2.6 F_2 + \epsilon,$$

where F_1 and F_2 are mutually correlated risk factors and ϵ is another risk factor, which is independent from the other two. All risk factors are assumed normally distributed with expected value zero. The standard deviations of F_1 and F_2 are 0.1 and 0.12, respectively, and their correlation coefficient is 0.48; the standard deviation of the third risk factor is 0.2.

- Assume that the annual risk-free return is $r_f = 2.5\%$ (annual compounding). What is the probability that the risky portfolio outperforms the risk-free investment?

- Assume that you have invested €1,000,000 in the risky portfolio. Find the annual V@R at 95% confidence level. Note: In this case, since the time horizon is one year, you cannot neglect the expected value of return (unlike daily V@R).

2.6 Prove that, under the equivalent martingale measure $\mathbb{Q}_n$ of Section 2.3.4, the ratio f_t/B_t is a martingale.

2.7 Let us consider a market on which three assets, indexed by $i = 1, 2, 3$, are traded, with current price

$$S_1(0) = S_2(0) = S_3(0) = 1.$$

The asset values in the future, at time $t = T$, depend on which state will be realized. We consider three possible scenarios ω_1, ω_2, and ω_3, with probabilities 55%, 30%, and 15%, respectively. The corresponding asset values are given in the following table:

State	$S_1(T,\omega)$	$S_2(T,\omega)$	$S_3(T,\omega)$
ω_1	1	3	1.2
ω_2	3	1	1.2
ω_3	0	0	1.2

We note that asset 3 is risk-free, and that state ω_3 is a "bad" state. Imagine an insurance contract against the occurrence of the bad state, whose payoff is 0 if states ω_1 or ω_2 occur, and 1 if state ω_3 occurs. What is the fair price of this insurance contract? Do you need further information about risk aversion?

Further reading

- In this chapter, we have outlined a few mathematical programming models; a simple introduction to such models may be found in [2].
- A good reference for a general introduction on mathematical finance, covering a lot of ground, is [7].
- For an overview of value-at-risk, you may refer to [8].
- Example 2.5 on volume risk has been inspired by the Harvard Business School case [5].
- Section 2.4 follows the treatment by [10]. You may also see [6] or [9].
- An overview of multiobjective optimization is given, e.g., in [4].
- To see more on LP and LP duality you may refer, e.g., to [1] or [11].

Bibliography

1 M.S. Bazaraa, J.J. Jarvis, and H.D. Sherali. *Linear Programming and Network Flows* (4th ed.). Wiley, Hoboken, NJ, 2010.

2. P. Brandimarte. *Quantitative Methods: An Introduction for Business Management.* Wiley, Hoboken, NJ, 2011.
3. J.Y. Campbell and L.M. Viceira. *Strategic Asset Allocation.* Oxford University Press, Oxford, 2002.
4. Y. Collette and P. Siarry. *Multiobjective Optimization: Principles and Case Studies.* Springer, Heidelberg, 2004.
5. M.A. Desai, A. Sjoman, and V. Dessain. *Hedging Currency Risks at AIFS.* Harvard Business School Publishing, 2007. Case no. 9-205-026.
6. R.J. Elliott and P. Ekkehard Kopp. *Mathematics of Financial Markets* (2nd ed.). Springer, New York, 2005.
7. C. Fries. *Mathematical Finance: Theory, Modeling, Implementation.* Wiley, Hoboken, NJ, 2007.
8. P. Jorion. *Value at Risk: The New Benchmark for Controlling Derivatives Risk* (3rd ed.). McGraw-Hill, New York, 2006.
9. S.F. Leroy and J. Werner. *Principles of Financial Economics* (2nd ed.). Cambridge University Press, Cambridge, 2014.
10. S.R. Pliska. *Introduction to Mathematical Finance: Discrete Time Models.* Blackwell Publishers, Malden, MA, 1997.
11. R.J. Vanderbei. *Linear Programming: Foundations and Extensions* (3rd ed.). Springer, Heidelberg, 2010.

Part Two

Fixed-income assets

Chapter Three

Elementary Theory of Interest Rates

The time value of money is one of the key ingredients in finance. We need a way to move cash flows backward and forward in time, in order to analyze and compare investment opportunities, as well as to come up with financial plans. In this chapter, we introduce the fundamental concepts related to interest rates, such as compounding frequencies, discount factors, the term structure of interest rates, and forward rates.

Interest rates are a key risk factor in the pricing of fixed-income assets, which include a multitude of securities, ranging from plain bonds to rather complicated interest rate derivatives. In this chapter, we only deal with elementary bond pricing, which can be accomplished without the need for dynamic models accounting for the uncertainty about interest rates in the future. Such advanced models shall be introduced in Chapter 11, whereas we rely here on a static picture of interest rates. Despite its (apparent) simplicity, this enables us to tackle some quite relevant problems:

- Given two bonds, how can we compare their return?
- Given a set of bond prices, how can we check whether there are arbitrage opportunities?
- How can we estimate the amount of money that we need to save each year, during our working life, in order to achieve a given target wealth at retirement?
- How can we measure the interest rate risk of a plain bond?

As an introduction to the issues involved in comparing investment opportunities, let us consider the following simple example.

Example 3.1 Two investment opportunities

A friend of ours needs to borrow money, say, $10,000. He will give that money back in one year, and to compensate us for our help, he will top $400 on it. Should we accept his proposal? Of course, we might do so just because he is a very dear friend of ours, but let us analyze the situation rationally as an investment analysis problem. Analyzing an investment requires some sort of relative comparison between comparable alternatives. Hence, let us say that our bank offers an interest rate of 3% for a deposit of one year. This means that if we lend our money to the bank, we will receive $10,300 in one year, which is less than the $10,400 offered by our friend. Another way to compare the two alternatives is to consider the return on the loan,

$$\frac{10{,}400 - 10{,}000}{10{,}000} = 4\%,$$

a return larger than the 3% offered by the bank. Thus, it would seem that we might be better off by lending to our friend.

Actually, comparing the two proposals may not be so trivial. We should also consider the possibility of not getting the money back at all, if a default occurs. Arguably, the bank should be financially more reliable than our friend, so the two interest rates might not be quite comparable, and we need a way to bring risk into the picture. Furthermore, taxes and more or less hidden fees or transaction costs may play a role as well. In this chapter, we do not consider additional complications like market frictions or uncertainty in cash flows, due to default and other risk factors. We assume that the interest rates that we analyze are, in a sense to be made clear, risk-free. Last but not least, in this simple example, we are comparing two opportunities resulting in cash flows that do not differ in their timing. What if we have to compare sequences of cash flows occurring at different times?

Interest rates have to do with the time value of money: $10,000 now is not the same as $10,000 in one year. To cope with more realistic and interesting problems than Example 3.1, we need the ability to shift money back and forth in time in order to compare different cash flow sequences on a common ground. The time value of money is the subject of the first two sections of this chapter. In Section 3.1, we show how interest rates are used to shift cash flows and money forward in time. There, we also introduce the fundamental concept of compounding, as well as the difference between quoted and effective rates. In Section 3.2, we consider shifting cash flows and money backward in time by discount factors. We justify discounting by the no-arbitrage principle and show the connection between discount factors and interest rates. Different ways of measuring rates can be adopted, and it is essential to pay attention to how rates are expressed in terms of compounding frequency. However, it is important to realize that interest rates may be quoted in different ways, according to convenience, but they are just different expressions of the same thing.

In Section, 3.3, we briefly discuss the fundamental difference between nominal and real interest rates, accounting for inflation. In Section, 3.4, we move on to a fundamental feature of interest rates: They differ as a function of maturity. Even a cursory look at rates quoted on a newspaper shows that an interest rate for a time period of six months is not the same as an interest rate for a longer period, say, five years. Usually, rates are larger for longer maturities, but this is not always the case, and theories have been proposed to explain these patterns.

Armed with these elementary concepts, we shall then examine the foundations of elementary bond pricing in Section 3.5. By "elementary" we mean that, as we have anticipated, we rely only on a deterministic analysis of cash flows; more advanced bond pricing models take into account the stochastic nature of interest rates and will be discussed in Chapter 14. Nevertheless, we will show that pricing floating-rate bonds, which feature stochastic cash flows, may boil down to a surprisingly simple formula. We also introduce a commonly used measure of bond return, yield-to-maturity (YTM), as well as a simplified view of interest rate risk. Elementary bond pricing and YTM are related to fundamental concepts in investment analysis, like net present value (NPV) and internal rate of return (IRR). NPV and IRR are discussed in more detail within the framework of *corporate* finance. Hence, in Section 3.6, we shall just have a brief discussion of NPV and IRR. We also hint at more sophisticated analysis tools based on real options.

We close the chapter with Section 3.7, where we discuss another essential concept related with the term structure of interest rates, namely, the relationship between spot and forward rates. This paves the way for the analysis of interest rate risk management strategies and a few simple interest rate derivatives, which will be introduced in Chapter 4. In that section, we also consider a possible explanation of the term structure of interest rates.

It is worth mentioning that in this chapter, as in the rest of the book, we take for granted that it is perfectly legitimate to lend money and expect a reasonable compensation for it. We just rule out the application of unreasonable interest rates, which amounts to usury and is in fact forbidden by the law. However, there are cultures in which this is simply forbidden, and different arrangements are required; a notable example is Islamic finance. We will not consider such issues.

Remark. As we shall see, interest rates are applied to time periods of quite different length. For comparison purposes, they are always quoted on a common annual basis. A rate shall always refer to one year, whereas a return may refer to an arbitrary holding period.

CHAPTER 3 Elementary Theory of Interest Rates

3.1 The time value of money: Shifting money forward in time

Suppose that we deposit an amount L in a bank account for one year. If we part ways with hard-earned money, even though for a limited span of time, we may require a compensation in the form of an annual interest rate r,[1] meaning that after one year the deposit will grow to

$$W_1 = L(1+r).$$

A little thought raises a few questions:

1. What if we want to borrow, rather than lend money? Should we expect to pay the same interest rate that we earn from a deposit?
2. Are we sure that we will get our money back, or is default an unpleasing possibility?
3. What if we want to invest L only for a fraction of a year?
4. What if we want to invest L for more than one year?

As we may expect, the first issue is reflected by a spread between the interest rate that is bid when we invest money and the larger rate that is asked when we borrow money. In general, whatever asset we are dealing with, we face a bid–ask spread. This applies, for instance, to currency exchange rates quoted by a dealer, as well as to stock prices traded through a broker. If a retail bank steps in as an intermediary between savers and borrowers, it has to make a living by applying a spread between the two rates. For instance, the bank will collect deposits that are rewarded at a given interest rate, but it will require a larger rate on loans and mortgages.

The second issue is related to credit risk, i.e., the risk that a default occurs. If we lend money to a bank, we should consider the possibility that the bank goes bankrupt.[2] When the bank lends us money, it takes into account the possibility that we will not be able to repay the debt, for reasons that might be independent of our good will. The default on the part of a client is normally more likely than the default on the part of a bank, and this will also contribute to a spread between lending and borrowing rates, since common sense suggests that credit risk will imply a larger interest rate. The prime rate quoted by banks is the interest rate offered to their best clients when they need financing, but this is definitely not the rate offered to normal clients. By the same token, large institutional investors may afford borrowing funds with a minimal spread.

In this chapter, we shall neglect bid–ask spreads and credit risk; hence, we only deal with risk-free rates and their connection with *time*. As we may

---

[1] In this section, we denote by r the interest rate applying to a holding period corresponding to one year; we will introduce a more precise and useful notation later.

[2] In many countries, bank deposits are guaranteed by the government, but only up to some limit. At the time of writing, there is strong political pressure to eliminate or at least reduce any such protection.

imagine, in normal market conditions, a larger rate is associated with longer-term investments. If we lock the money for five years rather than one, more often than not we will be compensated by a larger annual interest rate. If, on the other hand, we have a bank account from which we may withdraw money whenever we need or feel like it, the rate that we should expect to earn from our deposits will be much lower (possibly zero). Time is also essential when we need to shift money forward and backward in time, among other things, in order to compare investment opportunities. We must do so both for long and short time intervals, in order to deal with issues 3 and 4 above.

3.1.1 SIMPLE VS. COMPOUNDED RATES

As we have observed, when an annual interest rate r applies to a single year, an amount L now is equivalent to an amount $L(1 + r)$ in one year. If we invest only for a fraction $\alpha \in (0, 1)$ of a year, one possible idea is prorating the rate, i.e., to apply the formula

$$L(1 + \alpha r). \tag{3.1}$$

For instance, if the annual rate is 5% and we invest for six months, according to Eq. (3.1) we will earn 2.5% of L. In principle, we may apply the same concept when $\alpha > 1$. However, when the investment spans a long time interval, quite often interest is paid periodically along the way, and not only at the end of the time horizon. The interest we receive at the end of each time period can be immediately reinvested, so that we can earn interest on interest. For instance, if the money is invested for n years, and we assume that the rate will not change over time (no reinvestment risk), capital will grow as follows:

$$L \cdot \underbrace{(1+r) \cdot (1+r) \cdots (1+r)}_{n \text{ times}} = L \cdot (1+r)^n. \tag{3.2}$$

The mechanism underlying Eq. (3.1) is called a **simple interest rate**. The alternative of Eq. (3.2) corresponds to a **compounded interest rate**.

Example 3.2 Simple vs. compounded rates

Say that the annual rate is 5%, and we invest $1000 for two years. The wealth at the end of the holding period depends on how the interest rate is applied. Under the simple interest rate rule of Eq. (3.1), wealth after two years is

$$L \cdot (1 + 2r) = 1000 \times (1 + 2 \times 0.05) = \$1100.$$

If interest is paid annually and it may be immediately reinvested, wealth after two years will stem from the application of Eq. (3.2),

$$L \cdot (1+r) \cdot (1+r) = 1000 \times (1 + 0.05)^2 = \$1102.50.$$

> The slight difference between the two amounts is due to the compounding mechanism, since
>
> $$(1+r) \cdot (1+r) = 1 + 2r + r^2 > 1 + 2r,$$
>
> and the term r^2 corresponds to interest earned on interest.

Clearly, if compounding is applied over a large number of years, it implies an exponential increase of wealth. The impact is not so remarkable in Example 3.2, where the time horizon is rather short, but it may be quite relevant for the long-term investments that are associated with pension funds.

Example 3.3 Building pension capital

> Suppose that we are going to work for the next T years, and that at the beginning of each year we contribute an amount L to a pension fund, which is invested at an annual rate r for the future time periods (years) until retirement. If annual compounding applies, what is our wealth at retirement?
>
> To formalize the problem, let us introduce time instants (epochs) $t = 0, 1, \ldots, T$. We invest money at epochs $t = 0$ through $t = T - 1$, for a total of T contributions, and we need to evaluate wealth at epoch $t = T$. The key is that what we contribute at time t is invested for $T-t$ time periods. As a result, wealth at retirement is
>
> $$W_T = \sum_{t=0}^{T-1} L \cdot (1+r)^{T-t} = L \cdot (1+r)^T \cdot \sum_{t=0}^{T-1} \left(\frac{1}{1+r}\right)^t.$$
>
> To figure out the sum, we recall a property of the geometric series,
>
> $$\sum_{k=0}^{+\infty} \alpha^k = \frac{1}{1-\alpha},$$
>
> for $|\alpha| < 1$. Moreover, we may express a finite sum as the difference between two infinite sums:
>
> $$\sum_{k=0}^{\tau} \alpha^k = \sum_{k=0}^{+\infty} \alpha^k - \sum_{k=\tau+1}^{+\infty} \alpha^k$$

3.1 The time value of money: Shifting money forward in time

$$= \sum_{k=0}^{+\infty} \alpha^k - \alpha^{\tau+1} \cdot \sum_{k=0}^{+\infty} \alpha^k$$

$$= \frac{1 - \alpha^{\tau+1}}{1 - \alpha}.$$

In our case,

$$\alpha = \frac{1}{1+r} < 1, \qquad \tau = T - 1.$$

Therefore, we find

$$\sum_{t=0}^{T-1} \left(\frac{1}{1+r}\right)^t = \frac{1 - \dfrac{1}{(1+r)^T}}{1 - \dfrac{1}{1+r}}$$

$$= \frac{1+r}{(1+r)^T} \cdot \frac{(1+r)^T - 1}{(1+r) - 1} = \frac{(1+r)^T - 1}{r(1+r)^{T-1}}.$$

Hence,

$$W_T = L \cdot \frac{1+r}{r} \cdot \left[(1+r)^T - 1\right]. \tag{3.3}$$

As a quick check, observe that the formula yields $W_1 = L \cdot (1+r)$ for $T = 1$. For instance, if $L = \$10{,}000$, $r = 5\%$, and $T = 30$ years,

$$W_{30} = 10{,}000 \times \frac{1.05}{0.05} \times \left(1.05^{30} - 1\right) = \$697{,}607.90.$$

The interest rate has a remarkable impact. If $r = 4\%$, the above amount drops to \$583,283.40.

If simple interest applies, wealth at retirement is

$$W_T = \sum_{k=1}^{T}(1+kr)L = LT + rL \cdot \sum_{k=1}^{T} k.$$

It is easy to see that

$$\sum_{k=1}^{T} k = \frac{T(T+1)}{2},$$

which implies

$$W_T = LT \cdot \left(1 + \frac{r \cdot (1+T)}{2}\right).$$

> If the above 5% rate is applied with no compounding, wealth at retirement is only
>
> $$W_{30} = 10{,}000 \times 30 \times \left(1 + \frac{0.05 \times 31}{2}\right) = \$532{,}500,$$
>
> which is less than what we would obtain with 4% and compounding.

Example 3.3 aims at showing how remarkable the impact of compounding may be, a concept that we further elaborate on in the following. However, there are many issues that we did not take into due account:

1. We have assumed that the rate r does not change over time. When we roll investment over time, rates may move unfavorably, and we face reinvestment risk.

2. We did not consider inflation. Over a long planning horizon, we should note the difference between real and nominal rates, which we briefly discuss in Section 3.3.

3. We did not consider taxation. Taxes may be applied immediately, when money is contributed to a fund, or later, when we collect the terminal wealth. Deferred taxation may make a remarkable difference.

4. We only considered the *accumulation* phase, i.e., when capital is built, but not the *decumulation* phase, when capital is used for periodic pension payments. Randomness in the residual lifetime after retirement plays a key role; longevity risk is dealt with in actuarial mathematics.

3.1.2 QUOTED VS. EFFECTIVE RATES: COMPOUNDING FREQUENCIES

A further relevant point is that, so far, we have only considered annual compounding, i.e., interest is earned at the end of each year. But what if interest is compounded at a higher frequency? If interest is earned semiannually or quarterly, we may reinvest it earlier, and this should yield some advantage.

Interest rates are *always* quoted in annual terms, even though they may apply over quite different time periods. In this way, we may compare interest rates across different maturities. The quoted rate is referred to as the **annual percentage rate**. However, this rate may apply to a smaller time interval, like a semester or a quarter, and interest may be earned n times a year and immediately reinvested. Let us denote the annual percentage rate by APR_n, where n refers to the compounding frequency, i.e., the number of compounding periods within a single year: With semiannual compounding, $n = 2$, and with quarterly com-

3.1 The time value of money: Shifting money forward in time

pounding, $n = 4$. As the following example illustrates, a higher compounding frequency implies a larger **effective annual rate**, which we denote by EAR_n.

Example 3.4 The effect of compounding frequency

Let us consider again an annual percentage rate of 5%. If we invest $1000, with no compounding, wealth after one year is

$$W_1 = 1000 \times 1.05 = \$1050.$$

Now, what if interest is earned semiannually? The typical convention is that if the annual rate APR_2 is compounded semiannually, it means that a rate $\text{APR}_2/2$ applies to each semester. Hence,

$$W_1 = 1000 \times 1.025^2 = 1050.625.$$

While the quoted rate APR_2 is 5%, the equivalent effective annual rate EAR_2, with semiannual compounding, is a bit larger and can be found as follows:

$$1000 \times (1 + \text{EAR}_2) = 1050.625 \quad \Rightarrow \quad \text{EAR}_2 = 5.0625\%.$$

By a similar token, with quarterly compounding we find

$$W_1 = 1000 \times (1.0125)^4 = 1050.945,$$

which corresponds to an effective annual rate $\text{EAR}_4 = 5.0945\%$.

Example 3.4 shows that, given a quoted APR_n, a higher compounding frequency implies a larger EAR_n. The general formulas are obtained by the following equality, which relates wealth after one year using the two rates:

$$1 + \text{EAR}_n = \left(1 + \frac{\text{APR}_n}{n}\right)^n,$$

which implies

$$\text{EAR}_n = \left(1 + \frac{\text{APR}_n}{n}\right)^n - 1, \tag{3.4}$$

and

$$\text{APR}_n = n \cdot \left[(1 + \text{EAR}_n)^{1/n} - 1\right]. \tag{3.5}$$

An obvious question is: What happens when the compounding frequency is taken to the limit, i.e., $n \to +\infty$? This limit is referred to as **continuous compounding**, and in order to find the answer we have just to recall what we

know from basic calculus:
$$\lim_{n \to +\infty} \left(1 + \frac{1}{n}\right)^n = e,$$

where $e \approx 2.71828$, the Euler number. This implies
$$\lim_{n \to +\infty} \left(1 + \frac{r}{n}\right)^n = e^r.$$

More generally, if we invest L at a continuously compounded rate r for τ years,
$$W_\tau = L e^{r \cdot \tau},$$

where τ need not be an integer number. When compounding in continuous time, the relationship between the quoted and the effective rates, denoted by APR_∞ and EAR_∞, respectively, is found by observing that
$$1 + \text{EAR}_\infty = e^{\text{APR}_\infty},$$

which in turn implies[3]
$$\text{EAR}_\infty = e^{\text{APR}_\infty} - 1, \tag{3.6}$$
$$\text{APR}_\infty = \log(1 + \text{EAR}_\infty). \tag{3.7}$$

To get a feeling for the impact of the compounding frequency, in Table 3.1 we fix an effective rate of 5% and calculate the corresponding APR_n, for a few standard values of n, using Eqs. (3.5) and (3.7). We observe that, when n increases, a lower APR_n suffices to obtain the target EAR. Furthermore, at the displayed precision level, daily and continuous compounding are practically equivalent. In Table 3.2, we reverse the roles of the two rates, and for a quoted rate of 5% we show the corresponding EAR_n obtained by applying Eqs. (3.4) and (3.6). We observe that by increasing n we increase EAR_n, and that daily and continuous compounding are quite close again.

Continuously compounded rates may look like a mathematical abstraction and, indeed, they are not really available. Nevertheless, we observe what follows:

1. Daily compounding, i.e., $n = 365$, yields essentially the same rates as continuous compounding, as we have observed in Tables 3.1 and 3.2 .

2. The use of continuous compounding streamlines many calculations, as we shall see later in this chapter, e.g., when dealing with forward rates.

3. Continuous compounding provides us with a powerful modeling framework in continuous time, based on stochastic differential equations, which are quite useful to represent uncertainty in interest rates and to price interest rate derivatives, as we shall see in later chapters.

---

[3] We always use log to denote the natural logarithm with base e, rather than ln, since we never use decimal or binary logarithms.

Table 3.1 Calculating the APR_n that yields a given effective rate of 5%, for different compounding frequencies n.

Period	Frequency n	APR_n
1 year	1	0.05000
6 months	2	0.04939
1 quarter	4	0.04909
1 month	12	0.04889
1 week	52	0.04881
1 day	365	0.04879
Continuous	∞	0.04879

Table 3.2 Calculating the EPR_n obtained by applying a fixed annual percentage rate of 5%, with different compounding frequencies n.

Period	Frequency n	EAR_n
1 year	1	0.05000
6 months	2	0.05063
1 quarter	4	0.05095
1 month	12	0.05117
1 week	52	0.05125
1 day	365	0.05127
Continuous	∞	0.05127

3.2 The time value of money: Shifting money backward in time

When planning our financial future, we need to shift money into the future, possibly by making some educated guess about future interest rates. By transforming a sequence of cash flows spread over time into a single equivalent cash flow at one time instant, we may also compare different prospects. When analyzing investment opportunities, we do so by shifting cash flows *back* to time $t = 0$, i.e., now. On the one hand, this is clearly convenient, as this provides us with an idea of the present value of a sequence of cash flows; indeed, the net present value (NPV) is a cornerstone in investment analysis.[4] On the other hand, moving money forward in time may require some hypothesis about uncertain interest rates in the future, which is subject to forecast error. On the contrary, when we shift money back in time, we only use given rates applying

---
[4] See Section 3.6.1.

from now to some maturity.[5] Since we transform money now into money in the future multiplying by a growth factor, it stands to good reason that when money is shifted backward, we should do the opposite, i.e., divide by a growth factor. For instance, one question we might want to answer is: What is the value now of $10,000 in one year? If we apply a 5% interest rate, the answer is

$$\frac{10,000}{1.05} = \$9523.81.$$

This the amount of money that, if invested now at 5% for one year, would give us exactly $10,000 at the end of the time horizon. We also know that we should consider some form of compounding when dealing with multiple periods. Hence, the present value of $10,000 in two years is

$$\frac{10,000}{(1.05)^2} = \$9070.295.$$

Not surprisingly, this value is considerably smaller than the previous one. More generally, the present value of an amount L in n years, when an annual rate r is applied, is

$$\frac{L}{(1+r)^n}. \qquad (3.8)$$

This fundamental operation is called **discounting** and the factor $1/(1+r)^n$ is called **discount factor**. Equation (3.8) assumes that cash flows occur at times corresponding to integer multiples of one year and that annual compounding is adopted. We need a way to generalize this formula, and we shall see how in Section 3.2.2. Before doing so, it is quite instructive to formally *motivate* the use of discount factors. As it turns out, discounting is a pervasive concept in finance, and it plays a key role in asset pricing, where it may be justified by no-arbitrage arguments. We shall illustrate the idea by pricing a simple asset, a riskless zero-coupon bonds.

3.2.1 DISCOUNT FACTORS AND PRICING A ZERO-COUPON BOND

Consider a riskless zero-coupon bond maturing in one year, with face value $1000. We should clarify that by "riskless" we mean that there is no default risk. The face value will be certainly redeemed at maturity. What is the fair price of such a bond now? As we have already hinted at in Section 2.2.1, in order to find the answer, we have to ask what risk-free interest rate applies to a time horizon of one year. Let us assume that it is 4%. Then, the fair bond price now should be

$$\frac{1000}{1.04} = 961.5385,$$

---

[5]Things are not so simple, actually, as we may have to use rates that reflect the risk of the investment.

reasonably rounded to $961.54. More generally, the fair price of a zero-coupon bond with face value F, maturing in n years, is

$$\frac{F}{(1+r)^n},$$

where r is the annual interest rate, applying to an investment horizon of n years. We may use a risk-free rate, if we assume that default risk is irrelevant for that bond; otherwise, the discount factor should incorporate a risk premium. When referring to bonds, the interest rate r plays the role of an annual **yield** and is typically denoted by y. Also note that the fair price does not consider the impact of transactions costs and the presence of a bid–ask spread.

Since cash flow discounting is a common sense operation, this pricing formula may look deceptively obvious. Actually, there is a strong justification for the use of discounting in pricing assets featuring deterministic cash flows: The no-arbitrage argument. Any other price would lead, under some idealized assumptions about financial markets, to an arbitrage opportunity. To see this, let us assume that the bond price is lower than the one given above, say, $940.00. In such a case, an arbitrageur could step in and apply the following strategy:

- She may borrow $940.00 at 4% and use this amount to buy the bond.
- In one year, she will collect $1000, i.e., the face value of the bond.
- She will also have to repay the debt, which will amount to

$$940 \times 1.04 = 977.6.$$

To this aim, she will use part of the face value of the bond. The rest yields a sure profit of

$$1000 - 977.6 = \$22.4.$$

Note that this profit is risk-free and does *not* require any initial capital. If a profit of $22.4 does not look quite impressive, imagine scaling up the trading strategy by a large multiplicative factor. The point is that, if the bond price is $940.00, then the implied annual yield y of the bond is found by solving the simple equation

$$940 = \frac{1000}{1+y} \quad \Rightarrow \quad y = \frac{1000}{940} - 1 \approx 6{,}383\%. \tag{3.9}$$

This yield is larger than the risk-free rate of 4%, and such a misalignment cannot persist for long, as the above strategy would be applied by many arbitrageurs trading large volumes of the zero, pushing prices and rates back to a set of consistent levels.

On the other hand, what if the price is larger than the fair price, say, $990? In this case, the arbitrageur should reverse the above trade and pursue the following strategy:

- She may sell the bond short and invest the proceeds, $990, at 4% for one year.

- In one year, she will collect

$$990 \times 1.04 = \$1029.60,$$

which is sufficient to pay the bond face value back to the legitimate bondholder, earning a risk-free profit of $29.60.

As usual, this simple kind of arbitrage strategy amounts to selling an expensive asset and buying a cheap one. In this case, the asset value is related to a risk-free interest rate, and we cannot have two different risk-free rates in an economy. Actually, despite the simplicity and appeal of the reasoning, we must be aware of all of the hidden assumptions behind it. Indeed, we are assuming an idealized, frictionless market in which:

- There is no distortion due to taxes
- There is no limit to borrowing
- There is no limit to short-selling
- There are neither commissions nor bid–ask spreads
- There is no difference between borrowing and lending rates

Clearly, these assumptions do not match reality exactly, and they may be approximately true only for large institutional investors. This leaves room for some misalignment in prices, but the argument is approximately valid. Incidentally, in Section 2.4.1, we have seen that pricing functionals should be linear. Since a coupon-bearing bond may be considered as a portfolio of zeros, we may use a set of discount factors, with different maturities, to price a coupon-bearing bonds, too, as we shall see in Section 3.5.

We close this section by an example illustrating how we may use the concepts that we have just introduced, most notably the annual bond yield, to compare assets.

Example 3.5 Comparing zeros

Let us consider the prices of three zero-coupon bonds with face value $1000 and different maturities, as reported in Table 3.3. Given the bond prices, we may compute their annual holding period return, or yield, in other words. For instance, the holding period return for the zero Z_{30} maturing in 30 years is calculated as

$$\frac{1000 - 231.38}{231.38} \approx 332.19\%.$$

When compared with this seemingly stellar return, the holding period return for the bond $Z_{0.5}$, maturing in six months, just pales away. However, it is clear that such a comparison makes no sense at all. We should come up with a common ground for a comparison, and a

3.2 The time value of money: Shifting money backward in time

Table 3.3 Comparing three zero-coupon bonds with three maturities.

Bond	Maturity	Price ($)	Holding period return	Annual yield
$Z_{0.5}$	6 months	980.58	1.98%	4.0%
Z_2	2 years	915.73	9.20%	4.5%
Z_{30}	30 years	231.38	332.19%	5.0%

natural choice is expressing the return on an *annual* basis. We may invert the pricing formula to find the annual yield on each bond,

$$P_z(T) = \frac{F}{(1+y)^T} \quad \Rightarrow \quad y = \left(\frac{F}{P_z(T)}\right)^{1/T} - 1, \qquad (3.10)$$

where $P_z(T)$ is the price of a zero-coupon bond maturing in T. Carrying out this calculation gives the last column in Table 3.3. We observe that, indeed, the annualized return on longer maturity bonds is larger, which is usually the case. Plotting interest rates for different maturities provides us with a picture of the term structure of interest rates, as we shall see in Section 3.4, and an increasing structure may be justified by risk considerations. Thus, the return of the long-term bond in Table 3.3 is not really surprising (and just a bit less exciting, when compared with the yields of the other two bonds).

Example 3.5 is, in a sense, closer to reality, since information about interest rates and yields is squeezed out of market price data. Then, we may check whether a set of bond prices is consistent or not. It is important to observe that, in order to get a meaningful comparison, we should analyze a set of bonds with similar features in terms of default risk and liquidity.

The careful reader may wonder about the application of Eq. (3.10) when maturity T is not an integer number. For instance, for the bond maturing in six months, we set $T = 0.5$ and find

$$P_z(0.5) = 980.58 = \frac{1000}{(1+y)^{0.5}} \quad \Rightarrow \quad y = \left(\frac{1000}{980.58}\right)^2 - 1 = 0.04. \qquad (3.11)$$

We shall dig the issue in more depth in Section 3.2.2.

3.2.2 DISCOUNT FACTORS VS. INTEREST RATES

In Example 3.5, we have considered the prices of three zeros. On the one hand, the price of a zero implies a well-defined discount factor. On the other hand, to compare the three bonds, we have to find an annualized yield, which is essentially an annual interest rate implied by the bond price. Note that we assume that bonds are free from default risk, and that the corresponding rates are risk-free. However, we have also seen that there are different ways of compounding interest rates. Furthermore, we have to cope with the fact that all of these quantities are subject to random changes over time.

Before carrying out any further investigation, we need to introduce a suitable notation:

- $Z(t,T)$ is the price at time t of a zero-coupon bond with face value of $1 and maturing at time T, where time is measured in years. Actual bonds do not have a face value F of $1, but, by linearity of pricing, we may immediately find the price at time t of a bond maturing at time T as follows:

$$P_z(t,T) = F \cdot Z(t,T). \qquad (3.12)$$

Thus, $Z(t,T)$ is the discount factor at time t for deterministic cash flows at time T.

- $r_n(t,T)$ is the annual interest rate for the time interval (t,T), where interest is compounded n times per year. Note that the relevant time span need not be one year, but we always quote annualized rates.

- $r(t,T)$ is the annual interest rate applying to the time interval (t,T), where interest is compounded in continuous time.

The choice of compounding may be dictated by opportunity. If the interest rate applies to a time period of six months, it is convenient to quote it as $r_2(0,0.5)$, which is the annual rate, whereas the actual rate on the time period will be $r_2(0,0.5)/2$.

Example 3.6 From annual rates to cash flows

Let us consider investing €10,000 for six months, at an annual percentage rate $r_2(0,0.5) = 5\%$, with semiannual compounding. After six months, we will obtain

$$€10{,}000 \times \left(1 + \frac{r_2(0,0.5)}{2}\right) = €10{,}250.$$

Quoting a rate with annual compounding is not quite convenient in this case. To see why, let us recall that we may convert a rate with semiannual compounding to a rate with annual compounding, and

> vice versa, using the relationship
>
> $$\left(1 + \frac{r_2}{2}\right)^2 = 1 + r_1.$$
>
> In principle, for the above investment, we might quote an annual rate
>
> $$r_1(0, 0.5) = \left(1 + \frac{0.05}{2}\right)^2 - 1 = 0.050625.$$
>
> Then, the cash flow after six months should be calculated as
>
> $$\begin{aligned}\text{€}10{,}000 \times \sqrt{1 + r_1(0, 0.5)} &= \text{€}10{,}000 \times \sqrt{1.050625} \\ &= \text{€}10{,}250. \end{aligned} \quad (3.13)$$
>
> We find the same result as before, but in a somewhat twisted way.

Equation (3.13) provides us with a justification for Eq. (3.11), where we have used a discount factor

$$Z(0, 0.5) = \frac{1}{\sqrt{1 + r_1(0, 0.5)}}.$$

This does not look quite natural, but it is fine if we just need a common ground for a comparison. Indeed, in Example 3.5, we have compared the prices of three zeros, $P_z(0, 0.5)$, $P_z(0, 2)$, and $P_z(0, 30)$, with face value $F = 1000$, in terms of the annually compounded interest rates $r_1(0, 0.5)$, $r_1(0, 2)$, and $r_1(0, 30)$. We may do the same using semiannually compounded rates. The only essential requirement is that we must be consistent in the rates we use.

Often, semiannual compounding is assumed in practice, as coupon-bearing bonds usually pay semiannual coupons. As we shall see, continuous compounding is also very convenient from a mathematical viewpoint. It is important to understand, though, that the way we measure interest rates has no impact whatsoever on bond prices and discount factors. Our choice will be merely dictated by convenience or by adherence to market practice. What really matters, in bond pricing, is the set of discount factors. We may use different ways of quoting an annual interest rate, but the discount factor is always the same. A good way to understand this is to realize that we may measure space in kilometers or miles, but the distance between Boston and Los Angeles is what it is.

Since we find it much easier to deal with annualized interest rates, it is important to relate rates and discount factors. We have already used a simple relationship between the price of a zero-coupon bond, maturing in exactly m years, and an interest rate with annual compounding, which we may now express in a more precise way:

$$P_z(t, t+m) = \frac{F}{[1 + r_1(t, t+m)]^m} = F \cdot Z(t, t+m). \qquad (3.14)$$

In Section 3.1.2, we have learned that we may shift money forward in time using different compounding rules. By the same token, we may adopt different compounding rules to move money backward in time, based on rates with different compounding. The relationship of Eq. (3.14) may be stated more generally by using a discount factor depending on a discretely compounded rate $r_n(t, T)$ as follows:

$$Z(t, T) = \frac{1}{\left[1 + \frac{r_n(t,T)}{n}\right]^{n \cdot (T-t)}}. \qquad (3.15)$$

Note that we raise the denominator to a power related to time-to-maturity, which is $(T - t)$ years, but is expressed as a number of compounding periods (n per year). We may also go the other way around:

$$r_n(t, T) = n \cdot \left[\frac{1}{Z(t,T)^{1/n \cdot (T-t)}} - 1\right]. \qquad (3.16)$$

If we resort to continuous compounding, the equivalent relationships look much nicer:

$$Z(t, T) = e^{-r(t,T) \cdot (T-t)}, \qquad (3.17)$$

$$r(t, T) = -\frac{\log Z(t, T)}{T - t}. \qquad (3.18)$$

Note that, in order to go from discount factors to rates, we need natural logarithms. Since $Z(t, T) \leq 1$, the logarithm is negative, and the minus sign in front of it yields a non-negative interest rate in Eq. (3.18).

It is also important to notice that time-to-maturity is $T - t$, and *not* T; sometimes we will use $\tau = T - t$ to denote time-to-maturity. We insist on using a generic current time t, rather than $t = 0$; in this way, we will be able to study the stochastic evolution of prices and rates in time. Since interest rates are not constant, for a given maturity T, the zero prices $P_z(t, T)$ and the corresponding discount factors $Z(t, T)$ are stochastic processes with respect to t. By a similar token, if we fix time-to-maturity τ and we consider the rate $r(t, t + \tau)$ as a function of t, we obtain another stochastic process. This will be essential when we consider randomness in interest rates. Since changes in interest rates are numerically small, it may be useful to introduce a suitable unit to measure them.

DEFINITION 3.1 (Basis point) *A change of one basis point corresponds to a change of 0.0001 (i.e., 1% of 1%) in an interest rate.*

For instance, if an interest rate increases from 4% to 4.6%, we say that it has increased by 60 basis points. One hundred basis points correspond to a change of 1%.

3.3 Nominal vs. real interest rates

In Table 3.3, we may notice how a zero maturing in 30 years offers a remarkable holding period return. The return may look less remarkable when annualized, but there is a further reason of concern, when holding a security with such a long maturity: What about the *real* value of the face value that we will redeem at maturity? We should keep in mind that one of the essential functions of financial markets is to allow for consumption shifts over time. The nominal value of the bond is measured by a monetary amount, but its real value should be measured in terms of the provided ability to consume, i.e., in terms of purchasing power. The purchasing power of money is typically eroded over time, a fact that is measured by an inflation rate. Inflation may be hard to measure, as we have to define what we purchase exactly. Monetary authorities define a basket of goods and services that yields an index, whose composition is updated to reflect consumption trends and technological innovation. At the time of writing, inflation rates are rather low in Europe, but in the past there have been periods of very high inflation, and the issue is quite relevant, e.g., to a pension fund.

Thus, even if the monetary amount of our invested wealth increases by a nominal interest rate, we might be concerned by its real increase, net of inflation effects. To figure out the relationship between real, nominal, and inflation rates, let us consider the monetary price S_t of a financial asset. The nominal rate of return r_t at time t is related to the asset price:

$$r_t = \frac{S_t - S_{t-1}}{S_{t-1}} \quad \Rightarrow \quad 1 + r_t = \frac{S_t}{S_{t-1}}.$$

Now, let us consider the monetary price F_t of a consumption good, which we may identify as the reference basket for computing inflation. The inflation rate i_t and the price level F_t are related by

$$1 + i_t = \frac{F_t}{F_{t-1}}.$$

Now, let us consider the *real* value of the financial asset in term of consumption, i.e., purchasing power, which is given by S_t/F_t.[6] Hence, the real rate of return R_t is related to the real asset value as follows:

$$1 + R_t = \frac{S_t/F_t}{S_{t-1}/F_{t-1}} = \frac{1 + r_t}{1 + i_t}. \tag{3.19}$$

For the sake of simplicity, let us streamline notation and eliminate dependency on time. Therefore, let i_1 be the observed annual inflation rate with annual compounding. This means that the average of prices has increased by a factor $1 + i_1$ over the last year. Note that here we are talking about the observed inflation rate for the past, not the expected one for the future. Given the nominal

---

[6]We may say that the asset price is measured in units of the basket of goods, which plays the role of a **numeraire**. As we shall see, using suitable numeraires plays a key role in asset pricing.

rate r_1 over the same period, the real interest rate R_1, with annual compounding, is given by rewriting Eq. (3.19) and solving for R_1:

$$1 + R_1 = \frac{1 + r_1}{1 + i_1} \quad \Rightarrow \quad R_1 = \frac{r_1 - i_1}{1 + i_1}. \tag{3.20}$$

Note that this is the exact relationship, whereas the rule of thumb $R_1 \approx r_1 - i_1$ is often used. This can be justified, when rates are small enough, as follows:

$$1 + r_1 = (1 + R_1)(1 + i_1) = 1 + R_1 + i_1 + R_1 i_1 \approx 1 + R_1 + i_1$$
$$\Rightarrow \quad R_1 \approx r_1 - i_1. \tag{3.21}$$

If we use continuously compounded rates, it turns out that the rule of thumb of Eq. (3.21) is actually exact:

$$e^R = \frac{e^r}{e^i} \quad \Rightarrow \quad R = r - i. \tag{3.22}$$

As we shall see a few times in the following, formulas involving continuously compounded rates are indeed often simpler than the corresponding formulas for discretely compounded rates.

Example 3.7 The joint impact of inflation and tax rates

> In this book, we do not consider taxes and their impact on return in detail, but let us assume that we are subject to a tax rate t on the capital growth. This means that the nominal after-tax return rate is $r_1(1-t)$. Using the approximation of Eq. (3.21), the real after-tax return is
>
> $$r_1(1-t) - i_1 = (R_1 + i_1)(1-t) - i_1 = R_1(1-t) - i_1 t.$$
>
> The product term $i_1 t$ shows how tax and inflation rates compound in reducing the real increase of wealth.

The inflation index measures the past impact of inflation, but what about the impact of the *expected* future inflation? Economic common sense suggests that long-term interest rates should be somehow affected by the expectation about future inflation. A very simple formula expressing this view is **Fisher's equation**,

$$r_1(0, 1) = R_1(0, 1) + \mathrm{E}_0[i_1],$$

where we use a more careful notation [rates are given at time $t = 0$ and will apply to the time period $(0, 1)$] to insist on the fact that we are looking forward into the future and everything is conditional on information at time $t = 0$. As we have already observed, more often than not, rates for long maturities are larger than short-term interest rates. Inflation risk is a contributing factor, but not the only one, and certainly not in a simple way as suggested by Fisher's equation. We will refrain from discussing the impact of inflation any further, but we emphasize that a long-maturity bond, even if held until maturity, need not be a perfectly risk-free asset. It may be argued that indexing by inflation is required to make the asset truly risk-free. We will briefly consider inflation-indexed bonds in Chapter 5.

3.4 The term structure of interest rates

Let us consider two individuals, who want to borrow money from a bank. The first borrower will repay her debt in one year; the second one will repay his debt in ten years. All other things being equal, we would expect that the annual rate required for the longer-term loan will be higher. There could be different reasons behind this difference, since, in the long-term:

- There is a larger default risk.
- Inflation may have a larger impact.
- The money is locked for a longer period, with a corresponding reduction in liquidity, which means that more favorable investment opportunities could be lost.

It is a matter of fact that if we look at the set of rates $r(t, t+\tau)$ at a fixed time instant t, for different times-to-maturity τ, we observe a nonconstant function of τ. This function defines the **term structure of interest rates**, also called the **zero curve** or **spot rate curve**.[7] Sometimes, we will use the notation $r(t, \cdot)$ to emphasize that we refer to the full term structure observed at time t. The term structure may have different shapes, as depicted in Fig. 3.1. More often than not, the structure is indeed increasing, but there are cases in which this is not true. Decreasing curves are sometimes observed, and humps may also be observed during the transition from an increasing to a decreasing structure and vice versa. Note that here we are considering $r(t, t + \tau)$ as a *deterministic* function of τ for a fixed t. If we fix τ and let time t move forward, we will observe a quite jagged *stochastic process* like the one depicted in Fig. 3.2.[8] If we consider the rate as a joint function of both t and τ, we obtain a **random field**.

Now, two questions are in order:

1. How can we estimate the term structure?
2. How can we explain the shape of the term structure?

To answer the first question, we can take advantage of the link between the whole term structure and the price of coupon-bearing bonds. In Section 3.5.2,

---

[7]Sometimes, the term *yield* curve is used, but we will avoid it, as there is some possibility of confusion between yield of a zero and yield-to-maturity of a possibly coupon-bearing bond. Yield-to-maturity for bonds with the same maturity will differ if their coupon rates are different, as we shall see in Section 3.5.4.
[8]The picture has been obtained by simulating one of the stochastic short-rate models described in Chapter 14.

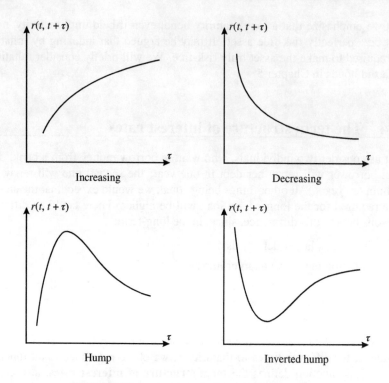

FIGURE 3.1 Different shapes of term structures.

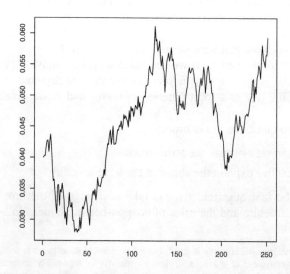

FIGURE 3.2 Evolution of an interest rate over time, for a given time-to-maturity. Time is measured in days on the horizontal axis.

we describe how bond prices may be used to estimate the term structure. The answer to the second question is definitely trickier. In fact, different theories have been proposed to explain the term structure:

- The expectation theory
- The market segmentation and preferred habitat theories
- The liquidity preference theory

We will investigate in a little more detail the last one, which may be thought as a generalization of the expectation theory, in Section 3.7, after introducing forward rates.

3.5 Elementary bond pricing

We have seen that the price at time t of a risk-free zero-coupon bond maturing at time T is related to a discount factor $Z(t, T)$ by Eq. (3.12), where the discount factor in turn may be expressed in terms of interest rate, as in Eqs. (3.15) or (3.17). The application of discounting is a consequence of the no-arbitrage principle. The same principle may be applied to price a more complex coupon-bearing bond by a simple decomposition approach. By taking advantage of the linearity of pricing, each individual future cash flow may be regarded as a zero-coupon bond, as we show in Section 3.5.1. To do so, we need a whole set of discount factors and, unless the term structure is flat, they correspond to different interest rates. In Section 3.5.2, we show how the link between bond prices and rates may be used to estimate the term structure.

To further complicate the matter, the set of underlying interest rates may also reflect the risk in the bond. Common sense suggests that, in presence of default risk, a larger interest rate will be commanded. For the sake of simplicity, we will just assume that the bond is free of default risk, and we will deal with risk-free interest rates. However, we should also realize that, in concrete, defining and measuring a risk-free rate is not as easy as it might seem, and the very concept is subject to some misunderstanding; we elaborate on this theme in Section 3.5.3.

This is an extensive section, where we shall also introduce essential concepts like yield-to-maturity and duration. We will show their relevance in interest rate risk management. To close the section, we will consider the pricing of a simple floating-rate bond, which leads to some possibly counterintuitive findings.

3.5.1 PRICING COUPON-BEARING BONDS

Let us consider a risk-free bond with face value F, maturing at time T, paying a coupon with constant rate c every six months; we will denote the price of this bond at time t by $P_c(t, T)$, to distinguish it from the price $P_z(t, T)$ of a zero with the same face value. Semiannual coupons are the most common case, but this need not be an absolute rule. Note that, consistently with market conventions, the coupon is expressed in annual terms, but the actual coupon paid every six months is $F \cdot c/2$. The current time is denoted by t as usual, and we assume that m coupons will be paid, at times T_i, $i = 1, \ldots, m$. The last time instant coincides with maturity, $T_m = T$, where the face value is also repaid. Hence, the bond is essentially a stream of cash flows timed as follows, assuming $F = 100$:

$$\left(100 \times \frac{c}{2}, T_1\right), \ldots, \left(100 \times \frac{c}{2}, T_{m-1}\right), \left(100 \times \left(1 + \frac{c}{2}\right), T_m\right).$$

By linearity of pricing,[9] the bond may be decomposed as a portfolio of $m - 1$ zero-coupon bonds with face value $100 \times c/2$, maturing at times T_i, $i = 1, \ldots, m-1$, respectively, and a zero-coupon bond with face value $100 \times (1 + c/2)$, maturing at time T_m. To find the price of the bond, we have just to price the individual zeros and add everything up:

$$\begin{aligned} P_c(t,T) &= \frac{100 \times c}{2} \cdot Z(t, T_1) + \frac{100 \times c}{2} \cdot Z(t, T_2) + \cdots \\ &\quad + \frac{100 \times c}{2} \cdot Z(t, T_{m-1}) + 100 \times \left(1 + \frac{c}{2}\right) \cdot Z(t, T_m) \\ &= \frac{c}{2} \sum_{i=1}^{m} P_z(t, T_i) + P_z(t, T_m). \end{aligned} \qquad (3.23)$$

We observe that the bond price depends on an array of discount factors, i.e., on the whole set of underlying risk-free rates. If we express discount factors in terms of discretely compounded rates, we obtain

$$P_c(t,T) = F \cdot \left[\frac{c}{2} \sum_{i=1}^{m} \frac{1}{[1 + r_n(t, T_i)/n]^{n \cdot (T_i - t)}} + \frac{1}{[1 + r_n(t, T_m)/n]^{n \cdot (T_m - t)}} \right]. \qquad (3.24)$$

A similar formula applies if we use continuous compounding:

$$P_c(t,T) = F \cdot \left[\frac{c}{2} \sum_{i=1}^{m} e^{-r(t,T_i) \cdot (T_i - t)} + e^{-r(t,T_m) \cdot (T_m - t)} \right]. \qquad (3.25)$$

We insist again: There is no conceptual difference between Eqs. (3.24) and (3.25), as the discount factors are just the same. They are only expressed in two different ways, and the latter one will prove to be more convenient, but this amounts to measuring the same objects using different units of measurement.

---

[9] See Section 2.4.1.

3.5 Elementary bond pricing

▣ Example 3.8 Bond prices and coupon rates

Let us consider a bond with face value $1,000, paying semiannual coupons with rate 4%, and maturing in two years. Each coupon payment amounts to $20. We assume a term structure consisting of the following continuously compounded rates:

$$r(0, 0.5) = 3.7\%, \quad r(0, 1) = 4.0\%,$$
$$r(0, 1.5) = 4.2\%, \quad r(0, 2) = 4.3\%.$$

The bond price is

$$P_c(0, 2) = 20 \times e^{-0.037 \times 0.5} + 20 \times e^{-0.04 \times 1} + 20 \times e^{-0.042 \times 1.5}$$
$$+ 1020 \times e^{-0.037 \times 2} = \$993.57.$$

If the coupon rate is 6%, the price increases to

$$P_c(0, 2) = 30 \times e^{-0.037 \times 0.5} + 30 \times e^{-0.04 \times 1} + 30 \times e^{-0.042 \times 1.5}$$
$$+ 1030 \times e^{-0.037 \times 2} = \$1031.56.$$

The bond prices in Example 3.8 reflect the face value of $1,000. However, the face value is not quite relevant, and usually bond prices are quoted as a percentage of the face value. For the two bonds of Example 3.8, this would correspond to 99.357 and 103.156, respectively.[10] We observe that, depending on the relationship between interest rates and the bond coupon rate, the price may be below or above the face value. This is expressed as follows.

DEFINITION 3.2 (Trading at premium and at discount) *If the bond price is larger than the face value, we say that the bond trades* **at premium**. *If the bond price is smaller than the face value, we say that the bond trades* **at discount**. *If the bond price corresponds to the face value, we say that the bond trades* **at par**.

Clearly, a zero always trades at discount and, if there is no change in the interest rates, its value will increase over time. When a bond trades at premium, its price will decrease over time, reflecting the fact that valuable coupons are detached. When a bond is issued, the coupon rate is usually set in such a way that the bond initially trades approximately at par.

---

[10] We shall discuss the practicalities of bond price quoting in Section 5.2.2.

3.5.2 FROM BOND PRICES TO TERM STRUCTURES, AND VICE VERSA

Equations (3.24) and (3.25) link bond prices to the term structure of interest rates and may used in two ways:

1. To use observed[11] bond prices to estimate a term structure of interest rates.
2. To find the fair value of a bond, given a term structure; a discrepancy between the fair value and the observed price might indicate that some bonds are either over- or underpriced, relative to other bonds, which may suggest an arbitrage opportunity.

Estimating the term structure at time $t = 0$ amounts to finding a set of risk-free rates $r(0, T_i)$, for a set of maturities T_i, $i = 1, \ldots, m$, such that a pricing model matches observed asset prices. The full term structure may then be recovered by a suitable interpolation strategy. If we had a broad set of zero-coupon bonds, issued by an extremely creditworthy issuer, we would have a rich set of discount factors $Z(t, T_i)$ from which rates could be easily obtained. There are a few problems, however.

- A T-bill would be a suitable zero-coupon bond as far as rates on USD are concerned, but such bonds are not available for long maturities. Long-term zeros are traded, but they result from stripping coupons from long-term bonds (a process called cash flow unbundling). Such bonds are actually issued by banks and, therefore, they are subject to some credit risk that would affect our estimate.
- Another issue is liquidity: Market prices are not necessarily the same as fair prices, a difference that may be due to liquidity and other factors, like the occasional flight to quality.
- It has been argued that even T-bills may not be the best choice since their price may be affected by bank regulations, which require banks to hold a stock of T-bills.

This is why further assets are often brought into the picture, i.e., interest rate derivatives like interest rate futures and swaps. These are extremely liquid assets and are sold with a wide range of maturities. The presence of a clearinghouse for futures and the limited credit exposure associated with a swap make counterparty risk almost irrelevant.

For the sake of illustration, let us consider basic procedures to estimate the term structure with a set of generic bonds. Let us assume that we have selected m bonds, indexed by $k = 1, \ldots, m$, with cash flows C_{ki} at times T_i, $i = 1, \ldots, m$. Note that the number of bonds is the same as the number of time

---

[11] Here, we use the term *observed* rather than *quoted* bond price. The reason is that, as we shall see in Chapter 5, the quoted bond price is not the cash price, as it does not consider accrued interest from the next coupon.

3.5 Elementary bond pricing

Table 3.4 **Data for bootstrapping in Example 3.9.**

Maturity	0.5	1	1.5	2
Coupon rate	0%	6%	3%	5%
Price	984.62	1023.12	985.13	1014.69

instants. If a bond matures at time $T_j < T_m$, all cash flows for $i > j$ are zero. The cash flows should be discounted by discount factors $Z(t, T_i)$, denoted by Z_i for the sake of simplicity, to yield the observed price P_k^o. Thus, we just have to solve the following system of linear equations:

$$
\begin{aligned}
P_1^o &= C_{11}Z_1 + C_{12}Z_2 + C_{13}Z_3 + \cdots C_{1m}Z_m \\
P_2^o &= C_{21}Z_1 + C_{22}Z_2 + C_{23}Z_3 + \cdots C_{2m}Z_m \\
P_3^o &= C_{31}Z_1 + C_{32}Z_2 + C_{33}Z_3 + \cdots C_{3m}Z_m \\
\vdots &= \vdots \\
P_m^o &= C_{m1}Z_1 + C_{m2}Z_2 + C_{m3}Z_3 + \cdots C_{mm}Z_m.
\end{aligned}
\tag{3.26}
$$

Note that, since we have a set of m unknown discount factors and a set of m equations, assuming that bonds are linearly independent, we will find exactly one set of discount factors, from which we may deduce a set of interest rates.

A particular case of this procedure is obtained when cash flows have a staircase structure, i.e., bond k has exactly k cash flows at times $T_1, \ldots, T_k$. This means that the first bond is a zero (or a bond with just one coupon left) maturing at T_1, that the second bond pays a coupon at time T_1 and matures at time T_2, etc. We find a system of linear equation, whose matrix is lower triangular:

$$
\begin{aligned}
P_1^o &= C_{11}Z_1 \\
P_2^o &= C_{21}Z_1 + C_{22}Z_2 \\
P_3^o &= C_{31}Z_1 + C_{32}Z_2 + C_{33}Z_3 \\
\vdots &= \vdots \\
P_m^o &= C_{m1}Z_1 + C_{m2}Z_2 + C_{m3}Z_3 + \cdots C_{mm}Z_m.
\end{aligned}
\tag{3.27}
$$

This system is solved by forward substitution, finding one discount factor at each step, as shown in the following example. This textbook approach is known as **bootstrapping the zero curve**. Actually, what we do, when solving the system of Eq. (3.26) or Eq. (3.27), is finding the **discount curve**, i.e., the curve of discount factors $Z(t, T)$, which is then converted to the zero curve of interest rates.

Example 3.9 Bootstrapping a term structure

Let us consider the bond prices of Table 3.4, where we assume that all face values are 1000 and coupons are semiannual, and find the implied continuously compounded rates. The price of the first zero maturing in six months yields the first discount factor immediately:

$$Z(0, 0.5) = \frac{984.62}{1000} = 0.98462$$

$$\Rightarrow \quad r(0, 0.5) = -\frac{\log Z(0, 0.5)}{0.5} = 0.031.$$

The second bond has two cash flows, 30 and 1030, in six months and one year, respectively. Hence

$$1023.12 = 30 \cdot Z(0, 0.5) + 1030 \cdot Z(0, 1)$$

$$\Rightarrow \quad Z(0, 1) = \frac{1023.12 - 30 \times 0.98462}{1030} = 0.96464$$

$$\Rightarrow \quad r(0, 1) = -\frac{\log Z(0, 1)}{1} = 0.036.$$

By a similar token,

$$985.13 = 15 \cdot Z(0, 0.5) + 15 \cdot Z(0, 1) + 1015 \cdot Z(0, 1.5)$$

$$\Rightarrow \quad Z(0, 1.5) = \frac{985.13 - 15 \times 0.98462 - 15 \times 0.96464}{1015} = 0.94176$$

$$\Rightarrow \quad r(0, 1.5) = -\frac{\log Z(0, 1.5)}{1.5} = 0.04.$$

The last step yields $r(0, 2) = 0.042$.

There are few issues with the above procedures, as it may be difficult to find a good set of risk-comparable bonds that are not affected by liquidity issues and feature a synchronized sequence of cash flows. In general, it may be better to use a larger number of securities. This leads to an overdetermined system, where we have more equations than unknown variables, but we can find a solution in the least-squares sense. Let $\widehat{P}_k(Z_1, \ldots, Z_m)$ be the price predicted for bond k as a function of the discount factors. We would like to find a set of discount factors such that the predicted prices are as close as possible to the observed prices. Hence, given a set of $n > m$ bonds, we may solve the optimization problem:

$$\min_{Z_1, \ldots, Z_m} \sum_{k=1}^{n} \left[P_k^0 - \widehat{P}_k(Z_1, \ldots, Z_m) \right]^2. \quad (3.28)$$

3.5 Elementary bond pricing

If $\widehat{P}_k(Z_1, \ldots, Z_m)$ is a linear function of the discount factors, this is a simple linear least-squares problem, which is quite easy to solve. In a more general setting, this is a nonlinear optimization problem, typically a nonconvex one, which requires numerical methods for its solution.[12] This would be the case, if we calibrate directly in terms of interest rates. This kind of model calibration has a wide scope of applicability and can be applied to quite sophisticated pricing models for derivatives. Note that, in solving Problem (3.28), we are still assuming a set of synchronized cash flows. A more flexible procedure should rely on a set of arbitrarily timed cash flows; a suitable interpolation approach may be adopted to find discount factors at generic time instants as a function of the subset $\{Z_1, \ldots, Z_m\}$. Last, but not least, we have assumed a *non-parametric* approach whereby we directly solve for the discount factors. An alternative approach would be to parameterize the zero or the discount curve and estimate the parameters of the curve. Note again that we may express prices in terms of discount factors or interest rates. A parametric model of the curve of interest rates, which we do not illustrate in detail, is the **Nelson–Siegel model**:

$$r(0,T) = \beta_0 + (\beta_1 + \beta_2) \cdot \frac{\tau_1}{T} \cdot \left(1 - e^{-T/\tau_1}\right) - \beta_2 e^{-T/\tau_1}.$$

The model depends on parameters β_0, β_1, β_2, and τ_1, and it is clearly more parsimonious than a model relying on several interest rates. This may sacrifice fit in favor of more robustness to liquidity and other issues. If the model may look a bit peculiar, it is because the Nelson–Siegel model, as many others, is actually a model for the *forward* rates, which we describe later in Section 3.7. Modeling the forward rates often turns out to be a more convenient approach than dealing with the spot rates directly. Given the forward rates, it is easy to find the spot rates.

3.5.3 WHAT IS A RISK-FREE RATE, ANYWAY?

In the rest of this book, we will often refer to risk-free rates or risk-free assets. However, since the term might be somewhat misleading, it is necessary to clarify what we really mean by risk-free rate and risk-free asset.

Consider, for instance, a T-bill, i.e., a short-maturity zero issued by the US treasury. In portfolio theory, when talking about a risk-free asset, the T-bill is quite often given as a concrete example, and its yield is proposed as a risk-free return over a holding period corresponding to its maturity. But is a T-bill a really safe asset? To find an answer, we must list and comment on the potentially relevant sources of risk for its holder:

- **Currency risk**. A T-bill may be a safe asset for an investor whose currency is the US dollar, but not for all investors.

---

[12] We deal with nonconvex optimization in Section 16.2. Further issues with model calibration are discussed in Section 14.4.

- **Inflation risk**. Since the T-bill has very short maturity (say, three or six months), inflation risk is not likely to be that relevant. Nevertheless, it would be relevant for a zero with longer maturity.
- **Interest rate risk**. From the bond pricing formulas, it is clear that bond prices are subject to change if interest rates change. Later, we will see that the impact may be large or small, depending on bond maturity and coupon rates. For a T-bill, the impact is limited and, probably, an investor will hold the bond until maturity. However, if a long-term zero is sold along the way, a possibly consistent loss may be incurred.
- **Default risk**. A T-bill is considered virtually risk-free from this viewpoint. Bonds issued by other governments are not so safe, and the same applies to corporate bonds.

Therefore, if we assume the US dollar as the reference currency and we rule out default risk, we may say that a T-bill is a reasonably safe asset. We can buy the asset now for a price $F \cdot Z(0,T)$, and we will receive the face value F at maturity. Thus, the corresponding holding period return is deterministic, rather than stochastic, as is the case with stock shares.

The price of a zero is related to a discount factor $Z(t,T)$, which is related in turn to an interest rate $r(t,T)$ (let us use the one with continuous compounding). Typically, when we talk about risk-free rates, we do not want to bring inflation and currency risk into the picture, and we assume that the investment is held for the whole time horizon to which the interest rate applies. So, can we say that the rate that we may calculate from a T-bill price is a risk-free rate? As we have pointed out before, some practitioners could object that the price of T-bills is somewhat affected by regulations requiring banks to hold some T-bills, which may have an impact on its price. Hence, they suggest that risk-free rates should be estimated on the basis of other quite liquid securities, like certain interest rate derivatives. We might disregard this issue, but it still remains a fact that a risk-free rate is defined with reference to a given time horizon. Using estimation procedures that we have outlined in Section 3.5.2, we may estimate the term structure of risk-free rates $r(t,T)$ at time t, which for fixed t is a deterministic function of T. However, the interest rate as a function of t is *not* deterministic. Indeed, when we will build continuous-time models for interest rates, we will see that they are based on stochastic processes, just like stock share prices. The idea of modeling a risk-free rate by a stochastic process may sound confusing, and some clarification is in order. If we invest at rate $r(t,T)$ at time t and we just collect the reward at time T, indeed, there is no risk involved. However, if we consider a process defined as

$$r(t, t + \tau)$$

for a *fixed* time-to-maturity τ, as a function of t, this is in fact a stochastic process. Risk-free rates do move and we are subject to not only the aforementioned interest rate risk, but also to **reinvestment risk**. The second kind of risk is associated with rolling a short-term investment forward in time. Consider a strategy whereby we invest in T-bills maturing in three months. When a T-bill matures,

3.5 Elementary bond pricing

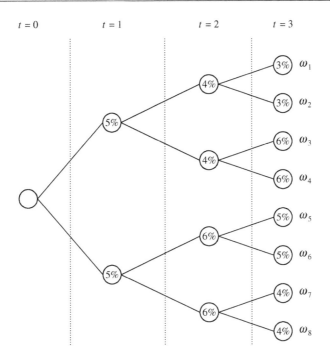

FIGURE 3.3 A scenario tree for a risk-free rate.

we reinvest its face value in a freshly issued T-bill. When we buy a T-bill, we know how much wealth we will have when it matures in three months, but we do *not* know the future prices of new T-bills, because the future interest rates, as well as the corresponding discount factors, are random.

Another way to understand reinvestment risk is to compare a bank account, whose interest rate will be reset at times $T_1, T_2, \ldots, T_m$, to a zero-coupon bond maturing at T_m. Assuming continuously compounded rates, the money invested in the zero will grow by a known multiplicative gain factor,

$$\exp\left[r(t, T_m) \cdot (T_m - t)\right].$$

The gain factor for the bank account is

$$\exp\left[r(t, T_1) \cdot (T_1 - t) + r(T_1, T_2) \cdot (T_2 - T_1) + \right.$$
$$\left. \cdots + r(T_{m-1}, T_m) \cdot (T_m - T_{m-1})\right].$$

In this expression, only $r(t, T_1)$ is known at time t, even though each rate $r(T_{i-1}, T_i)$ will be given at the beginning of the corresponding time interval.

From a formal viewpoint, all of this is related to the concept of a **predictable stochastic process** and it may be visualized by the scenario tree of Fig. 3.3. This tree is not meant to be realistic in any way, but the key message is how interest rates are associated with nodes. The return of a stock share over the time period $(t, t+\delta t)$,

$$\frac{S(t+\delta) - S(t)}{S(t)},$$

is only known at time $t+\delta$, the *end* of the time interval. However, the interest rate $r(t, t+\delta t)$ is known at time t, the *beginning* of the time interval. Hence, in the scenario tree of Fig. 3.3, the rates for all of the successors of any node are the same.[13] However, if we keep rolling the investment over time, we face reinvestment risk, and we cannot predict the final outcome exactly. For instance, the scenario ω_1 corresponds to a sample path in which the rate is 5% on the time interval (0,1), 4% on the time interval (1,2), and 3% on the time interval (2,3), Thus, the holding period return for this scenario is

$$R_H(\omega_1) = 1.05 \times 1.04 \times 1.03 - 1 = 12.476\%.$$

We note that this is the same result we obtain in scenario ω_2. However, the corresponding return for scenarios ω_7 and ω_8 is

$$R_H(\omega_7) = R_H(\omega_8) = 1.05 \times 1.06 \times 1.04 - 1 = 15.752\%.$$

The concept of a predictable stochastic process is easy to grasp in discrete time, whereas more technicalities are involved in the case of continuous time.

3.5.4 YIELD-TO-MATURITY

Pricing a bond at time t requires the knowledge of a full term structure $r(t, \cdot)$, since different interest rates enter the pricing formula. If we like to be picky, we could denote the price of a coupon-bearing bond maturing at T by

$$P_c(t; T, r(t, \cdot)).$$

But what can we say about the return of the bond? It should be clear that there is no reason to believe that any one of the rates involved in the pricing formula defines the holding period return. It should be even clearer that the coupon rate should not be confused with the bond return. To be more concrete, let us consider the following two bonds:

- Bond B_1 pays a 3% coupon and sells for 96.08
- Bond B_2 pays a 9% coupon and sells for 132.18

Clearly, in comparing the two bonds, maturity plays a role, but even if maturity is the same, how can we compare the two assets? The second bond offers a very palatable coupon, but it is very expensive. By the way, we notice that a coupon-bearing bond may have a price significantly larger than the face value,

---

[13]This may be expressed in terms of the filtration to which the process is adapted, as we will see in Chapter 11.

3.5 Elementary bond pricing

which is impossible for zeros. It stands to reason that when coupons are paid, the overall value of the second bond will be reduced,[14] since we strip a cash flow out. The first bond is much cheaper and its value will arguably increase, when maturity is approached; however, its coupon is much less attractive. We clearly see the need of a single number giving us a feeling for the relative value of the two investments.

Actually, assessing the return from holding a bond is not trivial at all. A first point is: Why are we holding that bond? If we are holding the bond as a way to meet a stream of future liabilities, the bond return, per se, might not be very relevant. Assuming that we are interested in an asset-only portfolio, what is the holding period over which we want to assess the return? Are we holding the bond until maturity, or are we planning to sell it along the way? In the former case, we do not have any uncertainty about the cash flows, but in the latter one, we face some uncertainty about the price at which we will sell the bond, as the interest rates are stochastic. Uncertainty in future interest rates has an impact even if we hold the bond until maturity. In fact, we will reinvest the cash flows from coupons along the way, but at which interest rates? Thus, for a given time horizon, the bond return is a random variable, and its characterization requires a specification of a stochastic model of interest rates. The answer depends on our assumptions and will not be the same for other market participants with different expectations. Clearly, if we want to find a simple and manageable answer, we must adopt some drastic simplification.

A simple, even though limited, answer is provided by **yield-to-maturity**, or **YTM** for short. YTM is a single interest rate that, when used to define discount factors in a bond pricing formula, matches the observed bond price $P_c(t,T)$. YTM is a feature of a specific bond, hence we will use a simplified notation, whereby y_n and y are the yields with discrete and continuous compounding, respectively, without reference to time. For instance, if we use semiannual compounding, which is a common practice, since most bonds pay semiannual coupons, we define the semiannually compounded yield y_2 as the solution of the following equation:

$$P_c(t,T) = F \cdot \left[\frac{c}{2} \sum_{i=1}^{m} \frac{1}{\left(1 + \frac{y_2}{2}\right)^{2\cdot(T_i - t)}} + \frac{1}{\left(1 + \frac{y_2}{2}\right)^{2\cdot(T_m - t)}} \right]. \qquad (3.29)$$

The idea can be extended to any compounding period in order to define y_n. In the case of continuous compounding, we have

$$P_c(t,T) = F \cdot \left[\frac{c}{2} \sum_{i=1}^{m} e^{-y\cdot(T_i - t)} + e^{-y\cdot(T_m - t)} \right]. \qquad (3.30)$$

Note that we just associate *one* yield y_n or y with the whole sequence of cash flows. We may think of YTM as a sort of average between interest rates in the

---

[14] A full picture is given in Fig. 5.1.

term structure. Indeed, it is the interest rate that would yield the observed bond price in the case of a flat term structure. Computing YTM requires numerical methods, as shown in the following example.

Example 3.10 The link between YTM and interest rates

Let us assume that the following term structure prevails on markets:
$$r_1(0,1) = 4\%, \quad r_1(0,2) = 4.5\%, \quad r_1(0,3) = 5\%.$$

A bond maturing in three years, paying a 3% annual coupon, has fair price
$$P_{3\%} = \frac{3}{1.04} + \frac{3}{1.045^2} + \frac{103}{1.05} = 94.6071.$$

Note that the coupon rate is lower than all of the relevant rates, and the bond sells at discount. To find YTM, with annual compounding, we have to solve the nonlinear equation
$$\frac{3}{1+y_1} + \frac{3}{(1+y_1)^2} + \frac{103}{(1+y_1)^3} = 94.6071,$$

which can be transformed into the polynomial equation
$$103x^3 + 3x^2 + 3x - 94.6071 = 0,$$

where $x = 1/(1+y_1)$. This equation has a single real root, as well as two complex conjugates that we ignore,
$$x = 0.9526 \quad \Rightarrow \quad y_1 = \frac{1-x}{x} = 0.0498.$$

Note that the "average" is tilted toward the largest rate, corresponding to the last cash flow, which includes the face value and is much larger than those consisting of coupons only. A similar bond, with 9% coupon, has fair price
$$P_{9\%} = \frac{9}{1.04} + \frac{9}{1.045^2} + \frac{109}{1.05} = 111.0537.$$

This bond sells at premium, and its yield is 0.0495. This is a bit smaller, as the first and second cash flows are relatively larger.

The difference in YTM may be quite significant. We omit the details, but if we consider similar bonds, paying one annual coupon with rates 3% and 9%, respectively, maturing in 20 years, and we assume that the term structure consists of annually compounded rates, increasing linearly over time from 2% to 5%, the two bond prices are
$$P_{3\%} = 79.12, \qquad P_{9\%} = 161.97,$$

with yields
$$y_{1,3\%} = 4.62\%, \qquad y_{1,9\%} = 4.31\%.$$

3.5 Elementary bond pricing

The careful reader might wonder whether there is any guarantee that, in general, we find a single YTM, if any. Indeed, a general polynomial equation of degree n may have up to n real roots, not necessarily positive (complex conjugate roots are of no use to us). We discuss this matter later in Section 3.6.2, when dealing with the internal rate of return of a cash flow sequence. We can anticipate that, in the case of a bond, it can be shown that there is exactly one real and positive root, so that there is no ambiguity in calculating YTM.

Example 3.10 clearly shows that, unlike interest rates, we *cannot* associate YTM with a specific maturity, as it depends on the peculiarities of each bond, such as the coupon rate and, possibly, liquidity. Despite this observation, it is instructive to see the impact that YTM may have on bond prices if we consider it as a single, catch-all risk factor. We shall do so in Section 3.5.5. Furthermore, while pricing a bond using YTM is a crude simplification with respect to a full-fledged term structure, it may help intuition building by providing us with simple bond price formulas.

Example 3.11 Pricing annuities

Annuities are assets that provide a stream of periodic payments over a period of time. This may correspond to the buyer's lifetime in the case of life insurances and pension funds in the decumulation phase (when the accumulated wealth is depleted in order to provide pension payments). Pricing an annuity under longevity risk requires tools from actuarial mathematics. Furthermore, the long time span involved implies considerable uncertainty about future interest rates. The picture may be further complicated if the annuity is inflation-indexed.

Here, we consider a simple annuity providing fixed payments over a given time horizon, disregarding interest rate risk. Such an annuity is just a bond whereby no face value is redeemed, and it may be priced given a term structure of interest rates. It is useful to find an explicit formula, under the further simplification of a flat term structure, e.g., for a given annually compounded yield:

$$r_1(t, T_i) = y_1,$$

where T_i, $i = 1, \ldots, m$, is the set of time instants at which a payment is made. Using y_1 makes sense if we consider annual payments (in practice, typical annuities involve monthly payments). The price of a unit annuity, paying \$1 at each relevant epoch, is

$$A = \sum_{i=1}^{m} \frac{1}{(1+y_1)^i}.$$

We may find a compact expression for this value, by relying on the geometric series and using the same trick as in Example 3.3. If we consider $\alpha \in (0,1)$,

$$\sum_{i=1}^{m}\alpha^i = \sum_{i=1}^{+\infty}\alpha^i - \sum_{i=m+1}^{+\infty}\alpha^i = \alpha\cdot\left[\sum_{i=0}^{+\infty}\alpha^i - \sum_{i=m}^{+\infty}\alpha^i\right]$$

$$= \alpha\cdot\left[\frac{1}{1-\alpha} - \frac{\alpha^m}{1-\alpha}\right] = \frac{\alpha(1-\alpha^m)}{1-\alpha}.$$

Plugging $\alpha = 1/(1+y_1)$, we find

$$A = \frac{\frac{1}{1+y_1}\cdot\left[1 - \frac{1}{(1+y_1)^m}\right]}{1 - \frac{1}{1+y_1}} = \frac{1}{y_1}\cdot\left[1 - \frac{1}{(1+y_1)^m}\right]. \qquad (3.31)$$

It is important to see the connection between this formula and Eq. (3.3). In that case, we have cash flows L at times $t = 0, \ldots, T-1$, and we are evaluating the terminal wealth W_T at time T. Here, we have cash flows $C = 1$ at times $t = 1, \ldots, T$, and we are evaluating the annuity A at time $t = 0$. To see the equivalence, we can shift W_T backward in time by $T+1$ time periods, to time $t = 0$. This requires dividing Eq. (3.3) by $(1+y_1)^{T+1}$:

$$\frac{W_T}{(1+y_1)^{T+1}} = \frac{L}{(1+y_1)^{T+1}}\cdot\frac{1+y_1}{y_1}\cdot\left[(1+y_1)^T - 1\right]$$

$$= \frac{L}{y_1}\cdot\left[1 - \frac{1}{(1+y_1)^T}\right],$$

which is consistent with Eq. (3.31).

The formula for an annuity immediately yields a formula to price a bond as a function of YTM. We just have to add the discounted cash flow corresponding to the face value. If we consider a hypothetical bond maturing in T years and paying a single coupon per year, at rate c, the bond price is

$$P_c(0,T) = \frac{cF}{y_1}\cdot\left[1 - \frac{1}{(1+y_1)^T}\right] + \frac{F}{(1+y_1)^T}. \qquad (3.32)$$

Note that this formula applies only when the bond is issued or just after the payment of a coupon. In the more realistic case of semiannual coupons, we find

$$P_c(0,T) = \frac{c/2\cdot F}{y_2/2}\cdot\left[1 - \frac{1}{(1+y_2/2)^{2T}}\right] + \frac{F}{(1+y_2/2)^{2T}}.$$

Note that, in this case, the number of coupons, i.e., the number of time periods is $2T$. We may also come up with a formula for a continuously compounded yield

3.5 Elementary bond pricing

y, but since all of these variations do not contribute much to intuition building, let us stick with the annually compounded yield y_1 and assume that coupons are paid annually, for the sake of simplicity.

A question that we should address is: Why is yield-to-maturity called that way? To find the answer, imagine that we buy a bond when it is issued and keep it until maturity, for exactly T years, reinvesting the coupons at the risk-free rate. In practice, we do not really know the future rates at which coupons will be reinvested, but let us assume that all of them are just given by y_1. Then, wealth at maturity T, just after collecting the last coupon plus the bond face value, can be found by shifting cash flows forward in time:

$$W_T = \sum_{t=1}^{T} cF \cdot (1+y_1)^{T-t} + F$$

$$= (1+y_1)^T \cdot \left[\sum_{t=1}^{T} \frac{cF}{(1+y_1)^t} + \frac{F}{(1+y_1)^T} \right]$$

$$= (1+y_1)^T \cdot P_c(0,T).$$

Thus, we see that y_1 gives the holding period return, assuming that the bond is kept until maturity, and that the term structure of interest rates is flat and constant over time. Clearly, these assumptions do not match the real world, but YTM may provide us with a rough-cut estimate of how much a bond yields, which is good enough for a comparison.

Furthermore, YTM is very useful to build some fundamental intuition. To see how, let us observe that Eq. (3.32) may be rewritten in two ways:

$$P_c(0,T) = \frac{cF}{y_1} + \frac{F \cdot (1 - c/y_1)}{(1+y_1)^T}, \qquad (3.33)$$

and

$$\frac{P_c(0,T)}{F} = \frac{c}{y_1} \cdot \left[1 - \frac{1}{(1+y_1)^T} \right] + \frac{1}{(1+y_1)^T}. \qquad (3.34)$$

Note that the second expression shows that the bond price, relative to its par value, is a weighted average of the ratio between coupon rate and yield, c/y_1, and 1.

Example 3.12 Pricing a perpetuity

By using Eq. (3.33) and taking the limit for $T \to +\infty$, it is easy to find the price of a perpetuity, i.e., an annuity where $T \to +\infty$, paying an annual amount C, which may be thought as a fraction c of a virtual nominal F:

$$P_c(0,\infty) = \frac{C}{y_1}. \qquad (3.35)$$

> The notation suggests the interpretation of this security as a coupon-bearing bond with infinite maturity. For instance, if $y_1 = 5\%$ and $C = 10{,}000$, we have
>
> $$P_c(0, \infty) = \frac{10{,}000}{0.05} = 2{,}000{,}000.$$
>
> Note that 5% of this value is exactly the annual payment, i.e., what is required to pay the annual coupon while keeping the capital intact, assuming that it will be reinvested at a rate $y_1 = 0.05$ forever. Given the stochastic nature of interest rates, this will be hardly the case.

A real-life example of a perpetuity was the British consol, a kind of perpetual bond. Equation (3.34) is extremely useful to investigate the relationships between YTM, coupon rate c, and bond price.

Example 3.13 A key result

> What happens if the value of YTM and the coupon rate c are the same? By applying Eq. (3.33), we find
>
> $$P_c(0, T) = \frac{cF}{c} + \frac{F \cdot (1 - c/c)}{(1+c)^T} = F.$$
>
> Thus, when $c = y_1$, the bond trades at par.
>
> By the same token if $c > y_1$, by using Eq. (3.34), we see that the bond price is an average between 1 and a number larger than 1. Hence, $P_c(0, T) > F$ and the bond trades at premium. On the contrary, if $c < y_1$, then $P_c(0, T) < F$ and the bond trades at discount. We also notice that the longer the maturity, the smaller the weight of 1 in Eq. (3.34), and the larger/smaller the bond price as a function of the coupon rate. A long-term bond with a large coupon rate is, in fact, quite expensive.

Usually, when a bond is issued, the coupon rate is chosen in such a way that the bond trades approximately at par. This means that the coupon rate reflects the "general" level of interest rates, i.e., a sort of average provided by YTM.

3.5.5 INTEREST RATE RISK

Since a bond price is the sum of discounted cash flows, it is clear that there is an inverse relationship between interest rates and bond prices. The impact of

3.5 Elementary bond pricing

Table 3.5 The interaction of coupon rates and maturities when yield is increased.

T (years)	$y_1 = 4\%$			$y_1 = 5\%$		
	3	10	30	3	10	30
$c = 0\%$	88.90	67.56	30.83	86.38	61.39	23.14
$c = 3\%$	97.22	91.89	82.71	94.55	84.56	69.26
$c = 9\%$	113.88	140.55	186.46	110.89	130.89	161.49

	% loss		
T (years)	3	10	30
$c = 0\%$	−2.83	−9.13	−24.96
$c = 3\%$	−2.75	−7.98	−16.27
$c = 9\%$	−2.62	−6.88	−13.39

a change in the term structure depends on the exact kind of change, which is not so trivial to analyze, as it may involve a vertical shift, a change in slope, or a twist in curvature. Here, we analyze interest rate risk with reference to a simplified setting, where we only consider an uncertain YTM with annual compounding. In Chapter 6, we will see that this is essentially equivalent to considering a parallel shift in the term structure, which is indeed a limited view. However, a simplified analysis is a good starting point to build intuition and get acquainted with a few essential concepts.

Let us consider the bond prices given in Table 3.5. Prices refer to bonds differing in coupon rate and maturity: (a) the coupon rates are zero, 3% or 9%, and a single coupon is paid per year; (b) for each possible coupon rate, we consider three bonds maturing in 3, 10, or 30 years. The resulting nine bonds are priced for two different values of YTM, 4% and 5%, in order to assess how the two bond features interact with changes in yield. We also give the percentage loss associated with the increase in YTM.

The lower part of the table shows that loss may be considerable when YTM increases by 100 basis points. The table also suggests that impact is:

- More significant for long maturities
- Less significant for large coupon rates

From a financial viewpoint, these observations may be explained by considering that, for a zero-coupon bond, an increase in maturity is just bad news. There is just one cash flow at maturity, and it is more heavily discounted when yield is increased. However, an increase in YTM has a partially positive effect on a coupon-bearing bond, if held up to maturity: Coupons can be reinvested at a larger rate. Clearly, such good news are more relevant for a bond with a large coupon rate.

The intuition may be reinforced and made more precise by introducing an important measure of interest rate risk: **duration**. Here we give a classical definition of duration that, as we shall see in Chapter 6, is rather limited. For the sake of simplicity, we will assume that one coupon is paid per year. Let us consider the bond price as a function of the annually compounded yield y_1,

$$P(y_1) = \sum_{t=1}^{T} \frac{C_t}{(1+y_1)^t}.$$

Note that C_t denotes the generic cash flow at the end of year t. Let us take the first-order derivative of this function:

$$\frac{dP}{dy_1}(y_1) = -\sum_{t=1}^{T} \frac{tC_t}{(1+y_1)^{t+1}} = -\frac{1}{1+y_1} \sum_{t=1}^{T} \frac{tC_t}{(1+y_1)^t}. \quad (3.36)$$

This formula measures the first-order sensitivity of the bond price with respect to y_1, and it looks much like the bond price formula with two differences:

1. There is a leading coefficient multiplying the sum, whose negative sign makes good sense, as bond price is decreased if yield is increased.[15]
2. The sum consists of terms in which each discounted cash flow is multiplied by its time of payment.

In real life, changes in yield may be relatively small, but not infinitesimal. Nevertheless, a simple measure of sensitivity, allowing us to write useful approximations, may come in handy. Let δy_1 be a small change in the annual yield, and let δP be the corresponding change in the bond price. Using Eq. (3.36), we may write

$$\frac{\delta P}{\delta y_1} \approx -\frac{1}{1+y_1} \sum_{t=1}^{T} \frac{tC_t}{(1+y_1)^t} = -\frac{P}{1+y_1} \cdot \frac{\sum_{t=1}^{T} \frac{tC_t}{(1+y_1)^t}}{P}$$

$$= -\frac{P}{1+y_1} \cdot \frac{\sum_{t=1}^{T} \frac{tC_t}{(1+y_1)^t}}{\sum_{k=1}^{T} \frac{C_k}{(1+y_1)^k}} = -\frac{P}{1+y_1} \cdot \sum_{t=1}^{T} w_t t, \quad (3.37)$$

---

[15]The fraction $1/(1+y_1)$ is a bit annoying, but it is the result of discrete compounding. If we use continuous compounding, we take derivatives of a exponential functions, which are much nicer.

3.5 Elementary bond pricing

where we define weights

$$w_t \doteq \frac{\dfrac{C_t}{(1+y_1)^t}}{\displaystyle\sum_{k=1}^{T} \dfrac{C_k}{(1+y_1)^k}}.$$

It is easy to see that these weights indeed add up to 1, as they consist of discounted cash flows divided by their total sum, which is just the bond price P.

Now we may rewrite Eq. (3.37) as follows:

$$\frac{\delta P}{P} \approx -\frac{1}{1+y_1} \cdot D_{\mathsf{mac}} \cdot \delta y_1, \tag{3.38}$$

where we define the **Macauley duration**, as

$$D_{\mathsf{mac}} \doteq \frac{\displaystyle\sum_{t=1}^{T} \dfrac{tC_t}{(1+y_1)^t}}{\displaystyle\sum_{k=1}^{T} \dfrac{C_k}{(1+y_1)^k}} = \sum_{t=1}^{T} w_t t. \tag{3.39}$$

The definition of Macauley duration involves a weighted sum of time instants, where weights are related to discounted cash flows, and is dimensionally measured in years. It is called duration, since it provides us with a sort of maturity taking cash flows into account. It is easy to see that the Macauley duration for a zero is just its maturity, whereas duration is smaller than maturity for coupon-bearing bonds.

To get rid of the leading fraction in Eq. (3.38), we may introduce the **modified duration**, defined as

$$D_{\mathsf{mod}} \doteq \frac{1}{1+y_1} \cdot D_{\mathsf{mac}}, \tag{3.40}$$

leading us to the following first-order approximation, linking a change in yield to the percentage change in bond price:

$$\frac{\delta P}{P} \approx -D_{\mathsf{mod}} \cdot \delta y_1. \tag{3.41}$$

Example 3.14 A numerical illustration of duration

> Let us check the application of duration to the bonds that we considered in Table 3.5. The duration of each zero corresponds to its maturity. So, using Eq. (3.41), the prediction of the price of the zero

maturing in three years is, after the increase of yield,

$$P(3;5\%) \approx P(3;4\%) \cdot (1 - D_{\text{mod}} \cdot \delta y_1)$$
$$= 88.90 \times \left(1 - \frac{1}{1+0.04} \times 3 \times 0.01\right) = 86.3356.$$

This is fairly close to the exact price, which is 86.38. We notice that the duration-based prediction is somewhat pessimistic, as the actual bond price after the increase in yield is larger. The same calculation for the bond maturing in 30 years gives

$$P(3;5\%) \approx 30.83 \times \left(1 - \frac{1}{1+0.04} \times 30 \times 0.01\right) = 21.9367.$$

In this case, the prediction is definitely pessimistic with respect to the actual price, which is 23.14.

Table 3.6 also shows the Macauley duration of the six coupon-bearing bonds of Table 3.5, featuring different maturities and coupon rates. These values may be obtained by direct application of the definition, which is somewhat inconvenient when 30 cash flows are involved, or by an analytical formula that we shall prove later. Let us check the accuracy for the 9% bonds maturing in 3 and 30 years:

$$P_{9\%}(3;5\%) \approx 113.88 \times \left(1 - \frac{1}{1+0.04} \times 2.77 \times 0.01\right)$$
$$= 110.8468$$
$$P_{9\%}(30;5\%) \approx 186.46 \times \left(1 - \frac{1}{1+0.04} \times 15.50 \times 0.01\right)$$
$$= 158.6703.$$

Again, by comparing the approximations with the exact prices in Table 3.5, we see that the approximation is pessimistic, but pretty accurate for a short maturity, a bit less for a long maturity. We will discuss this matter further in Example 6.1.

Example 3.14 shows that the approximate price predicted by duration is pessimistic, in the sense that it overestimates the drop in the bond price when yield is increased. By the same token, when there is a drop in yield, the actual bond price will be larger than what we predict using duration. This is a consequence of the convexity of the price–yield relationship. In fact, if we assume a continuously compounded yield y, the bond price is a sum of negative exponentials, like $C_t e^{-yt}$, which are convex in t. The same applies to discount factors

3.5 Elementary bond pricing

Table 3.6 Macauley duration for the bonds of Table 3.5.

T (years)	3	10	30
$c = 0\%$	3	10	30
$c = 3\%$	2.91	8.72	19.10
$c = 9\%$	2.77	7.50	15.50

involving an annually compounded yield y_1.[16] A linear approximation always underestimates a convex function (see Fig. 6.1). Needless to say, the practical relevance of duration has nothing to do with the approximation per se, as we may easily reprice the bond. The approximation, as we shall see, is relevant to define hedging strategies against interest rate risk. A look at Table 3.6 seems to suggest that duration is smaller for larger coupon rates. Furthermore, we might also guess that duration is increased when time-to-maturity is increased. Actually, the first guess is correct, but the second one is not. We investigate qualitative properties of duration in Section 3.5.5.1.

Equations (3.39) and (3.40) illustrate the traditional definitions of duration, which may be easily adapted to semiannual or continuously compounded yields, y_2 and y. The annoying distinction between Macauley and modified duration disappears when using y, as exponential functions yield nicer derivatives than rational functions, as we shall see later.

Duration is not only useful as a risk *measurement* tool, but also as a concept leading us to risk *management* strategies for fixed-income portfolios. We shall not elaborate too much on the traditional definition, though, since it is subject to significant limitations:

- We have defined duration with reference to YTM, which is equivalent to considering a flat term structure. What about a more realistic term structure? We will see that what we are doing amounts to assuming that the term structure is only subject to parallel shifts, but this does not account for changes in slope or curvature, which may be observed in practice.

- Duration is a first-oder sensitivity measure providing us with a first-order approximation. Indeed, Example 3.14 shows that the approximation may be not quite satisfactory. We may improve the approximation by introducing a second-order sensitivity measure, bond convexity.

- If we define duration using cash flows, we are in trouble when these are uncertain. A simple case is a floating-rate bond, discussed in Section 3.5.6, and the observation also applies to some derivatives such as vanilla interest rate swaps. Luckily, there is an easy way to redefine duration in a more general way.

---

[16]Convexity is further discussed in Section 15.1.

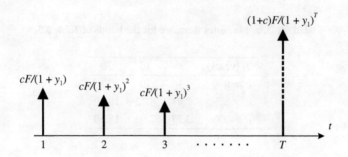

FIGURE 3.4 Duration as a center of gravity of discounted cash flows.

We shall pursue these further developments in Chapter 6.

3.5.5.1 Qualitative properties of duration

To investigate the properties of duration, it may be useful to find an analytical expression.[17] We do so for the Macauley duration in the case of annually compounded yield. The following analytical formula for Macauley duration is proved in Supplement S3.1:

$$D_{\mathsf{mac}} = 1 + \frac{1}{y_1} + \frac{T(y_1 - c) - (1 + y_1)}{c \cdot \left[(1 + y_1)^T - 1\right] + y_1}, \tag{3.42}$$

where c is the coupon rate of the bond and T is time-to-maturity. It is important to realize that this formula may be of limited practical use as it disregards the term structure and assumes that time-to-maturity is an *integer* number of years (or periods, if we do not consider annual yield, but rather a semiannual one). In other words, it applies when the bond is issued or immediately after the payment of a coupon. Nevertheless, it is useful for a qualitative investigation.

From Eq. (3.42), we immediately see that, in fact, the sensitivity of duration with respect to the coupon rate is negative. The coupon rate c occurs with a negative sign in the numerator of the ratio, and a positive sign in the denominator. Hence, an increase in c will decrease duration. This is actually intuitive, and we may understand why by looking at Fig. 3.4. If we interpret duration as a center of gravity of time instants, weighted by cash flows, increasing the coupon rate c has a large effect on cash flows on the left, much less on the last cash flow on the right, corresponding to maturity. Financially, if yield is increased, the loss on the bond value is partially offset by the opportunity of reinvesting coupons at a larger rate. The larger the coupon rate, the less interest rate risk we observe.

Now what about time-to-maturity? For a zero-coupon bond, duration is just time-to-maturity; hence, increasing T also increases duration. Intuition might suggest that the same applies to coupon-bearing bonds, but this is not necessar-

---

[17]The treatment in this section follows [2, Chapter 4].

3.5 Elementary bond pricing

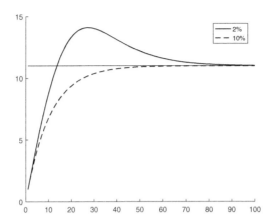

FIGURE 3.5 Duration as a function of time-to-maturity (measured in years), for bonds with coupon rates of 2% and 10%.

ily true. One starting observation is that when T goes to infinity, duration tends to a limit which is independent of the coupon rate:

$$\lim_{T \to +\infty} D_{\text{mac}} = 1 + \frac{1}{y_1}.$$

To see this, note that T occurs linearly in the numerator of the ratio in Eq. (3.42), but exponentially in the denominator. Thus, the limit of the ratio is zero. As a reality check, let us consider the price of a perpetuity, as given in Eq. (3.35):

$$P_c(0, \infty) = \frac{C}{y_1} \Rightarrow \frac{dP_c(0, \infty)}{dy_1} = -\frac{C}{y_1^2}$$

$$\Rightarrow D_{\text{mod}} = \frac{y_1}{C} \cdot \frac{C}{y_1^2} = \frac{1}{y_1}$$

$$\Rightarrow D_{\text{mac}} = (1 + y_1) \cdot \frac{1}{y_1} = 1 + \frac{1}{y_1}.$$

Duration will always tend to this limit as T increases, but convergence need not be monotonic from below, as shown in Fig. 3.5. The plot illustrates convergence to the limit ($D_{\text{mac}} = 11$) when $y_1 = 0.1$. Two bonds are considered, one with small coupon rate $c = 0.02$, and one with large coupon rate $c = 0.1$. We see that when c is smaller than y_1, we may have nonmonotonic convergence. To get some intuition about this counterintuitive effect, we may have a look at Fig. 3.4 again. When time-to-maturity is increased from T to $T + 1$, we add a new discounted cash flow $(1 + c)F/(1 + y_1)^{T+1}$, adding mass to the right end of the cash flow sequence, but we reduce the mass at T, which is also located at the right end, by $F/(1 + y_1)^T$. If c is small with respect to y_1, the net effect may well be a shift to the left of the center of gravity. Hence, a reduction of duration may result.

3.5.6 PRICING FLOATING RATE BONDS

Pricing a fixed-coupon bond looks like a rather simple affair, if we do not consider default risk. We have a sequence of deterministic cash flows, and all we need is a term structure of interest rates to discount them. However, what about pricing a floater, i.e., a floating-rate bond? As a starting point, let us clarify how a floating-rate bond works.[18] Let us consider a bond, issued at time $t = T_0$, paying semiannual coupons at times T_i, $i = 1, \ldots, m$, where T_m is maturity, when the face value F is also redeemed. At time T_0 the semiannually compounded spot rate $r_2(T_0, T_0 + 0.5) \equiv r_2(T_0, T_1)$, applying to the first semester, is observed and used to define the first coupon rate. Thus, the first coupon payment will be $F \cdot r_2(T_0, T_0 + 0.5)/2$. At time $t = 0.5$, the first coupon is paid and the new rate $r_2(T_0 + 0.5, T_0 + 1) \equiv r_2(T_1, T_2)$ is observed and used to set the next coupon rate. More generally, at time T_i a coupon C_i is paid, given by the rate observed six months before:

$$C_i = F \cdot r_2(T_{i-1}, T_i)/2,$$

where $T_i = T_{i-1} + 0.5$.[19] Thus, coupon dates are also **reset dates**, and we always know the amount of the next coupon in advance, even though the whole sequence is uncertain.

Let us denote by $P_f(t, T)$ the price at time t of the floater maturing at time T. Pricing a floater seems like a complicated affair involving stochastic cash flows, and the following questions arise:

- Should we define a stochastic model describing the evolution of interest rates over time?
- Is a floater more or less risky than a fixed-coupon bond?

The answer to first question is, luckily, in the negative. Rather surprisingly, pricing a floater is much easier than pricing a fixed-coupon bond. To see why, let us consider the last cash flow at maturity $T = T_m$,

$$C_m + F = F \cdot \left[1 + \frac{r_2(T_{m-1}, T_m)}{2}\right],$$

which consists of the face value of the bond and a last coupon, determined at T_{m-1} and paid six months later. The value of the bond at time T_{m-1} results from discounting this cash flow:

$$P_f(T_{m-1}, T) = F \cdot \frac{1 + r_2(T_{m-1}, T_m)/2}{1 + r_2(T_{m-1}, T_m)/2} = F. \qquad (3.43)$$

Thus, when the rate is reset for the last time at time T_{m-1}, the bond price is exactly the face value F, since the same random rate is used to both define and

---

[18] See also the discussion in Section 3.5.3.
[19] For the sake of simplicity, we assume that semesters consist of the same number of days, which is not really true.

3.5 Elementary bond pricing

discount the cash flow. If we step back to time T_{m-2}, the bond price may be found by considering the bond as a portfolio consisting of:

- A zero-coupon bond with face value

$$C_{m-2} = F \cdot r_2(T_{m-2}, T_{m-1})/2,$$

corresponding to next (second-to-last) coupon.

- An asset, the bond itself, after stripping the second-to-last coupon, whose value at time T_{m-1} will be F, no matter what, as we have seen in Eq. (3.43).

Thus,

$$P_f(T_{m-2}, T) = F \cdot \frac{1 + r_2(T_{m-2}, T_{m-1})/2}{1 + r_2(T_{m-2}, T_{m-1})/2} = F.$$

Unfolding the recursion, we see that at each reset date, just after the previously determined coupon has been paid, the bond price is exactly F. In particular, the bond trades at par when it is issued. Thus, we see that the bond price is known at reset dates, even though the future cash flows are not. This is a consequence of no-arbitrage, and it also implies that a floating-rate bond is not affected by interest rate risk, at least at reset dates.

In order to price the bond at a generic epoch t between two reset dates, $T_{i-1} < t < T_i$, we have just to discount the next coupon, fixed at the last reset time T_{i-1} and paid at time T_i, plus the bond value at the next reset time:

$$P_f(t, T) = Z(t, T_i) \cdot F \cdot \left[1 + \frac{r_2(T_{i-1}, T_i)}{2}\right]. \tag{3.44}$$

Thus, between two reset dates, the bond price is related to a discount factor $Z(t, T_i)$ that changes with time, and there is some interest rate risk. However, risk is essentially related to the price of a zero maturing in less than six months.

Now, what about the duration of a floating-rate bond? Clearly, the classical definition of Eq. (3.39) cannot be applied as it requires knowledge of a sequence of random cash flows. However, if we interpret the value in Eq. (3.44) as the price of a zero with time-to-maturity $T_i - t$, we may suspect that duration for such a bond is the time to the next coupon, rather than time-to-maturity. We will see in Chapter 6 that this is indeed the case.

Example 3.15 The risk of a floating-rate bond

It is interesting to compare the risk of a fixed- and a floating-rate bond. Let us consider a bond with face value $F = \$1000$, paying semiannual coupons, maturing at time $T_m = 4.75$ (four years and nine months), and let us assume that the term structure is flat and given by a semiannually compounded yield $y_2 = 4\%$. Given the time-to-maturity, the next reset date is $T_1 = 0.25$, i.e., three months

> from now, and ten coupons will be paid over the bond life. Then, the coupon rate was reset three months ago, and let us assume that the observed rate was 3%. Hence, the next coupon amounts to
>
> $$\$1000 \times \frac{0.03}{2} = \$15.$$
>
> Note that the relevant discount factor, using the semiannual yield over 0.25 years, is
>
> $$Z(0, 0.25) = \frac{1}{\sqrt{1 + 0.04/2}},$$
>
> and so the price of the floater is
>
> $$P_f(0, 4.75; 4\%) = \frac{1015}{\sqrt{1.02}} = \$1005.00.$$
>
> If the term structure is shifted up to 5%, the new bond price is
>
> $$P_f(0, 4.75; 5\%) = \frac{1015}{\sqrt{1.025}} = \$1002.55,$$
>
> with a very limited loss:
>
> $$\frac{1002.55 - 1005.00}{1005.00} = -0.24\%.$$
>
> It is easy to see that this loss would be the same for a bond maturing in 100 years! The reader is invited to compare these values with the corresponding ones for a fixed-coupon bond.

A comparison with Table 3.5 shows that, somewhat paradoxically, a floating-rate bond featuring stochastic cash flows may be less risky than a fixed-coupon bond, with deterministic cash flows. But is this really the case? The answer is a bit more complicated and depends on the intended use of bonds. If we buy a bond and plan to sell it shortly, there is no doubt that a floater is less risky. However, if we plan to hold the bond until maturity and use coupons to finance a stream of fixed liabilities, the picture may be different. Thus, the message is that risk must always be analyzed within a context.

3.6 A digression: Elementary investment analysis

The elementary approach to pricing a coupon-bearing bond and the concept of yield-to-maturity are related to a couple of fundamental tools in investment analysis, namely, the net present value and the internal rate of return, respec-

3.6 A digression: Elementary investment analysis

tively. Any textbook on corporate finance spends some pages discussing the pros and cons of these approaches in the context of capital budgeting. Since this is a book on financial markets, we will steer away from these discussions. Nevertheless, seeing the connection between investment analysis and asset pricing is quite useful.

3.6.1 NET PRESENT VALUE

Consider an investment project characterized by a stream of cash flows C_t at epochs $t = 0, 1, \ldots, T$. Cash flows may also be negative, corresponding to cash outflows related to investing money in the project. If we consider building a plant or designing a new product or service, it is quite likely that $C_0 < 0$, as this is the initial capital outlay but, in a complex project, there may be staged investments along the planning horizon, before revenue (hopefully) turns some cash flows into the positive.

If the cash flows were certain, analogy with bond pricing suggests that we could evaluate the investment by calculating its **net present value** (NPV) as follows:

$$\text{NPV} = C_0 + \frac{C_1}{1 + r_1(0,1)} + \frac{C_2}{[1 + r_1(0,2)]^2} + \cdots + \frac{C_T}{[1 + r_1(0,T)]^T}$$

$$= \sum_{t=0}^{T} \frac{C_t}{[1 + r_1(0,t)]^t},$$

where we are using a term structure of risk-free rates with annual compounding. When pricing a bond, $C_0 = -P_c(0,T) < 0$ corresponds to the cash outflow to buy the bond at its current price. The next cash flows for $t = 1, \ldots, T-1$ correspond to coupon payments $C_t = cF > 0$, and the final cash flow at bond maturity includes the face value, $C_T = cF + F > 0$. If NPV > 0, then the investment is worth pursuing; otherwise, it is better to use the required capital in some other way.

However, such ventures are very rarely risk-free and there may be considerable uncertainty about future cash flows, which are actually stochastic. One approach to deal with this issue is to consider the expected value of the future cash flows, and to account for uncertainty by discounting cash flows using a rate that includes a risk premium. To avoid further difficulties, a single **hurdle rate** R is used to discount all of the expected cash flows:

$$\text{NPV} = \sum_{t=0}^{T} \frac{\text{E}[C_t]}{(1+R)^t}.$$

Clearly, the difficulty in applying the approach lies in the estimation of expected cash flows and the choice of an appropriate hurdle rate. One idea to estimate an appropriate risk premium is related to the capital asset pricing model, as we shall see in Chapter 10.

3.6.2 INTERNAL RATE OF RETURN

When a valuation model, like the NPV, requires uncertain inputs, such as the hurdle rate R, a good idea is to check the impact of uncertainty on decisions by sensitivity analysis. In particular, useful information is provided by finding a limit value marking the difference between two different courses of action. Since the sign of the NPV depends on R, we may find the critical rate R such that the resulting NPV is zero. Such a rate is called **internal rate of return**, or IRR for short. This is obtained by solving the nonlinear equation

$$\text{NPV}(R) = \sum_{t=0}^{T} \frac{\text{E}[C_t]}{(1+R)^t} = 0.$$

This is actually a polynomial equation, since we may use substitution of variables,

$$z = \frac{1}{1+R},$$

and solve[20]

$$\sum_{t=0}^{T} \text{E}[C_t] z^t = 0.$$

Then, given a root z, we find the corresponding IRR

$$R = \frac{1-z}{z}.$$

We immediately observe that YTM is just the IRR for a bond. When the IRR exceeds a critical value, corresponding to an investment of comparable risk, the investment that we are analyzing is worth pursuing. Otherwise, we will be better off by considering an alternative investment.

Clearly, we are interested in roots leading to real and positive values of IRR, but what if such roots are not unique? Indeed, we know from the fundamental theorem of algebra that a polynomial of degree T has T roots, possibly complex conjugates. It may well be the case that there are multiple IRRs, typically when the sign of cash flows alternates, and this is why many theorists in corporate finance claim the definite superiority of NPV over IRR (some practitioners disagree). Luckily, the situation is much easier when dealing with YTM of a bond. In such a case, there is one negative leading cash flow $C_0 < 0$, followed by a stream of positive cash flows, and it can be shown that there is a unique IRR > 0.

---

[20] Any numerical computing environment, like MATLAB, provides us with tools to solve polynomial equations. We should not use generic procedures for nonlinear equations, as these are meant to find one root of the equation near an initial point provided by the user, whereas more specific procedures for polynomial equations find all of them.

3.6.3 REAL OPTIONS

As we have mentioned, in the corporate finance literature there is some controversy surrounding the use of NPV and IRR and their relative advantages and disadvantages. The situation is further complicated by the fact that, in a real-life capital budgeting problem, we should not consider single investments, but sets of competing ones, possibly under a budget constraint. The uncertainty in their cash flows may also be affected by correlations.

However, both approaches suffer from a quite important limitation. They consider cash flows as exogenously given, whereas in real life they depend on our decisions. To see the point, imagine a project consisting of a set of interrelated activities. It may be possible to execute them in some sequence over time, and condition our decisions on the observation of relevant risk factors. If a project is turning into a disaster, a wise course of action could be to just cut our losses and abandon it.[21] We may also scale an investment up or down, depending on the unfolding of uncertainty over time. In other cases, it may be worth delaying the project, in order to gather more information and reduce the level of uncertainty.

Traditionally, these planning problems under uncertainty have been analyzed using decision trees.[22] But after the considerable success of quantitative methods for financial option pricing, the name **real options** has been coined, in order to reflect their link with the real economy. There are some standard real options that are used to analyze flexible investment strategies, like delay options, abandonment options, growth options, etc. From a methodological viewpoint, there is an interesting difference with respect to the more traditional valuation approach, where we discount expected cash flows using a hurdle rate that reflects a risk premium. As we shall see, when dealing with option pricing, the approach is to use the risk-free rate for discounting, but to adopt a different probability measure in order to compute expectations.

3.7 Spot vs. forward interest rates

In Section 1.2.6.1, we have introduced forward contracts, which allow us to buy or sell an underlying asset in the future at a fixed delivery price, rather than facing uncertainty of future spot prices. A similar concept may be introduced for interest rates, even though they are not tradable assets. The current term structure $r(0, \cdot)$ at time $t = 0$ consists of an array of *spot* interest rates, which apply to time periods starting immediately. In this section, we show how to find

---

[21] A well-known behavioral bias is the sunk-cost syndrome: We tend to insist on an unfortunate endeavor, since we have already paid some cost that we cannot recover (irreversible investment). However, from a rational viewpoint, this should be regarded as a sunk cost that should not influence future decisions. Most of us experiment this syndrome, when we insist on watching a horrible movie, because we have paid the ticket, rather than just walk away.

[22] See, e.g., [1, Chapter 13].

rates that apply to a time period starting somewhere in the future. Such rates are called **forward rates**. In order to understand the nature and the role of forward rates, let us consider the following hypothetical situation:

- We are at time $t = 0$, and we will receive a payment in six months, at time $t = 0.5$.
- We will need that money only in one year, at time $t = 1$. Hence, we would like to invest it for the six-month period $(0.5, 1)$.
- The problem is that we know the spot rates $r(0, 0.5)$ and $r(0, 1)$, but there is some uncertainty about the future spot rate[23] $r(0.5, 1)$.

In general, uncertainty in the future spot prices of commodities, indexes, and other assets may be hedged away by resorting to forward or futures contracts. By the same token, we can use interest rate derivatives to manage interest rate risk.

It is easy to see that, if we assume that all rates are free from default risk, there *must be* a well-defined forward interest rate for time intervals in the future, in order to rule out arbitrage opportunities. Let us denote by $f(t, T_1, T_2)$ the **continuously compounded forward rate** observed at time t for a future time interval (T_1, T_2), where $t \leq T_1 \leq T_2$. We already know that, when the maturity of a forward contract is approached, there is a convergence between spot and forward prices. By the same token, when $T_1 = t$, we must have

$$r(T_1, T_2) = f(T_1, T_1, T_2).$$

Forward rates are indeed quoted and offered in real-life markets, in the form of forward rate agreements (FRAs), which we shall discuss in Chapter 4. For now, let us just discuss how to relate spot and forward rates on the basis of financial theory.

In the above hypothetical situation, we can show that the knowledge of spot rates $r(0, 0.5)$ and $r(0, 1)$ implies knowledge of the forward rate $f(0, 0.5, 1)$. To see how this results from application of the no-arbitrage principle, let us consider the following two strategies to invest money for one year, on the interval $(0, 1)$:

1. The straightforward possibility is to invest an arbitrary sum L at the spot rate $r(0, 1)$. At time $t = 1$, our wealth will be

$$L \cdot e^{r(0,1) \times 1}.$$

2. Alternatively, we might invest L for the initial six months at the spot rate $r(0, 0.5)$, and then on the next semester $(0.5, 1)$ at the forward rate $f(0, 0.5, 1)$. Wealth after one year will be

$$L \cdot e^{r(0,0.5) \times 0.5} \cdot e^{f(0,0.5,1) \times 0.5}.$$

---

[23]The *future spot rate* should not be confused with the *futures rate*, which underlies interest rate futures contracts. Futures contracts on interest rates will be discussed later, in Section 4.3.

3.7 Spot vs. forward interest rates

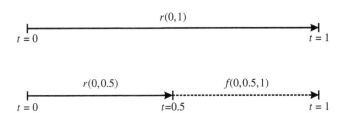

FIGURE 3.6 Comparing two investment paths.

Since both strategies start with the same money L and are riskless, wealth in one year must be the same:

$$L \cdot e^{r(0,1) \times 1} = L \cdot e^{r(0,0.5) \times 0.5} \cdot e^{f(0,0.5,1) \times 0.5},$$

which implies

$$r(0,1) = \frac{r(0,0.5) + f(0,0.5,1)}{2} \tag{3.45}$$

and

$$f(0,0.5,1) = \frac{r(0,1) - 0.5 \cdot r(0,0.5)}{0.5}. \tag{3.46}$$

It is useful to visualize the two strategies as paths over time, as shown in Fig. 3.6. If we multiply the capital growth along any path, we implicitly define a path rate. If the capital growth were larger along one of the two paths, it would be possible (under the usual somewhat idealized market conditions) to create an arbitrage strategy by borrowing money at the smaller path rate and immediately investing it at the larger path rate.

Equation (3.45) seems to suggest that the spot rate for the longer maturity is related to an arithmetic average of the spot rate for the shorter maturity and the forward rate. Indeed, we may generalize it as follows, if we consider time intervals of length τ_1 and τ_2 corresponding to maturities $T_1 = \tau_1$ and $T_2 = \tau_1 + \tau_2$, respectively. In this case, no-arbitrage requires

$$e^{r(0,T_2) \times T_2} = e^{r(0,T_1) \times T_1} \cdot e^{f(0,T_1,T_2) \times (T_2 - T_1)},$$

which implies

$$r(0,T_2) = \frac{r(0,T_1) \cdot T_1 + f(0,T_1,T_2) \cdot (T_2 - T_1)}{T_2}$$
$$= \frac{r(0,\tau_1) \cdot \tau_1 + f(0,\tau_1,\tau_1+\tau_2) \cdot \tau_2}{\tau_1 + \tau_2},$$

and

$$f(0,T_1,T_2) = \frac{r(0,T_2) \cdot T_2 - r(0,\tau_1) \cdot T_1}{T_2 - T_1}.$$

The equivalence of returns over time paths may be generalized to arbitrary pairs of paths over time, involving arbitrary time intervals, along with their forward rates. If we consider time instants $T_0, T_1, \ldots, T_n$, where $T_0 = 0$, corresponding to time intervals of length $\tau_i = T_i - T_{i-1}$, $i = 1, \ldots, n$, by no-arbitrage we find

$$r(0, T_n) = \frac{\sum_{i=1}^{n} \tau_i \cdot f(0, \tau_{i-1}, \tau_i)}{\sum_{i=1}^{n} \tau_i}, \qquad (3.47)$$

where, in order to streamline the expression, we use the identity

$$r(0, T_1) \equiv f(0, 0, T_1)$$

for the first time interval. Indeed, if we use continuous compounding, spot rates may be considered as weighted averages of forward rates. Clearly, if we have the set of spot rates $r(0, T_i)$, we may find the set of forward rates $f(0, T_i, T_j)$, $T_i \leq T_j$, and vice versa.

Example 3.16 Spot and forward curves

Assume that we are given a set of forward rates for annual investments:

$$f(0, 0, 1) = 2.0\%, \quad f(0, 1, 2) = 2.2\%,$$
$$f(0, 2, 3) = 2.4\%, \quad f(0, 3, 4) = 2.3\%.$$

Using Eq. (3.47) we may directly find the corresponding spot rates:

$$r(0, 1) = 2.0\%,$$
$$r(0, 2) = \frac{2.0\% + 2.2\%}{2} = 2.1\%,$$
$$r(0, 3) = \frac{2.0\% + 2.2\% + 2.4\%}{3} = 2.2\%,$$
$$r(0, 4) = \frac{2.0\% + 2.2\% + 2.4\% + 2.3\%}{4} = 2.225\%.$$

It is interesting to notice that spot rates are increasing, even though the forward rates are not. We will discuss this matter further in Section 3.7.4.

Going the other way around requires a little more work, as it involves a triangular system of linear equations, much in the same vein as the bootstrapping procedure of Example 3.9.

3.7 Spot vs. forward interest rates

The example suggests that the information provided by the spot rate curve may also be captured by a forward rate curve.

3.7.1 THE FORWARD AND THE SPOT RATE CURVES

Just like we define the spot rate curve $r(t, T)$ at time t, as a function of T, we may define a forward curve $f(t, T, T + \Delta)$ at time t, for a fixed Δ, as a function of T. In Section 3.4, we have observed that the term structure of (spot) interest rates is more often than not increasing, but it may take different shapes. It is interesting to investigate the relationships between the spot rate and the forward rate curves. We know that no-arbitrage enforces the condition

$$r(t, T) \cdot T + f(t, T, T + \Delta) \cdot \Delta = r(t, T + \Delta) \cdot (T + \Delta),$$

which implies

$$\begin{aligned} f(t, T, T + \Delta) &= \frac{r(t, T + \Delta) \cdot (T + \Delta) - r(t, T) \cdot T}{\Delta} \\ &= \frac{r(t, T + \Delta) \cdot (T + \Delta) - r(t, T) \cdot T + r(t, T) \cdot \Delta - r(t, T) \cdot \Delta}{\Delta} \\ &= r(t, T) + (T + \Delta) \cdot \frac{r(t, T + \Delta) - r(t, T)}{\Delta}. \end{aligned} \quad (3.48)$$

Equation (3.48) shows that the forward rate can be expressed as a spot rate plus a term involving an increment ratio. We observe that, for a given epoch T and time increment Δ, if the spot curve is increasing, the forward curve is above the spot curve. If the spot curve is decreasing, then the forward curve is below the spot curve. If we take the limit of the increment ratio for $\Delta \to 0$, we find an expression involving the partial derivative of the spot rate with respect to maturity:

$$\lim_{\Delta \to 0} f(t, T, T + \Delta) = r(t, T) + T \cdot \frac{\partial r(t, T)}{\partial T}. \quad (3.49)$$

This expression involves the *instantaneous* forward rate, i.e., a rate for a very small time interval. This concept is related to a short rate, which we will explore in detail in the context of continuous-time stochastic models for interest rates. Here, we observe that when the partial derivative is zero, i.e., when the spot rate has a maximum or a minimum, forward and spot rates take the same value. This is illustrated in Fig. 3.7 for the case of a maximum in a humped term structure.

It is worth noting that, in the context of dynamic models of interest rates under uncertainty, building a realistic and tractable model of a set of spot rates is not quite trivial. Some modeling approaches take an indirect route and represent forward rates directly, from which spot rates may be obtained.

3.7.2 DISCRETELY COMPOUNDED FORWARD RATES

We have used continuously compounded forward rates but, by the same token, we may define discretely compounded forward rates $f_n(t, T_1, T_2)$. Actually,

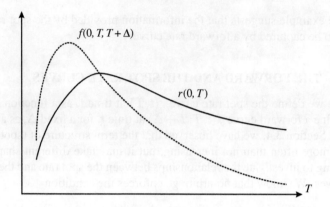

FIGURE 3.7 Comparing spot and forward curves.

discretely compounded rates are used in the definition of interest rate derivatives like, e.g., forward rate agreements. However, they are much less convenient to work with, since all of the calculations become more intricate. As an illustration, let us consider two consecutive time spans of i and j years, respectively, and rates with annual compounding, denoted by r_1 and f_1. Then, the no-arbitrage condition becomes

$$\left[1 + r_1(0, N)\right]^{i+j} = \left[1 + r_1(0, i)\right]^i \left[1 + f_1(0, i, i + j)\right]^j,$$

where $N = i+j$. If we consider a collection of N annual forward rates $f_1(0, i-1, i)$, $i = 1, \ldots, N$, we find

$$\left[1 + r_1(0, N)\right]^N = \left[1 + f_1(0, 0, 1)\right] \cdot \left[1 + f_1(0, 1, 2)\right] \cdots \left[1 + f_1(0, N-1, N)\right],$$

where $f_1(0, 0, 1) \equiv r_1(0, 1)$. This implies

$$r_1(0, N) = \left[\prod_{i=1}^{N} \left(1 + f_1(0, i-1, i)\right)\right]^{1/N} - 1.$$

We notice that the annually compounded spot rate for N years is related to a *geometric* average of forward rates, rather than being directly given by an arithmetic average as in the continuously compounded case. By using no-arbitrage, we may find any spot rate given forward rates, and any forward rate given spot rates. However, the calculations get a bit messy, especially if we consider semi-annual or shorter compounding. In this book, we will mostly use continuous compounding for the sake of simplicity, unless we have to stick to market conventions.

3.7.3 FORWARD DISCOUNT FACTORS

We emphasize once again that compounding issues arise when we want to quote annualized rates in a convenient way, but we are just applying different units to

3.7 Spot vs. forward interest rates

measure the same thing. What actually matters is the discount factor, which may be written in different ways as a function of spot rates:

$$Z(t,T) = e^{-r(t,T)\cdot(T-t)} = \frac{1}{[1+r_1(t,T)]^{T-t}} = \frac{1}{[1+r_2(t,T)/2]^{2\cdot(T-t)}}.$$

By the same token, we may find useful a **forward discount factor**, denoted by $F(t,T_1,T_2)$. No-arbitrage implies an interesting relationship among forward and spot discount factors:

$$F(t,T_1,T_2) = \frac{Z(t,T_2)}{Z(t,T_1)}. \tag{3.50}$$

As a quick reality check, notice that the forward discount factor must be less than 1, and in fact, under ordinary economic conditions, we have $Z(t,T_2) < Z(t,T_1)$, for $T_1 < T_2$. Equation (3.50) may look more familiar when rewritten in terms of rates:

$$\begin{aligned} Z(t,T_2) &= e^{-r(t,T_2)\cdot(T_2-t)} = e^{-r(t,T_1)\cdot(T_1-t)-f(t,T_1,T_2)\cdot(T_2-T_1)} \\ &= Z(t,T_1)\cdot F(t,T_1,T_2). \end{aligned}$$

3.7.4 THE EXPECTATION HYPOTHESIS

In this section, we investigate in slight more detail the expectation hypothesis as a theory explaining the term structure of spot rates. We analyze this hypothesis, disregarding alternatives, because it has an interesting connection with forward rates. Let us consider continuously compounded rates and a time period consisting of two consecutive years. We know that the spot rate for a time interval of two years is an average of spot and forward rates along a time path:

$$r(0,2) = \frac{r(0,1) + f(0,1,2)}{2}. \tag{3.51}$$

Now, the *pure* expectation hypothesis relates the spot rate over two years to the expectation of the spot rate for the second year, i.e.,

$$r(0,2) \stackrel{?}{=} \frac{r(0,1) + \mathrm{E}[r(1,2)]}{2}, \tag{3.52}$$

where the question mark emphasizes that this is just a hypothesis. A comparison between Eqs. (3.51) and (3.52) would suggest that the forward rate for a time interval is an expectation of the future spot rate on the same time interval, in this case:

$$f(0,1,2) \stackrel{?}{=} \mathrm{E}[r(1,2)]. \tag{3.53}$$

This would suggest that forward rates should be predictors of future spot rates,[24] but is this consistent with what we know about forward rates?

---

[24] A similar question may be asked when dealing with forward contracts on stock shares or foreign currencies. Evidence shows that, in general, forward prices are not really good predictors of spot prices.

If the structure is increasing, i.e., $r(0,2) > r(0,1)$, Eq. (3.48) implies that the forward rate $f(0,1,2)$ is larger than the spot rate $r(0,1)$, but then Eq. (3.53), in turn, would imply

$$r(0,1) < \mathrm{E}[r(1,2)],$$

i.e., an increase in the future one-year spot rate is expected next year. However, we have observed that the term structure is usually increasing, and how can it be the case that we expect an increase in spot rates most of the time? This does not sound quite reasonable. Furthermore, Example 3.16 shows that we may have increasing spot rates, even when forward rates are not increasing.

Thus, the pure expectation hypothesis expressed by the equality (3.53) does not seem quite plausible, and it must rather be the case that there is a gap between the forward rate and the expected spot rate,

$$f(0,1,2) > \mathrm{E}[r(1,2)].$$

Usually, an inequality like this may be explained in terms of a risk premium. The **liquidity preference** theory assumes that a risk premium is required in order to give up liquidity. By introducing a risk premium $\pi(0,1,2)$, we may write

$$f(0,1,2) = \mathrm{E}[r(1,2)] + \pi(0,1,2).$$

If there is a positive risk premium, an increasing term structure may result, even though there is no expectation of an increase in the spot rates.

Example 3.17 The effect of a liquidity premium

Let us consider an array of continuously compounded spot interest rates for time periods of one year:

$$r(0,1), \quad r(1,2), \quad r(2,3), \quad r(3,4), \ldots$$

Say that no increase is expected and

$$r(0,1) = 4\% = \mathrm{E}[r(1,2)] = \mathrm{E}[r(2,3)] = \mathrm{E}[r(3,4)] = \cdots$$

If the risk premium is

$$\pi(0,k,k+1) = 1\%, \quad k = 1,2,3,\ldots,$$

the one-year forward rates are

$$f(0,0,1) \equiv r(0,1) = 4\%,$$
$$f(0,k,k+1) = \mathrm{E}[r(k,k+1)] + \pi(0,k,k+1)$$
$$= 5\%, \quad k = 1,2,3,\ldots$$

3.7 Spot vs. forward interest rates

Then, the term structure is an average of forward rates,

$$r(0,\tau) = \frac{1}{\tau}\sum_{k=0}^{\tau-1} f(0,k,k+1), \qquad \tau = 1,2,3,\ldots,$$

which in this specific case yields

$$r(0,1) = 4\%,$$
$$r(0,2) = \frac{4\% + 5\%}{2} = 4.5\%,$$
$$r(0,3) = \frac{4\% + 5\% + 5\%}{3} = 4.67\%,$$
$$r(0,4) = \frac{4\% + 5\% + 5\% + 5\%}{4} = 4.75\%.$$

Thus, we observe an increasing term structure, even though there is no expected increase in the spot rates.

It may even happen that an increasing term structure results from decreasing expected spot rates, if the liquidity premium is increasing. For instance, let us consider

$$r(0,1) = 4\%, \quad \mathrm{E}[r(1,2)] = 3.75\%, \quad \mathrm{E}[r(2,3)] = 3.5\%,$$

and

$$\pi(0,1,2) = 0.5\%, \quad \pi(0,2,3) = 1.0\%.$$

Then, in this case, we find

$$f(0,0,1) = 4\%, \quad f(0,1,2) = 4.25\%, \quad f(0,2,3) = 4.5\%,$$

and

$$r(0,1) = 4\%,$$
$$r(0,2) = \frac{4\% + 4.25\%}{2} = 4.125\%,$$
$$r(0,3) = \frac{4\% + 4.25\% + 4.5\%}{3} = 4.25\%.$$

The resulting term structure is increasing, even though rates are expected to drop.

Example 3.17 shows that an increasing term structure need not imply that spot rates are expected to increase. However, if the term structure is decreasing, we may say that a decrease in the spot rates is expected, unless we assume negative liquidity premia. This may occur when an economic slowdown is anticipated.

3.7.5 A WORD OF CAUTION: MODEL RISK AND HIDDEN ASSUMPTIONS

The relationship between spot and forward rates looks rather obvious. The no-arbitrage argument sounds compelling, and no sophisticated mathematical model is actually used. In particular, there seems to be no place for any stochastic modeling issue. However, hidden assumptions may creep in without we realizing it. We have assumed that interest rates are risk-free, but when we step into the real world, things may be different, and we must refer to real-life quoted rates. As we shall see in later chapters, a widely used interest rate is LIBOR, which is an average of interbank rates. To be precise, it is a trimmed average of rates applying to interbank *unsecured* loans. Since a bank may default, this rate cannot be really considered risk-free, but under normal conditions the default risk was considered negligible. Indeed, only trustworthy banks may be included in the panel defining a LIBOR rate, for the very reason that it refers to unsecured loans. Something changed when the credit crisis erupted in 2007 and led, among other things, to the Lehman Brothers collapse in 2008. Due to the ensuing credit crunch, the LIBOR rates skyrocketed abruptly. While a risk-free rate essentially captures the time value of money and does not depend on the counterparties involved in a transaction, when credit risk is involved, the specific counterparty matters. Under an extreme stress, even the panel of banks involved in the definition of LIBOR rates may change.

Real-life forward rates play a role in forward rate agreements, as well as in interest rate swaps, which we will introduce in Chapter 4. The relevant feature of these derivative contracts is that the payoff is related to the difference between a floating and a fixed rate, applied to a notional amount. The notional amount is not really exchanged and, since the actual payoff is related to a difference of cash flows, default risk is less relevant than in other transactions. In a risk-free setting, a forward rate agreement may be replicated by transactions in spot rates, just like the forward rate may be related with spot rates. However, under market stress, an increasing gap was observed between the forward rates implied by spot LIBOR rates and the actual market rates of forward rate agreements, so that the standard equations that are proposed in classical textbook treatments, the present one included, broke down.

It is important to note that, before 2007, the standard replication argument had been consistent with market data for years, but the situation changed abruptly, creating a difficult puzzle to solve. The message is that we may rely on implicit model assumptions, even when we are not using sophisticated modeling frameworks, and this may result in model risk. More details can be found, e.g., in [8] or [9, Chapter 4].

S3.1 Proof of Equation (3.42)

Let us consider the price of a coupon-bearing bond, paying one coupon per year at rate c, when time-to-maturity is an integer number T of years: We know that, after a coupon payment, when there are T years to maturity, the bond price is

$$P_c(y_1) = \frac{cF}{y_1}\left[1 - \frac{1}{(1+y_1)^T}\right] + \frac{1}{(1+y_1)^T}.$$

The face value F is irrelevant if we want to find Macauley duration,

$$D_{\text{mac}} = -\frac{1+y_1}{P_c(y_1)} \cdot \frac{dP_c(y_1)}{dy_1}.$$

Hence, we may set $F = 1$ and rewrite the bond price as

$$P_c(y_1) = \frac{1}{y_1}\left[c \cdot \left(1 - (1+y_1)^{-T}\right) + y_1(1+y_1)^{-T}\right].$$

The form of the duration definition suggests the opportunity of taking the derivative of the *logarithm* of the bond price. In fact, using the chain rule for composite functions, we have

$$\frac{d \log P_c(y_1)}{dy_1} = \frac{1}{P_c(y_1)} \cdot \frac{dP_c(y_1)}{dy_1}.$$

Now we have

$$\log P_c(y_1) = -\log y_1 + \log\left\{c \cdot \left[1 - (1+y_1)^{-T}\right] + y_1(1+y_1)^{-T}\right\},$$

and by taking its derivative we find

$$\frac{d \log P_c(y_1)}{dy_1} = -\frac{1}{y_1} + \frac{cT \cdot (1+y_1)^{-T-1} + (1+y_1)^{-T} - Ty_1(1+y_1)^{-T-1}}{c \cdot \left[1 - (1+y_1)^{-T}\right] + y_1(1+y_1)^{-T}}$$

$$= -\frac{1}{y_1} + \frac{cT \cdot (1+y_1)^{-1} + 1 - Ty_1(1+y_1)^{-1}}{c \cdot \left[(1+y_1)^T - 1\right] + y_1}.$$

Now, to find Macauley duration, we multiply by $-(1+y_1)$ and simplify:

$$D_{\text{mac}} = \frac{1+y_1}{y_1} - \frac{cT + (1+y_1) - Ty_1}{c \cdot \left[(1+y_1)^T - 1\right] + y_1}$$

$$= 1 + \frac{1}{y_1} + \frac{T(y_1 - c) - (1+y_1)}{c \cdot \left[(1+y_1)^T - 1\right] + y_1},$$

which proves Eq. (3.42).

Problems

3.1 Assume that the prices of zero-coupon bonds, with face value 1000, maturing in 1, 2, 3, and 4 years are

$$947.87, \quad 885.81, \quad 815.15, \quad 757.22,$$

respectively.

- Find the term structure of interest rates.
- Find the one-year forward rates.
- Let us consider the zero maturing in two years and assume that, on a very short time interval, the change in the corresponding rate is modeled by a normal random variable with expected value 0 and standard deviation 0.01. Find the 95% V@R, if you have invested €100,000 in that bond. *Hint:* You may take advantage of a first-order approximation based on duration, and then compare the result against the exact one.

3.2 Consider a riskless bond paying one coupon per year with coupon rate 5%, maturing in three years, and with face value $1000. The forward rates with continuous compounding are:

$$f(0,0,1) = 3.7\%, \quad f(0,1,2) = 4.5\%, \quad f(0,2,3) = 5.1\%.$$

Find the bond price.

3.3 A bond with face value 100 matures in two years and has just paid a coupon. The bond pays one coupon per year and the coupon rate is 6%. If the bond price is 102, what is its yield-to-maturity?

3.4 A corporate bond (subject to default risk) matures in three years, pays one coupon per year, at rate 9% of a face value €1000, and trades at €960. The term structure for risk-free rates is flat at 7% (annual compounding). A bank offers an insurance against default, for a price of €200. This insurance covers both future coupons and the repayment of whole face value (for the sake of simplicity, we do not consider partial default). Should we accept the offer?

3.5 A bond portfolio consists of two bonds: A zero-coupon bond maturing in three years and a coupon bond with a single (annual) coupon of 4%, maturing in two years. Both bonds have a face value of €1000, and we hold 10 bonds of the first kind and 20 of the second one. Interest rates are subject to uncertainty, and we consider the following three term structure scenarios:

Scenario	Probability	One year	Two years	Three years
ω_1	0.2	3.1%	3.8%	4.3%
ω_2	0.5	3.2%	3.3%	3.5%
ω_3	0.3	3.0%	2.9%	2.8%

The three scenarios consist of annually compounded spot rates for maturities of 1, 2, and 3 years (note that, in general, making sure that scenarios are realistic and arbitrage-free is not trivial). We neglect the passage of time, i.e., we assume that these scenarios apply to the immediate future and are based on an instantaneous change in the term structure. Find the expected value of the portfolio wealth after the realization of the random scenario.

3.6 We hold a fixed-income portfolio including two bonds: A zero maturing in three years, and a coupon-bearing bond paying one coupon per year with coupon rate 4%, maturing in two years. The face value is €1000 for both bonds, and we have invested €53,000 and €93,000 in the two bonds, respectively (let us assume infinitely divisible assets, i.e., we may buy fractions of a bond). We are given the following risk-free forward rates, with annual compounding:

$$f_1(0,0,1) = 3\%, \quad f_1(0,1,2) = 4\%, \quad f_1(0,2,3) = 5\%.$$

The price of the two bonds is also related to sovereign risk, i.e., all of the interest rates used in pricing are incremented by a spread that is currently 2.3%. This rate reflects default risk on sovereign debt. Let us assume that the spread is subject to a random shock on the very short term, which is uniformly distributed between -1% and $+2\%$. Find V@R at 99% confidence level on the short term (in other words, we do not consider the effect of time on the bond prices).

3.7 How would you price a floater with a spread δ on observed interest rates?

3.8 How would you price a reverse floater, neglecting the possibility of negative rates?

Further reading

- We have not covered some institutional issues in bond trading, which are dealt with in [13].
- For a treatment of elementary interest rate and bond mathematics, see [3] or [11].
- A more extensive coverage of interest rates and related issues can be found in [4], [7], [12], and [14].
- A more detailed investigation on the term structure of interest rates is described in [2, Chapter 9].
- A specific book on interest rate risk management is [5].
- Dynamic stochastic models for interest rate risk management are covered in [6] and [10].

Bibliography

1 P. Brandimarte. *Quantitative Methods: An Introduction for Business Management*. Wiley, Hoboken, NJ, 2011.

2 O. de La Grandville. *Bond Pricing and Portfolio Analysis: Protecting Investors in the Long Run.* MIT Press, Cambridge, MA, 2001.

3 F.J. Fabozzi. *Fixed Income Mathematics: Analytical and Statistical Techniques* (3rd ed.). McGraw-Hill, New York, 1997.

4 F.J. Fabozzi, editor. *Fixed Income Analysis* (2nd ed.). Wiley/CFA Institute, Hoboken, NJ, 2007.

5 B.E. Gup and R. Brooks. *Interest Rate Risk Management: The Banker's Guide to Using Futures, Options, Swaps, and Other Derivative Instruments.* Irwin Professional Publishing, New York, 1993.

6 J. James and N. Webber. *Interest Rate Modelling.* Wiley, Chichester, 2000.

7 L. Martellini, P. Priaulet, and S. Priaulet. *Fixed-Income Securities: Valuation, Risk Management, and Portfolio Strategies.* Wiley, Chichester, 2003.

8 M. Morini. Solving the puzzle in the interest rate market. SSRN Paper 1506046, 2009. Available at SSRN: https://ssrn.com/abstract=1506046 or http://dx.doi.org/10.2139/ssrn.1506046.

9 M. Morini. *Understanding and Managing Model Risk: A Practical Guide for Quants, Traders and Validators.* Wiley, Chichester, 2011.

10 S.K. Nawalkha, G.M. Soto, and N.A. Beliaeva. *Interest Rate Risk Modeling.* Wiley, Hoboken, NJ, 2005.

11 D.J. Smith. *Bond Math: The Theory behind the Formulas.* Wiley, Hoboken, NJ, 2011.

12 S.M. Sundaresan. *Fixed Income Markets and Their Derivatives* (2nd ed.). South Western College Publishing, Cincinnati, OH, 2002.

13 S.R. Veale, editor. *Stocks, Bonds, Options, Futures: Investments and their Markets* (2nd ed.). Prentice Hall, Paramus, NJ, 2001.

14 P. Veronesi. *Fixed Income Securities: Valuation, Risk, and Risk Management.* Wiley, Hoboken, NJ, 2010.

Chapter Four

Forward Rate Agreements, Interest Rate Futures, and Vanilla Swaps

In this chapter, we consider a few simple interest rate derivatives, which are the natural counterparts of the forward and futures contracts introduced in Section 1.2.6, when the underlying variable is an interest rate. We will also appreciate a relationship with forward interest rates, introduced in Section 3.7.

In Chapter 3, we have repeatedly used the concept of risk-free rate and have shown how risk-free rates can be estimated from bond prices. However, if we want to trade derivatives whose underlying variable is a risk-free rate, we need a very precise reference rate, on which financial institutions may formally and legally agree, not just an estimate. In Section 4.1, we introduce two such rates, the LIBOR and EURIBOR rates. These are essential rates, as they are the underlying variables of several derivatives, but they cannot be considered risk-free, as the painful experience during the 2008 credit crunch has shown.

Then, we move on to consider three simple families of interest rate derivatives:

1. Forward rate agreements, in Section 4.2
2. Eurodollar futures, in Section 4.3
3. Vanilla interest rate swaps, in Section 4.4

A stylized presentation of simple forward rate agreements and vanilla swaps may rely only on no-arbitrage arguments, without the need for dynamic stochastic models, which are necessary to analyze options. We just use concepts like bond pricing and forward rates, which have been introduced in Chapter 3. The resulting pricing models are quite useful, but we have to keep in mind that we are implicitly assuming that everything is risk-free.[1]

---

[1] The assumption that there are no issues with credit and liquidity risk is important and must be made explicit to grasp the limitations of the reasoning lines that we shall pursue. Recent trends in derivative valuation pay due attention to the cost of funding a transaction, which are affected by credit issues. See, e.g., [5, Chapter 4].

The contracts that we describe are quite simple, but extensively traded on both regulated financial markets and OTC. Interest rate swaps are actually OTC derivatives, even though swap rates are readily available, and according to statistics provided by the Bank for International Settlements, the notional amount of outstanding contracts in December 2014 was a staggering $381 trillion. As we shall see, the notional amount of swaps grossly overstates the actual market value, but it gives a clue about the relevance of this kind of market. We should also mention that, in real life, there are plenty of institutional details that we will neglect in the mathematical presentation here; some of these details are covered in Chapter 5. The use of simple interest rate derivatives for risk management is discussed in Chapter 6.

4.1 LIBOR and EURIBOR rates

The seemingly intuitive concept of a "risk-free" rate is actually an elusive one.[2] The risk-free rate is not really constant in time, and when we try to squeeze risk-free rates out of bond prices or other securities, we may face difficulties due to liquidity and credit risk. Hence, when derivatives are written on an interest rate, which is supposed to be risk-free, it is important to specify what rate is used exactly and who is in charge of quoting it.

A widely used set of interest rates is defined by considering interbank offered rates:

- **LIBOR** (*London interbank offered rate*) rates result from a trimmed average of a set of rates offered by banks to other banks operating in London, in need of liquidity.
- **EURIBOR** (*Euro interbank offered rate*) rates are defined similarly, by averaging rates offered by banks in the eurozone.

The LIBOR is quoted for a set of currencies, and available maturities are in the range up to one year.

Rates are related to *unsecured* loans, i.e., not backed by any collateral. However, they were supposed to be almost risk-free, since they refer to relatively short-term loans among solid institutions. Actually, during the credit crunch crisis ensuing after the subprime mortgage crisis, those rates increased considerably, because of the lack of trust among banks. The price was paid, among others, by homeowners whose floating mortgage rates were related to LIBOR or EURIBOR plus a spread. We remark that the plain textbook treatment that we offer here is traditional and assumes that no credit or liquidity issue affects rates, but in stressed conditions this is not quite true. We should also mention the possibility that rates are manipulated by bank cartels. That this is not a remote possibility is shown by the fines that have been paid by banks found guilty of such manipulations. Alternative rates have been proposed to

---

[2] See Section 3.5.3.

overcome issues with LIBOR/EURIBOR, but the message does not change: In order to trade interest rate derivatives, we need a well-defined and official rate.

4.2 Forward rate agreements

A **forward rate agreement**, or **FRA** for short, is an OTC agreement between two counterparties, which is stipulated at time $t = 0$ as follows:

- The two parties agree on a **notional** amount of money N and a time period (T_1, T_2) in the future. Let $\Delta = T_2 - T_1$ be the length of the time interval (measured in years, also called **tenor**) and let $n = 1/\Delta$ be the corresponding compounding period. Usually, Δ is a fraction of a year, like a quarter or a semester, corresponding to $n = 4$ and $n = 2$, respectively.
- One party will pay a *fixed* interest rate on the notional, for the time period (T_1, T_2). This rate is agreed at $t = 0$ and we will denote it by K_n, to reflect the compounding frequency. Hence, the party paying fixed should pay
$$N \cdot \Delta \cdot K_n. \tag{4.1}$$
- The other party will pay a *floating* interest rate on the notional, for the time period (T_1, T_2). This rate is floating in the sense that it is unknown at $t = 0$ and will be set at T_1 by observing the future spot rate $r_n(T_1, T_2)$. Thus, the floating payer should pay
$$N \cdot \Delta \cdot r_n(T_1, T_2). \tag{4.2}$$

Note that we are using discretely compounded rates to reflect market practice, even though continuously compounded rates may be more convenient mathematically. We should also note that we are disregarding an important real-life complication: How many days are included in a quarter or a semester? This really depends on which months happen to be included, as a month may have 30, 31, or even 28 days (29 in leap years). Such day count issues are discussed in Section 5.1.

The two cash flows in Eqs. (4.1) and (4.2) are figurative and do not occur in practice, as only the actual net difference will be paid. From the viewpoint of the party paying fixed, the cash flow at time T_2 will be

$$V_{\text{fixed}}(T_2) = N \cdot \Delta \cdot [K_n - r_n(T_1, T_2)],$$

which is reversed for the party paying floating,

$$V_{\text{float}}(T_2) = N \cdot \Delta \cdot [r_n(T_1, T_2) - K_n] = -V_{\text{fixed}}(T_2).$$

These expressions give the payoff of contract, i.e., its value at time T_2. However, the payoff is actually determined at time T_1, when the spot rate $r_n(T_1, T_2)$ is observed. Hence, a contract may also be arranged so that the cash flow takes place at time T_1, in which case the payoff will be discounted using the spot rate $r_n(T_1, T_2)$.

Now, there are two related questions, which are common with forward contracts on other assets:

- How can we define a fixed rate K_n such that the fair value of the contract at time $t = 0$ is zero?
- What is the value of the contract at a generic epoch $0 < t < T_1$?

As we show in the following, we already know where to look for the first answer: We just use forward rates defined in Section 3.7 and set $K_n = f_n(0, T_1, T_2)$. However, we need something more to answer the second question. We illustrate two views. The first one is based on hedging/replication arguments and will also prove essential when dealing with options. The second one decomposes the FRA into two bonds. We will meet again both views when dealing with interest rate swaps later in this chapter.

The value of an FRA may be needed to assess the market value of a portfolio including an FRA, according to marking-to-market principles. Furthermore, it may be useful, for instance, to determine the fair value at which the two counterparties could agree to cancel the contract, or to assess the loss in case of bankruptcy of a counterparty. We should keep in mind that FRAs, as well as swaps, are of a different nature from eurodollar futures. The latter are liquid futures contracts, easily closed out by reversing the position. FRAs and swaps, just like forward contracts written on other underlying assets, are OTC contracts.

4.2.1 A HEDGING VIEW OF FORWARD RATES

In Section 2.3, we have used a replication argument to price an option. The idea is related to hedging risk away by synthesizing a position that eliminates a given risk exposure. We may apply the same principle to deal with an FRA. Consider a firm that will have to borrow a given amount of money N for a time interval $[T_1, T_2]$ in the future. As before, let $\Delta = T_2 - T_1$ be the length of the time interval and $n = 1/\Delta$ be the corresponding compounding frequency per year. At time $t = 0$, the future spot rate $r_n(T_1, T_2)$ is not known and the firm faces interest rate risk, since the current spot rate can increase. The firm may hedge this risk away by contracting an FRA, whose rate is locked now, with a bank. But what is the fair rate K_n that the firm and the bank should agree on (if we do not consider credit risk)?

Let us take the viewpoint of the bank. The bank is subject to interest rate risk, because the actual spot rate $r_n(T_1, T_2)$ may well be different from the agreed rate. What the bank does not want to do is collecting the necessary amount N at time $t = T_1$ from another bank, as this may occur at a bad moment (i.e., when the spot rates are larger than the contracted fixed rate). And even if the bank has that money, it might lose more valuable investment opportunities

4.2 Forward rate agreements

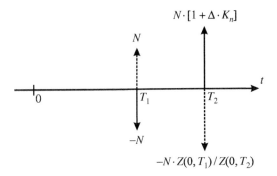

FIGURE 4.1 Hedging a forward contract.

if interest rates rise. From the bank's viewpoint, two cash flows will occur in the future:

- At time $t = T_1$ the bank will see a negative cash flow $-N$, i.e., it will lend N to the firm.
- At time $t = T_2$ the bank will see a positive cash flow

$$N \cdot \left[1 + \Delta \cdot K_n\right], \qquad (4.3)$$

i.e., it will receive the capital back, plus earned interest.

These cash flows are depicted in Fig. 4.1 as continuous arrows. In order to hedge those two cash flows *now*, in such a way that there is no uncertainty at all, the bank should synthesize opposite cash flows (the dashed arrows in Fig. 4.1), offsetting those of the trade with the firm. How can it achieve this aim?

To begin with, if the bank wants to synthesize a positive cash flow N at time T_1, it may buy a zero-coupon bond with face value N maturing at T_1. If we assume, for the sake of convenience, zeros with unit face value, we may think of buying N unit zeros. The price of these bonds now is

$$N \cdot Z(0, T_1),$$

which corresponds to a negative cash flow for the bank at $t = 0$. Now the bank should offset this negative cash flow at time $t = 0$, so that the initial net cash flow is zero. Furthermore, it has to offset the positive cash flow at time T_2, resulting from the trade with the firm. The solution is selling a zero maturing at time T_2. Since the price now of \$1 at time T_2 is the discount factor $Z(0, T_2)$, and the cash flow at time zero (to buy the bond maturing at T_1) is $-N \cdot Z(0, T_1)$, the bank may sell

$$\frac{N \cdot Z(0, T_1)}{Z(0, T_2)} \qquad (4.4)$$

units of the zero maturing in T_2. Note that this amount is nondimensional, as it measures a number of unit zeros traded. If we imagine multiplying the amount in Eq. (4.4) by the bond face value, $1, this is the money that the bank will have to pay at time T_2 to the bondholders, a negative cash flow. If the hedge has to work, this must exactly offset the cash flow given in Eq. (4.3), i.e., the positive cash flow at time T_2 resulting from the trade with the firm. The hedge is perfect if

$$N \cdot \left[1 + \Delta \cdot K_n\right] = N \cdot \frac{Z(0, T_1)}{Z(0, T_2)},$$

which implies

$$\frac{1}{1 + \Delta \cdot K_n} = \frac{Z(0, T_2)}{Z(0, T_1)}. \qquad (4.5)$$

As a quick reality check, observe that the ratio on the right-hand side of Eq. (4.5) should be less than 1, and in fact $Z(0, T_2) < Z(0, T_1)$, since $T_1 < T_2$. Furthermore, the ratio on the left-hand side has the typical form of a discount factor. If we rearrange Eq. (4.5) slightly,

$$Z(0, T_1) \cdot \frac{1}{1 + \Delta \cdot K_n} = Z(0, T_2),$$

we see a clear message: If we discount from T_2 to T_1, and then from T_1 to now, $t = 0$, we must get the same value that we find when discounting directly from T_2 to now. This sounds suspiciously like the kind of argument that we have seen when we introduced forward rates in Section 3.7. The only difference is that now we are discounting cash flows, moving them backward in time rather than moving them forward to see capital growth, but this is inconsequential. Indeed, in Section 3.7.3 we introduced a forward discount factor, $F(t, T_1, T_2)$, and by comparing Eq. (3.50) with Eq. (4.5), we see that the fixed rate K_n must be such that

$$\frac{1}{1 + \Delta \cdot K_n} = F(0, T_1, T_2).$$

Note that discount factors, unlike rates, do not depend on whatever compounding we use. If we use discrete compounding with frequency n, we find

$$K_n = f_n(0, T_1, T_2),$$

i.e., the fixed rate must be the current forward rate for the tenor of the FRA. If it is convenient, we may also use the equivalent rate $f(0, T_1, T_2)$ with continuous compounding. Indeed, we find ourselves again on familiar ground, if we express discount factors in terms of continuously compounded rates,

$$F(0, T_1, T_2) = e^{-(T_2 - T_1) \cdot f(0, T_1, T_2)} = \frac{e^{-T_2 \cdot r(0, T_2)}}{e^{-T_1 \cdot r(0, T_1)}} = \frac{Z(0, T_2)}{Z(0, T_1)},$$

which implies the known relationship between spot and forward rates,

$$f(0, T_1, T_2) = \frac{T_2 \cdot r(0, T_2) - T_1 \cdot r(0, T_1)}{T_2 - T_1}. \qquad (4.6)$$

Using this kind of hedging argument may seem a useless complication. After all, we find something that was easily obtained by linking spot rates with

4.2 Forward rate agreements

forward rates by no-arbitrage. However, this approach allows us to answer the second question: What is the value of an FRA contract after its inception? The two bonds form a hedging portfolio with zero value at $t = 0$, as the bank buys and sells zeros in such way that the net cash flow at times $t = 0$ is zero. Neglecting the nominal value N, the value of the portfolio at time t is related to a long position in one zero maturing at T_1 and a short position in $Z(0, T_1)/Z(0, T_2)$ bonds maturing at T_2:

$$1 \cdot Z(t, T_1) - \frac{Z(0, T_1)}{Z(0, T_2)} \cdot Z(t, T_2), \tag{4.7}$$

where we write "1" explicitly to point out that we are using one bond with price $Z(t, T_1)$ and $Z(0, T_1)/Z(0, T_2)$ bonds with price $Z(t, T_2)$. These numbers of bonds in the hedge do not change over time, but the bond prices do change and depend on t. The initial value is (somewhat trivially) zero when $t = 0$, but it will change as time goes by and interest rates move randomly.

It is interesting to rewrite Eq. (4.7) in terms of discretely compounded forward rates, in order to get a better feeling, consistent with market practice:

$$\begin{aligned}
& 1 \cdot Z(t, T_1) - \frac{Z(0, T_1)}{Z(0, T_2)} \cdot Z(t, T_2) \\
&= Z(t, T_2) \cdot \left[\frac{Z(t, T_1)}{Z(t, T_2)} - \frac{Z(0, T_1)}{Z(0, T_2)} \right] \\
&= Z(t, T_2) \cdot \left[\frac{1}{F(t, T_1, T_2)} - \frac{1}{F(0, T_1, T_2)} \right] \\
&= Z(t, T_2) \cdot [1 + \Delta \cdot f_n(t, T_1, T_2) - 1 - \Delta \cdot f_n(0, T_1, T_2)] \\
&= Z(t, T_2) \cdot \Delta \cdot [f_n(t, T_1, T_2) - f_n(0, T_1, T_2)].
\end{aligned} \tag{4.8}$$

Equation (4.8) gives the value of an FRA with unit nominal value from the viewpoint of the *floating* payer, at a time $t < T_1$.[3] Remember that, in our motivating example, the bank pays floating to the firm. The value for the fixed payer is obtained by changing the sign, and for a generic nominal we just multiply by N. To interpret the result, we note that the value is obtained by evaluating the future payoff, which is uncertain and will be known only at time T_1, by replacing the future spot rate $r_n(T_1, T_2)$ with the forward rate $f_n(t, T_1, T_2)$; the payoff is then discounted from time T_2 back to time t. Thus, we are using the forward rate *as if* this were a predictor of the future spot rate, which is a nontrivial result. The discussion about the pure expectation hypothesis in Section 3.7.4 showed that forward rates do not really predict spot rates; here, we use them in this manner, but only for valuation purposes.

---

[3] The value of the contract for $t \in [T_1, T_2]$ is trivially found by discounting the payoff, which is set at time T_1 and received at time T_2.

4.2.2 FRAs AS BOND TRADES

To find the value of an FRA, we may also use another clever trick, which we will meet again when dealing with swaps. The idea is to add the payment of the notional to both legs of the FRA. Clearly, this does not change the net payment in any way, but it allows us to recast the FRA as a difference of two bonds, a fixed- and a floating-rate one. The payment on the fixed leg, for a fixed rate K_n, is

$$N \cdot \Big[1 + \Delta \cdot K_n\Big],$$

and its value at time $t \leq T_2$ is obtained by straightforward discounting,

$$V_{\text{fixed}}(t) = Z(t, T_2) \cdot N \cdot \Big[1 + \Delta \cdot K_n\Big].$$

The payment on the floating leg is only known at $t = T_1$,

$$N \cdot \Big[1 + \Delta \cdot r_n(T_1, T_2)\Big],$$

but we may find its value at time $t < T_1$, by using the same approach that we used in Section 3.5.6 to price a floating-rate bond. Its value at T_1 is just the face value,

$$Z(T_1, T_2) \cdot N \cdot \Big[1 + \Delta \cdot r_n(T_1, T_2)\Big] = N,$$

since, on the left-hand side, we have the product of two random terms canceling each other. Thus, the value of the floating leg at time $t < T_1$ is

$$V_{\text{float}}(t) = Z(t, T_1) \cdot N.$$

The value of the FRA, from the viewpoint of the floating payer, is $V_{\text{fixed}}(t) - V_{\text{float}}(t)$. By using, once more, the link between discount factors and forward rates,[4]

$$1 + \Delta \cdot f_n(t, T_1, T_2) = \frac{Z(t, T_1)}{Z(t, T_2)}, \tag{4.9}$$

the FRA value may be rewritten as follows:

$$\begin{aligned}
V_{\text{fixed}}(t) &- V_{\text{float}}(t) \\
&= N \cdot [Z(t, T_2) \cdot (1 + \Delta \cdot K_n) - Z(t, T_1)] \\
&= N \cdot Z(t, T_2) \cdot \left[(1 + \Delta \cdot K_n) - \frac{Z(t, T_1)}{Z(t, T_2)}\right] \\
&= N \cdot Z(t, T_2) \cdot [(1 + \Delta \cdot K_n) - (1 + \Delta \cdot f_n(t, T_1, T_2))] \\
&= N \cdot Z(t, T_2) \cdot \Delta \cdot [K_n - f_n(t, T_1, T_2)]. \tag{4.10}
\end{aligned}$$

By setting the value of this contract to zero at $t = 0$, we find again the condition $K_n = f_n(0, T_1, T_2)$. For a later time $0 < t < T_1$, disregarding the nominal value N, we find the same result as Eq. (4.8): We have to replace the unknown future spot rate $r_n(T_1, T_2)$, in the formula for the payoff of a forward contract, by the current forward rate $f_n(t, T_1, T_2)$.

---

[4] See Section 3.7.3 on forward discount factors.

4.2 Forward rate agreements

Table 4.1 Sample evolution of a term structure.

Time to maturity τ	$r(0,\tau)$	$r(0.25, 0.25+\tau)$
$\tau = 0.25$	3.0%	3.1%
$\tau = 0.50$	3.2%	3.3%
$\tau = 0.75$	3.5%	3.7%
$\tau = 1.00$	4.0%	4.1%

4.2.3 A NUMERICAL EXAMPLE

An FRA with a nominal value of $100 million is agreed at $t=0$ for the semester $(0.5, 1)$. A set of continuously compounded rates for various maturities is given in Table 4.1. The table shows the known term structure at time $t=0$, as well as one possible future scenario at time $t=0.25$. We want to answer the following questions:

1. What is the payment on the fixed leg?
2. If the term structure after three months is the one shown in the rightmost column in Table 4.1, what is the value of the FRA for the fixed payer?

The first step is finding the fixed payment, which means finding the semiannually compounded forward rate $f_2(0, 0.5, 1)$, using Eq. (4.6). Then, in order to apply Eq. (4.10), we also have to find the relevant forward rate after three months, i.e., $f_2(0.25, 0.5, 1)$.

The current continuously compounded forward rate for the time interval $[0.5, 1]$ is

$$f(0, 0.5, 1) = \frac{1 \times r(0,1) - 0.5 \times r(0, 0.5)}{1 - 0.5} = \frac{0.04 - 0.5 \times 0.032}{0.5} = 4.8\%,$$

which corresponds to $f_2(0, 0.5, 1) = 2 \times \left(e^{0.5 \times 0.048} - 1\right) = 0.04858064$. The fixed payment is therefore determined as

$$100 \cdot 10^6 \times 0.5 \times 0.04858064 = \$2{,}429{,}032.$$

At time $t = 0.25$, we need to recalculate the new forward rate $f_2(0.25, 0.5, 1)$ for the tenor $[0.5, 1]$. Note that three months have elapsed, and now the relevant spot rates are

$$r(0.25, 0.5) = 3.1\% \quad \text{and} \quad r(0.25, 1) = 3.7\%,$$

which implies

$$\begin{aligned} f(0.25, 0.5, 1) &= \frac{0.75 \times r(0.25, 1) - 0.25 \times r(0.25, 0.5)}{0.75 - 0.25} \\ &= \frac{0.75 \times 0.037 - 0.25 \times 0.031}{0.5} = 4.0\%, \end{aligned}$$

which corresponds to
$$f_2(0.25, 0.5, 1) = 2 \times \left(e^{0.5 \times 0.04} - 1\right) = 0.04040268.$$
Now, from the viewpoint of the fixed payer, the FRA value after three months is

$$V_{\text{float}}(0.25) - V_{\text{fixed}}(0.25)$$
$$= N \times Z(0.25, 1) \times (0.75 - 0.25) \times [f_2(0.25, 0.5, 1) - f_2(0, 0.5, 1)]$$
$$= \$100 \cdot 10^6 \times e^{-0.75 \times 0.037} \times 0.5 \times [0.04040268 - 0.04858064]$$
$$= -\$397{,}706.86.$$

The result we find may look counterintuitive: There is an increase in the whole term structure, which should be good news to the fixed payer, who receives the floating rate. Nevertheless, her position is losing value. The puzzle may be explained by considering that in the pricing equation, the forward rates are used *as if* they were predictors of future spot rates. The forward rates implied by the initial term structure suggest an increase in the spot rates, which does indeed take place after three months, but by a lower amount than predicted. If we denote by $\widehat{r}_0(T_1, T_2)$ the "forecast" at time $t = 0$ of the future spot rates at time T_1 with maturity T_2, we see that

$$\widehat{r}_0(0.25, 0.5) = f(0, 0.25, 0.5) = \frac{0.5 \times 0.032 - 0.25 \times 0.03}{0.5 - 0.25}$$
$$= 0.034 > 0.031,$$
$$\widehat{r}_0(0.25, 0.75) = f(0, 0.25, 0.75) = \frac{0.75 \times 0.035 - 0.25 \times 0.03}{0.75 - 0.25}$$
$$= 0.0375 > 0.037,$$
$$\widehat{r}_0(0.25, 1) = f(0, 0.25, 1) = \frac{1 \times 0.04 - 0.25 \times 0.03}{1 - 0.25}$$
$$= 0.04643 > 0.041.$$

Furthermore, the forward rate for the time period $[0.5, 1]$ goes from 4.858% at time $t = 0$ down to 4.0403% at time $t = 0.25$.

The reader may also note that, in this case, we do not really need the semiannually compounded forward rates, since we could use the continuously compounded ones to find the fixed payment directly:

$$100 \cdot 10^6 \times \left(e^{0.5 \times 0.048} - 1\right) = \$2{,}429{,}032.$$

The message is that, if we want to find the value of an FRA where the fixed rate is quoted according to market practice, we must be careful about compounding.

4.3 Eurodollar futures

In this section, we briefly outline one kind of futures contract written on interest rates. We defer a discussion of futures contract on bonds to Section 5.3.2, as

4.3 Eurodollar futures

they involve some important institutional details. A bond futures depends in a nonlinear way on interest rates through the bond price, i.e., the price of a traded asset; eurodollar futures are an example of a linear contract directly defined on interest rates, which are not traded assets. We consider eurodollar futures because they are widely traded and practically relevant, and they also give us the opportunity of discussing a fundamental point: the difference between futures and forward rates.[5]

The term **eurodollar** may be misleading, as it has nothing to do with forex markets. By eurodollars we mean US dollars deposited on a non-US bank account. Eurodollar futures are very liquid and actively traded futures, whose underlying variable is the three-month LIBOR on dollar deposits. Interest rate futures are available for other currencies as well, based, e.g., on EURIBOR or Euro LIBOR. They can be used:

- To lock interest rates for future time periods
- To change the risk exposure of a fixed-income portfolio
- To speculate on interest rate movements

Contract maturities are standardized (March, June, September, December) up to ten years in the future.[6] Contracts are also available for the other months, but only short-term ones (i.e., within the current year). The nominal contract size is $1,000,000, to which the futures rate is applied over a three month period (0.25 years).

Let $L_4^f(t, T, T + 0.25)$ be the LIBOR futures rate at time t, with quarterly compounding, for the time interval $[T, T + 0.25]$. This futures rate, not to be confused with the forward rate, is implicit in the quoted futures price. In fact, the quote does not give the rate directly, but implicitly. The futures quote at time t is given as

$$100 \times [1 - L_4^f(t, T, T + 0.25)].$$

For instance, if the quoted futures price when taking a position in the contract is 97.22, the related futures rate is

$$L_4^f = \frac{100 - 97.22}{100} = 2.78\%.$$

Also note that a change of 1 basis point in the underlying futures rate implies a price change of $25, since

$$1,000,000 \times 0.0001 \times 0.25 = 25.$$

Let us illustrate this point with an example.

---

[5]This is a more general point concerning the relationship between forward and futures prices. As we shall see later, it can be shown that they should be the same under the assumption of constant interest rates. Clearly, this assumption is nonsensical when dealing with interest rate derivatives.

[6]For further details, see http://www.cmegroup.com/trading/interest-rates/stir/eurodollar_contract_specifications.html

218   CHAPTER 4   FRAs, Interest Rate Futures, and Vanilla Swaps

Example 4.1 Daily cash flows in eurodollar futures

> Let us consider a scenario in which the settlement price at the end of day k is 99.28. Suppose that the settlement price at the end of day $k+1$ was 99.33, which implies a drop of 5 basis points in the underlying futures rate. Then, at the end of day $k + 1$, the long position gains $125.

We recall that, in forward and futures contracts, the long position gains from an increase in the forward/futures price. Thus, if the quote increases, as in Example 4.1, the long position makes a profit, which means that the long position gains when interest rates fall. Hence, a long position may hedge against a fall in interest rates. The short position has an opposite exposure.

Example 4.2 Locking rates by eurodollar futures

> To see how, in principle, an investor can lock an interest rate for a future time interval $[T, T + 0.25]$, imagine that she takes a long position in one contract when the quote is 97.22 (futures rate is 2.78%). If the spot rate at maturity is 2.5%, the final settlement futures price will be 97.50, and if we disregard the time value of the daily cash flows, the net gain will be
>
> $$\$25 \times (97.50 - 97.22) = \$700.$$
>
> She will invest $1,000,000 at the spot rate 2.5% for three months, earning
>
> $$\$1{,}000{,}000 \times 0.025 \times 0.25 = \$6250.$$
>
> Adding the $700 profit from the futures position, she earns a total of $6950 and, apparently, she has the same gain as with a locked rate of 2.78%,
>
> $$\$1{,}000{,}000 \times 0.25 \times 0.0278 = \$6950.$$
>
> However, we have disregarded the possibility of investing daily profits (and the need of financing daily losses) due to marking-to-market of futures contracts. In other words, we are considering the futures contract as a forward.

For short time periods, it may be the case that the difference between futures and forward rates is not quite remarkable but, when dealing with an extended period of time, the difference cannot be disregarded. To see why, let us consider a contract with a payoff related to the difference $r(T_1, T_2) - K$, where K is a contracted rate (forward or futures).

4.3 Eurodollar futures

1. When marking-to-market is applied, intuition suggests that positive cash flows will be realized when interest rates rise; these are daily cash flows, and they can be immediately reinvested at higher rates. On the other hand, if interest rates drop, negative cash flows will occur; however, they may be financed at a lower rate. This makes marking-to-market a welcome feature with respect to a forward. Hence, due to demand–offer mechanism, the contracted rate K for a forward contract will tend to be smaller in order to make the forward contract more attractive and compensate for the lack of marking-to-market.

2. Furthermore, a forward contract is essentially settled at time T_2 (even if the contract is settled before, at time T_1, the payoff is just the discounted payoff that one would earn at time T_2). A futures contract with the above payoff, which is settled at time T_1, is preferable to a contract settled at T_2. This has the effect of reducing the forward rate, too.

The above argument is purely intuitive, but the difference between forward and futures rates (as well as between forward and futures prices for contracts on an underlying asset or commodity) may be analyzed by using the tools of stochastic calculus that we shall cover later. We just reinforce the intuition by mentioning a correction that can be used to recover forward rates from observed futures rates. The corresponding forward rate may be obtained by a "convexity correction." One such correction (related to the Ho–Lee model, which is a short-rate model based on stochastic differential equations) is

$$f(0, T_1, T_2) = r_{\text{fut}}(0, T_1, T_2) - \frac{1}{2}\sigma^2 T_1 T_2,$$

where rates are continuously compounded and σ is the volatility of short-term rates. We refrain from discussing the correction in any detail, but we point out the intuition: The forward rate tends to be smaller than the futures rate, and the difference is significant when volatility is large and for long maturities. On the contrary, when volatility is zero, there is no difference between forward and futures rates. Indeed, as we mentioned before, forward and futures prices should be the same in the case of constant interest rates, as we shall prove in Section 12.2.

A further consequence of daily marking-to-market is that, in general, a hedge based on futures contracts is more difficult to set up than a hedge based on forward contracts. For long maturities, the strategy should be dynamically adjusted by *tailing* the hedge (this will be further discussed later in Section 12.3.4).

We should also keep in mind the institutional side of the coin: Futures rate may be easily recovered by observing quite liquid exchange-traded securities. On the contrary, FRAs are over-the-counter contracts. Forward rates are implied by the term structure, which may be estimated on the basis of observed fixed-income asset prices, as we have discussed in Section 3.5.2. However, we recall that liquidity issues and other distortions may complicate this task.

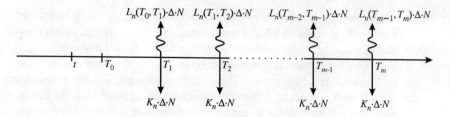

FIGURE 4.2 An illustration of swap cash flows.

4.4 Vanilla interest rate swaps

A vanilla interest rate swap is an agreement between two counterparties that will exchange periodical cash flows expressed as an interest on a nominal amount N. There are two legs in the swap:

- The *fixed leg*, associated with cash flows related to a fixed interest rate
- The *floating leg*, associated with cash flows related to a floating interest rate

To understand the swap mechanics, please refer to Fig. 4.2:

- Interest rates will apply to time periods $[T_{i-1}, T_i]$, $i = 1, \ldots, m$. We assume, for the sake of simplicity, that $\Delta = T_i - T_{i-1}$ is constant and independent from i, but this is not the necessarily case in real life. Rates are assumed to be compounded $n = 1/\Delta$ times per year, where usually $n = 2$ or $n = 4$.
- At time $t \leq T_0$ the contract is arranged and a fixed rate K_n, the **swap rate**, is established. If $t < T_0$, we speak of a *forward-start* swap contract. If $t = T_0$, we have a *spot-start* swap, which is the only case we consider here.
- The floating rate will be reset at time instants $T_0, T_1, \ldots, T_{m-1}$, where, for instance, the LIBOR rate $L_n(T_{i-1}, T_i)$ is observed.
- Payments will occur at times $T_1, T_2, \ldots, T_m$, when the floating rate is also reset (with the exception of T_m). On the floating leg, a payment $L_n(T_{i-1}, T_i) \cdot \Delta \cdot N$ is made at time T_i. On the fixed leg, a payment $K_n \cdot \Delta \cdot N$ is made at time T_i.

The two cash flow sequences are illustrated in Fig. 4.2. Needless to say, for the sake of convenience, only the net difference of the two cash flows is paid. Here we only consider this extremely simple swap agreement, but in practice there are many variations on the theme.[7] For instance, the frequencies of the fixed and the floating leg need not be the same; we assume so for the sake of simplicity. As a matter of fact, there is a huge market of interest rate swaps, which, like any

---

[7] See Section 5.3.1.

4.4 Vanilla interest rate swaps

derivative, may be used to change the nature of assets or liabilities, to speculate on interest rates, or to hedge interest rate risk.

There are two basic questions, which we have already considered when dealing with FRAs: How is the fixed rate K_n chosen at the inception of the contract, and how can we find the fair value of the contract at a later time? We may answer both questions by finding a way to value a swap contract, and then enforcing an initial value zero, just like we did with FRAs. There are two equivalent ways to value a vanilla interest rate swap:

1. As the difference of a fixed- and a floating-rate bond (which requires including a fictitious payment of notionals at maturity T_m).
2. As a portfolio of FRAs with a range of maturities, $T_1, \ldots, T_m$.

The first approach is, in a sense, a horizontal decomposition of the cash flows in Fig. 4.2, where floating-rate flows are drawn on the upper side of the figure, and fixed-rate flows are on the lower side. On the contrary, using FRAs is a vertical decomposition with respect to time, resulting in pairs of cash flows occurring at the same time instant. These two approaches mirror those we have discussed in Sections 4.2.1 and 4.2.2, respectively, and have advantages and disadvantages:

- If the payment frequency is not the same on the two legs, we cannot use the approach based on a sequence of FRAs.
- However, when dealing with forward-start swaps, using FRAs may be more natural.

4.4.1 SWAP VALUATION: APPROACH 1

Let us assume that we want to value the swap at time t, $T_0 \le t \le T_1$, when m payments are still due. The value of the fixed leg is the value of a fixed-coupon bond,

$$P_{\text{fixed}}(t, T) = N \cdot \left[\sum_{i=1}^{m} K_n \cdot \Delta \cdot Z(t, T_i) + Z(t, T_m) \right].$$

Note that we include a fictitious payment of the notional amount at maturity $T \equiv T_m$, which is not going to occur, as it is exactly offset by the same payment on the floating leg. The value of the floating leg is the value of a floating-coupon bond, where the first coupon is related to the already observed rate $L_n(T_0, T_1)$:

$$P_{\text{float}}(t, T) = Z(t, T_1) \cdot N \cdot [1 + L_n(T_0, T_1) \cdot \Delta].$$

The swap value, from the fixed payer viewpoint, is

$$P_{\text{float}}(t, T) - P_{\text{fixed}}(t, T).$$

If we consider the initial swap value, at $t = T_0$, the swap rate is chosen in such a way that the initial swap value is zero:

$$P_{\text{fixed}}(T_0, T) = P_{\text{float}}(T_0, T) = N.$$

Table 4.2 Term structure for Example 4.3.

Time to maturity	Spot rate $r(0,\tau)$
$\tau = 0.5$	2.3%
$\tau = 1.0$	2.6%
$\tau = 1.5$	2.8%
$\tau = 2.0$	3.0%

The swap rate is then found as a solution of the following equation:

$$N \cdot \left[\sum_{i=1}^{m} K_n \cdot \Delta \cdot Z(T_0, T_i) + Z(T_0, T_m)\right] = N,$$

which yields

$$K_n = \frac{1}{\Delta} \cdot \frac{1 - Z(T_0, T_m)}{\sum_{i=1}^{m} Z(T_0, T_i)}. \tag{4.11}$$

The swap rate is essentially a coupon rate such that the fixed-coupon bond sells at par. This rate is also called the **par yield**. This swap rate is quoted by dealers (with a bid–ask spread) for various maturities.

Example 4.3 Finding the swap rate

The current term structure of (continuously compounded) interest rates is given in Table 4.2. We want to find the swap rate for a contract maturing in two years with semiannual payments.

To find the swap rate, let us assume a nominal value of 1 (which is actually irrelevant) and solve the equation

$$\frac{K_2}{2} \times \left(e^{-0.5 \times 0.023} + e^{-1 \times 0.026} + e^{-1.5 \times 0.028} + e^{-2 \times 0.03}\right)$$
$$+ e^{-2 \times 0.03} = 1,$$

which yields

$$K_2 \approx 3.015\%.$$

Clearly, we may just apply Eq. (4.11) and find the same result. The swap rate K_2 with semiannual compounding corresponds to $K = 0.0299$ with continuous compounding.

4.4 Vanilla interest rate swaps

Table 4.3 Term structure for swap valuation in Examples 4.4 and 4.5.

Time to maturity	Spot rate $r(0,\tau)$
$\tau = 0.25$	4.3%
$\tau = 0.75$	4.7%
$\tau = 1.25$	5.0%

Example 4.4 Swap valuation

A swap was agreed in the past, with swap rate $K_2 = 6\%$ and a notional of $100 million. A payment occurred three months ago, and the LIBOR rate was reset to 5%, with semiannual compounding. Payments will occur in 3, 9, and 15 months, and the current term structure of (continuously compounded) rates is given by Table 4.3. We want to find the swap value from the fixed payer viewpoint.

Let us price the fixed-rate bond first:

$$P_{\text{fixed}} = 3 \times e^{-0.25 \times 0.043} + 3 \times e^{-0.75 \times 0.047} + 103 \times e^{-1.25 \times 0.05}$$
$$= 102.6236,$$

where we express value in millions of dollars. The floating-rate bond price is

$$P_{\text{float}} = (100 + 2.5) \times e^{-0.25 \times 0.043} = 101.404.$$

Then, the swap value for the fixed payer is

$$P_{\text{float}} - P_{\text{fixed}} = 101.404 - 102.6236 \Rightarrow -\$1,219,534.$$

4.4.2 SWAP VALUATION: APPROACH 2

If we regard the swap as a portfolio of FRAs, we know that, for valuation purposes, we may replace the random future spot rates by the corresponding forward rates, as we have seen in Section 4.2.1. The value of the swap, from the fixed payer's viewpoint, is

$$N \cdot \Delta \cdot \sum_{i=1}^{m} \left[f_n(t, T_{i-1}, T_i) - K_n \right] \cdot Z(t, T_i). \qquad (4.12)$$

224   CHAPTER 4   FRAs, Interest Rate Futures, and Vanilla Swaps

To show that we find the same swap value as before, let us set $t = T_0$, decompose the sum, and express the forward rate using discount factors:

$$N \cdot \sum_{i=1}^{m} \left[f_n(T_0, T_{i-1}, T_i) - K_n \right] \cdot \Delta \cdot Z(T_0, T_i)$$

$$= N \cdot \left[\sum_{i=1}^{m} \left(\frac{Z(T_0, T_{i-1})}{Z(T_0, T_i)} - 1 \right) \cdot Z(T_0, T_i) - \sum_{i=1}^{m} K_n \cdot \Delta \cdot Z(T_0, T_i) \right]$$

$$= N \cdot \left[\sum_{i=1}^{m} \left(Z(T_0, T_{i-1}) - Z(T_0, T_i) \right) - \sum_{i=1}^{m} K_n \cdot \Delta \cdot Z(T_0, T_i) \right]$$

$$= N \cdot \left[\left(1 - Z(T_0, T_m) \right) - \sum_{i=1}^{m} K_n \cdot \Delta \cdot Z(T_0, T_i) \right].$$

By setting this to zero, we find the same result as Eq. (4.11). In the second line we use Eq. (4.9) to link forward rates and discount factors. We get to the last line by using a telescoping sum and the identity $Z(T_0, T_0) \equiv 1$.

▨ Example 4.5   Swap valuation (continued from Example 4.4)

We use the data in Table 4.3 once more, but now we need the forward rates for the relevant maturities in order to "predict" the cash flows on the floating leg for the second and third payment:

$$f(0, 0.25, 0.75) = \frac{0.75 \times 0.047 - 0.25 \times 0.043}{0.5} = 0.049,$$

$$f_2(0, 0.25, 0.75) = 2 \times (e^{0.049 \times 0.5} - 1) = 0.04960518,$$

$$f(0, 0.75, 1.25) = \frac{1.25 \times 0.05 - 0.75 \times 0.047}{0.5} = 0.0545,$$

$$f_2(0, 0.75, 1.25) = 2 \times (e^{0.0545 \times 0.5} - 1) = 0.05524935.$$

The first cash flow is known, since it is related to the last observed LIBOR rate.

Then, we calculate and discount the net "forecasted" cash flows:

$$V_{\text{swap}} = \frac{100 \cdot 10^6}{2} \times \left[e^{-0.043 \times 0.25} \times (0.05 - 0.06) \right.$$
$$+ e^{-0.047 \times 0.75} \times (0.04960518 - 0.06)$$
$$\left. + e^{-0.05 \times 1.25} \times (0.05524935 - 0.06) \right] = -\$1{,}219{,}534.$$

The result, of course, is the same as with approach 1.

4.4.3 THE SWAP CURVE AND THE TERM STRUCTURE

If we consider a range of swap rates for different maturities and we plot them, we find the **swap curve**. As we mentioned in Section 4.1, LIBOR rates are not available for long maturities. If one wants to extend the LIBOR curve to long maturities, swap rates may be used. Actually, other fixed-income securities may be used. One possible advantage of using swap rates is that they feature good liquidity properties and limited credit risk, the factors that may hinder the possibility of estimating a term structure of truly *risk-free* rates by using plain bonds.

Example 4.6 Extending the LIBOR curve by using swap rates

> Let us assume that LIBOR rates for 6 and 12 months are 3% and 3.3%, respectively, with continuous compounding. We would like to find the LIBOR rate for 18 months, and we consider the quoted swap rate for a contract maturing in 18 months with semiannual payments. The quoted rates are
>
> $$\text{bid } 3.5\%, \quad \text{ask } 3.7\%,$$
>
> which reflect the bid–ask spreads observed in real-life markets (see Section 5.3.1). Let us take the average, 3.6%, as the swap rate. Note that these rates are to be considered as semiannually compounded and define *cash flows*, rather than discount factors. The swap rate is such that a fixed-coupon bond with coupon rate 3.6%, maturing in 18 months, trades at par. Assuming an arbitrary face value of \$1, we have to solve the following equation:
>
> $$1 = \frac{0.036}{2} \times e^{-0.03 \times 0.5} + \frac{0.036}{2} \times e^{-0.033 \times 1} + \left(1 + \frac{0.036}{2}\right) \times e^{-L(0,1.5) \times 1.5},$$
>
> which yields $L(0, 1.5) \approx 3.575\%$, with continuous compounding.

Once again, we stress the fact that, in Example 4.6, we have mixed LIBOR and swap rates as if they are both risk-free. After the credit crunch of 2008, this practice is a bit questionable. Nevertheless, the example shows that swaps may be used to estimate a term structure of interest rates. Indeed, even more complicated derivatives, namely, swaptions (options on swaps), are used to this purpose.

Problems

4.1 The current LIBOR rates for maturities of three and six months are 4.3% and 4.7%, respectively (with continuous compounding). We also have the following eurodollar futures market quotes:

Maturity	Price
6 months	94.9
9 months	94.5
12 months	94.2

Price a risk-free bond maturing in one year, paying semiannual coupons, with coupon rate 6%. For the sake of simplicity, we ignore day counts and we assume that all months consist of 30 days. Furthermore, we assume that there is no significant difference between forward and futures rates.

4.2 An investment bank agrees on an interest swap contract with a firm. Semiannual payments are arranged: the firm pays six-month LIBOR, whereas the bank pays a fixed rate of 4% with semiannual compounding. The notional value is €20 million and the maturity is four years. After exactly 28 months the firm goes bankrupt and defaults. Let us assume that, when default occurs, the term structure is flat, at 5% with continuous compounding, and that the last relevant LIBOR observation was 7%. What is the profit/loss for the bank?

Further reading

- More information on the simple interest rate derivatives that we have discussed in this chapter can be found in general books on fixed-income assets, like [4] and [6].
- A detailed treatment of eurodollar futures can be found in [1].
- For swaps and their variants, the reader may refer, e.g., to [2] or [3].

Bibliography

1 G. Burghardt. *The Eurodollar Futures and Options Handbook*. McGraw-Hill, New York, 2003.

2 H. Corb. *Interest Rate Swaps and Other Derivatives*. Columbia Business School Publishing, New York, 2012.

3 R.R. Flavell. *Swaps and Other Derivatives* (2nd ed.). Wiley, Chichester, 2010.

4 L. Martellini, P. Priaulet, and S. Priaulet. *Fixed-Income Securities: Valuation, Risk Management, and Portfolio Strategies*. Wiley, Chichester, 2003.

5 M. Morini. *Understanding and Managing Model Risk: A Practical Guide for Quants, Traders and Validators*. Wiley, Chichester, 2011.

6 P. Veronesi. *Fixed Income Securities: Valuation, Risk, and Risk Management*. Wiley, Hoboken, NJ, 2010.

Chapter Five

Fixed-Income Markets

We have introduced the fundamental concepts related to interest rates and fixed-income assets in Chapter 3. Then, in Chapter 4, we have introduced simple interest rate derivatives that may be used, among other things, to manage interest rate risk, as we shall illustrate in Chapter 6. In this chapter, we take a short break to deal with some topics that are quite relevant for the profession. Mathematically inclined readers may not be interested in certain nasty details of real markets, and indeed they can skip this chapter if they are only interested in the intellectual pleasure of quantitative models. However, there is little value in overly sophisticated and fragile models, without any understanding of the pitfalls and issues that are so pervasive in financial markets, especially in the fixed-income case. Due to space constraints, we will not be able to present an extensive picture, but it is important to get at least a feeling for some issues that do play an important role in practice. The choice of topics is somewhat arbitrary, and it has been made to illustrate just a few among the most essential issues.

- When calculating interest in elementary treatments, we deal with time measured in years or months. For the sake of simplicity, we pretend that every month is just the same and consists of 30 days, but this is not really the case. We have to consider the difference between months consisting of 30 or 31 days, and possibly 28 (or 29 in leap years).[1] Market practice may be rather peculiar and this leads to day count conventions, which are discussed in Section 5.1.

- Bonds may look like simple assets, whose prices can be obtained by straightforward discounting of cash flows. However, both market conventions and available assets are a bit more complex. Some additional details on bond markets are given in Section 5.2, where we deal with actual bond price quotes and bonds with embedded options, namely, callable and convertible bonds. We also hint at the issues related to bond ratings.

---

[1] When dealing with interest accrual, Saturdays and Sundays do matter. The picture is different if we consider equity markets, and specifically trading-induced volatility. It is common to consider a year consisting of 250 or 252 active trading days.

- In Section 5.3, we extend the treatment of elementary interest rate derivatives by considering additional features of swaps and bond futures/options.
- We should not disregard money markets, i.e., fixed-income markets with very short maturities. A relevant example is the market of repurchase agreements (repo for short) which we outline in Section 5.4, along with other securities that firms may use for short-term liquidity needs.
- Finally, in Section 5.5 we consider the securitization of illiquid assets to create fixed-income securities. We outline mortgage-backed securities, since they are a prominent example of this process and allow us to understand tranching issues.

The last item in the list is quite relevant, as it opens the door to the dangerous world of credit risk, which is beyond the scope of this book, yet cannot be ignored.

5.1 Day count conventions

When dealing with interest rates and bond mathematics, we usually treat time in a very straightforward fashion. For instance, if we have to discount a cash flow with a continuously compounded rate of, say, 4%, over a three-month period, we just use a discount factor like $e^{-0.04 \times 0.25}$, since three months amount to a quarter of a year. However, what if the quarter includes February? Is this quarter the same, in terms of how many days are included, as a quarter including August and July? In real life, whenever we have to discount or calculate cash flows, e.g., when analyzing an interest rate swap, care is needed.

This kind of question is also relevant when dealing with **accrued interest** in bond trading. Imagine that we want to buy a bond with a face value of $10,000 and paying semiannual coupons at rate 5%. The settlement date is April 20th, year t, and the bond matures in two years, on September 15th, year $t + 2$. Assuming semiannual coupons, the last coupon (2.5% of the face value, i.e., $250) was paid on March 15th, and the next one will be paid on September 15th. Hence, the current bondholder is about to sell the bond, but she has kept the bond in her portfolio for more than one month: How much of the next coupon payment is she entitled to? The easy answer is that we should prorate the interest rate of the whole period, a semester, in proportion to the fraction of the period that has elapsed from the last coupon payment. Using simple-minded approach, one month over six should amount to $1/6 \approx 0.1667$, but by doing so we are ignoring exact day counts. The actual number of days elapsed from the last coupon payment is 36 (the last 16 days of March, plus the first 20 days of April). The number of days between March 15th and September 15th is 184, hence, one possible answer is

$$\$250 \times \frac{36}{184} \approx \$48.91. \qquad (5.1)$$

In doing so, we have used one of the possible **day count** conventions, which is based on the ratio of actual number of days. However, there are other possibilities, as one may consider a year consisting of 12 months of 30 days, a practice that made life much simpler when computers were not available. Among the possible day count conventions, we mention:

1. Actual/Actual, which yields the result of Eq. (5.1).
2. 30/360, in which case the above calculation would be

$$\$250 \times \frac{35}{180} \approx \$48.61.$$

In this case, we consider equal months consisting of 30 days and a year consisting of 360 days.

3. Actual/360, in which case the above calculation would be

$$\$250 \times \frac{36}{180} = \$50.00.$$

As we notice, the impact of different conventions is not quite negligible. Actually, there are other possibilities, and we must also pay attention to leap years. Hence, we must be careful and keep in mind that different kinds of bonds, e.g., treasury vs. corporate, are subject to different conventions.

The careful reader might object that computing accrued interest is not really needed. If the bond is traded between two coupon payments, all we have to do is to properly discount cash flows by the selected day count convention. However, we should consider the fact that the market price is determined by factors, such as liquidity, which are not quite addressed by simple pricing formulas. Furthermore, as we discuss in Section 5.2, quoting bond prices requires setting the accrued interest apart from the actual bond price.

Last but not least, day count conventions play an important role in several interest rate derivatives, as they contribute to determining the payoff. Strange as it may sound, the misspecification of the payoff in interest rate derivatives is a potential danger and is one of the facets of model risk.

5.2 Bond markets

Bonds, like stock shares, are first issued on primary markets, possibly with the support of investment banks acting as underwriters. They are then traded on secondary markets, on which they may be more or less liquid. The procedure to issue bonds on primary markets is typically based on auctions, whose details may differ, depending on the nature of the bond (e.g., treasury vs. corporate). While bonds may look quite simple, when compared with exotic derivatives, they come in a variety of forms and may be classified according to the following criteria:

Issuer. As we have already mentioned, the issuer may be a central government (for treasury/sovereign bonds) or a corporation (corporate bonds), but there are alternative issuers, like municipalities or other public agencies. The kind of issuer may have an impact on taxation rules, as well as day count conventions.

Maturity. Short-term bonds are usually zero-coupon bonds, whereas long-term bonds are coupon-bearing bonds. Long-term zeros may be synthetically created by stripping coupons, however. US treasury bonds are classified according to time to maturity as follows: T-bills (maturity up to a year), T-notes (maturities up to ten years), and T-bonds (longer maturities, possibly thirty years). The market for short-term securities is called **money market**, whereas the term **capital market** refers to securities with longer maturities. There are other money market instruments, like banks' certificates of deposit, banker's acceptances, and repos, which we discuss in Section 5.4.

Rules to determine cash flows. In a plain bond, cash flows are linked to a fixed coupon rate applied to the face value. However, floaters and linkers are traded as well. In **floaters** (floating-rate bonds), coupons depend on the current level of interest rates. Usually, the coupon rate increases with interest rates, as it may be given by the six-month LIBOR $L_{0.5}^f$, observed at each reset date, plus a spread. In **reverse floaters**, the coupon rate is *reduced* by an increase of rates and it may be given as

$$\max\{0, K - L_{0.5}^f\},$$

to preserve non-negativity. With **linkers**, the coupon rate may depend on some other reference quantity, possibly linked with equity markets. A quite relevant example is an inflation-indexed bond, where the face value (and, as a consequence, the coupons) is indexed by the inflation rate. A TIPS (treasury inflation-protected security) is an example of inflation-indexed bond.

Embedded options. As we discuss in Section 5.2.3, bonds may have embedded options, like the possibility of early repayment of face value by the issuer (callable bonds) or the possibility of conversion to equity (convertible bonds). These should not be confused with options written on bonds.

Collateral. The price of a bond depends not only on interest rates, but also on the credit rating of the issuer. When default is a possibility, we cannot discount cash flows using a risk-free rate. However, the rating does not only depend on the issuer, as bonds may have different provisions for a collateral, as well as the specification of a "pecking order" in case of default. Senior bonds are relatively protected against default, whereas subordinated debentures have a lower degree of protection for the investor. Issuers' assets are first used to pay senior bondholders in case of default, and they may not be sufficient to pay holders of subordinated bonds. These features are specified in **bond indentures**. Needless to say, subordinated bonds offer a higher yield, as they trade at a lower price than senior bonds.

5.2.1 BOND CREDIT RATINGS

Corporate and sovereign bonds are debt instruments, subject to default risk. In the event of default, the bond issuer (often referred to as **obligor**) will fail to repay a part or the whole of his debt. Statistical modeling of loss given default (LGD) is an active research area. In order to help investors in assessing the creditworthiness of a bond issuer, rating agencies like Moody's and Standard & Poor's associate a credit rating with each bond, assessing the ability of obligors to meet their repayment obligations.

Each rating agency has a rating scale. For instance, the Standard & Poor's scale includes credit ratings like the following:

- AAA, which corresponds to the prime grade.
- AA+, AA, AA−, the high grade bonds.
- A+, A, A−, and BBB+, BBB, BBB−, medium grade.

These ratings correspond to investment grade bonds. Down the scale we meet ratings from BB+ to B−, which are flagged as non-investment grade and speculative. The rating CCC flags risky bonds, and ratings beginning with a D are reserved to bonds in default.

Non-investment grade bonds are also referred to as high yield, since their relatively low price is associated with higher yields. Junk bonds are quite speculative, and some institutional investors are forbidden to include them in their portfolio. Credit derivatives, like credit default swaps may be used to insure bonds against defaults.

5.2.2 QUOTING BOND PRICES

Imagine that the (ask) price of a bond with face value $1000 is quoted as 112.08, which must be interpreted as a percentage of face value. Does it mean that we have to pay $1120.80 to buy that bond? The answer is not quite so simple. To begin with, we must be aware of market conventions. For instance, US treasury bonds are quoted in 32nds, so that 0.08 does not really mean 8 cents, but rather $8/32 = \$0.25$. Leaving this issue aside, the quoted price does *not* include the **accrued interest** related to the next coupon to be paid. The current bondholder is entitled to some fraction of the next coupon, as she held the bond for the corresponding fraction of the time period on which interest accrues. The quoted price is just the **clean price**, whereas the actual cash price is called **dirty price** and is obtained by taking into account the time elapsed from the last coupon payment.

Example 5.1 Dirty vs. clean bond price

Let us consider a bond paying semiannual coupons with 6% coupon rate on a face value of $1000, which means that the bond pays $30

> every six months. Let us ignore day count issues, for the sake of simplicity, and assume that the last coupon was paid two months ago. Accrued interest is obtained by prorating the next coupon as follows:
>
> $$\$30 \times \frac{2}{6} = \$10.$$
>
> If yield is lower than the coupon rate, the bond shall trade at premium, say, at a quoted price of 112.08. Thus, the cash price would be
>
> Clean price + Accrued interest = 1120.80 + 10.00 = \$1130.80.
>
> In practice, this calculation should be carried out according to the relevant day count convention pertaining to the specific bond at hand.

Setting the accrued interest component apart makes sense because it is money that the bondholder is entitled to. However, one could just discount cash flows properly and this would account for everything, so why should we quote the clean price? The plots in Fig. 5.1 help to explain why this is convenient. Whenever a coupon is paid, we observe a jump in the bond price, since a cash flow is eliminated from the cash flow sequence that gives the bond price. The figure shows the effect for two bonds trading at premium and at discount, respectively. When the bond trades at premium (the case on the left in Fig. 5.1, where the coupon rate is larger than yield-to-maturity), we observe a decreasing lower envelope, to which a jagged price path is superimposed. When the bond trades at a discount, the lower envelope is increasing. Bond price jumps may also be observed because of shocks related with interest rate and credit risk, which may be a good reason for concern, whereas a jump due to a coupon payment is physiological. The clean price eliminates these "natural" jumps, so that they are not confounded with the effect of true risk factors.

When dealing with short-term zeros, like T-bills, still another convention may be used. Suppose that the quote for a T-bill maturing in three months (90 days) is 4.90. Clearly, this is not the bond price. Short maturity bonds may be quoted in terms of a **bank discount** with respect to the face value. The bill's discount is annualized on the basis of a 360-day convention, and this is then reported as a percentage of par value. This means that the actual bank discount for this T-bill is

$$4.90\% \times \frac{90}{360} = 1.225\%,$$

and a bond with \$10,000 par value could be purchased for

$$\$10,000 \times (100\% - 1.225\%) = \$9877.50.$$

In real life, a dealer may quote two discounts, reflecting bid–ask spreads. For instance, the relevant discount, if we want to sell a bond, could be 4.91 (which is

5.2 Bond markets

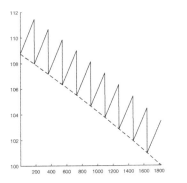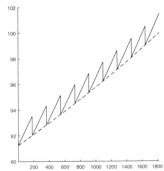

FIGURE 5.1 Plot of clean (dashed line) and dirty (continuous line) bond prices. Time-to-maturity is five years, yield is 5%, and coupon rate is 7% (left) and 3% (right). On the horizontal axis, time is expressed in days.

larger). This quoting approach is a traditional method, with a few shortcomings. In fact, it is based on a 360-day convention, which does not reflect the exact return for the investor. For the T-bill we are considering, the return over 90 days is

$$\frac{\$10,000}{\$9877.50} - 1 = 1.01240 - 1 = 1.24\%,$$

which can be annualized as

$$1.24\% \times \frac{365}{90} = 5.03\%.$$

In this book, we will not consider all of these difficulties, but they are relevant in the real life. Furthermore, a possible ambiguity arises when dealing, e.g., with options on bonds. Does the strike price refer to the clean or the dirty price of the bond? The exact terms of the agreement must be carefully checked and specified.

5.2.3 BONDS WITH EMBEDDED OPTIONS

Some bonds come with packaged options. This is the case with structured bonds, where the repayment of the face value is promised, and the possibility of a coupon depending on some other factor, like the return on an equity portfolio, is offered. The guarantee of a non-negative coupon may be engineered by bundling an option with the underlying bond.[2] Structured bonds were used to circumvent regulation forbidding certain funds to invest in options directly. Here, we illustrate two more examples of bonds with embedded options.

---

[2] See Example 1.12.

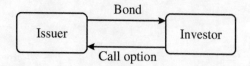

FIGURE 5.2 Decomposing the trade on a callable bond.

5.2.3.1 Callable bonds

Callable bonds are bonds that can be repurchased by the issuer before maturity. To understand the rationale behind callable bonds, imagine a firm issuing long-term debt when yield is relatively large, say, 6%. Coupon rates will reflect the general level of interest rates and credit spreads. Imagine that yield drops. This is extremely good news for the bondholder, as she could sell the bond for a large price (at premium). However, the firm will regret issuing the bond, as now it could collect capital at a considerably smaller cost. If a call provision is embedded in the bond, the issuer may indeed refinance the debt by repurchasing the old bonds and issuing brand new ones, featuring a smaller coupon rate.

Hence, the bondholder is exposed to reinvestment risk and must be compensated for that. Indeed, the price for a callable bond is smaller than a noncallable one. We may understand the point by thinking that the holder of a callable bond has a long position in the bond and a short position in a call option on the bond itself. In other words, the bondholder implicitly sells a call option to the issuer, as shown in Fig. 5.2.

▣ Example 5.2 Finding the implicit call option price

> The continuously compounded risk-free rates with maturities of 6, 12, 18, and 24 months are, respectively, 2.2%, 2.7%, 3.1%, and 3.49%. A callable bond, with no default risk, following the usual market conditions, with coupon rate 9%, maturing in two years, trades for €101.12 (face value is 100). What is the value of a call option on the corresponding noncallable bond?
>
> We notice that the bond has a large coupon rate, 9%, with respect to prevailing rates. Indeed, the bond sells at premium, but 101.12 does not seem large enough. Indeed, the price difference between a plain bond and a callable bond is just the value of the call (we assume that there is no default risk). The noncallable bond price is:
>
> $$4.5e^{-0.022 \times 0.5} + 4.5e^{-0.027 \times 1} + 4.5e^{-0.03 \times 1.5}$$
> $$+ 104.5e^{-0.0349 \times 2} = 110.59.$$
>
> Then, the value of the call option is $110.59 - 101.12 = \$9.47$.

In Example 5.2, we have taken a simple approach to find the value of a callable bond, assuming that there is no mispricing on markets. However, we need pricing models to detect arbitrage opportunities and to measure and manage risk. As we shall see, pricing options linked with interest rates is not quite trivial.

With a callable bond, the bondholder sells a call option; with puttable bonds, the bondholder also buys a put option, allowing her to sell the bond back and to collect the face value earlier. Clearly, a puttable bond is more expensive than a plain bond. A similar risk is associated with prepayment risk on mortgages. This kind of risk affects mortgage-backed securities and requires careful modeling.

5.2.3.2 Convertible bonds

Convertible bonds can only be issued by corporations, as they offer the possibility of converting the bond to a prespecified number of stock shares of the issuing firm, at a given price. Thus, a convertible bond includes an equity option. To be more precise, what is embedded is typically a **warrant**, rather than a call option. The peculiar feature of a warrant is that a *brand new* stock share is created when the bondholder exercises the conversion option. Hence, it is immediately understood that a warrant can only be issued by the corporation itself, rather than by an investment bank.

A convertible bond may be appealing to the issuer, as it is a way to issue debt at a somewhat lower price, as a convertible bond is less expensive than a plain bond. It may also be a more palatable way to raise equity since markets, under certain circumstances, may perceive the issuing of new equity as a bad signal. From the investors' viewpoint, convertible bonds offer the possibility of taking advantage of the firm upside potential, without incurring the risk of stock trading.

5.3 Interest rate derivatives

In Chapter 4, we have considered simple interest rate derivatives, like vanilla swaps and Eurodollar futures. In this section, we introduce more complicated swaps, and point out important institutional details about derivatives written on bonds.

5.3.1 SWAP MARKETS

We have seen in Section 4.4 that a swap rate may be determined in such a way that the current value of a swap contract is zero. Swap contracts are actually quoted by dealers (market makers), and a realistic quote on markets could be

$$\text{Bid} : 4.99, \qquad \text{Ask} : 5.03,$$

for maturity of, say, seven years. As usual, for any security or contract, there is a bid–ask spread. This means that if we are floating-rate payers, we will receive a fixed-rate of 4.99%, and we will have to pay 5.03% if we are floating-rate receivers. An alternative way to quote a swap rate may be in terms of spread with respect to a reference bond yield. For instance, a quote like 48–51 implies that the dealer is willing to pay 48 basis points above the bond yield, when receiving the floating rate, and is asking 51 basis points above, when paying the floating rate. Another practical complication is that the two payments need not be synchronized. Payment on the fixed leg may occur every six months, whereas payments on the floating leg may be linked to the three-month LIBOR, in which case they occur every three months. Day count conventions must also be specified and may be different on the two legs.

Furthermore, non-vanilla swaps are actively traded:

- The notional amount may change in time. In **amortizing swaps**, the notional is reduced over time, whereas it is increased in **accrediting swaps**.
- The swap may be based on floating-for-floating payments, where the two legs are linked to different markets and/or different maturities. This is the case with **basis swaps**.
- Another example of floating-for-floating swap is a **constant maturity swap** (CMS), whereby a floating rate is exchanged for a particular swap rate. For instance, the three-month LIBOR could be exchanged for an eight-year swap rate.

Another common kind of derivative is a **swaption**, i.e., an option giving the holder the possibility of entering into a swap contract with a predetermined swap rate at some later time.

5.3.2 BOND FUTURES AND OPTIONS

There are two broad families of derivatives related to interest rates.

- Some contracts, like Eurodollar futures, are directly written on a quoted interest rate. Since an interest rate is not a traded asset, such contracts specify a notional to which the interest rate is applied and are settled in cash. Other examples of such contracts are interest rate **caps** and **floors**. Caps and floors are portfolios of caplets and floorlets, respectively, much like a swap is a portfolio of forwards. A caplet has payoff

$$N \cdot \Delta \cdot \max\left\{L_n(T_{i-1}, T_i) - K_n, 0\right\},$$

where N is a notional amount and $L_n(T_{i-1}, T_i)$ is an interest rate, possibly LIBOR for tenor $[T_i, T_{i-1}]$, with the appropriate compounding frequency, where $\Delta = T_i - T_{i-1} = 1/n$. The time interval up to maturity T_m is partitioned in subintervals indexed by $i = 1, \ldots, m$. A cap has the effect of limiting the interest rate exposure on debt, as the derivative will

pay any interest in excess of the cap rate K_n. The payoff of a floorlet is
$$N \cdot \Delta \cdot \max \{K_n - L_n(T_{i-1}, T_i), 0\}.$$

- On the contrary, the underlying asset of bond futures and options is an actual bond. Hence, the dependence on the interest rate is mediated in a nonlinear way by the bond price, which may be also sensitive to other risk factors. Furthermore, the contract prescribes the actual delivery of the underlying asset, rather than being settled in cash.

The last point has important consequences, as the short position has to actually deliver the bond, unless the contract is closed out before maturity. This raises a set of important issues related to the risk of cornering the short positions. The contract could be written on a specific bond, say, a certain T-bond maturing in 15 years. However, especially for relatively illiquid bonds, speculators could purchase large amounts of the underlying assets before delivery, cornering the holders of short positions and forcing them to buy at high prices in order to comply with their obligations. Note that, given the variety of bonds in terms of coupon rates and maturities, cornering would be easier to carry out in this case than in equity markets.

Thus, rather than requiring a specific bond, these contracts specify a *range* of acceptable bonds. For instance, a contract may prescribe the delivery of a treasury bond that, when the futures matures, will have at least 15 years to maturity and is not callable before, say, 10 years. Clearly, such bonds may have different coupon rates and quite different prices. The short position has the possibility of choosing the **cheapest-to-deliver** bond. Since not all bonds are the same, a **conversion factor** is prescribed to convert the futures price into the bond delivery price when the contract is finally settled.

5.4 The repo market and other money market instruments

The term "money market" refers to securities and trades aimed at satisfying the need of financial and nonfinancial firms for short-term funding. Banks in need of short-term liquidity for treasury management may use the interbank market, which leads to the definition of important reference rates such as LIBOR and EURIBOR. The shortest maturity loans are related to overnight rates, which may follow dynamics that are not quite related with the term structure of interest rates for longer maturities.

A bank may also issue a **certificate of deposit**, which offers clients an interest rate for a possibly short-term deposit (say, three months). A firm may issue a **banker's acceptance**, which is a debt instrument guaranteed by a bank, issued as a part of a commercial transaction, and then possibly sold on secondary markets much like a T-bill.

These securities are fairly safe, given the short time horizon, but they cannot be considered completely risk-free, which is reflected in the interest rate underlying the transaction. One possibility that a borrower may use in order to reduce the interest rate is to provide some collateral. Good collateral is liquid and risk-free. Clearly, a corporation may offer real estate and machinery as a collateral, but we cannot really say that these are liquid assets. Now imagine a bank or a corporation holding some safe treasury bond. They could sell the bond to raise liquidity, but an alternative is to use it as a collateral. A convenient way for doing so is a **repurchase agreement**. Repurchase agreements are so common that a whole market, referred to as **repo** market, has grown. The underlying idea is fairly simple: The borrower sells a security to the lender, under the agreement to repurchase the same security at some later time for a slightly larger price. The difference between the two prices implies the payment of interest at a rate, called the **repo rate**. Since there is a collateral, the transaction is fairly safe, and the borrower may raise short-term liquidity without the need to actually liquidate assets. To be on the safe side, the lender may request a **haircut**, which is a reduction of the value of the collateral asset, to protect against possible loss on the guarantee. The effect of the haircut to the borrower is to increase the cost of raising short-term cash.

5.5 Securitization

Securitization is a way to engineer new assets by converting illiquid assets, such as a pool of mortgages, into a tradable security. The institutional arrangements are beyond the scope of the book,[3] but the general idea is that **asset-backed securities** (ABS) collect the cash flows from a pool of assets and are sold as bonds. In the specific case of a **mortgage-backed security** (MBS), there are two risks that the investor is subject to:

- The prepayment risk: If interest rates drop, the homeowner may find it convenient to terminate the old mortgage and open a new one at a lower rate. This kind of risk is similar to the reinvestment risk of a callable bond.

- The default risk, as homeowners may fail to comply with periodic payments.

Default risk can be diversified away by pooling mortgages, unless risks are strongly correlated. In normal economic conditions, defaults on mortgages are supposed to be uncorrelated. Furthermore, a default may be covered by the house itself, which is a collateral on the loan. Unfortunately, what happened during the subprime mortgage crisis is that economic downturn resulted in a large number of defaults, which turned out to be correlated. Furthermore,

---

[3] For instance, the bank has to set up a special purpose vehicle, SPV, to manage the cash flows from the pool of mortgages to owners of mortgage-backed securities.

5.5 Securitization

house prices dropped. This made default a good option to homeowners,[4] as they could default, give up the old home, and buy a new one at reduced prices. Thus, loss on the collateral was sustained. Because of securitization, losses were not sustained by reckless banks originating risky mortgages, but by the investors holding the MBSs.

It is worth emphasizing that ABSs are sometimes improperly referred to as *derivatives*. It is true that the value of such a security depends on something else, but this applies to whatever security we trade, including stock shares and bonds. ABSs should *not* be considered as derivatives, since this link is not formalized by a precise mathematical formula written into a contract.

An important concept related to securitization is **tranching**, a mechanism by which different securities, with different risks, are issued. The idea is that losses are sustained by different tranches in a well-defined sequence. To keep it simple, we may imagine that an ABS is tranched into the following three levels:

- The **equity tranche**, which is the first one to face any loss. For instance, the equity tranche may have to cover the first, say, 5% of loss. The equity tranche consists of cheap, but speculative-level securities.
- The **mezzanine tranche**, which has to sustain, say, the next 15% of loss.
- The **senior tranche**, which is supposed to be fairly safe, as it has to sustain only the loss in excess of the previous levels.

This kind of arrangement is common to other securities, like **collateralized debt obligations** (CDO), or credit default swaps (CDS), which are a credit derivative. A set of names (debtors) is pooled, and a security is created that will refund its holder in case of default. A counterintuitive feature of this kind of assets is related to default correlation.

Example 5.3 Correlation risk and tranching

> Common wisdom suggests that increasing correlation between risk sources has an adverse effect on investors. Actually, there may be somewhat paradoxical results. Imagine an ABS (or a CDO), where loss is related to default on the part of 100 debtors, and all potential losses are the same in monetary terms. The probability of default is 2% for each debtor. Consider an equity tranche liable to pay the first 5% of defaults. What is the probability of losing the whole value of the asset?
>
> If default events are both equiprobable and independent, the probability distribution of the number X of defaults is a binomial random variable, with probability 0.02 and size 100. The probability of a total

---
[4] Some homeowners were, in fact, speculators taking advantage of a prolonged period of increased real estate value.

loss for the equity tranche is

$$P\{X \geq 5\} = 1 - P\{X \leq 4\}$$
$$= 1 - \sum_{k=0}^{4} \binom{100}{k} \times 0.02^k \times 0.98^{100-k}$$
$$= 1 - (0.133 + 0.271 + 0.273 + 0.182 + 0.090)$$
$$\approx 5.08\%.$$

If correlation is increased to 1, i.e., in the case of perfect correlation, there are only two scenarios:

- A name defaults, hence, all of them do the same, with probability 2%, which is the probability of total loss.
- No one defaults, with probability 98%.

Therefore, we notice that increasing the correlation reduces the risk of losing the whole value of an equity tranche asset, contrary to common wisdom.

Remark. We notice that we are using the term "correlation" in a somewhat improper way, as the correlation coefficient refers to values of numerical random variables, whereas we are talking about correlation of events. The meaning can be made more precise, but we disregard these technical issues.

On the contrary, a senior tranche covering the last 80% of defaults is fairly safe with independent defaults, as the probability of observing more than 20 defaults is negligible. However, total loss has probability 2% in the case of perfect correlation. Hence, an increasing correlation increases risk in this case, as one would expect.

Clearly, this example is a limit case, but it shows that the effect of correlation risk may be counterintuitive.

A second example may be useful to understand the hidden risks in securitization, as well as what happened during the financial crisis in 2007–2009. Let us consider Fig. 5.3, where a second-level securitization of the original mezzanine tranche is illustrated. Since the assets in the mezzanine tranche were fairly risky, they were not easily sold to investors. This suggested the possibility of yet another level of securitization, whereby assets in the mezzanine tranche were pooled, repackaged, tranched again, and sold. As before, let us assume that in the original ABS created by the first-level securitization, the first 5% of loss is sustained by the equity tranche and the next 15% by the mezzanine. As shown in Fig. 5.3, this 15% may be further tranched in CDOs, created by

5.5 Securitization

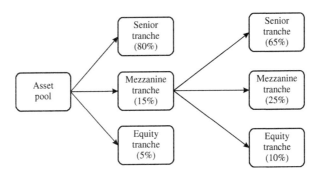

FIGURE 5.3 An illustration of second-level tranching.

a second-level securitization. Let us assume that the percentages for the three new tranches are 10%, 25%, and 65%, respectively.[5] One might assume that the senior tranche of the second-level security was fairly safe, but let us analyze a few scenarios.

1. If the loss on the original ABS is 10%, the first 5% is sustained by the equity tranche, and the second 5% is sustained by the mezzanine tranche. This 5% actually amounts to 33.3% (one third, 5% out of 15%) of the total potential loss of the original mezzanine tranche. This means that the equity tranche at the second level sustains a 10% of loss and is wiped out, and that the mezzanine tranche in the second-level CDO has to sustain $33.3\% - 10\% = 23.3\%$ of the loss deriving from the original mezzanine tranche, which amounts to $23.3/25 = 93.2\%$ of the second-level potential loss. The senior tranche of the CDO is safe in this scenario.

2. A slight increase in loss on the original assets, say, from 10% to 13%, has a significant impact. Now the original mezzanine tranche loses
$$\frac{13-5}{15} = 53.3\%$$
of its value, which means that the mezzanine tranche of the CDO is wiped out, too, while the senior tranche loses
$$\frac{53.3 - (10+25)}{65} = 28.2\%$$
of its value.

3. If the loss on the first-level ABS is 15%, the original mezzanine tranche loses $10/15 = 66.7\%$ of its value. Hence, the senior tranche of the second-level CDO loses
$$\frac{66.7 - (10+25)}{65} = 48.8\%$$
of its value.

---

[5] We use the same numbers as [2, Chapter 8], which is the basis for the treatment here.

Thus, we realize that senior tranches at the second level of securitization were actually quite risky. Despite this fact, these second-level CDOs received quite good ratings from specialized agencies, possibly due to a conflict of interest, since the agencies in charge of rating these securities were paid for this service by the investment banks originating them.

Problems

5.1 The risk-free rates with maturities of 6, 12, and 18 months are, respectively, 2.3%, 2.8%, and 3.2% (with continuous compounding). We also have the following swap rates:

Maturity	Bid	Offer
2 years	3.4	3.6
2.5 years	3.6	3.9
3 years	3.9	4.2

We assume semiannual payments, and the swap rates are semiannually compounded (this is consistent with market conventions). For the sake of simplicity, we neglect day count issues and take for granted that all months consist of 30 days. A callable bond with no default risk has coupon rate 10% and matures in two years. The bond trades for €97.12. What is the value of a call option on the corresponding noncallable bond?

5.2 Using the same logic as Problem 5.1, how could you price the put option within a puttable bond?

Further reading

- A general introduction to fixed-income markets can be found in [5], where institutional details about auction markets are also given.
- A more quantitative introduction can be found in [3] or [6].
- Conversion factors for bond derivatives are described in [2, Chapter 6].
- More information on interest rate swaps can be found, e.g., in [4].
- For an introduction to credit risk modeling see, e.g., [1].

Bibliography

1 C. Bluhm, L. Overbeck, and C. Wagner. *Introduction to Credit Risk Modeling* (2nd ed.). Chapman and Hall/CRC, Boca Raton, FL, 2010.

2 J.C. Hull. *Options, Futures, and Other Derivatives* (8th ed.). Prentice Hall, Upper Saddle River, NJ, 2011.

3 L. Martellini, P. Priaulet, and S. Priaulet. *Fixed-Income Securities: Valuation, Risk Management, and Portfolio Strategies.* Wiley, Chichester, 2003.

4 A. Sadr. *Interest Rate Swaps and Their Derivatives: A Practitioner's Guide.* Wiley, Hoboken, NJ, 2009.

5 S.M. Sundaresan. *Fixed Income Markets and Their Derivatives* (2nd ed.). South Western College Publishing, Cincinnati, OH, 2002.

6 P. Veronesi. *Fixed Income Securities: Valuation, Risk, and Risk Management.* Wiley, Hoboken, NJ, 2010.

Chapter Six

Interest Rate Risk Management

This is the final chapter of a sequence dealing with the elementary mathematics of interest rates and basic fixed-income securities, and it kind of summarizes all we have seen from Chapter 3 on. We had a taste of what interest risk is in Section 3.5.5, where we have seen that a shift in the level of interest rates may have a significant impact on the value of a bond. There, we have considered the bond price as a function of one risk factor, yield-to-maturity (YTM), and we have introduced the classical concept of duration for fixed-coupon bonds. However, this is just an approximation, since a bond price actually depends on the whole term structure of interest rates. From Section 2.2.3.3, we also know how first-order immunization against multiple risk factors may be achieved by using a set of hedging instruments and by measuring the first-order portfolio sensitivities to each risk factor. In this chapter, we analyze duration in more depth and extend it in order to deal with a broader set of securities, most notably interest rate swaps, and show in more detail how risk factor sensitivities may be used to *measure* and *manage* interest rate risk. Later, we shall need more sophisticated stochastic models in order to cope with interest rate options, which provide additional flexibility to the risk management toolkit.

The classical definition of duration has several limitations, and a more flexible one is provided in Section 6.1, allowing us to cope, e.g., with floating-rate bonds and swaps. We also introduce dollar duration to deal with securities whose value is zero. We dig deeper into the concept of duration in Section 6.2, where we interpret it as an investment time horizon and provide some connection between this sensitivity measure and proper risk measures. We deal with the application of duration to classical immunization in Section 6.3, pointing out some of its limitations. We also outline an alternative approach, based on cash flow matching, which may be further refined as a full-fledged optimization model. Then, in Section 6.4, we show how interest rate derivatives like swaps can be used to change the interest rate exposure of a fixed-income portfolio. Since duration is a first-order sensitivity measure, immunization may be refined by the introduction of a second-order sensitivity, convexity, which is defined in Section 6.5. Adding convexity to the picture may improve the quality of our risk management approaches, but it nevertheless deals with the exposure to a single risk factor. Immunization may be improved if we consider multiple risk factors, as we outline in Section 6.6. The resulting approaches rely, e.g., on factor durations, and still aim at achieving perfect immunization for small perturbations. It might be argued that it is better to be imperfectly hedged against larger perturbations, but this requires more sophisticated optimization models, like those we cover in Chapter 15.

6.1 Duration as a first-order sensitivity measure

The price of a bond depends on a whole array of interest rates, associated with maturities corresponding to cash flow times, i.e., periodic coupon payments as well as repayment of the face value at bond maturity. We denote, as usual, the continuously compounded rate at time t, with maturity T, by $r(t,T)$.[1] We use $r(t,\cdot)$ to denote the whole term structure at time t. We denote the price at time t of a generic asset depending on the term structure by $P(t, r(t,\cdot))$. This asset may be a plain bond with fixed coupons, of course, but also a more complicated bond, like a floater, or even an interest rate derivative. We will refer to this asset as a generic fixed-income security. The shape of the term structure can change in an array of ways, including a parallel shift, as well as a change in slope or curvature. Let us consider an instantaneous parallel shift δr, transforming the structure as follows:

$$r(t,\cdot) \quad \Rightarrow \quad \bar{r}(t,\cdot) = r(t,\cdot) + \delta r. \tag{6.1}$$

Let δP be the corresponding change in the price of a given fixed-income security. We define **duration** as follows, by taking the limit for $\delta r \to 0$,

$$D_P \doteq -\frac{1}{P}\frac{dP}{dr}, \quad \text{where} \quad \frac{dP}{dr} = \lim_{\delta r \to 0} \frac{\delta P}{\delta r}. \tag{6.2}$$

This immediately gives the approximate relationship

$$\frac{\delta P}{P} \approx -D_P \cdot \delta r, \tag{6.3}$$

which is just a first-order Taylor expansion.

We notice that this definition of duration is different from the classical definition of Macauley duration that we have given in Section 3.5.5. We recall the definition of Eq. (3.39) for the sake of convenience:

$$D_{\text{mac}} \doteq \frac{\sum_{t=1}^{T} \frac{tC_t}{(1+y_1)^t}}{\sum_{k=1}^{T} \frac{C_k}{(1+y_1)^k}}.$$

---

[1] In this chapter, we mostly use continuously compounded rates for the sake of mathematical convenience. When it is essential to conform to market practice, we use semiannually compounded rates.

6.1 Duration as a first-order sensitivity measure

The key point is that the definition we use here does *not* involve cash flows C_t at time t. We also recall that classical Macauley and modified durations differ because of the use of a discretely compounded yield. Here, we adopt continuously compounded rates, so that the difference disappears, and consider a parallel shift in the rates themselves, rather than a change in yield-to-maturity, which is a somewhat fictitious quantity. Nevertheless, we may easily reconcile all of the above concepts in *simple* cases, as we show below.

Let us consider a bond with cash flows C_i at times T_i, $i = 1, \ldots, m$, where T_m is the bond maturity. The bond price is

$$P(t, r(t, \cdot)) = \sum_{i=1}^{m} C_i Z(t, T_i) = \sum_{i=1}^{m} C_i e^{-(T_i - t) \cdot r(t, T_i)},$$

where the discount factors $Z(t, T_i)$ are expressed as a function of continuously compounded rates. Given the shift of Eq. (6.1), the bond price will change to

$$P(t, \bar{r}(t, \cdot)) = \sum_{i=1}^{m} C_i e^{-(T_i - t) \cdot [r(t, T_i) + \delta r]}.$$

Let us consider the bond price as a function of a single variable, $s \equiv \delta r$, corresponding to the shift:

$$F(s) \doteq \sum_{i=1}^{m} C_i e^{-(T_i - t) \cdot [r(t, T_i) + s]}.$$

Since we are interested in a small shift $\delta r \to 0$, let us evaluate the derivative of the bond price with respect to s, for $s = 0$:

$$\left. \frac{dF}{ds} \right|_{s=0} = \sum_{i=1}^{m} \left[-(T_i - t) C_i e^{-(T_i - t) \cdot [r(t, T_i) + s]} \right] \bigg|_{s=0}$$

$$= -\sum_{i=1}^{m} (T_i - t) \cdot C_i \cdot Z(t, T_i)$$

If we divide the last expression by $-P$, we may see that the definition of Eq. (6.2) is consistent with the definition of Macauley duration, which in this case is

$$D_{\mathsf{mac}} = \frac{\sum_{i=1}^{m} (T_i - t) \cdot C_i \cdot Z(t, T_i)}{\sum_{i=1}^{m} C_i \cdot Z(t, T_i)}.$$

However, there are a few critical differences that we may summarize as follows:

- As we have already noted, the classical definition of Macauley duration differs from the definition of modified duration in the case of discrete compounding, possibly leading to confusion. The use of continuously compounded rates simplifies all of the matter considerably[2] and is more consistent with the continuous-time stochastic models that we will have to introduce in order to cope with interest rate options.

- Usually, duration is defined with respect to yield-to-maturity, whereas here we make a connection with *one* possible change in the term structure. Duration is a single-factor sensitivity measure, but we may have to cope with multiple factors, or with more complex changes in the term structure. The new definition makes all of this more explicit.
- Defining duration in terms of cash flows is not feasible, when these are stochastic, as in the case of a floating-rate bond. However, if we define duration directly as a sensitivity of the bond price, we can apply it to fixed-income assets featuring stochastic cash flows, too. An even less obvious point is: What if the bond is callable and the implicit option is exercised, so that the bond does not get to mature?

The last observations are essential if we want to broaden the range of fixed-income securities that we use in interest rate risk management, in order to include interest rate derivatives. A notable example is an interest rate swap. The classical definition of duration cannot be applied to swaps, since cash flows are not deterministic. Furthermore, neither the definition in Eq. (6.2) can be applied, as it may involve division by zero. Indeed, at its inception, the value of a swap is zero, which makes the above definition useless. This issue will be solved by introducing the concept of dollar duration. In the rest of the section, we find the duration of simple securities, as well as the dollar duration of vanilla swaps.

6.1.1 DURATION OF FIXED-COUPON BONDS

Let us find the duration of a zero-coupon bond with face value $F = 100$. Its price is
$$P_z(t, r; T) = 100 \cdot Z(t, T) = 100 \cdot e^{-r(t,T) \cdot (T-t)},$$
where we may use the streamlined notation $P_z(t, r; T)$ to denote its price, which depends only on the rate for a single maturity, $r(t, T)$, which we may just denote by r for the sake of simplicity. The first-order derivative with respect to r is
$$\frac{dP_z}{dr} = 100 \cdot \left[-(T-t) \cdot e^{-r \cdot (T-t)} \right] = -(T-t) \cdot P_z(t, T),$$
which implies
$$D_z = -\frac{1}{P_z} \frac{dP_z}{dr} = (T-t).$$

---

[2] Also the expression of forward rates as a function of spot rates is simpler with continuous compounding.

6.1 Duration as a first-order sensitivity measure

Hence, the duration of a zero is just its time-to-maturity, which is coherent with the classical definition, if we use continuous compounding. Let us reconsider the kind of calculations that we carried out in Example 3.14.

Example 6.1 A numerical check

Let us check the accuracy of the approximation in Eq. (6.3) for a couple of zeros maturing in 3 and 20 years, respectively. If the continuously compounded yield r is 3%, we have

$$P_z(0, 0.03; 3) = 100 \times e^{-0.03 \times 3} = 91.39312,$$
$$P_z(0, 0.03; 20) = 100 \times e^{-0.03 \times 20} = 54.88116.$$

Note that, in practice, prices should be rounded to cents, but we refrain from doing so, in order to better illustrate numerical accuracy. If there is an upshift by ten basis points,

$$P_z(0, 0.031; 3) = 100 \times e^{-0.031 \times 3} = 91.11935,$$
$$P_z(0, 0.031; 20) = 100 \times e^{-0.031 \times 20} = 53.79444.$$

The approximate prices $\widehat{P}_z$ predicted by the duration approximation are:

$$\widehat{P}_z(0, 0.031; 3) = P_z(0, 0.03; 3) \cdot (1 - 3 \times 0.001) = 91.11894,$$
$$\widehat{P}_z(0, 0.031; 20) = P_z(0, 0.03; 20) \cdot (1 - 20 \times 0.001) = 53.78354.$$

We observe that the approximation seems a bit less accurate for the longer maturity zero, and that the duration approximation is somewhat "pessimistic," in the sense that it gives a lower price than exact repricing. A similar pattern occurs if there is a downshift by ten basis points:

$$P_z(0, 0.029; 3) = 100 \times e^{-0.029 \times 3} = 91.66771,$$
$$P_z(0, 0.029; 20) = 100 \times e^{-0.029 \times 20} = 55.98984,$$
$$\widehat{P}_z(0, 0.029; 3) = P_z(0, 0.03; 3) \cdot (1 + 3 \times 0.001) = 91.6673,$$
$$\widehat{P}_z(0, 0.029; 20) = P_z(0, 0.03; 20) \cdot (1 + 20 \times 0.001) = 55.97879.$$

Again, we observe that duration predicts a lower price than the exact one. When a larger shift occurs, accuracy is less impressive. For instance, if there is an increase by 100 basis points, we have

$$P_z(0, 0.04; 3) = 100 \times e^{-0.04 \times 3} = 88.69204,$$
$$P_z(0, 0.04; 20) = 100 \times e^{-0.04 \times 20} = 44.9329,$$

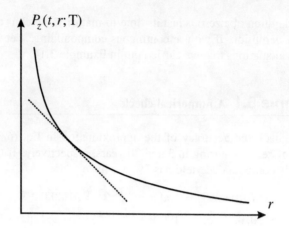

FIGURE 6.1 The price $P_z(t,T;r) = F \cdot \exp\left(-r \cdot (T-t)\right)$ of a zero is a convex function of the rate r, and it is globally underestimated by the tangent line at any point.

but the first-order approximation based on duration yields

$$\widehat{P}_z(0, 0.04; 3) = P_z(0, 0.03; 3) \cdot (1 - 3 \times 0.01) = 88.65132,$$
$$\widehat{P}_z(0, 0.04; 20) = P_z(0, 0.03; 20) \cdot (1 - 20 \times 0.01) = 43.90493.$$

The last approximation, in particular, is rather inaccurate and overly pessimistic, with a percentage error of

$$\frac{43.90493 - 44.9329}{44.9329} = -2.28\%.$$

In Example 6.1, we have observed that the linear approximation based on duration is pessimistic, i.e., it underestimates the actual bond price. Actually, this is not quite surprising, since the relationship between the risk factor and the bond price involves an exponential, which is a (differentiable) convex function and is globally underestimated by the tangent line at any point (see Fig. 6.1). More generally, the price–yield relationship is convex. We may better account for this nonlinearity by introducing bond convexity, as we shall see later, which involves a second-order approximation.

To find the duration of a coupon-bearing bond, whose price will be denoted by $P_c(t, r(t, \cdot))$, it may be useful to decompose it into a portfolio of zeros and take advantage of the linearity of derivative as an operator. Let us consider a portfolio consisting of N_1 bonds with price P_1 and N_2 bonds of price P_2. The portfolio value is

$$W = N_1 \cdot P_1 + N_2 \cdot P_2.$$

6.1 Duration as a first-order sensitivity measure

The duration of the portfolio is just

$$\begin{aligned}
D_W &= -\frac{1}{W}\frac{dW}{dr} \\
&= -\frac{1}{W}\left[N_1\frac{dP_1}{dr} + N_2\frac{dP_2}{dr}\right] \\
&= \frac{1}{W}\left[N_1 P_1 \cdot \left(-\frac{1}{P_1}\frac{dP_1}{dr}\right) + N_2 P_2 \cdot \left(-\frac{1}{P_2}\frac{dP_2}{dr}\right)\right] \\
&= \frac{N_1 P_1}{W}D_1 + \frac{N_2 P_2}{W}D_2 \\
&= w_1 D_1 + w_2 D_2,
\end{aligned} \qquad (6.4)$$

where D_1, D_2 are the two bond durations and w_1, w_2 are the weights of the two bonds in the portfolio. Thus, we observe that the duration of a bond portfolio is just the weighted combination of the individual bond durations. This can be easily generalized to a portfolio consisting of any number of bonds.

If we decompose a coupon bond maturing at T into a portfolio of m zeros maturing at times T_i, $i = 1, \ldots, m$, where $T_m = T$, we have

$$P_c\bigl(t; r(t, \cdot)\bigr) = \sum_{i=1}^{m-1} \frac{c}{2} P_z(t, T_i) + \left(1 + \frac{c}{2}\right) P_z(t, T_m),$$

which may be considered as a portfolio of m zeros with weights

$$\begin{aligned}
w_i &= \frac{\frac{c}{2} \cdot P_z(t, T_i)}{P_c\bigl(t; r(t, \cdot)\bigr)}, \qquad i = 1, \ldots, m-1, \\
w_m &= \frac{\left(1 + \frac{c}{2}\right) \cdot P_z(t, T_m)}{P_c\bigl(t; r(t, \cdot)\bigr)}.
\end{aligned}$$

Thus, the duration of a coupon-bearing bond is

$$\begin{aligned}
D_c &= \sum_{i=1}^{m} w_i \cdot (T_i - t) \\
&= \frac{1}{P_c(t; r(t, \cdot))} \cdot \left[\sum_{i=1}^{m-1} \frac{c}{2} P_z(t, T_i) \cdot (T_i - t) + \left(1 + \frac{c}{2}\right) P_z(t, T_m) \cdot (T_m - t)\right],
\end{aligned}$$

which is again consistent with the classical definition.

Example 6.2 Duration of a coupon-bearing bond

Consider a bond maturing in 18 months, paying semiannual coupons at annual rate 5%, with face value $10,000. The term structure is flat, and the continuously compounded rate is 3%. The current bond price is

$$P_c = 250 \times e^{-0.5 \times 0.03} + 250 \times e^{-1 \times 0.03} + 10{,}250 \times e^{-1.5 \times 0.03}$$
$$= 10{,}287.86.$$

Its duration is

$$D_P = \frac{1}{10{,}287.86} \times \left(0.5 \times 250 \times e^{-0.5 \times 0.03} + 1 \times 250 \times e^{-1 \times 0.03} \right.$$
$$\left. + 1.5 \times 10{,}250 \times e^{-1.5 \times 0.03}\right) = 1.4643.$$

We note that duration is fairly close to bond maturity, since there is a large terminal cash flow with respect to coupon payments.

6.1.2 DURATION OF A FLOATER

In Section 3.5.6, we have learned how pricing of a floating-rate bond is surprisingly simple. We need the following information, assuming semiannual coupons:

- The interest rate at the last reset time, which determines the amount of the next coupon payment.
- The time of the next reset time, when the next coupon will be paid.
- The current interest rate for a maturity corresponding to the next reset time.

If we denote the last and the next reset times by T_{i-1} and T_i, respectively, the next coupon payment will be

$$C_i = F \cdot \frac{r_2(T_{i-1}, T_i)}{2},$$

where we use a semiannually compounded rate to conform to market practice. We assume that T_i is not the bond maturity, as in that case no future uncertain cash flow is involved. Since we know that the bond trades at par at the reset times, the floating bond price is

$$P_f(t; r(t, \cdot)) = (C_i + F) \cdot e^{-(T_i - t) \cdot r(t, T_i)}.$$

Essentially, this is the price of a zero maturing at the next reset time. Thus, duration is just

$$D_f = T_i - t.$$

6.1 Duration as a first-order sensitivity measure

We note that duration does not depend on bond maturity, and that the duration of a floater is smaller than the duration of a fixed-coupon bond with the same maturity. Hence, the interest rate risk of a floating-rate bond is rather limited (see Example 3.15).

6.1.3 DOLLAR DURATION AND INTEREST RATE SWAPS

We have seen, in Section 4.4, that the value of a vanilla interest rate swap can be expressed as the difference between the prices of a fixed- and a floating-rate bonds. Hence, we might argue that the duration of a swap is related to the difference of the two bond durations. Unfortunately, this idea does not work in general, as the value of an interest rate swap may well be zero. By a similar token, we may hold a bond portfolio with long and short positions, possibly amounting to a current value of zero.

To overcome this issue, we define the **dollar duration** as

$$D_P^\$ \doteq -\frac{dP}{dr}. \tag{6.5}$$

Clearly, if the value P of the security (or portfolio of securities) is not zero, duration and dollar duration are related as follows:

$$D_P^\$ = P \cdot D_P.$$

For instance, in the case of a zero,

$$\frac{dP_z}{dr} = \frac{d}{dr}\left[F \cdot e^{-r \cdot (T-t)}\right] = -(T-t) \cdot F \cdot e^{-r \cdot (T-t)},$$

and so

$$D_z^\$ = P_z \cdot (T-t).$$

Assume that we hold a portfolio of fixed-income securities, indexed by $i = 1, \ldots, m$, and let N_i and $D_i^\$$ denote the number of units and the dollar duration of each security, respectively. Then, using linearity of the derivative operator again, it is easy to see that the dollar duration of the portfolio is

$$D_P^\$ = \sum_{i=1}^{m} N_i \cdot D_i^\$.$$

An immediate consequence is that the dollar duration of a swap is

$$D_{\text{swap}}^\$ = D_{\text{float}}^\$ - D_{\text{fixed}}^\$$$

for the fixed-rate payer, and

$$D_{\text{swap}}^\$ = D_{\text{fixed}}^\$ - D_{\text{float}}^\$$$

for the floating-rate payer. Since the duration of a floater is smaller than a fixed-coupon bond with corresponding maturity, we notice that the dollar duration for the fixed-rate payer will be typically negative. Hence, by entering into a swap agreement, an investor may reduce the duration of a fixed-income portfolio, without the need of additional capital, since the initial value of the swap is zero. Another advantage of swaps is that they relieve us from the possibly difficult task of taking extended short positions in bonds.

Example 6.3 Dollar duration of a swap

Let us consider a swap with the following features:

- Nominal amount, $100,000
- Maturity, 14 months, so that cash flows will occur in 2, 8, and 14 months
- Swap rate, 3.4% with semiannual compounding, so that the fixed payment is

$$100{,}000 \times \frac{0.034}{2} = \$1700$$

The current term structure, with continuous compounding, is

$$r(0, 2/12) = 3\%, \quad r(0, 8/12) = 3.5\%, \quad r(0, 14/12) = 4\%.$$

At the last reset time, four months ago, the six-month rate was 3.8% with semiannual compounding, so that the next floating payment will be

$$100{,}000 \times \frac{0.038}{2} = \$1900.$$

The value of the fixed-rate bond is

$$P_{\text{fixed}} = 1700 \times e^{-0.03 \times 2/12} + 1700 \times e^{-0.035 \times 8/12}$$
$$+ 101{,}700 \times e^{-0.04 \times 14/12} = \$100{,}415.35,$$

and its dollar duration is

$$D^{\$}_{\text{fixed}} = \frac{2}{12} \times 1700 \times e^{-0.03 \times 2/12} + \frac{8}{12} \times 1700 \times e^{-0.035 \times 8/12}$$
$$+ \frac{14}{12} \times 101{,}700 \times e^{-0.04 \times 14/12} = \$114{,}629.33.$$

The value of the floating-rate bond is

$$P_{\text{float}} = 101{,}900 \times e^{-0.03 \times 2/12} = \$101{,}391.77,$$

and its dollar duration is

$$D^{\$}_{\text{float}} = \frac{2}{12} \times 101{,}900 \times e^{-0.03 \times 2/12} = \$16{,}898.63.$$

> Hence, taking the viewpoint of the fixed payer, the value of the swap is
> $$P_{\text{swap}} = P_{\text{float}} - P_{\text{fixed}} = \$976.42,$$
> and its dollar duration is
> $$D^\$_{\text{swap}} = D^\$_{\text{float}} - D^\$_{\text{fixed}} = -\$97{,}730.70.$$
> Let us consider an instantaneous parallel upshift of ten basis points in the term structure. According to the dollar duration approximation, the new swap value can be approximated by
> $$\widehat{P}_{\text{swap}} \approx P_{\text{swap}} - D^\$_{\text{swap}} \cdot 0.001 = \$1074.15.$$
> If we reprice the swap exactly, we find the following values after the shift:
> $$\bar{P}_{\text{fixed}} = 1700 \times e^{-0.031 \times 2/12} + 1700 \times e^{-0.036 \times 8/12}$$
> $$+ 101{,}700 \times e^{-0.041 \times 14/12} = \$100{,}300.79,$$
> $$\bar{P}_{\text{float}} = 101{,}900 \times e^{-0.031 \times 2/12} = \$101{,}374.87,$$
> $$\bar{P}_{\text{swap}} = \bar{P}_{\text{float}} - \bar{P}_{\text{fixed}} = \$1074.09,$$
> showing the accuracy of the approximation for a small shift.
>
> We should note that, unlike plain bonds, the value of the swap may be increased by a rise in the the interest rates, which shows the potential of swaps for hedging interest rate risk. The increase in the swap value is slightly overestimated by the first-order approximation, whereas the *loss* on a bond is overestimated when using this kind of approximation. The explanation is provided by convexity vs. concavity (sometimes referred to as negative convexity) of the nonlinear functions that we are approximating to the first order.

A related concept is the **price value of a basis point**, denoted as PV01 or PVBP, which is the dollar loss for a perturbation $\delta r = 0.0001$, i.e., an increase of one basis point, which is just

$$\text{PV01} = -D^\$_P \times 0.0001.$$

6.2 Further interpretations of duration

A deeper understanding of the meaning of duration, as well as its limitations, may be acquired by linking it with interest rate risk in different ways. In this section, we disregard the full term structure and consider yield-to-maturity (annually or continuously compounded, according to convenience) as the single risk factor. This clearly limits the practical applicability of our findings, but not their conceptual usefulness.

6.2.1 DURATION AND INVESTMENT HORIZONS

Consider a plain zero maturing in five years, which is the bond duration, too. If our investment horizon is exactly five years, we do not care about a possible increase in yield. The bond price will drop if yield rises, but we will just recover the loss under the increased yield, and we will receive the face value anyway, at maturity. This is not true if the investment horizon is smaller than the bond duration, i.e., if we plan to sell the zero before its maturity. Now what if we are considering a coupon-bearing bond? In this case, we know that an increase in yield implies a drop in the bond price, but this may be somehow mitigated by the possibility of reinvesting coupons at a larger rate. The contrary happens when yield drops: We have a welcome capital gain, but we also suffer from coupon reinvestment risk. In both cases, we have contrasting effects, and the final outcome may also depend on our investment horizon.

As it turns out, Macauley duration is an investment horizon such that we are indifferent to small changes in yield. To see this, let us consider a bond maturing at time T, with price $P_c(y_1)$ depending on the annually compounded yield y_1. Say that we plan to hold the bond for a time period of length $H < T$. If the yield does not change, and we reinvest coupons at a rate y_1 and sell the bond at $t = H$, wealth will be

$$P_c(y_1) \cdot (1 + y_1)^H. \tag{6.6}$$

If yield changes instantaneously by an amount δy_1 and then remains constant, wealth at time $t = H$ will be

$$P_c(y_1 + \delta_1) \cdot (1 + y_1 + \delta y_1)^H. \tag{6.7}$$

By equating Eqs. (6.6) and (6.7), we may find the indifference horizon H. This is much easier if we take logarithms:

$$\log P_c(y_1) + H \cdot \log(1 + y_1) = \log P_c(y_1 + \delta_1) + H \cdot \log(1 + y_1 + \delta y_1).$$

Let us rearrange this equation and divide by δy_1:

$$\frac{\log P_c(y_1 + \delta_1) - \log P_c(y_1)}{\delta y_1} = -H \cdot \frac{\log(1 + y_1 + \delta_1) - \log(1 + y_1)}{\delta y_1}.$$

For a small perturbation δy_1, we may replace these ratios by the derivative of a logarithm. We recall that, by using the chain rule for the derivative of a composite function, we find

$$\frac{d \log f(x)}{dx} = \frac{f'(x)}{f(x)}.$$

6.2 Further interpretations of duration

Hence, we obtain

$$\frac{1}{P_c(y_1)} \cdot P_c'(y_1) = -H \cdot \frac{1}{1+y_1},$$

where the right-hand side includes the ratio between the derivative of the function $1+y_1$, which is just 1, and the function itself. A slight rearrangement shows that H is just the Macauley duration:

$$H = -(1+y_1) \cdot \frac{1}{P_c(y_1)} \cdot P_c'(y_1) \equiv D_{\text{mac}}.$$

Let us illustrate this finding with a simple example.

Example 6.4 Duration and investment horizons

Let us consider a bond with face value $10,000, maturing in three years, and paying an annual coupon at rate 6%. If annual yield is 4%, the bond price is

$$P_c(4\%) = \frac{600}{1.04} + \frac{600}{1.04^2} + \frac{10{,}600}{1.04^3} = \$10{,}555.02,$$

and its Macauley duration is

$$D_{\text{mac}} = \frac{1}{10{,}555.02} \times \left[1 \times \frac{600}{1.04} + 2 \times \frac{600}{1.04^2} + 3 \times \frac{10{,}600}{1.04^3} \right] = 2.84.$$

If yield does not change, we reinvest coupons at 4%, and sell the bond at time $H = 2.84$, wealth will be

$$W_H(4\%) = 600 \times 1.04^{2.84-1} + 600 \times 1.04^{2.84-2} + \frac{10{,}600}{1.04^{3-2.84}}$$
$$= \$11{,}797.82.$$

To understand this expression, note that the first two cash flows are reinvested up to time $t = 2.84$, whereas the third cash flow is discounted from time $t = 3$ to $t = 2.84$. If yield is increased by 50 basis points, wealth will be

$$W_H(4.5\%) = 600 \times 1.045^{2.84-1} + 600 \times 1.045^{2.84-2} + \frac{10{,}600}{1.045^{3-2.84}}$$
$$= \$11{,}797.85.$$

Indeed, up to an approximation error, future wealth at the right time horizon is insensitive to small changes in yield.

6.2.2 DURATION AND YIELD VOLATILITY

Duration may be considered a measure of interest rate sensitivity for fixed-income securities, but not a risk measure, unless we link it with an uncertainty model. If we consider the continuously compounded yield y as a risk factor, then the approximate relationship

$$\frac{\delta P}{P} \approx -D \cdot \delta y$$

implies that the standard deviation of bond return σ is proportional to the standard deviation of yield:

$$\text{Var}\left(\frac{\delta P}{P}\right) \approx D^2 \cdot \text{Var}(\delta y) \quad \Rightarrow \quad \sigma \approx D\sigma_y,$$

where σ_y is the standard deviation of the perturbation δy, i.e., yield volatility. Thus, duration is a factor determining volatility, which is a symmetric risk measure. In general, when we have a short time horizon and a low risk tolerance, a wise suggestion is to invest in short-duration money market securities.

6.2.3 DURATION AND QUANTILE-BASED RISK MEASURES

In order to overcome some limitations of volatility, which is a symmetric risk measure, we have introduced quantile-based risk measures like value-at-risk (V@R) in Section 2.2.2.[3] Duration may be used to find an approximation of such measures. The idea is pretty simple, if we assume a one-factor risk model under a normal distribution. If $\delta y \sim \mathsf{N}(\mu_y, \sigma_y^2)$, from the relationship

$$\delta P \approx -P \cdot D_P \cdot \delta y,$$

we immediately find that the loss $L_P = -\delta P$ is normally distributed, too, with parameters

$$\mu_L = P \cdot D_P \cdot \mu_y, \qquad \sigma_L = P \cdot D_P \cdot \sigma_y.$$

Example 6.5 Bond V@R

> Consider a bond maturing in 18 months, paying semiannual coupons at rate 5%, with face value $10,000. The term structure is flat, so we identify interest rate r and yield y, which is 3% with continuous compounding. We assume that this rate is subject to an instantaneous shock, characterized by a normal distribution with $\mu_r = 0\%$ and $\sigma_r = 0.5\%$, and we want to find V@R at confidence level 99%.

---

[3] See Section 7.4 for a more in-depth treatment.

> The current bond price is
>
> $$P_c = 250 \times e^{-0.5 \times 0.03} + 250 \times e^{-1 \times 0.03} + 10{,}250 \times e^{-1.5 \times 0.03}$$
> $$= \$10{,}287.86.$$
>
> Its duration is
>
> $$D_P = \Big(0.5 \times 250 \times e^{-0.5 \times 0.03} + 1 \times 250 \times e^{-1 \times 0.03}$$
> $$+ 1.5 \times 10{,}250 \times e^{-1.5 \times 0.03}\Big)/10{,}287.86$$
> $$= 1.4643.$$
>
> The approximated V@R is given by
>
> $$z_{0.99} \cdot \sigma_p \cdot P \cdot D_P = 2.3263 \times 0.005 \times 10{,}287.86 \times 1.4643$$
> $$= \$175.22.$$
>
> The exact V@R, in this simple case, can be found by just repricing the bond with the worst rate at 99% confidence level,
>
> $$r_{\text{worst99\%}} = 0.03 + 2.3263 \times 0.005 = 0.0416,$$
>
> which gives
>
> $$P_{c,\text{worst99\%}} = 250 \times e^{-0.5 \times 0.0416} + 250 \times e^{-1 \times 0.0416}$$
> $$+ 10{,}250 \times e^{-1.5 \times 0.0416} = \$10{,}114.14.$$
>
> Hence, the exact value-at-risk is
>
> $$\text{V@R} = 10{,}287.86 - 10{,}114.14 = \$173.72$$
>
> which is fairly close to the approximation.

We notice that the duration approximation does a good job in Example 6.5, where the estimate is slightly pessimistic because of a convexity effect. The actual problem lies in the normal approximation itself, which is pretty debatable. We should also notice that we have not considered the effect of time, as we have assumed an instantaneous shock.

6.3 Classical duration-based immunization

Duration is not only a way to measure interest rate sensitivity, but also a tool to immunize a portfolio against interest rate risk. The idea is just an application of the general approach introduced in Section 2.2.3.3, and it is best illustrated in the context of asset–liability management (ALM) problems. For the sake of simplicity, we consider a stream of deterministic liabilities L_j, to be paid at times T_j, $j = 1, \ldots, m$. On the asset side, we have to choose a fixed-income portfolio able to provide cash flows sufficient to cover the liabilities. This kind of problem, with some more realistic twists, is common to insurance companies and pension funds, among others. In this section, we pursue two ideas for setting up the asset portfolio:

- Matching the whole stream cash outflows over time.
- Matching the present value of assets and liabilities, making sure that some immunization guarantee is met.

6.3.1 CASH FLOW MATCHING

A portfolio exactly matching the liabilities would be easy to build if we had a rich set of zeros, with maturities corresponding to the time instants at which liabilities are to be met. Denoting by N_j the holding of each zero maturing at time T_j, with face value F_j, we would just set

$$N_j = \frac{L_j}{F_j}, \qquad j = 1, \ldots, m,$$

assuming asset divisibility. The current value of the asset portfolio would just match the present value of the liabilities:

$$\mathrm{PV}_A\big(r(0,\cdot)\big) \doteq \sum_{j=1}^{m} N_j F_j \cdot Z(0, T_j) = \sum_{j=1}^{m} L_j \cdot Z(0, T_j) \doteq \mathrm{PV}_L\big(r(0,\cdot)\big),$$

where we use discount factors $Z(0, T_j)$ associated with the current term structure $r(0,\cdot)$. Assuming that bonds are free from default risk, any change in the term structure would be irrelevant.

In practice, such a rich set of zeros, including long maturities, may not be available, or it may be expensive, not to mention default risk on the long term. Hence, we could consider a set of n bonds, possibly coupon-bearing ones, with current prices P_i, $i = 1, \ldots, n$, paying at time j a cash flow denoted by F_{ij}. Let us assume, for the sake of simplicity, that bond cash flows and liabilities are synchronized. Then, we may consider the following **cash flow matching** model:

$$\min \quad \sum_{i=1}^{n} P_i x_i$$

$$\text{s.t.} \quad \sum_{i=1}^{n} F_{ij} x_i \geq L_j \qquad j = 1, \ldots, m$$

$$x_i \geq 0,$$

6.3 Classical duration-based immunization

where x_i is the amount of each bond that we buy. This is a simple linear programming model, which may be easily solved by commercial software. A little thought, however, suggests that this model is too naive to be of any practical use:

- The cash flow timings need not be perfectly synchronized, and we should manage liquidity by short-term investing or borrowing.
- The model is static and does not consider the possibility of buying or selling bonds along the way.
- The solution is likely to be too expensive, as the approach is essentially based on a superhedging strategy, where any cash surplus is of no use.

A more sophisticated modeling approach, allowing for dynamic trading, should be considered, as we shall discuss in Chapter 15. Unfortunately, this model should account for the stochastic nature of interest rates and bond prices, and this may result in a quite challenging stochastic programming problem. A more down-to-earth idea is to rely on first-order immunization.

6.3.2 DURATION MATCHING

In asset–liability management, the first and foremost concern is *solvency*, which is to say that equity,

$$E \doteq \mathrm{PV}_A\big(r(0,\cdot)\big) - \mathrm{PV}_L\big(r(0,\cdot)\big),$$

should never be negative. In insurance management, some safety buffer (technically speaking, a reserve) is maintained, so that equity is strictly positive. For the sake of simplicity, let us assume that we wish to find a minimum cost portfolio, where equity is zero, and that the only relevant risk factor is the annually compounded yield y_1. Hence, we are satisfied when

$$\mathrm{PV}_A(y_1) = \mathrm{PV}_L(y_1). \tag{6.8}$$

Unfortunately, a change in yield may have a different impact on assets and liabilities, so that equity may get negative. If there is a *small* change δy_1 in yield, solvency will not be affected if

$$\frac{\delta \mathrm{PV}_A}{\delta y_1} = \frac{\delta \mathrm{PV}_L}{\delta y_1}, \tag{6.9}$$

which means that the two dollar durations are the same. Using the relationships between dollar duration, duration, and classical Macauley duration, we find:

$$D_A^\$ = D_L^\$ \;\Rightarrow\; \mathrm{PV}_A(y_1) \cdot D_A = \mathrm{PV}_L(y_1) \cdot D_L$$
$$\Rightarrow\; \mathrm{PV}_A(y_1) \cdot \frac{D_{A,\mathrm{mac}}}{1+y_1} = \mathrm{PV}_L(y_1) \cdot \frac{D_{L,\mathrm{mac}}}{1+y_1},$$

which, given Eq. (6.8), boils down to saying that the two Macauley durations must be the same. This is also consistent with the interpretation of Macauley duration as an investment horizon such that small changes in yield have no effect on terminal wealth.

In order to match $D_{L,\text{mac}}$, we may use two bonds, with prices P_1 and P_2, and Macauley durations $D_{1,\text{mac}}$ and $D_{2,\text{mac}}$, respectively. In order to match the value of assets and liabilities, we must hold bond amounts N_1 and N_2, such that

$$N_1 \cdot P_1(y_1) + N_2 \cdot P_2(y_1) = \text{PV}_L(y_1).$$

By dividing both sides of the equality by $\text{PV}_A(y_1)$ and recalling that the duration of a bond portfolio is the weighted average of the individual durations,[4] we end up with the following system of linear equations:

$$w_1 + w_2 = 1$$
$$w_1 \cdot D_{1,\text{mac}} + w_2 \cdot D_{2,\text{mac}} = D_{L,\text{mac}},$$

where w_1 and w_2 are the weights of the two bonds in the portfolio. Using the substitution $w_2 = 1 - w_1$, the second equation immediately gives

$$w_1 = \frac{D_{L,\text{mac}} - D_{2,\text{mac}}}{D_{1,\text{mac}} - D_{2,\text{mac}}}.$$

Expressing the whole thing in terms of bond holdings, we finally obtain

$$N_1 = \frac{\text{PV}_A}{P_1} \cdot \frac{D_{L,\text{mac}} - D_{2,\text{mac}}}{D_{1,\text{mac}} - D_{2,\text{mac}}}, \quad N_2 = \frac{\text{PV}_A}{P_2} \cdot \frac{D_{1,\text{mac}} - D_{L,\text{mac}}}{D_{1,\text{mac}} - D_{2,\text{mac}}}. \quad (6.10)$$

This classical approach looks simple enough, but we should mention a few difficulties:

- In principle, we might find a negative bond holding, which means that we should sell a bond short. This is not easily arranged for an extended period of time.
- The duration will change over time, and we need to periodically rebalance the asset portfolio, incurring transaction costs.
- Another reason to rebalance the asset portfolio is that some bonds will mature along the way. If we want to bracket a target duration with positive weights, we shall need a bond with a smaller duration and another bond with a larger duration than the liability, which means that a bond will mature early and will need to be replaced.
- We are only immunized to the first order against a single risk factor (or, equivalently, only against small parallel shifts in the term structure).

These difficulties may be eased by using more than two bonds, and by extensions that we discuss in the following: (a) taking advantage of the flexibility of interest rate derivatives; (b) introducing second-order sensitivities (bond convexity); (c) introducing multifactor models.

---

[4]See Eq. (6.4).

6.4 Immunization by interest rate derivatives

Using bonds as hedging instruments has some disadvantages in terms of cost, liquidity, and limits to short positions. In this section, we apply the generic framework introduced in Section 2.2.3.3 to interest rate risk immunization by simple interest rate derivatives. First-order immunization relies on duration or dollar duration. We consider here one risk factor, which may be a parallel shift in the term structure or a change in YTM. We should note that, when we consider YTM, different assets will have different yields, but they are affected in the same way by a parallel shift.

Let us consider a fixed-income portfolio, whose value at time t is $P(t; r(t, \cdot))$, where we emphasize dependence on the full term structure $r(t, \cdot)$ at time t. If we include ϕ units of a hedging instrument H, the value of the hedged portfolio is

$$P_h\big(t; r(t, \cdot)\big) = P\big(t; r(t, \cdot)\big) + \phi H\big(t; r(t, \cdot)\big).$$

Given an instantaneous shock δr, we find

$$\delta P_h \doteq P_h\big(t; r(t, \cdot) + \delta r\big) - P_h\big(t; r(t, \cdot)\big) \approx \frac{dP}{dr} \delta r + \phi \frac{dH}{dr} \delta r.$$

By setting $\delta P_h = 0$, we find the hedging ratio ϕ in terms of durations,

$$D_P \cdot P \cdot \delta r + \phi D_H \cdot H \cdot \delta r = 0 \quad \Rightarrow \quad \phi = -\frac{D_P \cdot P}{D_H \cdot H}, \quad (6.11)$$

which works with bonds, but not with a swap or a futures with initial value $H = 0$. In terms of dollar duration we have

$$D_P^\$ \cdot \delta r + \phi D_H^\$ \cdot \delta r = 0 \quad \Rightarrow \quad \phi = -\frac{D_P^\$}{D_H^\$}. \quad (6.12)$$

▣ Example 6.6 Hedging interest rate risk with bond futures

> Futures contracts that are sensitive to interest rates may be used as hedging instruments against interest rate risk. However, if we use *bond* futures, a complication arises, as we should consider the duration $D_{\text{ctd}}^\$$ of the bond that is likely to be actually delivered (the cheapest-to-deliver bond that we mentioned in Section 5.3.2) as well as its conversion factor CF. In this case, Eq. (6.12) reads
>
> $$\phi = -\frac{D_P^\$}{D_{\text{ctd}}^\$} \cdot \text{CF}.$$
>
> Using eurodollar futures or similar contracts avoids this trouble. In any case, hedging over a long time horizon using future contracts may require tailing the hedge, in order to account for daily marking-to-market.

6.4.1 USING INTEREST RATE SWAPS IN ASSET–LIABILITY MANAGEMENT

As we have seen in Section 6.3.2, a concern in asset–liability management problems is to keep equity,

$$E = A - L,$$

at a safe level. Interest rate swaps are quite convenient, since they do not require initial cash outlays and allow to modify the interest rate risk exposure in a very flexible way by taking the fixed- or floating-payer positions with a range of maturities.

Let us denote the dollar duration of assets and liabilities by $D_A^\$$ and $D_L^\$$, respectively. If we do not hedge, the dollar duration of equity,

$$D_E^\$ = D_A^\$ - D_L^\$,$$

may be far from zero, exposing us to interest rate risk. As we have seen in Section 6.1.3, the dollar duration of a swap is

$$D_{\text{swap}}^\$ = D_{\text{float}}^\$ - D_{\text{fixed}}^\$.$$

The exact sign of the difference does not matter, as we may easily reverse the swap by taking the opposite position. We may choose the notional N in such a way that the hedged portfolio has zero dollar duration:

$$D_E^\$ = D_A^\$ + N \cdot D_{\text{swap}}^\$ - D_L^\$ = 0 \quad \Rightarrow \quad N = \frac{D_L^\$ - D_A^\$}{D_{\text{swap}}^\$}.$$

Note again that we do not change the original value of equity, as the initial value of the swap is zero.

6.5 A second-order refinement: Convexity

Duration-based hedging relies on first-order sensitivities. One way to improve the accuracy and the performance of hedging strategies is to adopt a second-order approximation. To this aim, we may introduce (bond) **convexity**,

$$C_P \doteq \frac{1}{P} \frac{d^2 P}{dr^2}. \tag{6.13}$$

Convexity may be applied to any fixed-income security or portfolio, even though it was originally introduced for bonds. A look at Fig. 6.1, where we observe the convex relationship between bond price and yield, suggests that convexity is positive for a plain bond. By using convexity, we can write a second-order Taylor expansion of the relative change in P:

$$\frac{\delta P}{P} \approx -D_P \cdot \delta r + \tfrac{1}{2} \cdot C_P \cdot (\delta r)^2.$$

6.5 A second-order refinement: Convexity

It is important to notice that bond convexity is *not* the derivative of duration. Just as with duration, it is also convenient to introduce **dollar convexity**[5]:

$$C_P^\$ \doteq \frac{d^2 P}{dr^2}. \tag{6.14}$$

Just like we have done with duration, we may find the convexity C_z of a zero-coupon bond first. Since

$$P_z(t, T) = F \cdot e^{-r(t,T) \cdot (T-t)},$$

we have

$$\frac{d^2 P_z}{dr^2} = (T-t)^2 \cdot F \cdot e^{-r(t,T) \cdot (T-t)} = (T-t)^2 \cdot P_z,$$

and

$$C_z = (T-t)^2. \tag{6.15}$$

It is also easy to see that the convexity of a fixed-income portfolio is just the weighted combination of the individual convexities. More precisely, if the value of the portfolio is

$$V = \sum_{i=1}^{m} N_i P_i,$$

where N_i and P_i are the holdings (how many units) and the price of security i, respectively, then

$$C = \sum_{i=1}^{m} w_i C_i,$$

where C_i is the convexity of asset i and the weights are

$$w_i = \frac{N_i \cdot P_i}{V}.$$

As an immediate consequence, the convexity of a coupon-bearing bond is

$$C_c = \frac{1}{P_c(t, r(t, \cdot))} \cdot \left[\sum_{i=1}^{m-1} \frac{c}{2} \cdot P_z(t, T_i) \cdot (T_i - t)^2 \right.$$
$$\left. + \left(1 + \frac{c}{2}\right) \cdot P_z(t, T_m) \cdot (T_m - t)^2 \right]. \tag{6.16}$$

In Eq. (6.16), we find a formal verification that convexity is positive for a plain bond. As we noticed in Example 6.1, the convexity effect explains why the first-order approximation by duration is somewhat pessimistic with respect to the true changes in bond prices when yield is shifted. Now, let us see how the approximation is improved by adding a second-order term.

---

[5] In Section 13.5, dealing with option price sensitivities, we will appreciate the similarity of dollar duration and convexity with option delta and gamma, respectively.

Example 6.7 A numerical check (continued)

Let us check if and by how much using convexity improves the approximation of the bond price changes that we have considered in Example 6.1. As we have seen there, the prices of two zeros maturing in 3 and 20 years, respectively, are

$$P_z(0, 0.03; 3) = 100 \times e^{-0.03 \times 3} = 91.39312,$$
$$P_z(0, 0.03; 20) = 100 \times e^{-0.03 \times 20} = 54.88116,$$

when yield is 3%, and they drop to

$$P_z(0, 0.04; 3) = 100 \times e^{-0.04 \times 3} = 88.69204,$$
$$P_z(0, 0.04; 20) = 100 \times e^{-0.04 \times 20} = 44.9329,$$

when there is an increase by 100 basis points. If we use both duration and convexity, we find the approximations

$$\widehat{P}_z(0, 0.04; 3) = P_z(0, 0.03; 3) \cdot \left(1 - 3 \times 0.01 + \tfrac{1}{2} \times 3^2 \times 0.01^2\right)$$
$$= 88.69245,$$
$$\widehat{P}_z(0, 0.04; 20) = P_z(0, 0.03; 20) \cdot \left(1 - 20 \times 0.01 + \tfrac{1}{2} \times 20^2 \times 0.01^2\right)$$
$$= 45.00255.$$

These approximations are definitely more accurate than those provided by a first-order expansion. For the second zero, the percentage error drops (in absolute value) from 2.28% in Example 6.1 to

$$\frac{45.00255 - 44.9329}{44.9329} = 0.155\%,$$

when adding the second-order term.

We should note the following:

- Convexity is a nice property of a bond, since a larger convexity means a larger profit when yield drops and a smaller loss when yield rises. This is too good to be true, and one should expect that this is paid somehow in terms of bond price.[6]
- We may improve the immunization of an asset–liability portfolio by setting both duration and convexity of equity to zero. This requires addi-

---
[6] When dealing with option price sensitivities in Section 13.5.3, we shall see the related link between option gamma (convexity) and option theta (option price decay with time). A large gamma is associated with a faster option price decay.

tional hedging instruments, but it does not increase computational complexity significantly. However, it is still true that we are considering only a single risk-factor, and that we are perfectly hedged against *small* perturbations. When dealing with multiple risk factors, we have to cope with second-order cross-sensitivities, which may actually increase complexity.

6.6 Multifactor models in interest rate risk management

Duration-based immunization is a first-order approach aimed at hedging against a single risk factor. Introducing convexity does not really change the picture, since it introduces a second-order approximation that still copes with a single risk factor. Since the term structure involves multiple risk factors, we may have to take a different approach. In the following example, we illustrate how duration fails to cope with nonparallel shifts in the term structure, which is typically affected by changes in slope and curvature, too.

Example 6.8 The effect of nonparallel shifts

Let us consider again the coupon-bearing bond of Example 6.2. If the term structure is flat and the continuously compounded rate is 3%, we have seen that the bond price and durations are $P_c(0; 0.03) = \$10{,}287.86$ and $D_P = 1.4643$, respectively. Thus, the dollar duration of this bond is

$$D_P^\$ = 10{,}287.86 \times 1.4643 = \$15{,}064.21.$$

Suppose that we hedge interest rate risk with a short position in a zero maturing in six months, with a face value of $10,000, like the coupon-bearing bond. The price of this hedging instrument is

$$H(0; 0.03) = 10{,}000 \times e^{-0.03 \times 0.5} = \$9851.12,$$

its duration is 0.5, and its dollar duration is

$$D_H^\$ = 9851.12 \times 0.5 = \$4925.56.$$

Hence, the short position should consist of

$$\phi = -\frac{15{,}064.21}{4925.56} = -3.06$$

units of the zero. We neglect rounding issues, and we assume that the short sale is feasible, possibly through the repo market. No initial cash outlay is needed (we ignore haircuts, margins, etc.). Hence, the current value of the hedged portfolio is just the bond price:

$$V_H(0; 0.03) = \$10{,}287.86.$$

Let us assume that there is a parallel upshift by 100 basis point, i.e., the new term structure is flat at 4%. This is assumed to be instantaneous and takes place at time $t = 0$. The new bond prices are easily computed,

$$P_c(0; 0.04) = \$10{,}138.33, \qquad H(0; 0.04) = \$9801.99,$$

and the value of the hedged portfolio changes to

$$\begin{aligned} V_H(0; 0.04) &= P_c(0; 0.04) + \phi \cdot \left[H(0; 0.04) - H(0; 0.03) \right] \\ &= 10{,}138.33 - 3.06 \times (9801.99 - 9851.12) \\ &= \$10{,}288.67, \end{aligned}$$

which is very close to the initial one, with some difference due to convexity effect. The loss on the coupon-bearing bond is compensated by the profit from the short position on the zero.

However, what if the shift is nonparallel? Let us assume that the new term structure is not flat anymore:

$$\bar{r}(0, 0.5) = 3.8\%, \quad \bar{r}(0, 1) = 4.0\%, \quad \bar{r}(0, 1.5) = 4.2\%.$$

Note that, on the average, the new interest rate is 4%, as with the parallel shift. The new bond prices are

$$P_c\big(0; \bar{r}(0, \cdot)\big) = 10{,}109.66, \qquad H\big(0; \bar{r}(0, \cdot)\big) = 9811.79,$$

and the value of the hedged portfolio is only

$$\begin{aligned} V_H\big(0; \bar{r}(0, \cdot)\big) &= 10{,}109.66 - 3.06 \times (9811.79 - 9851.12) \\ &= \$10{,}230.01. \end{aligned}$$

In this case, the drop in the price of the zero is not enough to compensate the loss on the bond and we end up being under-hedged. The overall loss (−0.56%) does not look quite impressive, but this happens because the maturity of the coupon-bearing bond is rather short (see Problem 6.5). Probably, we would be better off with a zero with a longer maturity, but the general point is that, since we are dealing with three risk factors, we must introduce additional hedging instruments and assess individual sensitivities for each risk factor.

In order to deal with multiple risk factors, we have to introduce a multifactor model. Unfortunately, the whole term structure $r(t, \cdot)$ consists of a virtually infinite set of risk factors. The resulting complexity may be reduced in a few ways:

- One approach is to introduce factor durations, possibly durations corresponding to rates at carefully selected maturities. Then, first-order immunization may be carried out as suggested in Section 2.2.3.3.
- A slightly different approach relies on an alternative way of finding factors. One idea is to take linear combination of rates, e.g., by principal component analysis. Then again, factor durations are put to good use.
- These two approaches have one thing in common: They aim at *perfect* immunization for *small* changes in the risk factors. It has been argued that it may be preferable to be *approximately* hedged against *large* changes in the risk factors. To achieve this objective, one may resort to scenario-based optimization models, like those we introduce in Chapter 15. In that case, too, we need a suitable multifactor model to generate scenarios.

We note that the idea of factor models is not limited to fixed-income assets. Models in this vein for equity portfolios are discussed in Chapter 9. In that case, the structure is somewhat different. Generally, multifactor models for fixed-income portfolios deal with similar factors, i.e., rates or combination of rates, and perhaps credit spreads. In the equity case, the factors are substantially different, as we shall see, and range from macroeconomic factors like inflation rate or oil price, to financial factors like the amount of financial leverage of a specific firm or a broad market index. They may also include behavioral factors like market momentum.

Problems

6.1 In five years, you will have to pay a single and deterministic liability, amounting to $10,000. The only asset you may use is a bond paying a 7% annual coupon and maturing in six years. At present, yield is 6% with annual compounding.

- Assuming asset divisibility, how much should you buy of this bond?
- What is the performance of the resulting ALM policy if yield goes up or down by 100 basis points? How do you explain the result? Assume that the change is immediate and instantaneous.

6.2 In five years, you are going to pay €20,000 to purchase a machine for your firm. Consider a portfolio consisting of a zero-coupon bond maturing in seven years and a coupon bond with coupon rate 5% (the bond pays one coupon per year), maturing in three years. The term structure is flat, and the rate is now 4% with annual compounding. Assume a face value of €1000 for both bonds.

- Build an immunized portfolio. What is the problem with your choice of bonds?
- Repeat the procedure, but now consider the same zero and a similar coupon bond maturing in six years. Is this portfolio easy to implement?

6.3 In six years, you will have to pay a single, deterministic liability for an amount of €30,000. Consider a portfolio consisting of a zero maturing in eight years and a bond paying a single annual coupon with rate 4%, maturing in four years. The term structure is flat at a rate of 3.5%, with annual compounding. Assume a face value of €1000 for both bonds.

- Build a first-order immunized portfolio and check its performance for an immediate shift of ±50 basis points.
- Repeat the procedure, but now include another zero maturing in three years, in order to match both duration and convexity of the assets and the liability. Compare the performance against the previous portfolio.

6.4 Your portfolio consists of two sovereign bonds: A zero maturing in three years and a coupon bond maturing in two years, paying a single annual coupon at rate 4%. Assume a face value of €1000 for both bonds. The one-year risk-free forward rates are 3%, 4%, and 5%, respectively, with annual compounding (Actually, the first rate is the annual spot rate, and the last one applies to an investment over the time interval $[2, 3]$). The amounts invested in the two bonds are €53,000 and €93,000, respectively (assume asset divisibility). The two bonds have been issued by the same government, and the price is influenced by a spread due to specific country risk. The spread is 2.3% at present (applying uniformly to every maturity), and it is subject to a random shock, which we assume uniformly distributed between -1% e $+2\%$ (hence, the new spread will be in the range between 1.3% and 4.3%). Neglecting the passage of time, find value-at-risk at probability level 97%.

6.5 Repeat the analysis of Example 6.8, but now consider a bond maturing in five years. For the nonparallel shift, assume a new term structure of rates linearly increasing from 3% to 5%, so that the average is 4% in both cases, parallel and nonparallel shifts. In the second case, the rates for the ten maturities (six months, one year, one year and a half, all the way up to five years) are

$$3\%,\ 3.22\%,\ 3.44\%,\ 3.67\%,\ \ldots,\ 4.56\%,\ 4.78\%,\ 5\%.$$

Do you still observe a small loss as in Example 6.8? What if you increase the maturity of the zero to three or five years?

Further reading

- General textbooks on fixed-income securities, like [3] and [5], include extensive sections on interest rate risk management, dealing in more depth with all of the topics that we have outlined in this chapter. Example 6.8 is a simplified version of a case discussed in [5, Chapter 4].
- For a more specific treatment, you may consult [2] and [4].
- For a comprehensive reference, also covering the use of interest rate derivatives in interest rate risk management, see [1].

Bibliography

1. F.J. Fabozzi, editor. *Fixed Income Analysis* (2nd ed.). Wiley/CFA Institute, Hoboken, NJ, 2007.

2. B.E. Gup and R. Brooks. *Interest Rate Risk Management: The Banker's Guide to Using Futures, Options, Swaps, and Other Derivative Instruments.* Irwin Professional Publishing, New York, 1993.

3. L. Martellini, P. Priaulet, and S. Priaulet. *Fixed-Income Securities: Valuation, Risk Management, and Portfolio Strategies.* Wiley, Chichester, 2003.

4. S.K. Nawalkha, G.M. Soto, and N.A. Beliaeva. *Interest Rate Risk Modeling.* Wiley, Hoboken, NJ, 2005.

5. P. Veronesi. *Fixed Income Securities: Valuation, Risk, and Risk Management.* Wiley, Hoboken, NJ, 2010.

Part Three

Equity portfolios

Chapter Seven

Decision-Making under Uncertainty: The Static Case

Uncertainty is the rule in most financial decision-making problems. The prototypical case is the allocation of wealth to a set of assets with uncertain returns. If we make a here-and-now decision and observe the return of the portfolio after a given holding period, we are considering a **static** decision problem, since we disregard the possibility of adjusting our decisions along the way, when we observe the actual unfolding of uncertain risk factors. This is not to say that, in reality, the portfolio will not be adjusted after a while, possibly by solving the same model again; the point is that this is not explicitly considered in the decision model itself. On the contrary, **multistage** decision models take into account the possibility of updating decisions, depending on the incoming information flow over time. It is important to avoid a potential confusion between *multistage* and *multiperiod* models. A multiperiod problem requires the planning of decisions to be executed over a sequence of time instants. However, if the plan is specified here and now, once for all, the problem is actually static, as there is no dynamic adaptation. The solution of a multiperiod problem is a sequence of numbers, representing the decisions that are supposed to be implemented, no matter what. On the contrary, the solution of a multistage problem consists of a set of random variables, since decisions will be contingent on the realization of uncertain states. We may also explicitly express decisions as functions of the uncertain states or, alternatively, as functions of the realization of random risk factors.

In this chapter, we lay down the conceptual foundations of decision-making under uncertainty in the static, single-period case. It is useful to consider portfolio decisions as an application framework to understand the related issues; however, what we describe here is also relevant for asset pricing. Here, we do not consider either model building or solution algorithms.[1] In Chapter 8, we discuss a specific relevant case in detail, mean–variance portfolio optimization, whereas in Chapter 15 we outline more advanced models, including multistage

---

[1] A very simple introduction to deterministic optimization models and solution algorithms is given in Chapter 12 of [4]. Chapter 13 therein describes some models for decision-making under uncertainty, including, but not limited to financial problems.

ones. Here, we introduce three possible approaches to decision-making under uncertainty that are relevant to finance:

- Utility functions
- Mean–risk models
- Stochastic dominance

We start with a few simple introductory examples in Section 7.1. Then, in Section 7.2, we show that financial decision-making cannot rely on simple maximization of expected wealth or expected return. Risk should be carefully accounted for. One way for doing so, albeit not quite a practical one, is by introducing expected utility, as we illustrate in Section 7.3. A more practical approach is based on the definition of suitable risk measures and the solution of a mean–risk optimization problem. Mean–risk models, as we show in Section 7.4, are the foundation of the ubiquitous mean–variance portfolio optimization framework. However, there is no reason why we could not replace standard deviation (or variance) of return by an alternative risk measure. We have already introduced value-at-risk in Section 2.2.2. Here, we discuss some basic properties that a *coherent* risk measure should satisfy. As it turns out, value-at-risk is not quite satisfactory in this respect. Then, in Section 7.5, we outline a third approach, stochastic dominance. This last section is included for the sake of completeness, but it is not needed for the remainder of this book and may be safely skipped. We also include a couple of theorem proofs in Supplement S7.1, which may be safely skipped, too. Usually, we do not include complete and overly rigorous proofs, given the introductory nature of this book. However, some of them may be instructive and useful to the interested reader.

7.1 Introductory examples

A couple of simple examples may help in framing the kind of problems that we want to tackle in this chapter.

Example 7.1 A choice among lotteries

Consider the choice among the four lotteries depicted in Table 7.1. These lotteries are characterized by uncertain payoffs, which we model by four discrete random variables $L_i(\omega)$, $i = 1, 2, 3, 4$, taking values corresponding to three equally likely outcomes ω_1, ω_2, and ω_3. For each $L_i(\omega)$, in the table we also report its expected value μ_i and standard deviation σ_i. Which lottery should we choose?

It is easy to see that lottery L_4 would not be chosen, since its payoff is dominated by L_1 (as well as by L_3):

$$L_4(\omega_k) \leq L_1(\omega_k), \qquad k = 1, 2, 3,$$

7.1 Introductory examples

Table 7.1 Choice among four lotteries.

Lottery	ω_1	ω_2	ω_3	μ_i	σ_i
$L_1(\omega)$	100	200	300	200	81.65
$L_2(\omega)$	−800	200	1200	200	816.50
$L_3(\omega)$	150	200	244	198	38.40
$L_4(\omega)$	100	200	150	150	40.82

with strict inequality in scenario ω_3. Lottery L_2 is obtained from L_1 by shifting a payoff of 900 units from scenario ω_1 to ω_3. Thus, we do not change the expected value, but the payoff has a much larger variability, as measured by the standard deviation. Many would agree that, since the expected value is the same and there is less uncertainty, lottery L_1 should be preferred to L_2.

Actually, this is a matter of individual taste and depends on how much we like or dislike taking risk. If we are risk-averse, chances are that we may even like L_3 the most. This lottery is obtained from L_1 by increasing the payoff for event ω_1 by 50 and decreasing the payoff for ω_3 by 56. Thus, the expected value μ_3 is only 198, but the standard deviation is considerably reduced.

In Example 7.1, we have only considered expected value and standard deviation of a lottery. Indeed, there is a large body of knowledge, broadly referred to as **modern portfolio theory**, which revolves around this view. However, this may not quite enough. As we said, if we compare the payoffs of lotteries L_1 and L_4 in Table 7.1, state by state, the latter is clearly dominated. However, we cannot reach a clear conclusion by just considering expected value and standard deviation of the two payoffs, since $\mu_4 < \mu_1$ and $\sigma_1 > \sigma_4$. One issue is that standard deviation does not capture the features of a very skewed random variable, associated with an asymmetric probability distribution. Example 7.2 below further illustrates this point.

Example 7.2 A dominated lottery

Let us consider the two lotteries described in Table 7.2. Note that the states of nature (outcomes) are not equiprobable. We find the expected value and the standard deviation of the payoff of lottery $L_1(\omega)$

Table 7.2 A dominated lottery.

State	ω_1	ω_2	ω_3
Probability	0.4	0.4	0.2
Payoff $L_1(\omega)$	10	50	100
Payoff $L_2(\omega)$	10	50	500

as follows:

$$\mu_1 = 0.4 \times 10 + 0.4 \times 50 + 0.2 \times 100 = 44$$
$$\sigma_1 = \sqrt{0.4 \times 10^2 + 0.4 \times 50^2 + 0.2 \times 100^2 - 44^2} \approx 33.23$$

By the same token, for lottery $L_2(\omega)$ we find $\mu_2 = 124$ and $\sigma_2 \approx 188.85$. If we compare the two alternatives in terms of expected value and standard deviation, there is an unclear tradeoff between the two lotteries, as the second one is more attractive in terms of expected payoff, but it looks riskier. However, if we compare the payoffs state by state, $L_1(\omega)$ is clearly dominated by $L_2(\omega)$. The problem is that the large payoff of lottery $L_2(\omega)$ in state ω_3 increases not only the expected value, but also standard deviation. Its distribution is positively skewed, and a symmetric deviation measure, like standard deviation, does not properly account for this feature. We should also notice that, if we introduce a negative skew, standard deviation will not tell the difference with respect to a corresponding positive skew.

We have considered simple lotteries that may be represented by a *discrete* random variable X that takes values x_j with probabilities p_j, corresponding to scenarios (also called outcomes or states of the world) $\omega_j, j = 1, \ldots, m$. In risk management, this random variable usually represents loss, rather than profit, return, or payoff. We may also consider *continuous* random variables, as is common in asset allocation problems.

Example 7.3 Static asset allocation

We are endowed with wealth W_0 that we should allocate among a set of n assets with current price S_{i0}, $i = 1, \ldots, n$. At the end of a holding period of length T, the prices of the assets are represented by continuous random variables $S_{iT}(\omega)$. If we assume that assets are in-

7.1 Introductory examples

finitely divisible and short-selling is not allowed, our decision can be represented by decision variables $h_i \geq 0$, $i = 1, \ldots, n$, corresponding to the holding of each asset, i.e., the number of stock shares of firm i included in the portfolio.

Decision variables are subject to a budget constraint,

$$\sum_{i=1}^{n} h_i S_{i0} = W_0,$$

and define a random variable,

$$W_T(\omega) = \sum_{i=1}^{n} h_i S_{iT}(\omega),$$

which is the random terminal wealth for each outcome $\omega \in \Omega$.

In this case, the problem does not just require ranking a few simple lotteries. By choosing the portfolio holdings we define a continuous probability distribution of terminal wealth, and we might choose the most preferred one by defining and optimizing a suitable **functional** $F(\cdot)$, mapping a random variable into the set of real numbers:

$$\max_{h_1,\ldots,h_n} F\big[W_T(\omega)\big].$$

We are talking about a functional rather than a function, since we are mapping random variables (which are function themselves, and not just numerical variables), to real numbers. If we can find a suitable functional $F(\cdot)$, we may map a possibly complicated preference structure into the simple ordering of real numbers.

Throughout the chapter, we assume that we have a credible stochastic characterization of the probability distribution of uncertain risk factors. The distribution may be considered as an **objective** assessment of uncertainty, but it is most likely to be at least partially **subjective**. The difference is that, ideally, all market participants should agree on a truly objective representation of uncertainty. On the contrary, market views are to some extent subjective. In more sophisticated models, we explicitly consider distributional ambiguity and look for a *robust* solution. In such a case, we could be uncertain about a set of plausible probability distributions, or we might even take a radical view and give up the idea of a stochastic representation of uncertainty. In this chapter, for the sake of simplicity, we assume that a reliable stochastic representation of uncertainty is available.

7.2 Should we just consider expected values of returns and monetary outcomes?

Whenever we bet money on a lottery or invest wealth in risky assets, we pay due attention to the expected value of the payoff, i.e., a monetary outcome, or to expected return. The expected value is quite likely to be the first feature we consider, when dealing with a probability distribution. However, let us ask the following questions:

- Given a set of assets or alternative financial portfolios, should we just select the one with the largest expected return? No doubt, this would make life much easier when dealing with decision-making under uncertainty. However, as we show below, this does not take risk into account and may lead to quite unreasonable decisions. As a general rule, larger expected returns come with a larger exposure to risk, and this leads to the need of assessing difficult risk–return tradeoffs.

- Should we consider an asset with *negative* expected return for inclusion within a portfolio? Even if we suspect that expected return does not tell the whole story, one is tempted to think that there is little good to be expected from such an asset, unless short-selling is allowed. However, this simplistic view does not consider the correlations between returns. An asset with a negative expected return may be negatively correlated with other assets and contribute to reducing risk. Derivatives such as futures and forward contracts are in fact included in an asset portfolio (possibly a nonfinancial one, involving commodities) to reduce risk by exploiting a negative correlation. In real life, indeed, we often purchase insurance, which is an asset with (hopefully) negative expected return, as we expect to pay the insurance premium but hope that a severe accident will not occur.[2]

- Given a financial asset with an array of random payoffs, can we just consider the expected value of the payoff to price the asset fairly? This is a relevant question, when dealing with derivatives and insurance contracts. If an insurance company faces a large set of small-scale and independent risks, it may be argued that finding the *actuarially fair* price of an insurance policy, by estimating the expected cash outflow for the company, may be a good strategy. However, this need not apply in general, and the insurance business can get quite dangerous when risks turn out to be correlated.[3]

Among other things, these questions show the link between the three basic problems of asset allocation, risk management, and asset pricing.

---

[2] As a further, but quite different example, lottery tickets have a negative expected payoff, since it is unlikely that we will win. So, it seems that we may be risk lovers, at least in the small.

[3] A good lesson in this respect comes from the default risks on mortgages in 2008, leading to the subprime crisis and the ultimate demise of Lehman Brothers.

7.2.1 FORMALIZING STATIC DECISION-MAKING UNDER UNCERTAINTY

In this section, we consider possible ways of formalizing a static problem under uncertainty. A generic optimization model may be written as

$$\min_{\mathbf{x} \in S} f(\mathbf{x}),$$

where:

- $\mathbf{x} \in \mathbb{R}^n$ is the vector of decision variables
- $S \subseteq \mathbb{R}^n$ is the set of feasible solutions
- $f(\cdot)$ is the objective function, mapping solutions (vectors in $\mathbb{R}^n$) into a numerical evaluation of their quality (a number in $\mathbb{R}$)

In finance, the objective function is likely to be related to a monetary outcome, like profit/loss, or to a return. Depending on the choice of the objective function, the problem may be a minimization or a maximization one. In most fields of practical interest, some data or parameters of the optimization model are uncertain. One way of stating this is by considering a vector of random variables $\boldsymbol{\xi}(\omega)$, where $\omega \in \Omega$ is a random outcome, corresponding to a scenario, within the sample space Ω. Then, the objective function becomes a function $f(\mathbf{x}, \boldsymbol{\xi}(\omega))$ of both controllable and uncontrollable variables, and the feasible set may be random, too. This may have two consequences:

- The quality of the solution that we find is random and may turn out to be not quite what we expect.
- Possibly worse, the solution may even turn out to be infeasible for some realizations of the random data.

As we have pointed out before, in Example 7.3, a specific choice $\mathbf{x}_0$ of the decision variables defines the distribution of a random variable $Y_0 = f(\mathbf{x}_0, \boldsymbol{\xi}(\omega))$, and we need a way to rank probability distributions. The simplest choice is to rank distributions by the corresponding expected value. Thus, we might consider an optimization problem like

$$\min_{\mathbf{x} \in S} \mathrm{E}\big[f(\mathbf{x}, \boldsymbol{\xi}(\omega))\big].$$

However, just taking the expected value of an objective like cost or profit may not account for different attitudes toward risk. Thus, in general, we may consider a transformation of the random performance measure $f(\mathbf{x}, \boldsymbol{\xi}(\omega))$, say, $\mathcal{R}_0\big[f(\mathbf{x}, \boldsymbol{\xi}(\omega))\big]$, which should be considered as a **risk functional**. In concrete, as we shall see later, we may consider utility functions or mean–risk models.

Actually, stating an optimization problem under uncertainty in a precise way is not quite trivial, as different approaches may be pursued to model the interplay of decisions and observations, i.e., how to define a dynamic decision strategy, as well as how to cope with potential infeasibility of decisions made before knowing the values of uncertain data. To be more concrete, let us assume

that the feasible set is explicitly described by a set of inequalities:
$$g_i(\mathbf{x}, \boldsymbol{\xi}(\omega)) \leq 0, \qquad i = 1, \ldots, m.$$
Clearly, for a given **x**, we cannot be sure that the inequality will be satisfied for every value of $\boldsymbol{\xi}$. If we insist on guaranteed feasibility in every scenario,[4] an overly fat solution may be obtained. Here, too, we may introduce functionals $\mathcal{R}_i$, $i = 1, \ldots, m$, and require
$$\mathcal{R}_i[g_i(\mathbf{x}, \boldsymbol{\xi}(\omega))] \leq 0, \qquad i = 1, \ldots, m.$$
A naive approach would be to require that the expected value of the constraint function g_i is negative or zero, but this would be a very weak statement of an uncertain constraint. To see why, consider a standard normal distribution, where the expected value is zero, but the probability of a strictly positive value is 50%. As an alternative, we may settle for a probabilistic satisfaction of the constraints. We may introduce a set of **individual** chance constraints,
$$P\{g_i(\mathbf{x}, \boldsymbol{\xi}(\omega)) \leq 0\} \geq 1 - \alpha_i, \qquad i = 1, \ldots, m,$$
or a **joint** chance constraint,
$$P\{g_i(\mathbf{x}, \boldsymbol{\xi}(\omega)) \leq 0, \quad i = 1, \ldots, m\} \geq 1 - \alpha.$$
We should ask whether chance constraints are a suitable modeling framework, which means: (a) whether they allow us to express a financial decision-making problem in a sensible way, and (b) whether they lead to model formulations that may be efficiently solved. We will discuss this matter in Section 15.6.1.

We note again that, in a static decision problem under uncertainty, the solution is not dynamically adapted according to contingencies. Furthermore, it is practically impossible to find feasible solutions to problems involving random equality constraints. To this aim, we may take advantage of a more flexible modeling framework, stochastic programming with recourse, which will be introduced in Chapter 15.

7.2.2 THE FLAW OF AVERAGES

In common wisdom, we often consider loose statements of the law of large numbers, which is typically referred to as the "law of averages." Here, we rather consider the *flaw* of averages.[5] A comparison of the lotteries in Table 7.1 suggests that ranking alternatives on the basis of the expected value is probably neither safe nor sensible. Let us consider a few further examples reinforcing the point.

---

[4]Technically, we say that constraints are satisfied *almost surely*, i.e., with the exception of a set of null measure. Alternatively, we say that constraints are satisfied with probability one. When dealing with a finite set Ω of discrete outcomes, this boils down to the satisfaction of constraints in every discrete scenario.

[5]See [16].

7.2 Should we just consider expected values?

▣ Example 7.4 A single bet vs. multiple repeated bets

Consider a simple lottery based on the flip of a fair coin: If it lands tails, we win €10, otherwise we lose €5. Should we play this lottery? The expected payoff is €2.5, and most people answer that they would be willing to take the gamble. If we spice things up and scale the payoff by a factor of one *million*, the answer turns probably negative. Sure, an expected payoff of €2.5 million is quite palatable, but the considerable risk of losing €5 million makes the gamble not attractive to most people.

However, imagine playing the gamble repeatedly many times, say, one thousand times. Our answer could change if we are allowed to settle the score at the end of the game. Let X_i be the payoff of flip number i, $i = 1, \ldots, n$, where n is the number of independent and identically distributed flips. Thus, the variables X_i are i.i.d. random variables. Let $Y = \sum_{i=1}^{n} X_i$ be the total payoff, and let us denote the common expected value and standard deviation of the variables X_i by μ_X and σ_X, respectively. Then, the **coefficient of variation** of Y, under the hypothesis of independent flips, is

$$C_Y \doteq \frac{\sigma_Y}{|\mu_Y|} = \frac{\sqrt{n}\sigma_X}{n\mu_X} = \frac{C_X}{\sqrt{n}}, \tag{7.1}$$

where we assume $\mu_X > 0$. If n is large, the expected overall payoff becomes virtually certain (this is an informal glimpse of the law of large numbers). However, if we may go bankrupt along the way (i.e., we settle each flip of the coin individually, rather than assessing the overall profit/loss at the end of the game) or risks are correlated, this is not true anymore.

The natural interpretation of Example 7.4 is in terms of a bet repeated over time. An alternative view, which is relevant to insurers, concerns multiple bets taken at the same time. Indeed, Eq. (7.1) shows why an insurer providing coverage for a large number of uncorrelated risks may rely of expectations, plus some fudge consisting of reserves. However, correlated risks are much more dangerous. For an insurer, the random variables X_i correspond to losses. Let us assume that losses are pairwise correlated in the same way and that the common correlation coefficient is ρ. Then, the variance of the total loss becomes

$$\begin{aligned} \mathrm{Var}(Y) &= \sum_{i=1}^{n} \mathrm{Var}(X_i) + \sum_{i=1}^{n}\sum_{\substack{j=1 \\ j \neq i}}^{n} \mathrm{Cov}(X_i, X_j) \\ &= n\sigma_X^2 + n \cdot (n-1)\rho\sigma_X^2 \\ &= n\sigma_X^2 \cdot \left[1 + (n-1)\rho\right]. \end{aligned}$$

In the limit case $\rho = 1$, we have $\text{Var}(Y) = n^2 \sigma_X^2 = \text{Var}(nX)$ and there is no diversification of risk, in the sense that the coefficient of variation becomes

$$C_Y = \frac{n \sigma_X}{n \mu_X} = C_X.$$

As an example of correlated risks, we may think of home insurance in a region prone to earthquakes, or mortgage defaults under economic recession, as it happened during the subprime mortgage crisis.

Example 7.5 Putting all of our eggs in one basket

Consider an investor who must allocate her wealth to n assets. The return of each asset, indexed by $i = 1, \ldots, n$, is a random variable R_i with expected value $\mu_i = \text{E}[R_i]$. Asset allocations may be expressed by decision variables w_i, representing the fraction of wealth invested in asset i. If we rule out short-selling, these decision variables are naturally bounded by $0 \leq w_i \leq 1$. If we assume that the investor should just maximize expected return, she should solve the problem

$$\max \quad \sum_{i=1}^{n} \mu_i w_i$$
$$\text{s.t.} \quad \sum_{i=1}^{n} w_i = 1$$
$$w_i \geq 0.$$

This is a simple model that we have already met in Section 2.1.1, Eq. (2.1), and we know its quite trivial solution: Just pick the asset with maximum expected return, $i^* = \arg\max_{i=1,\ldots,n} \mu_i$, and set $w_{i^*} = 1$. It is easy to see that this concentrated portfolio is a very dangerous bet. In practice, portfolios are diversified, which means that decisions depend on something beyond expected values. Furthermore, one would also include additional constraints on portfolio composition, bounding exposure to certain geographic areas or types of industry, and they would render the above trivial solution infeasible. However, it may be necessary to add many such additional constraints to find a sensible solution; this means that the solution is basically shaped by the constraints that the decision maker enforces in order to rule out blatantly inadequate portfolios. Incidentally, if short-selling is allowed, the decision variables are unrestricted, and the expected value of future wealth goes to infinity. In fact, one would short-sell assets with low expected return, to raise money to be invested in the most promising asset. This is clearly risky and should be carefully disciplined.

7.2 Should we just consider expected values?

The next example is more akin to pricing a risky asset. It provides good evidence that pricing by the expected value of the payoff (possibly discounted, in order to take time value of money into account) does not seem a plausible approach.

Example 7.6 St. Petersburg paradox

Consider the following proposal. We are offered a lottery, whose outcome is determined by flipping a fair and memoryless coin. The coin is flipped until it lands tails. Let k be the number of times the coin lands heads; then, the payoff we get is \$$2^k$. Now, how much should we be willing to pay for this lottery? Even if we are unlucky and the game stops at the first flip, so that $k = 0$, we will get \$1, so we should be willing to pay at least this amount.

We may consider this as an asset pricing problem and set the expected value of the payoff as the fair price for this rather peculiar asset. The probability of winning \$$2^k$ is the probability of observing k consecutive heads followed by the tails that stops the game, after $k+1$ flips of the coin. Given the independence of events, the probability of this sequence is $1/2^{k+1}$, i.e., the product of $k+1$ individual event probabilities. Then, the expected value of the payoff is

$$\sum_{k=0}^{\infty} \frac{1}{2^{k+1}} 2^k = \tfrac{1}{2} \times 1 + \tfrac{1}{4} \times 2 + \tfrac{1}{8} \times 4 + \cdots$$
$$= \tfrac{1}{2} + \tfrac{1}{2} + \tfrac{1}{2} + \cdots$$
$$= +\infty.$$

This game looks so beautiful that we should be willing to pay any amount of money to play it! No one would probably do so. True, the game offers huge payoffs, but with vanishing probabilities. Again, we conclude that expected values do not tell the whole story.

The idea that most decision makers are risk-averse is intuitively clear, but what does **risk aversion** really mean in formal terms? To get a clue, let us compare two simple lotteries:

1. Lottery a_1, which is actually deterministic and guarantees a sure payoff μ
2. Lottery a_2, which offers two equally likely payoffs $\mu + \delta$ and $\mu - \delta$

The two lotteries are clearly equivalent in terms of expected payoff, but a risk-averse agent will arguably select lottery a_1. More generally, if we consider a random variable X, representing a payoff, and we add a **mean-preserving**

spread, i.e., an independent random variable $\tilde{\epsilon}$ with $\mathrm{E}[\tilde{\epsilon}] = 0$,[6] this addition is not welcome by a risk-averse decision maker and the lottery X is preferred to $X + \tilde{\epsilon}$. This idea may be further formalized and made operational by using different approaches that are discussed in the following.

7.3 A conceptual tool: The utility function

Given a set of lotteries, a decision maker should be able to pick the preferred one; or, given any pair of lotteries, the decision maker should be able to tell which one she prefers or state that she is indifferent between them. If so, she has a well-defined preference relationship among lotteries. Since preference relationships are a bit cumbersome and difficult to deal with, we could map each lottery to a real number measuring the attractiveness of that lottery to the decision maker, and then use the standard ordering of real numbers to rank lotteries. Such a function cannot be just the expectation, as this disregards risk aversion. A theoretical answer, commonly put forward in economic theory, can be found by assuming that decision makers order uncertain outcomes by a suitably chosen functional, rather than by straightforward expected monetary values. For an arbitrary preference relationship, a functional representing it may not exist but, under a set of more or less reasonable assumptions,[7] such a mapping does exist and can be represented by an **expected utility**. A particularly simple form of expected utility functional, which looks reasonable, but it is only justified by specific hypotheses on the preference relationship that it represents, is the **Von Neumann–Morgenstern expected utility**, defined as

$$U(X) = \mathrm{E}\big[u(X)\big],$$

for a suitably chosen function $u(\cdot)$. For a simple lottery a represented by a discrete random variable with n outcomes x_i and probabilities p_i, this boils down to

$$U(a) = \sum_{i=1}^{n} p_i u(x_i).$$

To be precise, we refer to function $u(\cdot)$ as the utility function, which is related to a certain payoff. On the contrary, $U(\cdot)$ is the expected utility *functional*, as it maps random variables to the real line. If $u(x) \equiv x$, then the expected utility functional boils down to the expected value of the payoff. Alternative choices of the utility function $u(\cdot)$ model different attitudes toward risk. For financial

---

[6]For the sake of convenience, when using Greek letters we denote by $\tilde{\epsilon}$ a random variable and by ϵ a realization of that variable. This notation is common in economics. In statistics, one typically uses X and x with the corresponding pair of meanings, but this is not quite convenient with Greek letters.

[7]The discussion of these assumptions is best left to books on microeconomics or decision theory; we should mention that most of them seem rather innocent and reasonable, under most circumstances, but they may lead to surprising effects in paradoxical cases.

7.3 A conceptual tool: The utility function

problems, it is reasonable to assume that utility $u(\cdot)$ is a strictly increasing function, since we prefer more wealth to less. Formally, this property is referred to as **non-satiation**.

Beside the requirement of increasing monotonicity, the utility function is typically assumed to be concave. It is easy to see that concavity may express risk aversion. For the sake of convenience, we recall that a function f is said to be concave on a domain $S \subseteq \mathbb{R}^n$, if

$$f(\lambda \mathbf{x} + (1-\lambda)\mathbf{y}) \geq \lambda f(\mathbf{x}) + (1-\lambda) f(\mathbf{y}), \quad \forall \mathbf{x}, \mathbf{y} \in S, \lambda \in [0,1]. \quad (7.2)$$

In words, the value of the function for a convex combination of points in the domain is larger than the corresponding convex combination of the function values.[8] Since a convex combination is a linear combination with non-negative weights adding up to one, we immediately see the link with expected values. If we consider a lottery featuring two possible outcomes, x_1 and x_2, with probabilities $p_1 = p$ and $p_2 = 1-p$, respectively, a risk-averse decision maker would prefer not taking chances:

$$u(\mathrm{E}[X]) = u(px_1 + (1-p)x_2) \geq pu(x_1) + (1-p)u(x_2) = \mathrm{E}[u(X)]. \quad (7.3)$$

This may be generalized to a generic, possibly continuous random variable by recalling **Jensen's inequality** for a concave function u of a random variable X:

$$u(\mathrm{E}[X]) \geq \mathrm{E}[u(X)]. \quad (7.4)$$

Example 7.7 Concavity and risk aversion

Let us consider again the sure lottery a_1, which guarantees a payoff μ with probability one, and lottery a_2, obtained by the mean-preserving spread $\tilde{\epsilon}$, featuring equally likely outcomes $-\delta$ and δ. Concavity implies risk aversion, since

$$U(a_1) = u(\mu) \geq \tfrac{1}{2} u(\mu - \delta) + \tfrac{1}{2} u(\mu + \delta) = U(a_2).$$

Since the inequality is not strict, we should say that lottery a_1 is at least as preferred as a_2, and the decision maker could be indifferent between the two.

As a numerical illustration, let us consider the logarithmic utility $u(x) = \log x$, and $\mu = 10$, $\delta = 5$:

$$U(a_1) = \log 10 = 2.3026,$$
$$U(a_2) = \tfrac{1}{2} \log 5 + \tfrac{1}{2} \log 15 = 2.1587.$$

Figure 7.1 illustrates the role of concavity in describing risk aversion.

---

[8] See Section 15.1 for more details on convex and concave functions.

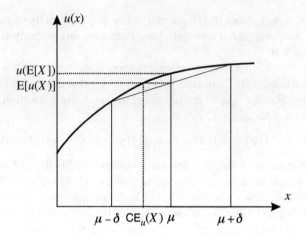

FIGURE 7.1 How concave utility functions imply risk aversion; the certainty equivalent is also shown.

It is fundamental to observe that the specific numerical value that the utility function assigns to a lottery is irrelevant per se; only the *relative* ordering of alternatives is essential. In fact, we speak of **ordinal** rather than *cardinal* utility. Given the linearity of expectation, we also see that an affine transformation of the utility function $u(\cdot)$ has no effect, provided it is increasing. To see this, let us consider $\bar{u}(x) \doteq au(x) + b$ instead of $u(x)$, where $a > 0$. Then, the ranking of alternatives according to u is clearly preserved by $\bar{u}$, since

$$\bar{U}(X) = \mathrm{E}\big[\bar{u}(X)\big] = \mathrm{E}\big[au(X) + b\big] = aU(X) + b.$$

Concavity implies risk aversion, from a qualitative viewpoint, but we would also like to come up with some *quantitative* way to measure risk aversion. We have said that a risk-averse decision maker would prefer a certain payoff to an uncertain one, when the expected values are the same. She would take the gamble only if the expected value of the risky lottery were suitably larger than the certain payoff. In other words, she requires a **risk premium**. The risk premium depends partly on the risk attitude of the decision maker, and partly on the uncertainty of the gamble itself. We will denote the risk premium by $\rho_u(X)$[9]; note that this is a number that a decision maker with utility $u(\cdot)$ associates with a random variable X. The risk premium is implicitly defined by the condition

$$u\big(\mathrm{E}[X] - \rho_u(X)\big) = U(X). \tag{7.5}$$

The risk premium also defines a **certainty equivalent**, i.e., a sure and guaranteed payoff $\mathrm{CE}_u(X)$, such that the agent would be indifferent between this certain amount and the uncertain lottery:

$$\mathrm{CE}_u(X) = \mathrm{E}[X] - \rho_u(X).$$

---

[9] Hopefully, no confusion will arise with the usual notation for the correlation coefficient ρ.

7.3 A conceptual tool: The utility function

Note that the certainty equivalent is smaller than the expected value, and the difference is larger when the risk premium is larger. These concepts may be better grasped by looking again at Fig. 7.1.

▣ Example 7.8 Certainty equivalent and risk premium

In Example 7.7, we have seen that the sure lottery a_1 is preferred to a_2 by a decision maker characterized by a logarithmic utility. Let us find the corresponding certainty equivalent for lottery a_2. We need a sure amount $x = \text{CE}_{\log}(a_2)$, such that

$$u(x) = \log x = U(a_2) = 2.1587.$$

Hence,

$$\text{CE}_{\log}(a_2) = e^{2.1587} = 8.6603,$$

and the risk premium is

$$\rho_{\log}(a_2) = 10 - 8.6603 = 1.3397.$$

We may interpret the risk premium as the additional expected payoff that a risk-averse decision maker requires to switch from the risk-free alternative a_1 to the risky alternative a_2, or the amount that she is willing to give up in order to get rid of the risk of a_2.

Example 7.8 points out a difficulty with the risk premium concept: It mixes the intrinsic risk of a lottery[10] with the subjective risk attitude of the decision maker. Thus, we might wish to separate the two sides of the coin. Consider a lottery $X = x + \tilde{\epsilon}$, where x is a given number and $\tilde{\epsilon}$ is a random variable with $\text{E}[\tilde{\epsilon}] = 0$ and $\text{Var}(\tilde{\epsilon}) = \sigma^2$. Hence,

$$\text{E}[X] = x, \quad \text{Var}(X) = \sigma^2.$$

Let us assume that the random variable $\tilde{\epsilon}$ is a "small" perturbation, in the sense that any possible realization ϵ is a relatively small number. Hence, we may approximate both sides of Eq. (7.5) by Taylor expansions. Consider, for instance, the expression $u(x + \epsilon)$. Since only numbers are involved here, we may write

$$u(x + \epsilon) \approx u(x) + \epsilon u'(x) + \tfrac{1}{2}\epsilon^2 u''(x).$$

By using this approximation for the random variable $\tilde{\epsilon}$, under the assumption that its realizations are small enough, and taking expected values, we may ap-

---

[10] Here we assume that the risk is related to objective probabilities, but the same concept would apply in the case of a subjective assessment of probabilities, if we disregard distributional ambiguity.

proximate the right-hand side of Eq. (7.5) as follows:

$$\begin{aligned} U(X) \doteq \mathrm{E}\big[u(x+\tilde{\epsilon})\big] &\approx \mathrm{E}\big[u(x) + \tilde{\epsilon}u'(x) + \tfrac{1}{2}\tilde{\epsilon}^2 u''(x)\big] \\ &= u(x) + \mathrm{E}[\tilde{\epsilon}]u'(x) + \tfrac{1}{2}\mathrm{E}[\tilde{\epsilon}^2]u''(x) \\ &= u(x) + 0 \cdot u'(x) + \tfrac{1}{2}\mathrm{Var}(\tilde{\epsilon})u''(x) \\ &= u(x) + \tfrac{1}{2}\sigma^2 u''(x). \end{aligned} \qquad (7.6)$$

In the second-to-last line, we have used the well-known identity $\mathrm{Var}(\tilde{\epsilon}) = \mathrm{E}[\tilde{\epsilon}^2] - \mathrm{E}^2[\tilde{\epsilon}] = \mathrm{E}[\tilde{\epsilon}^2] - 0$. We may also approximate the left-hand side of Eq. (7.5), which involves only numbers, by a first-order expansion around $\mathrm{E}[X] = x$:

$$u\big(\mathrm{E}[X] - \rho_u(X)\big) \approx u(x) - \rho_u(X)u'(x). \qquad (7.7)$$

By equating the two approximations (7.6) and (7.7) and rearranging, we find

$$\rho_u(X) = -\frac{1}{2}\frac{u''(x)}{u'(x)}\sigma^2. \qquad (7.8)$$

Since we assume that the utility function is concave and strictly increasing, the right-hand side of Eq. (7.8) is well-defined and positive.[11] We observe that the risk premium is factored as the product of a term depending on the agent's subjective risk aversion, represented by the utility function $u(\cdot)$, and another one depending on the intrinsic uncertainty of the lottery, represented by the standard deviation σ. This justifies the following definition of the **coefficient of absolute risk aversion**:

$$R_u^a(x) \doteq -\frac{u''(x)}{u'(x)}. \qquad (7.9)$$

The more concave the utility function, i.e., the larger $u''(x)$ in absolute value, the larger the risk aversion. We have observed that, given the linearity of the expectation operator, transforming the utility function $u(x)$ by an increasing affine transformation is inconsequential. Indeed, the definition of the risk aversion coefficient is consistent with this observation, as it is easy to see that the coefficients for $u(x)$ and $\bar{u}(x) = au(x) + b$ are the same.

We should also note that the coefficient $R_u^a(x)$ may change considerably as a function of x. If we consider the asset allocation problem of Example 7.3, we may use expected utility as the functional of terminal wealth $W_T(\omega)$, which we should maximize with respect to the vector **h** of the asset holdings. Let us denote by $R_\mathbf{h}$ the corresponding holding period return of the portfolio. Then, we should maximize

$$U(W_T) = \mathrm{E}\big[u\big(W_0 \cdot (1 + R_\mathbf{h})\big)\big].$$

In general, the solution may change as a function of W_0. From an investor's perspective, in fact, risk aversion may depend on the current level of wealth.

---

[11] We recall that, for a differentiable concave function of one variable, we have $u''(x) \leq 0$.

7.3 A conceptual tool: The utility function

By a similar token, we may define the **coefficient of relative risk aversion**. This is motivated by considering a *multiplicative*, rather than additive, shock on an expected value x: $X = x \cdot (1 + \tilde{\epsilon})$. Here $\mathrm{E}[\tilde{\epsilon}] = 0$ and $\mathrm{Var}(\tilde{\epsilon}) = \sigma^2$, as before, but

$$\mathrm{E}[X] = x, \quad \mathrm{Var}(X) = x^2 \sigma^2.$$

The mean is preserved again, but the random variable $\tilde{\epsilon}$ is related to a return in this case. Then, we may consider a *relative* risk premium $\pi_u(X)$ as the *fraction* of wealth that the decision maker is willing to give up in order to avoid taking chances,

$$\pi_u(X) \doteq \frac{x - \mathsf{CE}_u(X)}{x},$$

which implies

$$\rho_u(X) = x - \mathsf{CE}_u(X) = \pi_u(X) \cdot x.$$

Now, using a first-order Taylor approximation as before, we may write

$$\begin{aligned} u\big(\mathsf{CE}_u(X)\big) = u\big(\mathrm{E}[X] - \rho_u(X)\big) &\approx u(x) - \rho_u(X) u'(x) \\ &= u(x) - \pi_u(X) \cdot x u'(x). \end{aligned} \quad (7.10)$$

The utility of X can be approximated by a second-order expansion, for a realization ϵ:

$$u(X) = u(x + x\epsilon) \approx u(x) + u'(x) x \epsilon + \frac{1}{2} u''(x) x^2 \epsilon^2.$$

By taking expectations and observing that $\mathrm{E}[\epsilon^2] = \sigma^2$, we find

$$\mathrm{E}\big[u(X)\big] \approx u(x) + \frac{1}{2} u''(x) x^2 \sigma^2. \quad (7.11)$$

Putting Eqs. (7.10) and (7.11) together and rearranging yield

$$\pi_u(X) = -\frac{1}{2} \frac{u''(x)}{u'(x)} x \sigma^2,$$

which suggests the definition of the relative risk aversion coefficient,

$$R_u^r(x) \doteq -\frac{u''(x) x}{u'(x)}. \quad (7.12)$$

The only difference with respect to the absolute coefficient is the multiplication by x.

7.3.1 A FEW STANDARD UTILITY FUNCTIONS

Beside listing some common utility functions, in this section, we want to illustrate how to classify them according to some relevant criteria. This is best illustrated by a simple example.

Example 7.9 Logarithmic utility

A typical utility function is the logarithmic utility:

$$u(x) = \log(x). \tag{7.13}$$

Clearly this makes sense only for positive values of wealth. It is easy to check that, for the logarithmic utility, we have

$$R_u^a(x) = \frac{1}{x}, \qquad R_u^r(x) = 1.$$

Hence, logarithmic utility has decreasing absolute risk aversion, but constant relative risk aversion.

The coefficients of absolute and relative aversion may be decreasing, constant, or increasing with respect to their argument. Hence, utility functions may belong to one of the following families:

- Decreasing, or constant, or increasing absolute risk aversion, denoted by **DARA**, **CARA**, and **IARA**, respectively.
- Decreasing, or constant, or increasing relative risk aversion, denoted by **DRRA**, **CRRA**, and **IRRA**, respectively.

Thus, logarithmic utility is DARA and CRRA. Furthermore, it may be thought of as a limit case of the more general family of power utility functions:

$$u(x) = \frac{x^{1-\gamma} - 1}{1 - \gamma}, \qquad \gamma > 1. \tag{7.14}$$

To understand the reasons behind the parameterization with respect to γ, let us find the coefficient of relative risk aversion of power utility:

$$u'(x) = x^{-\gamma},$$
$$u''(x) = -\gamma x^{-(\gamma+1)},$$
$$R_u^r(x) = x \cdot \gamma \cdot \frac{x^\gamma}{x^{\gamma+1}} = \gamma.$$

Furthermore, using L'Hôpital's rule,[12] we find[13]

$$\lim_{\gamma \to 1} \frac{x^{1-\gamma} - 1}{1 - \gamma} = \lim_{\gamma \to 1} \frac{-\log(x) \cdot x^{1-\gamma}}{-1} = \log(x).$$

---

[12]L'Hôpital's rule is used to find the limit $\lim_{x \to x_0} f(x)/g(x)$, in the case where both functions $f(\cdot)$ and $g(\cdot)$ tend to zero. Subject to technical conditions, the limit is $\lim_{x \to x_0} f'(x)/g'(x)$.
[13]Here, we are also using the derivative of the function $f(x) = a^x = e^{x \log a}$, which is $f'(x) = a^x \cdot \log a$.

7.3 A conceptual tool: The utility function

We may also consider the exponential utility function

$$u(x) = -e^{-\alpha x}, \tag{7.15}$$

for $\alpha > 0$. Note that this is an increasing function, and it is easy to interpret the parameter α:

$$R_u^a(x) = -\frac{-\alpha^2 e^{-\alpha x}}{\alpha e^{-\alpha x}} = \alpha.$$

Hence, we conclude that the exponential utility is CARA. This feature may be somewhat at odds with intuition, as one might expect that wealthier individuals are less averse to risk. It is important to remark that some utility functions have been used in the academic literature, because they are easy to manipulate, but this does not imply that they always model realistic investors' behavior.[14]

Another common utility function is **quadratic utility**:

$$u(x) = x - \frac{\lambda}{2}x^2. \tag{7.16}$$

Note that this function is not monotonically increasing and makes sense only for $x \in [0, 1/\lambda]$. Another odd property of quadratic utility is that it is IARA:

$$R_u^a(x) = \frac{\lambda}{1 - \lambda x} \quad \Rightarrow \quad \frac{dR_u^a(x)}{dx} = \frac{\lambda^2}{(1 - \lambda x)^2} > 0.$$

This implies, for instance, that an investor becomes more risk-averse if her wealth increases, which is usually considered at odds with standard investors' behavior. Nevertheless, it may be argued that, since any concave utility function may be locally approximated by a quadratic utility function, this provides a useful tool anyway. Furthermore, quadratic utility emphasizes the role of variance, since we have

$$U(X) = \mathrm{E}\left[X - \frac{\lambda}{2}X^2\right] = \mathrm{E}[X] - \frac{\lambda}{2}\left(\mathrm{Var}(X) + \mathrm{E}^2[X]\right). \tag{7.17}$$

A decision maker with quadratic utility is basically concerned only with the expected value and the variance of an uncertain outcome. In chapter 8, we will discuss the connection with mean–variance portfolio optimization (see Section 8.5).

Example 7.10 Logarithmic utility and portfolio choice

Consider the following stylized portfolio optimization problem:

- We represent uncertainty in asset return by a binomial model: There are two possible states of the world in the future, the up and down states, with probabilities p and $q = 1 - p$, respectively.

---

[14] See Problem 7.1 for an example concerning the odd behavior of the exponential utility function.

- There are two assets: one is risk-free, the other one is risky.
- The risk-free asset has gain R_f in both states (recall that multiplicative gain is one plus holding period return; in other words, \$1 grows to \$$R_f$).
- Current price for the risky asset is S_0 and its gain is u in the up state and d in the down state. Hence, the two possible risky asset prices are uS_0 and dS_0. We use gain, rather than holding period return, to streamline notation.
- Initial wealth is W_0 and the investor has logarithmic utility.

In this problem, there is actually one decision variable, which we may take as δ, the number of stock shares purchased by the investor. To get rid of the budget constraint, we observe that δS_0 is the wealth invested in the risky asset, and $W_0 - \delta S_0$ is invested in the risk-free asset. Then, future wealth will be, for each of the two possible states:

$$W_u = \delta S_0 u + (W_0 - \delta S_0)R_f = \delta S_0(u - R_f) + W_0 R_f,$$
$$W_d = \delta S_0 d + (W_0 - \delta S_0)R_f = \delta S_0(d - R_f) + W_0 R_f,$$

and expected utility is $p \log(W_u) + q \log(W_d)$. The problem is then

$$\max_{\delta} \; p \log \{\delta S_0(u - R_f) + W_0 R_f\}$$
$$+ q \log \{\delta S_0(d - R_f) + W_0 R_f\}.$$

Let us write the first-order (stationarity) condition for optimality:

$$p \cdot \frac{S_0(u - R_f)}{\delta S_0(u - R_f) + W_0 R_f} + q \cdot \frac{S_0(d - R_f)}{\delta S_0(d - R_f) + W_0 R_f} = 0.$$

In order to solve for δ, we may rearrange the equation a bit:

$$\frac{\delta S_0(u - R_f) + W_0 R_f}{pS_0(u - R_f)} = -\frac{\delta S_0(d - R_f) + W_0 R_f}{qS_0(d - R_f)}.$$

Straightforward manipulations yield

$$\frac{\delta}{p} + \frac{W_0 R_f}{pS_0(u - R_f)} = -\frac{\delta}{q} - \frac{W_0 R_f}{qS_0(d - R_f)}$$

and

$$\delta \left[\frac{1}{p} + \frac{1}{q}\right] = -\frac{W_0 R_f \left[q(d - R_f) + p(u - R_f)\right]}{pqS_0(u - R_f)(d - R_f)}.$$

Then, one last step yields

$$\frac{\delta S_0}{W_0} = \frac{R_f \left[up + dq - R_f\right]}{(u - R_f)(R_f - d)}.$$

7.3 A conceptual tool: The utility function

> This relationship implies that the *fraction* of initial wealth invested in the risky asset does not depend on the initial wealth itself. We have derived this property in a simplified setting, but it holds more generally for logarithmic utility and is essentially due to its CRRA feature.

Example 7.10 shows how the features of each utility function may affect the solution of decision problems. One must be aware of the implied behavior, when choosing a specific utility function. Once again, we recall that, in this chapter, we are dealing with static decision problems. The definition of a utility function gets much more complicated in the case of multistage problems, as intertemporal issues arise.

7.3.2 LIMITATIONS OF UTILITY FUNCTIONS

Utility functions have been subjected to much criticism over the years:

- They rely on critical assumptions about the underlying preference relationships and may lead to paradoxes.
- They assume a significant degree of rationality in decision makers, who may be affected in real life by lack of information and cognitive limitations, leading to behavioral anomalies that are not explained within the standard utility framework. Some experiments shows that the observed behavior of decision makers may contradict the expected utility paradigm, as we discuss in Section 10.5.
- They aim at modeling subjective risk aversion, but a portfolio manager has to cope with multiple clients, and she should certainly *not* make decisions according to her own degree of risk aversion. Objective risk measures may be preferable.
- It is difficult to *elicit* a specific utility function from a decision maker.

In Section 7.4, we resort to an alternative approach, based on mean–risk models. The idea is to introduce an objective risk measure, which is a functional mapping random variables into real numbers, and trade expected profit/return against risk. This leads to a multiobjective optimization problem. As we have seen in Supplement S2.1, one possibility to cope with multiple objectives is to form a linear combination of two objective functions. For instance, when dealing with a random return R, a natural idea is to define a risk-adjusted expected return,

$$\mathrm{E}[R] - \tfrac{1}{2}\lambda \mathrm{Var}(R). \tag{7.18}$$

This mean–risk objective looks much like an expected quadratic utility, even though a comparison with Eq. (7.17) shows that they are not exactly the same. We shall introduce alternative risk measures to cope with asymmetric risks.

Before doing so, we may take advantage of the streamlined form of Eq. (7.18) to show how we might try to estimate the risk aversion coefficient λ in a simple case.[15]

Example 7.11 Estimating risk aversion

Say that we own a piece of real estate and we want to insure it against a disaster that may occur with probability p. If disaster strikes, our loss is 100% of the property value. Risk may be represented by a Bernoulli random variable:

- With probability p, return is -1 (we lose 100% of the property).
- With probability $1 - p$, return is 0.

Then,
$$E[R] = p \times (-1) + (1-p) \times 0 = -p,$$
$$\text{Var}(R) = p \times (-1)^2 + (1-p) \times 0^2 - p^2 = p(1-p).$$

Note that the expected return is negative, as we are facing a potential loss. By abusing proper quadratic utility a little bit, let us consider the mean–risk form of Eq. (7.18),
$$U(R) = E[R] - \tfrac{1}{2}\lambda \text{Var}(R).$$

In this specific case, the utility score is, for a given risk aversion coefficient λ,
$$U = -p - \tfrac{1}{2}\lambda p(1-p).$$

We may consider insuring the property for a given premium. The more we are willing to pay, the more risk-averse we are. If we are willing to pay at most ν, then the utility of the certain equivalent loss of $-\nu$ is equal to the above utility score:
$$U = -\nu \quad \Rightarrow \quad \nu = p + \tfrac{1}{2}\lambda p(1-p).$$

As a reality check, observe that a risk-neutral investor ($\lambda = 0$) would just pay p, the expected loss. Given the insurance premium ν that we are willing to pay, this relationship allows to figure out a sensible value of λ, since
$$\lambda = \frac{2(\nu - p)}{p(1-p)}.$$

To get a more intuitive feeling, imagine that p is small ($1 - p \approx 1$), so that
$$\nu \approx p + \tfrac{1}{2}\lambda p.$$

---

[15] The example is borrowed from [3, Chapter 6].

> Let us try a few values of λ:
>
> $$\lambda = 0 \;\Rightarrow\; \nu = p,$$
> $$\lambda = 1 \;\Rightarrow\; \nu \approx 1.5p,$$
> $$\lambda = 2 \;\Rightarrow\; \nu \approx 2p,$$
> $$\lambda = 3 \;\Rightarrow\; \nu \approx 2.5p.$$
>
> Therefore, for each unit increment in the risk aversion coefficient, we should be willing to pay another 50% of the expected loss.
>
> In portfolio optimization, it is commonly agreed that λ ranges between 2 and 4.

7.4 Mean–risk models

The framework of expected utility suffers from the limitations that we have outlined in Section 7.3.2. Arguably, the most critical one is that a utility function mixes objective risk measurement and subjective risk aversion in decision-making. This is quite evident in the concept of risk premium. Hence, practitioners in financial industry prefer to rely on the concept of a risk measure. From a mathematical viewpoint, we should arguably talk of a risk *functional*, since what we need is a way to map a random variable $X(\omega)$, which is itself a function, to a real number:

$$\xi : X(\omega) \to \mathbb{R}.$$

We will use both terms interchangeably. Armed with a risk measure, we may tackle the problem of finding a satisfactory risk–reward tradeoff by using concepts of multiobjective optimization, as discussed in Section S2.1. This results in mean–risk optimization models.

If we choose variance or standard deviation as risk measures, we end up with the mean–variance portfolio optimization model that we have introduced in Section 2.1.1. Mean–variance optimization relies on variance for the sake of computational convenience, as this choice leads to a simple quadratic programming model. However, the underlying idea is actually using standard deviation as a risk measure. Standard deviation can be considered as a risk measure: the smaller, the better. However, while standard deviation captures the dispersion of a probability distribution, is it really a good risk measure? Example 7.2 clearly shows that symmetric risk measures, like standard deviation or variance, may fail with skewed distributions. As an alternative, we have considered value-at-risk, which is an asymmetric risk measure, in Section 2.2.2.

Value-at-risk is an example of asymmetric risk measure based on quantiles. However, we may easily define an asymmetric risk measure based on variance, namely, **semivariance**. If X is a random variable modeling profit or return, its semivariance is defined as

$$\mathrm{E}\left[\left(\max\{0, \mu_X - X\}\right)^2\right]. \tag{7.19}$$

In practice, we consider only negative deviations with respect to the expected value. The idea can be generalized and made more flexible, if we introduce negative deviations with respect to a minimum target that we wish to achieve, i.e., shortfall amounts. Let us denote the random terminal wealth associated with a portfolio by W_T. If we choose a target wealth $W_{\min}$, we may be interested in evaluating the portfolio performance in terms of **shortfall probability**,

$$\mathrm{P}\{W_T < W_{\min}\},$$

or **expected shortfall**,

$$\mathrm{E}\left[\max\{0, W_{\min} - W_T\}\right].$$

Shortfall is zero if we achieve or exceed the target, so we are penalizing underachievement in an asymmetric way. Expected shortfall, when used within portfolio optimization modeling, may result in simple linear programming problems.[16] To this aim, we should discretize the expectation by generating a finite set of scenarios, as customary in stochastic programming. If we wish to penalize large shortfalls more heavily, we may consider the expected squared shortfall,

$$\mathrm{E}\left[\left(\max\{0, W_{\min} - W_T\}\right)^2\right],$$

which may be tackled by quadratic programming.

How do these measures compare against each other? In order to provide a sensible answer, we must clarify the desirable properties of a risk measure.

7.4.1 COHERENT RISK MEASURES

A **single-period risk measure** is a functional $\xi(\cdot)$ mapping a random variable $X(\omega)$ to the real line. The random variable might be interpreted as the value of a portfolio, or a profit or loss, i.e., a change in value. Furthermore, loss might be relative with respect to an expected future target, or an absolute loss. In this section, we list some desirable properties of a risk measure. In the literature, different statements of these properties may be found, depending on the interpretation of $X(\omega)$. Here, we assume that the random variable represents a profit or the value of a portfolio. Hence, the larger the random variable, the better, but the risk measure is defined in such a way that it should be minimized.

The following set of properties characterizes a **coherent** risk measure:

---

[16] See Chapter 15.

7.4 Mean–risk models

- **Normalization.** Consider a random variable that is identically zero, $X \equiv 0$. It is reasonable to set $\xi(0) = 0$; if we do not hold any portfolio, we are not exposed to any risk.
- **Monotonicity.** If $X_1 \leq X_2$,[17] then $\xi(X_1) \geq \xi(X_2)$. In plain English, if the value of portfolio 1 is never larger than the value of portfolio 2, then portfolio 1 is at least as risky as portfolio 2.
- **Translation invariance.** If we add a fixed amount a to the portfolio, the risk measure is affected: $\xi(X + a) = \xi(X) - a$. If $a > 0$, risk is reduced.
- **Positive homogeneity.** Intuitively, if we double the amount invested in a portfolio, we double risk. Formally: $\xi(bX) = b\xi(X)$, for $b \geq 0$.
- **Subadditivity.** Diversification is expected to decrease risk; at the very least, diversification cannot increase risk. Hence, it makes sense to assume that the risk of the sum of two random variables should not exceed the sum of the respective risks: $\xi(X + Y) \leq \xi(X) + \xi(Y)$.

We are dealing only with a single-period problem; tackling multiperiod problems may complicate the matter further, introducing issues related to time consistency, which we do not consider in this book.[18]

Remark. An interesting implication of translation invariance is

$$\xi\big(X + \xi(X)\big) = \xi(X) - \xi(X) = 0.$$

Thus, the risk measure of a portfolio with random value X may be interpreted as the minimum amount of additional capital that is needed to make the portfolio *acceptable*, where a portfolio X is said to be acceptable if its risk measure is $\xi(X) \leq 0$. In fact, risk measures (functionals) may also be interpreted as **acceptability functionals**.

We have listed theoretical requirements of a risk measure, but what about the practical ones?

- Clearly, a risk measure should not be overly difficult to compute. Unfortunately, computational effort may be an issue, if we deal with financial derivatives whose pricing itself requires intensive computation.
- When solving a portfolio optimization model, convexity is a quite important feature. Positive homogeneity and subadditivity may be combined into a convexity condition:

$$\xi\big(\lambda X + (1-\lambda)Y\big) \leq \lambda \xi(X) + (1-\lambda)\xi(Y), \qquad \forall \lambda \in [0,1].$$

Thus, apart from theoretical considerations, a coherent risk measure may be practically preferable from a computational viewpoint.

---

[17] Since we are comparing random variables, the inequality should be qualified as holding almost surely, i.e., for all of the possible outcomes, with the exception of a set of measure zero. The unfamiliar reader may consider this as a technicality.

[18] The essence of time consistency of a multiperiod risk measure is that if a portfolio is riskier than another portfolio at time horizon τ, then it is riskier at time horizons $t < \tau$ as well. See, e.g., [2].

- Another requirement is that the risk measure should be easily communicated to top management. A statistically motivated measure, characterizing a feature of a probability distribution, may be fine for the initiated, but a risk measure expressed in hard monetary terms can be easier to grasp. We also note that specific sensitivity measures, like bond duration (and the option Greeks that we shall meet later), do not enable us to summarize all risk contributions, irrespectively of the nature of the different positions held in the portfolio. These difficulties led to the development of value-at-risk.

7.4.2 STANDARD DEVIATION AND VARIANCE AS RISK MEASURES

We are aware that a major limitation of standard deviation and variance is their symmetry, since they measure dispersion without paying attention to direction of variability. Let us run a more formal check by asking whether they meet the coherence requirements.

- The normalization requirement is met, but we know that, for any real number a,
$$\text{Var}(X + a) = \text{Var}(X).$$

Hence, variance and standard deviation are not translation invariant. We find a translation invariant measure, however, if we consider

$$\xi(X) = -\text{E}[X] + \lambda\sqrt{\text{Var}(X)},$$

since

$$\xi(X + a) = -\text{E}[X + a] + \lambda\sqrt{\text{Var}(X + a)} = \xi(X) - a.$$

- Monotonicity fails, as we have seen in Example 7.2. More generally, if we have a random variable bounded by a constant,

$$X_1(\omega) \leq \alpha,$$

and we consider $X_2(\omega) = \alpha$, the monotonicity condition fails since $\text{Var}(X_1) \geq 0$ and $\text{Var}(X_2) = 0$.
- Let us consider positive homogeneity. Since

$$\text{Var}(bX) = b^2\text{Var}(X),$$

this condition fails for variance, but it is met by standard deviation (just take the square root).
- Let us complete the picture with subadditivity. Since

$$\text{Var}(X_1 + X_2) = \text{Var}(X_1) + \text{Var}(X_2) + 2\text{Cov}(X_1, X_2),$$

variance fails to meet subadditivity when covariance is positive. However, if we consider standard deviation and we express covariance using the correlation coefficient $\rho_{12} \leq 1$, we see that standard deviation is subadditive:

$$\sigma_{X_1+X_2} = \sqrt{\sigma_{X_1}^2 + \sigma_{X_2}^2 + 2\rho_{12}\sigma_{X_1}\sigma_{X_2}}$$
$$\leq \sqrt{\sigma_{X_1}^2 + \sigma_{X_2}^2 + 2\sigma_{X_1}\sigma_{X_2}} = \sigma_{X_1} + \sigma_{X_2}.$$

Hence, the picture is not quite encouraging for standard deviation and variance as risk measures, but standard deviation looks a bit better. From a practical viewpoint, when dealing with the return of a simple portfolio, variance may result in simple optimization problems, i.e., convex quadratic programs. However, this is not necessarily true when considering more complicated optimization models, where scenarios in terms of underlying risk factors are generated and mapped to asset prices by a nonlinear pricing model, and a stochastic programming model is solved.[19] Furthermore, while standard deviation or return or wealth may make sense, variance of wealth, which is measured in squared monetary units, cannot be really be interpreted. Even standard deviation of wealth may fail to convey a precise perception of directional risk. Nevertheless, these measures are broadly used in the context of modern portfolio theory, which relies on mean–variance optimization. As we shall see, this provides us with useful insights, like the capital asset pricing model, and it may be sometimes justified, since quadratic utility can approximate a generic concave utility function locally.

7.4.3 QUANTILE-BASED RISK MEASURES: V@R AND CV@R

We have introduced value-at-risk, in Section 2.2.2.1, as a quantile of the probability distribution of loss. There, we have considered typical textbook examples relying on normality, in order give a simple picture. However, in practice, estimating V@R is far from trivial for a complex trading book involving exotic derivatives, as well as equity or fixed-income assets. Whatever approach we use for its computation, V@R is not free from some fundamental flaws, which depend on its definition as a quantile. We should be well aware of them, especially when using sophisticated computational tools that may lure us into a false sense of security. The following example shows how a quantile cannot distinguish between different tail shapes.

◼ Example 7.12 Different shapes of a tail

> Consider the two loss densities in Fig. 7.2. In Fig. 7.2(a), we observe a normally distributed loss and its 95% V@R, which is just its quantile

---

[19] See Chapter 15.

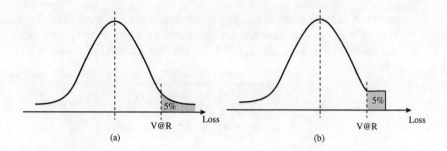

FIGURE 7.2 Value-at-risk can be the same in quite different situations.

> at probability level 95%; the area of the right tail is 5%. In Fig. 7.2(b), we observe a sort of truncated distribution, obtained by replacing the tail of the normal PDF with a uniform density. The tail accounts for 5% of the total probability. By construction, V@R is the same in both cases, since the areas of the right tails are identical. However, we might not associate the same risk with the two distributions. In the case of the normal distribution, there is no upper bound to loss; in the second case, there is a clearly defined worst-case loss. Whether the risk for density (a) is larger than density (b) or not, it depends on how we measure risk exactly; the point is that V@R cannot tell the difference between them.

One way to overcome the limitations of a straightforward quantile, while retaining some of its desirable features, is to resort to a conditional expectation on the tail. This observation has led to the definition of alternative risk measures, such as **conditional value-at-risk** (CV@R), which (informally) is the expected value of loss, conditional on being to the right of V@R. For instance, the conditional (tail) expectation yields the midpoint of the uniform tail in the truncated density of Fig. 7.2(b); the tail expectation may be larger in the normal case of Fig. 7.2(a), because of its unbounded support. In this section, we investigate the properties of both V@R and CV@R in terms of coherence and computational viability.

7.4.3.1 A remark on quantiles

When defining quantile-based risk measures, there is no particular difficulty with standard continuous distributions featuring a continuous and strictly increasing CDF

$$F_X(x) \doteq \mathrm{P}\{X \leq x\}.$$

7.4 Mean–risk models

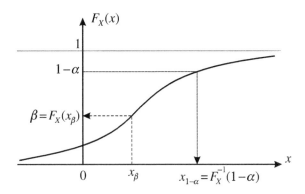

FIGURE 7.3 The link between quantiles and the CDF.

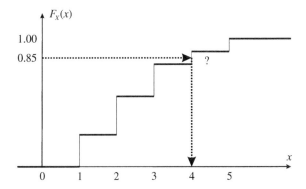

FIGURE 7.4 A noninvertible CDF of a discrete random variable.

In such a case, the CDF is invertible and the quantile $x_{1-\alpha}$ at probability level $1 - \alpha$ is easily found:

$$F_X(x_{1-\alpha}) = 1 - \alpha \quad \Rightarrow \quad x_{1-\alpha} = F_X^{-1}(1 - \alpha), \tag{7.20}$$

where $\alpha \in [0, 1]$ is the probability mass on the right tail (which is supposed to be small if we are considering a loss). This is illustrated in Fig. 7.3. Given a numerical value x_β, the CDF $F_X(x_\beta)$ gives the corresponding probability $\beta = P\{X \leq x_\beta\}$. Going the other way around, given the probability $1 - \alpha$, inversion of the CDF yields the corresponding quantile $x_{1-\alpha}$.

However, the case of a discrete random variable is more involved, as the CDF is piecewise constant and not invertible. Figure 7.4 shows the CDF for a discrete random variable with the following PMF:

x	0	1	2	3	4	5
$p_X(x)$	0.05	0.20	0.30	0.25	0.10	0.10

Here,

$$F_X(3) = 0.8, \qquad F_X(4) = 0.9,$$

and we are in trouble when looking for the quantile $x_{0.85}$. Then, we may define the quantile as the smallest number $x_{1-\alpha}$ such that $F_X(x_{1-\alpha}) \geq 1 - \alpha$. This relies on the definition of a **generalized inverse function**:

$$x_{1-\alpha} = \min \{x : F_X(x) \geq 1 - \alpha\}.$$

The generalized inverse function boils down to the standard inverse, when the CDF is continuous and strictly increasing. In the numerical case that we are considering, we find

$$x_{0.85} = 4,$$

which makes sense in terms of "staying on the safe side." The intuitive idea is that the value 4 "covers" loss with a 90% guarantee, which is larger than necessary, but the value 3 offers a guarantee of only 80%. This looks innocent enough, but we should wonder about the possibility of defining quantiles using a *strict* inequality, as in

$$\inf \{x : F_X(x) > 1 - \alpha\},$$

where we really have to use inf, since the strict inequality does not guarantee the existence of a minimum. Actually, the question is not trivial and leads to alternative definitions of quantiles and risk measures, which may differ in terms of coherence. For the sake of simplicity, we will cut a few corners as usual.[20]

7.4.3.2 Is value-at-risk coherent?

In this section, we consider a continuous random variable L_T, modeling loss over the time horizon T for which we want to evaluate V@R. We assume that its CDF is invertible, so that value-at-risk with confidence level $1 - \alpha$, for the given time horizon T, is the usual quantile V@R$_{1-\alpha,T}$, such that

$$P\{L_T \leq \text{V@R}_{1-\alpha,T}\} = 1 - \alpha, \qquad (7.21)$$

and we may disregard technical complications. If we consider an affine transformation $aL_T + b$ of loss, with $a \geq 0$, we may manipulate Eq. (7.21) and find

$$P\{aL_T + b \leq a\text{V@R}_{1-\alpha,T} + b\} = 1 - \alpha,$$

showing translation invariance[21] and positive homogeneity. Value-at-risk is clearly normalized and monotonic, but what about subadditivity? If we restrict our attention to specific classes of distributions, such as the normal, V@R is subadditive (see Problem 7.4). However, this depends on the fact that quantiles of a normal distribution are related to the standard deviation, which is subadditive. The following counterexample is often used to show that V@R is not subadditive in general.

---

[20]For a deeper analysis, see, e.g., [6].

[21]In this case, the constant b is added, rather than subtracted, which seems at odds with the previous definition of translation invariance. The point is that the random variable L_T represents a loss, rather than a profit.

Example 7.13 V@R is not subadditive

Let us consider two zero-coupon bonds, whose issuers may default with probability 4% (over some time horizon that we leave implicit). Say that, in the case of default, we lose the full face value, $100 (in practice, we might partially recover the face value of the bond). Let us compute the V@R of each bond with confidence level 95%. We represent the loss for the two bonds by random variables X and Y, respectively, which take values in the set $\{0, 100\}$. Since loss has a discrete distribution in this example, we should use the more general definition of V@R provided by the generalized inverse. The probability of default is 4%, and $1 - 0.04 = 0.96 > 0.95$; therefore, we find

$$\text{V@R}_{0.95}(X) = \text{V@R}_{0.95}(Y) = \$0$$
$$\Rightarrow \text{V@R}_{0.95}(X) + \text{V@R}_{0.95}(Y) = \$0.$$

Now what happens if we hold both bonds and assume independent defaults? We will suffer:

- A loss of $0, with probability $0.96^2 = 0.9216$
- A loss of $100, with probability $2 \times 0.96 \times 0.04 = 0.0768$
- A loss of $200, with probability $0.04^2 = 0.0016$

Now the probability of losing $0 is smaller than 95%, and

$$P\{X + Y \leq 100\} = 0.9216 + 0.0768 > 0.95.$$

Hence, with that confidence level,

$$\text{V@R}_{0.95}(X + Y) = 100 > \text{V@R}_{0.95}(X) + \text{V@R}_{0.95}(Y),$$

which means that risk, as measured by V@R, may be increased by diversification.

The lack of subadditivity also implies that minimization of V@R may not result in convex portfolio optimization problems. When uncertainty is represented by sampled scenarios, it turns out that it is not even a differentiable function of portfolio weights.[22] This is not to say that V@R is not useful and relevant. In fact, it is used to assess capital requirements for banks, i.e., to determine the liquidity needed as a buffer against short-term loss. However, it must be used with care, and alternative measures have been proposed for the same purpose.

---

[22] See, e.g., [5, pp. 615–618].

7.4.3.3 Conditional value-at-risk

Conditional value-at-risk, CV@R, is an asymmetric risk measure related to tail expectations and, as such, bears some similarity with expected shortfall (in fact, the two concepts are sometimes confused). However, in expected shortfall, we fix a target a priori; here, the threshold is given by V@R. Informally, CV@R is defined as a conditional tail expectation[23] of loss over a time horizon T, where the threshold is V@R with probability level $1 - \alpha$:

$$\text{CV@R}_{1-\alpha,T} \doteq \text{E}\big[L_T \mid L_T > \text{V@R}_{1-\alpha,T}\big]. \qquad (7.22)$$

Since CV@R looks like a complication of V@R, it seems reasonable to expect that it is an even more difficult beast to tame. On the contrary, CV@R is much better behaved:

- It can be shown that CV@R is a coherent risk measure.
- A consequence of coherence is that CV@R is a convex risk measure.

The last point is quite relevant in terms of optimization modeling, as it suggests that minimization of CV@R and optimization subject to an upper bound on CV@R may result in relatively simple convex problems. We will consider CV@R optimization later, in Section 15.6.2.1. For now, let us consider an example in which, rather unsurprisingly, CV@R is easy to find.

Example 7.14 CV@R in the normal case

In the case of a normally distributed loss, $L \sim \mathsf{N}(\mu_L, \sigma_L^2)$, we may find an explicit expression for CV@R. Let us consider a standard normal loss $Z \sim \mathsf{N}(0,1)$ first, where $\text{V@R}_{1-\alpha} = z_{1-\alpha}$, the familiar quantile for the standard normal distribution. We have

$$\text{E}[Z \mid Z > z_{1-\alpha}]$$
$$= \frac{1}{\alpha} \int_{z_{1-\alpha}}^{+\infty} x \cdot \frac{1}{\sqrt{2\pi}} e^{-x^2/2} dx$$
$$= \frac{1}{\alpha} \frac{1}{\sqrt{2\pi}} \int_{z_{1-\alpha}^2/2}^{+\infty} e^{-y} dy \quad \text{(variable substitution } y = x^2/2\text{)}$$
$$= \frac{1}{\alpha} \frac{1}{\sqrt{2\pi}} e^{-y} \Big|_{+\infty}^{z_{1-\alpha}^2}$$

---

[23] As we have already noted, there are some critical issues in the careful definition of quantile-based risk measures, especially when dealing with discrete distributions. We disregard such subtleties. We should also mention that the term "average value-at-risk" is also used to refer to CV@R.

$$= \frac{1}{\alpha} \frac{1}{\sqrt{2\pi}} e^{-z_{1-\alpha}^2/2} = \frac{1}{\alpha} \phi(z_{1-\alpha}),$$

where $\phi(z)$ is the PDF of the standard normal, as usual. For instance, if $\alpha = 0.05$ (note that this is the small area on the right tail of the loss distribution),

$$\mathrm{E}[Z \mid Z > z_{0.95}] = \frac{1}{0.05} \times \phi(1.6449) = 2.0627.$$

In the case of a generic normal loss $L \sim \mathsf{N}(\mu, \sigma^2)$, we just destandardize by considering

$$L = \mu + \sigma Z, \quad q_{1-\alpha} = \mu + \sigma z_{1-\alpha}.$$

Hence,

$$\begin{aligned}
\mathrm{E}[L \mid L > q_{1-\alpha}] &= \mathrm{E}[\mu + \sigma Z \mid Z > z_{1-\alpha}] \\
&= \mu + \sigma \cdot \mathrm{E}[Z \mid Z > z_{1-\alpha}] \\
&= \mu + \frac{\sigma}{\alpha} \cdot \phi(z_{1-\alpha}).
\end{aligned} \tag{7.23}$$

For instance, if $L \sim \mathsf{N}(-50, 200^2)$, where the negative expected loss corresponds to a positive expected profit of 50, we find

$$\mathrm{CV@R}_{0.95} = -50 + \frac{200}{0.05} \times \phi(1.6449) = 362.54.$$

7.4.4 FORMULATION OF MEAN–RISK MODELS

The exact formulation of a mean–risk model depends on both modeling and computational convenience. As we have seen in Supplement S2.1, there are different scalarization strategies to boil a multiobjective problem down to a sequence of single-objective problems. In the mean–risk case, we are dealing with a model where:

- We represent a portfolio by the vector **x** of decision variables, constrained by a feasible set $S \subset \mathbb{R}^n$.
- We want to maximize an expected profit/return $\pi(\mathbf{x})$.
- We want to minimize a risk measure $\xi(\mathbf{x})$.

In practice, we must select a scalarized model, which may be obtained by defining a risk-adjusted mean,

$$\max \quad \pi(\mathbf{x}) - \lambda \xi(\mathbf{x})$$
$$\text{s.t.} \quad \mathbf{x} \in S,$$

or by requiring a minimum expected reward,

$$\min \quad \xi(\mathbf{x})$$
$$\text{s.t.} \quad \mathbf{x} \in S,$$
$$\pi(\mathbf{x}) \geq \beta,$$

or by defining a risk budget,

$$\max \quad \pi(\mathbf{x})$$
$$\text{s.t.} \quad \mathbf{x} \in S,$$
$$\xi(\mathbf{x}) \leq \gamma.$$

Depending on the selected risk measure and the adopted scalarization, we formulate one of the mathematical programming problems to be discussed later, in Chapter 15. They range from manageable linear programming models to difficult nonconvex problems. Furthermore, the scalarizations involve a choice of parameters λ, β, or γ, which may have a more or less intuitive meaning to the decision maker. Efficient solvers are available for a wide class of (convex) optimization problems, enabling us to tackle many practically significant problems.

7.5 Stochastic dominance

In principle, the framework of utility functions allows to find a complete ordering of portfolios. However, utility functions are difficult to elicit, and an investor might be reluctant to commit to a specific utility. The mean–risk framework may provide us with a partial ordering of alternatives, as well as a set of efficient portfolios. The stochastic dominance framework is a third alternative framework, resulting in a partial ordering that may be related to broad families of utility functions.

To get the intuition and a possible motivation, let us consider again the two lotteries of Example 7.2. The example shows a limitation of mean–variance analysis, since one lottery is clearly dominated by the other one, yet, we have an unclear tradeoff in terms of mean and variance. We may introduce a concept of dominance between random returns/payoffs X and Y fairly easily. We say that X dominates Y if[24]

$$Y(\omega) \leq X(\omega), \quad \forall \omega \in \Omega, \tag{7.24}$$

---

[24] As usual, the condition should be better qualified, as it applies with the possible exception of a subset of the sample space Ω with null measure. If the random variables are discrete, this is not relevant.

7.5 Stochastic dominance

and
$$P\{Y < X\} > 0.$$

In other words, X is never worse than Y, and X is strictly better than Y in some scenarios. Note that there is no clear relationship between this concept of dominance and efficiency. Nevertheless, assuming that investors are nonsatiated, i.e., they prefer more to less, no one would prefer Y to X. This concept of (strict) dominance is quite simple and intuitive, but it is not likely to be very useful in practice. It is unlikely that it will establish a rich preference relationship between portfolios. Actually, under a no-arbitrage assumption, we should expect that we *never* detect this kind of dominance.[25]

Hence, we must weaken the idea of strict dominance in order to find a more useful concept. To get a further clue, let us fix a target payoff/return β and assume that
$$P\{X \leq \beta\} \leq P\{Y \leq \beta\}.$$

What does this condition suggest about the choice between X and Y? Since $P\{X \leq \beta\} = 1 - P\{X > \beta\}$, we may rephrase the condition in terms of complementary probabilities as follows:
$$P\{X > \beta\} \geq P\{Y > \beta\}.$$

If we consider β as a target performance, we see that the probability of exceeding the target is larger for X than for Y. This may suggest that X is a better investment than Y, but actually this conclusion is not warranted, as the relationship could be reversed for other values of the target β. However, if we assume that this relationship holds for *every* possible target, we come up with the following definition.

DEFINITION 7.1 (First-order stochastic dominance) *Consider random variables X and Y. We say that X has first-order stochastic dominance over Y if*
$$P\{X \leq \beta\} \leq P\{Y \leq \beta\}, \qquad \forall \beta \in \mathbb{R}$$

and
$$P\{X \leq \gamma\} < P\{Y \leq \gamma\}, \qquad \text{for some } \gamma \in \mathbb{R}.$$

Note that the condition in Definition 7.1 may be restated in terms of the CDF of the two random variables:
$$F_X(\beta) \leq F_Y(\beta), \qquad \forall \beta \in \mathbb{R}$$
$$F_X(\gamma) < F_Y(\gamma), \qquad \text{for some } \gamma \in \mathbb{R}.$$

In plain English, if we plot the two CDFs, F_X is never above F_Y, and it is strictly less somewhere.

---

[25] See Section 2.4 for the link between dominance and arbitrage opportunities.

Table 7.3 An example of first-order stochastic dominance.

State	ω_1	ω_2	ω_3	ω_4	ω_5	
Probability	0.2	0.2	0.2	0.2	0.2	
Return r_X (%)	3	4	5	6	7	
Return r_Y (%)	7	6	5	3	3	

Return β (%)	2	3	4	5	6	7	8
$\mathrm{P}\{r_X \leq \beta\}$	0.0	0.2	0.4	0.6	0.8	1.0	1.0
$\mathrm{P}\{r_Y \leq \beta\}$	0.0	0.4	0.4	0.6	0.8	1.0	1.0

▣ Example 7.15 First-order stochastic dominance

> Let us consider the two investments described in Table 7.3. The first table gives the percentage return of the two investments in five states of the world. Clearly, there is no state-by-state dominance between the returns of the two alternatives. The second table shows the two CDFs for relevant values of return. Note that the CDF does not bear any relationship with the states of the world. The CDF of r_X never exceeds the CDF of r_Y and is strictly less at one point (return 3%). Hence, r_X first-order stochastically dominates r_Y.

In Definition 7.1, we are essentially assuming that investors prefer more to less, which is expressed by a strictly increasing utility function. This fact is formalized as follows.

THEOREM 7.2 *If X and Y satisfy the condition in Definition 7.1, then*

$$\mathrm{E}\big[u(X)\big] > \mathrm{E}\big[u(Y)\big],$$

for every utility function u satisfying the condition $u'(x) > 0$ for all x (u is differentiable and strictly increasing).

Actually, it turns out that the condition is necessary and sufficient, and we shall just sketch a proof in Supplement S7.1. We may get a glimpse of intuition by considering the following relationship *in distribution* between random variables X and Y:

$$Y \stackrel{d}{=} X + \xi, \tag{7.25}$$

where ξ is a nonpositive random variable. We should carefully note the fundamental difference between Eq. (7.24) and Eq. (7.25). In the latter case, we are *not* requiring a strong state-by-state condition,

$$Y(\omega) = X(\omega) + \xi(\omega), \qquad \forall \omega \in \Omega,$$

7.5 Stochastic dominance

with $\xi(\omega) \leq 0$, but only a weaker condition in terms of distribution, which is actually a way to rephrase first-order stochastic dominance. Then, if the utility function u is strictly increasing, we find

$$\mathrm{E}\big[u(Y)\big] = \mathrm{E}\big[u(X+\xi)\big] < \mathrm{E}\big[u(X)\big].$$

First-order stochastic dominance is easier to observe in the real world than an unreasonable state-by-state dominance, but it is still too strong and may not allow to compare alternatives in many significant cases. To see why, let us consider the specific case $u(x) = x$, i.e., the utility function is the identity function, which is to say that the investor prefers more to less but is risk-neutral. We clearly see that the condition in Theorem 7.2 implies

$$\mathrm{E}[X] > \mathrm{E}[Y].$$

This means that we cannot compare distributions with the same expected value, which is a significant limitation. To overcome this difficulty, a weaker condition has been introduced.

DEFINITION 7.3 (Second-order stochastic dominance) *Let us consider random variables X and Y. We say that X has second-order stochastic dominance over Y if*

$$\int_{-\infty}^{\beta} \mathrm{P}\{X \leq s\}\, ds \leq \int_{-\infty}^{\beta} \mathrm{P}\{Y \leq s\}\, ds, \qquad \forall \beta \in \mathbb{R},$$

and

$$\int_{-\infty}^{\gamma} \mathrm{P}\{X \leq s\}\, ds < \int_{-\infty}^{\gamma} \mathrm{P}\{Y \leq s\}\, ds, \qquad \text{for some } \gamma \in \mathbb{R}.$$

Definition 7.3 involves integrals of the CDF of random variables, which we may denote by

$$\widetilde{F}_X(x) \doteq \int_{-\infty}^{x} F_X(s)\, ds \equiv \int_{-\infty}^{x} \mathrm{P}\{X \leq s\}\, ds. \qquad (7.26)$$

Hence, the condition of second-order stochastic dominance may be restated as follows:

$$\widetilde{F}_X(\beta) \leq \widetilde{F}_Y(\beta), \qquad \forall \beta \in \mathbb{R}$$
$$\widetilde{F}_X(\gamma) < \widetilde{F}_Y(\gamma), \qquad \text{for some } \gamma \in \mathbb{R}.$$

First-order stochastic dominance implies second-order dominance; hence, it is a stronger concept. This is reflected in a weakened version of Theorem 7.2, whereby we add a condition related to risk aversion.

THEOREM 7.4 *If X and Y satisfy the condition in Definition 7.3, then*

$$\mathrm{E}\big[u(X)\big] > \mathrm{E}\big[u(Y)\big],$$

for every utility function u satisfying the conditions $u'(x) > 0$ and $u''(x) < 0$ for all x (u is differentiable, strictly increasing, and concave).

Stochastic dominance is an interesting concept, allowing us to establish a partial ordering between portfolios, which applies to a large range of sensible utility functions. Unfortunately, it is not quite trivial to translate the concept into computational terms, in order to make it suitable to portfolio optimization. Nevertheless, it is possible to build optimization models including stochastic dominance constraints with respect to a benchmark portfolio (see the chapter references).

S7.1 Theorem proofs

S7.1.1 PROOF OF THEOREM 7.2

The proof that we sketch here is rather limited, as we only deal with the case of random variables with a common bounded support $[a, b]$, for finite $a, b \in \mathbb{R}$. Nevertheless, it is simple enough and rather instructive. We assume that the random variables X and Y are continuous with densities (PDFs) $f_X(x)$ and $f_Y(y)$, related with the CDF as usual:

$$F_X(x) \doteq \mathrm{P}\{X \leq x\} \quad \text{and} \quad f_X(x) = F_X'(x).$$

We assume differentiability throughout. We should consider the difference of the expected utilities, which may be written as follows:

$$\mathrm{E}[u(X)] - \mathrm{E}[u(Y)] = \int_a^b u(x) f_X(x)\, dx - \int_a^b u(y) f_Y(y)\, dy$$
$$= \int_a^b u(x) F_X'(x)\, dx - \int_a^b u(y) F_Y'(y)\, dy.$$

Now we use integration by parts for both integrals. For instance,

$$\int_a^b u(x) F_X'(x)\, dx = u(b) F_X(b) - u(a) F_X(a) - \int_a^b u'(x) F_X(x)\, dx$$
$$= u(b) - \int_a^b u'(x) F_X(x)\, dx,$$

since the assumption of bounded support implies $F_X(b) = 1$ and $F_X(a) = 0$. A similar relationship applies to Y, and we find

$$\mathrm{E}[u(X)] - \mathrm{E}[u(Y)] = \int_a^b u'(y) F_Y(y)\, dy - \int_a^b u'(x) F_X(x)\, dx$$
$$= \int_a^b u'(z) \Big[F_Y(z) - F_X(z) \Big]\, dz. \qquad (7.27)$$

Now we observe that, for every z, $u'(z) > 0$, by the assumption of increasing monotonicity, and $F_Y(z) - F_X(z) \geq 0$, by the assumption of first-order dominance. Hence, the integral is positive, which proves the theorem.

S7.1.2 PROOF OF THEOREM 7.4

The proof, under similar assumptions about bounded support and differentiability of the involved functions, is quite similar to that of Theorem 7.2. We use the function $\widetilde{F}_X(x)$, i.e., the integral of the CDF that we have introduced in Eq. (7.26), so that

$$F_X(x) = \widetilde{F}'_X(x).$$

We start from Eq. (7.27) and, since risk aversion involves the second-order derivative $u''(x)$, we integrate by parts once more as follows:

$$\begin{aligned} \mathrm{E}[u(X)] - \mathrm{E}[u(Y)] &= \int_a^b u'(z)\Big[F_Y(z) - F_X(z)\Big] dz \\ &= u'(b)\Big[\widetilde{F}_Y(b) - \widetilde{F}_X(b)\Big] - \underbrace{u'(a)\Big[\widetilde{F}_Y(a) - \widetilde{F}_X(a)\Big]}_{=0} \\ &\quad - \int_a^b u''(z)\Big[\widetilde{F}_Y(z) - \widetilde{F}_X(z)\Big] dz, \end{aligned}$$

where the second term vanishes, since $\widetilde{F}_X(a)$ and $\widetilde{F}_X(a)$ are integrals on an interval $[a,a]$ with zero measure. Hence,

$$\widetilde{F}_Y(a) = \widetilde{F}_X(a) = 0.$$

The result now follows from the assumptions $u'(z) > 0$, $u''(z) < 0$, and the definition of second-order stochastic dominance.

Problems

7.1 Consider an exponential utility function $u(x) = -e^{-\alpha x}$, with a strictly positive α. An investor characterized by this exponential utility has to allocate an initial wealth W_0 between a risk-free and a risky asset. We assume a binomial uncertainty model, so that the risky asset has two possible gains (not returns) R_u and R_d, with probabilities π_u and π_d, respectively. Let q be the wealth allocated to the risky asset; it is possible to borrow cash as well to short-sell the risky asset. How does q change as a function of initial wealth W_0? Do you think that your utility function is exponential?

7.2 An investor endowed with an initial wealth $W_0 = 1$ (e.g., in euro) maximizes the expected value of the quadratic utility function $u(x) = ax - bx^2/2$, where x is the terminal wealth obtained by investing in n risky assets. Accordingly, the investor chooses a portfolio. Another investor, with a different initial wealth of $W_0 = K$, by optimizing the *same* utility function, chooses a different portfolio (in the sense that the asset weights are different).

- How do you explain the difference?

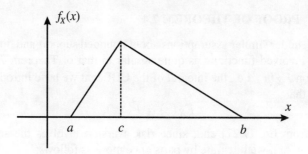

FIGURE 7.5 PDF of a triangular distribution.

- If the second investor changes the coefficient b to b', he finds the same portfolio as the first investor. What is the relationship between b and b'?

Note: This problem is borrowed from [13].

7.3 The value of your real estate property is $100,000. In case of a fire, your property may be lost or damaged, depending on how severe the accident is. Let us assume that the following scenarios give the residual value of your property in the future, depending on the possible occurrence of a fire:

State	Residual value	Probability
ω_1	$100,000	0.95
ω_2	$50,000	0.04
ω_3	$1	0.01

State ω_1 means that no accident occurred. Assume that your preferences are represented by a logarithmic utility depending on wealth, which is why we do not consider a residual value of $0, but $1. What is the maximum price that you would be willing to pay for an insurance guaranteeing coverage of any loss? Note: In the three states, the insurance will pay $0, $50,000, and $99,999, respectively, so that the value of your property is preserved.

7.4 Consider random variables L_1 and L_2, modeling loss from two portfolios, and assume that they are jointly normal. Show that, in this case, value-at-risk is subadditive.

7.5 Figure 7.5 shows the probability density function (PDF) of a generic triangular distribution with support (a, b) and mode c. For such a distribution, expected value and variance are given by the following formulas:

$$E[X] = \frac{a+b+c}{3},$$
$$\text{Var}(x) = \frac{a^2 + b^2 + c^2 - ab - ac - bc}{18}.$$

Say that the profit from a financial portfolio, with a holding period of a few weeks, has a triangular distribution with parameters (in €) $a = -75{,}000$, $b =$

55,000, and $c = 40,000$, so that the maximum possible loss is €75,000. Find V@R at level 95%. Note: The drawing of Fig. 7.5 is not in scale and is just meant as a qualitative hint.

7.6 Consider the following payoff distributions for two independent investment opportunities:[26]

Investment A		Investment B	
Payoff	Probability	Payoff	Probability
4	0.25	1	0.33
5	0.50	6	0.33
12	0.25	8	0.33

Compare the two alternatives in terms of stochastic dominance. *Hint:* You may plot the CDF, which is piecewise constant, and its integral, which is piecewise linear, for the two alternatives.

Further reading

- Decision-making under uncertainty is a topic of general interest, which is treated in different ways by different academic and practitioner communities. A thorough treatment of utility theory can be found, e.g., in [12], which is a treatment with a more economic flavor.
- There is an array of excellent books offering a treatment in a more financial vein, dealing with both utility theory and stochastic dominance. A concise, yet quite broad coverage of portfolio theory is offered in [11]. You may also see [7]. A more extensive treatment is offered in [9] or [10].
- The concept of coherent risk measure was introduced in [1].
- Risk measures are dealt with extensively in books with a more computational twist, especially stochastic programming. You may see [15], as well as [14]. A quite readable chapter on risk measures can also be found in [6].
- We have defined the concept of stochastic dominance, but portfolio optimization using this framework is more challenging than using utility functions or risk measures. See, e.g., [8] or [17].

Bibliography

1 P. Artzner, F. Delbaen, J.-M. Eber, and D. Heath. Coherent measures of risk. *Mathematical Finance*, 9:203–228, 1999.

---

[26]This is a numerical example borrowed from [7].

2. P. Artzner, F. Delbaen, J.-M. Eber, D. Heath, and H. Ku. Coherent multiperiod risk adjusted values and Bellman's principle. *Annals of Operations Research*, 152:5–22, 2007.

3. Z. Bodie, A. Kane, and A. Marcus. *Investments* (9th ed.). McGraw-Hill, New York, 2010.

4. P. Brandimarte. *Quantitative Methods: An Introduction for Business Management*. Wiley, Hoboken, NJ, 2011.

5. P. Brandimarte. *Handbook in Monte Carlo Simulation: Applications in Financial Engineering, Risk Management, and Economics*. Wiley, Hoboken, NJ, 2014.

6. M. Capiński and E. Kopp. *Portfolio Theory and Risk Management*. Cambridge University Press, Cambridge, 2014.

7. J.-P. Danthine and J.B. Donaldson. *Intermediate Financial Theory* (3rd ed.). Academic Press, San Diego, CA, 2015.

8. D. Dentcheva and A. Ruszczyński. Portfolio optimization with stochastic dominance constraints. *Journal of Banking & Finance*, 30:433–451, 2006.

9. C. Gollier. *The Economics of Risk and Time*. MIT Press, Cambridge, MA, 2001.

10. J.E. Ingersoll, Jr. *Theory of Financial Decision Making*. Rowman & Littlefield, Totowa, NJ, 1987.

11. M.S. Joshi and J.M. Paterson. *Introduction to Mathematical Portfolio Theory*. Cambridge University Press, Cambridge, 2013.

12. D.M. Kreps. *Notes on the Theory of Choice*. Westview Press, Boulder, CO, 1988.

13. D.G. Luenberger. *Investment Science* (2nd ed.). Oxford University Press, New York, 2014.

14. G.C. Pflug and A. Pichler. *Multistage Stochastic Optimization*. Springer, Heidelberg, 2014.

15. G.C. Pflug and W. Römisch. *Modeling, Measuring and Managing Risk*. World Scientific, London, 2007.

16. S.L. Savage. *The Flaw of Averages: Why We Underestimate Risk in the Face of Uncertainty*. Wiley, Hoboken, NJ, 2012.

17. A. Shapiro, D. Dentcheva, and A. Ruszczyński. *Lectures on Stochastic Programming: Modeling and Theory*. SIAM, Philadelphia, PA, 2009.

Chapter Eight

Mean–Variance Efficient Portfolios

This chapter is fairly technical and is meant to be a bridge between the general framework of mean–risk models, which we introduced in Section 7.4, and Chapters 9 and 10, where we describe factor and equilibrium models. Here, we adopt standard deviation as a risk measure, momentarily setting aside the critical remarks that we made in Section 7.4.1, in order to develop the theory of mean–variance efficient portfolios, which is the foundation of a body of knowledge broadly known as **modern portfolio theory** (MPT). In portfolio optimization, variance is typically used, rather than standard deviation, but this is just a matter of computational convenience. Despite its deceptive simplicity and the limitation of symmetric risk measures, MPT provides us with useful insights. Everything hinges on the determination of an efficient frontier of risky portfolios and the selection of an optimal portfolio mixing risky assets with a risk-free asset. The risk-free asset may be thought as a safe zero-coupon bond with maturity corresponding to the portfolio holding period, or a safe bank account offering a constant interest rate.

The theory, in its basic form, only deals with a single-period decision problem. To fix ideas, we will essentially consider equity portfolios, even though, in principle, any asset would do, since we may consider the holding period return from whatever asset, including a bonds and commodities. A further limitation of our treatment is that we do not consider transaction costs, taxes, etc., but the basic model can be extended to account for these and additional problem features. The resulting optimization models are quadratic programming problems that, per se, are certainly not hard to solve with numerical optimization methods.[1] A deeper limitation is of a statistical nature and concerns the reliability of the input data and of the resulting portfolio. We feed the optimization model with estimates of expected values and covariances of asset returns, but they are noisy estimates. Estimation errors may be magnified by the optimization process, resulting in unreliable and possibly weird solutions. Remedies have been proposed, including the Black–Litterman approach (Section 10.3) and robust optimization (Section 15.9).

---

[1] See Chapter 16.

Despite all of its limitations, the MPT framework sheds light on the fundamental decomposition of risk into systematic and idiosyncratic (specific) components. Another useful idea that we derive from MPT is the decomposition of the overall portfolio problem into independent subproblems: (1) the selection of a risky portfolio, and (2) its mixing with the risk-free asset. We will take a step-by-step, top-down process, whereby in Section 8.1 we first consider the simple problem of how wealth should be allocated between the risk-free asset and a generic risky portfolio. This leads to the definition of the capital allocation line (CAL). Then, in section 8.2, we consider in detail the problem of tracing the mean–variance efficient frontier, which leads us to an important separation property, as we shall see in Section 8.3. The property suggests that if all investors have the same view about the probability distribution of returns, then they should invest in the same risky portfolio, irrespective of their subjective risk aversion. Risk aversion should come into play only in the capital allocation between the optimal risky portfolio and the risk-free asset. The optimal risky portfolio can be found by maximization of a measure, the Sharpe ratio, trading off risk and expected reward, which we tackle in Section 8.4. In Section 8.5, we discuss whether and to what extent the theory of mean–variance efficient portfolios may be reconciled with the theory of expected utility. We close the chapter with some considerations about the stability of the portfolios generated by mean–variance optimization. As we illustrate in Section 8.6, a naive mean–variance approach may yield unreliable and unstable solutions, which would be hardly trusted by any portfolio manager.

We close the chapter with two technical sections, Supplements S8.1 and S8.2, where we prove some properties of the efficient portfolio frontier and give an explicit solution of a simplified portfolio optimization problem. These supplements may be safely skipped.

In this chapter we use some basic concepts in multiobjective optimization. The unfamiliar reader may refer to Supplement S2.1.

8.1 Risk aversion and capital allocation to risky assets

In this section, we consider a simple introductory problem, the capital allocation between a single risky asset and a risk-free one, over a given holding horizon. We will denote the risk-free return over the holding period as r_f,[2] and the return of a risky portfolio by the random variable $\tilde{r}$,[3] with expected value and variance

---

[2] We use the term risk-free *return*, rather than *rate*, as r_f is not annualized and may refer to an arbitrary holding period.

[3] There are two main conventions to denote random variables. We may use X for random variables and x for their realizations. As an alternative, we may use $\tilde{x}$ for random variables and x for their realizations. We mostly use the first convention in this book but, since in this chapter we shall use r for return and R for *excess* return, here we adopt the second one to avoid any ambiguity.

8.1 Risk aversion and capital allocation to risky assets

denoted by
$$\mathrm{E}[\tilde{r}] = \mu \quad \text{and} \quad \mathrm{Var}(\tilde{r}) = \sigma^2,$$
respectively. We will denote the fraction of wealth allocated to the risky asset by x, so that the corresponding fraction allocated to the risk-free asset is $1 - x$. Hence, the random return of the resulting portfolio is
$$\tilde{r}_p(x) = x\tilde{r} + (1-x)r_f,$$
with expected return and standard deviation given by
$$\mu_p(x) = \mathrm{E}[\tilde{r}_p(x)] = x\mu + (1-x)r_f = r_f + x \cdot (\mu - r_f), \quad (8.1)$$
$$\sigma_p(x) = \sqrt{\mathrm{Var}[\tilde{r}_p(x)]} = |x| \cdot \sigma, \quad (8.2)$$
respectively. Please note that if we allow a negative value of x, which corresponds to short-selling the risky portfolio, we have to use an absolute value in Eq. (8.2). It is also useful to introduce the following fundamental concepts.

DEFINITION 8.1 (Excess return) *The excess return $\tilde{R}$ of a risky asset (or a portfolio) is its return in excess of the risk-free return:*
$$\tilde{R} \doteq \tilde{r} - r_f.$$

DEFINITION 8.2 (Risk premium) *The risk premium π of a risky asset (or a portfolio) is its expected return in excess of the risk-free return:*
$$\pi \doteq \mathrm{E}[\tilde{R}] = \mu - r_f,$$
i.e., the expected excess return.

Remark. The concept of risk premium of definition 8.2 should not be confused with the utility theoretic concept of Eq. (7.5), which is related with a specific utility function and a certainty equivalent. Nevertheless, we may see some similarity between the two concepts, since the risk premium that we consider here does measure a compensation required to take a risk. To better grasp the idea, we may consider the ratio
$$\lambda = \frac{\mu - r_f}{\sigma}$$
as a **market price of risk**, i.e., the expected compensation that is expected by the market (rather than by an individual decision maker) per unit of risk. The relationship may be rewritten as
$$\mu = r_f + \lambda\sigma,$$
which sheds further light on the matter. If the market price of risk is zero, then $\mu = r_f$. This kind of reasoning will play a role in the risk-neutral pricing of options, as we shall see in Chapter 14. However, we should be careful about an interpretation that relies on the standard deviation of a single asset, rather than its contribution to the overall risk of a portfolio.

Table 8.1 Data for Example 8.1. The risk-free return is $r_f = 3\%$.

Portfolio	Risk premium	Expected return	Risk (St. dev.)
$P1$	4%	7%	10%
$P2$	7%	10%	20%
$P3$	2%	5%	30%

In order to find the most preferred portfolio, we have to trade off expected return and risk. Given the background concepts introduced in Chapter 7, we may use a mean–risk approach, adjusting expected return and building a sort of expected utility function as follows:

$$U(x) = \mu_p(x) - \tfrac{1}{2}\lambda \sigma_p^2(x), \tag{8.3}$$

where λ corresponds to the degree of risk aversion. Such a function is essentially a quadratic utility function and disregards higher-order moments. We may consider this function as a sort of utility function,[4] with all of the pitfalls that we have outlined in Chapter 7, or as the scalarized objective function of a mean–risk model.

Example 8.1 A numerical example

> Let us consider three portfolios, with expected return and standard deviation given in Table 8.1. Comparing $P1$ and $P2$, we observe that the former is less risky than the latter, but it features a lower expected return. The choice between the two depends on the degree of risk aversion. In Figure 8.1, we plot the expected utility of Eq. (8.3) for a range of values of λ, which results in a plain straight line for each portfolio. We observe that there is a critical value for which a decision maker would be indifferent between $P1$ and $P2$. On the contrary, $P3$ looks always worse than the other two alternatives. Indeed, we observe from Table 8.1 that it features the smallest expected return and the largest risk.

In general, the values of an expected utility can be used to rank alternatives, but they have no concrete financial meaning; indeed, we speak of *ordinal*, rather than *cardinal* utility. In this specific case, however, the utility score can be interpreted as a certainty equivalent return, i.e., a risk-free return that would

---

[4] We should note that the interpretation of Eq. (8.3) as a quadratic utility function is a bit imprecise. See Eq. (7.17) and the discussion in Section 8.5.

8.1 Risk aversion and capital allocation to risky assets

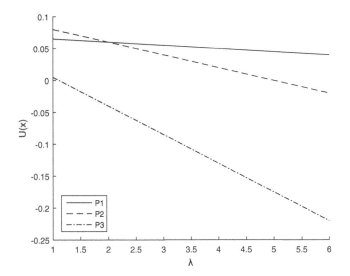

FIGURE 8.1 Plots of expected utility as a function of risk aversion in Example 8.1.

provide the investor with the same utility as the risky portfolio. In practice, sensible values for λ are usually considered to lie in the range between 2 and 4.

Now, as we said, the problem is to find an allocation between the risky portfolio and the risk-free asset. We may get a feeling for the involved tradeoff by plotting the range of portfolios that we can form as a function of x, on a mean–risk plane with coordinates $\sigma_p(x)$ and $\mu_p(x)$. To this aim, let us assume $x \geq 0$ [to get rid of the absolute value in the expression of $\sigma_p(x)$] and eliminate x between Eqs. (8.1) and (8.2). We solve the latter equation for x and plug it into the former one, which yields

$$\mu_p(x) = r_f + \frac{\mu - r_f}{\sigma} \cdot \sigma_p(x).$$

Thus, on the mean–risk plane, the portfolios trace the line shown in Fig. 8.2, which is called the **capital allocation line** (CAL). The CAL has an intercept r_f and a slope given by a ratio trading off risk and reward:

$$S_p = \frac{\mu_p - r_f}{\sigma_p}. \tag{8.4}$$

This ratio, called **Sharpe ratio**, relates the risk premium of the portfolio with its volatility, and it is also known as **reward-to-volatility** ratio. In the case of a single risky asset, the Sharpe ratio of the portfolio does not really depend on x, the decision variable defining the portfolio. In fact, if we assume $x \geq 0$ and use Eqs. (8.1) and (8.2), we find

$$S_p = \frac{\mu_p(x) - r_f}{\sigma_p(x)} = \frac{r_f + x \cdot (\mu - r_f) - r_f}{x\sigma} = \frac{\mu - r_f}{\sigma}.$$

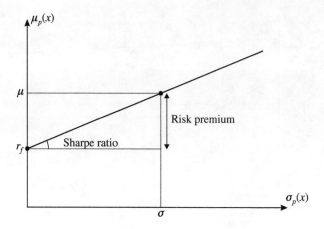

FIGURE 8.2 The capital allocation line.

Hence, the Sharpe ratio of the portfolio boils down to the Sharpe ratio of the risky asset. In Fig. 8.2, the two bullets correspond to the two assets that we are considering, and the CAL has positive slope, since we assume that the risky asset has positive risk premium. The portion of line connecting the two bullets is spanned by x in the range $[0, 1]$. For $x > 1$ we have a *leveraged* portfolio, in which expected return is boosted by borrowing money at the risk-free rate in order to increase the investment in the risky asset.[5] For $x < 0$ we are selling the risky asset short. The picture does not show the corresponding line, which has the same intercept as the CAL, but a negative slope. Since, in this case, the risk premium is positive, portfolios with $x < 0$ are clearly dominated and not efficient.

8.1.1 THE ROLE OF RISK AVERSION

The CAL defines the universe of possible portfolios, whereas risk aversion determines which one is selected. To this aim, let us maximize the expected utility of Eq. (8.3),

$$U(x) = r_f + x(\mu - r_f) - \frac{1}{2}\lambda x^2 \sigma^2,$$

with respect to x. Since this function is differentiable and concave, it suffices to write the first-order optimality condition,

$$U'(x) = (\mu - r_f) - \lambda x \sigma^2 = 0,$$

which yields the optimal allocation to the risky asset,

$$x^* = \frac{\mu - r_f}{\lambda \sigma^2}. \tag{8.5}$$

---

[5] Here, we are assuming that cash may be borrowed or lent at the same rate. See Problem 8.1 to find out what happens in the case of different rates.

8.2 The mean–variance efficient frontier with risky assets

If the risk premium is positive, x^* is positive as well. We also notice that, quite sensibly, this allocation is increasing with respect to the risk premium $\mu - r_f$, and decreasing with respect to risk σ and risk aversion λ. The expression of the optimal portfolio depends on a ratio trading off risk premium and risk, even though it is not really a Sharpe ratio, as it involves variance rather than standard deviation.

8.2 The mean–variance efficient frontier with risky assets

In capital allocation, we consider a given portfolio of risky assets as the only risky asset available to investors. But how can we choose that portfolio of risky assets? Here, we stick to measuring risk by standard deviation of return, within a mean–variance framework. This leads us to tracing the efficient frontier of portfolios of risky assets on the mean–risk plane. We deal with the case of two risky assets first, and then we move on to the case of n assets. Before doing so, we should ask why we should include multiple assets in a portfolio. Combining n risky assets should allow us to strike a better balance between risk and reward. However, as we show below, there is a limit to the amount of risk reduction that can be achieved by pure diversification.

8.2.1 DIVERSIFICATION AND PORTFOLIO RISK

Common sense suggests that holding a diversified portfolio should reduce risk. Indeed, we know from inferential statistics that, given a sample of i.i.d. (independent and identically distributed) variables, the variance of the sample mean $\overline{X}$ goes to zero when the sample size goes to infinity. In fact,

$$\text{Var}(\overline{X}) = \text{Var}\left(\frac{1}{n}\sum_{i=1}^{n} X_i\right) = \frac{1}{n^2}\sum_{i=1}^{n} \text{Var}(X_i) = \frac{\sigma^2}{n},$$

which vanishes for an increasing sample size n. However, this holds under an assumption of *independence*.[6] Such an assumption is not quite realistic in a financial context. If we consider an equally weighted portfolio ($w_i = 1/n$), the variance of its return includes not only contributions from individual variances, but also from the whole array of covariances:

$$\text{Var}\left(\sum_{i=1}^{n} \frac{1}{n} \cdot \tilde{r}_i\right) = \frac{1}{n^2}\sum_{i=1}^{n} \sigma_i^2 + \frac{1}{n^2}\sum_{i=1}^{n}\sum_{\substack{j=1 \\ j \neq i}}^{n} \sigma_{ij},$$

where $\sigma_i^2 \doteq \text{Var}(\tilde{r}_i)$ and $\sigma_{ij} \doteq \text{Cov}(\tilde{r}_i, \tilde{r}_j)$.

---
[6] Actually, assuming that correlation is zero would be sufficient.

In order to figure out what happens when we push on diversification, by letting $n \to +\infty$, let us define an "average variance,"

$$\overline{\sigma}^2 \doteq \frac{1}{n} \sum_{i=1}^n \sigma_i^2,$$

and an "average covariance,"

$$\overline{\text{Cov}} \doteq \frac{1}{n(n-1)} \sum_{i=1}^n \sum_{\substack{j=1 \\ j \neq i}}^n \sigma_{ij}.$$

Then, the return variance of the equally weighted portfolio is

$$\sigma_p^2 = \frac{1}{n} \cdot \overline{\sigma}^2 + \frac{n-1}{n} \cdot \overline{\text{Cov}}.$$

We immediately notice that, for an increasing value of n, the first component of variance does indeed go to zero, but the second one does not. There is a *nondiversifiable* component of risk. In financial markets, we have to distinguish:

- An array of **idiosyncratic** risk factors, associated with specific firms. These specific components of risk are nonsystematic and may be eliminated by diversification.
- A **market risk** component, which is a systematic and nondiversifiable risk factor.

This will be more apparent when dealing with factor models in Chapter 9, where we shall also see how we might hedge systematic risk by using long–short portfolios. This, however, assumes that short-selling is allowed.[7]

8.2.2 THE EFFICIENT FRONTIER IN THE CASE OF TWO RISKY ASSETS

To build intuition, let us consider first the simple case of only two risky assets, say, A_1 and A_2, with expected returns μ_1 and μ_2, and standard deviations of return σ_1 and σ_2, respectively. The returns are not necessarily independent, and we denote their coefficient of correlation by ρ_{12}. The asset weights in the portfolio are denoted by w_1 and w_2, subject to the constraint

$$w_1 + w_2 = 1. \tag{8.6}$$

Hence, the expected return of the portfolio is a linear (affine) combination of the two expected returns:

$$\mu_p = w_1 \mu_1 + w_2 \mu_2.$$

---

[7] To hedge market risk, we may also use derivatives like index futures or options. In particular, a short position in an index futures may be used to emulate a short position in the market portfolio, as we show in Chapter 12.

8.2 The mean–variance efficient frontier with risky assets

Variance is
$$\sigma_p^2 = w_1^2 \sigma_1^2 + 2\rho_{12} w_1 w_2 \sigma_1 \sigma_2 + w_2^2 \sigma_2^2.$$

The minimum variance portfolio can be found by straightforward minimization[8] and is, for two assets:

$$w_1 = \frac{\sigma_2^2 - \rho_{12}\sigma_1\sigma_2}{\sigma_1^2 - 2\rho_{12}\sigma_1\sigma_2 + \sigma_2^2}, \qquad w_2 = 1 - w_1. \tag{8.7}$$

As we have seen in Section 2.1.1, we might trace the efficient frontier of portfolios by setting a range of target portfolio returns $\mu_{\min}$. In the case of two risky assets, the resulting optimization problem is a bit dull, since the additional constraint
$$w_1\mu_1 + w_2\mu_2 = \mu_{\min},$$
together with Eq. (8.6), immediately give the resulting portfolio:
$$w_1 = \frac{\mu_{\min} - \mu_2}{\mu_1 - \mu_2}, \qquad w_2 = \frac{\mu_1 - \mu_{\min}}{\mu_1 - \mu_2}.$$

To figure out the qualitative shape of the efficient frontier, let us consider expected return and variance as a function of $w = w_1$:

$$\mu_p = w(\mu_1 - \mu_2) + \mu_2, \tag{8.8}$$
$$\sigma_p^2 = w^2\sigma_1^2 + 2\rho_{12}w(1-w)\sigma_1\sigma_2 + (1-w)^2\sigma_2^2. \tag{8.9}$$

By letting w range over the whole real line, i.e., assuming that short-selling is possible, we may plot a curve on a mean–risk plane. In this case, we find a curve corresponding to the whole set of attainable portfolios, which we may call the **attainable set**. The qualitative shape of this set is outlined in Fig. 8.3 for three specific values of correlation, $\rho_{12} \in \{-1, 0, 1\}$, assuming that short-selling is not allowed, so that $w \in [0, 1]$. In order to rule out pathological cases, we assume $\mu_1 \neq \mu_2$ and $\sigma_1 \neq \sigma_2$. If we allow short-selling, the plots stretch beyond the extreme points corresponding to assets A_1 and A_2. Let us consider the two limit cases first.

Case $\rho_{12} = 1$. In this case,
$$\sigma_p = \sqrt{w^2\sigma_1^2 + 2w(1-w)\sigma_1\sigma_2 + (1-w)^2\sigma_2^2} = \sqrt{\big(w\sigma_1 + (1-w)\sigma_2\big)^2}$$
$$= |w(\sigma_1 - \sigma_2) + \sigma_2|.$$

Note that, to be fully general, we have to consider the absolute value. If we rule out short-selling and use Eq. (8.8) to eliminate the portfolio weight, we find
$$\sigma_p = \frac{\mu_p - \mu_2}{\mu_1 - \mu_2} \cdot (\sigma_1 - \sigma_2) + \sigma_2,$$

---

[8] One possibility is to use a Lagrange multiplier to deal with constraint (8.6), as we shall see in the general case of n assets. In this simple case, we may just eliminate w_2 and solve an unconstrained problem with respect to w_1, by enforcing the usual first-order optimality condition.

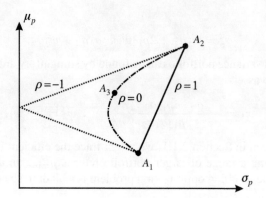

FIGURE 8.3 Plotting the mean–risk tradeoff curve for portfolios including two risky assets.

which is a linear relationship between standard deviation and expected return. In the case of Fig. 8.3, $\mu_1 < \mu_2$ and $\sigma_1 < \sigma_2$, so the slope is positive. Also note that we may find a portfolio with zero risk by choosing

$$w = \frac{\sigma_2}{\sigma_2 - \sigma_1} > 1,$$

which implies short-selling asset A_2.

Case $\rho_{12} = -1$. By a similar token, in this case we find

$$\sigma_p = \sqrt{\big(w\sigma_1 - (1-w)\sigma_2\big)^2} = |w(\sigma_1 + \sigma_2) - \sigma_2|.$$

Now, we attain zero risk by setting

$$w = \frac{\sigma_2}{\sigma_1 + \sigma_2}.$$

In this case, $w \in [0, 1]$ and we may get rid of risk without resorting to short-selling. The set of attainable portfolios corresponds to *two* line segments, even if we require $w \geq 0$.

For intermediate values of correlation, we have a nonlinear curve featuring a minimum variance portfolio. In Supplement S8.1, we show that the exact shape is a hyperbola. Note that the plot corresponding to all of the attainable portfolios is *not* the efficient frontier, as only the portion of the attainable set curve above the minimum variance portfolio is efficient.

An interesting observation is that a sensible portfolio may include a dominated asset. To see this, let us consider asset A_3, which happens to lie on the mean–risk curve in Fig. 8.3, and imagine tracing the frontier with only the two assets A_1 and A_3. Since A_3 is on the previous attainable set, it means that its expected return and standard deviation may be replicated by a portfolio including A_1 and A_2, which implies that we get exactly the same attainable set by using A_1 and A_3. Clearly, asset A_1 is dominated by A_3, and in the case of perfect positive correlation, $\rho_{13} = 1$, the set of attainable portfolios with no short-selling would be a line segment joining A_1 and A_3, with a negative slope, and it would consist of nonefficient portfolios. However, Fig. 8.3 shows that, for certain values of correlation, it might be worthwhile to include a dominated asset into a portfolio, as this could reduce the overall portfolio risk. This is especially true for a negative correlation. Thus, we should not analyze the individual risks of assets, but their contribution to the overall risk within a well-diversified portfolio.[9]

8.2.3 THE EFFICIENT FRONTIER IN THE CASE OF N RISKY ASSETS

Given the basic intuition provided by the case of two risky assets, let us now consider a portfolio including n risky assets with expected value $\boldsymbol{\mu}$ and covariance matrix $\boldsymbol{\Sigma}$. As we have said,[10] the case of n risky assets may be tackled by two alternative scalarization approaches (assuming fully invested portfolios):

- One possibility is enforcing a constraint on a target expected return $\mu_{\min}$ and letting it range over a suitable interval:

$$\min \quad \tfrac{1}{2}\mathbf{w}^\mathsf{T} \boldsymbol{\Sigma} \mathbf{w}$$
$$\text{s.t.} \quad \mathbf{i}^\mathsf{T} \mathbf{w} = 1,$$
$$\boldsymbol{\mu}^\mathsf{T} \mathbf{w} = \mu_{\min},$$

where $\mathbf{i} = [1, 1, \ldots, 1]^\mathsf{T}$.

- An alternative is setting up a (sort of) utility function characterized by a risk aversion parameter λ:

$$\max \quad \boldsymbol{\mu}^\mathsf{T} \mathbf{w} - \frac{\lambda}{2} \mathbf{w}^\mathsf{T} \boldsymbol{\Sigma} \mathbf{w} \qquad (8.10)$$
$$\text{s.t.} \quad \mathbf{i}^\mathsf{T} \mathbf{w} = 1. \qquad (8.11)$$

We may interpret this scalarized objective as a risk-adjusted expected return. By letting λ range from 0 to $+\infty$, we trace the efficient frontier.

Quite often, these two approaches are equivalent, but they are not in general. As a counterexample, let us consider tracing the mean–variance efficient frontier subject to a cardinality constraint. This means that we limit the number of assets that may have a nonzero weight in the portfolio. In this case, the second scalarization approach does not necessarily work. To see why, consider Fig. 8.4. Here, we show the possible shape of an efficient frontier involving three assets,

---

[9]This concept will be more evident in Section 10.2, where we deal with the role of beta in the capital asset pricing model and introduce the security market line.
[10]See Supplement S2.1.

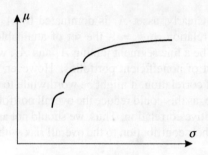

FIGURE 8.4 Qualitative sketch of a cardinality-constrained efficient frontier.

where the maximum cardinality is two. We may find a sort of "union" of three frontiers, each one corresponding to a pair of assets. This frontier does not look like the boundary of a convex set, and the risk-adjusted objective of Eq. (8.10) may fail to generate the whole frontier.[11]

In our simple setting, we may work with either scalarization approach but, for the sake of convenience, we pursue the second approach in this section, leaving the first one to Supplement S8.2. As a first step, let us find the minimum variance portfolio. This portfolio is obtained by letting $\lambda \to +\infty$ in the maximization problem (8.10), which is easily solved by using the Lagrange multiplier method. We plug the equality constraint (8.11) into the Lagrangian function, associated with a multiplier ν:

$$\mathcal{L}(\mathbf{w}, \nu) = \tfrac{1}{2}\mathbf{w}^T \boldsymbol{\Sigma} \mathbf{w} + \nu \cdot \left(1 - \mathbf{i}^T \mathbf{w}\right).$$

Stationarity with respect to portfolio weights yields a system of linear equations, which is solved under the assumption of a full-rank (invertible) covariance matrix:

$$\boldsymbol{\Sigma} \mathbf{w} - \nu \mathbf{i} = \mathbf{0} \quad \Rightarrow \quad \mathbf{w} = \nu \boldsymbol{\Sigma}^{-1} \mathbf{i}.$$

Plugging this vector of weights into constraint (8.11) immediately yields the value of the multiplier,

$$\nu = \frac{1}{\mathbf{i}^T \boldsymbol{\Sigma}^{-1} \mathbf{i}} \quad \Rightarrow \quad \mathbf{w}^*_{\min} = \frac{\boldsymbol{\Sigma}^{-1} \mathbf{i}}{\mathbf{i}^T \boldsymbol{\Sigma}^{-1} \mathbf{i}}. \tag{8.12}$$

Then, the minimum variance of the resulting portfolio is

$$\sigma^2_{\min} = (\mathbf{w}^*_{\min})^T \boldsymbol{\Sigma} \mathbf{w}^*_{\min} = \frac{1}{\mathbf{i}^T \boldsymbol{\Sigma}^{-1} \mathbf{i}},$$

which is just the Lagrange multiplier.

---

[11] We face a similar issue in Fig. 2.9. The problem with a cardinality constraint may be solved using the mixed-integer modeling tools that we shall discuss in Section 15.4.1.

8.2 The mean–variance efficient frontier with risky assets

The case of finite risk aversion can be tackled along the same lines, and it is a useful exercise to prove that:

$$\mathbf{w}^* = \frac{\Sigma^{-1}\mathbf{i}}{\mathbf{i}^T\Sigma^{-1}\mathbf{i}} + \frac{1}{\lambda} \frac{\left(\mathbf{i}^T\Sigma^{-1}\mathbf{i}\right)\Sigma^{-1}\mu - \left(\mathbf{i}^T\Sigma^{-1}\mu\right)\Sigma^{-1}\mathbf{i}}{\mathbf{i}^T\Sigma^{-1}\mathbf{i}}, \quad (8.13)$$

$$\mu^* = \mu^T\mathbf{w}^* = \frac{\mathbf{i}^T\Sigma^{-1}\mu}{\mathbf{i}^T\Sigma^{-1}\mathbf{i}} + \frac{1}{\lambda} \frac{\left(\mathbf{i}^T\Sigma^{-1}\mathbf{i}\right)\left(\mu^T\Sigma^{-1}\mu\right) - \left(\mathbf{i}^T\Sigma^{-1}\mu\right)^2}{\mathbf{i}^T\Sigma^{-1}\mathbf{i}}, \quad (8.14)$$

$$(\sigma^*)^2 = (\mathbf{w}^*)^T\Sigma\mathbf{w} = \frac{1}{\mathbf{i}^T\Sigma^{-1}\mathbf{i}} + \frac{1}{\lambda^2} \frac{\left(\mathbf{i}^T\Sigma^{-1}\mathbf{i}\right)\left(\mu^T\Sigma^{-1}\mu\right) - \left(\mathbf{i}^T\Sigma^{-1}\mu\right)^2}{\mathbf{i}^T\Sigma^{-1}\mathbf{i}}. \quad (8.15)$$

As a reality check, we may note that if λ goes to infinity, the portfolio of Eq. (8.13) boils down to the minimum variance portfolio.

We may streamline Eqs. (8.14) and (8.15) by introducing suitable constants

$$A \equiv \frac{\mathbf{i}^T\Sigma^{-1}\mu}{\mathbf{i}^T\Sigma^{-1}\mathbf{i}},$$

$$B \equiv \frac{\left(\mathbf{i}^T\Sigma^{-1}\mathbf{i}\right)\left(\mu^T\Sigma^{-1}\mu\right) - \left(\mathbf{i}^T\Sigma^{-1}\mu\right)^2}{\mathbf{i}^T\Sigma^{-1}\mathbf{i}},$$

$$C \equiv \frac{1}{\mathbf{i}^T\Sigma^{-1}\mathbf{i}},$$

and rewriting the two equations as follows:

$$\mu^* = A + \frac{B}{\lambda} \quad \Rightarrow \quad \frac{B}{\lambda^2} = \frac{(\mu^* - A)^2}{B}$$

$$(\sigma^*)^2 = C + \frac{B}{\lambda^2} = C + \frac{(\mu^* - A)^2}{B}.$$

The last equation can be rearranged as

$$\frac{(\sigma^*)^2}{C} - \frac{(\mu^* - A)^2}{B/C} = 1,$$

which is the equation of a hyperbola,[12] assuming that $B > 0$ and $C > 0$.[13] To be precise, since risk aversion λ is supposed to be positive, we only draw a portion of a hyperbola. The resulting curve is depicted in Fig. 8.5. Unlike the case of two assets, this curve does not correspond to the whole set of attainable portfolios, but only to the portfolios minimizing variance for given target expected return. Thus, we should talk of the **minimum variance curve** (or min-variance

---
[12] Supplement S8.1 describes a similar analysis for the case of two assets.
[13] The inequality $C > 0$ is a consequence of the positive definiteness of the inverse of a positive definite matrix. Hence, it holds if we assume that the covariance matrix Σ is positive definite, not only semidefinite, so that $\mathbf{i}^T\Sigma^{-1}\mathbf{i} > 0$. The inequality $B > 0$ is a consequence of the Cauchy–Schwartz inequality applied to vectors μ and $\mathbf{i}$, where we consider an inner product involving matrix Σ^{-1}; therefore, $\langle \mathbf{i}, \mu \rangle = \mathbf{i}^T\Sigma^{-1}\mu$ and $\|\mu\|^2 = \langle \mu, \mu \rangle = \mu^T\Sigma^{-1}\mu$.

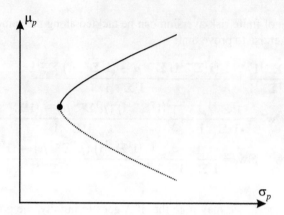

FIGURE 8.5 The set of minimum variance portfolios, for given target expected returns, with n risky assets. The portion above the overall minimum variance portfolio is the efficient frontier.

set). The dotted portion of the min-variance set, below the point corresponding to the minimum variance portfolio, is not efficient. The efficient frontier is the portion of the plot above the minimum variance point.

8.3 Mean–variance efficiency with a risk-free asset: The separation property

In this section, we examine the structure of the efficient frontier when a risk-free asset is introduced. As it turns out, the efficient frontier is just a straight line, mixing a risky portfolio and the risk-free asset, just like in the CAL. Furthermore, all investors should hold the *same* risky portfolio.

First, let us tackle the problem formally. Let w_0 be the weight of the risk-free asset, earning the risk-free return r_f. The problem

$$\max \quad w_0 r_f + \boldsymbol{\mu}^T \mathbf{w} - \frac{\lambda}{2} \mathbf{w}^T \boldsymbol{\Sigma} \mathbf{w}$$
$$\text{s.t.} \quad w_0 + \mathbf{i}^T \mathbf{w} = 1,$$

can be rewritten as an unconstrained problem by eliminating w_0:

$$\max \quad \mathbf{w}^T \boldsymbol{\pi} - \frac{\lambda}{2} \mathbf{w}^T \boldsymbol{\Sigma} \mathbf{w},$$

where $\boldsymbol{\pi} = \boldsymbol{\mu} - r_f \mathbf{i}$ is the vector of risk premia. We apply the first-order optimality condition to the unconstrained maximization problem,

$$\boldsymbol{\pi} - \lambda \boldsymbol{\Sigma} \mathbf{w} = \mathbf{0},$$

8.3 Mean–variance efficiency with a risk-free asset: The separation property

which immediately yields

$$\mathbf{w}^* = \frac{1}{\lambda}\Sigma^{-1}\boldsymbol{\pi}, \tag{8.16}$$

$$w_0^* = 1 - \mathbf{w}^\mathsf{T}\mathbf{i} = 1 - \frac{1}{\lambda}\mathbf{i}^\mathsf{T}\Sigma^{-1}\boldsymbol{\pi}. \tag{8.17}$$

As before, we assume a full-rank, invertible covariance matrix. As a reality check, we observe that when $\lambda \to +\infty$ the whole wealth is allocated to the risk-free asset. We may build further intuition by considering a couple of simple examples.

Example 8.2 The case of uncorrelated assets

Suppose that asset returns are uncorrelated, so that Σ is a diagonal matrix, with entries $\sigma_{ii} \equiv \sigma_i^2$. Then, the weight for each risky asset i in the optimal portfolio, which also includes the risk-free asset, is

$$w_i^* = \frac{1}{\lambda}\frac{\mu_i - r_f}{\sigma_i^2} = \frac{1}{\lambda}\frac{\pi_i}{\sigma_i^2}.$$

In this simplified case, portfolio weights depend in an obvious way on the risk aversion coefficient λ, the risk premium π_i, and risk σ_i. We also observe that this solution is formally identical to the case of simple capital allocation; see Eq. (8.5).

Example 8.3 The case of two correlated assets

The covariance matrix for two correlated assets is

$$\Sigma = \begin{bmatrix} \sigma_1^2 & \rho\sigma_1\sigma_2 \\ \rho\sigma_1\sigma_2 & \sigma_2^2 \end{bmatrix},$$

whose inverse is

$$\Sigma^{-1} = \frac{1}{1-\rho^2}\begin{bmatrix} \dfrac{1}{\sigma_1^2} & -\dfrac{\rho}{\sigma_1\sigma_2} \\ -\dfrac{\rho}{\sigma_1\sigma_2} & \dfrac{1}{\sigma_2^2} \end{bmatrix}.$$

Clearly, the inverse exists if we rule out the case of perfect correlation ($\rho \neq \pm 1$). Then, the weights in the optimal portfolio are

$$w_1^* = \frac{1}{\lambda(1-\rho^2)}\left(\frac{\pi_1}{\sigma_1^2} - \rho\frac{\pi_2}{\sigma_1\sigma_2}\right),$$

$$w_2^* = \frac{1}{\lambda(1-\rho^2)}\left(\frac{\pi_2}{\sigma_2^2} - \rho\frac{\pi_1}{\sigma_1\sigma_2}\right),$$

$$w_0^* = 1 - w_1^* - w_2^*.$$

Note that, since we do not constrain the portfolio in any way, the weight of an asset can be positive or negative, depending on the sign and the size of the risk premia, on the volatilities, and on the correlation. For instance, let us rewrite the weight of asset 1 in terms of the Sharpe ratios $S_1 \doteq \pi_1/\sigma_1$ and $S_2 \doteq \pi_2/\sigma_2$:

$$w_1^* = \frac{1}{\lambda(1-\rho^2)\sigma_1}(S_1 - \rho S_2). \tag{8.18}$$

With a sufficiently large correlation, so that there is no diversification effect, if S_2 is sufficiently larger than S_1, we should sell the first asset short.

These expressions are useful to check the sensitivity of the portfolio composition to the input data and to challenge our intuition. What about the sensitivity to risk aversion? We see that

$$\frac{\partial w_1^*}{\partial \lambda} = -\frac{1}{\lambda}w_1^*,$$

which may be positive or negative. If we hold a long (positive) position in the risky asset 1, increasing risk aversion will decrease its weight. However, if asset 1 is sold short, increasing λ will increase the portfolio weight, in the sense that it is shrunk towards zero (we reduce the amount of short-selling).

An easy finding is

$$\frac{\partial w_1^*}{\partial \pi_1} = \frac{1}{\lambda(1-\rho^2)\sigma_1^2} > 0,$$

which makes good sense: The larger the risk premium of an asset, the larger the corresponding portfolio weight. One would expect that increasing the risk premium of the other asset will drive the portfolio weight down, which is not necessarily true. In fact,

$$\frac{\partial w_1^*}{\partial \pi_2} = \frac{-\rho}{\lambda(1-\rho^2)\sigma_1\sigma_2}.$$

8.3 Mean–variance efficiency with a risk-free asset: The separation property

> In the case of *negative* correlation, increasing π_2 will increase both w_1^* and w_2^*, because of a diversification effect. This effect is less strong when risk aversion and volatilities are large, and it is increasing with the absolute value of the correlation. By a similar token, the effect of increasing the volatility of the other asset depends on the sign of the correlation:
> $$\frac{\partial w_1^*}{\partial \sigma_2} = \frac{\rho \pi_2}{\lambda(1-\rho^2)\sigma_1 \sigma_2^2}.$$
> When correlation is positive, increasing the volatility of the second asset increases the weight of the first one, but the contrary applies with negative correlation.
>
> The sensitivity of w_1^* with respect to σ_1 is trickier:
> $$\frac{\partial w_1^*}{\partial \sigma_1} = \frac{1}{\lambda(1-\rho^2)} \left(-2 \frac{\pi_1}{\sigma_1^3} + \rho \frac{\pi_2}{\sigma_1^2 \sigma_2} \right)$$
> $$= \frac{1}{\lambda(1-\rho^2)\sigma_1^2} \left(-2 S_1 + \rho S_2 \right).$$
>
> If the two Sharpe ratios are positive and close enough, this sensitivity will be negative, which corresponds to our intuition. However, the sensitivity can be positive. For instance, this may happen if risk premia are both positive, as well as correlation, but asset 1 has a smaller Sharpe ratio, so that it is sold short. An increase in volatility will increase its weight, in the sense that we should reduce the amount of short-selling in this case.
>
> The analysis of sensitivity with respect to correlation is left as an exercise. We only observe that if risk premia are positive and we change a positive correlation into a negative one, without changing its absolute value, we will increase the exposure to both risky assets. This is clearly due to a diversification effect.
>
> The bottom line of this example, with respect to the uncorrelated case of Example 8.2, is that introducing correlation makes the portfolio behavior less intuitive. In practice, we should also be aware that uncertainty in parameters may be interpreted as an effect of statistical estimation noise, which may have an adverse effect on the stability of the mean–variance optimal portfolio (see Section 8.6).

Equation (8.16) has an important implication: Within a mean–variance framework, the relative composition of the risky portfolio does not change as a function of risk aversion. The coefficient λ scales the weights of the risky assets in the portfolio uniformly. Hence, the overall portfolio does depend on risk aversion, but only in the way the risk-free asset is mixed with a risky port-

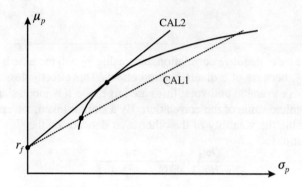

FIGURE 8.6 CALs and the tangency portfolio.

folio. The risky portfolio only depends on the covariance matrix and the vector of risk premia. Let us explore this finding in a more intuitive way by comparing different portfolios that we may build by mixing the risk-free asset with a risky portfolio.

To begin with, it is *never* optimal to consider a nonefficient risky portfolio. To see this, let us consider a nonefficient portfolio w_{ne}. If this portfolio is not efficient, there must exist another, efficient portfolio w_e dominating it. Two possibilities should be considered, in terms of expected return and risk:

1. $\mu_{ne} < \mu_e$ and $\sigma_{ne} \geq \sigma_e$
2. $\mu_{ne} \leq \mu_e$ and $\sigma_{ne} > \sigma_e$

In the first case, let w_0 be the weight of the risk-free asset. The expected return of the portfolio including the efficient portfolio is

$$w_0 r_f + (1 - w_0)\mu_e > w_0 r_f + (1 - w_0)\mu_{ne},$$

whereas variance is

$$(1 - w_0)^2 \sigma_e^2 \leq (1 - w_0)^2 \sigma_{ne}^2.$$

Hence, we are better off mixing the efficient portfolio with the risk-free asset. The second possibility is treated similarly, leading us to the conclusion that we should never mix a nonefficient portfolio with the risk-free asset.

Now, let us consider different CALs mixing the risk-free asset with an efficient risky portfolio, as depicted in Fig. 8.6. It is immediately apparent that:

- All of the portfolios on the line CAL1 are dominated by the portfolios on the line CAL2.
- The line CAL2 is associated with a tangency portfolio, which is the efficient portfolio corresponding to the maximum CAL slope.
- Since the slope of the CAL is a Sharpe ratio, in order to find the optimal risky portfolio, we have to maximize the Sharpe ratio itself. We will address this task in Section 8.4.

The bottom line of the reasoning is that we have found a **separation property**, telling us that the portfolio choice problem may be decomposed into two independent tasks:

- Determination of the optimal risky portfolio (purely technical),
- Allocation of available wealth between the risk-free asset and the risky portfolio (this depends on subjective preferences).

The fact that all investors should hold the same portfolio, independently from their risk attitude, will have important consequences, leading us to the capital asset pricing model in Chapter 10. However, this conclusion relies on some critical assumptions. Investors are assumed to be mean–variance optimizers,[14] and they are supposed to share a common view about risk premia, volatilities, and correlations, i.e., they use the same data in the same way. As we shall see in Chapter 9, issues in the statistical estimation of expected returns and covariances are not quite negligible, not to mention the fact that we should use forecasts for the *future*, rather than estimates based on past data.

8.4 Maximizing the Sharpe ratio

The tangency portfolio is found by maximizing the slope of the CAL, which in turn requires finding a risky portfolio maximizing the Sharpe ratio, trading off risk premium and standard deviation. The corresponding optimization model is

$$\max \quad \frac{\sum_i w_i (\mu_i - r_f)}{\sqrt{\mathbf{w}^\mathsf{T} \Sigma \mathbf{w}}} = \frac{\boldsymbol{\pi}^\mathsf{T} \mathbf{w}}{\sqrt{\mathbf{w}^\mathsf{T} \Sigma \mathbf{w}}} \doteq S_p(\mathbf{w})$$

$$\text{s.t.} \quad \sum_i w_i = 1,$$

where the vector $\boldsymbol{\pi}$ collects the asset risk premia. We allow short sales, as we have done so far, so there is only one constraint normalizing portfolio weights. We can tackle the problem by introducing Lagrange multipliers, but we may also take advantage of a clever trick. In fact, it is easy to see that the Sharpe ratio function $S_p(\mathbf{w})$ is a homogeneous function of degree zero, i.e.,

$$S_p(\alpha \mathbf{w}) = S_p(\mathbf{w}), \qquad \forall \alpha > 0.$$

Geometrically, this means that the Sharpe ratio is constant along rays emanating from the origin, on which it is not defined, because of division by zero (see Fig. 8.7). Hence, we can disregard the weight normalization constraint and solve an unconstrained optimization problem, where we maximize the Sharpe ratio as a function $S_p(\mathbf{x})$. The vector $\mathbf{x}$ collects a set of *pseudoweights* x_i, which will be

---

[14] See Section 8.5 for a connection with the theory of utility functions.

normalized after solving the unconstrained problem, in order to find weights:

$$w_i = \frac{x_i}{\sum_k x_k}.$$

To find the optimal portfolio, we have to write the usual first-order optimality conditions for the function

$$S_p(\mathbf{x}) = \frac{\boldsymbol{\pi}^T \mathbf{x}}{(\mathbf{x}^T \boldsymbol{\Sigma} \mathbf{x})^{1/2}}.$$

Hence, we should solve the system:

$$\frac{\partial S_p}{\partial x_1} = 0, \quad \frac{\partial S_p}{\partial x_2} = 0, \quad \ldots, \quad \frac{\partial S_p}{\partial x_n} = 0.$$

To figure out the solution, let us generalize the reasoning and consider the maximization of a function like

$$\theta(\mathbf{x}) = \frac{f(\mathbf{x})}{\sqrt{g(\mathbf{x})}}.$$

The first-order conditions are

$$\frac{\partial f}{\partial x_k} \cdot g^{-\frac{1}{2}}(\mathbf{x}) - \frac{1}{2} f(\mathbf{x}) \cdot g^{-\frac{3}{2}}(\mathbf{x}) \cdot \frac{\partial g}{\partial x_k} = 0, \quad k = 1, \ldots, n,$$

which, assuming $g(\mathbf{x}) \neq 0$, can be simplified and rewritten as

$$\frac{\partial f}{\partial x_k} = \frac{1}{2} \frac{f(\mathbf{x})}{g(\mathbf{x})} \frac{\partial g}{\partial x_k}, \quad k = 1, \ldots, n.$$

If we use a more compact vector form, we end up with the equation

$$\nabla f(\mathbf{x}) = \frac{1}{2} \frac{f(\mathbf{x})}{g(\mathbf{x})} \cdot \nabla g(\mathbf{x}). \tag{8.19}$$

In our specific case, we may take advantage of the specific form of functions $f(\cdot)$ and $g(\cdot)$ to solve Eq. (8.19):

- $g(\mathbf{x})$ is a quadratic form, and we know from matrix theory that the gradient of the quadratic form $\mathbf{x}^T \boldsymbol{\Sigma} \mathbf{x}$ is $2 \boldsymbol{\Sigma} \mathbf{x}$.
- The gradient of the linear function $f(\mathbf{x}) = \boldsymbol{\pi}^T \mathbf{x}$ is just vector $\boldsymbol{\pi}$.
- Finally, the term $f(\mathbf{x})/g(\mathbf{x})$ is just a number, which can be safely be disregarded. In fact, we may think that it is included in vector $\mathbf{x}$ in the resulting equation, whose solution must be normalized anyway.

So, the first-order optimality conditions boil down to a system of linear equations, which is readily solved under the assumption of a full-rank covariance matrix:

$$\boldsymbol{\Sigma} \mathbf{x} = \boldsymbol{\pi} \quad \Rightarrow \quad \mathbf{x} = \boldsymbol{\Sigma}^{-1} \boldsymbol{\pi}. \tag{8.20}$$

8.4 Maximizing the Sharpe ratio

It is interesting to note that the shape of this solution is essentially the same as that given in Eq. (8.16), where the portfolio is scaled by the risk aversion coefficient λ to give the mix with the risk-free asset. Finally, the solution in terms of pseudoweights must be normalized, in order to obtain the weights of a fully invested risky portfolio.

Example 8.4 Maximizing the Sharpe ratio: A numerical example

The following example is taken from [4, Chapter 6]. Consider three assets with expected returns, standard deviations, and correlation matrix given by:

$$\boldsymbol{\mu} = \begin{bmatrix} 0.14 \\ 0.08 \\ 0.20 \end{bmatrix}, \quad \boldsymbol{\sigma} = \begin{bmatrix} 0.06 \\ 0.03 \\ 0.15 \end{bmatrix}, \quad \mathbf{R} = \begin{bmatrix} 1.0 & 0.5 & 0.2 \\ 0.5 & 1.0 & 0.4 \\ 0.2 & 0.4 & 1.0 \end{bmatrix}.$$

Assume further that the risk-free return is 5%. Then, we should solve the following system of linear equations (Note: There is an inconsistency, as returns have been multiplied by 100 and covariances by 100×100, but this is inconsequential; why?):

$$14 - 5 = 36x_1 + 0.5 \times 6 \times 3x_2 + 0.2 \times 6 \times 15x_3$$
$$8 - 5 = 0.5 \times 6 \times 3x_1 + 9x_2 + 0.4 \times 3 \times 15x_3$$
$$20 - 5 = 0.2 \times 6 \times 15x_1 + 0.4 \times 3 \times 15x_2 + 225x_3.$$

The system can be simplified to

$$1 = 4x_1 + x_2 + 2x_3$$
$$1 = 3x_1 + 3x_2 + 6x_3$$
$$5 = 6x_1 + 6x_2 + 75x_3,$$

whose solution is

$$x_1 = \frac{14}{63}, \quad x_2 = \frac{1}{63}, \quad x_3 = \frac{3}{63}.$$

The sum of the three variables is 18/63. Dividing the pseudoweights by this normalization factor, we get the actual portfolio weights

$$w_1 = \frac{14}{18}, \quad w_2 = \frac{1}{18}, \quad w_3 = \frac{3}{18}.$$

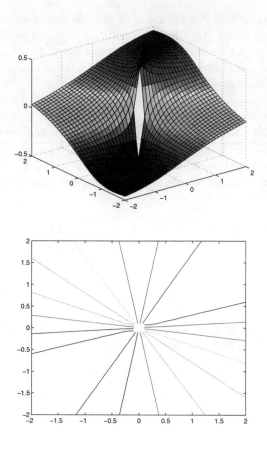

FIGURE 8.7 Plotting the Sharpe ratio $S_p(\mathbf{x})$, as a function of pseudoweights x_1 and x_2. We show a surface plot and the corresponding contour plot.

8.4.1 TECHNICAL ISSUES IN SHARPE RATIO MAXIMIZATION

We have maximized the Sharpe ratio by enforcing the first-order optimality condition, but how can we be sure that this is correct? Indeed, a more careful analysis is needed, as the Sharpe ratio is *not* a concave function. To see this, let us consider the case in Fig. 8.7, which shows the surface and the contour plots in a case involving two risky assets. In the surface plot, we notice the singularity at the origin, where the Sharpe ratio is not defined. The plot shows that the function is not concave at all. The contour plot shows that the Sharpe ratio is constant along rays emanating from the origin, which is expected, as it is a homogeneous function of degree zero. Unfortunately, this questions the validity of Eq. (8.20), since it is not at all obvious that the first-order conditions are sufficient for optimality. However, we are only interested in the Sharpe ratio for a normalized vector of weights. So, we may consider a section of the surface plot, corresponding to the line $x_1 + x_2 = 1$, where normalization is enforced. On that

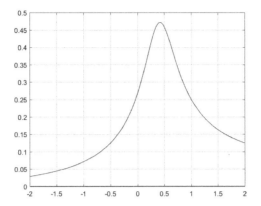

FIGURE 8.8 A section of the Sharpe ratio surface plot, obtained by enforcing the normalization condition $x_1 + x_2 = 1$ on the pseudoweights. The Sharpe ratio is displayed as a function of $x_1 \equiv w_1$.

line, we set $S_p(x_1, x_2) = S_p(w_1, 1 - w_1)$ and plot the Sharpe ratio as a function of $x_1 \equiv w_1$, as shown in Fig. 8.8, we see that the first-order stationarity condition looks indeed necessary and sufficient for optimality. The function is clearly not convex, but it is indeed unimodal, as it is first increasing, then decreasing. Technically, a function like this is *quasiconcave*.[15] In Section 15.3.1, we shall see that, in fact, the maximization of the Sharpe ratio can be recast as a convex optimization model. The resulting formulation is a convex quadratic program, where we may include additional constraints on the portfolio composition.

8.5 Mean–variance efficiency vs. expected utility

In this chapter, we have often used a kind of utility function involving mean and variance of return. This function may be interpreted as a risk-adjusted expected return that bears some resemblance to expected quadratic utility, which we know has some peculiar properties,[16] since it is not monotonically increasing and is IARA. Hence, we could and should ask whether the mean–variance framework is consistent with the expected utility framework. The question can be recast as follows: If we maximize expected utility, do we find a mean–variance efficient portfolio? The answer is "not necessarily." In fact, this is no big surprise, since in MPT we consider a symmetric risk measure and disregard higher order moments (e.g., skewness and kurtosis).

---

[15] See Section 15.1 for a discussion of convexity and concavity, as well as quasiconvex/quasiconcave functions.
[16] See Section 7.3.1.

Let us consider a quadratic utility function of end-of-horizon wealth,

$$u(W) = W - \frac{\lambda}{2}W^2,$$

with $\lambda > 0$, which is increasing on the range $W < 1/\lambda$. If we assume that initial wealth is unitary, terminal wealth boils down to one plus return, i.e., a multiplicative gain. Let us consider a random terminal wealth $\widetilde{W}$, with

$$\mu = \mathrm{E}\big[\widetilde{W}\big], \qquad \sigma^2 = \mathrm{Var}\big(\widetilde{W}\big).$$

The expected utility of $\widetilde{W}$ is

$$\mathrm{E}\big[u\big(\widetilde{W}\big)\big] = \mathrm{E}\big[\widetilde{W}\big] - \frac{\lambda}{2}\mathrm{E}\big[\widetilde{W}^2\big] = \mu - \frac{\lambda}{2}\big(\mu^2 + \sigma^2\big) \doteq U\big(\mu, \sigma^2\big), \quad (8.21)$$

which only depends on expected return μ and variance σ^2. We should notice the difference between the legit expected quadratic utility of Eq. (8.21) and the the risk-adjusted expected return,

$$\mu - \frac{\lambda}{2}\sigma^2,$$

which is used in mean–variance optimization. They differ by a term involving the squared expected return μ^2. Nevertheless, we also observe that

$$\frac{\partial U}{\partial \sigma^2} = -\lambda < 0,$$

hence, only minimum variance portfolios can be optimal, for a given μ. Furthermore

$$\frac{\partial U}{\partial \mu} = 1 - \lambda \mu,$$

which is positive on the range for which quadratic utility is increasing. Hence, a portfolio maximizing expected quadratic utility will also be mean–variance efficient. In this sense, we may say that mean–variance optimization is justified if we assume a quadratic utility.

It can also be shown[17] that, for a generic utility function, the optimal portfolio is mean–variance efficient in the case of a multivariate normal distribution of return. The result depends on the shape of the density function of a multivariate distribution,

$$f_{\mathbf{X}}(\mathbf{x}) = \frac{1}{\sqrt{(2\pi)^n \det(\mathbf{\Sigma})}} \exp\left\{-\frac{1}{2}(\mathbf{x} - \boldsymbol{\mu})^\mathsf{T} \mathbf{\Sigma}^{-1}(\mathbf{x} - \boldsymbol{\mu})\right\}. \quad (8.22)$$

The level curves of this density are ellipses determined by the quadratic form involved in the argument of the exponential function, which depends on the

---

[17] See [7, p. 97].

vector of expected values μ and the covariance matrix Σ. The result can be generalized to the family of elliptical distributions. This family includes the multivariate normal, as well as the multivariate version of Student's t. In both cases, the density depends on the quadratic form inside Eq. (8.22),[18] so that its level curves are elliptical.

It has been argued that mean–variance optimization relies on questionable assumptions, since elliptical distributions are symmetric and quadratic utility has some odd features. Supporters of the approach, in turn, argue that quadratic utility may be considered as a local, second-order approximation of a more generic utility function.[19]

8.6 Instability in mean–variance portfolio optimization

The output of any portfolio optimization procedure depends on the input data, which may be affected by statistical estimation noise. A robust procedure should not be too sensitive to such noise. Furthermore, one may wish to choose the input data in order to reflect possible views about the future. For instance, there may be little point in estimating expected returns or risk premia by a sample average of recent data. The sensitivity analysis of Example 8.3 shows that perturbations in the input data may have quite different consequences, depending on the specific case.

To further investigate the issue, let us consider a numerical example originally proposed in [6]. The investment opportunities consist of seven assets, as well as the risk-free asset. The risk premia are as follows:

$$\pi = \begin{bmatrix} 3.9\%, & 6.9\%, & 8.4\%, & 9\%, & 4.3\%, & 6.8\%, & 7.6\% \end{bmatrix}^\mathsf{T}.$$

The correlations among the excess returns, over an investment horizon, are given by the following symmetric matrix:

$$\mathbf{R} = \begin{bmatrix} 1.000 & 0.488 & 0.478 & 0.515 & 0.439 & 0.512 & 0.491 \\ \cdot & 1.000 & 0.664 & 0.655 & 0.310 & 0.608 & 0.779 \\ \cdot & \cdot & 1.000 & 0.861 & 0.355 & 0.783 & 0.668 \\ \cdot & \cdot & \cdot & 1.000 & 0.354 & 0.777 & 0.653 \\ \cdot & \cdot & \cdot & \cdot & 1.000 & 0.405 & 0.306 \\ \cdot & \cdot & \cdot & \cdot & \cdot & 1.000 & 0.652 \\ \cdot & \cdot & \cdot & \cdot & \cdot & \cdot & 1.000 \end{bmatrix}$$

and the vector of volatilities is

$$\sigma = \begin{bmatrix} 16\%, & 20.3\%, & 24.8\%, & 27.1\%, & 21\%, & 20\%, & 18.7\% \end{bmatrix}^\mathsf{T}.$$

---

[18] See, e.g., [1, pp. 124–128].
[19] See the discussion in [9] and [10].

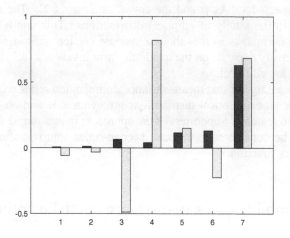

FIGURE 8.9 Effect of a perturbation in the risk premia on asset allocation.

Based on these data, we may easily build the covariance matrix Σ. The risk aversion coefficient is set to $\lambda = 2.5$, and short-selling is allowed.

The resulting portfolio weights are

$$\mathbf{w} = [0.0064, \ 0.0114, \ 0.0650, \ 0.0406, \ 0.1165, \ 0.1312, \ 0.6295]^\mathsf{T}.$$

We observe that the largest allocation is not to the asset with the largest risk premium, as volatilities and correlations have an impact on the solution.

Now, let us introduce the following perturbation on the risk premia:

$$\delta\pi = [0\%, \ 0\%, \ -0.8\%, \ 2.4\%, \ 0\%, \ -0.8\%, \ 0\%]^\mathsf{T}.$$

As we shall see later, when discussing the Black–Litterman model in Section 10.3, this could be supposed to reflect a change in the portfolio manager's view with respect to equilibrium risk premia. Essentially, one asset is expected to outperform two other assets, whereas the premia for the four remaining assets are not perturbed. The resulting portfolio now is

$$\mathbf{w} = [-0.0582, \ -0.0316, \ -0.4879, \ 0.8185, \ 0.1502, \ -0.2254, \ 0.6837]^\mathsf{T},$$

which is dramatically different from the previous one. The sum of these weights is not 1, as the difference is allocated to the risk-free asset. We notice an impressive shift in the weights of the assets involved in the perturbation, which results in extreme positions and heavy short-selling. The weights of the other assets are affected as well. The difference is visualized in Fig. 8.9. Increasing the risk aversion coefficient will have an impact the absolute values of the weights, but not on their relative change, as the effect of increasing λ is just to scale weights uniformly down and to increase the weight of the risk-free asset.

The example shows the swings and the extreme positions that may arise as a consequence of estimation noise or investors' views. Furthermore, there seems to be no sensible way to express the *confidence* in these subjective views. One possible remedy would be to control this behavior by enforcing constraints on the portfolio composition but, by doing so, the portfolio would be defined by our way of defining constraints, rather than by translating data and views to decisions.

There are alternative strategies to deal with such issues. One possible approach, due to Black and Litterman, is to take advantage of an equilibrium model, the capital asset pricing model, to integrate subjective views. The approach, which is discussed in Section 10.3, may be regarded as a shrinkage estimation or a Bayesian estimation approach. Alternative approaches have been proposed along similar lines, and the overall idea is to reduce the noise in the estimates that we feed into the optimization model. The Treynor–Black model of Section 9.3 is also similar in spirit. Essentially, we should follow the market, adding a little twist on a few selected assets, for which we trust our views or private information. A more recent stream of research tackles estimation uncertainty *within* the optimization model, by applying robust optimization approaches. See Section 15.9.

S8.1 The attainable set for two risky assets is a hyperbola

Let us rewrite, for the sake of convenience, the equations describing the attainable set on the mean–risk plane,

$$y = w(\mu_1 - \mu_2) + \mu_2, \tag{8.23}$$

$$x^2 = w^2 \sigma_1^2 + 2w(1-w)\sigma_{12} + (1-w)^2 \sigma_2^2, \tag{8.24}$$

where the coordinates of each point are $y = \mu_p$ and $x = \sigma_p$.

The equation of a hyperbola is usually given in the standard form

$$\frac{(x-h)^2}{a^2} - \frac{(y-k)^2}{b^2} = 1. \tag{8.25}$$

We should eliminate the parameter w in Eqs. (8.23) and (8.24) and recast the result in the form of Eq. (8.25). To this aim, let us solve for w in Eq. (8.23)

$$w = \frac{y - \mu_2}{\mu_1 - \mu_2},$$

and plug this result into Eq. (8.24) to find

$$x^2 = \frac{(y-\mu_2)^2 \sigma_1^2 + 2(y-\mu_2)(\mu_1-y)\sigma_{12} + (\mu_1-y)^2 \sigma_2^2}{(\mu_1-\mu_2)^2}.$$

This expression may be rearranged as

$$x^2 = \frac{By^2 - 2Cy + D}{A}, \qquad (8.26)$$

where

$$A = (\mu_1 - \mu_2)^2,$$
$$B = \sigma_1^2 - 2\sigma_{12} + \sigma_2^2,$$
$$C = \sigma_1^2 \mu_2 + \sigma_2^2 \mu_1 - \sigma_{12}(\mu_1 + \mu_2),$$
$$D = \sigma_1^2 \mu_2^2 + \sigma_2^2 \mu_1^2 - 2\sigma_{12}\mu_1\mu_2.$$

Since we assume $\mu_1 \neq \mu_2$, we have $A > 0$. We also see that $B > 0$, since we do not consider perfect correlation, i.e., we assume $\rho_{12} \in (-1, +1)$. Now we may use the standard trick of completing the square:

$$By^2 - 2Cy + D = B\left[\left(y - \frac{C}{B}\right)^2 - \frac{C^2}{B^2} + \frac{D}{B}\right] = B(y-k)^2 + H,$$

where

$$k = \frac{C}{B}, \qquad H = \frac{BD - C^2}{B}.$$

Therefore, we may recast Eq. (8.26) as

$$x^2 = \frac{B(y-k)^2 + H}{A} \quad \Rightarrow \quad \frac{x^2}{H/A} - \frac{(y-k)^2}{H/B} = 1.$$

We already know that $A, B > 0$, and in order to make sure that this is a hyperbola we need to prove that $H > 0$ as well. This implies a rather lengthy (and boring) calculation, involving plenty of trivial algebra and simplifications, showing that

$$BD - C^2 = \left(\sigma_1^2 - 2\sigma_{12} + \sigma_2^2\right) \cdot \left(\sigma_1^2\mu_2^2 + \sigma_2^2\mu_1^2 - 2\sigma_{12}\mu_1\mu_2\right)$$
$$- \left[\sigma_1^2\mu_2 + \sigma_2^2\mu_1 - \sigma_{12}(\mu_1 + \mu_2)\right]^2$$
$$= A\sigma_1^2\sigma_2^2\left(1 - \rho_{12}^2\right) > 0,$$

under the assumption $-1 < \rho_{12} < 1$. We note that the center of the hyperbola is on the vertical axis. For the limit cases $\rho_{12} = \pm 1$, the hyperbola degenerates to a pair of lines, as we have seen in Section 8.2.2.

S8.2 Explicit solution of mean–variance optimization in matrix form

In Section 8.2.3, we have solved a multiobjective optimization problem by relying on a scalarized mean–risk objective function, which plays the role of a

S8.2 Explicit solution of mean–variance optimization in matrix form

risk-adjusted expected return. Here, we take the alternative scalarization route, where we minimize risk, subject to a constraint on expected return. We have to solve the quadratic program

$$\min \tfrac{1}{2}\mathbf{w}^\mathsf{T} \mathbf{\Sigma} \mathbf{w} \tag{8.27}$$

$$\text{s.t.} \quad \boldsymbol{\mu}^\mathsf{T} \mathbf{w} = \mu_{\min}, \tag{8.28}$$

$$\mathbf{i}^\mathsf{T} \mathbf{w} = 1, \tag{8.29}$$

where $\mathbf{i} = [1, 1, \ldots, 1]^\mathsf{T}$. When inequality constraints are added to this model, possibly including constraints forbidding short-selling, we have to resort to numerical solution methods, but in this case the problem may be solved in closed form. Let us introduce Lagrange multipliers[20] γ_1 and γ_2, associated with constraints (8.28) and (8.29), respectively, and build the Lagrangian function

$$\mathcal{L}(\mathbf{w}, \gamma_1, \gamma_2) = \tfrac{1}{2}\mathbf{w}^\mathsf{T} \mathbf{\Sigma} \mathbf{w} + \gamma_1 \cdot (\mu_{\min} - \boldsymbol{\mu}^\mathsf{T} \mathbf{w}) + \gamma_2 \cdot (1 - \mathbf{i}^\mathsf{T} \mathbf{w}).$$

The first-order optimality condition for the Lagrangian yields a system of linear equations,

$$\nabla_\mathbf{w} \mathcal{L}(\mathbf{w}, \gamma_1, \gamma_2) = \mathbf{\Sigma} \mathbf{w} - \gamma_1 \boldsymbol{\mu} - \gamma_2 \mathbf{i} = \mathbf{0},$$

which is solved assuming that the covariance matrix has full rank:

$$\mathbf{w} = \gamma_1 \mathbf{\Sigma}^{-1} \boldsymbol{\mu} + \gamma_2 \mathbf{\Sigma}^{-1} \mathbf{i}. \tag{8.30}$$

If we premultiply Eq. (8.30) by $\boldsymbol{\mu}^\mathsf{T}$ and use Eq. (8.28), we find the scalar equation

$$\boldsymbol{\mu}^\mathsf{T} \mathbf{w} = \gamma_1 \boldsymbol{\mu}^\mathsf{T} \mathbf{\Sigma}^{-1} \boldsymbol{\mu} + \gamma_2 \boldsymbol{\mu}^\mathsf{T} \mathbf{\Sigma}^{-1} \mathbf{i} = \mu_{\min}. \tag{8.31}$$

By a similar token, if we premultiply Eq. (8.30) by $\mathbf{i}^\mathsf{T}$ and use Eq. (8.29), we find

$$\mathbf{i}^\mathsf{T} \mathbf{w} = \gamma_1 \mathbf{i}^\mathsf{T} \mathbf{\Sigma}^{-1} \boldsymbol{\mu} + \gamma_2 \mathbf{i}^\mathsf{T} \mathbf{\Sigma}^{-1} \mathbf{i} = 1. \tag{8.32}$$

Therefore, in order to find the multipliers, we have to solve a system consisting of the two linear equations (8.31) and (8.32). Let us streamline notation by introducing

$$A \doteq \boldsymbol{\mu}^\mathsf{T} \mathbf{\Sigma}^{-1} \boldsymbol{\mu}, \quad B \doteq \boldsymbol{\mu}^\mathsf{T} \mathbf{\Sigma}^{-1} \mathbf{i} = \mathbf{i}^\mathsf{T} \mathbf{\Sigma}^{-1} \boldsymbol{\mu}, \quad C \doteq \mathbf{i}^\mathsf{T} \mathbf{\Sigma}^{-1} \mathbf{i}.$$

The Cramer's (determinant) method applied to the system

$$A\gamma_1 + B\gamma_2 = \mu_{\min}$$
$$B\gamma_1 + C\gamma_2 = 1$$

yields

$$\gamma_1 = \frac{C\mu_{\min} - B}{AC - B^2}, \quad \gamma_2 = \frac{A - B\mu_{\min}}{AC - B^2},$$

---

[20] See Section 16.1.3 for details about the Lagrange multiplier method.

which may be plugged into Eq. (8.30) to find the portfolio weights of the minimum variance portfolio corresponding to the target return $\mu_{\min}$. It is also easy to see that, by introducing suitable vectors **g** and **h** depending on A, B, and C, we may write the portfolio weights as a linear function of the target return:

$$\mathbf{w} = \mathbf{g} + \mathbf{h}\mu_{\min}. \tag{8.33}$$

When $\mu_{\min}$ ranges over the real line, we trace the set of minimum variance portfolios for a given target, whose upper portion, above the minimum variance portfolio, is the efficient frontier. Equation (8.33) has a further implication: Every portfolio on the minimum variance set may be expressed as a linear function of the target expected return. Let us consider two portfolios on the curve,

$$\mathbf{w}_1 = \mathbf{g} + \mathbf{h}\mu_{\min,1}, \quad \mathbf{w}_2 = \mathbf{g} + \mathbf{h}\mu_{\min,2},$$

and take an affine combination

$$\alpha \mathbf{w}_1 + (1-\alpha)\mathbf{w}_2 = \mathbf{g} + \mathbf{h}[\alpha\mu_{\min,1} + (1-\alpha)\mu_{\min,2}],$$

for an arbitrary value of α. The resulting portfolio must be the minimum variance portfolio for the target expected return $\alpha\mu_{\min,1} + (1-\alpha)\mu_{\min,2}$, and this shows that we may find any portfolio on the minimum variance set by taking combinations of just *two* portfolios on the set. This result is known as **two-fund separation theorem**. Clearly, this applies under the rather unrealistic assumption that there are no additional constraints on the portfolio composition. A further consequence is that, since we know from Supplement S8.1 that the set of attainable portfolios built by combining two assets is a hyperbola, the minimum variance set is a hyperbola in the case of $n > 2$ assets, too.

Problems

8.1 Let us consider the CAL with different borrowing and lending rates. In practice, you may borrow money only at a rate $r_f^B > r_f$. What is the shape of the CAL in this setting?

8.2 Show that the general formula of Eq. (8.12), for the minimum variance portfolio, boils down to Eq. (8.7) in the case of two risky assets.

8.3 Analyze the sensitivity of the optimal portfolio in Example 8.3 with respect to the correlation coefficient ρ.

8.4 Prove Eqs. (8.13–8.15).

8.5 Find the optimal risky portfolio maximizing the Sharpe ratio analytically, for the case of two assets with parameters μ_1, μ_2, σ_1, σ_2, and ρ.

8.6 The efficient market hypothesis (EMH), in its strong form, implies that asset returns over time are independent random variables. An investor does not believe the EMH and thinks that there is some degree of persistence in return. She estimates the following relationship between the random returns on an index at time periods $t-1$ and t:

$$\tilde{r}_t = a + b\tilde{r}_{t-1} + \epsilon_t.$$

Note that, if b is positive, we have persistence in return. Assume that we know the parameters of the model, as well as the variance of the random shock term ϵ_t (whose expected value is zero). We use this statistical model to make portfolio choice on the basis of a mean–variance function

$$\mathrm{E}[\tilde{r}_p] - \lambda \cdot \mathrm{Var}(\tilde{r}_p),$$

where $\tilde{r}_p$ is the holding period return for a time period consisting of two consecutive time periods t and $t+1$. In the portfolio, we mix the index and a risk-free asset with constant return r_f for each time period. Find the optimal portfolio weights.
Hint: Which assumption could we make, in order to make the problem manageable, if the returns are small enough?

Further reading

- Mean–variance optimization is a traditional topic, and an account of modern portfolio theory can be found in [4].
- A succinct, yet accurate treatment is offered in [8], whereas [5] offers a more extensive coverage.
- For the mathematically inclined reader, a good reference is [7].
- The mathematics of the efficient frontier is also dealt with by [2].
- The effect of distributional ambiguity is best investigated by considering out-of-sample performance. The experiments carried out in [3] show that a naive, equally weighted portfolio may yield remarkably robust performance.

Bibliography

1 P. Brandimarte. *Handbook in Monte Carlo Simulation: Applications in Financial Engineering, Risk Management, and Economics*. Wiley, Hoboken, NJ, 2014.

2 M. Capiński and E. Kopp. *Portfolio Theory and Risk Management*. Cambridge University Press, Cambridge, 2014.

3 V. DeMiguel, L. Garlappi, and R. Uppal. Optimal versus naive diversification: How inefficient is the $1/N$ portfolio strategy? *The Review of Financial Studies*, 22:1915–1953, 2009.

4 E.J. Elton, M.J. Gruber, S.J. Brown, and W.N. Goetzman. *Modern Portfolio Theory and Investment Analysis* (8th ed.). Wiley, Hoboken, NJ, 2011.

5 J.C. Francis and D. Kim. *Modern Portfolio Theory: Foundations, Analysis, and New Developments*. Wiley, Hoboken, NJ, 2013.

6 G. He and R. Litterman. The intuition behind Black–Litterman model portfolios. Technical report, *Investment Management Research*, Goldman & Sachs Company, 1999. A more recent version can be downloaded from http://www.ssrn.org.

7 J.E. Ingersoll, Jr. *Theory of Financial Decision Making*. Rowman & Littlefield, Totowa, NJ, 1987.

8 M.S. Joshi and J.M. Paterson. *Introduction to Mathematical Portfolio Theory*. Cambridge University Press, Cambridge, 2013.

9 Y. Kroll, H. Levy, and H.M. Markowitz. Mean–variance versus direct utility maximization. *The Journal of Finance*, 39:47–61, 1984.

10 H.M. Markowitz. Mean-variance approximations to expected utility. *European Journal of Operational Research*, 234:346–355, 2014.

… Chapter Nine

Factor Models

In Chapter 8, we have investigated the mathematics of mean–variance portfolio optimization. From a computational viewpoint, this leads to rather easy optimization models, but things are far from trivial from a financial perspective. One could question the use of a symmetric risk measure, as well as a model neglecting multiperiod dynamics, transaction costs, etc. Some of these issues may be addressed by introducing more sophisticated optimization models, but an essential question remains: How can we provide the optimization model with suitable inputs? The mean–variance model, in its basic form, does not seem to require much: A vector of expected returns and a covariance matrix. Apparently, all we need is simple inferential statistics to estimate these parameters. Reality, unfortunately, is a tad more complicated. To begin with, we would be better off with *forecasts* rather than estimates based on past history. Moreover, a huge amount of data would be needed to estimate a covariance matrix reliably, and these data are simply not available. Last but not least, the solution of the optimization problem critically depends on the reliability of the estimates, leaving all of the mean–variance optimization framework to rest on shaky foundations. As we have seen in Section 8.6, the resulting portfolio may be quite sensitive to perturbations in the data. In this chapter, we consider factor models as a possible remedy. As we shall see, factor models have deep practical and theoretical implications since, augmented with additional assumptions, they lead to equilibrium models like CAPM and APT, discussed in Chapter 10.

We start by discussing statistical estimation issues for portfolio optimization in Section 9.1. Then, we introduce the single-index model in Section 9.2, showing how a single factor may ease these difficulties. We also show how to maximize the portfolio Sharpe ratio within the single-index model. Then, before generalizing the idea to multifactor models, we explore the implications of the single-index model in terms of practical portfolio management. The model, despite its simplicity, points out the role of systematic and specific risk factors, and it sheds light on the contrasting views associated with passive and active portfolio management. Active and passive strategies may be actually blended, in order to tilt the portfolio away from a passive benchmark in a limited way, depending on a set of views that we feel reliable enough. In Section 9.3, we illustrate one way of doing so, the Treynor–Black model. Later, in Section 10.3 we will describe an alternative approach, the Black–Litterman model. In practice, models relying on multiple factors are used, and they are outlined in

Section 9.4. Section 9.5 shows how factor models may be used to shape the risk exposure of a portfolio.

This chapter revolves about equity portfolios, but factor models have a much larger applicability. As we have outlined in Section 6.6, multiple risk factors, related to a set of interest rates with different maturities, should be taken into account for fixed-income portfolio management. This also applies to portfolios including derivatives, where not only underlying asset prices or interest rates, but also model parameters, such as volatility, are relevant risk factors. The main difference is that we consider here statistical *linear* models, whose aim is to explain the return of a primary security, a stock share, whereas fixed-income and derivative securities involve *nonlinear* pricing models.

In this chapter, returns refer to an arbitrary holding period, not necessarily one year. Thus, we shall never use the term "rate," and return should be interpreted as a holding period return, not annualized.

9.1 Statistical issues in mean–variance portfolio optimization

The Markowitz approach to mean–variance efficiency and its variations need two crucial inputs: The vector of expected returns μ and the covariance matrix Σ. Naive thinking may suggest that all we have to do is collect a sufficient amount of data and estimate these inputs by sample mean and sample variances/covariances. Given a sample of realized returns r_{it}, for assets indexed by $i = 1, \ldots, n$, over time periods $t = 1, \ldots, T$, sample means and sample covariances are given by

$$\bar{R}_i = \frac{1}{T} \sum_{t=1}^{T} r_{it}, \quad S_{ij} = \frac{1}{T-1} \sum_{t=1}^{T} (r_{it} - \bar{r}_i)(r_{jt} - \bar{r}_j),$$

respectively. However, as we may expect (pun intended), we need *expected* returns for the future, rather than sample averages from the past. Clearly, some subjective assessment would be needed, and it is difficult to think that accurate predictions may be found for a large number of assets. For the moment, let us pretend that the data generating process of returns is constant over time, so that estimates based on past data make sense. How reliable are our estimates?

To get a clue, let us consider a universe of $n = 500$ firms. The covariance matrix consists of $500 \times 500 = 250,000$ entries. Actually, since $\sigma_{ij} = \sigma_{ji}$, the matrix is symmetric. Therefore, about half of that amount, 125,000 parameters, is really needed. A more accurate assessment is that we need

- 500 variances $\sigma_{ii} \equiv \sigma_i^2$,
- $n \times (n-1)/2 = 124{,}750$ covariances $\sigma_{ij}, i > j$,

amounting to a total of 125,250 entries. Clearly, estimating such a huge number of parameters reliably is a hopeless endeavor. We would need an extremely long time series, which is certainly not available for new firms (say, Facebook). Moreover, even when have a long time series for firms like, say, IBM, these data are hardly all relevant, since the market conditions, and possibly the firms themselves, have changed considerably over time. One way to simplify the task is to reduce dimensionality by spotting common risk factors, such as market return, inflation, economic growth, oil price, etc., which explain the return of a stock share together with firm-specific factors. This simplification may also be useful for forecasting purposes, as it is certainly easier to predict a few key variables than a large array of returns.

9.2 The single-index model

The simplest factor model consists of a linear relationship between the return of each asset, represented by the random variables $\tilde{r}_i$, $i = 1, \ldots, n$, and a single common risk factor, related to market return, represented by the random variable $\tilde{r}_M$. The market return may be proxied by a relevant index, such as S&P500, which is why such a model is referred to as **single-index model**. It may be convenient to express the single-index model in terms of excess returns of each asset, $\tilde{R}_i = \tilde{r}_i - r_f$, and of the market index, $\tilde{R}_M = \tilde{r}_M - r_f$. Then, the single-index model is given by the linear regression model

$$\tilde{R}_i = \alpha_i + \beta_i \tilde{R}_M + \tilde{\epsilon}_i, \qquad i = 1, \ldots, n, \tag{9.1}$$

where the error term $\tilde{\epsilon}_i$ is interpreted as a **specific risk factor** of asset i. We also use terms like **systematic** and **idiosyncratic** risk factors to refer to common and specific factors, respectively. We make the following assumptions:

- The specific risks are random variables with expected value $\mathrm{E}[\tilde{\epsilon}_i] = 0$. This makes sense, as any predictable component should be included in the intercept α_i.
- The systematic and specific risks are uncorrelated. We know from the theory of linear regression that, after estimating the model using ordinary least-squares, we indeed have $\mathrm{Cov}(\tilde{R}_M, \tilde{\epsilon}_i) = 0$.
- Last but not least, a further important assumption is the lack of correlation among specific risks,

$$\mathrm{Cov}(\tilde{\epsilon}_i, \tilde{\epsilon}_j) = 0, \qquad i \neq j.$$

This condition characterizes a **diagonal** model, since the covariance matrix for the errors is diagonal. Indeed, if idiosyncratic factors are truly specific, this assumption should hold. However, this is a debatable condition, and it holds only if the single common factor really accounts for all of the common risk. When multiple common risk factors, possibly related to macroeconomic conditions, are at play, a single factor will not account for the whole correlation, and this may be reflected in correlated errors. Nevertheless, the diagonality assumption will prove to be quite convenient.

Before delving into statistical issues, it is important to interpret the single-index model in terms of contribution to risk premium. According to the index model,
$$\pi_i = \mathrm{E}[\tilde{R}_i] = \alpha_i + \beta_i \mathrm{E}[\tilde{R}_M] = \alpha_i + \beta_i \pi_M,$$
i.e., the risk premium[1] of asset i depends on:

- A specific risk premium α_i, depending on the exposure to specific risk. Intuitively, securities with a negative alpha are overpriced, and their weight should be reduced; they could even be sold short.
- The degree of exposure to the systematic risk, which is rewarded according to the coefficient β_i, measuring the sensitivity of each individual asset to the common risk factor.

From the theory of linear regression by least-squares, we recall that
$$\beta_i = \frac{\mathrm{Cov}(\tilde{R}_i, \tilde{R}_M)}{\mathrm{Var}(\tilde{R}_M)} \equiv \frac{\sigma_{iM}}{\sigma_M^2}.$$
Thus, the exposure of an asset to systematic risk is related to its covariance with the market return.[2] The total risk of an asset is the sum of systematic and specific risks:
$$\sigma_i^2 = \beta_i^2 \sigma_M^2 + \sigma_{\epsilon i}^2, \tag{9.2}$$
where σ_M^2 and $\sigma_{\epsilon i}^2$ are the variances of $\tilde{R}_M$ and $\tilde{\epsilon}_i$, respectively. We recall from Section 8.2.1 that there is a limit to how much risk may be diversified away. Now we may get a deeper understanding of that finding: Specific risks can be reduced by straightforward diversification in long-only portfolios. To reduce systematic risk, we may have to short-sell assets (or use derivatives), unless we find a suitable set of stock shares with positive and negative betas, offsetting each other. We should also observe that estimating β_i by a linear regression model, based on a sample of past observations, may make sense if we do not expect too many changes with respect to the past. On the contrary, we should not expect good results by estimating the specific risk premium α_i using past data. Forecasting α_i requires considerable skills in security analysis, and the "quest for alpha" is a typical endeavor of active portfolio managers.

9.2.1 ESTIMATING A FACTOR MODEL

It is easy to figure out why the factor model is much easier to deal with than estimation of a full covariance matrix. Variances are given by Eq. (9.2), which

---

[1] In order to avoid confusion, let us recall that we denote by $\mu_i = \mathrm{E}[\tilde{r}_i]$ and $\mu_M = \mathrm{E}[\tilde{r}_M]$ the expected returns of the individual assets and the index, respectively, and by $\pi_i = \mathrm{E}[\tilde{R}_i] = \mu_i - r_f$ and $\pi_M = \mathrm{E}[\tilde{R}_M] = \mu_M - r_f$ the corresponding risk premia.
[2] We will meet this expression again in Chapter 10, when discussing the capital asset pricing model.

9.2 The single-index model

requires the betas and the variances of the systematic and specific risk factors. The covariance between assets returns, under the diagonal model, is the product of betas and the market risk:

$$\text{Cov}(\tilde{R}_i, \tilde{R}_j) = \beta_i \beta_j \sigma_M^2, \qquad i \neq j.$$

As to the correlation, it is the product of correlations with the common risk factor:

$$\rho_{ij} = \frac{\beta_i \beta_j \sigma_M^2}{\sigma_i \sigma_j} = \frac{(\beta_i \sigma_M^2) \cdot (\beta_j \sigma_M^2)}{(\sigma_i \sigma_M) \cdot (\sigma_j \sigma_M)} = \rho_{im} \cdot \rho_{jm}.$$

The expected returns may be found from Eq. (9.1), which requires an estimate of the alphas. Hence, in order to feed a mean–variance optimization model, we need to estimate:

- The $2n$ parameters α_i and β_i, $i = 1, \ldots, n$
- The expected value μ_M (or, equivalently, the risk premium π_M) and the variance σ_M^2 of the common risk factor
- The n variances $\sigma_{\epsilon i}^2$ of idiosyncratic risk factors

The total number of these parameters is $3n + 2$, rather than the

$$2n + \frac{n(n-1)}{2}$$

parameters required by the full covariance matrix model. In the case of $n = 500$ assets, the total is 1502 parameters, which should be compared with the 125,250 entries of the full covariance matrix (to which 500 expected returns should be added).

The reduction in estimation complexity achieved by a single-index model is certainly relevant, but we are faced with a daunting task anyway. In order to tackle it, the index model suggests a decomposition, as well as an organizational decentralization of the portfolio management task:

- Macroeconomic analysis may be used to estimate/forecast the risk premium and the risk of the market index.
- Statistical analysis may be used to estimate/forecast the beta coefficients of all securities and their residual variances.
- The alpha value distills the incremental risk premium attributable to (legitimate) private information developed from security analysis.

Different specialists may tackle each of these subtasks. An array of statistical techniques may be used, and here we just mention one.

Example 9.1 Shrinkage estimators and beta

In inferential statistics, we learn about the most obvious and desirable property of an estimator: The expected value of the estimator,

> which is a random variable, should just be the value of the unknown
> parameter. In such a case, we say that the estimator is **unbiased**. For
> instance, it is easy to show that the sample mean is an unbiased esti-
> mator of the expected value. However, another feature of an estimator
> is its variance. If the value of the estimator is too sensitive to the input
> data, the corresponding instability in estimates may adversely affect
> the quality of the resulting decisions.
>
> Sometimes, we may reduce variance by accepting a moderate
> amount of bias. In fact, a relevant issue in statistical modeling is
> the need to address the **bias–variance tradeoff**. One example of this
> idea is the introduction of shrinkage estimators, i.e., estimators that
> shrink variability by mixing the sample estimate with a fixed value.
> For instance, the following adjusted estimator of beta has been pro-
> posed:
>
> $$\beta_{\text{adj}} = \tfrac{2}{3}\beta_{\text{sample}} + \tfrac{1}{3}.$$
>
> The idea is to take a weighted average between the sample estimate
> and a fixed value, which in this case is 1. A unit beta is, in some sense,
> a standard beta, as it implies that the risk of the asset is just the same
> as the market portfolio. Furthermore, when firms grow and diversify
> their lines of business, there is an empirical tendency of beta to move
> towards 1.

Shrinkage estimators introduce an amount of bias, but reduce sampling variability. A similar concern is addressed by regularized regression models, which we discuss in Section 14.4.1. The Black–Litterman model of Section 10.3 may also be interpreted as a sort of shrinkage estimator, where we merge different knowledge sources.

9.2.2 PORTFOLIO OPTIMIZATION WITHIN THE SINGLE-INDEX MODEL

It is useful to see how the simple problem of maximizing the Sharpe ratio of an unconstrained portfolio may be tackled within the single-index framework. Let portfolio weights be denoted by w_i. Then, the portfolio excess return is

$$\tilde{R}_p = \sum_{i=1}^n w_i(\alpha_i + \beta_i \tilde{R}_M + \tilde{\epsilon}_i) = \sum_{i=1}^n w_i \alpha_i + \tilde{R}_M \sum_{i=1}^n w_i \beta_i + \sum_{i=1}^n w_i \tilde{\epsilon}_i$$
$$= \alpha_p + \beta_p \tilde{R}_M + \tilde{\epsilon}_p, \tag{9.3}$$

where

$$\alpha_p \doteq \sum_{i=1}^n w_i \alpha_i, \qquad \beta_p \doteq \sum_{i=1}^n w_i \beta_i \tag{9.4}$$

9.2 The single-index model

are the portfolio alpha and beta, respectively, and

$$\tilde{\epsilon}_p \doteq \sum_{i=1}^{n} w_i \tilde{\epsilon}_i$$

accounts for specific risks. Then, the portfolio risk premium is

$$\pi_p = \mathrm{E}[\tilde{R}_p] = \alpha_p + \beta_p \pi_M, \tag{9.5}$$

where π_M is expected excess return (risk premium) of the market portfolio (more generally, the expected value of whatever common risk factor we choose), and the portfolio variance is

$$\sigma_p^2 = \beta_p^2 \sigma_M^2 + \sum_i w_i^2 \sigma_{\epsilon i}^2 = \beta_p^2 \sigma_M^2 + \sigma_{\epsilon p}^2, \tag{9.6}$$

where $\sigma_{\epsilon p}^2 \doteq \sum_i w_i^2 \sigma_{\epsilon i}^2$ is the residual variance of the portfolio, i.e., the component associated with specific risks.

Finally, we may express the Sharpe ratio as

$$\frac{\pi_p}{\sigma_p} = \frac{\sum_{i=1}^{n} w_i \alpha_i + \pi_M \sum_{i=1}^{n} w_i \beta_i}{\left[\sigma_M^2 \left(\sum_{i=1}^{n} w_i \beta_i \right)^2 + \sum_{i=1}^{n} w_i^2 \sigma_{\epsilon i}^2 \right]^{1/2}}. \tag{9.7}$$

As we have seen in Section 8.4, we may maximize the Sharpe ratio by solving a system of linear equations in terms of asset pseudo-weights x_i, which have to be renormalized to yield true weights w_i. Using Eq. (8.20), we write the first-order optimality conditions:

$$\sigma_M^2 \beta_k \left(\sum_{i=1}^{n} x_i \beta_i \right) + x_k \sigma_{\epsilon k}^2 = \alpha_k + \pi_M \beta_k, \quad k = 1, \ldots, n, \tag{9.8}$$

or

$$\sum_{\substack{i=1 \\ i \neq k}}^{n} x_i \beta_i + x_k \left(\beta_k + \frac{\sigma_{\epsilon k}^2}{\sigma_M^2 \beta_k} \right) = \frac{\alpha_k + \pi_M \beta_k}{\sigma_M^2 \beta_k}, \quad k = 1, \ldots, n. \tag{9.9}$$

All we have to do is solve this system of linear equations and then normalize the pseudo-weights. Clearly, the solution is so easy because we allow short-sales and are essentially dealing with an unconstrained optimization problem.[3]

---

[3] See Section 15.3.1 for a general reformulation of Sharpe ratio maximization.

9.3 The Treynor–Black model

The single-index model sheds more light on the financial issues involved in the statistical estimation of risk premia and risk exposures. The estimation of the asset betas and the variances of the risk factors may be considered as the task of specifying a risk model, as they are essentially related with the estimation of covariances. The estimation of alphas, the specific risk premia, is a task of a different nature, as it implies the analysis of future perspectives of individual stock shares. This is a typical task of active portfolio managers, whereas passive portfolio management essentially requires tracking an index. In Chapter 10, we will discuss the capital asset pricing model (CAPM), which is an equilibrium model, rather than a statistical model. According to CAPM, all alphas should be zero in equilibrium, i.e., the only risk premium is related to the portfolio exposure to the undiversifiable systematic risk, since specific risks can be diversified away and are not rewarded. The practical consequence is that, according to CAPM, there is no point in pursuing active strategies based on stock-picking.

Even without taking such a radical view, there is no doubt that generating alpha is difficult and arguably feasible only when analyzing a limited number of assets. Hence, assuming that we trust our alpha generation skills and we want to tilt away from a market portfolio, we should do so in a limited manner. This is the idea underlying the Treynor–Black model, which considers a portfolio including n individual assets *and* the market (index) portfolio as asset $n+1$. All of the notation that we have introduced in Section 9.2.2 applies, changing the upper limit of sums from n to $n+1$, with the conditions:

$$\alpha_{n+1} \equiv \alpha_M = 0,$$
$$\beta_{n+1} \equiv \beta_M = 1,$$
$$\sigma_{\epsilon,n+1} \equiv \sigma_{\epsilon M} = 0.$$

It is important to understand that the last condition does *not* imply that the market portfolio is riskless. We are only saying that it has no residual (specific) risk.

Rather than just maximizing the Sharpe ratio numerically, we take a decomposition approach that helps in building intuition and gaining financial insights. The overall portfolio selection can be broken down as a top-down process:

- Allocate available wealth between the risk-free asset and an optimal risky portfolio.
- Decompose the optimal risky portfolio into a passive and an active component.
- Find the optimal active portfolio.

It is also convenient to change notation slightly, in order to better express the above decomposition:

- w_M is the weight of the passive component of the overall portfolio.
- w_A is the weight of the active component of the overall portfolio.

9.3 The Treynor–Black model

- w_i, $i = 1, \ldots, n$, are the weights of each individual asset *in the active portfolio*. These weights add up to 1 and are used to determine the composition of the active portfolio; the actual weights of each asset i in the overall portfolio depend on both w_A and w_i.

The capital allocation between the risky portfolio and the risk-free asset is accomplished as we discussed in Section 8.1. As a first step to find the risky portfolio itself, let us maximize the Sharpe ratio of a portfolio including the following two components:

- The passive (market) portfolio.
- The active portfolio, characterized by the following features:

$$\alpha_A = \sum_{i=1}^n w_i \alpha_i, \quad \beta_A = \sum_{i=1}^n w_i \beta_i, \quad \sigma_{\epsilon A}^2 = \sum_{i=1}^n w_i^2 \sigma_{\epsilon i}^2.$$

Note that $\sigma_{\epsilon A}^2$ is just the active risk component, *not* the overall variance of the active portfolio. Let us apply optimality condition (9.9), which in this case is a system of two linear equations:

$$\sigma_M^2 \beta_A (x_A \beta_A + x_M) + x_A \sigma_{\epsilon A}^2 = \alpha_A + \beta_A \pi_M \qquad (9.10)$$

$$\sigma_M^2 (x_A \beta_A + x_M) = \pi_M. \qquad (9.11)$$

By plugging Eq. (9.11) into Eq. (9.10), we immediately obtain

$$x_A = \frac{\alpha_A}{\sigma_{\epsilon A}^2}.$$

Then, from Eq. (9.11) we find

$$x_M = \frac{\pi_M}{\sigma_M^2} - \beta_A \frac{\alpha_A}{\sigma_{\epsilon A}^2}.$$

By normalizing pseudo-weights, we find the weight of the active component,

$$w_A = \frac{x_A}{x_A + x_M} = \frac{\dfrac{\alpha_A}{\sigma_{\epsilon A}^2}}{\dfrac{\pi_M}{\sigma_M^2} + (1 - \beta_A) \dfrac{\alpha_A}{\sigma_{\epsilon A}^2}}.$$

The weight of the passive component of the risky portfolio is just $w_M = 1 - w_A$. We may rewrite the weight of the active portfolio as

$$w_A = \frac{w_A^0}{1 + (1 - \beta_A) w_A^0}, \qquad (9.12)$$

where

$$w_A^0 \doteq \frac{\alpha_A / \sigma_{\epsilon A}^2}{\pi_M / \sigma_M^2} \qquad (9.13)$$

is the weight of the active portfolio when $\beta_A = 1$. In order to grasp the message behind Eq. (9.12), we observe that w_0 is a ratio expressing the relative reward–risk tradeoffs of the active and the passive components of the risky portfolio. The tradeoffs are not exactly expressed as Sharpe ratios, as they involve variances rather than standard deviations, but the interpretation is similar. Let us assume that the active and passive risk premia α_A and π_M are positive, so that $w_A^0 > 0$, and consider the sensitivity of w_A with respect to β_A:

$$\frac{\partial w_A}{\partial \beta_A} = \frac{(w_A^0)^2}{\left[1 + (1 - \beta_A) w_A^0\right]^2}.$$

This sensitivity is not defined for a critical value of β_A, where we divide by zero:

$$\beta_A^* = \frac{1 + w_A^0}{w_A^0} > 1.$$

Otherwise, the sensitivity is always positive, and w_A is an increasing function of β_A. In fact, w_A, as a function of β_A, is a hyperbola with horizontal and vertical asymptotes. We observe that

$$\beta_A = 0 \quad \Rightarrow \quad w_A = \frac{w_A^0}{1 + w_A^0} < 1,$$

which increases to w_A^0 when $\beta_A = 1$. This increases and goes to infinity when β_A approaches the critical value β_A^*. In this range, we may interpret this behavior by observing that the larger the systematic risk of the active portfolio, the less effective is the diversification obtained from the market portfolio, and hence the larger the weight of the active component. However, beyond the critical value, the weight of the active component would get negative, which means that the active component is sold short, and it is increasing toward the horizontal asymptote $w_A = -1$ for large β_A. This kind of behavior may require rather pathological assumptions about the problem data, but it illustrates the potential intricacies of unconstrained portfolio optimization.

To find the composition of the optimal active portfolio, we can analyze its contribution to the Sharpe ratio of the risky portfolio. The excess return of the risky portfolio, as a function of the excess returns of the active and the passive components, is

$$\begin{aligned}
\tilde{R}_p &= w_A \tilde{R}_A + (1 - w_A) \tilde{R}_M \\
&= w_A (\alpha_A + \beta_A \tilde{R}_M + \tilde{\epsilon}_A) + (1 - w_A) \tilde{R}_M \\
&= w_A \alpha_A + \left[1 - w_A (1 - \beta_A)\right] \tilde{R}_M + w_A \tilde{\epsilon}_A.
\end{aligned}$$

Then, we find the risk premium and variance of the risky portfolio:

$$\pi_p = w_A \alpha_A + \left[1 - w_A (1 - \beta_A)\right] \pi_M, \tag{9.14}$$

$$\sigma_p^2 = \left[1 - w_A (1 - \beta_A)\right]^2 \sigma_M^2 + w_A^2 \sigma_{\epsilon A}^2. \tag{9.15}$$

9.3 The Treynor–Black model

Hence, the squared Sharpe ratio of the overall portfolio is:

$$S_p^2 = \frac{\pi_p^2}{\sigma_p^2} = \frac{\{w_A \alpha_A + [1 - w_A(1-\beta_A)]\pi_M\}^2}{[1 - w_A(1-\beta_A)]^2 \sigma_M^2 + w_A^2 \sigma_{\epsilon A}^2}. \qquad (9.16)$$

By plugging the optimal weight w_A of the active portfolio, given by Eq. (9.12), into Eq. (9.16), and carrying out a bit of algebra, as shown in detail in Supplement S9.1, we find the following fundamental relationship:

$$S_p^2 = S_M^2 + \left[\frac{\alpha_A}{\sigma_{\epsilon A}}\right]^2, \qquad (9.17)$$

where S_M is the Sharpe ratio of the market portfolio and $\alpha_A/\sigma_{\epsilon A}$ is called the **information ratio** of the active portfolio. Clearly, the Sharpe ratio of the market portfolio is constant and, in order to maximize S_p, we should maximize the information ratio of the active component. Let us express the information ratio in more detail:

$$\frac{\alpha_A}{\sigma_{\epsilon A}} = \frac{\sum_{i=1}^{n} w_i \alpha_i}{\sqrt{\sum_{i=1}^{n} w_i^2 \sigma_{\epsilon i}^2}}.$$

Note that the passive component does not contribute anything, since both α_M and $\sigma_{\epsilon M}$ are zero. Using the optimality conditions of Eq. (9.9) again, to maximize the information ratio, it is easy to see that we obtain the equations

$$\alpha_i = \sigma_{\epsilon i}^2 x_i \qquad i = 1, \ldots, n,$$

since we are assuming a simple diagonal model, where specific risks are uncorrelated. Hence, the weights of individual assets in the active portfolio should be proportional to the ratio

$$x_i = \alpha_i / \sigma_{\epsilon i}^2,$$

which makes sense, as this ratio trades off the active risk premium α_i against active risk $\sigma_{\epsilon i}^2$. If we plug x_k into the squared information ratio, we obtain

$$\left[\frac{\alpha_A}{\sigma_{\epsilon A}}\right]^2 = \frac{(\sum_i x_i \alpha_i)^2}{\sum_i x_i^2 \sigma_{\epsilon i}^2} = \frac{(\sum_i \alpha_i^2/\sigma_{\epsilon i}^2)^2}{\sum_i \sigma_{\epsilon i}^2 \alpha_i^2/\sigma_{\epsilon i}^4} = \sum_i \left[\frac{\alpha_i}{\sigma_{\epsilon i}}\right]^2,$$

from which we observe that each asset in the active portfolio contributes its individual squared information ratio to the overall squared information ratio.

Taking into account the normalization within the active portfolio, and its weight in the overall portfolio, we find the true weight of each individual asset in the risky portfolio,

$$w_i^* = w_A \frac{\dfrac{\alpha_i}{\sigma_{\epsilon i}^2}}{\sum_k \dfrac{\alpha_k}{\sigma_{\epsilon k}^2}}.$$

9.3.1 A TOP-DOWN/BOTTOM-UP OPTIMIZATION PROCEDURE

The process of building the overall portfolio, according to the Treynor–Black model, may be thought as a conceptual top-down decomposition of portfolio optimization. To actually implement it, we have to go bottom-up, finding the active portfolio first. This may be summarized by the following procedure[4]:

1. Compute the pseudo-weights of each individual asset in the active portfolio,
$$w_i^0 = \frac{\alpha_i}{\sigma_{\epsilon i}^2}, \qquad i = 1, \ldots, n.$$

2. Scale pseudo-weights so that they add up to 1,
$$w_i = \frac{w_i^0}{\sum_{k=1}^{n} w_k^0}.$$

3. Compute alpha and residual variance of the active portfolio,
$$\alpha_A = \sum_{i=1}^{n} w_i \alpha_i, \qquad \sigma_{\epsilon A}^2 = \sum_{i=1}^{n} w_i^2 \sigma_{\epsilon i}^2.$$

4. Compute the "initial" position of the active component in the risky portfolio,
$$w_A^0 = \frac{\alpha_A / \sigma_{\epsilon A}^2}{\pi_M / \sigma_M^2}.$$

5. Compute beta of the active portfolio,
$$\beta_A = \sum_{i=1}^{n} w_i \beta_i.$$

6. Adjust the weight of the active portfolio,
$$w_A^* = \frac{w_A^0}{1 + (1 - \beta_A) w_A^0}.$$

7. Find the weights of the n individual assets and the passive component,
$$w_i^* = w_A^* w_i, \qquad w_M^* = 1 - w_A^*.$$

8. Compute the features of the risky portfolio,
$$\beta_p = w_M^* + \beta_A w_A^*,$$
$$\pi_p = \beta_p \pi_M + w_A^* \alpha_A,$$
$$\sigma_p^2 = (w_M^* + \beta_A w_A^*)^2 \sigma_M^2 + [w_A^* \sigma_{\epsilon A}]^2.$$

---
[4]Here, we are following [1].

9.3 The Treynor–Black model

To fully understand these expressions, we should think that the weight w_M^* of the passive component multiplies a unit beta. Also note that the weight w_A^* of the active component multiplies σ_M^2 in the formula for σ_p^2; therefore, the active portfolio has an impact on the overall contribution of systematic risk.

9. Solve the capital allocation problem, taking into account subjective risk aversion.

Example 9.2 Numerical illustration of the Treynor–Black model

Let us consider four assets with the following features:

α_i	5.6%	−0.4%	7.4%	0.0%
β_i	1.3	1.8	0.7	1.0
$\sigma_{\epsilon i}$	28%	21%	30%	24%

We also have:
$$r_f = 4\%, \quad \mu_M = 12\%, \quad \sigma_M = 24\%.$$

Therefore, the risk premia are

$$\pi_M = \mu_M - r_f = 8\%,$$
$$\pi_i = \alpha_i + \beta_i \pi_M \quad (i=1,2,3,4) \quad \Rightarrow \quad [16\%, 14\%, 13\%, 8\%]^\mathsf{T}$$

In Step 1, we find the unscaled pseudoweights in the initial portfolio, w_i^0,
$$[0.7143, -0.0907, 0.8222, 0]^\mathsf{T},$$
which, after normalization in Step 2, yield the weight w_i of each asset in the active portfolio,
$$[0.4940, -0.0627, 0.5687, 0]^\mathsf{T}.$$

We notice a small short position for the asset with negative alpha, and a zero position for the asset with no specific risk premium. These individual weights are large, but this does not necessarily imply a large final position. The alpha and the residual variance of the active portfolio are calculated as in Step 3:

$$\alpha_A = 7\%, \quad \sigma_{\epsilon A}^2 = 0.0484.$$

Step 4 yields the "initial" position of the active portfolio,
$$w_A^0 = 1.0410.$$

In Step 5 we find its beta,

$$\beta_A = 0.9274,$$

which is used in Step 6 to find the weight of the active portfolio,

$$w_A^* = 0.9678,$$

which is fairly large. Then, in the next steps, we find the weight of the index,

$$w_M^* = 0.0322$$

and the weight w_i^* of each asset within the optimal risky portfolio,

$$[0.4782, -0.0607, 0.5504, 0]^\mathsf{T}.$$

We notice that these weights have been only partially moderated by the index, because of the peculiarity of the data we are using. However, let us see what overall portfolio would an investor with risk aversion $\lambda = 4$ choose. We recall from Section 8.1 that the optimal allocation to the risky portfolio is

$$x^* = \frac{\pi_p}{\lambda \sigma_p^2}.$$

The risk premium and the variance for the risky portfolio are given by

$$\begin{aligned}
\pi_p &= w_A^* \alpha_A + \beta_p \pi_M \\
&= w_A^* \alpha_A + [(1 - w_A^*) \cdot 1 + w_A^* \cdot \beta_A] \pi_M \\
&= 0.9678 \times 0.07 + [0.0322 + 0.9678 \times 0.9274] \times 0.08 \\
&= 14.21\%,
\end{aligned}$$

and

$$\sigma_p^2 = \beta_p^2 \sigma_M^2 + w_A^{*2} \sigma_A^2 = 0.0951,$$

respectively. Therefore, the weights of the risky portfolio and of the risk-free asset are

$$x^* = 0.3735, \qquad 1 - x^* = 0.6265,$$

respectively. Then, the actual weight of the index is

$$x^* \cdot (1 - w_A^*) = 0.3735 \times 0.0322 = 1.2\%,$$

and the actual weights of individual assets are given by $x^* \cdot w_A^* \cdot w_i$, which yields

$$[0.1786, -0.0227, 0.2056, 0]^\mathsf{T}.$$

9.4 Multifactor models

The single-index model is quite useful conceptually, but it is unlikely that a single common factor may explain the systematic component of stock returns completely. One consequence is that the resulting model is not diagonal, i.e., we find specific residual risks that are not uncorrelated. The other consequence is that we are not able to assess risk and forecast alpha accurately. Multifactor models have been proposed as a generalization of the single-index model:

$$\tilde{R}_i = \alpha_i + \sum_{j=1}^{m} \beta_{ij} \tilde{F}_j + \tilde{\epsilon}_i, \qquad i = 1, \ldots, n,$$

where we include m common risk factors $\tilde{F}_j$, as well as a specific risk factor $\tilde{\epsilon}_i$ for each asset i. Specific risks are assumed uncorrelated, and the coefficients β_{ij} measure the exposure (sensitivity) of asset i to systematic risk factor j. The above model is written in terms of excess returns, but we are free to use returns as well.

There is considerable latitude in the selection of factors:

- We may use market-related financial factors such as different indexes or a set of given benchmark portfolios. In this case, it is quite natural to use returns as systematic factors.
- We may use financial factors that are related to accounting measures commonly used in corporate finance, like the book-to-market ratio. These are called **fundamental** factors. In this case, the interpretation of factors as return is less natural.
- We may use **macroeconomic** factors, like inflation or oil price. Again, factors like these are not interpreted as returns.
- It is also possible to use **behavioral** factors, like momentum, which are related to market anomalies.[5]

When a factor $\tilde{F}_j$ is not directly related to a return, we often estimate the model in such a way that $\mathrm{E}[\tilde{F}_j] = 0$, i.e., the expected value of the factor is included in the constant term. Thus, the factor should be interpreted as an *unanticipated surprise* with respect to an expectation. This is common for macroeconomic factors, like inflation, not for fundamental factors.

Example 9.3 The three-factor Fama–French model

A well-known factor model is the three-factor Fama–French model [2]. The model extends the single-index model, based on market risk,

---

[5] Indeed, there is no contradiction between behavioral finance and quantitative models. What is ruled out in behavioral finance is the transition from factor models to equilibrium models, as we discuss in Section 10.5.

by introducing two additional factors related to the company size and the company book-to-market ratio. Company size takes into account the empirical difference in return between small and large capitalization firms. The book-to-market ratio takes into account the difference between **value** and **growth** stocks. Value stocks are characterized by a high book-to-market ratio, i.e., the market price is small with respect to the book value of the firm, which suggests that the stock is underpriced. Value investing is a strategy based on investing on stock shares which are expected to outperform the market if their market value approaches their intrinsic value. Growth stocks are not cheap, but the rationale of investing in a growth stock is that the firm has a sustainable competitive advantage and is able to generate increasing cash flows over time.

The regression model is specified as follows:

$$R_{it} - R_{ft} = \alpha_i + \beta_i \cdot R_{Mt} + s_i \cdot \mathsf{SMB}_t + h_i \cdot \mathsf{HML}_t + e_{it},$$

where R_{it} is the return on asset i in month t, R_{ft} and R_{Mt} are the corresponding risk-free return and market return, respectively, SMB_t (small minus big) is the difference in return between diversified portfolios consisting of small and big cap stocks, and HML_t (high minus low) is the difference in return between diversified portfolios consisting of high and low book-to-market stocks.

The seminal three-factor Fama–French model has been extended in several ways, both by introducing additional factors and by changing the model structure. A significant pitfall of the linear models that we consider here is that the effect of a factor is proportional to its level and does not interact with other factors in any way. To get the point, consider a low value of book-to-market ratio. A reasonably small value might suggest an undervalued stock,[6] but a *very* small value does not necessarily imply an even better investment opportunity. This could be accounted for by a nonlinear function reaching its maximum at a sensible value that should be estimated. As another example, consider the amount of dividend payout. It may be difficult to assess its impact outside any context. A firm that does not pay dividends may be a growing firm, which invests net income in new products or services, or, on the contrary, a firm struggling with poor performance. A firm that pays rich dividends may be a good "cash cow," or a firm that is not investing anything in order to maintain its competitive position. These considerations suggest the opportunity of introducing factor interactions into the model.

---

[6] We should always bear in mind that the stated book value of a firm may rely on questionable asset valuations, not to mention the latitude in accounting practice.

9.5 Factor models in practice

As we shall see in Chapter 10, factor models, under suitable but controversial hypotheses, are the foundation of well-known equilibrium models like CAPM and APT. Even if one does not believe these assumptions, factor models may be used for practical purposes, including the following ones:

- Factor models may be used to spot potential for excess return in a portfolio, i.e., to "generate alpha."
- Factor models may be used for performance and risk attribution, i.e., to understand which factors contribute to the realized return of a given portfolio.
- By combining assets with known exposures to common risk factors, we may shape the portfolio risk and make it selectively insensitive to undesired risk factors. Otherwise, we may increase the exposure to a risk factor on which we feel like making a bet.

Example 9.4 A market-neutral long–short portfolio

> A **market-neutral** portfolio is a portfolio which is not exposed to systematic risk. The only source of return is specific risk. Such a portfolio is also referred to as **beta-neutral**, since its betas with respect to systematic risk factors are zero. A possible rationale behind such a portfolio is that we may have a view about the relative performance of stock shares, but we do not feel safe in making a bet on the direction of the market as a whole. The analysis may suggest that some stocks have positive alpha, and other stocks have negative alpha. However, investing in the positive alpha stocks may still result in a loss if the market takes a negative turn. We might find only a partial consolation in a portfolio losing less than the index. Thus, we may take a **long–short** strategy, whereby we short-sell the stocks with negative alpha in order to neutralize the overall beta, i.e., the exposure to portfolio risk.
>
> Let us consider a stylized example of the strategy. We have a subset of n assets, characterized by the single-index model
>
> $$\tilde{R}_i = \alpha + \beta \tilde{R}_M + \tilde{\epsilon}_i, \qquad i = 1, \ldots, n.$$
>
> We assume that $\alpha > 0$ and β are the same for all of these assets. We also have another subset of n assets, characterized by the single-index model
>
> $$\tilde{R}_i = -\alpha + \beta \tilde{R}_M + \tilde{\epsilon}_i, \qquad i = n+1, \ldots, 2n,$$
>
> where again we assume that the numerical values of the involved parameters are the same and identical to those of the first subset of as-

sets. Note that the assets in the second set have negative alpha and, as such, are natural candidates for short-selling.

Imagine that we go long for a total amount of $\$W$ in an equally weighted portfolio of the stocks in the first subset, and we short the same dollar amount in an equally weighted portfolio of the stocks in the second subset. Note that, if we do not consider transaction costs, the initial value of this long–short portfolio is zero. For this reason, such a portfolio is said to be **dollar-neutral**. We cannot define return for this portfolio, but its profit/loss is

$$\tilde{\Pi} = \frac{W}{n} \sum_{i=1}^{n} \left(\alpha + \beta \tilde{R}_M + \tilde{\epsilon}_i \right) - \frac{W}{n} \sum_{i=n+1}^{2n} \left(-\alpha + \beta \tilde{R}_M + \tilde{\epsilon}_i \right)$$
$$= 2\alpha W + \tilde{\epsilon}_p,$$

where

$$\tilde{\epsilon}_p \doteq \frac{W}{n} \left(\sum_{i=1}^{n} \tilde{\epsilon}_i - \sum_{i=n+1}^{2n} \tilde{\epsilon}_i \right)$$

accounts for specific risk. Note that the portfolio is, in fact, market-neutral, and that for large n the total contribution of specific risk is diversified away (see Problem 9.3).

Clearly, Example 9.4 is based on an oversimplified picture, but it illustrates a possible strategy to build a market-neutral, long–short portfolio. Long–short portfolios need not be dollar-neutral. A common strategy is 130–30, which means that 30% of the portfolio value is shorted, in order to increase investment in stocks that, we believe, will outperform. More realistic strategies in this vein are used by some hedge funds. When short-selling is not readily applicable, one way to get rid of systematic risk is by taking a position in suitable derivatives, such as a short position in index futures.[7]

S9.1   Proof of Equation (9.17)

As a preliminary step, let us solve Eq. (9.12) for w_A^0, which yields

$$w_A^0 = \frac{w_A}{1 - w_A(1 - \beta_A)}. \tag{9.18}$$

---
[7] See Section 12.3.3.

We also slightly rearrange Eq. (9.13) as

$$w_A^0 = \frac{\alpha_A \sigma_M^2}{\pi_M \sigma_{\epsilon A}^2}. \tag{9.19}$$

This allows us to rewrite the squared Sharpe ratio in Eq. (9.16) as follows:

$$S_p^2 = \frac{\left[\dfrac{w_A}{1 - w_A(1-\beta_A)}\alpha_A + \pi_M\right]^2}{\dfrac{w_A^2}{[1 - w_A(1-\beta_A)]^2}\sigma_{\epsilon A}^2 + \sigma_M^2} = \frac{(w_A^0 \alpha_A + \pi_M)^2}{(w_A^0)^2 \sigma_{\epsilon A}^2 + \sigma_M^2} \tag{9.20}$$

By plugging Eq. (9.19) into Eq. (9.20), we find

$$S_p^2 = \frac{\left(\dfrac{\alpha_A \sigma_M^2}{\pi_M \sigma_{\epsilon A}^2}\alpha_A + \pi_M\right)^2}{\dfrac{\alpha_A^2 \sigma_M^4}{\pi_M^2 \sigma_{\epsilon A}^4}\sigma_{\epsilon A}^2 + \sigma_M^2} = \frac{\left(\dfrac{\alpha_A^2}{\sigma_{\epsilon A}^2}\cdot\dfrac{\sigma_M^2}{\pi_M} + \dfrac{\pi_M^2}{\sigma_M^2}\cdot\dfrac{\sigma_M^2}{\pi_M}\right)^2}{\dfrac{\alpha_A^2}{\sigma_{\epsilon A}^2}\dfrac{\sigma_M^4}{\pi_M^2} + \sigma_M^2}$$

$$= \frac{\left(\dfrac{\alpha_A^2}{\sigma_{\epsilon A}^2} + \dfrac{\pi_M^2}{\sigma_M^2}\right)^2}{\dfrac{\alpha_A^2}{\sigma_{\epsilon A}^2} + \dfrac{\pi_M^2}{\sigma_M^2}} = \dfrac{\alpha_A^2}{\sigma_{\epsilon A}^2} + S_M^2,$$

which is the sum of squared information ratio and squared Sharpe ratio of the market portfolio, as in Eq. (9.17).

Problems

9.1 You have estimated the following single-index model for two asset returns:

$$\tilde{r}_a = 0.14 + 0.8\tilde{r}_M + \tilde{\epsilon}_a,$$
$$\tilde{r}_b = 0.08 + 1.2\tilde{r}_M + \tilde{\epsilon}_b,$$

where $\tilde{r}_M$, $\tilde{\epsilon}_a$, and $\tilde{\epsilon}_b$ are uncorrelated random variables with zero expected value (the expected value of the market return has been included into the constant term of the regression) and standard deviations 20%, 30%, and 25%, respectively. The model is expressed in terms of returns, not excess returns. Find the weights of the minimum variance portfolio (we are only considering risky assets, not the risk-free asset).

9.2 A pension fund manager has chosen a portfolio consisting of the risk-free asset, with annual return 3% (annual compounding), and two risky assets. The holding period return can be expressed by the following factor model:

$$\tilde{r}_i = \alpha_i + \beta_{i1}\tilde{F}_1 + \beta_{i2}\tilde{F}_2 + \tilde{\epsilon}_i, \qquad i = a, b,$$

with the following parameters:

Asset i	α_i	$\beta_{i,1}$	$\beta_{i,2}$
$i = a$	0.01	0.8	1.2
$i = b$	0.03	−0.4	0.3

All factors are normally distributed with the following parameters:

$$\tilde{F}_1 \sim N(0.03, 0.30^2), \quad \tilde{F}_2 \sim N(0.13, 0.40^2),$$
$$\tilde{\epsilon}_a \sim N(0, 0.40^2), \quad \tilde{\epsilon}_b \sim N(0, 0.50^2).$$

Each specific risk factor is uncorrelated with the other factors, whereas the correlation coefficient between the two systematic factors is 0.68. The portfolio weight of the risk-free asset is 40%, and the rest of the portfolio is equally allocated to the risky assets. The manager receives a bonus depending on the realized annual return: if it is at least 9%, the bonus is €200,000; if the return exceeds 12%, she will receive additional €100,000. What is the expected value of the bonus earned by the manager?

9.3 This problem is a numerical illustration of Example 9.4. Consider a single-index model, in which all assets have unit beta. The volatility of each specific risk is 30%, and you are considering a universe of 20 stocks, half of which have alpha $+2\%$ and half have alpha -2%. You go long $1 million with an equally weighted portfolio consisting of the stocks with positive alpha, and short $1 million of a similar portfolio of stocks with negative alpha. Note that the resulting portfolio is both dollar-neutral and beta-neutral. What are the expected profit and risk of the long–short portfolio? How does your answer change if you consider 50 or 100 stocks?

9.4 You have invested $100,000 in asset A_1 and $250,000 in asset A_2. The annual returns are represented by a single-index model with parameters

$$\alpha_1 = 0.7\%, \quad \alpha_2 = -0.3\%, \quad \beta_1 = 1.1, \quad \beta_2 = 0.8.$$

The annual return of the market portfolio has expected return 7% and standard deviation 37%. The volatilities of the specific risks are 22% and 31%, respectively, on an annual basis. Assuming that the risk factors are normally distributed, find the annual and the daily value-at-risks, with confidence level 99%.

9.5 Let us consider a multifactor model that we want to use in mean–variance portfolio optimization, using the risk-adjusted form of the objective function with risk-aversion coefficient λ. We want to build a long–short portfolio that is both dollar- and beta-neutral. The first requirement means that the net investment is zero. The second requirement implies that the resulting portfolio has no exposure to systematic risk. This kind of portfolio, in principle, can be used to generate *portable alpha*. Portable alpha is a strategy by which we may add alpha to a portfolio without changing its systematic risk exposure, and without the need for more capital.

- Write the optimization model, using portfolio weights as decision variables.

- Note that we do not have asset covariances, but we use the multifactor model parameters. How does beta-neutrality simplify the model?
- Write down the optimality conditions, using Lagrange multipliers.

Hint: For the second question, you may write the model in either compact matrix form or in the more explicit form. If you take the former route, you may use the following fact: Let $\mathbf{X}$ be an n-dimensional random vector with covariance matrix $\mathbf{\Sigma_x}$, and consider the transformed variable $\mathbf{Z} = \mathbf{AX}$, where $\mathbf{A} \in \mathbb{R}^{m \times n}$. Then, the covariance matrix of $\mathbf{Z}$ is $\mathbf{\Sigma_z} = \mathbf{A}\mathbf{\Sigma_x}\mathbf{A}^\mathsf{T}$.

Further reading

- A concrete illustration of how factor models may be used in quantitative portfolio management is given in [5], which also includes a chapter on nonlinear models. See also [4].
- To appreciate the several flavors of regression models, you may have a look at the introductory treatment in [3], where the bias–variance tradeoff is carefully discussed.
- The Treynor–Black model was originally introduced in [6]. Our presentation follows the one provided in [1].

Bibliography

1 Z. Bodie, A. Kane, and A. Marcus. *Investments* (9th ed.). McGraw-Hill, New York, 2010.

2 E.F. Fama and K.R. French. Common risk factors in the returns on stocks and bonds. *Journal of Financial Economics*, 33:3–56, 1993.

3 G. James, D. Witten, T. Hastie, and R. Tibshirani. *An Introduction to Statistical Learning: With Applications in R*. Springer, New York, 2013.

4 J. Knight and S. Satchell, editors. *Linear Factor Models in Finance*. Elsevier, Amsterdam, 2005.

5 E.E. Qian, R.H. Hua, and E.H. Sorensen. *Quantitative Equity Portfolio Management: Modern Techniques and Applications*. CRC Press, London, 2007.

6 J.L. Treynor and F. Black. How to use security analysis to improve portfolio selection. *The Journal of Business*, 46:66–86, 1973.

Chapter Ten

Equilibrium Models: CAPM and APT

This chapter might be regarded as an outgrowth of Chapter 9 on factor models, as there is a clear relationship between single and multiple factor models and the equilibrium models we treat here. The single-index model is related with the capital asset pricing model (CAPM), and multifactor models are related with arbitrage pricing theory (APT). However, equilibrium models require much more than a statistical model, as they rely on crucial assumptions about investors' behavior. If we accept these assumptions, we find quite drastic conclusions. For instance, CAPM implies that there is no point in pursuing active portfolio management, based on stock-picking or market-timing strategies. There is no specific risk premium, and we should just follow a passive strategy tracking the market portfolio. A fierce debate revolves around this conclusion, and it is related to conflicting views about market efficiency and investors' rationality. More generally, the view behind equilibrium models is challenged by the behavioral approach to finance. It is interesting to notice that the behavioral approach, per se, is not necessarily incompatible with the use of quantitative factor-based models.

Taken literally, the assumptions behind CAPM are rather hard to accept, as we shall see. APT is a bit less demanding than CAPM in terms of critical assumptions, but it is much less specific in its consequences, as it does not specify the factors that we should use in the model. Plenty of empirical investigations have been carried out to verify the validity of equilibrium models, with mixed evidence. Even if we do not fully believe in the validity of equilibrium models, they are conceptually useful and provide us with benchmarks and performance evaluation tools.

In Section 10.1, we briefly describe the nature of equilibrium models and contrast their requirements for critical assumptions against the practical simplicity of arbitrage-based approaches. Then, in Section 10.2, we derive CAPM and show some possible applications. A notable example of use of CAPM is the Black–Litterman model, described in Section 10.3. The idea is to mix investors' subjective beliefs with the CAPM view, in order to overcome some issues with mean–variance portfolio optimization, which we have raised in Section 8.6. The Black–Litterman approach may be regarded as a way to ease statistical estimation issues either by shrinkage estimation or by a Bayesian approach. Hence, we provide some background on Bayesian statistics in Supplement S10.1.[1] The APT model and some of its potential applications are discussed in Section 10.4. Finally, Section 10.5 closes the chapter with some critical remarks related to the behavioral view of financial markets.

10.1 What is an equilibrium model?

Equilibrium models are a mainstay of economic theory, more specifically, microeconomics, and they come in quite different forms. We have simple equilibria in basic game theory, and more complicated models involving dynamics, uncertainty, learning, and heterogeneous beliefs. We may investigate equilibria in terms of actions and strategies chosen by agents in competitive markets, or in terms of prices of commodities and assets. In deterministic price equilibrium models, agents are endowed with an amount of goods and are then engaged in trading activities, where goods are exchanged at some prices, in order to maximize their utility from consumption of bundles of goods. The model aims at finding a set of prices such that markets clear, i.e., aggregate demand and offer are perfectly matched. In a financial stochastic model, uncertainty comes into play, and we may consider wealth levels in different future states of the world as different goods. Clearly, we have to model the attitude toward risk of each agent, and the initial wealth endowments also play a role, as risk aversion may depend on wealth.

We will not delve into details of sophisticated equilibrium models, but it is clear that it is difficult to use them to obtain more than qualitative insights, possibly supporting a theory,[2] unless rather radical simplifications are applied. Apart from computational difficulties, how can we specify an array of different utility functions and endowments for a large number of market players, not to mention the possibility of differential information? As we shall see, in CAPM we assume that agents have the same information and the same decision-making procedure, possibly differing only in the subjective degree of risk aversion.

An alternative approach, which we have already seen in action in Chapter 2, is to assume the lack of arbitrage opportunities. Essentially, this requires that agents prefer more to less and are quite efficient in taking advantage of any opportunity for risk-free profit. It is intuitive that markets cannot be in equilibrium if they offer arbitrage opportunities. Hence, relying on no-arbitrage is less demanding and simpler from a practical viewpoint. Indeed, this is the cornerstone of what we shall apply in option pricing, and it is the foundation of APT. However, we should also be aware that no-arbitrage approaches, too, rely

---

[1] We will say something about shrinkage estimation later, in Section 14.4.1, where we discuss regularization issues in linear regression and the basic tradeoff between bias and variance in statistical estimation. See also Example 9.1.

[2] For instance, using the equilibrium modeling framework, interesting theoretical connections are found between market completeness and certain welfare theorems in microeconomics.

on key assumptions, although they may be less critical than those required by a full-fledged equilibrium model. For instance, we assume that short-selling is allowed, unlimited liquidity is available, and that there are no market frictions (such as taxes, transaction costs, and bid–ask spreads).

10.2 The capital asset pricing model

The capital asset pricing model (**CAPM**) is an equilibrium model, whose development in the mid-1960s is credited to Sharpe, Lintner, and Mossin. One way to regard CAPM is as a relationship between the risk premium of an individual asset i and the risk premium of the market portfolio. This is expressed as follows:

The CAPM formula:

$$E[\tilde{r}_i] - r_f = \beta_i \cdot (E[\tilde{r}_M] - r_f). \tag{10.1}$$

where r_f is the risk-free return over the selected holding period.

We shall prove this formula in Section 10.2.1, and we will see that the coefficient β_i is the same as in the single-index model. Equation (10.1) may be rewritten in terms of risk premia:

$$\pi_i = \beta_i \pi_M.$$

Unlike the single-index model, we do not observe any specific risk premium α_i in Eq. (10.1). Moreover, the single-index model involves a relationship among random variables, whereas CAPM is a statement about expectations. Hence, there is more to CAPM than the above formula. We should also understand its more general implications, in the form of a CAPM *principle*.

As we have pointed out, in order to build an equilibrium model that is sensible and goes beyond simple mathematical exercises, we need some drastically simplifying assumptions:

- Individual investors are price takers: They act as if security prices were not affected by their own trades. In other words, we assume very deep and liquid markets with no market impact from large trades.
- All investors plan their investment for a single-period time horizon.
- Investments are limited to publicly traded financial assets (we do not consider nontradable assets such as education, privately owned firms, etc.).
- There are no market frictions, hence, no taxes or transaction costs.
- Investors are rational mean–variance portfolio optimizers, i.e., they make their portfolio decisions using the machinery of Chapter 8. In particular, they mix the risk-free asset with the tangency portfolio, the risky portfolio maximizing the Sharpe ratio.

- We only consider asset allocation decisions, i.e., we disregard consumption decisions or the need to pay for liabilities.
- Information is costless and available to all investors, who have homogeneous expectations (i.e., they use the same expected returns and covariances to make their asset allocation decisions).

In CAPM extensions, some of these assumptions have been relaxed.[3] From our viewpoint, the most relevant assumption concerns the use of the mean–variance optimization model. This is certainly a strong assumption, which in turn implies that either all investors have a quadratic utility or they believe that returns follow an elliptic distribution.[4] What we do *not* assume is that investors have the same degree of risk aversion. However, as we have seen in Section 8.3, the separation property implies that mean–variance optimizers will invest in the tangency portfolio, irrespective from risk aversion. The latter comes into play only in how the tangency portfolio is mixed with the risk-free asset. If investors have different views, they will assume different expected returns and covariances, and they will hold different risky portfolios. However, what should happen if they have the same views, as CAPM assumes?

> **The CAPM principle.** Under the CAPM assumptions, all investors should hold the *same* risky portfolio. Hence, the tangency portfolio must be the *market* portfolio.

To see why, note that when we aggregate individual portfolios, lending and borrowing assets will cancel out, and the aggregate risky portfolio, obtained by putting all of the individual portfolios together, will equal the entire wealth of the economy. The fraction of each stock must correspond to its market value, i.e., the asset price times the number of shares outstanding. If there were a mismatch between the market and the aggregate portfolio, e.g., because investors do not want to include an asset in their portfolio, its price would drop, making it interesting for inclusion in the portfolio.

The CAPM principle has further and quite significant implications:

- The market portfolio is mean–variance efficient and maximizes the Sharpe ratio.
- The capital allocation line (CAL) goes through the tangency portfolio, which under CAPM is the market portfolio. Hence, the CAL becomes the **capital market line** (CML), as illustrated in Fig. 10.1.
- Since the optimal risky portfolio that investors should hold is just the market portfolio, which may be proxied by a broad market index, there is no point in pursuing an active portfolio strategy. This is also reflected in the absence of any specific risk premium in the CAPM formula of Eq.

---

[3] For instance, Consumption CAPM (CCAPM) is a multiperiod extension allowing for consumption.
[4] See the discussion in Section 8.5.

10.2 The capital asset pricing model

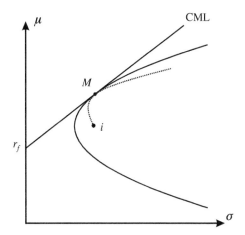

FIGURE 10.1 Proving the CAPM formula.

(10.1). Investors should just pursue a passive strategy, tracking a broad market index.

10.2.1 PROOF OF THE CAPM FORMULA

Here, we prove the CAPM formula,

$$\mathrm{E}[\tilde{r}_i] - r_f = \beta_i(\mathrm{E}[\tilde{r}_M] - r_f), \tag{10.2}$$

where the constant β_i is defined as

$$\beta_i \doteq \frac{\mathrm{Cov}(\tilde{r}_i, \tilde{r}_M)}{\mathrm{Var}(\tilde{r}_M)}. \tag{10.3}$$

The proof is instructive and not quite complicated, but readers who are not interested in it may safely skip this section. In Fig. 10.1, we observe that the market portfolio, with expected return $\mu_M \doteq \mathrm{E}[\tilde{r}_M]$ and volatility σ_M, is the tangency portfolio, under the CAPM assumptions. Now, let us consider a generic asset i, with expected return $\mu_i \doteq \mathrm{E}[\tilde{r}_i]$, volatility σ_i, and covariance $\sigma_{iM} \doteq \mathrm{Cov}(\tilde{r}_i, \tilde{r}_M)$ with the market portfolio, and form a portfolio with weight w for asset i and $1-w$ for the market portfolio. The portfolio return is a random variable

$$\tilde{r}_p(w) = w\tilde{r}_i + (1-w)\tilde{r}_M,$$

with the following expected value and standard deviation:

$$\mu_p(w) = w\mu_i + (1-w)\mu_M$$
$$\sigma_p(w) = \left[w^2\sigma_i^2 + 2w(1-w)\sigma_{iM} + (1-w)^2\sigma_M^2\right]^{1/2}.$$

When $w = 1$, the portfolio boils down to asset i; when $w = 0$, we just hold the market portfolio. By changing w, we generate a range of portfolios that describe a curve of attainable portfolios in the mean–risk plane (displayed as a dotted line in Fig. 10.1). For $w = 0$, the portfolio (dotted) curve must touch both the CML and the efficient frontier of risky portfolios. Since the market portfolio is efficient, however, these curves cannot cross, and a tangency condition must hold. The slope of the portfolio curve (i.e., the slope of its tangent line) at the point corresponding to $w = 0$ must be the same as the slope of the CML, i.e., the Sharpe ratio of the market portfolio. Let us write the tangency condition in explicit terms. We need the slope of the portfolio curve for $w = 0$. Hence, as a first step, let us find the derivatives of expected return and risk with respect to w:

$$\frac{d\mu_p(w)}{dw} = \mu_i - \mu_M,$$
$$\frac{d\sigma_p(w)}{dw} = \frac{w\sigma_i^2 + (1 - 2w)\sigma_{iM} + (w - 1)\sigma_M^2}{\sigma_p(w)}.$$

Then, taking the ratio of these equations, we find

$$\left.\frac{d\mu_p(w)}{d\sigma_p(w)}\right|_{w=0} = \left.\frac{d\mu_p(w)/dw}{d\sigma_p(w)/dw}\right|_{w=0} = \frac{(\mu_i - \mu_M)\sigma_M}{\sigma_{iM} - \sigma_M^2}.$$

This is the slope of the tangent line to the curve of attainable portfolios, which must be the same as the Sharpe ratio as the market portfolio. Therefore,

$$\frac{(\mu_i - \mu_M)\sigma_M}{\sigma_{iM} - \sigma_M^2} = \frac{\mu_M - r_f}{\sigma_M} \quad \Rightarrow \quad \mu_i = r_f + \frac{\sigma_{iM}}{\sigma_M^2}(\mu_M - r_f),$$

which proves Eqs. (10.2) and (10.3).

10.2.2 INTERPRETING CAPM

The CAPM formula can be restated as

$$\frac{\mathrm{E}[\tilde{r}_i] - r_f}{\sigma_{iM}} = \frac{\mathrm{E}[\tilde{r}_M] - r_f}{\sigma_M^2}.$$

These ratios can be interpreted as market prices of risk, i.e., risk premia divided by a risk measure. For the market portfolio, risk is represented by the variance of its return, rather than standard deviation. For the return of asset i, risk is represented by its covariance with the market portfolio, corresponding to the *systematic* component of risk. The important message is that, according to CAPM, specific risk can be diversified away, and the investor is not rewarded for bearing this risk. Only systematic risk is rewarded and priced. Also note that, at equilibrium, the market price of risk must be the same for each asset. Otherwise, the investor would be better off by tilting the portfolio toward assets with a better ratio.

We have already pointed out that the CAPM relationship is quite similar to a regression equation in terms of excess return, but the resulting α_i is zero.

10.2 The capital asset pricing model

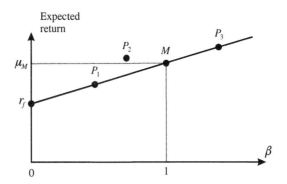

FIGURE 10.2 The security market line.

According to CAPM, assets have no alpha, the benefit of active portfolio management is an illusion, and investors should just pursue a passive strategy. Note that the beta of each asset measures the risk of each individual asset *within a well-diversified portfolio*. This risk measure is more relevant than the individual variance or standard deviation of asset return. In fact, even a dominated asset can be useful, if it has a certain correlation with other assets.[5] Thus, we need a measure of the contribution of each asset to the overall portfolio risk, which is provided by beta.

Based on this reasoning, we may position assets and portfolios within a mean–risk plane in which *beta*, rather than volatility, is used to represent risk. This results in the so-called **security market line** (SML), displayed in Fig. 10.2. The SML plots, for both an individual asset or a portfolio, the relationship between its risk and expected return. Note that the CML does the same for a *well-diversified* portfolio, for which variance may be a sensible risk measure. Geometrically, the SML goes through two key points that define it: The risk-free asset, corresponding to $\beta = 0$ (no risk premium) and the market portfolio, corresponding to $\beta = 1$. Points P_1 and P_3 in Fig. 10.2 correspond to portfolios (or individual assets) in line with CAPM. The former has less systematic risk and has an expected return lower than μ_M, whereas the latter is rewarded for its additional systematic risk. On the contrary, according to CAPM, portfolio P_2 cannot exist, since it features an additional risk premium that can only be attributed to specific risk, i.e., to a positive α.

An investor who believes CAPM literally would never look for a portfolio like P_2, trying to beat the market. An active manager, on the contrary, will try to outperform the market, by using additional information to generate a positive alpha. Even if one does not believe CAPM literally, it is certainly true that a portfolio like P_2 is hard to find *consistently*. Please note that CAPM is all about ex-ante expectations. Ex-post, we may well observe a portfolio outperforming the market.

---

[5] Consider asset A_3 in Fig. 8.3.

10.2.3 CAPM AS A PRICING FORMULA AND ITS PRACTICAL RELEVANCE

From the CAPM formula, it is not quite apparent why the term "CAPM" includes the word "pricing." Let S_0 be the current price of an asset, and let S_T be its (random) price at the end of the holding period. The CAPM states that the expected return from holding the asset is given by

$$\frac{\mathrm{E}[S_T] - S_0}{S_0} = r_f + \beta \cdot (\mu_M - r_f),$$

which implies the pricing relationship

$$S_0 = \frac{\mathrm{E}[S_T]}{1 + r_f + \beta \cdot \pi_M}. \tag{10.4}$$

Equation (10.4) gives the asset price as a discounted expectation. Note, however, that the expected future price is *not* discounted by using only the risk-free holding period return r_f, and that the market risk premium is involved, mediated by the asset exposure to systematic risk.

This idea may be used in capital budgeting decisions, when we evaluate the net present value of a stream of risky cash flow stream to assess the merit of an investment.[6] The CAPM may be used to find a suitable hurdle rate, which accounts for systematic risk. We should discount using the risk-free return only when the cash flows are riskless, or when the risk of the investment is uncorrelated with market risk (and it should not be rewarded). In practice, we should find a security, whose risk is comparable with the risk of the investment opportunity that we are evaluating, and use its beta to define the hurdle rate. It is also useful to observe that later, when dealing with option pricing, we shall take a "dual" approach. Here, we use expected cash flows under the true probability measure, adjusting the discount factor for risk. In risk-neutral option pricing, we will use the risk-free rate to discount, but expectations of cash flows will be taken under a risk-neutral (or risk-adjusted) risk measure.

The CAPM has been empirically tested, generating plenty of controversy. Empirical evidence does not fully support the model, although even defining *how* the test should be carried out is controversial. In fact, we have stated the CAPM in relationship with an index portfolio, but we should actually include any traded asset, not only stock shares, which is far from trivial. Nevertheless, even though we may not believe in the CAPM literally, this does not imply that it is useless. As we shall see in Section 10.3, the CAPM is the foundation of the Black–Litterman portfolio optimization model, which can be interpreted as an application of Bayesian statistics. Furthermore, the CAPM can provide us with a benchmark for security analysis ex-ante, and for performance evaluation ex-post. Let us illustrate how we may use the ex-post SML to evaluate realized performance, taking ex-ante risk into account. The left plot in Fig. 10.3

---

[6]See Section 3.6.

10.3 The Black–Litterman portfolio optimization model

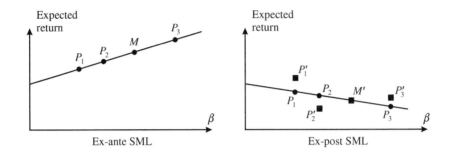

FIGURE 10.3 Using CAPM for ex-post performance evaluation.

shows the ex-ante SML. All portfolios P_1, P_2, and P_3 are on the SML, following the CAPM principle. We may increase the portfolio risk premium only by increasing its exposure to systematic, nondiversifiable risk. The CAPM is about ex-ante expectation, but let us stretch the model a bit and draw the SML ex-post. This means that we plot the actual return performance of each portfolio, while keeping the ex-ante assessment of risk. Let us assume that the market return falls well below its expectation, resulting in a loss. In the picture on the right, the ex-post SML features a negative slope, going through the risk-free asset and the point M' corresponding to the actual market performance. If the observed performance of the three portfolios were in line, the corresponding points P_1, P_2, and P_3 should lie on the SML. On the contrary, say that we observe points P_1', P_2', and P_3', where return is observed ex-post and risk is assessed ex-ante. Note that this approach is not quite correct, nor theoretically sound, as it is not supported by the CAPM. However, the idea is to use the ex-post SML to define a sensible benchmark. For instance, point P_3 corresponds to the benchmark performance of a portfolio, consistent with its exposure to market risk and the realized market performance. Point P_3', corresponding to the actual performance, is above P_3. This portfolio did not perform well in absolute terms, but it is above the line. Hence, in terms of relative performance, the portfolio manager did fairly well. The same can be said for portfolio P_1, which looks even better than P_3, whereas portfolio P_2 had an unsatisfactory performance, as it lies below the ex-post SML.

By a similar token, let us assume that the market performs according to or above expectations, so that the ex-post SML has positive slope. If an extremely risky portfolio yields a remarkable profit, before getting too excited, we should check if this profit is in line with the risk that the manager has taken. If observed return is positive, but below the SML, we might wonder if the performance has really been that good.

10.3 The Black–Litterman portfolio optimization model

We have observed, in Section 8.6, how the solution of a mean–variance portfolio optimization model is affected by estimates of risk premia (equivalently, of expected returns). Estimation errors may yield unacceptable swings in the portfolio, as well as unreasonable weights, and there seems to be no way to include subjective views along with a measure of how much we trust them. Furthermore, our views may be relative, in the sense that we may not feel safe in forecasting the return of a specific asset, but we may believe that one asset will outperform another one by some amount. It is not quite clear how this information may be exploited in a portfolio optimization model. A cynical view states that portfolio optimization is the best way to maximize the effect of estimation errors. This applies to errors in estimating covariances, but it is commonly agreed in the literature that the impact of perturbations in risk premia is definitely more severe. Furthermore, one may expect that a factor model should yield stable estimates of risk exposures in terms of betas, whereas the task of forecasting *future* expected returns is quite dangerous. Hence, here we assume that the covariance matrix is given and known, and we focus on risk premia.

According to CAPM, all of this effort is wasted, and we should pursue a purely passive strategy. In Section 10.5.1, we shall see that this view is reinforced by the efficient market hypothesis, which roughly states that beating the market is not possible, at least, not systematically. Here, we take a somewhat intermediate view. Even if we do not believe CAPM literally, pursuing an active strategy requires critical forecasts that we may only be comfortable with for a limited number of assets. Hence, we may use the market portfolio as a reference from which we may tilt away in a disciplined and limited manner. As we have seen in Section 9.3, the Treynor–Black model is one possible approach in this vein, and the Black–Litterman approach that we describe here is another one.

The Black–Litterman model may be regarded as a way to come up with more reliable estimates of risk premia, and it is somehow related to shrinkage estimation. Here, we connect the Black–Litterman model with the Bayesian approach to statistics, in order to emphasize the subjective nature of portfolio views. We provide some background on the Bayesian approach to statistics in Supplement S10.1. The main difference with respect to orthodox statistics is that, in the Bayesian approach, parameters of probability distributions are random variables themselves, and their distribution represents our uncertainty about them. Hence, rather than just relying on data, we may represent possibly subjective knowledge by a prior distribution of parameters, which is revised as we gather more evidence, resulting in a posterior distribution. Actually, this sequential learning interpretation is not essential. What is essential is that the Bayesian approach allows to merge two sources of knowledge, each one equipped with some measure of reliability. In our case, knowledge implied by market equilibrium is merged with subjective views.

10.3.1 BLACK–LITTERMAN MODEL: THE ROLE OF CAPM AND BAYESIAN STATISTICS

We recall that the mean–variance optimal portfolio weights are given by Eq. (8.16), which we repeat here for the sake of convenience,

$$\mathbf{w}^* = \frac{1}{\lambda} \Sigma^{-1} \pi, \tag{10.5}$$

10.3 The Black–Litterman portfolio optimization model

where λ is the coefficient of risk aversion. The remaining fraction of wealth is invested in the risk-free asset. While we assume that the covariance matrix Σ is estimated with sufficient accuracy, possibly by a factor model, the vector π of risk premia is a quite critical input. However, if we accept the validity of CAPM, we may reverse the optimization model and find the risk premia implied by the market portfolio.

Let us denote by w_{Mj} the weight of asset $j = 1, \ldots, n$ in the market portfolio. The market weights may be collected into vector $\mathbf{w}_M$. Then, the random market return is given by

$$\tilde{r}_M = \sum_{j=1}^{n} w_{Mj} \cdot \tilde{r}_j. \qquad (10.6)$$

According to CAPM, the following holds for each asset j:

$$\mu_j - r_f = \beta_j(\mu_M - r_f), \qquad (10.7)$$

where

$$\mu_M = \sum_{j=1}^{n} w_{Mj} \cdot \mu_j \qquad (10.8)$$

is the expected return of the market portfolio. Let us denote the risk premium implied by CAPM by

$$\pi_{ej} = \mu_j - r_f,$$

where the subscript e is used to emphasize that this view is implied by market equilibrium. Then, by using Eq. (10.6) and making β_j explicit, we may rewrite the CAPM Eq. (10.7) as follows:

$$\pi_{ei} = \frac{\text{Cov}(\tilde{r}_i, \tilde{r}_M)}{\sigma_M^2} \cdot (\mu_M - r_f)$$

$$= \frac{1}{\sigma_M^2} \cdot \text{Cov}\left(\tilde{r}_i, \sum_{j=1}^{n} w_{Mj} \cdot \tilde{r}_j\right) \cdot (\mu_M - r_f)$$

$$= \frac{\mu_M - r_f}{\sigma_M^2} \cdot \sum_{j=1}^{n} w_{Mj} \cdot \text{Cov}(\tilde{r}_i, \tilde{r}_j). \qquad (10.9)$$

By collecting the market-implied risk premia in vector π_e, Eq. (10.9) may be rewritten in matrix form,

$$\pi_e = \delta \Sigma \mathbf{w}_M, \qquad (10.10)$$

where we define the market price of risk

$$\delta \doteq \frac{\mu_M - r_f}{\sigma_M^2}. \qquad (10.11)$$

If we compare Eqs. (10.5) and (10.10), we see that we are, in fact, inverting the optimization problem, and the coefficient of risk aversion λ may be identified with δ, which plays the role of an aggregate coefficient of risk aversion for the market.

Remark. We use the notation $\boldsymbol{\pi}_e$ for the sake of simplicity. Within the Bayesian framework, the uncertain risk premia are considered as a vector $\tilde{\boldsymbol{\theta}}$ of *random variables*. The estimates implied by market equilibrium are based on background knowledge, which we may represent as a conditioning information $\mathcal{M}_e$. Hence, our notation may be interpreted as a shortcut for

$$\boldsymbol{\pi}_e = \mathrm{E}\big[\tilde{\boldsymbol{\theta}}\,\big|\,\mathcal{M}_e\big].$$

By the same token, below we use the notation $\boldsymbol{\pi}_s$ for the estimate implicit in subjective views, and $\boldsymbol{\pi}_{\mathrm{BL}}$ for the aggregate Black–Litterman estimate. Here, we assume that both returns and uncertain parameters are normally distributed. This simplifies calculations considerably, making an analytical solution possible.[7] Numerical methods may be used for different, possibly more realistic, distributional assumptions.

The market-implied risk premia may be considered as one ingredient of the estimate, which is not just backward-looking and based on historical data, but also forward-looking. But how can we include subjective views that an investor might have? For instance, an investor could believe that asset 2 will outperform asset 5 by, say, 2%. Then, we could write a condition such as

$$\pi_{s2} - \pi_{s5} = 0.02, \tag{10.12}$$

where we use the subscript s to emphasize the subjective nature of this view. If the subjective views about risk premia are expressed by linear relationships, we may collect all of them in the following matrix form:

$$\mathbf{P}\boldsymbol{\pi}_s = \mathbf{q}, \tag{10.13}$$

where $\mathbf{P} \in \mathbb{R}^{k \times n}$, $\mathbf{q} \in \mathbb{R}^k$, and k is the number of subjective views. The view of Eq. (10.12) would correspond to a row of matrix $\mathbf{P}$, with elements set to 1 and -1 in columns 2 and 5, respectively, and zero otherwise; the corresponding element in vector $\mathbf{q}$ would be 0.02. Of course, subjective views are uncertain, and we might express our confidence in them as follows:

$$\mathbf{P}\tilde{\boldsymbol{\theta}} \sim \mathsf{N}(\mathbf{q}, \boldsymbol{\Omega}). \tag{10.14}$$

We are claiming that a linear transformation of the uncertain parameter vector $\tilde{\boldsymbol{\theta}}$, which is again normal, has expected value $\mathbf{q}$, corresponding to subjective views, and covariance matrix $\boldsymbol{\Omega}$, typically a diagonal matrix whose elements are related to the confidence in subjective views. A view that we really believe in will be associated with a small variance.

---

[7] In Bayesian parlance, we say that using multivariate normal distributions for both the prior and the likelihood results in a *conjugate* pair, so that the posterior is normal, too.

10.3 The Black–Litterman portfolio optimization model

It is important to realize that Ω has nothing to do with the covariance matrix Σ between asset returns. However, Σ does play a role in the Black–Litterman model. In fact, the Bayesian framework suggests that we might merge the two sources of knowledge by using Eq. (10.37). We note that, unlike the usual Bayesian sequential learning framework, we cannot really speak of a "prior" and a "posterior." We are just merging two different sources of knowledge. The historical data are not used in estimating the risk premia, as past data may be not quite relevant to the future, but they do play a role in the estimation of Σ. The market implied view is expressed by the following distributional assumption about $\tilde{\boldsymbol{\theta}}$:

$$\tilde{\boldsymbol{\theta}} \sim \mathsf{N}(\boldsymbol{\pi}_e, \tau\boldsymbol{\Sigma}). \tag{10.15}$$

The expected value is given by the CAPM risk premia, whereas the covariance matrix of $\tilde{\boldsymbol{\theta}}$ is related to the covariance matrix $\boldsymbol{\Sigma}$ by the factor τ. One of the main difficulties of the Bayesian framework is specifying sensible multivariate priors. Here, we are using the parameter τ to fine tune the degree of confidence in the market view, without the need to specify covariances.

Putting everything together, we end up with the following **Black–Litterman estimator** of the risk premia:

$$\boldsymbol{\pi}_{\text{BL}} = \left[(\tau\boldsymbol{\Sigma})^{-1} + \mathbf{P}^\mathsf{T}\boldsymbol{\Omega}^{-1}\mathbf{P}\right]^{-1} \cdot \left[(\tau\boldsymbol{\Sigma})^{-1}\boldsymbol{\pi}_e + \mathbf{P}^\mathsf{T}\boldsymbol{\Omega}^{-1}\mathbf{q}\right]. \tag{10.16}$$

Technical remark. Equation (10.16) states that the estimate $\boldsymbol{\pi}_{\text{BL}}$ of the risk premia is a weighted average of the market-implied estimate $\boldsymbol{\pi}_e$ and the estimate $\boldsymbol{\pi}_s$ implied by the subjective views. The weights are related to the degree of uncertainty associated with these two sources of knowledge, and the background is illustrated in Section S10.1.3. In particular, Eq. (10.16) is related to Eq. (10.37). However, while the term $(\tau\boldsymbol{\Sigma})^{-1}$ is clearly understood as the inverse of the covariance matrix $\tau\boldsymbol{\Sigma}$ associated with the market-implied estimate of the risk premia, the role of matrix $\mathbf{P}$ is less obvious. Without going into too many calculation details, let us just recall that Bayesian updating with a normal distribution requires focusing on the essential terms in multivariate densities, sweeping all of the rest under the rug of a proportionality constant. In the case of market-implied risk premia, the multivariate normal density of the random vector $\tilde{\boldsymbol{\theta}}$ is proportional to the following expression:

$$\exp\left[\tfrac{1}{2}(\boldsymbol{\theta} - \boldsymbol{\pi}_e)^\mathsf{T}(\tau\boldsymbol{\Sigma})^{-1}(\boldsymbol{\theta} - \boldsymbol{\pi}_e)\right].$$

In the case of subjective views, we have a distributional information on a *linear transformation* of the random vector $\tilde{\boldsymbol{\theta}}$, which involves a multivariate normal density proportional to

$$\exp\left[\tfrac{1}{2}(\mathbf{P}\boldsymbol{\theta} - \mathbf{q})^\mathsf{T}\boldsymbol{\Omega}^{-1}(\mathbf{P}\boldsymbol{\theta} - \mathbf{q})\right]$$
$$= \exp\left[\tfrac{1}{2}\left(\boldsymbol{\theta}^\mathsf{T}\mathbf{P}^\mathsf{T}\boldsymbol{\Omega}^{-1}\mathbf{P}\boldsymbol{\theta} - \boldsymbol{\theta}^\mathsf{T}\mathbf{P}^\mathsf{T}\boldsymbol{\Omega}^{-1}\mathbf{q} - \mathbf{q}^\mathsf{T}\boldsymbol{\Omega}^{-1}\mathbf{P}\boldsymbol{\theta} + \mathbf{q}^\mathsf{T}\boldsymbol{\Omega}^{-1}\mathbf{q}\right)\right]$$
$$\propto \exp\left[\tfrac{1}{2}\boldsymbol{\theta}^\mathsf{T}\mathbf{P}^\mathsf{T}\boldsymbol{\Omega}^{-1}\mathbf{P}\boldsymbol{\theta} - \boldsymbol{\theta}^\mathsf{T}\mathbf{P}^\mathsf{T}\boldsymbol{\Omega}^{-1}\mathbf{q}\right],$$

where we use the symmetry of the inverted covariance matrix Ω^{-1} and keep only terms involving θ. This helps to explain the occurrence of the matrix products $\mathbf{P}^T\Omega^{-1}\mathbf{P}$ and $\mathbf{P}^T\Omega^{-1}\mathbf{q}$ in Eq. (10.16). A proper proof is beyond the scope of this book.

10.3.2 BLACK–LITTERMAN MODEL: A NUMERICAL EXAMPLE

Let us assume a very simple market on which two assets, A and B, are traded.[8] The market weights of the two assets are

$$w_A = 0.25, \quad w_B = 0.75.$$

Historical data yield the following estimates about the two volatilities and the correlation among assets:

$$\sigma_A = 0.08, \quad \sigma_B = 0.17, \quad \rho = 0.3,$$

which imply the following covariance matrix

$$\Sigma = \begin{bmatrix} 0.08^2 & 0.3 \times 0.08 \times 0.17 \\ 0.3 \times 0.08 \times 0.17 & 0.17^2 \end{bmatrix} = \begin{bmatrix} 0.0064 & 0.00408 \\ 0.00408 & 0.0289 \end{bmatrix}$$

As we have pointed out, we assume that this information can be trusted, and we may also find the variance of the market portfolio:

$$\sigma_M^2 = w_A^2\sigma_A^2 + w_B^2\sigma_B^2 + 2w_Aw_B\sigma_{AB}$$
$$= 0.25^2 \cdot 0.0064 + 0.75^2 \cdot 0.0289 + 2 \cdot 0.25 \cdot 0.75 \cdot 0.00408 = 0.018186.$$

Now, in order to proceed, we need a market-implied estimate of the risk premia π_{eA} and π_{eB}. To this aim, we may either come up with an estimate of the market risk premium and use CAPM, or we may directly use Eq. (10.10) along with an estimate of the aggregate risk aversion. It is commonly agreed that sensible values of the risk aversion coefficient, within a mean–variance framework, are in the range from 2 to 4. Assuming an aggregate risk aversion $\delta = 3$, we find

$$\boldsymbol{\pi}_e = 3 \times \begin{bmatrix} 0.0064 & 0.00408 \\ 0.00408 & 0.0289 \end{bmatrix} \times \begin{bmatrix} 0.25 \\ 0.75 \end{bmatrix} = \begin{bmatrix} 1.40\% \\ 6.81\% \end{bmatrix}. \quad (10.17)$$

As a reality check, the corresponding market risk premium is[9]

$$\pi_{eM} = \delta \cdot \sigma_M^2 = 3 \times 0.018186 = 5.46\%.$$

---

[8]The numerical inputs for this example are borrowed from [3, Chapter 27].
[9]Alternatively, we might estimate δ on the basis of forecasts of μ_M and σ_M, using Eq. (10.11). Forecasting expected return for the whole market is arguably less difficult than doing the same for an individual asset.

10.3 The Black–Litterman portfolio optimization model

To complete the market implied view, we have to choose a value for the coefficient τ, which is used to express the uncertainty about the market-implied risk premia. A sensible rule of thumb is to use a standard deviation for the uncertain parameter that is 10% of the standard deviation of return. Since τ multiplies variances and covariances, rather than standard deviations, this implies $\tau = (0.1)^2 = 0.01$:

$$\tau\Sigma = \begin{bmatrix} 0.000064 & 0.0000408 \\ 0.0000408 & 0.000289 \end{bmatrix}.$$

To get a feeling for this choice, let us recall that the variance of a sample mean is given by

$$\mathrm{Var}(\overline{X}) = \frac{\sigma_X^2}{m},$$

where m is the sample size. Therefore, choosing $\tau = 1/100$ is equivalent to the precision of an estimate based on 100 past observations.

Now, we must integrate market equilibrium with subjective views. Say that the portfolio manager expects asset A to outperform asset B by 0.5%,

$$\pi_{sA} - \pi_{sB} = 0.5\%.$$

This view can be expressed within our formalism as follows:

$$\mathbf{P} = \begin{bmatrix} 1 & -1 \end{bmatrix}, \quad \mathbf{q} = 0.5\%.$$

The matrix $\mathbf{P}$ is a row vector and the vector $\mathbf{q}$ is a scalar because we are expressing only one view. The matrix $\mathbf{\Omega}$ is a scalar as well, and in order to choose its value we might want to express that our confidence in the subjective view is roughly the same as the confidence in the market equilibrium view. If we represent the two uncertain risk premia by the random variables $\tilde{\theta}_A$ and $\tilde{\theta}_B$, respectively, the variance of their difference, conditional on the market equilibrium information $\mathcal{M}_e$, is

$$\mathrm{Var}(\tilde{\theta}_A - \tilde{\theta}_B \mid \mathcal{M}_e) = 0.000064 + 0.000289 - 2 \times 0.0000408$$
$$= 0.0002714 \approx 0.0003,$$

where we use the entries in matrix $\tau\Sigma$. If we want to assign roughly the same weight to market-implied and subjective views, we may choose $\Omega = 0.0003$. If we rewrite Eq. (10.14) as

$$\mathbf{P}\pi_s = \mathbf{q} + \tilde{\epsilon},$$

our choice corresponds to assuming a standard deviation $\sigma_{\tilde{\epsilon}} = \sqrt{0.0003} = 1.73\%$ for the noise term $\tilde{\epsilon}$, which is just a scalar in our example, since we are considering one view.

Now, we have to just apply Eq. (10.16), which requires a few straightforward matrix calculations. We just outline a couple of them, for illustration purposes:

$$(\tau\Sigma)^{-1} = 100 \times \begin{bmatrix} 171.7033 & -24.2405 \\ -24.2405 & 38.0243 \end{bmatrix}$$

$$\mathbf{P}^T \Omega^{-1} \mathbf{P} = \frac{1}{0.0003} \times \begin{bmatrix} 1 \\ -1 \end{bmatrix} \times \begin{bmatrix} 1 & -1 \end{bmatrix} = \frac{10{,}000}{3} \begin{bmatrix} 1 & -1 \\ -1 & 1 \end{bmatrix}.$$

The final result is

$$\pi_{\text{BL}} = \begin{bmatrix} 1.64\% \\ 4.24\% \end{bmatrix}.$$

We notice the difference with respect to the market-implied forecasts[10] of Eq. (10.17). The difference between the risk premia is

- 0.5% according to the subjective view,
- 1.40% − 6.81% = −5.41% according to the market view,
- 1.64% − 4.24% = −2.60% according to the integrated view, which is roughly half-way between the two views.

As it turns out, Black–Litterman portfolios are less "extreme" than portfolios obtained on the basis of naive estimates.

10.4 Arbitrage pricing theory

The arbitrage pricing theory (**APT**) model may be considered as an extension of the CAPM allowing for multiple factors. However, this is a partial view, as it disregards the difference between a statistical model and an equilibrium model. The hypotheses behind the APT equilibrium are less stringent than those required by CAPM. For instance, the critical assumption that market participants are all mean–variance optimizers is relaxed. Essentially, APT requires that they will not leave arbitrage opportunities on the table. At first sight, this seems a rather obvious assumption, but it still relies on the hypothesis that markets are liquid and frictionless, and that unlimited short-selling is possible.[11]

If we consider the required assumptions, APT looks like an extension and an improvement over CAPM. However, it is not quite clear how to choose of relevant factors in APT. Furthermore, as usual, relaxing the assumptions behind

---

[10] Here, we are using the terms "estimate" and "forecast" rather liberally. Within an orthodox framework, this would not make any sense. Within Bayesian statistics, we may be forgiven for the confusion, since parameters are random variables and we may use their expected values as forecasts.

[11] These assumptions also underlie the use of no-arbitrage principles in pricing derivatives. The difference between the concepts that we use in Chapter 13 and APT is that we deal here with portfolios of primary assets, rather than derivatives, and we do not consider dynamic modeling in any way.

10.4 Arbitrage pricing theory

a mathematical model, in order to strive for generality, comes with a price, which in this case is:

- From a technical viewpoint, the proof of the APT is much more involved than the proof of the CAPM. In fact, we will provide some intuition, rather than a mathematically rigorous proof.
- From a financial viewpoint, APT applies to well-diversified portfolios, rather than to individual assets. To be more precise, the possibility is left open that APT does not apply to *all* of the individual assets, and that some violation is possible.

10.4.1 THE INTUITION

Before stating the theorem formally, it is important to gain some financial intuition. To this aim, let us consider a multifactor linear model based on two systematic factors,

$$\tilde{r}_i = \alpha_i + \beta_{i1}\tilde{F}_1 + \beta_{i2}\tilde{F}_2 + \tilde{\epsilon}_i, \qquad i = 1,\ldots,n,$$

where i may refer to an asset or a portfolio. Note that we are writing the model in terms of return $\tilde{r}_i$, rather than excess return $\tilde{R}_i$, but this is not essential. If we assume a diagonal model and only consider *well-diversified* portfolios, rather than individual assets, we may assume that specific risks have been diversified away, so that the model can be rewritten as

$$\tilde{r}_i = \alpha_i + \beta_{i1}\tilde{F}_1 + \beta_{i2}\tilde{F}_2, \qquad i = 1,\ldots,n.$$

Now, leaving distributional knowledge about the common factors aside, how many features do we need to specify a portfolio p? In this case, we need three coefficients, one alpha and two betas. Equivalently, we may be concerned with the expected return, $\mu_p \doteq \mathrm{E}[\tilde{r}_p]$, and the two factor loadings β_{p1} and β_{p1}. Hence, geometrically, we may associate each portfolio with a point in a three-dimensional space with coordinates $(\mu_p, \beta_{p1}, \beta_{p2})$. We also know that, in such a space, three (linearly independent) points define a plane, whose equation may be written as

$$\mu_p = \lambda_0 + \lambda_1 \beta_{p1} + \lambda_2 \beta_{p2},$$

for some constants λ_0, λ_1, and λ_2.

For instance, let us assume that the three well-diversified portfolios displayed in Table 10.1 are available.[12] To find the plane corresponding to these three portfolios, we may solve the following system of linear equations:

$$\begin{cases} 15 = \lambda_0 + 1.0\lambda_1 + 0.6\lambda_2 \\ 14 = \lambda_0 + 0.5\lambda_1 + 1.0\lambda_2 \\ 10 = \lambda_0 + 0.3\lambda_1 + 0.2\lambda_2 \end{cases},$$

---

[12] The numerical example is borrowed from [8].

Table 10.1 Three well-diversified portfolios and their systematic factor exposures.

Portfolio i	Expected return μ_i	β_{i1}	β_{i2}
A	15%	1.0	0.6
B	14%	0.5	1.0
C	10%	0.3	0.2

where, for the sake of convenience, we express return as a percentage. This gives the equation

$$\mu_p = 7.75 + 5\beta_{p1} + 3.75\beta_{p2}. \tag{10.18}$$

If we combine the three portfolios, we just get another point on this plane (note that building a portfolio means taking a linear affine combination of other assets/portfolios, i.e., a linear combination where weights add up to 1).

Now, let us compare the following two portfolios:

- Portfolio D, which is obtained by an equally weighted combination of portfolios A, B, and C.
- Portfolio E, which has expected return 15%, $\beta_{E1} = 0.6$, and $\beta_{E2} = 0.6$.

The risk exposures of portfolio D are found by combining the betas of its three building blocks:

$$\beta_{D1} = \tfrac{1}{3} \times 1.0 + \tfrac{1}{3} \times 0.5 + \tfrac{1}{3} \times 0.3 = 0.6$$
$$\beta_{D2} = \tfrac{1}{3} \times 0.6 + \tfrac{1}{3} \times 1.0 + \tfrac{1}{3} \times 0.2 = 0.6.$$

Thus, we observe that portfolio D has the same risk exposure as portfolio E. However, its expected return is

$$\mu_D = \tfrac{1}{3} \times 15 + \tfrac{1}{3} \times 14 + \tfrac{1}{3} \times 10 = 13\% < \mu_E.$$

By construction, portfolio D lies on the plane of Eq. (10.18), whereas portfolio E does not. This introduces an inconsistency and, if both portfolios D and E exist, then we have a perfect money-making machine. All we have to do is to short the "ugly" portfolio D and buy the "nice" portfolio E, for an arbitrary initial capital, as shown in Table 10.2. Note that the betas of portfolio D are negative, since we are short-selling it, and they exactly offset the betas of portfolio E. We assume that we short $1 of D in order to buy $1 of E, but this is not essential, as we may scale the trade up arbitrarily. The resulting long–short portfolio is beta-neutral and riskless (we assume specific risk is negligible), and it is dollar-neutral (we do not need any initial capital to set up the trade). However, assuming that the multifactor model applies, it earns a positive risk-free return, $15\% - 13\% = 2\%$. This is an arbitrage opportunity, which should not exist in equilibrium.

If we have to rule out such arbitrage opportunities, all portfolios must lie on the same plane, which means that there must exist constants λ_0, λ_1, and λ_2,

10.4 Arbitrage pricing theory

Table 10.2 An arbitrage trade. We assume an investment of $1, but this may be scaled up at will.

	Initial cash flow	β_{p1}	β_{p2}	μ_p
Portfolio D	+$1	−0.6	−0.6	−13%
Portfolio E	−$1	0.6	0.6	15%
Arbitrage long–short portfolio	0	0.0	0.0	2%

such that
$$\mu_p = \lambda_0 + \lambda_1 \beta_{p1} + \lambda_2 \beta_{p2},$$
for *any* well-diversified portfolio p. By generalizing the result to an arbitrary number of factors, we obtain the statement of the APT theorem.

Remark. The careful reader will notice a difference with the arbitrage strategies that we have considered in Section 2.3. For instance, in Example 2.8, we did not rely on any model, but only the well-defined payoffs of derivatives. Here, we are relying on a multifactor model, whose parameters are *estimated*. If the betas of portfolios D and E are subject to estimation error, we cannot claim that we really have a clean and guaranteed arbitrage opportunity. Furthermore, we are not considering the uncertainty associated with residual specific risk.

10.4.2 A NOT-SO-RIGOROUS PROOF OF APT

One way to generalize our intuition and see why APT should hold is to exploit concepts from linear algebra, provided that we still assume that specific risks can be disregarded. First, let us state the APT theorem formally.

THEOREM 10.1 (APT pricing equation) *Consider the multifactor model*

$$\tilde{r}_i = \alpha_i + \sum_{j=1}^m \beta_{ij} \tilde{F}_j, \qquad i = 1, \ldots, n,$$

where no specific risk factor is involved. Then, assuming no-arbitrage, there exist constants $\lambda_0, \lambda_1, \ldots, \lambda_m$ *such that*

$$\mu_i = \lambda_0 + \sum_{j=1}^m \beta_{ij} \lambda_j, \qquad i = 1, \ldots, n. \tag{10.19}$$

To prove the claim, let us consider the case $m = 2$ and say that we invest an amount x_i (of any monetary unit) in each asset i, forming a portfolio such that:

$$\sum_{i=1}^n x_i = 0, \quad \sum_{i=1}^n x_i \beta_{i1} = 0, \quad \sum_{i=1}^n x_i \beta_{i2} = 0.$$

In plain English, we are building a long–short portfolio with no risk exposure (beta-neutral) and zero initial value (dollar-neutral). Since the portfolio does not require any initial investment and is riskless, if we rule out arbitrage opportunities, its expected profit (in monetary units) must be zero, too:

$$\sum_{i=1}^{n} \mu_i x_i = 0.$$

Let us translate this financial idea in terms of linear vector spaces, by introducing vectors $\mathbf{x}$, $\boldsymbol{\mu}$, $\boldsymbol{\beta}_1$, $\boldsymbol{\beta}_2$, which collect traded amounts and the features of each asset i, and $\mathbf{i} = [1, 1, \ldots, 1]^\mathsf{T}$. Then,

$$\mathbf{x}^\mathsf{T}\mathbf{i} = 0, \quad \mathbf{x}^\mathsf{T}\boldsymbol{\beta}_1 = 0, \quad \mathbf{x}^\mathsf{T}\boldsymbol{\beta}_2 = 0 \quad \Rightarrow \quad \mathbf{x}^\mathsf{T}\boldsymbol{\mu} = 0,$$

i.e., any vector $\mathbf{x}$ which is orthogonal to $\mathbf{i}$, $\boldsymbol{\beta}_1$, and $\boldsymbol{\beta}_2$ must also be orthogonal to $\boldsymbol{\mu}$. According to a theorem of linear algebra, this implies that $\boldsymbol{\mu}$ must be a linear combination of $\mathbf{i}$, $\boldsymbol{\beta}_1$, $\boldsymbol{\beta}_2$,

$$\boldsymbol{\mu} = \lambda_0 \mathbf{i} + \lambda_1 \boldsymbol{\beta}_1 + \lambda_2 \boldsymbol{\beta}_2,$$

which proves Eq. (10.19). Clearly, this may be generalized to an arbitrary number of factors.

10.4.3 APT FOR WELL-DIVERSIFIED PORTFOLIOS

So far, we took for granted that there was no need to consider specific risk. Let us try to justify the claim for well-diversified portfolios in a more general setting, where we include specific risks in the multifactor model

$$\tilde{r}_i = \alpha_i + \sum_{j=1}^{m} \beta_{ij} \tilde{F}_j + \tilde{\epsilon}_i, \qquad i = 1, \ldots, n.$$

If we build a portfolio p, with weights w_i, we have

$$\tilde{r}_p = \alpha_p + \sum_{j=1}^{m} \beta_{pj} \tilde{F}_j + \tilde{\epsilon}_p,$$

where

$$\alpha_p = \sum_{i=1}^{n} \alpha_i w_i, \quad \beta_{pj} = \sum_{i=1}^{n} \beta_{ij} w_i, \quad \text{Var}(\epsilon_p) = \sigma_{\epsilon p}^2 = \sum_{i=1}^{n} \sigma_{\epsilon i}^2 w_i^2,$$

where $\sigma_{\epsilon i}$ is the volatility of each specific risk factor. Now, let us assume the following:

1. Specific risks are bounded, in the sense that

$$\sigma_{\epsilon i}^2 \leq S^2, \qquad \forall i$$

for some constant S.

10.4 Arbitrage pricing theory

2. The portfolio is well diversified, in the sense that

$$w_i \leq W/n, \quad \forall i$$

for some constant $W \approx 1$, so that there is no overweighted asset.

Then,

$$\sigma_{\epsilon p}^2 \leq \frac{1}{n^2} \sum_{i=1}^{n} W^2 S^2 = \frac{1}{n} W^2 S^2. \tag{10.20}$$

If we let $n \to \infty$, Eq. (10.20) implies $\sigma_{\epsilon p}^2 \to 0$. Hence, we see that the previous results do apply to well-diversified portfolios, at least asymptotically, also if we include specific risks in the model.

10.4.4 APT FOR INDIVIDUAL ASSETS

What we have seen so far does not imply that APT holds for individual assets. Dealing with individual assets rigorously requires some more sophisticated technical machinery (see, e.g., [11]), but the net result is the following. Consider a universe of n assets, where a multifactor model with m common risk factors applies, and let $n \to \infty$. There exist numbers $\lambda_0, \lambda_1, \ldots, \lambda_m$ such that, if we define the error

$$\nu_i \doteq \mu_i - \lambda_0 - \sum_{j=1}^{m} \beta_{ij} \lambda_j,$$

we find

$$\lim_{n \to \infty} \frac{1}{n} \sum_{i=1}^{n} \nu_i^2 = 0.$$

This means that the mean square error of the model goes to zero. This, in turn, implies that most of the errors ν_i (i.e., all but a finite number of them) must be negligible. Hence, APT for individual assets holds as an asymptotic result, requiring an infinite universe of assets. Thus, most assets are correctly priced by the model, but APT might fail for a few of them.

A more intuitive argument is the following. Assume that APT does not apply to *any* individual asset. Nevertheless, we may build a well-diversified portfolio, for which we know that APT applies. Is it possible that APT does not apply to the individual assets, while it applies to the portfolio? In principle it may happen for a single portfolio, if we take some lucky combination of portfolio weights, so that errors cancel each other. It may also happen for a few portfolios, but the chances of finding that lucky combination of weights are diminishing. Since we assume an infinite universe of assets, we may build an infinite number of well-diversified portfolios. However, it is quite unlikely that APT applies to all such portfolios, but not to individual assets. Still, APT might well be violated for a finite set of assets.

10.4.5 INTERPRETING AND USING APT

Let us try to interpret the numbers λ_i in Theorem 10.1. It is convenient to assume a *centered* factor model, where common factors have expected value zero. In other words, the model is based on "unexpected" shocks on factors. Then, the constant term α_i in the multifactor model of return $\tilde{r}_i$ is just the expected return:

$$\mathrm{E}[\tilde{F}_j] = 0, \quad \forall j \quad \Longrightarrow \quad \mu_i = \mathrm{E}[\tilde{r}_i] = \alpha_i, \quad \forall i.$$

The APT pricing equation states that

$$\mu_i = \lambda_0 + \sum_{j=1}^{m} \beta_{ij} \lambda_j.$$

As a first step, consider a well-diversified portfolio p with no exposure to risk factors, i.e., $\beta_{pj} = 0$ for all risk factors $j = 1, \ldots, m$. Then, its return is deterministic (in this case, we do not include any specific risk). But if the portfolio is risk-free, no-arbitrage implies that

$$\mu_p = \lambda_0 = r_f,$$

where r_f is the risk-free return.

In order to interpret the coefficients λ_j, $j \geq 1$, imagine that we build a portfolio p_1 with unit exposure to factor F_1, and zero exposure to the remaining factors. In other words,

$$\beta_{p_1, 1} = 1; \qquad \beta_{p_1, j} = 0, \ j = 2, 3, \ldots$$

Then,

$$\mathrm{E}[\tilde{r}_{p_1}] = \lambda_0 + \lambda_1 \quad \Rightarrow \quad \lambda_1 = \mathrm{E}[\tilde{r}_{p_1}] - r_f.$$

We see that λ_1 is the expected excess return of a portfolio with unit exposure to factor F_1 only. The same reasoning applies to all other systematic factors. If we introduce, for each factor j, the parameter

$$\delta_j \doteq \mathrm{E}[\tilde{r}_{p_j}],$$

i.e., the expected return of a portfolio with unit exposure to $\tilde{F}_j$ only, we may rewrite the APT equation as

$$\mathrm{E}[\tilde{r}_i] = r_f + \sum_{j=1}^{m} \beta_{ij} \cdot [\delta_j - r_f].$$

The coefficients $\lambda_j \equiv \delta_j - r_f$ can be interpreted as risk premia, in the sense that they tell us by how much the expected return increases for each additional unit of exposure to systematic risk factor $\tilde{F}_j$. Hence, we may read APT as follows:

$$\mathrm{E}[\tilde{r}_i] = r_f + \sum_{j=1}^{m} \beta_{ij} \pi_j,$$

where π_j is the risk premium associated with systematic factor $\tilde{F}_j$.

10.4 Arbitrage pricing theory

Table 10.3 Well-diversified portfolios for Example 10.1.

Portfolio	β_{i1}	β_{i2}	Expected return
A	1.6	−0.8	9.5%
B	0.5	1.3	11.7%
C	−0.1	0.4	4.1%

Example 10.1 A toy numerical example

An APT model is based on two independent systematic factors $\tilde{F}_1$ and $\tilde{F}_2$, both with zero expected value (centered model). We assume that the holding period is one year, and the annual risk-free rate is 4%. We are given the three well-diversified portfolios displayed in Table 10.3. Let us find the coefficients of an APT model based on portfolios A and B, and check whether the introduction of portfolio C creates an arbitrage opportunity.

To find the first answer, we solve the system of linear equations

$$0.095 = 0.04 + 1.6\lambda_1 - 0.8\lambda_2,$$
$$0.117 = 0.04 + 0.5\lambda_1 + 1.3\lambda_2,$$

which yield $\lambda_1 = 0.0537$ and $\lambda_2 = 0.0386$. Note that the problem data give $\lambda_0 = 0.04$ directly.

Hence, according to APT, the equilibrium expected return of portfolio C should be

$$\mu_C = 0.04 - 0.1 \times 0.0537 + 0.4 \times 0.0386 = 0.0501, \qquad (10.21)$$

which is larger than the expected return in Table 10.3. *If* we believe the parameter estimate in the table, then we have an arbitrage opportunity. Using portfolios A and B, as well as the risk-free asset, we may generate a portfolio (let us call it portfolio D) with the same risk exposure of portfolio C, but a larger expected return. Hence, we should short portfolio C and buy portfolio D. To find a combination of A and B with the same risk exposure as C, we solve the following system:

$$1.6 w_A + 0.5 w_B = -0.1,$$
$$-0.8 w_A + 1.3 w_B = 0.4,$$
$$w_0 + w_A + w_B = 1,$$

where w_0 is the weight of the risk-free asset (which has zero beta by definition). Note that we need the third equation to obtain a set of normalized weights, which would not be obtained by using only the first two equations. Solving this system, we find $w_A = -0.1331$, $w_B = 0.2258$, $w_0 = 0.9073$.

Given these weights, the return of the resulting portfolio D is

$$\tilde{r}_D = w_A \tilde{r}_A + w_B \tilde{r}_B + w_0 r_f$$
$$= -0.1331 \times (0.095 + 1.6\tilde{F}_1 - 0.8\tilde{F}_2)$$
$$+ 0.2258 \times (0.117 + 0.5\tilde{F}_1 + 1.3\tilde{F}_2) + 0.9073 \times 0.04$$
$$= 0.0501 - 0.1\tilde{F}_1 + 0.4\tilde{F}_2.$$

The expected return of portfolio D is, in fact, 0.0501, as predicted by the consistency condition in Eq. (10.21). Finally, we build a long–short portfolio with $w_C = -1$ (short portfolio C) and $w_D = 1$ (buy portfolio D, which replicates the risk exposure of C). The portfolio is both dollar-neutral and beta-neutral. Assuming no specific risk (and no estimation errors in our models), its return is deterministic:

$$w_C \tilde{r}_C + w_D \tilde{r}_D$$
$$= -(0.041 - 0.1\tilde{F}_1 + 0.4\tilde{F}_2) + (0.0501 - 0.1\tilde{F}_1 + 0.4\tilde{F}_2)$$
$$= 0.0091.$$

If we short-sell $1,000,000 of portfolio C and buy $1,000,000 of portfolio D, we earn a sure profit of $9,100.

Example 10.1 is rather stylized, and we should wonder what can go wrong in real life:

- The impact of specific risk: Unless portfolios are really well-diversified, some residual risk can affect our trade.
- The impact of model risk: APT is not quite explicit in the selection of factors, and a wrong choice of factors will have a detrimental impact.
- The impact of estimation risk: We may base our strategy on wrong estimates of betas and expected returns. The arbitrage strategies of Examples 2.8 and 10.1 are quite different, as the first one is a true, model-free arbitrage opportunity.
- The impact of execution risk: We may plan a trade, but execution prices may not be in line with our plan.

In practice, terms like risk arbitrage and statistical arbitrage are used, to emphasize that certain arbitrage strategies do entail an amount of risk.

10.4 Arbitrage pricing theory

◼ Example 10.2 Hedging a single risk factor away

Here we consider a seemingly trivial problem: How to hedge a single systematic risk factor away, when there is no specific risk. Let us consider two well-diversified portfolios whose returns follow the single-factor model

$$\tilde{r}_1 = \alpha_1 + \beta_1 \tilde{F},$$
$$\tilde{r}_2 = \alpha_2 + \beta_2 \tilde{F},$$

where we have streamlined notation a bit to account for the presence of one risk factor. Without loss of generality, let us assume a centered model with $\mathrm{E}[\tilde{F}] = 0$, so that each alpha is the expected portfolio return. It is easy to find a portfolio with suitable weights, such that the common risk factor is hedged away. From

$$w\tilde{r}_1 + (1-w)\tilde{r}_2 = w\alpha_1 + (1-w)\alpha_2 + \left[w\beta_1 + (1-w)\beta_2\right] \cdot \tilde{F},$$

we see that the portfolio exposure to the risk factor is given by the term within brackets. By setting

$$w\beta_1 + (1-w)\beta_2 = 0$$

we find a weight

$$w^* = \frac{\beta_2}{\beta_2 - \beta_1},$$

such that the portfolio is beta-neutral. However, since the resulting portfolio is risk-free, no-arbitrage implies that it must earn the risk-free return, no more, no less. The return of the portfolio with weights w^* and $1 - w^*$ is

$$\frac{\beta_2}{\beta_2 - \beta_1} \cdot \alpha_1 - \frac{\beta_1}{\beta_2 - \beta_1} \alpha_2 = r_f,$$

which may be rewritten as

$$\frac{\alpha_1 - r_f}{\beta_1} = \frac{\alpha_2 - r_f}{\beta_2}. \qquad (10.22)$$

We have not made any assumption about the two portfolios we are considering. Hence, this equality must hold for *any* pair of well-diversified portfolios, and the ratios in Eq. 10.22 must be a constant depending only on the risk factor:

$$\frac{\alpha - r_f}{\beta} = \lambda \quad \Rightarrow \quad \alpha = r_f + \lambda \beta. \qquad (10.23)$$

Thus, the expected return α is just the risk-free rate plus the risk exposure β multiplied by a coefficient λ which may be interpreted as a

> *market price of risk.* Clearly, λ is just the coefficient associated with the systematic risk factor in the APT pricing equation. We shall meet this kind of reasoning again in Chapter 14.

In concrete, APT does not tell much about which factors we may use. As we have already pointed out in Section 9.4, there are plenty of factors we may choose from, including the following broad families:

- **Macroeconomic factors.** These are rather general factors, like market return, change in the short-term interest rate, change in industrial production level, change in inflation, change in oil price, etc.
- **Fundamental factors.** These are factors related to a specific firm, like size, book-to-price, dividend yield, earning variability, financial leverage, growth, etc.
- **Statistical factors.** These are obtained as a combination of factors, using multivariate statistic methods for data reduction, such as principal component analysis (PCA) and factor analysis. An interesting feature of PCA is that it yields uncorrelated factors; however, these may be hard to interpret. Exploratory factor analysis may be used when looking for latent, i.e., unobservable, factors.

APT may be applied ex-ante as well as ex-post with respect to portfolio choice (see, e.g., [5]). In active portfolio management, we may use APT in making factor bets, i.e., tilting the portfolio in order to take advantage of information about a risk factor. The estimation of risk exposures and risk premia may also be helpful in spotting under- or over-priced stocks. In passive portfolio management, we may build an index portfolio tracking a particular well-diversified benchmark by replicating its betas. APT has been proposed as a tool to evaluate performance of mutual funds, much like CAPM (see, e.g., [13]). A further application is *return attribution*, where ex-post return is factored into the following contributions:

1. Expected return, which is the reward for the risk taken.
2. Unexpected factor return, which arises from factor bets and factor surprises.
3. Alpha, which arises from stock selection.

10.5 The behavioral critique

We have considered simple and stylized equilibrium and no-arbitrage models in this chapter. The common underlying assumption is some form of rationality in the behavior of market participants. The rather drastic conclusion of CAPM

10.5 The behavioral critique

is that there is no point in trying to outperform the market. A similar conclusion is claimed by the **efficient market hypothesis** (EMH) that we outline below. In Chapter 11, we will explore its consequences in terms of dynamic modeling. The bottom line is that future asset prices are driven by stochastic processes, whose increments are unpredictable and independent from the past. As a consequence, any trading strategy trying to exploit patterns in stock prices is claimed to be futile.

Actually, there is a rich set of strategies, which we roughly place under the common label of **technical analysis**, trying to exploit different kinds of patterns. For instance, trend following strategies try to take advantage of the momentum in stock prices. Technical analysis may be contrasted with **fundamental analysis**, whereby one tries to take advantage of knowledge about the firm, possibly by a factor model. In technical analysis, we only use a time series model based on the prices themselves, assuming that there is little point in trying to explain erratic prices. Both of them may be contrasted with CAPM. An alternative view is proposed by behavioral finance, which admits that market participants are neither fully rational nor fully informed. These limitations, as well as behavioral biases related to decision makers' psychology, have been put forward to explain market anomalies and bubbles, as well as seemingly irrational patterns in both individual consumption–saving decisions and even corporate finance choices.

No-arbitrage models seem less demanding than CAPM in terms of critical assumptions. However, we should always keep in mind how they rely on implicit assumptions concerning liquid and frictionless markets.[13] In practice, we have to deal with:

- Limitations to short-selling (especially over extended time periods)
- Short-term losses and liquidity issues
- Implementation cost
- Model risk

These issues have more to do with market structure than with market anomalies due to investors' psychology. Needless to say, a carefully planned arbitrage strategy may prove to be quite dangerous in case of irrational patterns.

Given the rich variety of models, recently enriched by machine learning approaches, it is difficult to draw clear lines between competing strategies, and space does not allow a full treatment. In this section, we just hint at some alternative ideas that it is worth bearing in mind, when we feel the temptation of trusting any model too much, both in terms of market structure and investors' behavior. A lot of empirical work has been and is being carried out on these topics, with plenty of contradictory and confusing evidence. One thing is sure. Critical remarks about CAPM and related models do *not* necessarily imply that quantitative models are irrelevant. Quite the contrary. While CAPM suggests

---

[13] Also see the joke of Example 2.6.

the opportunity of a passive strategy, quantitative models may try to take advantage of market imperfections by pursuing active strategies. Factor models may rely on behavioral factors, too. Furthermore, even if one is pursuing a purely passive strategy, the problem remains of doing so efficiently, at minimum cost.[14]

10.5.1 THE EFFICIENT MARKET HYPOTHESIS

There are different forms in which we may state the essential idea of the EMH:

1. The **weak form** states that current asset prices reflect all of the information implicit in past prices and trades.
   Practical implication: Investors cannot outperform the market using only historical price and volume data.
2. The **semistrong form** states that current asset prices reflect all the publicly available information.
   Practical implication: Investors cannot outperform the market using only publicly available information, i.e., historical price, fundamental data, analysts' recommendations, etc.
3. The **strong form** states that prices (immediately) reflect all of the available information, public and private.
   Practical implication: Investors cannot outperform markets at all.

The EMH views markets as populated by knowledgeable and perfectly rational investors. Even if not all investors are like that, informed professional traders are supposed to keep prices in line, possibly by exploiting arbitrage opportunities, which are occasional and disappear quickly. If we assume that prices incorporate new information instantly, then there is no pattern to be exploited. If we assume that new, possibly positive, information is slowly incorporated in prices, then we may observe a trend pattern. This is a case of market under-reaction. By a similar token, if we believe EMH literally, we should not observe market over-reactions associated with bad news. The EMH view is clearly debatable. However, it is true that empirical evidence suggests how outperforming the market consistently is quite difficult.

10.5.2 THE PSYCHOLOGY OF CHOICE BY AGENTS WITH LIMITED RATIONALITY

Many models in economics assume a rational decision maker, with infinite memory and information processing ability. The decision-making paradigm based on utility functions is an expression of this unbounded rationality view. However, emotions (like disappointment and regret) play a major role in decision-making, influencing actual behavior. This applies to general decision-making under uncertainty, and to financial decision-making as well. Furthermore, due

---

[14] See Section 15.4.1 for an optimization model aimed at passive portfolio management.

to information processing limitations, heuristics are often used in real life. As a consequence, a few recurring "irrational" behavior patterns have been identified, based on mechanisms including:

Mental accounting. The portfolio is not considered in its entirety, but is partitioned. If we receive a large dividend, we may be tempted to spend it, whereas we do not sell a stock with a high capital gain.

Anchoring. We use past values as reference points or "anchors." Quite often, losing stocks are kept for too long (the price at which they are bought is an anchor), rather than admitting the mistake and selling them (aversion to sure loss). Often, the winning stocks are sold instead, which may not be rational also in terms of taxation.

Framing. The answer to a question may depend on how a problem is framed. This has been a topic of plenty of psychological research activity. For instance, when people are asked how much money would they be willing to pay to avoid a one-in-a-thousand chance of being killed, they give an answer. If the question is rephrased as how much money would they ask to accept that risk, the amount tends to be much larger, even though the two questions are equivalent.

Behavioral biases and cognitive limitations result in anomalous patterns and errors. Among the commonly cited examples, we mention the following tendencies:

- Overconfidence. It is well-known that 90% of drivers think they are above average drivers. By a similar token, one may be overconfident about his views about a stock share.
- Memory bias. Too much weight may be attributed to recent observations, which may result in forecasting errors.
- Conservatism. Investors may be too slow in revising their beliefs, which may justify certain momentum strategies.
- Sample size neglect and representativeness heuristics. This may explain the initial overreaction to earning reports.

Last, but not least, we mention again information asymmetry as one of the main features of competitive financial markets. There are informed market participants (possibly, participants who *believe* they are informed), and noise traders. The latter may contribute to market anomalies, as well. Furthermore, the actions of investors' who know or believe that other market participants are more informed may explain herding and imitative behavior, possibly generating market instability.

10.5.3 PROSPECT THEORY: THE AVERSION TO SURE LOSS

In Chapter 7, we have introduced the expected utility function as a possible paradigm for decision-making under uncertainty. However, apart from practical difficulties in eliciting individual utility functions, the concept itself relies on critical assumptions, and the literature is rich in examples pointing out the contradictions with empirically observed behavior.

Example 10.3 A decision-making paradox

When asked to make a choice, most people prefer $2400 with certainty to a lottery in which they may win $2500 with probability 0.33, $2400 with probability 0.66, and nothing with probability 0.01. Note that the expected value in the second case is $2409 > $2400, but the fear of painful regret in the case of a zero payoff is such that we prefer to avoid the risk. Let us assume a utility function such that $u(0) = 0$, with no loss of generality. Hence the above choice implies the following:

$$u(2400) > 0.33 \cdot u(2500) + 0.66 \cdot u(2400)$$
$$\Rightarrow \quad 0.34 \cdot u(2400) > 0.33 \cdot u(2500).$$

However, most people prefer $2500 with probability 0.33 and nothing with probability 0.67 to $2400 with probability 0.34 and nothing with probability 0.66. Arguably, this is due to the fact that we are not really able to perceive the small difference in probabilities, whereas the additional $100 have a clear meaning. However, according to utility theory, this second choice implies

$$0.34 \cdot u(2400) < 0.33 \cdot u(2500),$$

which contradicts the first one.

Prospect theory is a descriptive framework for decision-making under uncertainty that tries to explain some inconsistencies with expected utility theory. A common pattern is known as *certainty effect*, i.e., the tendency to overweight outcomes that are considered as certain.

Example 10.4 The aversion to sure loss

Consider the following two choices:

1. You have to choose between losing $7400 for sure, and a risky alternative, whereby you lose $10,000 with probability 0.75 and nothing with probability 0.25. What do you prefer?
2. You may take a gamble whereby you win $7400 with probability 0.25, and you lose $2600 with probability 0.75. Do you accept?

10.5 The behavioral critique

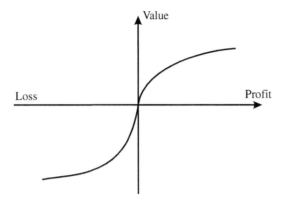

FIGURE 10.4 The value function in prospect theory.

> Most people take the risky alternative in the first case, but they do not in the second one. Note that, in the second case, the expected payoff of the gamble is
>
> $$0.25 \times 7400 - 0.75 \times 2600 = -100.$$
>
> Since we expect to lose $100, any risk-averse or risk-neutral decision maker would not take the chance. However, in the first case, the expected loss of the risky alternative is $7500, which is larger than the sure loss by $100. If we consider the sure loss of $7400 as a sunk cost, taking chances in the first choice is equivalent to adding the gamble of the second choice to the sure loss. Hence, it seems that, when loss is involved, we may behave as risk-lovers, contradicting the idea of concave utility functions.

More generally, most people are averse to sure losses, which may explain the reluctance to selling losing stocks or to cut losses when pursuing projects that are losing money. Furthermore, it seems that uncertain choices depend on reference values (0 or -7400 in Example 10.4). Prospect theory, as an alternative to utility functions, is based on **value functions** that are concave for positive values (gains) and convex for negative values (losses), as shown in Fig. 10.4. The value function represents risk aversion for positive values, whereas the opposite behavior is associated with losses. Also note that the curve is steeper for negative values. Unlike utility functions, here there is a reference point, with respect to which gains and losses are evaluated. The reference point may be current wealth, but this depends on how the prospect is formulated.

S10.1 Bayesian statistics

Parameter estimation plays a pivotal role in statistics and is central in finance, too. In **orthodox statistics**, parameters are *numbers* that we do not know. As such, there is no probability measure associated with parameters. For instance a typical trick question is the following one:

> We estimate a 95% confidence interval for an expected value, and the available sample yields the interval $(5.66, 6.13)$. Can we say that the expected value is contained in that interval with probability 0.95?

Any student in inferential statistics knows that this a wrong view, but a quite tempting one. The confidence level gives a measure of *coverage*, i.e., the probability that a random confidence interval includes the unknown true value. Therefore, a 95% confidence level tells us that if we sample 100 times, approximately 95 intervals should include the true value, but we cannot say anything about a single realized confidence interval. To understand the point, let us consider a random variable X with expected value μ. The probability $P\{X \le \mu\}$ is a sensible mathematical object, even though we cannot evaluate it exactly when μ or the underlying distribution are unknown. However, if we consider a realization $x = 10$ of the variable, we cannot even talk about the probability $P\{10 \le \mu\}$, since we are comparing two numbers. Either the inequality is satisfied or not, and if μ is unknown, we cannot say anything.

However, it would be nice to have a way to express how much we trust an estimate. This is certainly relevant in finance, when we express views about expected returns or risk premia, as well as critical but unobservable model parameters like volatility. In orthodox statistics, there is no such a thing as "probabilistic knowledge" about parameters, and data are the only source of information; subjective knowledge, more or less reliable, is disregarded.

Bayesian statistics provides a framework in which the above view about confidence intervals makes sense. Furthermore, within Bayesian statistics, subjective views may be stated in a precise way. The key shift in the paradigm is that parameters are regarded as random variables themselves. The following example shows some additional difficulties that we may encounter within the orthodox framework.[15]

Example 10.5 Difficulties with orthodox confidence intervals

> Let X be a uniformly distributed random variable, and let us assume that we do not know where the support of this distribution is located, but we know that its width is 1. Then, $X \sim \mathsf{U}[\mu - 0.5, \mu + 0.5]$, where

---

[15]This example is taken from [9], page 45, which in turn refers back to [7].

μ is the unknown expected value of X, as well as the midpoint of the support. To estimate μ we take a sample of $n = 2$ independent realizations X_1 and X_2 of the random variable. Now consider the order statistics

$$X_{(1)} = \min\{X_1, X_2\}, \qquad X_{(2)} = \max\{X_1, X_2\}$$

and the confidence interval

$$\mathcal{I} = [X_{(1)}, X_{(2)}]. \tag{10.24}$$

What is the confidence level of $\mathcal{I}$, i.e., the probability $\mathrm{P}\{\mu \in \mathcal{I}\}$? Both observations have a probability 0.5 of falling to the left or to the right of μ. The confidence interval will not contain μ if both observations fall on the same half of the support. Then, since X_1 and X_2 are independent, we have

$$\begin{aligned}\mathrm{P}\{\mu \notin \mathcal{I}\} &= \mathrm{P}\{X_1 < \mu, X_2 < \mu\} + \mathrm{P}\{X_1 > \mu, X_2 > \mu\} \\ &= 0.5 \times 0.5 + 0.5 \times 0.5 = 0.5.\end{aligned}$$

So, the confidence level for $\mathcal{I}$ is the complement of this probability, i.e., 50%.

Now suppose that we observe $X_1 = 0$ and $X_2 = 0.6$. What is the probability that μ is included in the confidence interval $\mathcal{I}$ resulting from Eq. (10.24), i.e., $\mathrm{P}\{0 \leq \mu \leq 0.6\}$? In general, this question does not make any sense, since μ is a number. But in this specific case, we have some additional knowledge, leading to the conclusion that the expected value is included in that interval for sure. Since the absolute deviation $|X - \mu|$ from the expected value is bounded by 0.5, a confidence interval of width 0.6 *must* contain μ. By a similar token, if we observe $X_1 = 0$ and $X_2 = 0.001$, we have some reason to argue that such a small interval is quite unlikely to include μ, but there is no way in which we can express this view properly, within the framework of orthodox statistics.

S10.1.1 BAYESIAN ESTIMATION

Say that we want to estimate an uncertain parameter $\tilde{\theta} \in \Theta \subseteq \mathbb{R}$, characterizing the probability distribution of a continuous random variable X. A natural possibility is to build an estimate $\hat{\theta} = \mathrm{E}[\tilde{\theta}]$, for which we need the distribution of $\tilde{\theta}$. We have some prior information about $\tilde{\theta}$, that we would like to express in a sensible way. Such a knowledge or subjective view may be expressed by a probability density $p(\theta)$, which is called the **prior** distribution of $\tilde{\theta}$. Here, we care about the difference between the random variable $\tilde{\theta}$ from its realized value θ, but will not always be so picky. One possible prior is a uniform distribution. Note that a uniform prior may seem uninformative, but this need not be the case, as a uniform prior bounds Θ.

Then, the prior should be somehow merged with experimental evidence. Experimental evidence consists of independent observations $X_1, \ldots, X_n$ from the unknown distribution. Let us denote the density of X by $f(x \mid \theta)$, to emphasize its dependence on the value θ of the parameter. Since a random sample consists of independent random variables, their joint distribution, *conditional* on $\tilde{\theta} = \theta$, is

$$f_n(x_1, \ldots, x_n \mid \theta) = f(x_1 \mid \theta) \cdot f(x_2 \mid \theta) \cdots f(x_n \mid \theta).$$

The conditional density $f_n(x_1, \ldots, x_n \mid \theta)$ is called the **likelihood function**.

Since we are dealing with the $n+1$ random variables $\tilde{\theta}, X_1, \ldots, X_n$, we could also consider their joint density $g(x_1, \ldots, x_n, \theta)$, but this will not be really necessary for what follows. Let us denote the marginal density of the n observations by

$$g_n(x_1, \ldots, x_n).$$

Given the likelihood $f_n(x_1, \ldots, x_n \mid \theta)$ and the prior $p(\theta)$, we can find the marginal density of $X_1, \ldots, X_n$ by applying the total probability theorem:

$$g_n(x_1, \ldots, x_n) = \int_\Theta f_n(x_1, \ldots, x_n \mid \theta) p(\theta) d\theta,$$

where we integrate over the domain Θ of $\tilde{\theta}$, i.e., the support of the prior distribution. Now we need to invert the conditioning, i.e., we would like to obtain the distribution of $\tilde{\theta}$ conditional on the observed values $X_i = x_i, i = 1, \ldots, n$, i.e.,

$$p_n(\theta \mid x_1, \ldots, x_n).$$

This **posterior** density should merge the prior and the density of observed data conditional on the parameter. This is obtained by applying Bayes' theorem to densities, which yields

$$p_n(\theta \mid x_1, \ldots, x_n) = \frac{g(x_1, \ldots, x_n, \theta)}{g_n(x_1, \ldots, x_n)} = \frac{f_n(x_1, \ldots, x_n \mid \theta) \cdot p(\theta)}{g_n(x_1, \ldots, x_n)}. \quad (10.25)$$

Note that the posterior density involves a term $g_n(x_1, \ldots, x_n)$, which does not really depend on θ. Its role is to normalize the posterior distribution, so that its integral is 1. Sometimes, it might be convenient to rewrite Eq. (10.25) as

$$p_n(\theta \mid x_1, \ldots, x_n) \propto f_n(x_1, \ldots, x_n \mid \theta) \cdot p(\theta), \quad (10.26)$$

i.e., p_n is *proportional* to the product of f_n and p. In plain English,

$$\text{posterior} \propto \text{likelihood} \times \text{prior}.$$

What we are saying is that, given some prior knowledge about the parameter and the distribution of observations, conditional on the parameter, we obtain an updated distribution of the parameter, conditional on the actually observed data.

S10.1.2 BAYESIAN LEARNING IN COIN FLIPPING

We tend to take for granted that coins are fair, and that the probability of getting heads is 1/2. Let us consider flipping a possibly unfair coin, with an unknown probability θ of getting heads. In order to learn this unknown value, we flip the coin repeatedly, i.e., we run a sequence of independent Bernoulli trials with unknown parameter $\tilde{\theta}$.[16] If we do not know anything about the coin, we might just assume a uniform prior

$$p(\theta) = 1, \quad 0 \leq \theta \leq 1.$$

If we flip the coin n times, we know that the probability of getting h heads is related to the binomial probability distribution

$$f_n(h \mid \theta) \propto \theta^h (1-\theta)^{n-h}. \tag{10.27}$$

This is our likelihood function. If we regard this expression as the probability of observing h heads, given θ, this should actually be the probability mass function (PMF) of a binomial variable with parameters θ and n, but we are disregarding a binomial coefficient, which does not depend on θ and just normalizes the distribution. In fact, if X is a binomial random variable with parameters θ and n, we should write the PMF as

$$\mathrm{P}\{X = k\} = \binom{n}{k} \theta^k (1-\theta)^{n-k}, \quad \text{where} \quad \binom{n}{k} \doteq \frac{n!}{(n-k)!k!},$$

but the leading binomial coefficient plays no role for what we need to accomplish. If we multiply the likelihood function by the prior, which is just 1, we obtain the posterior density for $\tilde{\theta}$, given the number of observed heads:

$$p_n(\theta \mid h) \propto \theta^h (1-\theta)^{n-h}, \quad 0 \leq \theta \leq 1. \tag{10.28}$$

Equations (10.27) and (10.28) look like the same thing, because we use a uniform prior, but they are very different in nature. Equation (10.27) should be interpreted as the PMF of a discrete random variable, the number of observed heads. On the contrary, Eq. (10.28) gives the posterior density of the continuous random variable θ, conditional on the fact that we observed h heads and $n - h$ tails. If we look at it this way, we recognize the density of a beta distribution.[17] To normalize the posterior, we should multiply it by the appropriate value of the beta function. Again, this normalization factor does not depend on θ and can be disregarded.

In Fig. 10.5, we display posterior densities, normalized in such a way that their maximum is 1, after flipping the coin n times and having observed h heads.

---

[16]This example is based on [22, Chapter 2].

[17]The beta distribution is a continuous distribution with support $[0, 1]$. It includes, as a special case, the uniform distribution on the unit interval, as well as some shapes of the triangular distribution. See, e.g., [4, Chapter 7].

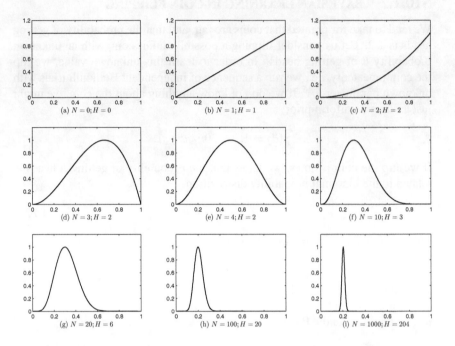

FIGURE 10.5 Updating the posterior density of the probability of observing heads in coin flipping. In each plot, H is the number of heads observed in N flips.

The plot in Fig. 10.5(a) is just the uniform prior. Now imagine that the first flip lands heads. After observing the first heads, we know for sure that $\theta \neq 0$; indeed, if θ were zero, we could not observe any heads. The posterior is now proportional to a triangle:

$$p_1(\theta \,|\, 1) \propto \theta^1 (1-\theta)^{1-1} = \theta, \qquad 0 \leq \theta \leq 1.$$

This triangle is shown in Fig. 10.5(b). If we observe another heads in the second flip, the updated posterior density is a portion of a parabola, as shown in Fig. 10.5(c):

$$p_2(\theta \,|\, 2) \propto \theta^2 (1-\theta)^{2-2} = \theta^2, \qquad 0 \leq \theta \leq 1.$$

If we get tails at the third flip, we rule out $\theta = 1$ as well. Proceeding this way, we get beta distributions, progressively concentrated around the true (unknown) value of θ. Incidentally, the figure has been obtained by Monte Carlo simulation of coin flipping with $\theta = 0.2$.

S10.1.3 THE EXPECTED VALUE OF A NORMAL DISTRIBUTION

Consider a sample $(X_1, \ldots, X_n)$ from a normal distribution with unknown expected value θ and known variance σ^2. It may seem odd that we assume knowl-

S10.1 Bayesian statistics

edge of variance, when the expected value is uncertain, but this may be relevant to finance, if we assume that we trust the estimate of the covariance matrix much more than the estimate of expected return. In terms of a single-index model, we assume that the estimates of betas are fairly reliable and stable over time, whereas the estimate of alphas is much more critical and related to subjective views about the future. Implicitly, we assume that returns are normally distributed, which we know to be subject to criticism. We do so for the sake of simplicity, but the Bayesian framework for portfolio management may be applied to other distributions as well. We treat the univariate case in detail. The multivariate extension is conceptually straightforward.

Given the mutual independence among observations, we have the following likelihood function:

$$f_n(x_1, \ldots, x_n \mid \theta) = \frac{1}{(2\pi)^{n/2} \sigma^n} \exp\left\{ -\sum_{i=1}^{n} \frac{(x_i - \theta)^2}{2\sigma^2} \right\}.$$

Let us assume that the prior distribution of $\tilde{\theta}$ is normal, too, with expected value μ_0 and standard deviation σ_0:

$$p(\theta) = \frac{1}{\sqrt{2\pi}\sigma_0} \exp\left\{ -\frac{(\theta - \mu_0)^2}{2\sigma_0^2} \right\}.$$

We should not confuse σ_0, the standard deviation in the prior, which represents our uncertainty about the expected value θ, with σ, which is the known standard deviation of the observations. Sometimes, we refer to (μ_0, σ_0^2) as a pair of *hyperparameters*, as they are the parameters of the distribution of a parameter.

To get the posterior, we may simplify our work by considering in each function only the part that involves θ, wrapping the rest within a proportionality constant. We may rewrite the prior as

$$p(\theta) \propto \exp\left\{ -\frac{(\theta - \mu_0)^2}{2\sigma_0^2} \right\}. \tag{10.29}$$

In order to simplify the expression of the likelihood, we observe that

$$\sum_{i=1}^{n}(x_i - \theta)^2 = \sum_{i=1}^{n}(x_i - \bar{x} + \bar{x} - \theta)^2$$

$$= \sum_{i=1}^{n}(x_i - \bar{x})^2 + \sum_{i=1}^{n}(\bar{x} - \theta)^2 + 2 \cdot \underbrace{\sum_{i=1}^{n}(x_i - \bar{x})}_{\equiv 0} \cdot \sum_{k=1}^{n}(\bar{x} - \theta)$$

$$= \sum_{i=1}^{n}(x_i - \bar{x})^2 + n(\theta - \bar{x})^2,$$

where $\bar{x}$ is the average of x_i, $i = 1, \ldots, n$. Then, we rewrite the likelihood as

$$f_n(x_1, \ldots, x_n \mid \theta) \propto \exp\left\{ -\frac{n}{2\sigma^2}(\theta - \bar{x})^2 \right\}. \tag{10.30}$$

By multiplying Eqs. (10.30) and (10.29), we obtain the posterior

$$p_n(\theta|x_1,\ldots,x_n) \propto \exp\left\{-\frac{1}{2}\left[\frac{n}{\sigma^2}(\theta-\bar{x})^2 + \frac{1}{\sigma_0^2}(\theta-\mu_0)^2\right]\right\}. \quad (10.31)$$

Finally, a bit of tedious algebra leads to

$$p_n(\theta|x_1,\ldots,x_n) \propto \exp\left\{-\frac{1}{2\xi^2}(\theta-\nu)^2\right\}, \quad (10.32)$$

where we define

$$\nu \doteq \frac{n\sigma_0^2 \bar{x} + \sigma^2 \mu_0}{n\sigma_0^2 + \sigma^2}, \quad (10.33)$$

$$\xi^2 \doteq \frac{\sigma_0^2 \sigma^2}{n\sigma_0^2 + \sigma^2}. \quad (10.34)$$

We recognize the familiar shape of a normal density, with expected value ν and variance ξ^2. Hence, given an observed sample mean $\bar{X}$ and a prior μ_0, Eq. (10.33) tells us that the Bayes' estimator of θ can be written as

$$\hat{\theta} = \mathrm{E}\left[\tilde{\theta}|X_1,\ldots,X_n\right] = \frac{n\sigma_0^2}{n\sigma_0^2 + \sigma^2}\bar{X} + \frac{\sigma^2}{n\sigma_0^2 + \sigma^2}\mu_0$$

$$= \frac{\frac{n}{\sigma^2}}{\frac{n}{\sigma^2} + \frac{1}{\sigma_0^2}}\bar{X} + \frac{\frac{1}{\sigma_0^2}}{\frac{n}{\sigma^2} + \frac{1}{\sigma_0^2}}\mu_0. \quad (10.35)$$

If we define the **precisions**

$$\tau_0 \doteq \frac{1}{\sigma_0^2}, \qquad \tau \doteq \frac{1}{\sigma^2},$$

we may recast Eq. (10.35) as

$$\mu_n = \mathrm{E}\left[\tilde{\theta}|X_1,\ldots,X_n\right] = \frac{n\tau \bar{X} + \tau_0 \mu_0}{n\tau + \tau_0}, \quad (10.36)$$

which has a quite natural interpretation: The posterior estimate μ_n is a weighted combination of the prior and the empirical evidence, where weights are given by the respective precisions. The variance of the normal posterior density is

$$\left(\frac{n}{\sigma^2} + \frac{1}{\sigma_0^2}\right)^{-1},$$

which may be more conveniently rewritten in terms of an updated precision:

$$\tau_n = n\tau + \tau_0.$$

PROBLEMS

The generalization to a multivariate normal distribution is conceptually straightforward. The unknown vector of expected values, $\boldsymbol{\theta}$, has a normal prior $\mathsf{N}(\boldsymbol{\mu}_0, \boldsymbol{\Sigma}_0)$. We observe n independent realizations

$$\mathbf{X}_i \sim \mathsf{N}(\boldsymbol{\theta}, \boldsymbol{\Sigma}), \quad i = 1, \ldots, n,$$

with sample mean $\overline{\mathbf{X}}$. In this case, the posterior estimate involves inverses of the covariance matrices,

$$\boldsymbol{\mu}_n = \mathrm{E}\left[\boldsymbol{\theta} \mid \mathbf{X}_1, \ldots, \mathbf{X}_n\right] = \left[n\boldsymbol{\Sigma}^{-1} + \boldsymbol{\Sigma}_0^{-1}\right]^{-1} \cdot \left[n\boldsymbol{\Sigma}^{-1}\overline{\mathbf{X}} + \boldsymbol{\Sigma}_0^{-1}\boldsymbol{\mu}_0\right], \quad (10.37)$$

and the updated covariance matrix is

$$\boldsymbol{\Sigma}_n = \left[n\boldsymbol{\Sigma}^{-1} + \boldsymbol{\Sigma}_0^{-1}\right]^{-1}. \quad (10.38)$$

Formally, Eqs. (10.37) and (10.38) are the same expressions as in the scalar case, where division by variance σ^2 is replaced by premultiplication by the inverse covariance matrix $\boldsymbol{\Sigma}^{-1}$. We insist again on the fact that $\boldsymbol{\Sigma}$ is the covariance matrix of the random vector $\mathbf{X}$, which we have assumed known, whereas $\boldsymbol{\Sigma}_n$ is related with the posterior uncertainty about the vector of expected values of $\mathbf{X}$.

Problems

10.1 Imagine that there are two stock markets in the world, whose relative capitalization weights are 25% and 75%, respectively. The expected returns of the two markets are 6% and 4%, and their volatilities are 15% and 10%, respectively. Kurtosis of return is 5 in the first market and 7 in the second one; skewness is zero for both of them.

- Consider a risk-averse investor X, with logarithmic utility function. If she wants a portfolio with expected return of 3%, is it possible to achieve her objective?
- What is the volatility of the above portfolio, assuming that the two markets are statistically independent?
- Now assume that all investors are exactly like investor X. Is investor X still able to achieve the above portfolio at market equilibrium? Why?
- Can the CAPM hold for these two markets, populated by investors like X? Why?

10.2 You are analyzing three stock shares: Joint, Eppon, and Peculiar Motors. Based on your analysis, the price of a Joint stock share should be the same as the sum of one Eppon share and one Peculiar Motors share. The expected price of a Joint share in one year, according to your forecast, is $100. The current price of a Peculiar Motors is $30, which is fair in your opinion. The annual risk-free rate is 5% (annual compounding), the market risk premium is 10%, the market volatility 38%, and the beta of Joint is 2. In equilibrium, what is the fair price of one Eppon share?

10.3 Consider an individual asset, with random holding period return $\tilde{r}_i$, and the market portfolio, with corresponding return $\tilde{r}_M$. We describe uncertainty by five discrete scenarios, as in the following table:

Scenario	Probability	$\tilde{r}_i(\omega)$	$\tilde{r}_M(\omega)$
ω_1	0.2	0.03	0.09
ω_2	0.2	0.17	0.16
ω_3	0.3	0.28	0.10
ω_4	0.2	0.05	0.02
ω_5	0.1	−0.04	0.16

- Assuming that CAPM holds, find the risk-free return.
- Does the result look sensible? If not, how can you explain the anomaly?

10.4 An APT model is based on three mutually independent systematic factors F_1, F_2, and F_3. The annual risk-free rate, with annual compounding, is 4% (below we assume annual returns). We consider three well-diversified portfolios, $i = A, B, C$, with the following features:

Portfolio	β_{i1}	β_{i2}	β_{i3}	$E[\tilde{r}_i]$
A	1.5	−0.9	2.0	8.5%
B	0.5	1.2	0.6	12.8%
C	−0.1	0.4	−0.3	4.9%

- Find the coefficients characterizing the APT model.
- Find the expected return of a portfolio with unit exposure to F_1, neutral to F_2 and F_3.

10.5 An APT model is based on two mutually independent systematic factors F_1 and F_2. The annual risk-free rate, with annual compounding, is 4% (below we assume annual returns). We consider three well-diversified portfolios, $i = A, B, C$, with the following features:

Portfolio	β_{i1}	β_{i2}	$E[\tilde{r}_i]$
A	1.6	−0.8	9.5%
B	0.5	1.3	11.7%
C	−0.1	0.4	4.1%

- Find the coefficients characterizing the APT model based on portfolios A and B.
- If the expected return of portfolio C is as given in the table, are there arbitrage opportunities? If so, devise an arbitrage strategy.

Further reading

- For a treatment of equilibrium models from the theoretical viewpoint of financial economics, see [1]. See also [6] for links with CAPM and arbitrage theory, as well as equilibrium with differential information.
- CAPM was introduced in [20], and APT was introduced in [18].
- Some of the concepts that we have used in proving or justifying CAPM and APT, as well as some examples, have been borrowed from [8], [11], and [16].
- The Black–Litterman portfolio management approach was introduced in [2]; see also [10] and [19].
- A general overview of Bayesian methods in finance can be found in [17].
- An introductory overview on market efficiency and behavioral finance can be found in [3], which has been the basis for part of our exposition. Fundamental concepts in behavioral finance are also illustrated in, e.g., [21]. See [12] for a key reference about prospect theory.
- Technical analysis is one of the most controversial topics in finance, and a non-negligible amount of related literature has probably little to do with a serious and objective investigation. Nevertheless, there are exceptions. An empirical analysis can be found in [15], and [14] is an interesting source of interviews with technical analysis practitioners.

Bibliography

1 E. Barucci. *Financial Markets Theory: Equilibrium, Efficiency and Information*. Springer, London, 2003.

2 F. Black and R. Litterman. Global portfolio optimization. *Financial Analysts Journal*, 48:28–43, 1992.

3 Z. Bodie, A. Kane, and A. Marcus. *Investments* (9th ed.). McGraw-Hill, New York, 2010.

4 P. Brandimarte. *Quantitative Methods: An Introduction for Business Management*. Wiley, Hoboken, NJ, 2011.

5 E. Burmeister, R. Roll, and S.A. Ross. A practitioner's guide to arbitrage pricing theory. In *A Practitioner's Guide to Factor Models*, pp. 1–30. The Research Foundation of the Institute of Chartered Financial Analysts, Charlottesville, VA, 1994.

6 J.-P. Danthine and J.B. Donaldson. *Intermediate Financial Theory* (3rd ed.). Academic Press, San Diego, CA, 2015.

7 M.H. DeGroot. *Probability and Statistics*. Addison-Wesley, Reading, MA, 1975.

8 E.J. Elton, M.J. Gruber, S.J. Brown, and W.N. Goetzman. *Modern Portfolio Theory and Investment Analysis* (8th ed.). Wiley, Hoboken, NJ, 2011.

9 I. Gilboa. *Theory of Decision under Uncertainty*. Cambridge University Press, New York, 2009.
10 G. He and R. Litterman. The intuition behind Black–Litterman model portfolios. Technical report, *Investment Management Research*, Goldman & Sachs Company, 1999. A more recent version can be downloaded from http://www.ssrn.org.
11 J.E. Ingersoll, Jr. *Theory of Financial Decision Making*. Rowman & Littlefield, Totowa, NJ, 1987.
12 D. Kahneman and A. Tversky. Prospect theory: An analysis of decision under risk. *Econometrica*, 47:263–292, 1979.
13 B.N. Lehmann and D.M. Modest. Mutual fund performance evaluation: A comparison of benchmarks and benchmark comparisons. *Journal of Finance*, 42:233–265, 1987.
14 A.W. Lo and J. Hasanhodzic. *The Heretics of Finance: Conversations with Leading Practitioners of Technical Analysis*. Wiley, Hoboken, NJ, 2009.
15 A.W. Lo, H. Mamaysky, and J. Wang. Foundations of technical analysis: Computational algorithms, statistical inference, and empirical implementation. *Journal of Finance*, 55:1705–1765, 2000.
16 D.G. Luenberger. *Investment Science* (2nd ed.). Oxford University Press, New York, 2014.
17 S.T. Rachev, J.S.J. Hsu, B.S. Bagasheva, and F.J. Fabozzi. *Bayesian Methods in Finance*. Wiley, Hoboken, NJ, 2008.
18 S.A. Ross. The arbitrage theory of capital asset pricing. *Journal of Economic Theory*, 13:341–360, 1976.
19 S. Satchell and A. Scowcroft. A demystification of the Black–Litterman model: Managing quantitative and traditional portfolio construction. *Journal of Asset Management*, 1:138–150, 2000.
20 W.F. Sharpe. Capital asset prices: A theory of market equilibrium under conditions of risk. *Journal of Finance*, 19:425–442, 1964.
21 H. Shefrin. *Behavioral Corporate Finance*. McGraw-Hill/Irwin, New York, 2005.
22 D.S. Sivia and J. Skilling. *Data Analysis: A Bayesian Tutorial* (2nd ed.). Oxford University Press, Oxford, 2006.

Part Four

Derivatives

Chapter Eleven

Modeling Dynamic Uncertainty

From Chapter 8 on, we have considered static portfolio management models, where we make a decision at time $t = 0$ and observe the result at time $t = T$, the end of a predefined holding period. Representing uncertainty in this context requires the characterization of a multivariate distribution of risk factors, which is not quite trivial. However, there are even more complicated problems, calling for a *dynamic* characterization of uncertainty.

- A first example is asset–liability management (ALM) problems, where we consider a sequence of time instants $t_i \in [0, T]$, $i = 1, \ldots, m$, at which we must meet a possibly uncertain liability L_i. We have considered simple approaches to interest rate risk management for ALM problems in Section 6.3. In these limited approaches, we actually solve a static decision problem. It may be the case that a better plan is obtained by a multistage decision model, but even if we do not want to pay the price of such a challenging optimization model,[1] we may need to check the performance of whatever plan on a set of random scenarios for both the assets and the liabilities. Thus, we must characterize the uncertain evolution of the underlying risk factors over time, in order to generate a rich and reliable set of scenarios.

- Another quite relevant example is provided by the need to hedge an option dynamically. In Section 2.3.4, we have considered a single-step binomial model for option pricing. Quite clearly, we need a more refined uncertainty model for both pricing and hedging purposes.[2] Furthermore, we certainly cannot use a static model when dealing with American-style options, which can be exercised at any time before expiration. Thus, we need a dynamic model, describing the evolution of the relevant factors for the option value, most notably the price of the underlying stock share in the case of an equity option.

---

[1] See Section 15.6.3.2.
[2] We will emphasize the tight relationship between option pricing and hedging later, in Section 13.3.1.

We are familiar with dynamic models of deterministic physical systems, where a collection of state variables $s_j(t)$, $j = 1, \ldots, n$, collected into a vector $\mathbf{s}(t) \in \mathbb{R}^n$, evolves over time.[3] Depending on the representation of time, we may choose between the following classes of models:

Continuous-time differential equations, where the derivative of the state variables with respect to time $t \in [0, T] \subset \mathbb{R}$ is given as a function of state and time:
$$\frac{d\mathbf{s}}{dt} = \mathbf{g}(\mathbf{s}(t), t).$$
The vector function $\mathbf{g}(\cdot, \cdot)$ is referred to as **transition function**, as it implicitly describes the transition from one state to another one. Furthermore, the transition function may change over time or not.

Discrete-time difference equations, where we describe the state transition in a more explicit form,
$$\mathbf{s}_{k+1} = \mathbf{g}_k(\mathbf{s}_k), \qquad k = 0, 1, 2, 3, \ldots$$
Here, we discretize time in uniform steps of length δt and consider time instants $t_k = k \cdot \delta t$. For the sake of notational simplicity, we typically write $\mathbf{s}_k$ as a shorthand for $\mathbf{s}(k \cdot \delta t)$. Again, the transition function may depend on time or not.

As we shall see, the choice between a continuous- and a discrete-time representation may be dictated by computational convenience, when we have to discretize the continuous time in order to apply a numerical method. However, perhaps surprisingly, it may be the case that a continuous-time model proves more convenient, if it provides us with an analytical solution. There are further variations on the theme of dynamic modeling, which are quite relevant for finance.

Continuous vs. discrete states. Differential equations assume continuous state variables. However, some state variables are intrinsically discrete, like the credit rating of a bond. Stock prices, strictly speaking, should be considered as a discrete state variable, as we use a limited number of decimals in actual markets. If the minimum tick is one cent of whatever monetary unit, then the stock price will be expressed as a number with two decimals, which is a discrete state, if we measure prices in cents. A similar consideration applies to interest rates. For modeling convenience, we may represent stock prices by real numbers. Sometimes, we go the other way around and discretize a continuous state space for computational purposes.

Stochastic transitions. Needless to say, whatever dynamic model we use in finance, it has to account for uncertainty. While, in probability theory, we deal with random variables like $X(\omega)$, here, we need to deal with a *collection* of random variables indexed by time, $X_t(\omega)$. Such a collection is

---

[3] We often denote dependence on time by a subscript, as in s_t, but when multiple subscripts might obscure notation, we shall also use $s(t)$.

called a **stochastic process**. As we shall see, we have to deal with *four* combinations of continuous- vs. discrete-time, as well as continuous- vs. discrete-state stochastic processes. The introduction of time adds a further layer of complication, as we need to take into due account the dependence (or lack thereof) of random variables over time.

Controlled transitions. If we deal with state variables like stock prices, we may assume that their evolution over time is driven by exogenous risk factors. However, a state variable like the funding level of a pension fund is influenced by decisions. Hence, we have to introduce control variables that partially affect state transitions. For instance, a discrete-time dynamic model may be written as

$$\mathbf{S}_{k+1} = \mathbf{g}_k(\mathbf{S}_k, \mathbf{x}_k, \epsilon_{k+1}), \qquad k = 0, 1, 2, 3, \ldots,$$

where we collect decision variables in vector $\mathbf{x}_k$. We use capital letters for state variables to insist on their random nature, whereas control variables are denoted by lowercase letters. Please also note the use of subscripts for the random factors ϵ_{k+1}. Especially in discrete time, it is important to figure out the sequence of events:[4] (1) We first observe the current state $\mathbf{S}_k$. (2) Then we make a decision $\mathbf{x}_k$. (3) Random risk factors ϵ_{k+1} are realized *after* the decision, leading to a new state $\mathbf{S}_{k+1}$ through the transition function $\mathbf{g}_k$.

After this quick overview, the reader should not be surprised by the considerable variety of dynamic models that are used in finance, including:

- Stochastic differential equations
- Time series models
- Scenario trees
- Recombining lattices
- Discrete- and continuous-time Markov chains

In this chapter, we describe some of this models, in a more or less detailed way. Most emphasis will be placed on stochastic differential equations, given their key role in option pricing. On the contrary, we shall not treat time series models as they deserve, and readers will be referred to some excellent books on financial econometrics.

In Section 11.1, we introduce stochastic processes formally, along with a few very simple examples. We also introduce important subclasses of stochastic processes, namely, Markov processes and martingales, as well as filtrations modeling the flow of information over time. In Section 11.2, we deal with continuous-time, continuous-state processes and introduce the key building block of many financial models, the standard Wiener process. Continuous-time financial models rely on stochastic differential equations, which are intro-

---

[4]This framework lends itself to optimization by stochastic dynamic programming, as we shall see in Section 15.7. See also the consumption–saving example of Section 2.1.2.

duced in Section 11.3. As it turns out, we should not really talk about differential equations, but rather about *integral* equations. In fact, in order to model uncertainty in a sensible way, we have to deal with processes featuring nondifferentiable or even discontinuous sample paths. Thus, we shall not deal with traditional derivatives, but with differentials, i.e., increments over infinitesimal time intervals, which are integrated over time. Hence, we shall need the introduction of a suitable concept of stochastic integral. Furthermore, we shall also find that the familiar rules of deterministic calculus may fail to work, within our stochastic context, and must be replaced by a suitable extension, namely, Itô's lemma. All of this material forms the core of stochastic calculus and is dealt with in Section 11.4. Then, armed with the necessary theoretical background, in Section 11.5, we outline some of the most common models adopted in financial modeling. In Section 11.6, we briefly describe how to tackle sample path and scenario generation based on dynamic uncertainty models, which is often required by numerical solution methods.

As we shall see, this chapter mostly revolves around stochastic calculus. A rigorous exposition would require considerable mathematical background, including machinery based on measure theory, quadratic variations, change of measure by the Girsanov theorem and Radon–Nikodym derivatives, and other wild beasts. Given the introductory nature of this book, and for the sake of brevity, we shall pursue a rather informal and heuristic approach. In order to not disappoint mathematically inclined readers too much, and to provide other readers with some intuition necessary to tackle the more advanced literature, in Supplement S11.1, we introduce a slightly more formal and measure-theoretic view, with emphasis on the link between measurability concepts and the evolution of information over time.

11.1 Stochastic processes

The simplest way to think of a stochastic process is as a collection of random variables indexed by time.[5] If time is a continuous variable, i.e., it is represented by a real number $t \in \mathbb{R}_+ \doteq [0, +\infty)$, we have a continuous-time stochastic process, denoted by X_t or $X(t)$. If time is a discrete variable, represented by a non-negative integer number, $k \in \mathbb{Z}_+ \doteq \{0, 1, 2, \ldots\}$, we have a discrete-time stochastic process, denoted by X_k. The random variable itself may take real or an integer values, but we may also consider arbitrary discrete sets, as is the case of credit ratings. In the following, we will be mostly dealing with stochastic processes taking numerical values. Beside continuous- and discrete-time processes, there is a third category, **discrete-event** stochastic processes. Actually, these are continuous-time processes, with the peculiar feature that the state of the system is piecewise constant, with occasional jumps. We shall

---

[5] It may be worthwhile to mention that, in other application domains, we might be interested in collection of random variables indexed by space.

11.1 Stochastic processes

consider one such case, the Poisson process, in Example 11.4. As a financially motivated example, imagine a pool of loans and a process $N(t)$ counting the number of defaults occurred in the time interval $[0,t]$.

We may also collect multiple random variables into a vector stochastic process $\mathbf{X}(t)$, with components $X_j(t)$, $j = 1,\ldots,m$. The usual terminology in econometrics is as follows:

- When we consider the realization of different random variables at the same time instant, we are analyzing **cross-sectional** data.
- When we consider the realization of a single variable at different time instants, we are analyzing **longitudinal** data.
- When we consider different random variables at different time instants, we are analyzing **panel** data.

On a more formal level, we know that random variables should be regarded as *functions* $X(\omega)$, mapping outcomes ω within a sample space Ω into numerical values, i.e., real or integer numbers. This more formal framework is needed to see precisely how a probability measure on the underlying sample space is translated to a probability measure for random variables. From a practical viewpoint, we often disregard the underlying framework and just settle for useful descriptions like a cumulative probability function (**CDF**), a probability density function (**PDF**, for continuous random variables), or a probability mass function (**PMF**, for discrete random variables). By the same token, we should consider a stochastic process in more formal terms, as a function of *two* arguments, $X(t,\omega)$. In this context, ω is associated with a possible history of the process. If we fix $\omega = \omega^*$ and consider $X(t,\omega^*)$ as a function of time, we have a **sample path**, also called **scenario**, of the stochastic process. If we fix time $t = t^*$, we observe a random variable $X(t^*,\omega)$. When unnecessary, we will avoid stressing the role of ω, but in certain cases, it is important to make it more explicit.

A concrete example of stochastic process that we are interested in is the price $S(t,\omega)$ of a stock share. However, we may be interested in more than one stock share, in which case we observe a set of stochastic processes $S_j(t,\omega)$, $j = 1,\ldots,n$, collected into a vector, $\mathbf{S}(t,\omega)$. In this case, if we fix $t = t^*$, we are considering a multivariate distribution $\mathbf{S}(t^*,\omega)$, and we need to represent the relationships among different stock shares. The usual approach is to specify their mutual correlations, but this may not be sufficient, especially when we consider risk management problems and the occurrence of extreme market scenarios.

Example 11.1 Random fields

> The dynamics of stock share prices over time may be difficult to model accurately, but interest rates are much more complex. In earlier chapters, we have considered continuously compounded interest rates

> $r(t, t+\tau)$ for an investment over the time interval $(t, t+\tau)$. If we introduce randomness, we have to cope with a mathematical object like $r(t, t+\tau, \omega)$. This object is called a **random field**. If we fix $\tau = \tau^*$, we obtain a stochastic process representing how an interest rate for a given time-to-maturity changes over time. If we fix $t = t^*$, we obtain a function of time-to-maturity τ, describing the term structure of rates at time t^*, for different scenarios ω. Clearly, there must be some connection between interest rates at different time instants, and rates for different maturities. Modeling interest rates in financially sensible, yet computationally tractable way is no easy task.

As we shall see, apart from considering the mutual dependence of different components in a vector stochastic process, the following related issues play a key role:

The flow of information. Consider a bounded time interval $[0, \tau]$. If we have observed a sample path $X(t, \omega)$ for $t \in [0, \tau]$, what can we say about the future evolution of the process, for $t > \tau$? The key point is that we usually cannot say which sample path ω we are following, because there will be a subset of sample paths which, up to time τ, cannot be distinguished from ω. We shall introduce filtrations as a formal way of modeling the information generation in a stochastic process.

The dependence over time. Sometimes, the process has no memory of the past, and knowledge of the sample path observed so far does not tell us anything about the future. At the other end of the spectrum, we may have a strongly path-dependent process whose future evolution depends on the whole sample path so far. In intermediate cases, only a limited amount of memory plays a role.

11.1.1 INTRODUCTORY EXAMPLES

In this section, we illustrate the differences between discrete and continuous features of stochastic processes in terms of time and state, by a set of simple examples. In doing so, we will streamline notation and avoid pointing out the dependence on the outcome (scenario) ω.

Example 11.2 A discrete-time random walk

> Consider the discrete-time stochastic process
>
> $$X_k = X_{k-1} + \epsilon_k, \quad k = 1, 2, 3, \ldots, \quad (11.1)$$

11.1 Stochastic processes

where the initial state is often set to $X_0 = 0$, and $\epsilon_k \sim \mathsf{N}(0,1)$ is an element of a sequence of i.i.d. standard normals. X_k is the state of the system at discrete time k, and it is a continuous random variable, since we add normal variables at each time step. Hence, we have a continuous-state, discrete-time stochastic process. The state is affected by a sequence of shocks ϵ_k, which are mutually independent, have zero expected value, and are also independent on the current state. Hence, the shocks are *unpredictable*. Note that we insist on adding a shock indexed by k to a state variable indexed by $k-1$, to emphasize the nature of the shocks, which are often referred to as *innovations* in econometric parlance. Their independence on the past is related to the efficient market hypothesis. This kind of process is called a **random walk**. As we shall see, there is a corresponding process in continuous time, called standard Wiener process, which plays a key role in financial modeling.

By unfolding Eq. (11.1) recursively, we find

$$X_k = \sum_{i=1}^{k} \epsilon_i.$$

Given the mutual independence and the normality of the driving shocks, we easily find the *unconditional* distribution of the state, $X_k \sim \mathsf{N}(0,k)$. Please note that the variance is k. Hence, the standard deviation is $\sqrt{k}$, i.e., it scales with the square root of (discrete) time. It is also easy to find *conditional* distributions. Conditional on the value $X_h = x$ at time $h < k$, we have

$$X_k = x + \sum_{i=h+1}^{k} \epsilon_i,$$

and

$$X_k \mid \{X_h = x\} \sim \mathsf{N}(x, k-h).$$

This is an example of a **Gaussian process**, since the joint distribution of the random variables X_k is normal.

In a random walk, it is important to observe that all we need to characterize the probability distribution of future states is the knowledge of the current state $X_k = x$. In fact, this is what state variables are all about, in any dynamic system. As we shall see, in the stochastic process parlance, this is related to a property characterizing Markov processes, to be formally defined later. The same property characterizes the next example.

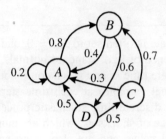

FIGURE 11.1 A discrete-time Markov chain.

▇ Example 11.3 A discrete-time Markov chain

A discrete-time Markov chain is a process with a state variable X_k, $k = 0, 1, 2, 3, \ldots$, taking values in a discrete set. The discrete state space may consist of an infinite, yet countable set, like the set of integer numbers, or a finite set, like the credit ratings of a bond. As we shall see, we may also consider a continuous-state Markov chain. The name *chain* is related to the nature of the state space. In fact, we may represent the process by a graph, as shown in Fig. 11.1. Here, the state space is the set $\{A, B, C, D\}$. Nodes correspond to states, and directed arcs represent possible transitions, labeled by the corresponding probabilities. For instance, if we are in state C now, at the next step we will be in state B, with probability 0.7, or in state A, with probability 0.3. Note that transition probabilities depend only on the current state, not on the whole past history. Thus, we may describe the chain in terms of the **conditional transition probabilities**,

$$P_{ij} \equiv \mathrm{P}\{X_{t+1} = j \mid X_t = i\},$$

which, in the case of a finite state space of cardinality n, may be collected into the single-step transition probability matrix $\mathbf{P} \in \mathbb{R}^{n \times n}$. For the chain in Fig. 11.1,

$$\mathbf{P} = \begin{bmatrix} 0.2 & 0.8 & 0 & 0 \\ 0.4 & 0 & 0 & 0.6 \\ 0.3 & 0.7 & 0 & 0 \\ 0.5 & 0 & 0.5 & 0 \end{bmatrix}.$$

Note that the matrix need not be symmetric, and that the current state is associated with a matrix row, whereas the next states are associated with columns. On the diagonal, we have the probability of staying in the current state. After a transition, we must land somewhere within

11.1 Stochastic processes

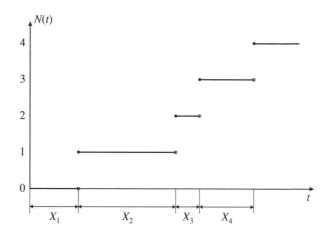

FIGURE 11.2 Sample paths of a Poisson process.

> the state space. Therefore, each and every row adds up to 1:
> $$\sum_{j=1}^{n} P_{ij} = 1, \qquad i = 1, \ldots, n.$$

In Example 11.3, we may imagine that we observe the credit rating associated with a bond with a given frequency, say, one month. Hence, we only observe state transitions at the beginning of each month. The sojourn time, i.e., the random time that the system spends at a given state, will be an integer number. However, in reality, state transitions might occur at arbitrary time instants. Hence, we may consider a continuous-time Markov chain, too, where the time that the process spends in a given state is modeled by a continuous random variable. As we shall see later, Markov processes are, in a sense, "memoryless." Hence, in discrete-time the actual distribution is given by a geometric random variable. The corresponding memoryless distribution in continuous time is an exponential random variable, which is at the heart of the following example.

Example 11.4 The Poisson process

> The Poisson process is a continuous-time stochastic process $N(t)$, counting the number of events that have occurred in the time interval $[0, t]$. A sample path is illustrated in Fig. 11.2, and it is important to understand the underlying dynamics:

> - The process starts at the initial state $N(0) = 0$.
> - After a random time X_1, the first event occurs, and the state of the system jumps to $N(X_1) = 1$.
> - After a random time X_2, the second event occurs, and $N(X_1 + X_2) = 2$.
>
> The striking feature of the Poisson process is that the sample path is piecewise constant and there is a jump corresponding to each event. This is a very simple example of a **discrete-event stochastic process**.
>
> There are other examples of counting processes. The Poisson process is obtained when we make a specific assumption about the random times elapsing between events. Let X_k, $k = 1, 2, 3, \ldots$, be the random time elapsing between events $k-1$ and k; by convention, X_1 is the epoch of the first event after the start time $t = 0$. We obtain a Poisson process, if we assume that variables X_k are independent and exponentially distributed with parameter λ, which is the event rate, i.e., the average number of events occurring per unit time. Note that the expected value of X_k is $1/\lambda$, and that the number of events occurring during a time interval of length τ is a discrete Poisson random variable with parameter $\lambda\tau$.

The Poisson process is a possible model for customer arrivals at a bank but, per se, it is not quite used in finance. However, it is a fundamental building block of other processes, so it is worth exploring in more detail.

- The process "jumps" whenever an event occurs, and the sample paths are piecewise constant. The jump introduces a discontinuity in the process, but it is important to understand what kind of discontinuity is involved exactly. If a jump occurs at time t^*, and there is a transition from state m to state $m+1$, we have a limit from the left

$$\lim_{t \uparrow t^*} N(t) = N(t^*_-) = m,$$

but the process is not continuous, as $N(t^*) = m + 1$. On the contrary, the process is continuous from the right, as the limit from the right is

$$\lim_{t \downarrow t^*} N(t) = N(t^*_+) = m + 1 = N(t^*).$$

We might even stumble on esoteric jargon like a *càdlàg function* when dealing with similar stochastic processes in finance. This is just a French acronym for "continue à droite, limitée à gauche," since the sample path is continuous from the right, and is limited (or bounded, i.e., it does not go to infinity) from the left.

11.1 Stochastic processes

In principle, we may also define a stochastic process with jumps, which is continuous from the *left*. From a financial viewpoint, the difference is quite relevant, since a process which is continuous from the right has an element of intrinsic unpredictability, which makes the corresponding risk factor difficult to hedge.

- If we assume an arbitrary distribution for the times elapsing between consecutive events in a counting process, we do not obtain a Poisson process, which requires a sequence of i.i.d. exponential variables. As a consequence of the lack of memory of the exponential distribution, the Poisson process may represent the random occurrence of events that have no mutual relationships at all. The Poisson process is, in fact, a Markov process, to be defined later.

Let us explore the consequence of the requirement on inter-event times X_k in more detail. If we consider a time interval $[t_1, t_2]$, with $t_1 < t_2$, then the number of events occurring in this interval, i.e., $N(t_2) - N(t_1)$, has Poisson distribution with parameter $\lambda \cdot (t_2 - t_1)$. Thus, the increment of the event count for intervals of equal length is the same, and where the intervals are located is irrelevant. Hence, we say that the Poisson process has **stationary increments**. Furthermore, if we consider another time interval $[t_3, t_4]$, where $t_3 < t_4$, which is disjoint from the previous one, i.e., $t_2 < t_3$, then the random variables $[N(t_2) - N(t_1)]$ and $[N(t_4) - N(t_3)]$ are independent. Hence, we say that the Poisson process has **independent increments**. As we shall see later, a wider class of stochastic processes, called Lévy processes, is characterized by stationary and independent increments. This class includes quite different processes, like the Poisson process and the Wiener process, which is a continuous process extending the Gaussian random walk of Example 11.2 to continuous time.

We mention a couple of possible extensions of the simple Poisson process:

- If we introduce a time-varying event rate $\lambda(t)$, we obtain the so-called **inhomogeneous** Poisson process.
- If we associate a random variable Y_k with each event, corresponding to the size of the jump, and the variables Y_k are an i.i.d. sequence, then we obtain the **compound** Poisson process

$$M(t) = \sum_{k=1}^{N(t)} Y_k.$$

Note that the sum includes only the events occurred up to time t.

Compound Poisson processes are relevant in finance, since we do observe jumps in stock prices, but they have a random, possibly negative, size. This jump component may be integrated with a continuous component, leading to jump–diffusion processes. In practice, we may need to use different kinds of stochastic processes jointly in a financial model. For instance, a bond price depends on interest rates, which may be modeled by continuous-state processes, but also on credit ratings, which are discrete-state. The reader has certainly noticed

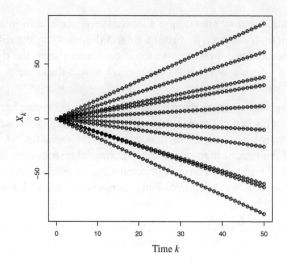

FIGURE 11.3 A few sample paths of the stochastic process of Eq. (11.2).

that, in this introductory set of examples, we did not consider continuous-state, continuous-time processes. We shall do so later, as this requires the more complicated machinery of stochastic differential equations.

11.1.2 MARGINALS DO NOT TELL THE WHOLE STORY

All we need to know about a single random variable X is encoded in its CDF, or alternatively in its PDF or PMF, for continuous and discrete variables, respectively. If we think of a stochastic process $X(t)$ or X_k as a collection of random variables, it is tempting to believe that all we need is a characterization of each individual variable for each time instant. To see how wrong this idea is, let us illustrate a simple counterexample. Consider the discrete-time stochastic process

$$X_k = k \cdot \epsilon, \qquad k = 0, 1, 2, 3, \ldots, \tag{11.2}$$

where $\epsilon \sim \mathsf{N}(0, 1)$. We immediately see that each X_k is normal with expected value 0 and variance k^2. This marginal distribution is the same as for the process

$$X_k = k \cdot \epsilon_k, \qquad k = 0, 1, 2, 3, \ldots, \tag{11.3}$$

where ϵ_k is again standard normal, but we have one such variable for each time instant. More precisely, ϵ_k is a sequence of i.i.d. standard normals. However, the two processes have little in common, despite the fact that their marginal PDFs are the same for every time instant. To see this, we may compare the sample paths of the two processes. Figure 11.3 shows a few sample paths of the process described by Eq. (11.2). In fact, this is a rather degenerate process, since uncertainty is linked to the realization of a *single* random variable. If we want to simulate a sample path of this process, we just need one sample from

11.1 Stochastic processes

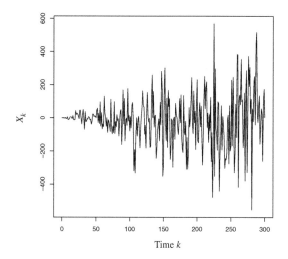

FIGURE 11.4 A single sample path of the stochastic process of Eq. (11.3).

the standard normal distribution, which yields the slope of a line. Indeed, the sample path is just a line, going through the origin, featuring that slope. If we know a *single* point on any sample path, we have full information about the whole sample path, and prediction is quite easy. Figure 11.4 shows one sample path of the process described by Eq. (11.3), which looks completely different and is quite unpredictable.

The random variables X_{k_1} and X_{k_2}, for any $k_1 \neq k_2$, are perfectly correlated in the first process, but independent in the second one. These two processes are, in a sense, at the opposite side of the spectrum. In the first process, if we know the value of the process at just one time instant, we know the whole sample path. In the second one, no information at all is provided by such knowledge. The random walk of Example 11.2 has a slightly different marginal distribution, $N(0,t)$ rather than $N(0,t^2)$, but what is really relevant is that it is somewhere between these two extremes, in terms of information and prediction. The random walk process is not as trivially predictable as the process of Fig. 11.3, but the current state has an impact on the future states, unlike the case of Fig. 11.4.

The key is the dependence between each variable in the collection $X(t)$ or X_k. On second thought, this is pretty obvious. Also in the static case of a multivariate distribution **X**, we may have *different* joint distributions featuring the *same* marginals.[6] In fact, a full characterization of a stochastic process, in general, would require the joint distribution of all of the variables for different times t. However, specifying the dependence by a multivariate density function is out of the question, especially in continuous time. We have to find some con-

---

[6]Technically speaking, the mutual dependence may be characterized by a copula.

venient way of describing how a stochastic process evolves, possibly limiting our interest to tractable cases.

One very interesting family of processes, to be described below, consists of the class of Markov processes, where the only information we need to characterize future evolution is the knowledge of its current state. This is the case of the discrete-time random walk: The path that leads to the current state does not matter for the future evolution. We may get to a state by quite different paths, but once we get to that state, any memory of the observed path so far is lost. The random walk also enjoys another important property:

$$\mathrm{E}\big[X(t)\,|\,X(\tau)=x\big]=x, \qquad \forall t \geq \tau.$$

In plain English, the expected value of the process, given its current value, is just its current value. There is no tendency to drift up or down. As will shall see, processes featuring this property are called martingales. In order to better understand these properties, we need a formal way to characterize the information generated by a stochastic process.

11.1.3 MODELING INFORMATION: FILTRATION GENERATED BY A STOCHASTIC PROCESS

In this introductory book, we avoid using quite formal approaches to probability, based on measure theory. However, as we have pointed out, a stochastic process should be regarded as a function $X(t,\omega)$ depending on two arguments, time t and an element ω of a sample space Ω. In the case of a random variable $X(\omega)$, we define a probability measure on Ω, which in turn allows to associate probabilities with the numerical values of the random variable. This requires the definition of an algebra of sets, based on subsets of the sample space, which may be combined by taking the usual set operations, like difference, union, and intersection. More formally, we should deal with a probability space,

$$\big\{\Omega, \mathcal{F}, P\big\},$$

where:

- Ω is the underlying **sample space**.
- $\mathcal{F}$ is a **field** field of events, i.e., a family of subsets of Ω, which is closed under the usual set-theoretic operations. This means that if we consider two events E_1 and E_2 in $\mathcal{F}$, their union, intersection, and difference are also in $\mathcal{F}$. A field is also called an **algebra**.
- $P(\cdot)$ is a **probability measure**, a function associating a probability with any event in $\mathcal{F}$. The nature of $\mathcal{F}$ ensures that, if we are able to associate a probability with any pair of events, we are also able to assign a sensible probability to their intersection (i.e., the joint event), their union, as well as the complement of a single event (related to the probability that an event does not happen).

11.1 Stochastic processes

Since this machinery is not really needed in this elementary book, we just outline these concepts in Supplement S11.1, which may be safely skipped by uninterested readers. The whole thing gets more complicated when time comes into play, as the underlying algebra of events evolves over time, reflecting the information generated by the stochastic process. The following example provides us with the basic intuition.

Example 11.5 Filtration generated by coin flipping

Let us consider three consecutive flips of a fair and memoryless coin. There are two outcomes H and T for each flip, and eight possible scenarios:

- $\omega_1 = (H, H, H)$
- $\omega_2 = (H, H, T)$
- $\omega_3 = (H, T, H)$
- $\omega_4 = (H, T, T)$
- $\omega_5 = (T, H, H)$
- $\omega_6 = (T, H, T)$
- $\omega_7 = (T, T, H)$
- $\omega_8 = (T, T, T)$

The sample space is $\Omega = \{\omega_1, \omega_2, \ldots, \omega_8\}$. After the three flips, we know exactly which scenario ω occurred. Hence, we observe events (subsets of Ω) of the form $\{\omega_i\}$, consisting of a single outcome. If we associate numerical values $X_k(\omega)$, $k = 1, 2, 3$, with the outcome, at the end of the time horizon we have observed a specific sample path. To be concrete, we may associate this process with a binomial tree, where an asset price goes up when we draw heads, and it goes down when we draw tails. On the contrary, before the first flip, we do not know anything. We can only deal with sets $\emptyset$, the empty set, and Ω. Considering the empty set may sound weird, but this is really needed to make sure that the field of events satisfies the technical conditions of closure with respect to set theoretic operations. The complement of Ω is, in fact, the empty set $\emptyset$. The probability measure of $\emptyset$ is zero, and there are technical reasons why, at any time, we should also consider events with probability zero. We will gloss over such technicalities.

Now, imagine that the first flip of the coin results in heads. We do not know which scenario ω will occur eventually, but we may rule out outcomes ω_5, ω_6, ω_7, and ω_8. By a similar token, we may rule out the first four outcomes if the first flip is T. Hence, after the first flip

> we deal with an algebra including events
>
> $$\Omega_H = \{\omega_1, \omega_2, \omega_3, \omega_4\} \quad \text{and} \quad \Omega_T = \{\omega_5, \omega_6, \omega_7, \omega_8\},$$
>
> completed with Ω and $\emptyset$, which may be obtained from Ω_H and Ω_T by unions and intersections. The two subsets Ω_H and Ω_T form a **partition** of the sample space. We note that these subsets are, in a sense, smaller (finer) than the whole sample space that we deal with before the first flip. By a similar token, after two flips, we know that one among the four events
>
> $$\Omega_{HH} = \{\omega_1, \omega_2\}, \ \Omega_{HT} = \{\omega_3, \omega_4\},$$
> $$\Omega_{TH} = \{\omega_5, \omega_6\}, \text{ and } \Omega_{TT} = \{\omega_7, \omega_8\},$$
>
> will occur. Again, the subsets Ω_{HH}, Ω_{HT}, Ω_{TH}, and Ω_{TT} are a partition of the sample space. These four subsets are smaller than the two subsets Ω_H and Ω_T that we deal with after the first flip. So, they generate a finer algebra. After the whole sequence of flips, we deal with singletons, the smallest possible events by which the sample space may be partitioned.

Example 11.5 shows that, when we accumulate information, we deal with successive partitions of the sample space, which involve smaller and smaller subsets, generating finer and finer algebras. Formally, this is represented by time-dependent algebras of events $\mathcal{F}_t$, such that the algebra gets finer and finer as time progresses. Technically, if $t_1 < t_2$,

$$\mathcal{F}_{t_1} \subseteq \mathcal{F}_{t_2},$$

i.e., any subset in $\mathcal{F}_{t_1}$ is in $\mathcal{F}_{t_2}$, but there may subsets generated by the richer structure in $\mathcal{F}_{t_2}$ that cannot be generated in $\mathcal{F}_{t_1}$. This increasing sequence of algebras is called a **filtration**.

From our viewpoint, we may just think of a filtration as the information generated by the stochastic process over time. The filtration $\mathcal{F}_t$ is the information generated by a given stochastic process on the time interval $[0, t]$. Clearly, information is added as time progresses, and never subtracted. Filtrations are important in defining conditional expectations. For instance, when we write

$$\mathrm{E}[X(t) \mid \mathcal{F}_\tau],$$

for $t \geq \tau$, we mean the expected value of $X(t)$ at a future time t, given (conditional on) the *whole* history of the process up to time τ. Needless to say,

$$\mathrm{E}[X(t) \mid \mathcal{F}_t] = X(t).$$

11.1 Stochastic processes

By the same token, we may consider a conditional probability like

$$P\{X_t \in A \mid \mathcal{F}_\tau\},$$

i.e., the probability that X_t will be within a subset A of the state space, conditional on the whole history up to time $\tau \leq t$.

11.1.4 MARKOV PROCESSES

We have repeatedly hinted at Markov processes, and it is finally time to provide a formal definition. The essential property of Markov processes is related to their limited memory: The only information that is needed to characterize the future system evolution is the current state. Formally, we may express this in terms of a conditional transition probability. A discrete-state, discrete-time Markov process is characterized by the following condition (**Markov property**):

$$\begin{aligned} P\{X_{t+1} = i_{t+1} \mid X_t = i_t, X_{t-1} = i_{t-1}, X_{t-2} = i_{t-2}, \ldots\} \\ = P\{X_{t+1} = i_{t+1} \mid X_t = i_t\}. \end{aligned} \quad (11.4)$$

This means that the conditional probability of being at discrete state i_{t+1} at time $t+1$, given the past history of the process, only depends on the state i_t visited at time t. Thus, any path dependency is ruled out, as the path that the process followed to reach state i_t is irrelevant. For instance, in Example 11.3, we only need to consider single-step transition probabilities. In the case of continuous states, the property stated in Eq. (11.4) does not make sense, as the probability of any single state is identically zero. However, we may consider transitions probabilities to *subsets* of the state space and state the Markov property as follows:

$$\begin{aligned} P\{X_{t+1} \in A \mid X_t = i_t, X_{t-1} = i_{t-1}, X_{t-2} = i_{t-2}, \ldots\} \\ = P\{X_{t+1} \in A \mid X_t = i_t\}, \end{aligned} \quad (11.5)$$

for every possible subset A. From a practical viewpoint, we may look for a **transition density**, which gives a transition probability when integrated over subsets of the state space.

From a conceptual viewpoint, a much more elegant solution relies on the filtration concept that we have introduced in Section 11.1.3. This allows to express the Markov property in full generality:

$$P\{X_t \in A \mid \mathcal{F}_\tau\} = P\{X_t \in A \mid X_\tau\}, \quad (11.6)$$

for every subset A of the state space and for any $\tau \leq t$. This condition can also be applied to continuous-time processes. Equation (11.6) states that the only relevant piece of information in the whole filtration $\mathcal{F}_\tau$, i.e., the whole history of the process on the time interval $[0, \tau]$, is the value of the process X_τ at the last time instant τ.

Example 11.6 The random walk, again

The discrete-time random walk

$$X_k = X_{k-1} + \epsilon_k, \quad k = 1, 2, 3, \ldots,$$

is an example of a Markov process that can be described by an explicit transition equation, which is an example of the more general case

$$\mathbf{X}_{t+1} = \mathbf{g}_t(\mathbf{X}_t, \boldsymbol{\epsilon}_{t+1}), \tag{11.7}$$

where a vector of state variables and a vector of shocks are considered. There is a strong connection between dynamic equations based on state variables and Markov processes. To be more precise, we should also require that the driving process ϵ_t consists of a sequence of i.i.d. variables.

Example 11.7 Redefining the state space

Lookback options are a family of exotic, path-dependent options where the payoff depends on the maximum or the minimum price of the underlying asset, observed along the sample path up to maturity. For instance, the option payoff could be

$$S_{\max} - S_T,$$

where S_T is the asset price at maturity and $S_{\max}$ is the maximum observed price on the time interval $[0, T]$. Let us assume a binomial, discrete-time process where

$$S_{k+1} = \begin{cases} uS_k, & \text{with probability } p_u, \\ dS_k, & \text{with probability } p_d. \end{cases}$$

The price dynamics is clearly Markovian, but the maximum price so far,

$$M_k \doteq \max_{j=0,1,2,\ldots,k} S_j,$$

seems to introduce an unavoidable path dependence, destroying the Markov property. Actually, we may notice that M_k can be defined recursively as

$$M_k = \max\{S_k, M_{k-1}\} \equiv S_k \vee M_{k-1},$$

where we introduce the common shorthand $a \vee b \doteq \max\{a, b\}$. This suggests the possibility of augmenting the state variable, which is now

11.1 Stochastic processes

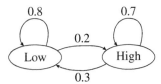

FIGURE 11.5 A simple regime switching model, with states corresponding to low and high volatility.

a vector with two components, whose state transition equation is

$$\begin{bmatrix} S_{k+1} \\ M_{k+1} \end{bmatrix} = \begin{cases} \begin{bmatrix} uS_k \\ M_k \vee uS_k \end{bmatrix}, & \text{with probability } p_u, \\ \begin{bmatrix} dS_k \\ M_k \vee dS_k \end{bmatrix}, & \text{with probability } p_d. \end{cases}$$

The trick of augmenting the state space is a general strategy that can be attempted, in order to transform a non-Markov process into a Markov one. In practice, needless to say, it can only be pursued when it leads to a moderate increase in the size of the state space.

Example 11.8 A regime-switching model

Financial markets are characterized by *volatility*, which is linked to the standard deviation of returns. One interesting feature of volatility is that we observe periods of relative calm, in which volatility is reasonable, followed by periods of nervousness, where volatility is quite large. Imagine that we want to build a model in which markets can be in one of two states, low and high; the time bucket that we consider is a single trading day. High volatility tends to persist; hence, we cannot just assign a probability that, on one day, markets will be in one of the two states. We may build a regime-switching model, accounting for the fact that each state tends to persist: After a day of high volatility, we are more likely to observe another day of high volatility; the same holds for a day with low volatility. A naive regime-switching model is illustrated in Fig. 11.5. Here we assume that if the last day was in the low state, the next day will feature the same level of volatility with probability 0.8. However, there is a probability 0.2 that we will observe a day with high volatility. If we get to the high state one day, the next day will feature high volatility again with probability 0.7, whereas we have a 0.3 probability of moving back to the low state. Clearly, this is an overly naive model, and we are not considering estimation issues arising when a state is hidden, i.e., not directly observable. Formally, we have the following transition probabilities:

$$P(\text{low} \mid \text{low}) = 0.8, \qquad P(\text{high} \mid \text{low}) = 0.2,$$
$$P(\text{high} \mid \text{high}) = 0.7, \qquad P(\text{low} \mid \text{high}) = 0.3.$$

A natural question is: If we are currently in a state, what is the expected number of time buckets that we will spend in that state? In other words, what is the expected **sojourn time**? The answer can be found by using the geometric distribution. We recall that the geometric distribution is the discrete counterpart of the exponential distribution, in terms of a memoryless property. The expected value of a geometric random variable with "failure" probability p is $1/p$, where we count the number of trials to obtain the first success, which in this case corresponds to a transition out of the current state. Hence, the expected sojourn time in the high state is

$$E[T_{\text{high}}] = \frac{1}{0.7} = 1.43.$$

In continuous-time Markov chains, the sojourn time in each state is exponentially distributed.

11.1.5 MARTINGALES

The discrete-time random walk of Eq. (11.1) is clearly a Markov process, but it is characterized by another relevant feature: The expected value of the process at any time $t > \tau$, conditional on $X_\tau = x$, is just x:

$$E[X_t \mid X_\tau = x] = x. \tag{11.8}$$

In plain English, the value of the process is not expected to increase or decrease. A process enjoying this characteristic is called a **martingale**. The defining property of martingales can be formalized as follows[7]:

$$E\{X_t \mid \mathcal{F}_\tau\} = X_\tau, \qquad t \geq \tau. \tag{11.9}$$

---

[7] In rigorous expositions of martingale processes, an integrability condition is also required. As usual, we disregard technical issues.

11.1 Stochastic processes

This property should *not* be confused with either Eq. (11.8), which is weaker, or with the Markov property stated by Eq. (11.6).

- On the one hand, Eq. (11.9) is about expectations, whereas Eq. (11.6) is about transition probabilities. It is true that probabilities may be expressed as expected values of indicator functions, but Markov processes may take non-numerical values, whereas martingales make only sense for processes taking numerical values.
- On the other hand, the martingale property (11.9) involves the whole history up to time τ, encoded in the filtration $\mathcal{F}_\tau$, and not only the state X_τ, as is the case of Eq. (11.8). The random walk is a Markov process, so the relevant information boils down to a single state, but a martingale need not be a Markov process, as it is possible to define martingales that are not Markov processes. Going the other way around, the Poisson process is a Markov process, but it is not a martingale.

From a historical viewpoint, martingales have their root in gambling strategies. In fact, a martingale may be interpreted as a fair game, in which wealth is not expected to increase or decrease. To see why martingales are relevant in finance, let us consider the value at time τ of a vanilla European-style derivative maturing at T. The payoff at maturity is a random variable f_T, depending on the value of the underlying asset at T. As we have already pointed out, intuition *might* suggest that the fair value at time τ should be given by the conditional expectation of the discounted expected payoff,

$$f_\tau \stackrel{?}{=} e^{-r \cdot (T-\tau)} \cdot \mathrm{E}[f_T \mid \mathcal{F}_\tau],$$

which may be rewritten as

$$\frac{f_\tau}{e^{r\tau}} \stackrel{?}{=} \mathrm{E}\left[\frac{f_T}{e^{rT}} \,\middle|\, \mathcal{F}_\tau\right].$$

Note the use of the question mark to stress that fact that we are just guessing. This amounts to saying that the stochastic process

$$X_t \doteq \frac{f_t}{e^{rt}},$$

defined as the ratio between the value of the derivative and a bank account process at time t, is a martingale. The bank account process gives the amount available at time t, if we start from an initial deposit of $1 and earn interest at the continuously compounded risk-free rate r. We already know, from Section 2.3.4, that this is not true, and financial intuition helps to understand why. Since the derivative is a risky asset, it stands to good reason that it should offer a risk premium to investors, i.e., its expected return should be larger than the risk-free rate. Formally, this implies that the process X_t should be expected to increase over the time interval $[\tau, T]$,

$$\frac{f_\tau}{e^{r\tau}} < \mathrm{E}\left[\frac{f_T}{e^{rT}} \,\middle|\, \mathcal{F}_\tau\right].$$

The process, however, is a martingale under a risk-neutral measure, which would apply in a risk-neutral world, where no risk premium is demanded by investors. We have shown that, for a simple single-step binomial model, the process is a martingale if we switch to risk-neutral probabilities:

$$\frac{f_\tau}{e^{r\tau}} = \mathrm{E}_{\mathbb{Q}_n}\!\left[\left.\frac{f_T}{e^{rT}}\,\right|\mathcal{F}_\tau\right].$$

An important point to be realized is that a process may be a martingale or not, depending on the selected probability measure. By a suitable change of measure, a process may be transformed into a martingale. As we shall see in Chapters 13 and 14, finding an equivalent martingale measure, under which the ratio of two stochastic processes is a martingale, provides a powerful machinery to price options.

11.2 Stochastic processes in continuous time

In this section, we start investigating continuous-time stochastic models, which are essential in pricing financial derivatives. To this aim, we shall need the more sophisticated machinery of stochastic calculus, which provides us with essential tools like stochastic integrals and stochastic differential equations. Before tackling these concepts, we need to get acquainted with an essential building block, the standard Wiener process, which is essentially the continuous-time counterpart of the discrete-time random walk of Example 11.2. As we shall see, the standard Wiener process is a Gaussian process, featuring continuous sample paths. Gaussian means that it involves normal distributions. Continuous sample paths means that we are ruling out jumps in asset prices or interest rates. Thus, using the standard Wiener process as a building block may fail to capture some essential features of financial markets, like excess kurtosis[8] and heavy-tailed distributions, which arise from both non-normality and jumps. The net result is that we may underestimate risk, by assuming a safer world than the real one, where some risks cannot be hedged away. Given the introductory nature of this book, we will stay within the safe domain of Wiener processes, but we also describe a more general building block, the class of Lévy processes, which subsumes the standard Wiener process.

11.2.1 A FUNDAMENTAL BUILDING BLOCK: STANDARD WIENER PROCESS

The **standard Wiener process** $W(t)$ is a continuous-time stochastic process that may be characterized by the following properties:

1. $W(0) = 0$ (this is actually a convention).

---

[8] We have excess kurtosis when this is larger than 3, the kurtosis of a normal distribution.

11.2 Stochastic processes in continuous time

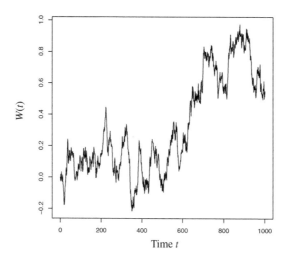

FIGURE 11.6 A sample path of the standard Wiener process.

2. Given any time interval $[s, t]$, the increment $W(t) - W(s)$ is distributed as $\mathsf{N}(0, t-s)$, a normal random variable with expected value zero and standard deviation $\sqrt{t-s}$.

3. Increments are stationary, in the sense that they do not depend on where the time interval is located, but only on its width.

4. Increments are independent: If we take time instants $t_1 < t_2 \leq t_3 < t_4$, defining two nonoverlapping time intervals, then $[W(t_2) - W(t_1)]$ and $[W(t_4) - W(t_3)]$ are independent random variables.

The second condition makes sure that standard deviation scales with the square root of time, which is just what happens with the discrete-time random walk. In fact, the standard Wiener process may be thought of as the limit of a discrete-time random walk, when the time step δt gets infinitesimal.

We may write increments of the Wiener process as follows:

$$\delta W(t) \doteq W(t + \delta t) - W(t) \stackrel{d}{=} \epsilon \sqrt{\delta t},$$

where ϵ is standard normal. We use the notation $\stackrel{d}{=}$ to point out that two random variables have the same distribution, even though they need not be the same mathematical object. We are just saying that $\delta W(t) \sim \mathsf{N}(0, \delta t)$, and that we may sample increments of the standard Wiener process by sampling and then scaling a standard normal variable. This is used in Monte Carlo simulation to generate sample paths of the standard Wiener process, which look like the one depicted in Fig. 11.6. In this figure, we observe that sample paths of the Wiener process look continuous, but not differentiable. This may be stated precisely, but introducing continuity and differentiability rigorously requires a precise definition of stochastic convergence. To get an intuitive feeling for this fact, let us consider the increment ratio

$$\frac{\delta W(t)}{\delta t} = \frac{W(t+\delta t) - W(t)}{\delta t}.$$

Given the defining properties of the Wiener process, it is easy to see that

$$\text{Var}\left[\frac{\delta W(t)}{\delta t}\right] = \frac{\text{Var}\left[W(t+\delta t) - W(t)\right]}{(\delta t)^2} = \frac{1}{\delta t}.$$

If we take the limit for $\delta t \to 0$, this variance goes to infinity. Strictly speaking, this is no proof of nondifferentiability of $W(t)$, but it does suggest that there is some trouble in using an object like $dW(t)/dt$. In fact, we will never take derivatives of stochastic processes, but we shall work with possibly infinitesimal increments, formally treated as differentials to be integrated.

Note that most of the properties defining the standard Wiener process are shared by the Poisson process, which also features stationary and independent increments. The difference is that the standard Wiener process is continuous (in some well-defined sense), whereas the Poisson process is not. Both belong to a more general class of stochastic processes, the Lévy processes, featuring stationary and independent increments.

11.2.2 A GENERALIZATION: LÉVY PROCESSES

Lévy processes are defined by the following properties:

- $X_0 = 0$ (technically, this is just required with probability 1).
- If $0 \leq t_1 < t_2 < t_3 < t_4$, then the increments $(X_{t_2} - X_{t_1})$ and $(X_{t_4} - X_{t_3})$ are independent random variables (independent increments).
- The increment $(X_t - X_\tau)$, for $\tau < t$, has the same distribution as $X_{t-\tau}$ (stationary increments).
- For any real number $\epsilon > 0$ and any time instant $t \geq 0$, we have the following *stochastic continuity* property:

$$\lim_{h \to 0} P\{|X_{t+h} - X_t| > \epsilon\} = 0.$$

The first three properties should ring some bells, since we have already met them when we defined the Poisson and the standard Wiener processes. The stochastic continuity property may be puzzling a bit, as it seems to exclude the jumps featured by Poisson processes. Actually, this is not true, as the limit is in probability terms. If the number of jumps is, in some sense, limited, stochastic continuity is preserved. In fact, the number of jumps of a Poisson process in a finite time interval has Poisson distribution. Since this is a discrete distribution with unbounded support, there is no upper bound to the number of jumps, but only a countable number of jumps is possible.

The standard Wiener and the Poisson processes are specific cases of a Lévy process. As we shall see, they can be generalized in different ways, and jump–diffusion processes merge their characteristics. The definition of a Lévy process leaves much room to different choices of the distribution of increments. If we choose a heavy-tailed distribution, we may circumvent some limitations of Gaussian processes based on the Wiener process. One such case is the Lévy flight, which should be contrasted against a random walk. However, not every distribution will do. It can be shown that Lévy processes involve infinitely divisible distributions. This essentially states that the random variable may be decomposed into a sum of an arbitrary number of i.i.d. random variables. The normal and Poisson distribution are infinitely divisible, but other distributions are not.

11.3 Stochastic differential equations

The standard Wiener process is a useful building block, but it cannot be directly used to model stock prices or interest rates, since (among other issues) it may take negative values. To that aim, we must introduce a powerful modeling framework to describe continuous-time, continuous-state processes. Informally, we shall talk about stochastic differential equations. As we shall see, the concept that we will be actually using is a form of stochastic integral.

To get started, let us consider a deterministic, ordinary differential equations that we are all familiar with from physics,

$$m\frac{d^2x}{dt^2}(t) = F(t), \qquad (11.10)$$

linking force, mass, and acceleration. Ordinary differential equations involve derivatives of some order, and Eq. (11.10) is a second-order equation. In finance, we essentially use first-order equations but, as we have already hinted at, a Wiener process features continuous, but nondifferentiable sample paths. Therefore, we shall never meet a notation like $dW(t)/dt$, in terms of derivatives. We only use the **differential** $dW(t)$ of the Wiener process. Informally, we may think of $dW(t)$ as a random variable with distribution $N(0, dt)$. Actually, we should think of this differential as an increment over a small time period δt, where the increment has distribution $N(0, \delta t)$ and we take the limit for $\delta t \to 0$. Cutting a few corners, the fundamental theorem of calculus states that

$$\int_s^t dF(\tau) = F(t) - F(s),$$

i.e., when we have an exact differential, possibly found by taking the antiderivative $F(\cdot)$ of an integrand function $f(\cdot) = F'(\cdot)$, the integral of the increments over a time interval $[s, t]$ is the difference between the values of $F(\cdot)$ at the extreme points t and s. By direct analogy, we may argue that the integration of the increments of the standard Wiener process yields

$$\int_s^t dW(\tau) = W(t) - W(s).$$

By integrating a deterministic function of time, we find a number, whereas by integrating a stochastic process, we find a random variable. This stochastic integral looks harmless enough, but assigning a proper meaning to it, as we shall see, is not quite trivial, and we need specific rules to work with it.

11.3.1 A DETERMINISTIC DIFFERENTIAL EQUATION: THE BANK ACCOUNT PROCESS

In order to get acquainted with the (rather informal) approach that we will use to cope with differential equations, it is best to consider an extremely simple example.[9] We do not take advantage of the general framework for linear ordinary differential equations, but we work with differentials directly. Consider the function $B(t)$, describing how the amount deposited on a bank account grows over time, when a continuously compounded and *constant* interest rate r is applied. The increase in wealth on a time interval dt is given by

$$dB(t) = rB(t) \cdot dt, \tag{11.11}$$

which may be interpreted as the limit of the increment of wealth over a small time interval δt,

$$\delta B(t) \doteq B(t + \delta t) - B(t) = rB(t) \cdot \delta t.$$

We may rewrite Eq. (11.11) as

$$\frac{dB(t)}{B(t)} = r \cdot dt. \tag{11.12}$$

If we integrate these differentials over the time interval $[0, T]$, we obtain

$$\int_0^T \frac{dB(t)}{B(t)} = r \int_0^T dt = r \cdot (T - 0),$$

where integration of the right-hand side is trivial. The left-hand side can be dealt with by recalling the derivative of the logarithm:

$$\frac{d \log B(t)}{dB(t)} = \frac{1}{B(t)}$$

$$\Rightarrow \quad d \log B(t) = \frac{dB(t)}{B(t)}$$

$$\Rightarrow \quad \int_0^T \frac{dB(t)}{B(t)} = \int_0^T d \log B(t) = \log B(T) - \log B(0).$$

---

[9] We shall see slightly more challenging examples of deterministic ordinary differential equations in Section 14.2.1.

11.3 Stochastic differential equations

Finally, we find a well-known equation,

$$\log \frac{B(T)}{B(0)} = rT \quad \Rightarrow \quad B(T) = B(0)e^{rT}. \tag{11.13}$$

If the risk-free rate is not constant, but given by a deterministic function of time $r(t)$, we may follow the same drill, and Eq. (11.13) is generalized to

$$B(T) = B(0) \cdot e^{\int_0^T r(t)\,dt}. \tag{11.14}$$

In practice, the rate $r(t)$ should be interpreted as a *short rate*, i.e., an interest rate that applies to a very small time interval, and is subject to random variations and reinvestment risk.[10] In this case, $r(t)$ is a stochastic process, which should be written as $r(t,\omega)$, for each outcome ω in a sample space Ω. The integral in Eq. (11.14) should be interpreted as a standard Riemann integral for each sample path $r(\cdot,\omega)$, which results in a random variable $B(T,\omega)$.

11.3.2 THE GENERALIZED WIENER PROCESS

Now let us turn to the task of building a stochastic differential equation, using the standard Wiener process $W(t)$ as a building block. A good starting point is the simplest differential equation we may think of,

$$dX(t) = a\,dt,$$

where a is a given constant, with initial condition $X(0) = x_0$. By integrating the differential, as we did in Section 11.3.1, we find that the solution is a straight line,

$$X(t) = x_0 + at.$$

We may introduce noise by adding the differential of a Wiener process, which yields the stochastic differential equation

$$dX(t) = a\,dt + b\,dW(t). \tag{11.15}$$

This equation defines a **generalized Wiener process**, and intuition suggests that it may be solved by straightforward integration over the time interval $[0,t]$,

$$X(t) - X(0) = \int_0^t dX(\tau) = \int_0^t a\,d\tau + \int_0^t b\,dW(\tau) = at + b\,[W(t) - W(0)], \tag{11.16}$$

which may be rewritten as

$$X(t) = x_0 + at + bW(t).$$

---

[10] We know the interest rate applying to a time period at the beginning of the period, but we do not know the interest rate that will apply to future time periods. See Section 3.5.3.

Hence, for the generalized Wiener process, we find the following distribution of the random variable $X(t)$:

$$X(t) \sim \mathsf{N}\Big(x_0 + at, b^2 t\Big).$$

Thus, sample paths are lines with slope a, to which a Gaussian noise is superimposed, whose impact depends on how large b is (in absolute value).

Example 11.9 Simulating the generalized Wiener process

We may use Eq. (11.16) to generate sample paths of the generalized Wiener process by Monte Carlo sampling. We rewrite the equation for a small time step δt,

$$X(t + \delta t) - X(t) = a \cdot \delta t + b\left[W(t + \delta t) - W(t)\right],$$

and express the increment of the Wiener process as

$$\delta W(t) \doteq W(t + \delta t) - W(t) \stackrel{d}{=} \epsilon_{t+\delta t}\sqrt{\delta t},$$

where $\epsilon_t \sim \mathsf{N}(0,1)$ is a sequence of i.i.d. variables. As before, we multiply the standard normal ϵ by the square root of the time step, to obtain the correct distribution of the increment. This boils down to the updating rule

$$X(t + \delta t) = X(t) + a \cdot \delta t + b\epsilon_{t+\delta t}\sqrt{\delta t},$$

which can be used to sample $X(t + \delta t)$, conditional on $X(t)$. For instance, if we set $\delta t = 1$, $X(0) = 20$, $a = 1$, and $b = 3$, the simulation of 200 time steps yields the sample path of Fig. 11.7. The dashed line corresponds to the **mean value function** $\mu(t) = 20 + t$, which gives the unconditional expected value of the process as a function of time. We will further discuss sample path generation in Section 11.6.

From a mathematical viewpoint, we see that solving a stochastic differential equation amounts to finding the distribution of the state $X(t)$ in the future, conditional the current state. We will steer away from theoretical issues concerning existence and uniqueness of a solution. From a financial viewpoint, the generalized Wiener process is a simple stochastic process, but can we use it for financial modeling? A first observation is that it may take negative values, which is not the case for stock prices, for instance. Furthermore, this process is based on additive increments which are independent on the current state. However, a stock price increment of $0.10 is not the same for a stock with price $1.00 or $100.00. Hence, we should consider multiplicative shocks or, equivalently, express increments in terms of returns.

11.3 Stochastic differential equations

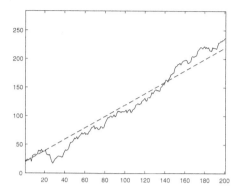

FIGURE 11.7 Sample path of a generalized Wiener process.

11.3.3 GEOMETRIC BROWNIAN MOTION AND ITÔ PROCESSES

If we want to write a stochastic differential equation modeling the random return of an asset, we might consider the following modification of the generalized Wiener process:

$$\frac{dX(t)}{X(t)} = \mu\, dt + \sigma\, dW(t).$$

This equation, in order to avoid trouble with division by zero, is usually written as

$$dX(t) = \mu X(t)\, dt + \sigma X(t)\, dW(t). \tag{11.17}$$

A process satisfying Eq. (11.17) is called **geometric Brownian motion**, or **GBM** for short. The constant μ is called **drift coefficient**, and it is related to expected return, whereas the constant σ, the **volatility coefficient**, is related to standard deviation of return. To see this, let us approximate the equation for a small time step δt:

$$\frac{\delta X(t)}{X(t)} \approx \mu\, \delta t + \sigma\, \delta W(t).$$

Since the increment $\delta W(t)$ of the Wiener process has normal distribution $\mathsf{N}(0, \delta t)$, this approximation *might* suggest that the stock return has a normal distribution $\mathsf{N}(\mu\, \delta t, \sigma^2\, \delta t)$. Note once again that the standard deviation, $\sigma\sqrt{\delta t}$, scales with the square root of time. This does not seem satisfactory, as a normal return may lead to negative prices. However, the above approximation is not quite correct and suffers from a discretization error. By exact integration of the corresponding stochastic differential equation, we will show that the GBM process cannot take negative values and that the true distribution of $X(t)$ is *lognormal*.

More generally, we may consider an equation like

$$dX(t) = a\bigl(X(t), t\bigr) dt + b\bigl(X(t), t\bigr) dW(t), \tag{11.18}$$

for given functions $a(\cdot,\cdot)$ and $b(\cdot,\cdot)$. This differential equation defines an **Itô process**. By integration of differentials, we could argue that the solution should be something like

$$X(t) = X(0) + \int_0^t a(X(s),s)\,ds + \int_0^t b(X(\tau,\tau))\,dW(\tau). \qquad (11.19)$$

Here, the first integral looks like a familiar Riemann integral of a function of time. To be precise, this applies to each sample path, which is a function of time for each outcome ω. But what about the second one? We need to assign a precise meaning to it, and this will lead us to the definition of a *stochastic* integral, which we illustrate later. Here we just notice that, unlike the generalized Wiener process, we do not have a numerical constant b, but a function $b(\cdot,\cdot)$ that cannot be moved outside the integral, and we do not have a differential that can be easily integrated.

Leaving a careful theoretical analysis aside, from a practical viewpoint, how can we *compute* a stochastic integral? Analogy with the example in Section 11.3.1 would suggest considering

$$\frac{dX(t)}{X(t)}$$

as the increment of the log-price process $\log X(t)$. If this *were* true, we *could* rewrite the equation defining GBM as

$$d\log X(t) \stackrel{?}{=} \mu\,dt + \sigma\,dW(t),$$

where we use again the notation $\stackrel{?}{=}$ to underline that this is just a possibly wrong conjecture, and not a proven fact. This is a generalized Wiener process in terms of the logarithm of $X(t)$, and straightforward integration would lead us to

$$\log X(t) - \log X(0) = \log \frac{X(t)}{X(0)} \stackrel{?}{=} \mu t + \sigma\big(W(t) - W(0)\big),$$

or

$$X(t) \stackrel{?}{=} X(0)\cdot e^{\mu t + \sigma W(t)}.$$

Given the properties of the standard Wiener process, we could rewrite the solution as

$$X(t) \stackrel{?}{=} X(0)\cdot e^{\mu t + \sigma\sqrt{t}\,\epsilon},$$

where ϵ is standard normal. Hence, $X(t)$ would be the exponential of a normal random variable, i.e., a lognormal random variable, which cannot take negative values. The solution looks like a simple and quite sensible generalization of Eq. (11.13), so that sample paths look like an exponential growth, on which Gaussian noise is superimposed.

Unfortunately, all of this relies on the application of calculus rules, specifically, the chain rule to take derivatives or differentials of composite functions. If we have a deterministic function $f(t)$ of time, provided that it is strictly positive and suitably well-behaved, we may apply the chain rule and write

$$\frac{d\log f(t)}{dt} = \frac{1}{f(t)} \cdot \frac{df(t)}{dt}.$$

However, there is really no guarantee that we may apply this idea to differentials of stochastic processes. In fact, as we shall see, all of the above is not terribly far from truth, but it is wrong nevertheless. The usual rules of deterministic calculus do not work in a stochastic calculus, and we need a more careful treatment of stochastic integrals.

11.4 Stochastic integration and Itô's lemma

The stochastic differential equation of Eq. (11.18) defines the Itô process $X(t)$, driven by the standard Wiener process $W(t)$. We will take for granted that the state $X(t)$ depends only on the history of $W(t)$ over the time interval from 0 to t. Given the past history, i.e., the filtration $\mathcal{F}_t$ generated by $W(t)$ up to time t, we know the value of the random variable $X(t)$. Technically speaking, we say that process $X(t)$ is **adapted** to process $W(t)$.[11] Now, let us consider a stochastic integral like

$$\int_0^T X(t)\, dW(t). \tag{11.20}$$

How can we assign a meaning to this expression?

11.4.1 A DIGRESSION: RIEMANN AND RIEMANN–STIELTJES INTEGRALS

Let us consider the definite integral of a deterministic (and well-behaved) function over a bounded interval,

$$I \doteq \int_a^b f(x)\, dx.$$

The proper way to define the integral starts with the definition of integrals for simple, i.e., piecewise constant functions, which is natural enough as it relies on areas of rectangles. Then, we approximate a more general function by lower and upper bounding simple functions and take a limit. We will pursue a less formal (and rigorous) route, which is just meant to be a useful refresher to grasp the intuition we need.[12] We will also take for granted that the integrand function

---

[11] In other words, the random variable $X(t)$ is measurable with respect to $\mathcal{F}_t$.
[12] See [5] for a clear treatment.

is continuous and all of the integrals we consider exist. The idea is to partition the integration interval $[a, b]$ by choosing a grid of $n + 1$ points,

$$a \equiv x_0 < x_1 < x_2 < \cdots < x_n \equiv b,$$

and defining n intervals $[x_k, x_{k+1}]$, $k = 0, 1, \ldots, n - 1$. Then, we approximate the integral by the sum

$$\widehat{I}_n \doteq \sum_{k=0}^{n-1} f(c_k) \cdot [x_{k+1} - x_k],$$

where we pick a "representative" point $c_k \in [x_k, x_{k+1}]$ for each subinterval in the partition. Then, we would expect that when $n \to +\infty$, i.e., when we take the limit for finer and finer partitions, the approximation $\widehat{I}_n$ will tend to the integral I in some sense, leading to the definition of the classical **Riemann integral**. A natural question, however, is how should we choose the point c_k within each subinterval. We might choose the right endpoint x_{k+1}, or the left endpoint x_k, or maybe the midpoint. When we take finer and finer partitions, we might argue that, if the function is well-behaved, the choice should make no difference, as the function will be approximately constant on a small subinterval. As we shall see, the choice is quite relevant in the case of a stochastic process, but it is not critical for a deterministic and continuous function.

However, the differential $dW(t)$ in the integral of Eq. (11.20) does not only involve an independent variable, like dx. In fact, we may see some similarity with the following extension of the Riemann integral, which is known as **Riemann–Stieltjes integral**:

$$\int_a^b f(x)\, dg(x),$$

where $g(\cdot)$ is called the **integrator function**. Following the above line of reasoning, we might partition the interval $[a, b]$ and use the approximation

$$\sum_{k=0}^{n-1} f(c_k) \cdot \left[g(x_{k+1}) - g(x_k)\right].$$

We notice that this is a sort of *path* integral, as it depends on the path of the integrator function. Again, we assume that the limit for $n \to +\infty$ exists, and that it does not depend on the choice of the representative point c_k.

11.4.2 STOCHASTIC INTEGRAL IN THE SENSE OF ITÔ

Analogy with the Riemann–Stieltjes integral suggests the following approximation of the stochastic integral (11.20):

$$\sum_{k=0}^{n-1} X(t_k) \left[W(t_{k+1}) - W(t_k)\right], \quad (11.21)$$

11.4 Stochastic integration and Itô's lemma

where the integration interval has been partitioned into subintervals defined by points $0 \equiv t_0, t_1, t_2, \ldots, t_n \equiv T$. Again, intuition suggests taking the limit for $n \to +\infty$ and hoping for the best, but there are some crucial points that we must notice.

- The approximation of Eq. (11.21) defines a random variable, not a number. Hence, when taking the limit, we have to rely on a stochastic convergence concept. Given the nature of this book, we steer away from the involved technicalities. As it turns out, convergence in mean square will do, but we should expect that additional conditions are required for convergence.
- What is arguably more important, we should pay attention to how we have chosen the time instants in the expression above: $X(t_k)$ is evaluated at the *left endpoint* of the subinterval (t_k, t_{k+1}). This is actually *one* possible choice: Why not take the value of $X(t)$ at the midpoint or the right extreme of each interval?

The choice of where we evaluate the process $X(t)$ is quite critical as it leads to different stochastic integrals. Here we are considering the stochastic integral **in the sense of Itô**, where the value of the integrand is taken at the left extreme of each interval, as in Eq. (11.21). Choosing the value at the midpoint defines the Stratonovich stochastic integral, which we will not consider in this book. Before discussing the mathematical implications of our choice, we should realize that it is well-motivated by the need to model the actual financial decision-making process.

Example 11.10 Financial motivation of Itô integrals

Consider a set of m assets, whose prices are modeled by stochastic processes $S_i(t)$, $i = 1, \ldots, m$, described by stochastic differential equations like (11.18). Let us assume that we pursue a portfolio strategy represented by functions $h_i(t)$, which give the number of stock shares of each asset i that we hold at time t. But which functions make sense? An obvious requirement is that functions $h_i(\cdot)$ should not be anticipative: $h_i(t)$ may depend on all the history so far, over the interval $[0, t]$, but clairvoyance should be ruled out. Furthermore, we should think of $h_i(t)$ as the number of shares we hold over a time interval of the form $[t, t + dt)$. Note that the interval is half-closed, to point out that we make a decision and rebalance the portfolio at time t; then, we keep the portfolio constant for a while and, at time $t + \delta t$, we will observe the result and make a new decision.

Now, assume that we are endowed with an initial wealth that we have to allocate among the m assets. The initial portfolio value, depending on the portfolio strategy represented by the functions $h_i(\cdot)$,

is
$$V_\mathbf{h}(0) = \sum_{i=1}^{m} h_i(0)S_i(0) = \mathbf{h}^\mathsf{T}(0)\mathbf{S}(0),$$

where we have collected the processes $h_i(t)$ and $S_i(t)$ into column vectors $\mathbf{h}$ and $\mathbf{S}$, respectively. What about the dynamics of the portfolio value? If the portfolio is self-financing, i.e., we can trade assets at any time, but we do not invest any additional cash after $t = 0$, and we never withdraw funds for consumption purposes, it can be shown that the portfolio value will satisfy the equation

$$dV_\mathbf{h}(t) = \sum_{i=1}^{m} h_i(t)\,dS_i(t) = \mathbf{h}^\mathsf{T}(t)\,d\mathbf{S}(t).$$

This looks fairly intuitive and convincing, but some careful analysis is needed to prove it (see, e.g., [1, Chapter 6]). Then, we may reasonably guess that the wealth at time $t = T$ will be

$$V_\mathbf{h}(T) = V_\mathbf{h}(0) + \int_0^T \mathbf{h}^\mathsf{T}(t)\,d\mathbf{S}(t).$$

However, it is fundamental to interpret the stochastic integral as the limit of an approximation like (11.21), i.e.,

$$\int_0^T \mathbf{h}^\mathsf{T}(t)\,d\mathbf{S}(t) \approx \sum_{k=0}^{n-1} \mathbf{h}^\mathsf{T}(t_k)\left[\mathbf{S}(t_{k+1}) - \mathbf{S}(t_k)\right].$$

The number of stock shares that we hold at time t_k does *not* depend on future prices $\mathbf{S}(t_{k+1})$. First, we allocate wealth at time t_k, and *then* we observe return over the time interval $[t_k, t_{k+1})$. This makes financial sense and is why Itô stochastic integrals are used for financial modeling.

Example 11.10 shows that, in a dynamic and stochastic setting, we have to pay due attention to the actual flow of information and decisions, whereas the choice of the representative point in each subinterval does not seem so critical in a deterministic setting. The choice has an important effect:

In Eq. (11.21), the random variable $X(t_k)$ and the random increment $\left[W(t_{k+1}) - W(t_k)\right]$ by which it is multiplied are independent.

Let us explore the consequences of this fact. To begin with, what is the expected value of the integral in Eq. (11.20)? This is a good starting question, because the stochastic integral is a random variable, and the first and foremost feature of

11.4 Stochastic integration and Itô's lemma

a random variable is its expected value. We may get a clue by considering the approximation (11.21) again and taking expectations:

$$\mathrm{E}\left[\int_0^T X(t)\,dW(t)\right] \approx \mathrm{E}\left\{\sum_{k=0}^{n-1} X(t_k)\left[W(t_{k+1}) - W(t_k)\right]\right\}$$

$$= \sum_{k=0}^{n-1} \mathrm{E}\left\{X(t_k)\left[W(t_{k+1}) - W(t_k)\right]\right\}$$

$$= \sum_{k=0}^{n-1} \mathrm{E}\left[X(t_k)\right] \cdot \mathrm{E}\left[W(t_{k+1}) - W(t_k)\right] = 0, \quad (11.22)$$

where we have used the independence between $X(t_k)$ and the increments of the Wiener process over the time interval $[t_k, t_{k+1}]$, along with the fact that the expected value of these increments is zero. This shows that the integral of Eq. (11.20) is a random variable with expected value zero, but can we say something more? One immediate consequence of Eq. (11.22) is the following.

THEOREM 11.1 *Let us consider a stochastic process $X(t)$, adapted[13] to the standard Wiener process $W(t)$. Then, the stochastic process*

$$\mathcal{I}(t) \doteq \int_0^t X(s)\,dW(s),$$

defined by a stochastic integral in the sense of Itô, is a martingale.

PROOF The proof is quite simple and it relies on the additivity property of integrals, which applies to stochastic integrals as well:

$$\int_0^t X(s)\,dW(s) = \int_0^\tau X(s)\,dW(s) + \int_\tau^t X(s)\,dW(s), \quad (11.23)$$

for $\tau < t$. If we define the stochastic process

$$\mathcal{I}(t) \doteq \int_0^t X(s)\,dW(s),$$

we may rewrite Eq. (11.23) as

$$\mathcal{I}(t) = \mathcal{I}(\tau) + \int_\tau^t X(s)\,dW(s).$$

Taking the expectation conditional on the filtration $\mathcal{F}_\tau$, i.e., given the history up to time τ, we have

$$\mathrm{E}\big[\mathcal{I}(t)\,|\,\mathcal{F}_\tau\big] = \mathrm{E}\big[\mathcal{I}(\tau)\,|\,\mathcal{F}_\tau\big] + \mathrm{E}\left[\int_\tau^t X(s)\,dW(s)\,\Big|\,\mathcal{F}_\tau\right] = \mathcal{I}(\tau) + 0,$$

---

[13] Formally, "adapted" means that the process $X(t)$ is measurable with respect to the filtration generated by $W(t)$. In plain English, it means that the value of X at time t depends only on the history of W on the interval $[0, t]$. To be concrete, we may interpret X as the price process of a stock share, depending on the path of an underlying risk factor W.

which proves the result. ∎

Note that we are not claiming that *any* stochastic integral is a martingale. For instance, the stochastic integral in Example 11.10 does not involve the differential of a standard Wiener process, which is itself a martingale. Hence, the wealth process $V_h(t)$ need not be a martingale. Another useful result is the following theorem, which we state without proof.

THEOREM 11.2 (Itô's isometry) *Let X_t be a stochastic process adapted to the standard Wiener process W_t. Then,*

$$\mathrm{E}\left[\left(\int_0^t X_\tau\, dW_\tau\right)^2\right] = \mathrm{E}\left[\int_0^t X_\tau^2\, d\tau\right].$$

We shall use Itô's isometry in Section 14.2.1, in order to find the variance of a random interest rate at time t.

The definition of stochastic integral does not yield a precise way to compute it practically. We may try, however, to consider a specific case to build some intuition. The following example illustrates one nasty consequence of the way the Itô stochastic integral is defined.

Example 11.11 Chain rule and stochastic differentials

Say that we want to "compute" the stochastic integral

$$\int_0^T W(t)\, dW(t).$$

Analogy with ordinary calculus would suggest using the chain rule for the differentiation of composite functions, in order to obtain a differential that can be integrated directly. Specifically, we *might guess* that

$$dW^2(t) \stackrel{?}{=} 2W(t)\, dW(t),$$

where we use $\stackrel{?}{=}$ again to underline that we are just making guesses. This in turn *would* suggest that

$$\int_0^T W(t)\, dW(t) \stackrel{?}{=} \frac{1}{2}\int_0^T dW^2(t) = \frac{1}{2}W^2(T).$$

Unfortunately, this *cannot* be the correct answer, as it contradicts our previous findings. We have just seen that the expected value of an

11.4 Stochastic integration and Itô's lemma

> integral of this kind is zero, but
>
> $$\mathrm{E}\left[\frac{1}{2}W^2(T)\right] = \frac{1}{2}\mathrm{E}\left[W^2(T)\right] = \frac{1}{2}\left\{\mathrm{Var}\left[W(T)\right] + \mathrm{E}^2\left[W(T)\right]\right\}$$
> $$= \frac{T}{2} \neq 0. \qquad (11.24)$$
>
> We see that the two expected values do not match at all, and there must be something wrong somewhere.

Example 11.11 shows that the chain rule does not work in Itô stochastic calculus. Hence, the guess we made in Section 11.3.3, about geometric Brownian motion, is quite likely wrong, too. To proceed further, we need to find the right rule to work with stochastic differentials, and the answer is provided by Itô's lemma.

Remark. It is interesting to note that we may choose a different definition of stochastic integral, whereby we take an average of the values of process $X(t)$ at the endpoints of each subinterval $[t_k, t_{k+1}]$. As we said, this leads to the stochastic integral in the sense of Stratonovich. This concept of integral has its applications in physics, but it is less natural in finance. Nevertheless, it turns out that, with the Stratonovich integral, we do not run into this kind of trouble with the chain rule.

11.4.3 ITÔ'S LEMMA

So far, we have tried guessing the differentials of functions of a stochastic process X_t,[14]

$$dX_t^2 \stackrel{?}{=} 2X_t\, dX_t, \quad d\log X_t \stackrel{?}{=} \frac{dX_t}{X_t},$$

but we have found evidence contrary to their validity. More generally, let us consider a financial option written on an underlying asset, whose price is described by an Itô process following the stochastic differential equation

$$dX_t = a(X_t, t)\, dt + b(X_t, t)\, dW_t. \qquad (11.25)$$

If the option is European-style, with a payoff depending only on the price X_T at maturity, we may argue that the its price will be a function $f(X_t, t)$ of the underlying asset price and time. How can we find a stochastic differential equation for the option price?

What we need is a chain rule to take differentials of functions of stochastic processes. In ordinary calculus, we use the chain rule to find derivatives

---

[14] For the sake of convenience, here we switch from notation $X(t)$ to X_t.

of composite functions. In stochastic calculus, this role is played by Itô's lemma. Since, in this introductory book, we do not pursue a rigorous approach to stochastic calculus, we may only give an informal, yet quite instructive argument.[15] The argument is instructive, as it provides us with valuable intuition explaining what went wrong in Example 11.11. As a starting point, we may consider Eq. (11.25) as the continuous limit of the discretized equation

$$\delta X_t = X_{t+\delta t} - X_t = a(X_t, t)\,\delta t + b(X_t, t)\epsilon_{t+\delta t}\sqrt{\delta t}, \qquad (11.26)$$

where $\epsilon_{t+\delta t} \sim N(0,1)$.

Let us take a little step back into the realm of deterministic calculus and consider what is the rationale behind the Taylor expansion of a function $g(x,y)$ of two variables. The key ingredient we need for our reasoning is the formula for the differential of such a function:

$$dg = \frac{\partial g}{\partial x}dx + \frac{\partial g}{\partial y}dy,$$

which indeed may be obtained from Taylor expansion:

$$\delta g = \frac{\partial g}{\partial x}\,\delta x + \frac{\partial g}{\partial y}\,\delta y + \frac{1}{2}\frac{\partial^2 g}{\partial x^2}(\delta x)^2 + \frac{1}{2}\frac{\partial^2 g}{\partial y^2}(\delta y)^2 + \frac{\partial^2 g}{\partial x\,\partial y}\,\delta x\,\delta y + \cdots,$$

for $\delta x, \delta y \to 0$. When taking this limit, only the first-order terms are relevant, as the second-order terms are negligible. Now, we may apply this Taylor expansion to the function $f(X_t, t)$ of the stochastic process X_t, limiting it to the leading terms. In doing so, it is important to notice that the term $\sqrt{\delta t}$ in Eq. (11.26) needs careful treatment when squared. In fact, streamlining notation a bit, we deal with something like

$$(\delta X)^2 = a(X,t)^2(\delta t)^2 + 2a(X,t)b(X,t)\epsilon(\delta t)^{3/2} + b(X,t)^2\epsilon^2\delta t$$
$$\approx b(X,t)^2\epsilon^2\delta t,$$

which implies that the term in $(\delta X)^2$ cannot be neglected in the approximation. In fact, when we square $\sqrt{\delta t}$, we get a *first-order* term, which must be accounted for in the differential, whereas higher-order terms in $(\delta t)^{3/2}$ and $(\delta t)^2$ are neglected.

Since ϵ is a standard normal variable, we have $E[\epsilon^2] = 1$ and $E[\epsilon^2\delta t] = \delta t$. A delicate point is the following: It can be shown that, as δt tends to zero, the term $\epsilon^2\,\delta t$ can be treated as nonstochastic, and it is equal to its expected value. An informal (far from rigorous) justification relies on the variance of this term:

$$\begin{aligned}\text{Var}(\epsilon^2\,\delta t) &= (\delta t)^2\left\{E[\epsilon^4] - E^2[\epsilon^2]\right\}\\ &= (\delta t)^2\left\{3 - \text{Var}^2[\epsilon]\right\}\\ &= (\delta t)^2\{3 - 1\} = 2(\delta t)^2,\end{aligned}$$

---

[15] We follow the treatment in [12, Chapter 13].

11.4 Stochastic integration and Itô's lemma

which, for $\delta t \to 0$, can be neglected with respect to first-order terms. Here we have used the fact $E[\epsilon^4] = 3$, which can be checked by using moment generating functions or, perhaps cheating a bit, by recalling that the kurtosis of any normal, including a standard one, is 3. A useful way to remember this point is the *formal* rule

$$(dW)^2 = dt. \tag{11.27}$$

Hence, when δt tends to zero, in the Taylor expansion, we have

$$(\delta X)^2 \to b(X,t)^2 \, dt.$$

Neglecting higher-order terms and taking the limit, as both δX and δt tend to zero, we end up with

$$df = \frac{\partial f}{\partial X} dX + \frac{\partial f}{\partial t} dt + \frac{1}{2} \frac{\partial^2 f}{\partial X^2} b(X,t)^2 \, dt,$$

which, by replacing dX with Eq. (11.25), becomes the celebrated **Itô's lemma**:

$$df(X_t, t) = \left[a(X_t, t) \frac{\partial f}{\partial X_t} + \frac{\partial f}{\partial t} + \frac{1}{2} b(X_t, t)^2 \frac{\partial^2 f}{\partial X_t^2} \right] dt + b(X_t, t) \frac{\partial f}{\partial X_t} dW_t. \tag{11.28}$$

Although this proof is far from rigorous, we see that all the trouble is due to the term of order $\sqrt{\delta t}$, which is introduced by the increment of the Wiener process.

It is instructive to see that if we set $b(X, t) = 0$ in the Itô differential equation, i.e., if we eliminate randomness, then we step back to familiar ground. In such a case, the stochastic differential equation is actually deterministic and can be rewritten as

$$dx = a(x, t) dt \quad \Rightarrow \quad \frac{dx}{dt} = a(x, t),$$

where we use the notation x rather than X to point out the deterministic nature of function $x(t)$. Then, Itô's lemma boils down to

$$df = a(x, t) \frac{\partial f}{\partial x} dt + \frac{\partial f}{\partial t} dt, \tag{11.29}$$

which can be rearranged to yield the following chain rule for a function $f(x, t)$:

$$\frac{df}{dt} = \frac{\partial f}{\partial x} \frac{dx}{dt} + \frac{\partial f}{\partial t}. \tag{11.30}$$

As a first application of Itô's lemma, let us consider Example 11.11 once more.

Example 11.12 Applying Itô's lemma to Example 11.11

In order to apply Itô's lemma to the computation of the stochastic integral

$$\int_0^T W_t \, dW_t,$$

we have to recast the problem a bit. Here we have $X_t \equiv W_t$, which implies $a(X_t, t) \equiv 0$ and $b(X_t, t) \equiv 1$. Furthermore, since our guess involved the differential of W_t^2, we may look for the differential of the function $f(X_t, t) = X_t^2$. To apply Itô's lemma, we need the following partial derivatives:

$$\frac{\partial f}{\partial t} = 0, \qquad (11.31)$$

$$\frac{\partial f}{\partial X_t} = 2X_t,$$

$$\frac{\partial^2 f}{\partial X_t^2} = 2.$$

These derivatives are trivial, but it is important to point out that the partial derivative with respect to time in Eq. (11.31) is zero. It is true that $f(X_t, t)$ depends on time through X_t, but here we have no direct dependence on t; thus, the *partial* derivative with respect to time vanishes. To put it another way, the dependence of X_t with respect to time does not play any role, when taking the partial derivative with respect to t, because X_t is held fixed. We may also have another look at Eq. (11.30), to see the relationship between the partial and the total derivative of $f(X_t, t)$ with respect to t.

Putting everything together, the application of Itô's lemma yields

$$df = d(W_t^2) = dt + 2W_t\, dW_t. \qquad (11.32)$$

It is essential to notice that dt is exactly the term that we would *not* expect by applying the usual chain rule. But this is the term that allows us to get the correct expected value of W_t^2. Indeed, by straightforward integration of Eq. (11.32), now we find

$$W_T^2 = W_0^2 + \int_0^T d(W_t^2)$$

$$= 0 + \int_0^T dt + 2\int_0^T W_t\, dW_t.$$

Hence, the required integral is

$$\int_0^T W_t\, dW_t = \frac{W_T^2}{2} - \frac{T}{2}.$$

By taking expectations and recalling Eq. (11.24), we find

$$\mathrm{E}\left[\int_0^T W_t\, dW_t\right] = \frac{\mathrm{E}[W_T^2]}{2} - \frac{T}{2} = 0,$$

> which is consistent with Eq. (11.22). Essentially, by subtracting a compensating factor $T/2$, we make this stochastic integral a martingale.

Let us summarize our findings: With respect to Eq. (11.29), Itô's lemma includes an extra term in dW, which is expected, given the form of the differential equation defining the stochastic process $X(t)$, but also an unexpected term

$$\frac{1}{2}b^2 \frac{\partial^2 f}{\partial X^2}.$$

In deterministic calculus, second-order derivatives occur in second-order terms linked to $(\delta t)^2$, which can be neglected. On the contrary, here we have a term of order $\sqrt{dt}$ which must be taken into account even though it is squared.

11.5 Stochastic processes in financial modeling

A wide class of models for financially relevant prices and risk factors may be built by choosing suitable functional forms for the drift function $a(\cdot,\cdot)$ and the volatility function $b(\cdot,\cdot)$ in the equation

$$dX_t = a(X_t, t)\, dt + b(X_t, t)\, dW_t.$$

This stochastic differential equation, actually, should be intended as a shorthand notation for the integral form of Eq. (11.19). Indeed, strictly speaking, stochastic differential equations make no sense, and only stochastic integral equations do. However, the differential form is much more manageable and does not obscure the financial intuition behind modeling choices. In this section, we describe some models that have been proposed to represent the uncertain dynamics in prices, interest rates, and other risk factors. We start with geometric Brownian motion, which is the foundation of the Black–Scholes–Merton option pricing model. Then, we will hint at some generalizations, including the possibility of adding a jump component.

11.5.1 GEOMETRIC BROWNIAN MOTION

Geometric Brownian motion is defined by a specific choice of the functions in Itô stochastic differential equation:

$$dS_t = \mu S_t\, dt + \sigma S_t\, dW_t,$$

where μ and σ are constant parameters referred to as drift and volatility coefficients, respectively. Intuition would suggest to rewrite the equation as

$$\frac{dS_t}{S_t} = \mu\, dt + \sigma\, dW_t,$$

and then consider this as the differential of log-price, $d \log S$, and integrate it directly. After all, this worked well for the deterministic bank account process in Section 11.3.1. However. we have learned that additional care is needed and we have to resort to Itô's lemma, in order to find the stochastic differential equation for $f(S_t, t) = \log S_t$. As a first step, we compute the partial derivatives:

$$\frac{\partial f}{\partial t} = 0, \qquad \frac{\partial f}{\partial S_t} = \frac{1}{S_t}, \qquad \frac{\partial^2 f}{\partial S_t^2} = -\frac{1}{S_t^2}.$$

Once again, we note that there is no direct dependence on time, so that the partial derivative with respect to t is zero. Then, putting all of it together, we find

$$\begin{aligned} df &= \left(\frac{\partial f}{\partial t} + \mu S_t \cdot \frac{\partial f}{\partial S_t} + \frac{1}{2}\sigma^2 S_t^2 \cdot \frac{\partial^2 f}{\partial S_t^2} \right) dt + \sigma S_t \frac{\partial f}{\partial S_t}\, dW_t \\ &= \left(\mu - \frac{\sigma^2}{2} \right) dt + \sigma\, dW_t. \end{aligned}$$

Now we see that our guess was not that bad, as this equation may be easily integrated and yields

$$\log S_t = \log S_0 + \left(\mu - \frac{\sigma^2}{2} \right) t + \sigma W_t.$$

Recalling that W_t has a normal distribution and can be written as $W_t = \epsilon\sqrt{t}$, where $\epsilon \sim N(0, 1)$, we conclude that the logarithm of price (log-price) is normally distributed:

$$\log S_t \sim N\left[\log S_0 + \left(\mu - \frac{\sigma^2}{2} \right) t,\ \sigma^2 t \right].$$

We can rewrite the solution in terms of S_t:

$$S_t = S_0 \cdot \exp\left\{ \left(\mu - \frac{\sigma^2}{2} \right) t + \sigma W_t \right\},$$

or

$$S_t = S_0 \cdot \exp\left\{ \left(\mu - \frac{\sigma^2}{2} \right) t + \sigma\sqrt{t}\epsilon \right\}. \tag{11.33}$$

On the one hand, this shows that prices, according to the geometric Brownian motion model, are lognormally distributed and cannot take negative values. On the other hand, now we have a way to generate sample paths, based on a discretization of a prescribed time horizon into time periods of length δt, like we

11.5 Stochastic processes in financial modeling

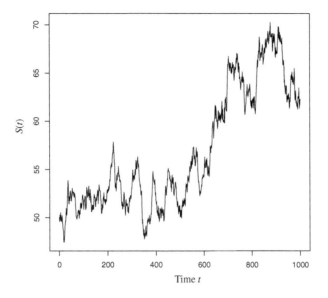

FIGURE 11.8 A sample path of geometric Brownian motion.

did with the generalized Wiener process.[16] This generates sample paths like the one depicted in Fig. 11.8.

Remark. We notice that sample paths of GBM essentially results from superimposing Gaussian noise to an exponential trendline. Then, one could wonder why stock share prices do not seem to grow indefinitely like an exponential function. Among other reasons, we should consider the fact that when dividends are paid, a down jump is expected, which limits the price growth.

By using properties of the lognormal distribution, we also find

$$E\left[\log \frac{S_t}{S_0}\right] = \left(\mu - \frac{\sigma^2}{2}\right)t,$$

$$\text{Var}\left[\log \frac{S_t}{S_0}\right] = \sigma^2 t,$$

$$E\left[\frac{S_t}{S_0}\right] = e^{\mu t}, \quad (11.34)$$

$$\text{Var}\left[\frac{S_t}{S_0}\right] = e^{2\mu t}(e^{\sigma^2 t} - 1), \quad (11.35)$$

from which we see that the drift parameter μ is linked to the continuously compounded return. The volatility parameter σ is related to standard deviation of the increment of logarithm of price. The roles of drift and volatility can also be

---

[16] See Example 11.9.

grasped intuitively by considering the following approximation of the differential equation:

$$\frac{\delta S_t}{S_t} \approx \mu\,\delta t + \sigma\,\delta W_t,$$

where $\delta S_t/S_t$ is the return of the asset over small time interval δt. According to this approximation, we see that return can be approximated by a normal variable with expected value $\mu\,\delta t$ and standard deviation $\sigma\sqrt{\delta t}$. Actually, this normal distribution is only a local approximation of the "true" (according to the model) lognormal distribution.

Remark. Equation (11.35) looks a bit intimidating, but it is related to the variance of a lognormal variable and may be obtained by resorting to moment generating functions. Equation (11.34), on the contrary, looks deceptively simple but it is often subject to a misunderstanding. Let us consider a lognormal variable $Y = e^X$, where $X \sim \mathsf{N}(\nu, \xi^2)$. We have to keep very clear in mind that the expected value of a function need not be the function of the expected value. In this specific case,

$$\mathrm{E}[e^X] = e^{\nu + \xi^2/2} \geq e^\nu = e^{\mathrm{E}[X]}.$$

This is a consequence of Jensen's inequality for convex functions like the exponential. To really grasp where Eq. (11.34) comes from, we have to go through the correct calculation:

$$\mathrm{E}[S_t] = S_0 \cdot \mathrm{E}\left[\exp\left\{\left(\mu - \frac{\sigma^2}{2}\right)t + \sigma\sqrt{t}\,\epsilon\right\}\right]$$
$$= S_0 \cdot \exp\left\{\mu t - \frac{\sigma^2 t}{2} - \frac{(\sigma\sqrt{t})^2}{2}\right\} = S_0 \cdot \exp\{\mu t\}.$$

11.5.2 GENERALIZATIONS

GBM is a simple diffusion process, admitting an analytical solution, but it is arguably not adequate to describe the dynamics of real-life risk factors. Nevertheless, there are different ways to extend GBM in order to build more realistic models.

Multidimensional extensions. When we have to cope with multiple risk factors, like different stock share prices or interest rates with different maturities, we need a system of differential equations. This raises the issue of accounting for correlations among risk factors.

Changing drift and volatility functions. The mean value function of GBM is $\mathrm{E}[S_t] = S_0 \cdot e^{\mu t}$. This exponential trendline makes no sense for interest rates or volatilities. By replacing the drift and the volatility functions in a diffusion process, we may model more realistic behavior.

Going beyond pure diffusions. In GBM, the diffusion component is given by the standard Wiener process, a Gaussian process with continuous sample paths. This component may be replaced in order to introduce heavier tails or jumps.

11.5.2.1 Correlated Wiener processes

The need to model the dynamics of a set of asset prices arises when we deal with a portfolio, or when we want to price a rainbow option depending on multiple underlying assets. The simplest model that we may adopt is a direct generalization of GBM. According to this approach, the prices $S_i(t)$ of n assets satisfy the equations

$$dS_i(t) = \mu_i S_i(t)\,dt + \sigma_i S_i(t)\,dW_i(t), \qquad i = 1,\ldots,n,$$

where the standard Wiener processes $W_i(t)$ are not necessarily independent. They are characterized by a set of instantaneous correlation coefficients ρ_{ij}, whose meaning can be grasped by an extension of the formal rule of Eq. (11.27):

$$dW_i \cdot dW_j = \rho_{ij}\,dt.$$

In terms of path generation, we require the generation of increments

$$\delta W_i = \epsilon_i \sqrt{\delta t}, \qquad \delta W_j = \epsilon_j \sqrt{\delta t},$$

where ϵ_i and ϵ_j are standard normals with correlation ρ_{ij}.

11.5.2.2 Ornstein–Uhlenbeck processes

This exponential growth of GBM makes no sense for several financial variables, which rather feature mean reversion. Then, we may resort to an Ornstein–Uhlenbeck process,

$$dX(t) = \gamma\bigl[\bar{X} - X(t)\bigr]dt + \sigma\,dW(t).$$

This is still a Gaussian diffusion, but the drift can change sign, pulling the process back to a long-run average $\bar{X}$, with mean reversion speed γ. The application of this model to a short-term interest rate $r(t)$ yields the **Vasicek model**:

$$dr(t) = \gamma\bigl[\bar{r} - r(t)\bigr]dt + \sigma\,dW(t).$$

11.5.2.3 Square-root diffusions

One drawback of Ornstein–Uhlenbeck processes is that they do not prevent negative values, which make no sense for stock prices and interest rates. A possible adjustment is the following:

$$dr(t) = \gamma\bigl[\bar{r} - r(t)\bigr]dt + \sigma\sqrt{r(t)}\,dW(t).$$

This is an example of a square-root diffusion, which in the context of short-term interest rates is known as the **Cox–Ingersoll–Ross model**. For a suitable choice of parameters, it can be shown that a square-root diffusion stays non-negative.

This changes the nature of the process, which is not Gaussian anymore.[17] Similar considerations hold when modeling a stochastic and time-varying volatility $\sigma(t)$. GBM assumes a constant volatility σ, whereas, in practice, we may observe time periods in which volatility is higher than usual. Square-root diffusions are used in a common stochastic volatility model, the **Heston model**, which consists of a pair of stochastic differential equations:

$$dS(t) = \mu S(t)\, dt + \sigma(t) S(t)\, dW_1(t),$$
$$dV(t) = \alpha \big[\bar{V} - V(t)\big]\, dt + \xi \sqrt{V(t)}\, dW_2(t),$$

where $V(t) = \sigma^2(t)$ is the variance process, $\bar{V}$ is its long-term average, and different assumptions can be made about the correlation between the two driving Wiener processes $W_1(t)$ and $W_2(t)$.

11.5.2.4 Jump–diffusions

In order to account for jumps, we may devise processes with both a diffusion and a jump component, such as

$$X(t) = \alpha t + \sigma W(t) + Y(t),$$

where $Y(t)$ is a compound Poisson process. From a formal point of view, we require the definition of a stochastic integral of a stochastic process $Y(t)$ with respect to a Poisson process $N(t)$:

$$\int_0^t Y(\tau)\, dN(\tau) = \sum_{i=1}^{N(t)} Y(t_i),$$

where $N(t)$ is the number of jumps occurred up to time t in the Poisson process, and t_i, $i = 1, \ldots, N(t)$, are the time instants at which jumps occur. We note that the Wiener and the Poisson process, which are the basic building blocks of jump–diffusions, are quite different, but they do have an important common feature: stationary and independent increments. In fact, the class of Lévy processes generalizes both of them by emphasizing this feature. A more radical departure from GBM is represented by *pure* jump processes. The rationale behind these models is that, at the market microstructure level, stock share prices do move by jumps, depending on the flow of buy/sell orders and the dynamics of the limit order book.

11.6 Sample path generation

In this introductory book, we do not discuss in depth numerical solution methods for optimization or pricing problems in finance. However, a cursory overview

---

[17] See Section 14.2.2.

of how continuous-time, continuous-state dynamic models are translated into a a discrete-time, discrete-state approximation amenable to computational techniques is useful. There are different kinds of discretized representations, including:

- **Linear scenarios**, i.e., independent sample path realizations that share only the initial state.
- **Scenario trees**, which are better suited to optimization modeling, as they represent the limited information on which we base multistage decisions. The main issue with trees is that they grow exponentially in size.
- **Recombining lattices**, which are similar to trees in some respect, but grow linearly in size. We will explore binomial lattices in the context of option pricing. A limitation of recombining lattices is that they do not allow to cope with path dependency.
- **Markov chains** and **stochastic meshes**, which we shall not discuss.

There is a wide array of methods that can be used to generate discretized representations, but they may be divided into two broad classes:

- Random sampling, which is the foundation of ubiquitous Monte Carlo methods. Conceptually, random sampling relies on the law of large numbers to estimate a quantity of interest, like an expected value, a probability, or a quantile.
- Deterministic methods, which aim at finding a discretization that is optimal in some sense. One possible idea is moment matching, where we generate a discrete distribution matching essential properties of a target continuous distribution. Since expectations are integrals, we may also borrow ideas from numerical integration, most notably clever Gaussian quadrature formulas. Another approach relies on low-discrepancy sequences, which yield a hybrid between Monte Carlo and deterministic methods, sometimes nicknamed quasi-Monte Carlo sampling.

All of these approaches have advantages and disadvantages, and the choice among them depends on the application at hand, as well as the problem size and the computational budget that we may afford. Hybrid strategies, based on the integration of the above ideas, have also been proposed.

11.6.1 MONTE CARLO SAMPLING

We have seen a hint of how Monte Carlo sampling may be used to generate sample paths of a stochastic process in Example 11.9. There, we have dealt with a generalized Wiener process and no difficulty was involved. In general, things are not that easy. To see why, let us consider an Itô process described by the equation

$$dX_t = a(X_t, t)dt + b(X_t, t)dW_t.$$

A natural discretization with time step δt is

$$X_{t+\delta t} - X_t = a(X_t, t)\delta t + b(X_t, t)\sqrt{\delta t} \cdot \epsilon,$$

where ϵ is an element of a sequence of i.i.d. standard normals. This simple scheme is known as the **Euler discretization scheme**, and it involves an obvious approximation: The drift and volatility functions, $a(\cdot, \cdot)$ and $b(\cdot, \cdot)$, are kept constant over the time step, whereas they change continuously. This results in a **discretization error**, which is a common issue in numerical methods for solving deterministic differential equations, too.

Example 11.13 Euler discretization of a GBM

Given the GBM

$$dS_t = \mu S_t \, dt + \sigma S_t \, dW_t,$$

the Euler scheme yields

$$\delta S_t = \mu S_t \, \delta t + \sigma S_t \, \delta W_t.$$

The resulting discretized process is

$$S_{t+\delta t} = (1 + \mu)S_t + \sigma S_t \sqrt{\delta t} \cdot \epsilon_{t+\delta}.$$

An obvious issue with this discretization is that it generates the wrong distribution. GBM is supposed to generate lognormal variables, but here we are generating normal (possibly negative) random variables.

The difficulties of Euler discretization may be partially eased by choosing a suitably small step, but this implies a corresponding increase in computational effort. The GBM is a lucky case, as Itô's lemma provides the exact discretization of Eq. (11.33),

$$S_{t+\delta t} = S_t e^{(\mu - \sigma^2/2)\delta t + \sigma\sqrt{\delta t} \cdot \epsilon_{t+\delta}}.$$

In other cases, the matter is a bit thornier, but sometimes we do find an exact discretization. A relevant example is a square-root diffusion, which requires sampling noncentral chi-square random variables, rather than standard normals. Given the wide range of methods to sample from probability distributions, this is easily accomplished. In other cases, the discretization error cannot be avoided, but it can be kept under control by either taking small steps in a simple discretization scheme, or by resorting to more accurate discretization schemes. Beside discretization issues, there is a more general difficulty with Monte Carlo methods, the **sampling error**, which is of a statistical nature. This is well-known from elementary inferential statistics, where we sample a sequence of n

11.6 Sample path generation

i.i.d. observations X_k, $k = 1, \ldots, n$, and use the sample mean,

$$\overline{X} \doteq \frac{1}{n} \sum_{k=1}^{n} X_k,$$

as an estimator of an unknown expected value μ. In the case of a large sample from a normal population, we find the confidence interval

$$\overline{X} \pm z_{1-\alpha/2} \cdot \frac{S}{\sqrt{n}},$$

where the sample standard deviation S is used to estimate the standard deviation of the sample mean,

$$\sigma_{\overline{X}} = \frac{\sigma_X}{\sqrt{n}}. \tag{11.36}$$

In finance, the sample size in a Monte Carlo simulation is large enough to warrant use of the quantile $z_{1-\alpha/2}$ of standard normal distribution. Equation (11.36) is a fundamental key to understand both the strength and the weakness of Monte Carlo methods, as it features both bad and good news. Good news is that, at least in principle, whatever the problem dimensionality is, the estimation uncertainty, relating to sampling error, goes to zero in the same way, depending on σ_X. To be fair, we could argue that both the computational effort and the sampling variability are likely to be affected by problem dimensionality, but this does not appear explicitly. The bad news, unfortunately, is that uncertainty is reduced at a slower and slower rate, when the sample size n is increased, because of the concavity of the square root function. To get a feeling for this, imagine that we want to reduce the width of the confidence interval by a factor $1/10$, which means gaining one order of magnitude in terms of precision. Then, because of the square root, the sample size should be multiplied by a factor 100, not 10. Linear behavior would be much preferred, and it can be attained by quasi-Monte Carlo methods, where, unfortunately, the problem dimensionality does play a role, limiting the applicability of the approach to intermediate problem dimensionalities. In the literature, however, there is an array of variance reduction strategies, where clever sampling is applied to reduce σ_X. All of this is beyond the scope of this book, and we refer the interested reader to the end-of-chapter bibliography.

11.6.2 SCENARIO TREES

If we have to approximate a continuous distribution by a discrete one, we may visualize the result by the scenario fan in Fig. 1.2. This is sufficient for a static decision problem under uncertainty, but when a multistage, dynamic problem must be tackled, we have to resort to a scenario tree, like the one in Fig. 1.3, which we repeat here in Fig. 11.9, for the sake of convenience. The main advantage of a scenario tree is that it allows to express nonanticipativity in dynamic decisions, i.e., the fact that we are not allowed to foresee the future. At

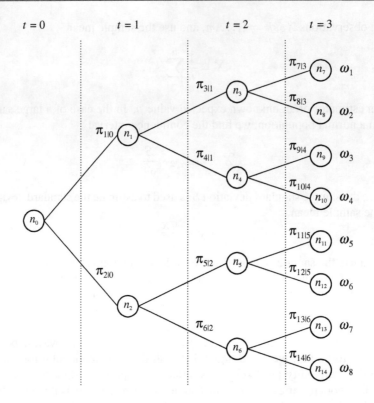

FIGURE 11.9 Uncertainty unfolds over time in a scenario tree.

node n_0, we have to make a single, here-and-now decision, without knowing which scenario will unfold. From the viewpoint of state n_0, the decision that we will make at time $t = 1$ is random, since it is contingent on the realization of random data. Probably, the decision in node n_1 will be different from the decision at node n_2. If we are at node n_1, we know that one of the scenarios in the set $\{\omega_1, \omega_2, \omega_3, \omega_4\}$ is going to occur, but we do not know which one. Hence, at node n_1 the decision will be the same for all of the scenarios in the set $\{\omega_1, \omega_2, \omega_3, \omega_4\}$, which at time $t = 1$ cannot be distinguished yet. When time goes on, we gather more information. For instance, if we are at node n_3, we know that one of the scenarios in the smaller set $\{\omega_1, \omega_2\}$ is going to occur. As we show in Supplement S11.1, this flow of information may be properly formalized by the introduction of filtrations. We will also see how scenario trees may be used in the challenging stochastic optimization models discussed in Chapter 15.

The issue now is: How can we *generate* a scenario tree? One possible answer is by Monte Carlo sampling, but we have observed that sampling error may be significant for a small sample size. As we shall see, if we have to price a European-style option, we have just to estimate the expected value of the payoff, under a risk-adjusted probability measure. Since there is no decision to be made along the way, we just have to generate a set of independent scenarios, evaluate the option payoff for each scenario, and then take a sample mean. We may easily afford a large number of scenarios, and this kind of computation lends itself to parallel processing. However, solving a multistage optimization problem is much more challenging: We need a scenario tree to represent nonanticipative decisions, and a tree is prone to exponential growth in size. If, in Fig. 11.9, we use a branching factor 100, rather than 2, we have 100 nodes at time 1, 10,000 nodes at time 2, and one million nodes at time 3. How can we represent uncertainty accurately without resorting to a huge tree? One possible answer is deterministic scenario generation, rather than brute force sampling. Let us illustrate one such strategy, **moment matching**, with a simple example.

Example 11.14 Moment matching

> Given a normal random variable $X \sim \mathsf{N}(\mu, \sigma^2)$, how can we approximate it by a discrete distribution with only two realizations x_1 and x_2?
>
> To begin with, given the symmetry of the normal distribution, a natural choice is to take two equiprobable realizations, $\mu \pm \delta$, symmetric with respect to the expected value. To find the displacement δ, we may match variance, which is equivalent to matching the second order moment:
>
> $$\mathrm{E}[X^2] = \tfrac{1}{2}x_1^2 + \tfrac{1}{2}x_2^2 = \tfrac{1}{2}(\mu+\delta)^2 + \tfrac{1}{2}(\mu-\delta)^2 = \sigma^2 + \mu^2,$$
>
> which boils down to $\delta = \sigma$. Note that, by symmetry, we also match skewness, i.e., the third-order central moment. We leave it as an exercise to prove that if we add a third point $x_3 = \mu$, we find $\delta = \sqrt{3/2} \cdot \sigma$. With five points, we may also afford to match the fourth-order moment, kurtosis, which is 3 for any normal variable.

Moment matching with only two realizations may sound overly silly, but we will appreciate it when dealing with option pricing in Section 13.3.2. This strategy is also related with deterministic numerical integration strategies like Gaussian quadrature. Unfortunately, when dealing with multivariate distributions, moment matching may not be quite up to the task, as the number of descendant nodes that must be branched out of a parent node may be too large. Furthermore, critics of the approach stress the fact that quite different distributions may share the first few moments, and advocate more sophisticated approaches. As always, there is no generally valid answer, and scenario tree generation is an active research topic, for which we refer to the literature.

We close this section by stressing that scenario tree generation for financial decisions has still an additional twist, with respect to other application areas. To get the message, let us consider a simple scenario fan, consisting of two nodes, representing the uncertainty in the holding period return for two assets. Imagine that we generate two nodes with the following realizations:

$$\begin{bmatrix} S_1(\omega_1) \\ S_2(\omega_1) \end{bmatrix} = \begin{bmatrix} 0.15 \\ 0.07 \end{bmatrix}, \quad \begin{bmatrix} S_1(\omega_2) \\ S_2(\omega_2) \end{bmatrix} = \begin{bmatrix} 0.09 \\ 0.03 \end{bmatrix}.$$

Does this make sense? A moment of reflection clearly shows that this choice makes no sense. Asset 1 dominates Asset 2, and this implies that there is an arbitrage opportunity, as we have discussed in Section 2.4. Detecting arbitrage opportunities in a more realistic case is not quite trivial, and generating arbitrage-free scenario trees requires sophisticated strategies.

S11.1 Probability spaces, measurability, and information

Successful investing in stock shares is typically deemed a risky and complex endeavor. However, the following piece of advice seems to offer a viable solution:[18]

> Buy a stock. If its price goes up, sell it. If it goes down, don't buy it.

In this section, we dig a little deeper into concepts related to measurability of random variables and their relationships with the flow of information in dynamic decision-making. Despite their more theoretical character,[19] the concepts that we consider here are often stumbled upon when reading books on quantitative finance, where it is common to meet filtrations and adapted processes. We will not attempt a full and rigorous treatment, which would require a quite sophisticated machinery. Nevertheless, we will be able to understand what is wrong with the above suggestion, from a probabilistic perspective, and the concepts that we illustrate should look less intimidating after getting an intuitive feel for them.

We pointed out that a random variable should be actually regarded as a function,

$$X : \Omega \to \mathbb{R},$$

mapping the set Ω of outcomes of a random experiment into the set of real numbers $\mathbb{R}$. Indeed, while we often denote random variables by capital letters like X, the more precise notation $X(\omega)$ emphasizes the true nature of random variables. However, not all conceivable mappings are legitimate random variables. To understand why, we need to clarify the concept of probability space.

---

[18] In his history of the Great Crash of 1929, John K. Galbraith attributes this fundamental piece of advice to an American comedian; see J.K. Galbraith, *The Great Crash 1929*, originally published in 1954, reprinted by Mariner Books, 2009.

[19] This supplement may be safely skipped. The only place in which we use related concepts is Section 15.6, where we illustrate multistage stochastic optimization models.

S11.1 Probability spaces, measurability, and information

DEFINITION 11.3 (Probability space) *A probability space is a triple, usually denoted by $(\Omega, \mathcal{F}, P)$, where Ω is the sample space, consisting of outcomes of a random experiment, $\mathcal{F}$ is a family of subsets of Ω, the events, with suitable closure properties (clarified below), and P is a probability measure mapping events into the interval $[0, 1]$.*

To fully get the message behind this definition, we should observe that for a given sample space Ω, consisting of a set of outcomes ω_i, we may not only define different probability measures, but also different families of events. Events are *subsets* of the sample space, and they need not be just singletons like $\{\omega_i\}$.

Example 11.15 The die has been cast

If we roll a die, the obvious sample space is $\Omega = \{1, 2, 3, 4, 5, 6\}$. If we can observe and are interested in the exact outcome, then we may consider singleton events consisting of a single outcome,

$$\{1\}, \{2\}, \{3\}, \{4\}, \{5\}, \text{ and } \{6\}.$$

The natural probability measure assigns $1/6$ with each outcome, but we would use a different measure in the case of an unfair die. However, we might also be interested in assigning a probability to other events as well. The probability that we observe 2 or 3 should be associated with the event $\{2, 3\}$, and intuition suggests that

$$P(\{2, 3\}) = P(\{2\} \cup \{3\}) = P(\{2\}) + P(\{3\}) = \tfrac{1}{3}.$$

In this case, since the two events are mutually exclusive, we just add probabilities. More generally, given events E_1 and E_2, we might be interested in the probabilities

$$P(E_1 \text{ or } E_2) \equiv P(E_1 \cup E_2),$$
$$P(E_1 \text{ and } E_2) \equiv P(E_1 \cap E_2),$$
$$P(\text{not } E_1) \equiv P(\Omega \setminus E_1),$$

where we see a natural connection with set operations like union, intersection, and difference. Note that or is related to an *inclusive* "or," rather than to the exclusive "either... or..." (but not both). By applying arbitrary combinations of these operations to singletons and composite events, we may generate a rather large family of all subsets of cardinality 1, 2, etc., also including Ω itself and its complement, the empty set $\emptyset$:

$$\mathcal{F}_1 = \{\emptyset, \{1\}, \{2\}, \ldots, \{6\}, \{1, 2\}, \{1, 3\}, \ldots$$
$$\ldots \{5, 6\}, \{1, 2, 3\}, \ldots, \Omega\}. \qquad (11.37)$$

The family $\mathcal{F}_1$ of all subsets of Ω is clearly closed with respect to the set operations, as by applying set operations to subsets in $\mathcal{F}_1$, we can only generate an element in $\mathcal{F}_1$.

However, we may constrain events a bit in order to reflect the possibly limited amount of information. For instance, we might consider the following family of events:

$$\mathcal{F}_2 = \{\Omega, \emptyset, \{1,3,5\}, \{2,4,6\}\}, \tag{11.38}$$

which makes sense when all we may observe (or are interested in) is whether the result is odd or even. This family of events, with respect to $\mathcal{F}_1$, is definitely less rich, and this reflects lack of information. However, it is easy to check that if we try taking complements and unions of elements in $\mathcal{F}_2$, we still get an element of $\mathcal{F}_2$.

The closure property of events with respect to elementary operations is important because we need to assign probability measures to events and their combinations. We do not want to generate a subset to which we cannot assign a probability measure. This may be expressed by requiring that $\mathcal{F}$ be a field.

DEFINITION 11.4 (Field) *A family $\mathcal{F}$ of subsets of Ω is called a **field** if the following conditions hold:*

1. $\Omega \in \mathcal{F}$
2. $E \in \mathcal{F} \Rightarrow (\Omega \setminus E) \in \mathcal{F}$
3. $E, G \in \mathcal{F} \Rightarrow (E \cup G) \in \mathcal{F}$

Thus, if a subset is in the field, its complement is, too, and the union of elements of the field is still in the field. Since set intersection may be expressed using complements and unions, a field is closed with respect to set intersection as well. Note that the first and second conditions imply that the empty set $\emptyset$ belongs to the field $\mathcal{F}$. A field is also called an **algebra of sets**.

Example 11.16 A family of subsets that is not a field

Given $\Omega = \{1,2,3,4\}$, consider the family of subsets

$$\mathcal{G} = \{\Omega, \emptyset, \{1\}, \{2,3,4\}, \{1,2\}, \{3,4\}\}.$$

This is not a field, since, for instance, $\{1\} \cup \{3,4\}, \notin \mathcal{G}$.

Actually, to cope with continuous random variables and, more generally with probability distributions with infinite support, a stronger concept is needed: a σ-**field**, also called a σ-**algebra**. To define this stronger concept, the third condition is extended to a countable union of events:

$$E_1, E_2, E_3, \ldots \in \mathcal{F} \Rightarrow \bigcup_{i=1}^{\infty} E_i \in \mathcal{F}.$$

Unfortunately, whenever the concept of infinity comes into play, pathological cases can occur. We will not be concerned with these anomalies, since we limit our treatment to a finite sample space.

Since describing a finite field by enumerating all of its subsets may be a daunting task, we may describe it implicitly by considering a finite *partition* $\mathcal{P}$ of Ω, i.e., a finite family of subsets E_i, $i = 1, \ldots, n$, such that

1. $E_i \cap E_j = \emptyset$, for $i \neq j$, and
2. $\bigcup_{i=1}^{n} E_i = \Omega$.

In plain English, a partition consists of mutually disjoint and collectively exhaustive subsets. Given a partition $\mathcal{P}$, we may generate the σ-field $\sigma(\mathcal{P})$ by combining subsets in the partition in any possible way. In the case of (11.37), the partition consists of all singleton sets, whereas in the case of Eq. (11.38), we have the two subsets of even and odd outcomes.

A probability space is fully described when we give the sample space Ω, a field of events $\mathcal{F}$, and a probability measure P, associating a real number with elements of $\mathcal{F}$ and satisfying the following rules of the game:

1. The probability measure is properly bounded: $P(E) \in [0, 1]$, for any event $E \in \mathcal{F}$.
2. $P(\Omega) = 1$, i.e., the whole sample space is, in some sense, the largest event, and "something has to happen."
3. Probability is additive for disjoint events:

$$P(E_1 \cup E_2) = P(E_1) + P(E_2),$$

for $E_1, E_2 \in \mathcal{F}$ such that $E_1 \cap E_2 = \emptyset$. This condition is extended to a countable union of events in the case of a σ-field.

Using these simple rules of the game, we may prove obvious and less obvious relationships and, above all, we can assign a probability with any event in the field $\mathcal{F}$.

Let us now turn our attention to **random variables**. Given a probability space, we may define random variables as mappings of outcomes in the sample space into real numbers, but we should clarify how we associate a probability measure with a random variable. In fact, we are able to associate a probability measure with the underlying events, but there must be some consistency between the way we combine events and associate probabilities with them, and the information that we obtain from observing the realization of a random variable.

To be specific, let us consider a discrete random variable $X(\omega)$, taking integer values, and an integer value a. How can we define the probability $P(X(\omega) = a)$? We should consider the subset of outcomes $\omega \in \Omega$ such that $X(\omega) = a$, which essentially amounts to inverting the function $X(\omega)$:[20]

$$X^{-1}(a) \equiv \{\omega \in \Omega : X(\omega) = a\}.$$

Then, the probability we seek is just the probability measure of the subset $X^{-1}(a)$. However, this is only possible if any such subset is an event in the σ-field $\mathcal{F}$.

Example 11.17 A nonmeasurable random variable

Consider the sample space $\Omega = \{1, 2, 3, 4\}$ and the partition

$$\mathcal{P} = \{\{1\}, \{2, 3, 4\}\}.$$

Let $\mathcal{F} = \sigma(\mathcal{P})$ be the field generated by this partition, and define the mapping $X(\omega)$ as follows:

$$X(\omega) = 1 + \omega.$$

This seemingly innocent mapping is *not* a random variable with respect to the field $\mathcal{F}$. In fact, we cannot assign the probability $P(X = 3)$, since

$$\{\omega \in \Omega : X(\omega) = 3\} = \{2\} \notin \mathcal{F}.$$

To be a random variable, the mapping should be *constant* for the three outcomes in $\{2, 3, 4\}$, which is an element of the field $\mathcal{F}$. For instance, the random variable $Y(\omega)$ defined as

$$Y(\{1\}) = -1, \qquad Y(\{2\}) = Y(\{3\}) = Y(\{4\}) = 1$$

is a legitimate random variable.

What is wrong with Example 11.17 is not the mapping $X(\omega)$ per se; it is its association with the field $\mathcal{F}$. If we had a richer field, generated by the partition of Ω into its singletons, there would be no issue. Technically speaking, we say that X is not $\mathcal{F}$-measurable.

---

[20] The careful reader will immediately guess that what we are saying works only for a discrete random variable, since, for continuous random variables, the probability of observing a specific value is zero. Indeed, in a rigorous treatment, we should consider events $\{\omega \in \Omega : X(\omega) \leq a\}$, but to keep things as intuitive as possible, we will refrain from doing so.

S11.1 Probability spaces, measurability, and information

DEFINITION 11.5 (Measurable random variable) *We say that a random variable $X(\omega)$ is $\mathcal{F}$-measurable if*

$$\{\omega \in \Omega : X(\omega) = x\} \in \mathcal{F}$$

for all values of x.

In other words, the inverse function for any value x must be an event in the field $\mathcal{F}$, so that we may associate a probability measure with it. This can be done or not, depending on the random variable and the richness of the field of events $\mathcal{F}$. If we go back to dice throwing, it is clear that if our field is given by Eq. (11.38), we can assign only one value to all even outcomes, and another value to all odd outcomes. The field is, in a sense, smaller than $\mathcal{F}_1$, since all of the events in $\mathcal{F}_2$ are events in $\mathcal{F}_1$, but the converse is not true; this represents a limitation in the available information.

Remark. Readers with some more background in integration theory will immediately see the link between measurability of random variables and the integral in the sense of Lebesgue. Unlike the classical Riemann integral that we have hinted at in Section 11.4.1, which is built by partitioning the *domain* of a function, the Lebesgue integral focuses on its *range*. In particular, we consider subsets within the range, on which the function is constant, and the corresponding subsets within the domain, which are found by function inversion. If we can associate a measure to these subsets in the domain, then we may define the integral by multiplying each value that the function takes by the measure of the corresponding subset in the domain, and adding everything up. Clearly, this can be done for simple (piecewise constant) functions only, but then we may go through the usual drill by taking limits. The Lebesgue integral extends the Riemann integral in the sense that they coincide when the latter is defined, but there are functions that are Lebesgue-integrable and not Riemann-integrable. Furthermore, we may also consider integrals defined on abstract spaces that cannot be partitioned in intervals, but may be associated with a measure.

The link between fields, measurability, and information can be further clarified if we consider a dynamic model. Let us consider a stochastic process in the form of the scenario tree depicted in Fig. 11.10. To be concrete, let us interpret this as a stochastic process describing the price of a stock share. At time $t = 0$, the stock price is $X_0 = 10$. Then, the price may go up or down, resulting in a stochastic process X_t, $t = 0, 1, 2, 3$. In this case, the sample space consists of outcomes ω_i, $i = 1, 2, \ldots, 8$, and each outcome corresponds to a scenario, i.e., a possible path of stock prices. For instance, outcome ω_3 is associated with the price scenario $(10, 12, 11, 13)$. If we are at any terminal node in the scenario tree, we know which scenario has occurred, since we can observe the whole history of stock prices. However, if we are, e.g., at node n_4, we do not know whether we are observing scenario ω_3 or ω_4, since they cannot be distinguished at time $t = 2$. Nevertheless, we do have some information, since by observing the past history of stock prices, we can rule out any other scenario. At the root node n_0 we have the least information, since any scenario is possible.

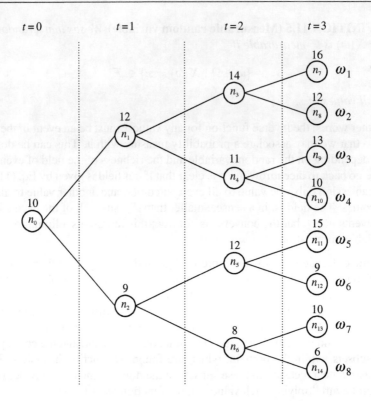

FIGURE 11.10 A scenario tree.

All of this is reflected in the event fields with which the random variables X_t, $t = 0, 1, 2, 3$, are associated. We can capture information by suitable partitions of the sample space

$$\Omega = \{\omega_1, \omega_2, \omega_3, \omega_4, \omega_5, \omega_6, \omega_7, \omega_8\}.$$

At time $t = 0$, we cannot say anything, and our field of events is

$$\mathcal{F}_0 = \{\emptyset, \Omega\} = \{\emptyset, \{1, 2, 3, 4, 5, 6, 7, 8\}\}.$$

At time $t = 1$, we can at least rule out half of the scenarios. This is reflected by the more refined partition

$$\mathcal{P}_1 = \{\{1, 2, 3, 4\}, \{5, 6, 7, 8\}\},$$

which generates the field

$$\mathcal{F}_1 = \{\emptyset, \{1, 2, 3, 4\}, \{5, 6, 7, 8\}, \Omega\}.$$

At time $t = 2$, there is a further branching, refining the partition

$$\mathcal{P}_2 = \{\{1, 2\}, \{3, 4\}, \{5, 6\}, \{7, 8\}\},$$

which generates an even richer field,

$$\mathcal{F}_2 = \{\ \emptyset, \{1,2\}, \{3,4\}, \{5,6\}, \{7,8\}, \{1,2,3,4\}, \{1,2,5,6\}, \ldots, \{5,6,7,8\}$$
$$\{1,2,3,4,5,6\}, \ldots, \{3,4,5,6,7,8\}, \Omega\ \}.$$

Finally, at time $t = 3$, we have the finest partition, consisting of singletons

$$\mathcal{P}_3 = \{\ \{1\}, \{2\}, \{3\}, \{4\}, \{5\}, \{6\}, \{7\}, \{8\}\ \},$$

which generates the richest field $\mathcal{F}_3$, consisting of all possible subsets of Ω. We note that, as time goes by, we get richer and finer fields. We say that this sequence of fields is increasing, in the sense that all of the events in $\mathcal{F}_t$ are included in $\mathcal{F}_{t+1}$. A partition with smaller subsets is associated with more information, and it generates a richer field.

DEFINITION 11.6 (Filtration) *An increasing sequence of σ-fields*

$$\mathcal{F}_0 \subseteq \mathcal{F}_1 \subseteq \mathcal{F}_2 \subseteq \cdots$$

defined on a common sample space Ω is called a **filtration**.

A filtration defines precisely how information is collected by observing a stochastic process. This concept can also be defined for continuous-time processes with a continuous state space, but this requires more sophisticated mathematical machinery. However, the essential message is quite simple:

> *The sequence of dynamic decisions that we make while observing the stochastic process at times $t = 0, 1, 2, \ldots$, must reflect the available information and cannot be anticipative.*

The piece of advice with which we have opened this section is clearly not implementable: It would require knowledge of the future. In that case, we have a sample space consisting of two outcomes,

$$\Omega = \{\omega_{\text{up}}, \omega_{\text{down}}\},$$

corresponding to bull and bear markets, respectively. The corresponding field, at the end of the holding period H is

$$\mathcal{F}_H = \{\emptyset, \{\omega_{\text{up}}\}, \{\omega_{\text{down}}\}, \Omega\},$$

However, at time $t = 0$, we do not know anything, and the corresponding field is just

$$\mathcal{F}_0 = \{\emptyset, \Omega\}.$$

Thus, the initial decision should be a "degenerate" random variable, measurable with respect to $\mathcal{F}_0$, which means that the decision has to be constant and the same for whatever future scenario. The piece of advice results in a nonmeasurable mapping, which is not a proper random variable and assumes clairvoyance. Let us generalize this idea to a truly dynamic case.

▣ Example 11.18 Nonanticipative decisions and measurability

Let us consider the scenario tree of Fig. 11.10 and define consider decision variables Z_t^b and Z_t^s, representing the number of stock shares that we buy and sell, respectively, at time t. At time $t = 0$, we have a unique decision, since we can just buy or sell, here and now. From the perspective at time $t = 0$, the decisions made at the next time instants are random variables, as they depend on the observed path of the stochastic process and the expectations about the future, which are encoded in the scenario tree. The random variables at time t must be $\mathcal{F}_t$-measurable. If we are at node n_4 in the scenario tree of Fig. 11.10, we cannot say "buy if scenario is ω_3" and "do not buy if scenario is ω_4." The decision, whatever it is, must be the same for the two scenarios ω_3 and ω_4. Otherwise, the random variable corresponding to the decision at that node would not be constant over the outcomes included the event $\{\omega_3, \omega_4\}$, and it would not be measurable; we would be in the same trouble as in Example 11.17.

Technically speaking, we say that decisions must be **adapted** to the filtration $\mathcal{F}_t$. Dynamic stochastic optimization models, as we shall see in Section 15.6, must comply with nonanticipativity.

Problems

11.1 The capital X_t of an insurance company, measured in millions of dollars, follows the generalized Wiener process

$$dX_t = 0.5\,dt + 2\,dW_t,$$

where unit time is one year. The insurance company goes bankrupt if capital gets negative. Find an initial capital, such that the probability that capital is negative at the end of a four-year period is no more than 5%.

Note: Actually, bankruptcy may well occur before the four years, and we should study the distribution of *hitting times*, i.e., the random time at which a stochastic process assumes a given value. Since this is beyond the scope of this book, we assume that capital is only checked at the end of the time horizon.

11.2 Consider a stochastic process described by the stochastic differential equation $dS_t = \mu_t\,dt + \sigma_t\,dW_t$, where drift and volatility are piecewise constant functions of time (measured in years). For the first three years, $\mu_t = 2$ and $\sigma_t = 3$; for the next three years, $\mu_t = 3$ and $\sigma_t = 4$. If the initial value of the process is $S_0 = 5$, what is the probability distribution of the value of the process at the end of the six-year period?

11.3 Assume that Y_t is a GBM with drift coefficient 3 and volatility coefficient 7.

- Find the stochastic differential equation of the process $X_t = e^{Y_t^2}$.
- Is X_t a GBM?

11.4 The stochastic process S_t is a GBM with drift coefficient 2 and volatility coefficient 3. Write the stochastic differential equation for the process
$$Y_t = (S_t - 10)^2 \cdot e^{-2t}.$$

11.5 Given a standard Wiener process W_t, find:

- The joint probability
$$P\{(W_5 - W_2 \geq 0) \cap (W_1 \leq 0)\}.$$

- The variance
$$\mathrm{Var}\big[(W_5 - W_2) \cdot W_1\big].$$

Note: The set theoretical notation $\cap$ refers to the intersection of events, and is equivalent to the logical and. We look for the probability of a joint event.

11.6 The stochastic processes $S_a(t)$ and $S_b(t)$ are two GBMs and represent two stock prices driven by the *same* sources of risk (i.e., they share the same driving Wiener process). The drift and volatility coefficients for the two stocks are denoted by μ_a, μ_b, σ_a, and σ_b, respectively. Given the two initial prices, $S_a(0)$ and $S_b(0)$, find the expected value of the product and ratio of prices, $\mathrm{E}[S_a(T) \cdot S_b(T)]$ and $\mathrm{E}[S_a(T)/S_b(T)]$, at time $t = T$.

11.7 Consider the stochastic process $Y_t = e^{W_t}$, where W_t is a standard Wiener process.

- Is Y_t a geometric Brownian motion? Is it a martingale? Why or why not?
- Compute the conditional probability $P\{Y_{10} > 150 \mid W_5 = 3\}$.

11.8 Consider the two GBMs
$$dS_1(t) = \mu_1 S_1(t)\, dt + \sigma_1 S_1(t)\, dW(t),$$
$$dS_2(t) = \mu_2 S_2(t)\, dt + \sigma_2 S_2(t)\, dW(t),$$

which are driven by the *same* standard Wiener process.

- Find the stochastic differential equation for the process $Y(t) = S_1(t) \cdot S_2(t)$. [*Hint:* Use logarithms to get rid of the product, as well as additivity of integrals, which extends to stochastic integrals, too.]
- Imagine a Finnish investor, whose home currency is EUR, investing on the US stock market. The stock price in EUR is the product of the stock price in USD and the exchange rate between the two currencies. Would you use the above model in this case? Why or why not?

11.9 Let y_t be the continuously compounded yield-to-maturity, at time t, of a zero-coupon bond maturing at time T (let us assume that the face value is \$1).

The yield-to-maturity is modeled by the following stochastic process with mean reversion:

$$dy_t = \alpha \cdot (\overline{y} - y_t)\,dt + \sigma y_t\,dW_t,$$

where α, $\overline{y}$, and σ are known parameters related to speed of mean reversion, long-term yield, and volatility of yield, respectively.

- Find the stochastic differential equation for the price of a zero-coupon bond, depending on yield y_t.
- Does the volatility of the bond price behave in a sensible way? Why?

Further reading

- All mathematically inclined textbooks on financial engineering cover the material in this chapter. See, e.g., [1] or [12].
- There is a huge literature on modeling with stochastic processes. For a generic textbook treatment, you might refer, e.g., to [16] or [18].
- Stochastic calculus for finance is dealt with quite thoroughly in [5] and [17]. A shorter treatment can be found in [15].
- Numerical solution of stochastic differential equation is treated in [13], where the very meaning of solution for this kind of equations is also discussed. A short overview is also given in [10].
- We did not deal with Lévy processes extensively. The interested reader may refer to [6].
- An extensive treatment of Monte Carlo methods, including the alternative low-discrepancy sequences, is given in [14]. An excellent treatment is also given in [7], which is more specifically oriented toward financial applications. An introductory, but less comprehensive treatment may be found in [2] or [3]; the former provides MATLAB code, whereas the latter is based on R.
- Scenario tree generation for stochastic optimization is covered, e.g., in [9] and [11].
- We did not cover discrete-time models based on financial time series. An elementary treatment geared toward finance is given in [4], whereas the monumental [8] deals with time series in general.

Bibliography

1 T. Björk. *Arbitrage Theory in Continuous Time* (2nd ed.). Oxford University Press, Oxford, 2004.

2 P. Brandimarte. *Numerical Methods in Finance and Economics: A MATLAB-Based Introduction* (2nd ed.). Wiley, Hoboken, NJ, 2006.

3 P. Brandimarte. *Handbook in Monte Carlo Simulation: Applications in Financial Engineering, Risk Management, and Economics*. Wiley, Hoboken, NJ, 2014.

4 C. Brooks. *Introductory Econometrics for Finance* (2nd ed.). Cambridge University Press, Cambridge, 2008.

5 G. Campolieti and R.N. Makarov. *Financial Mathematics: A Comprehensive Treatment*. CRC Press, Boca Raton, FL, 2014.

6 R. Cont and P. Tankov. *Financial Modeling with Jump Processes*. Chapman & Hall/CRC Press, Boca Raton, FL, 2004.

7 P. Glasserman. *Monte Carlo Methods in Financial Engineering*. Springer, New York, 2004.

8 J.D. Hamilton. *Time Series Analysis*. Princeton University Press, Princeton, NJ, 1994.

9 H. Heitsch and W. Roemisch. Scenario reduction algorithms in stochastic programming. *Computational Optimization and Applications*, 24:187–206, 2003.

10 D.J. Higham. An algorithmic introduction to numerical simulation of stochastic differential equations. *SIAM Review*, 43:525–546, 2001.

11 K. Hoyland and S.W. Wallace. Generating scenario trees for multistage decision problems. *Management Science*, 47:296–307, 2001.

12 J.C. Hull. *Options, Futures, and Other Derivatives* (8th ed.). Prentice Hall, Upper Saddle River, NJ, 2011.

13 P.E. Kloeden and E. Platen. *Numerical Solution of Stochastic Differential Equations*. Springer, Berlin, 1992.

14 D.P. Kroese, T. Taimre, and Z.I. Botev. *Handbook of Monte Carlo Methods*. Wiley, Hoboken, NJ, 2011.

15 T. Mikosch. *Elementary Stochastic Calculus with Finance in View*. World Scientific Publishing, Singapore, 1998.

16 S.M. Ross. *Stochastic Processes* (2nd ed.). Wiley, New York, 1996.

17 S. Shreve. *Stochastic Calculus for Finance (vol. II)*. Springer, New York, 2003.

18 H.C. Tijms. *A First Course in Stochastic Models*. Wiley, Chichester, 2003.

Chapter Twelve

Forward and Futures Contracts

Forward and futures contracts are classified as linear derivatives, unlike options, which are characterized by nonlinear payoffs. We have already introduced forward and futures contracts related with interest rates and fixed-income assets in Chapters 3 and 4. There, we used no-arbitrage arguments to find forward rates and to evaluate forward rate agreements. In Section 4.3, we have also pointed out that forward and futures rates need not be the same, because of the potential interaction of interest rate movements with daily marking-to-market of futures contracts. As we know from Section 1.2.6, a further difference between forward and futures is standardization, which has the effect of making perfect hedging impossible in practice.

In this chapter, we consider forward and futures contracts on equity and foreign currencies. Under stylized assumptions, we show in Section 12.1 how no-arbitrage arguments immediately lead to a fair forward price, which is the only price such that the initial value of the contract is zero. On the contrary, applying the idea to derivatives written on commodities is more troublesome, as commodities cannot be included in a financial portfolio, and they also raise issues with storage costs as well as limitations to short-selling. Finding the fair futures price is less trivial, due to the presence of daily cash flows. Nevertheless, as we show in Section 12.2, forward and futures prices would actually be the same, if interest rates were constant. Clearly, this assumption makes little sense for interest rate derivatives, but it might be considered as a reasonable approximation for short-maturity contracts on other assets. Even if we approximate futures prices with forward prices, hedging with futures is complicated by the limited availability of underlying assets and maturities. Nevertheless, it is a fact of life in risk management that most hedging is carried out with liquid, exchange-traded futures contracts, rather than over-the-counter forward contracts. We discuss hedging with linear derivatives and some of the related issues in Section 12.3. In particular, we consider hedging using index futures and how we may account for daily marking-to-market by tailing a hedge based on futures contracts.

Throughout the chapter, depending on notational convenience, we will liberally alternate notation like $S(t)$ and $F(t, T)$ with S_t and F_t, for spot and forward/futures prices at time t, respectively. In the second case, the contract maturity is left implicit.

12.1 Pricing forward contracts on equity and foreign currencies

In this section, we use a deceptively simple argument to find the price of forward contracts on stock shares and foreign currencies. The key concept is the spot–forward parity theorem, which we shall introduce for the case of a non-dividend-paying stock share. Then, we extend the idea to the case of assets providing the owner with income. This case includes stock shares paying dividends, as well as foreign currencies. Throughout this section, we will assume that interest rates are constant, and we will be able to appreciate the relevance of this assumption.

12.1.1 THE SPOT–FORWARD PARITY THEOREM

Let us denote the spot price of an asset at time t by $S(t)$, and its forward price for delivery in T by $F(t,T)$. At time $t=0$ we know the spot price $S(0)$, and we wonder about the fair forward price $F(0,T)$, which we would like to set in such a way that the value of the contract is zero, so that (in principle) it costs nothing to take a position in the forward. We assume that the continuously compounded risk-free rate r is constant, and that the asset does not provide any income.

Let us compare the two following strategies:

- Invest an amount $S(0)$ at the risk free rate over the time interval $[0,T]$. The cash flow at time $t=0$ is $-S(0)$, and the cash flow at time $t=T$ is $S(0)e^{rT}$, with no risk.
- Invest an amount $S(0)$ to buy one unit of the asset and enter into a short position to sell the asset at time T at the forward price $F(0,T)$. The cash flow at time $t=0$ is again $-S(0)$, and the cash flow at time $t=T$ is $F(0,T)$, with no risk.

Since both strategies require the same initial investment and are risk-free, there is only one forward price that leaves no room to arbitrage opportunities:

$$F(0,T) = S(0)e^{rT}. \qquad (12.1)$$

This condition is known as **spot–forward parity** condition.

An alternative, but equivalent and quite instructive argument is based on the hedging needs of the part holding the short position in the contract. The short position will have to deliver the item at the forward price, no matter what the spot price $S(T)$ is. Waiting until maturity to buy the asset in order to deliver it is quite risky. However, the short position may just borrow $S(0)$ in order to buy the asset, which is delivered at the forward price $F(0,T)$. The initial net cash

12.1 Pricing forward contracts on equity and foreign currencies

flow is zero and there is no risk. The cash flow at maturity is

$$F(0,T) - S(0)e^{rT},$$

when the short position sells the asset to the corresponding long position and repays debt. The no-arbitrage condition requires that this cash flow is zero, from which we find Eq. (12.1) again.[1] This kind of strategy is called **cash-and-carry**, and $S(0)e^{rT}$ is the cost of buying the underlying asset and holding it until maturity. The forward price should just be the cost of cash-and-carry, if there is no additional cost or income from holding the asset. If we want to hedge a long position, then it is easy to see that we should sell the asset short, in such a way that we hedge the risk of taking delivery when the future spot price is smaller than the agreed forward price. In this case, we invest the proceeds of the short sale at the risk-free rate r, and buy the asset back at the forward price to close the short sale.

These hedging arguments are useful to appreciate the hidden assumptions behind Eq. (12.1):

- When hedging the short position, we buy and hold the asset. On the one hand, we have implicitly assumed that the asset does not yield any income, as it would be the case with a stock share paying a dividend. Furthermore, carrying a stock share in inventory costs nothing.[2] However, with a physical commodity, storage cost would be an issue.

- When hedging a long position, we should short-sell the underlying asset, and we are assuming that there are no limit to short-selling. This is arguably not true for commodities that are not investment assets. In such a case, rather than a spot–forward parity, we can only claim that there is an *upper bound* on the forward price,

$$F(0,T) \leq S(0)e^{rT}.$$

If we can only buy and hold the asset (at no extra storage cost), it is easy to see that a forward price larger than $S(0)e^{rT}$, which is the cost of cash-and-carry, would lead to an arbitrage strategy. However, we may observe a lower price if we cannot take exploit the opportunity of short-selling and then buying the asset forward, to close the short position in the asset.

- We may borrow or lend money at the same constant risk-free rate r, and markets are perfectly liquid, so there are no frictions (no transaction costs, no taxes, no bid-ask spreads). Clearly, these assumptions are not strictly true, but they are approximately true for large institutional investors, which might be sufficient to enforce or get close to spot–forward parity.

---

[1] See Example 2.12 for an example of arbitrage strategy if the spot–forward parity is violated.

[2] We might say that the cost-of-carry is the risk-free rate, which is used to finance the purchase of the asset. The assumption that the transaction can be funded at the risk-free rate is clearly debatable, and recent derivative models pay due attention to the cost of funding.

The spot–forward parity Eq. (12.1) is somewhat counterintuitive, as one seemingly sensible guess for a fair forward price would be

$$F(0,T) \stackrel{?}{=} \mathrm{E}[S_T]. \qquad (12.2)$$

A bit of reflection shows that the expectation should be intended in a market-consensus sense, since individual investors may have different expectations, making the definition of a sensible expectation rather difficult in practice. Anyway, spot–forward parity seemingly rules out any role for expectations, but this is true if the assumption of idealized and frictionless markets is valid. In any case, we have seen in Section 2.3.4 that the guess of Eq. (12.2) would make sense under a risk-neutral measure $\mathbb{Q}_n$. In fact, Eq. (2.25) suggests the following guess:

$$F(0,T) \stackrel{?}{=} \mathrm{E}_{\mathbb{Q}_n}[S(T)] = S(0)e^{rT}. \qquad (12.3)$$

We introduced the risk-neutral measure $\mathbb{Q}_n$ in a very simple single-step binomial model. It turns out that our findings may be extended to continuous-time models. However, Eq. (12.3) relies on the risk-neutral measure, which is valid if we assume a *constant* interest rate. If interest rates are stochastic, Eq. (12.3) makes no sense. There exists a probability measure, called forward risk-neutral,[3] such that the forward price is indeed the expected spot price in the future, but this is a different measure, if interest rates are random.

Last, but not least, we observe that Eq. (12.1) may be generalized to

$$F(t,T) = S(t)e^{r \cdot (T-t)},$$

which is consistent with the spot–forward convergence condition $F(T,T) = S(T)$, i.e., at maturity, the forward price for immediate delivery is just the spot price. It is also important to realize that a parity condition does not imply any causality. It is tempting to think that, given the spot price, we find a fair forward price. This sounds at odds with the common claim that financial speculation on the huge derivatives market has an adverse effect on spot prices. In fact, the parity relationship is just a consistency condition, and it does not say anything about which price is the driving force. Indeed, the tail may wag the dog.

12.1.1.1 The value of a forward contract

The forward price is the only price, such that the initial value of the contract is zero. The same consideration applies to vanilla interest rate swaps.[4] There is only one swap rate, such that no money needs to be exchanged when the swap is agreed on. On the contrary, in the case of asymmetric contracts like options, a range of strike prices is available, associated with different (positive) option prices. However, after the forward contract is agreed at time 0, with delivery

---
[3] See Section 14.3.3.
[4] See Section 4.4 for how to value a swap agreement after its inception.

12.1 Pricing forward contracts on equity and foreign currencies

price $F(0,T)$, the spot price will move randomly, and the forward price will, too. At time t, a different forward price $F(t,T)$ would apply to delivery at time T. Hence, the old contract will have a positive or negative value, depending on whose position we are holding.

What we know is that, at time 0, the value of the forward contract is zero, which we may formally express by using a pricing operator $\Pi_t(\cdot)$ which maps a future (random) cash flow into its value at time t:

$$\Pi_0\big[S(T) - F(0,T)\big] = 0.$$

However, we are wondering about the value at a later time t, $\Pi_t\big[S(T)-F(0,T)\big]$. We take advantage of the availability of a *new* forward price $F(t,T)$ as follows:

$$\begin{aligned}\Pi_t\big[S(T)-F(0,T)\big] &= \Pi_t\big[S(T)-F(t,T)+F(t,T)-F(0,T)\big] \\ &= \underbrace{\Pi_t\big[S(T)-F(t,T)\big]}_{=0} + \Pi_t\big[F(t,T)-F(0,T)\big] \\ &= \big[F(t,T)-F(0,T)\big]\cdot e^{-r(T-t)},\end{aligned} \qquad (12.4)$$

where we use the linearity of pricing,[5] the fact that the value of the new forward contract is zero, and the fact that deterministic cash flows can be just discounted at the risk-free rate. We observe that the value of the contract is obtained by replacing, in the discounted payoff, the random future spot price $S(T)$ by its current forward price $F(t,T)$. Thus, we are using the forward price as a sort of forecast, which should not be taken for granted. We found a similar result in Section 4.2.1, when dealing with forward rate agreements.

The result in Eq. (12.4) applies to a long position and makes intuitive sense. If the current forward price increases, the old contract (whose delivery price is fixed to the old forward price) gets more appealing to the long position and assumes a positive value. Signs are reversed for a short position. We may also interpret a futures contract as a sort of forward in which the accumulated value is paid immediately at the end of each day, and the delivery price is reset, so that the new value of the contract is always zero.

12.1.2 THE SPOT–FORWARD PARITY THEOREM WITH DIVIDEND INCOME

Let us consider a forward contract written on a stock share that will distribute dividends before the maturity of the contract. If the underlying asset provides its owner with an income, with present value I, the spot–forward parity relationship becomes

$$F(0,T) = \big(S(0)-I\big)\cdot e^{rT}. \qquad (12.5)$$

To interpret this relationship, we may think of the cost of hedging a short position by a cash-and-carry strategy. Clearly, the income provided by the asset

---

[5] See Section 2.4.1.

reduces the hedging cost. The net cash flow at maturity is

$$F(0,T) + I \cdot e^{rT} - S(0)e^{rT}.$$

Setting this expression to zero yields Eq. (12.5). The hedge for a long position would be reversed and, in this case, when we close the short sale and give back the underlying asset to its legitimate owner, we should also compensate her for the missed dividends, paying an additional amount $I \cdot e^{rT}$.

In some cases, it may be convenient to think of a dividend flow that is paid continuously, rather than a sequence of lump payments. In the case of a continuous dividend yield q, we find

$$F(0,T) = S(0)e^{(r-q)T}. \tag{12.6}$$

We see, as with lump dividends, that the dividend yield lowers the forward price, but the formula should be clarified a bit. To this aim, we should think that the dividend flow is continuously reinvested in the stock itself. Imagine again that we hold a short position in the contract, but at time $t = 0$, we set up the hedge by buying only an amount

$$h(0) = e^{-qT} < 1 \tag{12.7}$$

of the underlying asset, financing this purchase by borrowing $S(0)e^{-qT}$ at the risk-free rate. We denote by $h(t)$ the amount of the underlying asset that we hold at time t as a hedge. To see how the hedge will change over time, let us consider a small time interval δt. Given price $S(t)$, over the time interval $[t, t + \delta t]$, we receive a cash amount

$$qh(t)S(t) \cdot \delta t,$$

which is reinvested by purchasing an additional amount

$$\frac{qh(t)S(t) \cdot \delta t}{S(t)} = qh(t) \cdot \delta t$$

of the stock share. This additional amount is the increment in the asset holding over the time interval:

$$\delta h(t) = h(t + \delta t) - h(t) = qh(t) \cdot \delta t.$$

This is actually an approximation, but if we let $\delta t \to 0$, we find a familiar differential equation

$$\frac{dh(t)}{dt} = qh(t),$$

whose solution, given the initial condition of Eq. (12.7), is

$$h(T) = h(0)e^{qT} = e^{-qT}e^{qT} = 1.$$

12.1 Pricing forward contracts on equity and foreign currencies

Thus, we end up holding exactly the stock share that we need to hedge the short position at maturity T. The net cash flow at time T will be

$$F(0,T) - S(0)e^{-qT}e^{rT}.$$

By setting this to zero, we find Eq. (12.6).

A continuous dividend yield may sound like a rather unrealistic assumption. However, it may be useful, for instance, to deal with derivatives written on stock market indexes. In that case, the index is affected by all stock shares paying dividends with different timings. Keep in mind that the index is not protected against dividends, which have the effect of reducing stock share prices. The overall effect of many lump payments may be approximated by a continuous yield. As we show in the next section, a forward on a foreign currency can be interpreted in this way, too.

12.1.3 FORWARD CONTRACTS ON CURRENCIES

Let $S(0)$ be the current spot price in our domestic currency (say, EUR) of one unit of foreign currency, and let $F(0,T)$ be the forward price for delivery at time $t = T$. The parity relationship is

$$F(0,T) = S(0)e^{(r_d - r_f)T}, \qquad (12.8)$$

where r_d is the continuously compounded *domestic* risk-free rate and r_f is the *foreign* risk-free rate. We may interpret the parity relationship as a condition involving an asset, the foreign currency, which provides a risk-free income at the continuous yield rate r_f. To prove Eq. (12.8) formally, let us compare the following two investment strategies, available, e.g., to an Italian investor (thus, the domestic rate is the rate on EUR, and the foreign rate is the rate on, say, USD):

1. She may invest €1 at rate r_d for a time interval of length T. Hence, terminal wealth is
$$\text{€} \, e^{r_d T}.$$

2. She may exchange €1 and buy $\$1/S(0)$, which is invested at rate r_f for a time interval of length T. Then, terminal wealth in USD is converted back to EUR, at no risk, by taking a short position in the forward contract. This yields the following wealth:
$$\text{€} \, \frac{F(0,T)e^{r_f T}}{S(0)}.$$

Since the two strategies are both riskless, no-arbitrage dictates that they must yield the same terminal wealth,

$$e^{r_d T} = \frac{F(0,T)e^{r_f T}}{S(0)}.$$

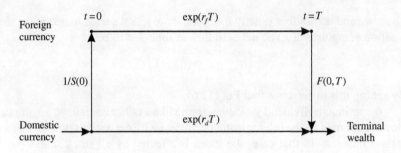

FIGURE 12.1 Alternative trades using a foreign currency forward.

A simple rearrangement yields Eq. (12.8). The argument can be visualized as shown in Fig. 12.1. The two paths, where arcs are labeled by a gain factor multiplying the flow of money, should yield the same terminal wealth. If a path yields a lower overall gain, we can reverse its flow and create wealth out of nothing. More specifically:

- If the quoted forward price is lower than the fair one, it is cheap to buy the currency forward in the future, which can be used to repay a debt in the foreign currency. Thus, we should borrow the foreign currency, convert it to domestic currency, which is invested at the domestic risk-free rate. Part of the terminal wealth is converted to the foreign currency, at the cheap forward price, to close the debt.

- If the quoted forward price is higher than the fair one, there is an advantage in selling the currency forward in the future. In this case, we should invest in the foreign currency, at the foreign risk-free rate. To do so, we borrow domestic currency at the domestic risk-free rate, convert it to the foreign currency which is invested and converted back to domestic currency at the overstated forward price. Part of the resulting wealth is used to pay the outstanding debt in the domestic currency.

Example 12.1 A currency arbitrage

To see how an arbitrage opportunity may arise if parity is violated, say that the two-year rates in Australia and the USA are 5% and 7%, respectively, and that the spot price of 1 AUD is 0.62 USD. Then, the two-year forward price should be

$$0.62 \times e^{(0.07-0.05) \times 2} = 0.6453.$$

Now assume that $F(0,2) = 0.63$. The forward price is lower than it should be, so it is cheap to *buy* AUD forward. Then, we can:

12.1 Pricing forward contracts on equity and foreign currencies

- Borrow 1000 AUD at 5% per annum for two years and convert them to 620 USD, which are invested at 7%.
- Enter a forward contract to buy $1000 \times e^{0.05 \times 2} = 1105.17$ AUD, which is the amount we shall need to repay our debt in AUD, for $1105.17 \times 0.63 = 696.26$ USD.

The 620 USD grow to $620 \times e^{0.07 \times 2} = 713.17$ USD; 696.26 are used to buy the AUD that we need to repay our debt, resulting in a risk-free profit of 16.91 USD.

If, on the contrary, the forward price is too large, say 0.66, we should *sell* AUD forward. Thus, the strategy is as follows:

- Borrow 1000 USD at 7% per annum for two years and convert them to $1000/0.62 = 1612.90$ AUD, which are invested at 5%. At maturity, we will have $1612.90 \times e^{0.05 \times 2} = 1782.53$ AUD.
- Enter a forward contract to sell 1782.53 AUD for $1782.53 \times 0.66 = 1176.47$ USD.

At maturity, we sell 1782.53 AUD for 1176.47 USD, but we need only $1000 \times e^{0.07 \times 2} = 1150.27$ USD to repay our debt, resulting in a risk-free profit of 26.20 USD.

As usual, forward and futures contracts on foreign currencies may be used for both speculation and hedging. Many transactions on currencies involve banks with good credit standing, and for this reason, this is a kind of underlying asset for which OTC forward contracts are more common, whereas in other cases futures contracts are the rule. We should also mention that vanilla swaps on foreign currencies are available as well. Just like vanilla interest rate swaps, currency swaps may be priced by regarding them as portfolios of simple forward contracts.

12.1.4 FORWARD CONTRACTS ON COMMODITIES OR ENERGY: CONTANGO AND BACKWARDATION

From the spot–forward parity formula,

$$F(t, T) = S(t)e^{r \cdot (T-t)},$$

where the cost-of-carry is just the risk-free rate, we may observe that the forward price is always above the spot price. Hence, the spot–forward convergence will be as in Fig. 12.2(a), where the forward price converges to the spot from above. This kind of pattern is called **contango**. However, when the asset is providing income, as is the case of a dividend yield q, the parity relationship becomes $F(t, T) = S(t)e^{(r-q) \cdot (T-t)}$, where the cost-of-carry is reduced. If

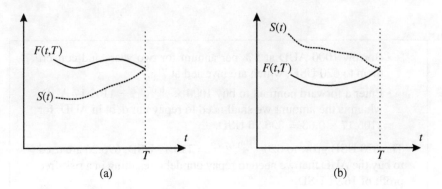

FIGURE 12.2 Different patterns of spot–forward convergence: (a) contango, (b) normal backwardation.

$q > r$, the cost-of-carry may even get negative, and in this case convergence will be from below, as in Fig. 12.2(b). This kind of pattern is called **normal backwardation**.

When dealing with contracts on commodities or energy, the picture gets more complicated. On the one hand, commodities may be difficult to store, and this increases the cost-of-carry. On the other hand, holding a commodity may in some cases provide its holder with a convenience yield. As a result, we may observe either a backwardation or a contango pattern, even for the same underlying asset at different times. We should also note that the no-arbitrage argument may break down in the case of assets that are not suitable as investment purposes. In particular, short sales may be not quite feasible. Hence, as we noted in Section 12.1.1, it may be the case that we only have an upper bound on the forward price, and backwardation may be observed.

In general, pricing contracts on nonfinancial assets requires dealing with strongly incomplete and possibly illiquid markets, where the idealized assumptions behind no-arbitrage arguments break down. We should note that, in such a context, the terms contango and backwardation may refer to the term structure of forward prices, where we do not observe a single price $F(t, T)$ over time for a fixed maturity, as in Fig. 12.2, but rather we fix time t and plot futures prices for a range of maturities. When forward prices are decreasing with maturity, we speak of backwardation. When forward prices are increasing with maturity, we speak of contango. In some cases, like contracts on energy, we may even observe seasonal patterns in forward prices.

12.2 Forward vs. futures contracts

Assuming that we have a sensible way to find the fair forward price, can we use it as a futures price as well? The answer is that by doing so we disregard the impact of marking-to-market. It is true that if we sum all daily cash flows

12.2 Forward vs. futures contracts

along the life of a futures contract, disregarding the time value of money, we obtain the same total net cash flow that we receive from a forward.[6] However, if we consider the possibility of investing profits and financing losses with debt, interest rates and their correlation with the underlying asset price come into play. This is especially critical for interest rate futures, maybe less so for other assets.

In general, we cannot take for granted that futures and forward prices are the same, especially for contracts with long maturities. For short maturities or assets whose prices are not correlated too much with interest rates, we may approximate the futures price by a forward price. There is a case in which the approximation is actually exact. The following clever argument shows that, in fact, forward and futures prices are the same if interest rates are constant.[7] Let us consider a time horizon consisting of T days. Time is indexed by $t = 0, 1, 2, \ldots, T$, and the following sequence of events takes place:

- We buy a certain number of futures contracts at time $t = 0$, during day 1, when the futures price is F_0 (here, we streamline the notation from $F(t, T)$ to F_t, for the sake of notational convenience). Thus, we hold a long position. As we shall see, this position is changed dynamically over time.

- Marking-to-market occurs at the end of each day before maturity, and the settlement futures prices are F_t, $t = 1, 2, \ldots, T - 1$.

- The first cash flow occurs at time $t = 1$, i.e., at the end of day 1, when the settlement futures price is F_1. Thus, the cash flow for *one* long position is $F_1 - F_0$, the difference between the first settlement price and the futures price at which we have bought the contract.

- We have daily cash flows corresponding to the differences $F_t - F_{t-1}$. The actual cash flow depends on the number of contracts we hold.

- The contract matures at the end of day $t = T$, when the futures price F_T converges to the spot price S_T. Depending on the underlying asset, the contract may be finally settled in cash, or actual delivery of the underlying asset takes place. For the sake of convenience, let us assume that the contract is settled in cash,[8] so that the last cash flow for one contract is $S_T - F_{T-1}$.

We consider a fixed, continuously compounded risk-free rate, and we denote the *daily* interest rate by δ. The key of the argument is that the trading strategy is dynamic, based on taking and increasing a long position: At the end of each day (before maturity), we increase the position by a factor e^δ, corresponding to capital growth over one day for a riskless investment. The additional positions are entered at the settlement price, so they do not contribute to the cash flow on

---

[6] See Eq. (1.9).
[7] Here we follow the treatment given by [4], in an online technical note.
[8] From a mathematical viewpoint, this is inconsequential.

that day, but they will at the end of the next day. Thus, our position consists of

$$e^{(k+1)\delta}, \qquad k = 0, 1, \ldots, T-1$$

units of the futures contract at each time instant. Note that the first position, when we enter at time $t = 0$, is e^δ. At the end of the first day it is increased to $e^{2\delta}$, and it is increased to $e^{T\delta}$ at the end of the day before maturity, at time $t = T - 1$. Positive cash flows are invested up to maturity and grow by a factor e^δ each day, whereas negative cash flows are financed by debt that will be repaid at maturity, growing by a factor e^δ each day. At maturity, the contract is finally settled and the final cash flow occurs.

The resulting sequence of cash flows, at the end of each day, is as follows:

$$(F_k - F_{k-1})e^{k\delta}, \qquad k = 1, \ldots, T.$$

Note that at time $t = 1$, we have a cash flow that results from holding the initial amount e^δ units,

$$(F_1 - F_0)e^\delta.$$

Keep in mind that, at the end of the day, we increase the position at the settlement futures price; hence, this will affect the cash flow only the day after. At the end of day $T - 1$, we finally increase the position to $e^{T\delta}$ units, and the last cash flow, at time $t = T$, is

$$(F_T - F_{T-1})e^{T\delta}.$$

Since we hold a long position, the cash flows will be positive when the futures price increases, and negative when it decreases. Cash flows are invested (or financed) up to maturity, at time $k = T$, and grow by a factor e^δ each day. Hence, each cash flow corresponds to the following amount of money at maturity:

$$(F_k - F_{k-1})e^{k\delta}e^{(T-k)\delta} = (F_k - F_{k-1})e^{T\delta}, \qquad k = 1, \ldots, T.$$

By summing everything up, we find a telescoping sum where almost every term is canceled, and the final profit/loss is

$$\sum_{t=1}^{T}(F_k - F_{k-1})e^{T\delta} = (F_T - F_0)e^{T\delta} = (S_T - F_0)e^{T\delta}.$$

Note that this amount is random, as it involves the price of the underlying asset at maturity. We may combine this trading strategy with a riskless investment of an amount F_0 to yield a terminal wealth

$$F_0 e^{T\delta} + (S_T - F_0)e^{T\delta} = S_T e^{T\delta}.$$

Now, imagine that the corresponding *forward* price, at time $k = 0$, is G_0. If we combine a riskless investment of an amount G_0 and a long position in $e^{T\delta}$ forward contracts, we end up with the terminal wealth

$$G_0 e^{T\delta} + (S_T - G_0)e^{T\delta} = S_T e^{T\delta},$$

where again we assume that the contract is settled in cash. These two investment strategies require an initial investment F_0 and G_0, respectively, and yield the same terminal wealth in any possible scenario. The consequence is that, by no-arbitrage arguments, we must have

$$F_0 = G_0,$$

i.e., the futures and the forward price must be the same. In the reasoning, as usual, we assume rather idealized markets, as well as no bid–ask spread in interest rates. Hence, we should not believe that the conclusion is literally true, but it might be a sensible approximation in some cases.

Remark. The relationship between forward and futures prices may be analyzed in more detail by using probabilistic arguments related to risk-neutral pricing. We have preferred an argument that emphasizes hedging, as this shows the need for dynamic hedge adjustments when using futures. We will say more on risk-neutral measures and the pricing of forward and futures contracts in Sections 13.7.3 and 14.3.3.

12.3 Hedging with linear contracts

In principle, by using OTC forward contracts, where we may choose at will underlying asset, contract size, and maturity, we might hope to set up a perfect hedge. However, as someone said, a perfect hedge can only be found in a Japanese garden. Some residual risk will remain, if the volume required is uncertain, and forward contracts are not completely free from counterparty risk, anyway. Furthermore, a hedge based on OTC contracts may be difficult and expensive to unwind if the hedging needs change. Therefore, futures contracts might be preferred for hedging purposes (and certainly for speculation purposes), but this raises a few issues that we outline in this section.

12.3.1 QUANTITY-BASED HEDGING

Suppose that we hold Q_A unit of an asset with spot price S_t, and that we plan to sell the asset at time T. We may hedge the risk exposure with a short position in N units of a linear contract (forward or futures) maturing at T, with contract size Q_F, written on the same asset. Assuming that the contract is settled in cash, the cash flow at maturity will be

$$Q_A S_T + N Q_F (F_0 - F_T),$$

where we neglect the time value of daily cash flows in the case of futures. Thanks to spot–forward convergence, $F_T = S_T$, we may rewrite this cash flow as

$$Q_A S_T + N Q_F (F_0 - S_T) = (Q_A - N Q_F) S_T + N Q_F F_0.$$

Thus, in order to eliminate randomness, we just have to eliminate dependence on the random variable S_T by choosing

$$N = \frac{Q_A}{Q_F}. \tag{12.9}$$

In practice, unless we use a customized OTC contract, a certain rounding will be involved, given the standardization of contract sizes. Therefore, we have to round N to the integer number of contracts that we will actually buy or sell, depending on the kind of hedge we need (long or short). If we consider a unit exposure in the asset, i.e., if we set $Q_A = 1$, we talk of a **hedging ratio**, $h = 1/Q_F$, the number of contracts we need per unit of the asset. Standardization of futures contracts also implies that we cannot choose the underlying asset and the maturity at will, which introduces further sources of risk.

12.3.2 BASIS RISK AND MINIMUM VARIANCE HEDGING

When we hedge with futures, the hedge may be less than perfect because of a maturity and/or an asset mismatch. A maturity mismatch may be of two kinds. The hedging horizon T^*, i.e., the horizon for which we want to hedge the risk exposure, may be longer or shorter than the maturity T of the hedging instrument. We discuss the related issues first. Then, we consider the more general case involving an asset mismatch, too.

12.3.2.1 Hedging with a maturity mismatch

Let us consider an asset with spot price S_t, and a relatively short hedging horizon T^*. We assume that futures contracts are available for that asset, with maturities larger than T^*. Then, we may choose a contract with maturity $T > T^*$, so that the cash flow at the hedging horizon is (for a short hedge)

$$S_{T^*} + h(F_0 - F_{T^*}), \tag{12.10}$$

where we ignore the time value of daily cash flows and use the hedging ratio h, the number of contracts per unit of the underlying asset. Here, h need not be 1, since, before maturity, we *cannot* claim that $S_{T^*} = F_{T^*}$. The difference between the future spot price and the futures price gives rise to **basis risk**. If we assume that futures and forward prices are close enough, and that spot–forward parity holds, we may rewrite Eq. (12.10) as follows:

$$S_{T^*} + h\left[F_0 - S_{T^*} e^{r \cdot (T-T^*)}\right] = h + S_{T^*}\left[1 - h e^{r \cdot (T-T^*)}\right]. \tag{12.11}$$

Hence, we may easily find the hedge ratio that will eliminate uncertainty:

$$h = e^{-r \cdot (T-T^*)}. \tag{12.12}$$

This formula has clear limitations: We ignore uncertainty in the risk-free rate, which is related to the assimilation of futures to forward prices, and take spot–forward parity for granted.

12.3 Hedging with linear contracts

If, on the contrary, we have a long-term exposure, and no long-term futures contract is available, we may consider the possibility of choosing short-term contracts, which are closed before their maturity and replaced by new contracts. This kind of strategy, where we roll the hedge forward in time, is called **stack-and-roll**. Clearly, when the time horizon is long, ignoring the effect of daily marking-to-market becomes questionable. Furthermore, liquidity issues may come into play, as the following well-known example illustrates.

Example 12.2 The Metallgesellschaft case

The following well-known and quite instructive real-life story is well suited to illustrate the dangers of stack-and-roll and the role of liquidity. At the beginning of the 1990s, Metallgesellschaft (MG) sold rather long-term contracts (5–10 years) for the supply of oil-related products, like heating oil, at a fixed delivery price. Hence, in order to hedge the risk of an increase in the price of oil and its derivatives, they took long positions in oil futures. Since oil futures were not available with a corresponding long maturity, the hedge was rolled forward. Then, after 1992, a drop in the oil price occurred. In terms of the contracts that MG were selling, this was great news. Unfortunately, this implied considerable losses in the hedge, and this had to be sustained immediately, given marking-to-market. These losses, in principle, would have been compensated in the future by the payoff of the contracts that MG had sold, but this did very little to alleviate the ensuing short-term liquidity issues. In the end, the hedging strategy had to be stopped and the outstanding contracts with their customers were canceled, with a loss that is estimated to be around $1.33 billion.

12.3.2.2 Minimum variance hedging

When using futures contracts for hedging, basis risk does not only arise from a maturity mismatch. It may well be the case that we want to hedge against adverse movements in the price S_t of an asset, for which no futures contract is available. Then, we may resort to a contract written on another asset, whose price P_t is correlated with S_t. This strategy is called **cross-hedging**. If the hedging horizon is $T^* < T$, where T is the maturity of the futures contracts, and we disregard cash flow timing, a short hedge with the naive ratio $h = 1$ would yield the following cash flow at time T^*:

$$S_{T^*} + (F_0 - F_{T^*}) = F_0 + \underbrace{(S_{T^*} - P_{T^*})}_{\text{asset mismatch}} + \underbrace{(P_{T^*} - F_{T^*})}_{\text{maturity mismatch}},$$

where we rewrite the expression in order to point out the two components of the hedging error. Hence, we must optimize the hedging ration h, in order to improve performance. As we have seen in Section 2.2.3.2, a simple strategy to reduce basis risk is minimum variance hedging, where choose the hedging ratio in order to minimize

$$\text{Var}\left[S_{T^*} + h \cdot (F_0 - F_{T^*})\right] = \sigma_S^2 + h^2 \sigma_{F_{T^*}}^2 - 2h\rho \sigma_{S_{T^*}} \sigma_{F_{T^*}},$$

where $\sigma_{S_{T^*}}^2$ and $\sigma_{F_{T^*}}^2$ are the variances of the spot and futures price at time T^*, respectively, and ρ is their correlation coefficient. Minimizing variance with respect to h yields the optimal hedging ratio,

$$h^* = \rho \cdot \frac{\sigma_{S_{T^*}}}{\sigma_{F_{T^*}}}. \qquad (12.13)$$

Note that this boils down to the perfect hedge if a contract is available with maturity T^* for the asset we have to sell (or a perfectly correlated asset).

It is interesting to note that the hedge ratio h^* is essentially the slope of the regression line when we regress the spot price on the futures price. This makes intuitive sense, as in linear regression we aim at reducing the variance of the residuals, whereas here we minimize the variance of the hedging error. We obtain the same result, if we consider the random variables,

$$\delta S = S_{T^*} - S_0, \qquad \delta F = F_{T^*} - F_0,$$

i.e., the random variations in the spot and futures prices. This is relevant if we do not really plan to sell the asset, but we consider its potential loss of value. Since S_0 does not contribute to variance, this is mathematically inconsequential and corresponds to shifting the axes in the regression model. It may also be interesting to consider what happens if we consider *returns*, i.e., relative rather than absolute variations. This will play a role in Section 12.3.3, where we deal with hedging stock portfolio risk by stock index futures and consider the role of the portfolio beta in finding the hedge ratio. Then, in Section 12.3.4, we deal with one more missing piece in the picture, i.e., how we might account for daily marking-to-market when hedging with futures.

12.3.3 HEDGING WITH INDEX FUTURES

When we need to hedge a stock portfolio, index futures come in handy. Needless to say, it would be impossible to trade a wide array of futures contracts on individual stock shares, as the market would not be liquid enough. Hence, we have to hedge the portfolio with stock index futures, which may eliminate the systematic risk component. As we have seen in Section 8.2.1, systematic risk cannot be easily eliminated by pure diversification, unlike specific risk. However, by taking a suitable position in stock index futures we may change the beta of the portfolio, possibly making it market-neutral. Even if we neglect specific risk, we have to settle for an imperfect hedge. On the one hand, even if we hold a well-diversified portfolio, it may differ from the index. In particular, its beta may be larger or smaller than 1. On the other hand, there may be a mismatch

12.3 Hedging with linear contracts

between the hedging horizon and the contract maturity. The optimal amount of futures contracts in the hedge may be found by minimizing the variance of the hedging error.

Futures contracts are available on market indexes like the S&P500, which is broad enough to be used to hedge stock portfolio risk for the US stock market. Other indexes must be used for other stock markets. As we have pointed out in Section 12.1.2, we may use a continuous dividend yield to account for the dividend income provided by a portfolio matching the index. Stock-index futures, by their very nature, must be settled in cash, as delivering a well-diversified portfolio consisting of several stock shares (500 in the case of the S&P500 index) would be impractical. Here, when we talk about the futures price, we do not really mean a price, but rather an index value which must be converted to a monetary value. In the case of S&P500 futures, the futures price is multiplied by a factor $M_F = 250$.

Example 12.3 Marking-to-market S&P500 index futures

> Suppose that we sell three S&P500 futures when the futures price is $1963. At end of the day, if the settlement price is $2001, we will observe a negative cash flow of
>
> $$\$250 \times 3 \times (1963 - 2001) = -\$\,28{,}500$$
>
> on our margin account.

Let us denote the current value of the portfolio by

$$V_A = Q_A S(0),$$

where, in this case, Q_A may be interpreted as the number of shares of a common fund that we hold, and $S(0)$ is the monetary value of each share, which is the underlying asset. The "face value" of each futures contract, with maturity T, is

$$V_F = M_F F(0, T),$$

where M_F is the multiplier transforming the index value to a monetary price (as we have said, $M_F = 250$ for S&P500). To find the number Q_F of required futures contracts, we may consider the random variable

$$Q_A S(T^*) - Q_F M_F \big[F(0, T) - F(T^*, T)\big], \tag{12.14}$$

where $T^* \leq T$ is the hedging horizon. For the sake of simplicity, let us neglect the effect of marking-to-market.[9] We may streamline Eq. (12.14) by consider-

---

[9] We shall discuss how we may to account for marking-to-market by applying a tailing correction later, in Section 12.3.4.

ing the equivalent expression

$$\delta S + h M_F \cdot \delta F, \tag{12.15}$$

where we define the (absolute) variations

$$\delta S \doteq S(T^*) - S(0), \qquad \delta F \doteq F(T^*, T) - F(0, T),$$

and $h = Q_F/Q_A$ is the hedging ratio. Note that the quantities in Eqs. (12.14) and (12.15) are not really the same, but they are equivalent in terms of variance, as they differ by a constant. If we introduce the standard deviations $\sigma_{\delta S}$ and $\sigma_{\delta F}$ of the two variations, and their correlation coefficient $\rho_{\delta S, \delta F}$, the familiar minimum variance hedging drill yields the optimal hedging ratio

$$h^* = -\frac{\rho_{\delta S, \delta F}}{M_F} \cdot \frac{\sigma_{\delta S}}{\sigma_{\delta F}}, \tag{12.16}$$

where the negative sign just underlines that we should take a short position. If we introduce the beta between the two absolute variations,

$$\beta_{\delta S, \delta F} \doteq \rho_{\delta S, \delta F} \cdot \frac{\sigma_{\delta S}}{\sigma_{\delta F}}, \tag{12.17}$$

we may write the optimal number of futures contract as

$$Q_F^* = -\beta_{\delta S, \delta F} \cdot \frac{Q_A}{M_F} = -\beta_{\delta S, \delta F} \cdot \frac{V_A}{M_F S(0)}, \tag{12.18}$$

which should be rounded to an integer number.

We may also find Eq. (12.18) by estimating a linear regression model,

$$\delta S = \beta_{\delta S, \delta F} \cdot \delta F + \epsilon.$$

Since we are regressing between variations, we do not include the intercept term in the regression. If we neglect the error term ϵ and plug the resulting approximation,

$$\delta S \approx \beta_{\delta S, \delta F} \cdot \delta F,$$

into Eq. (12.15), the hedging condition becomes

$$\beta_{\delta S, \delta F} \cdot \delta F + h M_F \cdot \delta F = 0.$$

Thus, we find

$$h^* = -\frac{\beta_{\delta S, \delta F}}{M_F},$$

which is equivalent to Eq. (12.18).

We should note that the beta in (12.17) involves *absolute* variations, and it should not be confused with the beta in the single-index model of Section 9.2, which involves returns, i.e., *relative* variations. Let us recall the index model in the following form:

$$r_p = \alpha_p + \beta_p r_M + \epsilon_p,$$

12.3 Hedging with linear contracts

which relates the portfolio return r_p with the return r_M on the index and the specific risk factor ϵ_p. In our case, we may write

$$r_p = \frac{\delta S}{S_0},$$

and, if we assume that relative changes in the index are tracked by relative changes in the index futures price, we also have

$$r_M \approx \frac{\delta F}{F_0}.$$

In order to relate the portfolio beta with the beta of absolute variations, we observe that

$$\text{Cov}\left(\frac{\delta S}{S_0}, \frac{\delta F}{F_0}\right) = \frac{\text{Cov}(\delta S, \delta F)}{S_0 F_0},$$

$$\text{Var}\left(\frac{\delta F}{F_0}\right) = \frac{\text{Var}(\delta F)}{F_0^2}.$$

Therefore,

$$\beta_p = \frac{\text{Cov}\left(\frac{\delta S}{S_0}, \frac{\delta F}{F_0}\right)}{\text{Var}\left(\frac{\delta F}{F_0}\right)} = \beta_{\delta S, \delta F} \cdot \frac{F_0}{S_0}.$$

Now, let us rewrite Eq. (12.18) by using the portfolio beta:

$$Q_F^* = -\beta_p \cdot \frac{S_0}{F_0} \cdot \frac{Q_A}{M_F} = -\beta_p \cdot \frac{V_A}{V_F}. \tag{12.19}$$

Hence, we find an expression based on a ratio of *values*, rather than a ratio of *quantities*.

Equation (12.19) makes intuitive sense. If the portfolio has $\beta > 1$, i.e., it is quite exposed to systematic risk, we should increase the hedging ratio with respect to the case of a unit beta. Otherwise, under-hedging would result. On the contrary, we should decrease the hedging ratio when $\beta < 1$, in order to avoid over-hedging.

If we are worried about short-term drops due to market turmoil, we may use index futures to temporary hedge systematic risk away, without the need of liquidating huge portfolio positions, with a potential adverse market impact. We may also change the portfolio beta according to market-timing strategies (see Problem 12.3).

12.3.4 TAILING THE HEDGE

The idea of tailing the hedge stems from the fact that positive cash flows from marking-to-market may be invested from each settlement date to the time at which the hedge is closed (possibly, but not necessarily, the maturity if the contract), whereas negative cash flows may be financed by debt. For the sake of simplicity, we assume that a single interest rate applies to both borrowing and lending. A simple example, borrowed from [7], will illustrate the idea.

Example 12.4 An illustration of tailing the hedge

Suppose that we will have to buy one unit of an asset in 1 year. To hedge risk, we go long one futures contract at price $100, which will be settled in cash at maturity. Note that, with a forward contract, we would lock a price of to be paid in one year, with a single cash flow, resulting in a perfect hedge. However, a futures contract is marked to market each day.

Assume that, at the end of the first day, at settlement, the futures price drops to $99 and then stays constant until maturity. Then, we incur a loss of $1 on day 1, which is financed by borrowing $1 at a continuously compounded rate of, say, 5% per year. After day 1, no additional cash flow will occur due to marking-to-market. At maturity, we will buy the asset at $99 on the spot market, and the overall cash flow stream is equivalent to a negative cash flow of

$$-99 - 1 \times e^{0.05 \times 1} = -\$100.0513.$$

This is a bit of bad news, as the total cost turns out to be more than anticipated. By a similar token, let us assume that, at the settlement on the first day, the futures price rises to $101 and then stays constant. In this case, we have an immediate profit of $1, which may be invested for one year. Again, no additional cash flow occurs, due to marking-to-market, and the equivalent cash flow at maturity is

$$-101 + 1 \times e^{0.05 \times 1} = -99.9487.$$

In this case, we receive good news, but the point is that the hedge is not perfect.

Essentially, we are *over-hedging*, and we should reduce the initial hedge by a factor $e^{-0.05 \times 1}$. In other words, rather than buying one futures contract, we should buy

$$e^{-0.05 \times 1} \approx 0.9512$$

contracts. In the two scenarios, the equivalent cash flow at maturity would be the same:

$$-99 - e^{0.05 \times 1} \cdot \times e^{0.05 \times 1} = -100,$$

and

$$-101 + e^{0.05 \times 1} \cdot \times e^{0.05 \times 1} = -100.$$

In Example 12.4, we are tailing the hedge once, but in principle we should do so every day. When maturity is approaching, since the interest rate is applied to shorter and shorter time periods, the degree of tailing should be reduced. The careful reader will see some similarity with the argument that we have used in Section 12.2 to show the equivalence of futures and forward prices when interest rates are constant.

We may extend the example by recalling the telescoping sum of Eq. (1.9), which we recall here for the sake of convenience:

$$\sum_{i=1}^{m} \left[F(t_1, T) - F(t_{i-1}, T) \right] = F(t_m, T) - F(t_0, T)$$
$$= S(T) - F(t_0, T), \qquad (12.20)$$

assuming that the contract is held until maturity $T \equiv t_m$. Here, the hedging ratio is 1, but we are neglecting the time value of money. If we assume a constant interest rate r, with continuous compounding, we should rewrite the sum as

$$\sum_{i=1}^{m} e^{r \cdot (T-t_i)} \cdot \left[F(t_1, T) - F(t_{i-1}, T) \right] \neq S(T) - F(t_0, T).$$

We do find the telescoping sum of Eq. (12.20) if we tail the hedge by a factor $e^{-r \cdot (T-t_i)}$ each day:

$$\sum_{i=1}^{m} \frac{e^{r \cdot (T-t_i)}}{e^{r \cdot (T-t_i)}} \cdot \left[F(t_1, T) - F(t_{i-1}, T) \right] = S(T) - F(t_0, T).$$

In real life, interest rates will change over time, and this may affect tailing. Another complicating factor is that we may be uncertain about the hedging horizon, i.e., we may wish to close the hedge earlier than anticipated. We should also consider the cost of implementing a nervous hedging strategy. As a result, the hedge is adjusted with a lower frequency, and this is worth doing only when a long time horizon is involved.

In order to account for tailing, if we assume that spot–forward parity in the form $F_0 = S_0 e^{rT}$ applies, we may write the number of futures contract we need by a slight modification of the quantity-based ratio of Eq. (12.9):

$$N_{\text{tailed}} = \frac{N}{e^{rT}} = \frac{Q_A}{Q_F \cdot e^{rT}} = \frac{Q_A S_0}{Q_F S_0 \cdot e^{rT}} = \frac{Q_A S_0}{Q_F F_0} = \frac{V_A}{V_F}, \qquad (12.21)$$

where $V_A = Q_A S_0$ is the total value of our assets, and $V_F = Q_F F_0$ is the dollar "face value" of a futures contract. Thus, we see that we may account for tailing by using a ratio of *values*, rather than the ratio of *quantities*.

Problems

12.1 The current spot price of one GBP is €1.2. The continuously compounded interest rate in the Eurozone is 2.4%, whereas the corresponding UK interest rate is 3.1%. If the forward price of one GBP, for delivery in six months, is €1.22, are there any arbitrage opportunities? If so, devise a suitable trading strategy to take advantage of them.

12.2 You are a German investor who enters into a futures contract to buy 150,000 GBP in four months. When you take this long position, the following data apply: Spot price of one GBP is €1.13, the interest rate in the Eurozone is 2%, and the interest rate in UK is 3% (both are continuously compounded and assumed constant over the following days). At market settlement, the same day, the spot price of GBP goes up to €1.15. The day after, the settlement spot price of GBP is €1.17. On the third day, when spot price of GBP is €1.11, you close the contract.

- Assuming that forward and futures prices are the same, what are your cash flows?
- Your broker requires that you deposit cash on a margin account, with a margin ratio of 25%. If we neglect the time value of money, what has been the return of your investment over the three days?

12.3 In Section 12.3.3, we have considered how index futures may be used to neutralize a stock portfolio with respect to systematic risk. However, we may also wish to *change* the portfolio beta, rather than setting it to zero. Let β^* be the target beta that we want to achieve, which may be smaller or larger than the current portfolio β, depending on our strategy. Find the number of futures contracts that we should use.

Further reading

- A standard reference on the topics of this chapter is [4], where you may also find information about derivatives written on commodities.
- Another extensive reference is [5], whereas [7] is one of the few texts explicitly covering the need for tailing in hedging with futures.
- For an extensive coverage of stock-index futures, see, e.g., [2, Chapter 3].
- A more mathematically inclined treatment may be found in [3]. An extensive collection of papers and surveys may be found in [6].
- The link between risk-neutral pricing and futures/forward contracts is discussed in [8, Chapter 5] and [1, Chapter 26].

Bibliography

1 T. Björk. *Arbitrage Theory in Continuous Time* (2nd ed.). Oxford University Press, Oxford, 2004.
2 K. Cuthbertson and D. Nitzsche. *Financial Engineering: Derivatives and Risk Management*. Wiley, Chichester, 2001.

3 D. Duffie. *Futures Markets*. Prentice Hall, Upper Saddle River, NJ, 1989.

4 J.C. Hull. *Options, Futures, and Other Derivatives* (8th ed.). Prentice Hall, Upper Saddle River, NJ, 2011.

5 R.W. Kolb and J.A. Overdahl. *Understanding Futures Markets* (6th ed.). Blackwell Publishing, Malden, MA, 2006.

6 A.G. Malliaris and W.T. Ziemba, editors. *The World Scientific Handbook of Futures Markets*. World Scientific, Singapore, 2015.

7 R.J. Rendleman. *Applied Derivatives: Options, Futures and Swaps*. Blackwell Publishing, Malden, MA, 2002.

8 S. Shreve. *Stochastic Calculus for Finance (vol. II)*. Springer, New York, 2003.

Chapter Thirteen

Option Pricing: Complete Markets

We have introduced vanilla options in Section 1.2.6.3 and, in Section 2.3.4, we have seen how the no-arbitrage principle can be used to find an option price in the simple single-step binomial setting. The approach relies on a replication argument: It is possible to replicate the option payoff in both states of the world by a portfolio consisting of two primary assets, since the market is complete. We have also seen that lack of arbitrage is related to the existence of a risk-neutral probability measure, and that this measure is unique in a complete market. Hence, we may use that measure for pricing purposes, and risk aversion does not play any role, since any payoff may be replicated exactly, state by state. When markets are not complete, the pricing measure is not unique anymore, and risk aversion cannot be disregarded. From a more practical viewpoint, market completeness is a somewhat paradoxical assumption. If markets were complete, options would be redundant assets, and it is hard to see why an option market should exist in the first place. Another paradoxical consequence of market completeness would be that we could always get rid of any risk, since we could synthesize any payoff we wish. Clearly, this is too good to be true and, in fact, markets are not complete. In real life, we have to cope with issues related to incompleteness and residual risk. Nevertheless, a reasonably deep understanding of the simpler case of complete markets is essential to get a grasp of both the essential financial concepts and the mathematical tools needed for option pricing in the more general setting of incomplete markets. We shall deal with incomplete markets in Chapter 14.

We follow common parlance and use the term option *pricing*, whereas *valuation* would be more correct. The fair option value, in a complete market, is just the cost of an exact hedging policy, which offsets the risk exposure of the option writer and allows her to break even in every scenario. Clearly, some markup will be added by the option writer, in order to earn profit and to cover for idealized assumptions and model risk. By the same token, the price demanded by the manufacturer of a good is not just the production cost. We must also bear in mind the true purpose of pricing models. Apparently, they are of no use, at least for actively traded and liquid vanilla options, since prices are continuously quoted and determined by demand and offer mechanisms. However, we do need pricing models for over-the-counter and illiquid securities, as well as to measure and manage risk with respect to an array of risk factors.

We introduce basic option terminology, as well as some examples of exotic options, in Section 13.1. Pricing an option requires choosing a model and assigning a numerical value to its parameters, and in doing so, we are exposing ourselves to model risk. Some useful model-free restrictions on option prices, however, can be found by just relying on the no-arbitrage assumption, as we show in Section 13.2. In Section 13.3, we extend the binomial model to multiple steps in order to price a vanilla option. We also show a different perspective on pricing, in terms of hedging cost, rather than in terms of portfolio replication. In complete markets, the two views boil down to the same approach, but the hedging view is instructive and may be more useful in incomplete markets, where we cannot get completely rid of risk. The chapter's highlight is Section 13.4, where we use the continuous-time modelling framework, which we have introduced in Section 11.3, to price a vanilla option. We will prove the celebrated and controversial Black–Scholes–Merton (**BSM**) formula for a European-style vanilla option. Having an explicit formula has obvious computational advantages. What may be less obvious is that it also enables us to calculate a set of option sensitivities, collectively known as the Greeks, which may be even more important than the option price itself. We introduce the basic Greeks in Section 13.5, and we illustrate the particular role of volatility in Section 13.6. In this chapter, unless otherwise noted, we will consider options written on a stock share that will not pay any dividend before the option maturity. We relax the assumption a bit in Section 13.7, where we hint at options on assets providing income, such as options on dividend-paying stocks, indexes, currencies, and futures. Then, we close the chapter with two sections covering essential material for practitioners. In Section 13.8, we show how options can be used in speculative or risk management strategies. We also hint at numerical methods for option pricing in Section 13.9. In fact, when we foray outside the safe BSM world, explicit formulas may be a rare commodity, and we find ourselves in need for numerical methods.

13.1 Option terminology

In this section, we introduce some pieces of essential terminology. In fact, there is often some confusion about basic terms like *vanilla* and *European-style* option. The former refers to the functional form of the payoff, whereas the latter refers to the lack of early exercise features. These two dimensions may be combined in any way and should not be confused. As strange as it may sound, we may find European- or American-style Asian options!

13.1.1 VANILLA OPTIONS

By **vanilla option**, we mean European- or American-style call and put options, written on a single underlying asset, featuring a very simple and path-independent payoff. **European-style** options can be exercised only at maturity, whereas **American-style** options may be exercised at any time before expiration.[1] An intermediate case, **Bermudan-style** options, occurs when there is a finite set of early exercise opportunities before expiration.

Let S_t be the market price of the underlying asset at a time instant t within the set $\mathcal{T}$ of exercise opportunities. The payoffs of vanilla calls and puts with strike K are

$$\max\{S_t - K, 0\}$$

and

$$\max\{K - S_t, 0\},$$

respectively. In the case of European-style options, which can only be exercised at maturity, $\mathcal{T} = \{T\}$. In the case of American-style options, $\mathcal{T} = [0, T]$. Bermudan-style options feature a discrete and finite set $\mathcal{T}$, but several variations are possible. We note that the above payoffs are path-independent and depend on the price of the underlying asset at just one point in time.

It is useful to introduce common terminology, referring to the potential advantage of exercising an option at any time. We say that an option is:

- **In-the-money** at time t, if exercising the option would be profitable. Hence, a call option is in-the-money when $S_t > K$, and a put is when $S_t < K$.
- **Out-of-the-money** at time t, if exercising the option would not be profitable. Hence, a call option is out-of-the-money when $S_t < K$, and a put is when $S_t > K$.
- **At-the-money** at time t, if exercise price and asset price are equal, $S_t = K$.

As one may expect, out-of-the-money options should be cheaper than in-the-money ones. A related concept is the **intrinsic value** of an option, i.e., the payoff that could be obtained by exercising the option immediately, i.e., $S_t - K$ for a call and $K - S_t$ for a put. Clearly, an option is in-the-money when the intrinsic value is positive, and it is out-of-the-money when the intrinsic value is negative. When a European-style option is in-the-money before maturity, we cannot exercise it anyway. When an American-style option is in-the-money, we may wonder whether it is wise to exercise it immediately. As we shall see, this need not be the case.[2] The **time value** is the difference between the option value and its intrinsic value. For an in-the-money American-style option, it is worthwhile to wait, if the time value is strictly positive. By no-arbitrage,

---

[1] In this case, the term *expiration* makes more sense than *maturity*.
[2] See Section 13.2.

options cannot have a negative price. Hence, even a deeply out-of-the-money European-style option may have some time value.

13.1.2 EXOTIC OPTIONS

Exotic options may differ from vanilla options in terms of both underlying assets, which may be more than one, and payoff, which is typically path-dependent. An example of option depending on several assets is a **spread option**, whose payoff is
$$\max\{S_1(T) - S_2(T) - K, 0\},$$
where $S_1(T)$ and $S_2(T)$ are the prices of two stock shares at maturity T. We may define payoffs depending on several asset prices at the same time instant, in which case, we speak of **rainbow** options. On the contrary, path-dependent options feature a payoff depending on the price of the same underlying asset at different time instants. We introduce below a few examples of simple path-dependent options, namely, barrier options, Asian options, and lookback options. In some lucky cases, despite their additional complexity, we have analytical pricing formulas for exotic options. However, as a rule, we have to resort to numerical methods. Unlike vanilla options, exotic options are not liquid securities traded on regulated exchanges, but rather OTC securities, whose fair price can only be found by a pricing model. As one may imagine, we may couple the two dimensions and define a rainbow and path-dependent option, as we have seen in Example 1.12.

13.1.2.1 Barrier options

In barrier options, a specific asset price S_b is selected as a barrier value. If we consider a barrier version of a vanilla call or put option, there are two classification criteria:

In vs. out. In **knock-out options**, the contract is canceled if the barrier value is crossed at any time during the whole option life. On the contrary, **knock-in options** are activated only if the barrier is crossed.

Up vs. down. The barrier S_b may be above or below the initial asset price S_0. If $S_b > S_0$, we have an up option, and if $S_b < S_0$, we have a down option.

Example 13.1 Down-and-out put options

A down-and-out put option is a put option that becomes void if the asset price falls below the barrier S_b. We must have $S_b < S_0$, otherwise the option has already been canceled, and $S_b < K$, otherwise, the option will be canceled before getting in-the-money. The rationale behind such an option is that the risk for the option writer is reduced. So, it is reasonable to expect that a down-and-out put option

> is cheaper than a vanilla one. Now, consider a down-and-in option. This option is activated only if the barrier level $S_b < S_0$ is crossed. Holding both a down-and-out and a down-and-in put option is equivalent to holding a vanilla put option. So, we have the following parity relationship:
>
> $$P = P_{\text{di}} + P_{\text{do}}, \tag{13.1}$$
>
> where P is the price of the vanilla put, and P_{di} and P_{do} are the prices of the down-and-in and the down-and-out options, respectively. Sometimes a rebate is paid to the option holder if the barrier is crossed and the option is canceled; in such a case the above parity relationship is not correct.

In principle, the barrier might be monitored continuously; in practice, periodic monitoring may be applied (e.g., the price could be checked each day at the close of trading). This may affect the price, as a lower monitoring frequency makes the detection of barrier crossing less likely. In Sections 13.5 and 13.6.2, we shall see that the price of a barrier option may feature complicated dependence patterns with respect to the current price of the underlying asset, its volatility, and the barrier location.

13.1.2.2 Asian options

Barrier options exhibit a weak degree of path dependency. A stronger degree of path dependency is typical of Asian options, whose payoff depends on the *average* asset price over the option life. Different Asian options may be devised, depending on how the average is computed. Sampling may be discrete or (in principle) continuous, and the average may be arithmetic or geometric. The discrete arithmetic average is

$$A_{\text{da}} = \frac{1}{n} \sum_{i=1}^{n} S(t_i),$$

where t_i, $i = 1, \ldots, n$, are the discrete sampling times. The discrete geometric average is

$$A_{\text{dg}} = \left[\prod_{i=1}^{n} S(t_i) \right]^{1/n}.$$

If continuous-time sampling is assumed, we have the following continuous arithmetic and geometric averages, respectively:

$$A_{\text{ca}} = \frac{1}{T} \int_0^T S(t)\, dt, \qquad A_{\text{cg}} = \exp\left[\frac{1}{T} \int_0^T \log S(t)\, dt \right].$$

As one can imagine, continuous time monitoring is not quite practical. Given a choice of the average A, we may use it to define a rate or a strike. An **average rate** call has a payoff given by

$$\max\{A - K, 0\},$$

whereas for an average strike call, we have

$$\max\{S(T) - A, 0\}.$$

By the same token, we may define an average rate put,

$$\max\{K - A, 0\},$$

or an average strike put,

$$\max\{A - S(T), 0\}.$$

Early exercise opportunities may also be specified in the contract, so that we may define European-, American-, or Bermudan-style Asian options.

13.1.2.3 Lookback options

Lookback options come in different forms, and their basic feature is that the maximum (or the minimum) underlying asset price is monitored during the option life. Assuming continuous monitoring, we may observe the maximum and the minimum asset price:

$$S_{\max} = \max_{t \in [0,T]} S(t),$$
$$S_{\min} = \min_{t \in [0,T]} S(t).$$

A European-style **lookback call** has a payoff given by

$$S(T) - S_{\min},$$

whereas in the case of a **lookback put**, we have

$$S_{\max} - S(T).$$

Just like with Asian options, we may also include early exercise features.

13.2 Model-free price restrictions

As we shall see later, in order to find an option price, we have to take our chances and choose a specific model for the dynamics of the underlying asset

13.2 Model-free price restrictions

price. Note that this is not the case with simple forward contracts, as spot–forward parity only relies on a no-arbitrage assumption.[3] In general, finding an option price does require a dynamic model, which will always be an approximation of reality. Therefore, model risk becomes an issue, and there is an unclear tradeoff between model sophistication, which is hopefully associated with increased realism, and robustness.

In this section, we find restrictions on option prices, such as bounds and parity relationships, requiring only the no-arbitrage condition. In principle, they may be useful to spot arbitrage opportunities, but they are mostly important tools for intuition building. We consider both European- and American-style vanilla options, written on a stock share with price process S_t. The underlying asset may or may not pay dividends. We denote the price of European- and American-style call options at time t by C_t^e or C_t^a, respectively. By the same token, we use P_t^e or P_t^a for put options.

It is easy to understand that an American-style option cannot be less expensive than the corresponding European-style contract, since it offers a larger set of exercise opportunities. We may wonder whether an American-style option should be immediately exercised when the intrinsic value is positive. However, the decision is not so trivial, as we may keep the option alive and wait for better opportunities. Indeed, pricing an option with early exercise opportunities is a stochastic dynamic optimization problem, belonging to the class of optimal stopping problems. We should keep in mind the following decomposition of the option value:

$$\text{Option value} = \text{Intrinsic value} + \text{Time value}.$$

When the intrinsic value is positive, we may exercise an American-style option and earn a profit. However, it is optimal to do so only when there is no time value, i.e., no reason to wait for better opportunities. When the time value is positive, we should keep the option alive and continue.

13.2.1 BOUNDS ON CALL OPTION PRICES

At maturity, the value of a European-style vanilla call is just given by the payoff,

$$C_T^e = \max\{S_T - K, 0\}.$$

Clearly, the payoff is never negative, since the option provides us with a right to exercise, without any obligation. Furthermore, the largest payoff occurs with a call option with strike zero. Hence, the option payoff is bounded as follows:

$$0 \leq C_T^e \leq S_T.$$

---

[3] To be precise, we also rely on some modeling assumptions concerning the market, like the absence of transaction costs and the possibility of unlimited short-selling. However, in simple cases, we do not need to specify a full-fledged dynamic model.

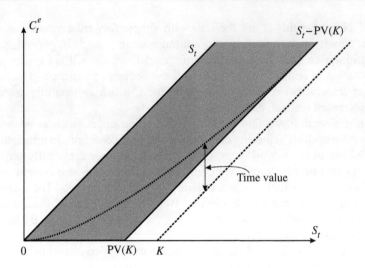

FIGURE 13.1 Bounding call option prices.

An immediate consequence is that the call option price at any time t cannot be negative, and it cannot be larger than the current stock share price:

$$0 \leq C_t^e \leq S_t, \qquad \forall t \in [0,T].$$

This is just a consequence of no-arbitrage. Intuitively, the right to buy the stock share at strike price K cannot be larger than the stock price itself. A slightly less trivial observation is that the payoff at maturity is always larger than the intrinsic value,

$$C_T^e = \max\{S_T - K, 0\} \geq S_T - K.$$

Note that the intrinsic value can be negative, and it may be thought as the value of a portfolio consisting of a long position in one stock share and a short position in a zero with face value K, maturing at time $t = T$. What is the value of this portfolio at time t? The no-arbitrage value of the stock share at time t, just like any security with a payoff S_T at maturity, is clearly S_t. Note that we *cannot* discount S_T back to time t, as this is a random variable. We may do so with the risk-free bond, whose current value is found by discounting its face value by the risk-free, continuously-compounded rate r. Therefore, we find a lower bound on the option value,

$$C_t^e \geq S_t - Ke^{-r \cdot (T-t)} = S_t - \mathsf{PV}(K). \tag{13.2}$$

where $\mathsf{PV}(K)$ is the present value of the strike price.

By putting all of these observations together, we see that the value of a call option must be contained in the shaded diagonal strip in Fig. 13.1. We may also make an *educated guess* about a qualitative plot of the option value at time t, shown in Fig. 13.1 as a dotted line, based on the following observations:

13.2 Model-free price restrictions

- The call option value should be an increasing function of the current stock price S_t.
- If the stock price is zero, it means that the firm went bankrupt, and the option will be out-of-the-money at maturity for sure. Hence,

$$\lim_{S_t \to 0} C_t^e = 0.$$

- By a similar token, if the stock price is very large, the call option will be in-the-money at maturity with a very high probability. Thus, with probability close to 1, the payoff will be $S_T - K$, which implies

$$\lim_{S_t \to +\infty} C_t^e = S_t - Ke^{-r \cdot (T-t)} = S_t - \mathsf{PV}(K).$$

The option value must be above the increasing line given by $S_t - \mathsf{PV}(K)$. When we approach maturity, this line, defining the lower bound, is shifted down and to the right, converging to the line $S_T - K$ corresponding to the option payoff, if the derivative is in-the-money. Thus, we may expect that the price of a vanilla call option, all other factors being constant, will be a *decreasing* function of time.[4]

A careful look at Fig. 13.1 also leads us to a possibly surprising finding. In the figure, we show the time value of the option, which is always *positive* before maturity. In fact, we may write

$$C_t^a \geq C_t^e \geq S_t - Ke^{-r \cdot (T-t)} > S_t - K, \qquad t < T. \qquad (13.3)$$

Hence, the value of an American-style call option is always larger than the intrinsic value, before the expiration. This proves the following theorem.

THEOREM 13.1 (Early exercise of American-style call options) *It is never optimal to exercise early an American-style call option written on a non-dividend-paying stock. As a consequence, $C_t^a = C_t^e$.*

The theorem may seem counterintuitive at first. In order to gain some intuition, let us examine two cases in which we may consider early exercise of an American-style, in-the-money call option. If we are not really interested in buying and keeping the stock share, Eq. (13.3) shows that it is more profitable to sell the call option, rather than exercising it early. Let us assume that, on the contrary, we really want to buy and hold the stock. In this case, the later we pay the exercise price, the better. There is little point in paying K before maturity. Furthermore, the stock price could fall below the strike price at a later time, and we would regret our decision to exercise early. However, if the underlying stock share pays a dividend before the option expiration, then it may be profitable to exercise just before the stock goes ex-dividend, in order to earn the right to the dividend.

---

[4] A quick peek at Fig. 13.10, which is based on the BSM model, shows that the intuition is reasonable. This need not be the case with other options.

To see more clearly how the situation may change, if we consider a stock distributing dividends, let us denote by D the value at maturity of the dividends paid from time t to the option maturity.[5] Now, let us compare two portfolios at time t:

Portfolio A consists of a long position in the European-style call and an amount of cash $(D + K)e^{-r(T-t)}$.

Portfolio B consists of a long position in the stock share.

We assume that the risk-free rate r, at which we may invest cash, will not change over time. At maturity, portfolio A has value

$$\max\{S_T - K, 0\} + D + K = \max\{S_T, K\} + D.$$

At maturity, portfolio A has value

$$S_T + D.$$

Since $\max\{S_T, K\} \geq S_T$, we find that the value of portfolio A at maturity cannot be less than the value of portfolio B, which implies

$$C_t^e \geq S_t - (D + K)e^{-r(T-t)}.$$

We see that, with respect to the case of no dividends, the lower bound is decreased by an amount $De^{-r(T-t)}$. It could be the case that this lower bound is smaller than the intrinsic value $S_t - K$. Hence, in the case of dividends, we cannot claim that the American-style option should never be exercised. As it turns out, it might be optimal to exercise just before the stock goes ex-dividend.

13.2.2 BOUNDS ON PUT OPTION PRICES: EARLY EXERCISE AND CONTINUATION REGIONS

By following the line of reasoning that we have used in Section 13.2.1 for a call option, we can find bounds on the value of a put option. Clearly, the put value cannot be negative. Furthermore, since the stock share price is non-negative as well, we immediately find the upper bounds

$$P_t^e \leq Ke^{-r(T-t)} = \mathsf{PV}(K),$$
$$P_t^a \leq K.$$

This bound corresponds to the case of a firm going bankrupt, so that $S_t = S_T = 0$. However, with a European-style option, the payoff K cannot be received now, but only at maturity; so, the amount K must be discounted. For a European-style put, it also easy to see that

$$\max\{K - S_T, 0\} \geq K - S_T,$$

---

[5] Long-term dividends are uncertain, but the amount of the next one is usually communicated in advance. Hence, unless we are dealing with an option with a long maturity, we may assume that D is known.

13.2 Model-free price restrictions

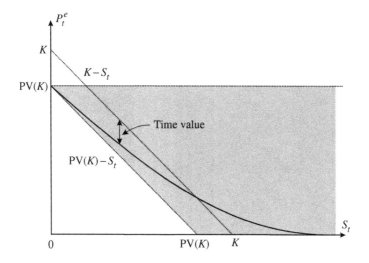

FIGURE 13.2 Bounding put option prices.

which implies the lower bound

$$P_t^e \geq Ke^{-r(T-t)} - S_T.$$

Furthermore, it stands to reason that

$$\lim_{S_t \to 0} P_t^e = Ke^{-r(T-t)}.$$

and

$$\lim_{S_t \to +\infty} P_t^e = 0.$$

Putting all of this together, we find the strip and the plausible put option price depicted in Fig. 13.2. Note that, unlike the case of a call, the downward sloping line $PV(K) - S_t$ will shift up and to the left with time. We also see that, for an in-the-money put, the time value can be *negative*. Thus, the value of a put could increase in time.[6]

This has a deep consequence for the corresponding American-style put, as the option value cannot be less than the intrinsic value,

$$P_t^a \geq K - S_t,$$

since the option can always be exercised immediately. If the intrinsic value is positive and the time value is zero, then the value of the option is just its immediate payoff; hence, there is no reason to wait and it is optimal to exercise the option immediately. In this case, the intrinsic value is larger than the **continuation value**, which is the value of keeping the option alive. This is illustrated

---

[6] In Section 13.5, when discussing option Greeks, we shall see that theta can be positive for an in-the-money put.

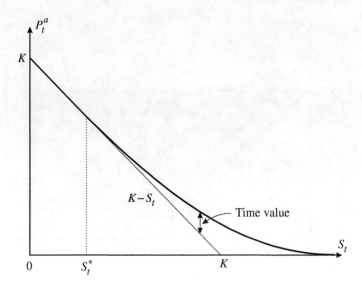

FIGURE 13.3 Bounds and time value for an American-style put option.

in Fig. 13.3, where a critical price S_t^* is shown, separating the **exercise region** from the **continuation region** at time t. If the stock price S_t is below the strike, the put is in-the-money, but it may be better to continue, without exercising. If $S_t \leq S_t^*$, then it is optimal to exercise. We note that the option value is given by a straight line in the exercise region, and it is larger than the intrinsic value in the continuation region, where time value is positive. In Fig. 13.3, we also observe that the option value is not only continuous, but continuously differentiable, too, where the linear and the nonlinear portions of the curve meet. The slope of the put price function is -1 in the exercise region, and there a "smooth pasting" between the two curves at the point corresponding to the critical price. This smooth pasting condition can be justified formally, but this is beyond the scope on an introductory book. In Fig. 13.4, we plot the critical stock price S_t^* as a function of time, which allows us to get a clearer picture of the boundary between the continuation and exercise (shaded) regions. This plot should be intended as an educated guess, as we did not show at all that the boundary is convex. However, it makes sense that S_t^* increases with time. When we get closer and closer to maturity, there are less and less opportunities to wait for, and we should exercise, settling for smaller and smaller intrinsic values.

The net consequence of all of these observations is that, unlike call options, American-style puts can be more expensive than the corresponding European-style counterparts. Finding this price is challenging as it requires finding the boundary of the exercise region by solving an optimization problem. In fact, as a general rule, pricing options with early exercise features calls for the application of numerical methods. We discuss a simple approach, based on binomial lattices, in Section 13.3.4; then, in Chapter 15, we generalize the idea within the framework of stochastic dynamic programming.

13.2 Model-free price restrictions

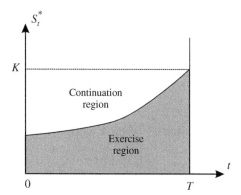

FIGURE 13.4 Exercise and continuation regions for an American-style put option.

13.2.3 PARITY RELATIONSHIPS

The bounds that we have considered in Sections 13.2.1 and 13.2.2 involve a single option. Parity relationships involve two derivatives. We have seen an example of arbitrage opportunity in Example 2.8, which involved trading both a vanilla call and a vanilla put. The arbitrage opportunity was a consequence of the violation of the following **put–call parity** relationship, which links European-style vanilla call and put options, written on the same asset, with the same maturity, and the same strike:

$$P_t^e + S_t = C_t^e + Ke^{-r \cdot (T-t)}. \qquad (13.4)$$

This is a model-free relationship that can be proven by considering the following two portfolios:

1. **Protective put**, i.e., a European-style put option plus the underlying asset.
2. **Call-plus-bills**, i.e., a European-style call option plus a risk-free zero-coupon bond with the same maturity and face value K.

The value of the protective put at maturity is

$$\max\{K - S_T, 0\} + S_T = \max\{K, S_T\}.$$

At maturity, the call-plus-bills portfolio has value

$$\max\{S_T - K, 0\} + K = \max\{S_T, K\}.$$

The two portfolios will have the same value in the future, whatever scenario is realized. By the law of one price, the two portfolios must have the same value now, in order to rule out arbitrage opportunities; this implies Eq. (13.4). A practical consequence is that we only need to find a formula for the price of the call option, as the price of the put is immediately obtained. This parity relationship may be adapted to the case of dividends.[7] In the case of American-style options, things are not so easy, but bounds may be obtained.

There are other parity relationships, and we have seen an example involving barrier options in Eq. (13.1). We also have parity-relationships involving more complicated derivatives, including interest rate derivatives.

Example 13.2 A parity relationship involving interest rate swaps

We have discussed the valuation of vanilla interest rate swaps in Section 4.4. The net cash flow received by the fixed-rate payer at time T_i, $i = 1, \ldots, m$, is

$$N \cdot \Delta \cdot \left[L_n(T_{i-1}, T_i) - K_n\right],$$

where we assume that $\Delta = T_i - T_{i-1}$ is the time elapsing between two consecutive dates (assumed constant for the sake of simplicity), $n = 1/\Delta$ is the corresponding compounding frequency, N is the notional value, $L_n(T_{i-1}, T_i)$ is the LIBOR rate with discrete compounding for the tenor $[L_{i-1}, L_i]$, and K_n the fixed swap rate. Time T_0 is the last time at which the LIBOR was reset, and we are typically interested in the swap value at time t, $T_0 < t < T_1$.

As we have see in Section 5.3.2, interest rate derivatives are traded, which look much like call and put options on interest rates. An interest rate cap consists of a portfolio of m caplets with payoff

$$N \cdot \Delta \cdot \max\left\{L_n(T_{i-1}, T_i) - K_n, 0\right\}.$$

By a similar token, an interest rate floor consists of a portfolio of m floorlets with payoff

$$N \cdot \Delta \cdot \max\left\{K_n - L_n(T_{i-1}, T_i), 0\right\}.$$

Clearly, for any random variable S and any given number K, we have

$$S - K = \max\{S - K, 0\} - \max\{K - S, 0\}.$$

Thus, assuming that cash flows have the same timing and that swap, cap, and floor rates are identical, by no-arbitrage, we find a parity relationship among the values of the three derivatives at time t:

$$V_{\text{swap}}(t) = V_{\text{cap}}(t) - V_{\text{floor}}(t).$$

---

[7] See Problem 13.6.

13.3 Binomial option pricing

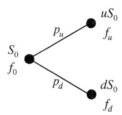

FIGURE 13.5 Single-step binomial model for option pricing.

13.3 Binomial option pricing

Model-free relationships are useful tools, but they provide us with limited information. To proceed further, we need to take our chances with a specific model. In Section 2.3.4, we have considered a simple pricing approach based on a single-step binomial model, whose structure is depicted again in Fig. 13.5, for the sake of convenience. We relied on replication arguments and the law of one price, to conclude that the current option value must be

$$f_0 = e^{-rT} \cdot \{\pi_u f_u + \pi_d f_d\}, \qquad (13.5)$$

which does *not* involve the real world probabilities p_u and p_d, but the risk-neutral probabilities

$$\pi_u = \frac{e^{rT} - d}{u - d}, \qquad \pi_d = \frac{u - e^{rT}}{u - d}.$$

Hence, we may express the option value in the following form:

$$f_0 = e^{-rT} \cdot \mathrm{E}_{\mathbb{Q}_n}[f_T], \qquad (13.6)$$

which is interpreted as the discounted expected value of the payoff under the risk-neutral measure $\mathbb{Q}_n$. Under the risk-neutral measure, the expected return of any security is the risk-free rate. For instance, the expected value of the underlying asset price at maturity is

$$\mathrm{E}_{\mathbb{Q}_n}[S_T] = \pi_u S_0 u + \pi_d S_0 d = S_0 e^{rT}.$$

This might happen only in a world where investors do not care about risk, but only about the expected return, and do not ask for any risk premium. In such a risk-neutral world, the expected return of all assets, at equilibrium, should be the same, and if there is a risk-free asset in the market, its return will be the expected return for all risky assets as well.

We shall extend our understanding of risk-neutral measures to martingale measures later, in Chapter 14. Here, we extend the single-step binomial model to multiple steps, but before doing so, let us repeat the argument in a different perspective, which shall prove useful later.

13.3.1 A HEDGING ARGUMENT

It is very instructive to recast the replication kind of argument in terms of hedging the risk borne by an option writer. Assume that we have *written* a European-style, vanilla call option on a stock. How can we *hedge* against risk? The writer might just cash the option premium in, cross her finger and wait. This corresponds to a "naked" position, as nothing is done to manage risk. We recall from Fig. 1.7 that the payoff for a call option writer may consist of an unbounded loss, showing the dangers of the naked position. An alternative is the "covered" position, where the writer buys one share of the underlying asset (for each option), just in case the option is exercised. However, this strategy is risky as well, as the price of the underlying asset may decrease considerably, so that the option expires worthless, but we have to liquidate the hedge at a low price. The naked position is under-hedged, and the covered position is over-hedged. We should try to find the "right" number of shares to hold, which should lie somewhere between 0 (naked position) and 1 (covered position). This should be a *long* position in the underlying asset, for the writer of a call option, as the gain from this position will compensate for the loss on the option payoff, when the underlying asset price goes up. Before proceeding further, it is important to ask what hedge should be adopted by the writer of a European-style, vanilla put option. A moment of reflection suggests that, in this case, we should take a *short* position in the underlying asset. If its price goes down and the option is exercised by its holder, the short hedge will compensate the negative payoff for the writer.

Say that we purchase Δ stock shares to cover the risk of writing a generic option with payoffs f_u and f_d. If we have written the option, the initial value of our portfolio is

$$\Pi_0 = \Delta S_0 - f_0. \tag{13.7}$$

Remark. It is important to understand that Eq. (13.7) gives the *value* of the portfolio at time $t = 0$, and that the value f_0 of the derivative occurs with a negative sign because the writer will lose money (corresponding to holder's profit) if the option value increases. Indeed, the writer has a short position in the option. We are interested in how the portfolio value $\Pi_t = \Delta S_t - f_t$ changes over time. In terms of *cash flows* at time $t = 0$, signs would be reversed, as the option premium is earned by the writer. It is quite common to see a confusion between values and cash flows. However, for instance, bond values are what they are; then, cash flow signs depend on whether we buy or sell the bonds.

In the binomial model, the two possible portfolio values at maturity T are

$$\Pi_u = \Delta u S_0 - f_u,$$
$$\Pi_d = \Delta d S_0 - f_d.$$

We can make the portfolio riskless by choosing Δ such that

$$\Pi_u = \Pi_d \quad \Rightarrow \quad \Delta = \frac{f_u - f_d}{S_0(u - d)}.$$

13.3 Binomial option pricing

If the portfolio is riskless, no-arbitrage dictates that it must earn the risk-free interest rate r, i.e.,

$$\Pi_u = \Pi_d = \Pi_0 e^{rT},$$

where we assume continuous interest compounding. Therefore, we may write

$$(\Delta S_0 - f_0)e^{rT} = \Delta u S_0 - f_u.$$

Solving for the current option value we find

$$f_0 = \Delta S_0 \left(1 - u e^{-rT}\right) + f_u e^{-rT},$$

and by plugging the expression for Δ and rearranging, we obtain Eq. (13.5) again.

Since Δ is the number of stock shares that we should hold, in order to hedge the risk of each option, it plays the role of a hedging ratio. Note that Δ is positive for a call option, but negative for a put option. It is also useful to interpret the increment ratio

$$\Delta = \frac{f_u - f_d}{S_0(u - d)} = \frac{f_u - f_d}{S_u - S_d} \tag{13.8}$$

as a discretized approximation of the partial derivative of the option value with respect to changes in the underlying price, i.e., $\Delta \approx \partial f / \partial S$. When we move on to a continuous-time model, we shall see that, in fact, the hedging ratio is exactly the sensitivity of the option value to the underlying asset price.

To get the full picture, let us recall Eq. (2.22),

$$\Psi = e^{-rT} \cdot \frac{u f_d - d f_u}{u - d},$$

which gives the amount of cash which is needed to hedge the option. If this amount is negative, it means that the option writer has to borrow money to set up the hedge, as the following example illustrates.

Example 13.3 A numerical example

Let us consider a call option with maturity $T = 1$ year and strike price $11. The current price of the stock share is $10, and the risk-free interest rate is 10%. The two possible returns of the stock share in one year are either 20% or -10%, which implies $u = 1.2$ and $d = 0.9$. The corresponding payoffs in the up and down states are:

$$f_u = \max\{10 \times 1.2 - 11, 0\} = 1,$$
$$f_d = \max\{10 \times 0.9 - 11, 0\} = 0,$$

respectively. To find the replicating portfolio, we need the number of stock shares Δ and the amount Ψ to invest in (or to borrow from) the bank account. To this aim, we have to solve the system of linear equations:

$$12\Delta + \Psi e^{0.1} = 1,$$
$$9\Delta + \Psi e^{0.1} = 0.$$

If we subtract the second equation from the first one, we get

$$(12-9)\Delta = 1 \quad \Rightarrow \quad \Delta = \tfrac{1}{3} \approx 0.3333.$$

Thus, for each call option, we should buy one third of a stock share. Plugging this value back into the first equation yields

$$\Psi = \frac{1 - \frac{12}{3}}{e^{0.1}} = -3 \cdot e^{-0.1} \approx -2.7145,$$

which means that the writer should borrow some cash. Note that the writer has to repay a debt that, at maturity, will amount to $3. By putting everything together, we obtain the fair option premium

$$f_0 = \Delta S_0 + \Psi = \tfrac{10}{3} - 2.7145 \approx \$0.6188.$$

This option premium and the additional borrowed cash are used to buy Δ shares.

Note that, with a naked position, when the stock price goes up, the writer will have to buy one share at $12 just to hand it over to the option holder for $11, losing $1. With a covered position, if the price goes down and the option is not exercised, the writer loses $1 since she has to sell for $9 the share that was purchased for $10.

Let us check that, on the contrary, risk is hedged away if the writer buys $\Delta = 1/3$ shares. If the stock price goes down to $9, the option writer will just unwind the hedge and sell 1/3 shares for $3. This is just what the writer needs in order to repay debt at maturity. Hence, the option writer breaks even in the "down" scenario. In the "up" scenario, 2/3 additional shares are purchased at the unit price of $12, in order to sell a whole share to the option holder. The option writer breaks even again, as the $11 cashed in from the option exercise are exactly what she needs, $8 for the additional shares plus $3 to repay debt.

13.3.2 LATTICE CALIBRATION

In order to make binomial lattices a working tool for option pricing, we need to pause and tackle a necessary task, **lattice calibration**. For now, we have really no idea about how we could choose the multiplicative shocks u and d, as well as the related probabilities. Here, we use the moment matching approach that was introduced in Section 11.6.2. The binomial model should be a good discretization of a continuous-time, continuous-state model. Let us use the simplest model, geometric Brownian motion (GBM),

$$dS_t = \mu S_t\, dt + \sigma S_t\, dW_t.$$

For pricing purposes, we should change the probability measure to the risk-neutral one. It turns out that this amounts to a simple change in the drift coefficient. Rather unsurprisingly, under the risk-neutral measure, the model is

$$dS_t = r S_t\, dt + \sigma S_t\, dW_t.$$

Indeed, the drift coefficient is related to the rate of return, which would be the risk-free rate in the risk-neutral world. We will better motivate this change later.[8] Hence, we should find the three parameters u, d, and π_u, in such a way that some essential properties of the continuous-time model are preserved by the binomial model. Since the GBM is a process with constant drift and volatility coefficients, the structure of the single-step binomial lattice is the same for each time step. By approximating the dynamics over a small time step of length δt, we may replicate the building block of Fig. 13.5 and generate the multistep recombining binomial lattice, as depicted later in Fig. 13.6. Therefore, we may focus on the calibration of a single step.

In Section 11.5.1, we have learned about the essential properties of GBM. Conditional on being in state S_t at time t, the new state at time $t+\delta t$ is a random variable $S_{t+\delta t}$, such that

$$\log\left(\frac{S_{t+\delta t}}{S_t}\right) \sim \mathsf{N}\left[\left(r - \frac{\sigma^2}{2}\right)\cdot \delta t,\ \sigma^2 \cdot \delta t\right].$$

Using properties of the lognormal distribution, we find

$$\mathrm{E}[S_{t+\delta t}\,|\,S_t] = S_t \cdot e^{r\cdot \delta t} \qquad (13.9)$$

and

$$\mathrm{Var}[S_{t+\delta t}\,|\,S_t] = S_t^2 \cdot e^{2r\cdot \delta t}\left(e^{\sigma^2 \cdot \delta t} - 1\right). \qquad (13.10)$$

Now, we may use the same idea that we applied, within a simpler setting, in Example 11.14. A reasonable requirement on the discretized dynamics is that it

---

[8] A rigorous approach to change of measure requires more sophisticated concepts from probability and stochastic calculus, like Radon–Nikodym derivatives and the Girsanov theorem, which are beyond the scope of this book.

should match the two above features of the continuous model. Note that these are two conditions, but we have three parameters: π_u, u, and d. Hence, we have one additional degree of freedom, which may be used by choosing $u = 1/d$. Note that this is just one possible choice, which is not required by the need for recombination. The lattice with multiplicative shocks is recombining anyway, since $S_t u d = S_t d u$, so that the two sequences of shocks, up–down and down–up, result in the same state. Our choice leads to the so-called CRR (Cox, Ross, and Rubinstein) lattice calibration.

On the lattice, we have

$$E[S_{t+\delta t} \mid S_t] = \pi_u u \cdot S_t + (1 - \pi_u) d \cdot S_t,$$

which, together with (13.9), yields[9]

$$\pi_u u \cdot S_t + (1 - \pi_u) d \cdot S_t = e^{r\,\delta t} S_t \quad \Rightarrow \quad \pi_u = \frac{e^{r\,\delta t} - d}{u - d}.$$

To match variance, we observe that, on the lattice,

$$\operatorname{Var}(S_{t+\delta t} \mid S_t) = \mathrm{E}[S_{t+\delta t}^2 \mid S_t] - \mathrm{E}^2[S_{t+\delta t} \mid S_t]$$
$$= S_t^2 \left[\pi_u u^2 + (1 - \pi_u) d^2 \right] - S_t^2 e^{2r\,\delta t},$$

which should be matched against Eq. (13.10). After a few tedious calculations,[10] involving a linear approximation of a nonlinear equation, we end up with the parameterization

$$u = e^{\sigma \sqrt{\delta t}}, \qquad d = e^{-\sigma \sqrt{\delta t}}, \qquad \pi_u = \frac{e^{r\,\delta t} - d}{u - d}. \tag{13.11}$$

We observe that the larger the volatility, the larger the gap between u and d, which makes intuitive sense.

13.3.3 GENERALIZATION TO MULTIPLE STEPS

Pricing by a single-step binomial lattice is clearly too crude to be of any practical use. In principle, we could rely on a single-step tree (i.e., a scenario fan), but the usual replication/hedging arguments would require a large number of spanning assets, which is not practical. Furthermore, the underlying asset price at maturity is not sufficient to price path-dependent and American-style derivatives, for which the whole path over multiple time steps is required. Thus, we need a multistep model allowing for trading at intermediate times. Since a full-fledged scenario tree is plagued by the curse of dimensionality, the recombining binomial lattice shown in Fig. 13.6 is a quite popular discrete-time, discrete-state model. Here, we consider a three-step lattice, where option maturity is

---

[9] We find a familiar formula for the risk-neutral probability π_u, which might be taken for granted. Actually, there are alternative calibrations. For instance, we might calibrate on the basis of log-

13.3 Binomial option pricing

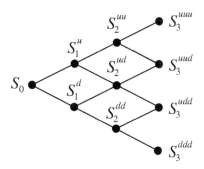

FIGURE 13.6 Multistep binomial model for option pricing.

$T = 3 \cdot \delta t$. Each up-step corresponds to multiplication by the shock u, and each down-step corresponds to multiplication by the shock d. For instance,

$$S_3^{uud} = S_0 uud.$$

The lattice is recombining because of the multiplicative nature of the shocks, $S_0 ud = S_0 du$. In fact, the exact sequence of up and down shocks does not really matter. What matters is the *number* of up and down shocks in the sequence. For instance, we find the same price S_3^{uud} for the following three paths:

up-up-down, up-down-up, down-up-up,

where a single down shock occurs. Furthermore, if we use the calibration of Eq. (13.11), where $ud = 1$, we find a limited set of different prices since, for instance, $S_3^{uud} = S_1^u$. Clearly, the recombining feature is convenient in computational terms: For the lattice in Fig. 13.6, we have 4 nodes after 3 steps. In general, with N time steps, we have $N + 1$ nodes in the last time layer of a recombining lattice. In a binomial tree, we would have an exponential growth, with 2^N nodes at the last time layer. This saving, however, has a price: We cannot price strongly path-dependent options like Asian options. For complex derivatives, alternative numerical methods can be used, which we hint at in Section 13.9.[11]

Assuming that the risk-free interest rate and the volatility are constant in time, the lattice calibration of Eq. (13.11) applies to the entire lattice of the underlying asset price, which is easy to generate. Then, we have to build and fill the corresponding lattice of option prices. This requires a backward calculation process, where we start from the terminal (rightmost) layer in the option lattice, which corresponds to the option payoff at maturity. Then, we should just apply

---

price, rather than price, and we could use the available degree of freedom to set $\pi_u = \pi_d = 1/2$, and then find the shocks accordingly.

[10] See, e.g., [4, Chapter 7] for details. There, it is also shown that the choice $ud = 1$ may yield some computational advantages in terms of memory requirements.

[11] In Problem 13.9, we give an illustration of how trees might be used to price path-dependent options.

Eq. (13.5) recursively, going backward one step at a time, until we reach the initial node. In order to formalize the approach, let f_{ij} be the option value at node (i, j), where j refers to the time instant $j\,\delta t$, $j = 0, \ldots, N$, and i is the ith node in period j. Here, N is the number of time steps we consider, and $N\delta t = T$ is the option maturity. Note that we start with $j = 0$, and N time steps correspond to $N + 1$ time instants indexed by $j = 0, \ldots, N$. We assume, by convention, that node numbers increase going up in the lattice; at time step j, we have $j + 1$ nodes labeled by (i, j), $i = 0, \ldots, j$. With these conventions, the price of the underlying asset at node (i, j) is $S_{ij} = S_0 u^i d^{j-i}$. We observe that i is nothing but the number of up-steps. For instance, if we consider a call option, we have the following option values at maturity:

$$f_{i,N} = \max\{0, S_0 u^i d^{N-i} - K\}, \qquad i = 0, 1, \ldots, N.$$

By going backward in time, decreasing time subscript j, we find

$$f_{ij} = e^{-r\cdot\delta t} \cdot \left[\pi_u f_{i+1,j+1} + (1 - \pi_u) f_{i,j+1}\right]. \tag{13.12}$$

All of this is best illustrated by a simple example.

▣ Example 13.4 Pricing a call option by a binomial lattice

Suppose that we want to find the price of a vanilla European-style call with $S_0 = K = 50$, $r = 0.1$, $\sigma = 0.4$, and $T = \frac{5}{12}$ (time-to-maturity is five months). We must first set up the lattice parameters. Suppose that each time step is one month. Then,

$$\delta t = \tfrac{1}{12} = 0.0833,$$
$$u = e^{\sigma\sqrt{\delta t}} = 1.1224,$$
$$d = 1/u = 0.8909,$$
$$\pi_u = \frac{e^{r\,\delta t} - d}{u - d} = 0.5073.$$

Figure 13.7 shows the resulting lattices for the stock price and the option value. To see how the lattice for the stock price is built, at node $(1, 1)$ we have

$$S_{1,1} = S_0 u = 50 \cdot 1.1224 \approx 56.12,$$

which is the same as $S_{3,5} = S_0 u^3 d^2$. The lattice of option values is initialized with the option payoffs. So, for instance,

$$f_{5,5} = \max\{S_{5,5} - K, 0\} = \max\{89.07 - 50, 0\} = 39.07.$$

The option value at the uppermost node in the second-to-last time layer, $f_{4,4}$, depends on the option values $f_{5,5}$ and $f_{4,5}$, and is obtained

13.3 Binomial option pricing

> as follows:
>
> $$e^{-r \cdot \delta t} \left[\pi_u \cdot 39.07 + (1 - \pi_u) \cdot 20.77 \right]$$
> $$= e^{-0.1 \cdot 0.0833} \left[0.5073 \cdot 39.07 + 0.4927 \cdot 20.77 \right] \approx 29.77.$$
>
> By going backward recursively, we find that the estimated option price is about 6.36. Later, in Example 13.6, we will see that the exact option price is 6.1165. Hence, the approximation we have found is not bad at all, considering that we have used a very crude discretization, where each time step is one month. In practice, something like 1000 steps is needed to find a satisfactory approximation.

We should note that, in principle, the above calculation is not really necessary to price a vanilla, path-independent, European-style option. We may just consider the last time layer, associate probabilities with each node, and compute the expected value. It is fairly easy to see that the probabilities of each terminal node depend on the number of up and down steps, as well as their respective risk-neutral probabilities, and all boils down to a familiar binomial distribution. There are different ways to approach binomial pricing, with tradeoffs in terms of time and numerical accuracy. However, we do need to move backward one step at a time, when pricing American-style derivatives.

13.3.4 BINOMIAL PRICING OF AMERICAN-STYLE OPTIONS

Binomial lattices are a simple and practical tool to price American-style options, which lack analytical pricing formulas. From a formal viewpoint, the price of an American-style option stems from the solution of the following optimization problem:

$$\max_{\tau} \, \mathrm{E}_{\mathbb{Q}_n}\!\left[e^{-r\tau} f(S_\tau) \right], \tag{13.13}$$

where the function $f(\cdot)$ is the option payoff, the expectation is taken under the risk-neutral measure $\mathbb{Q}_n$, and τ is a **stopping time**. In our context, a stopping time is a random variable representing the time at which the option is exercised, following a given rule. This random variable is adapted to the available information, which is the sample path observed so far; in other words, the stopping time is associated with a *nonanticipative* exercise policy. For instance, consider the boundary separating the continuation and early exercise regions in Fig. 13.4. Given any such boundary, we may define a stopping time corresponding to the first time instant at which the boundary is crossed. Hence, a boundary is associated with a decision rule and a stopping time, and solving problem (13.13) means finding the optimal rule. This is nontrivial in general, since in multidimensional cases, the exercise region need not be a connected subregion of the state space.

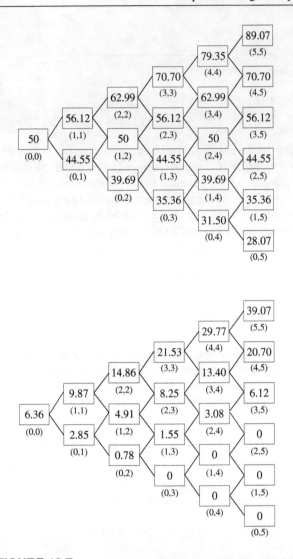

FIGURE 13.7 Numerical example of binomial option pricing.

Remark. It makes intuitive sense that Eq. (13.13) gives the fair value of an option with early exercise features, but it is not at all clear how to justify it theoretically. In fact, this would require the introduction of supermartingales, to show that the optimization problem yields a trading strategy majorizing the cost for the option writer, and that there is no "smaller" process with the same property. We rely on intuition, referring interested readers, e.g., to [15, Chapter 4] for a simple treatment within a binomial setting.

The formulation in terms of a stopping time is a bit too abstract, and we need to find a financially motivated exercise rule. As we have already discussed

13.3 Binomial option pricing

in Section 13.2.2, the early exercise decision should be made by comparing the intrinsic value of the option, i.e., the payoff obtained by exercising the option immediately, and the continuation value, i.e., the value of keeping the option alive and waiting for better opportunities. The difficulty is in the estimation of the continuation value. In general, we may do so by applying stochastic dynamic programming concepts, as we shall illustrate in Section 15.7.3. Within a binomial setting, this boils down to a straightforward extension of Eq. (13.12). Since, as we have seen, there is no point in pricing an American-style call option, as it is never optimal to exercise early, we illustrate the idea by pricing a vanilla American-style put option on a non-dividend paying stock.

Let us consider a point (i, N) on the last time layer of the binomial lattice. If the option is in-the-money at expiration, it is obviously optimal to exercise it. Hence, in the last time layer we have

$$f_{iN} = \max\{K - S_{iN}, 0\},$$

where $S_{iN} = S_0 u^i d^{N-i}$ is the underlying asset price at node (i, N). Now, consider a point in the second-to-last time layer. If the option is not in-the-money, i.e., if $S_{i,N-1} > K$, we do not exercise. But if the option is in the money, we should compare the intrinsic value, $K - S_{i,N-1}$, with the continuation value. Let us denote the continuation value at a generic node (i, j) by f_{ij}^c. If we continue and keep the option alive, we own an asset whose value, at node $(i, N-1)$, is

$$f_{i,N-1}^c = e^{-r \cdot \delta t} \cdot (\pi_u f_{i+1,N} + \pi_d f_{i,N}),$$

where $\pi_d = 1 - \pi_u$. This is the discounted risk-neutral expectation of the option payoff at the last time step, conditional on being at node $(i, N-1)$. We should exercise if the intrinsic value exceeds the continuation value. Hence, the option value at each node in the second-to-last time layer is

$$f_{i,N-1} = \max\left\{K - S_{i,N-1},\ e^{-r \cdot \delta t}(\pi_u f_{i+1,N} + \pi_d f_{i,N})\right\}.$$

The same argument may be repeated in a recursive fashion for any time layer. This means that we should start from the last time layer, where the option value is just the option payoff, and we should proceed backward in time using a slight modification of the discounted expectation scheme of Eq. (13.12):

$$f_{i,j} = \max\{K - S_{ij},\ e^{-r \cdot \delta t}(\pi_u f_{i+1,j+1} + \pi_d f_{i,j+1})\}. \tag{13.14}$$

By finding if and where it is optimal to exercise, we obtain an approximation of the early exercise boundary, as well as the related stopping time. Let us illustrate the idea with a numerical example.

Example 13.5 Pricing an American-style put by a binomial lattice

We want to find the price of a vanilla American-style put call with $S_0 = 60$, $K = 70$, $r = 0.02$, $\sigma = 0.45$, and $T = 1$, by using a three-step binomial lattice (each time step consists of four months). The lattice calibration is:

$$\delta t = \tfrac{4}{12} = \tfrac{1}{3},$$
$$u = e^{\sigma\sqrt{\delta t}} = 1.2967,$$
$$d = 1/u = 0.7712,$$
$$\pi_u = \frac{e^{r\,\delta t} - d}{u - d} = 0.4481.$$

The discount factor at each step is $e^{-0.02/3} = 0.9934$. Figure 13.8 shows the resulting lattices for the stock price and the option value. The stock price lattice is built like in Example 13.4, and the option price lattice is initialized by using the put option payoff (the option is in-the-money in the lower part of the lattice). The interesting node is $(0,2)$, the shaded one, where the continuation value is

$$f^c_{0,2} = e^{-r\cdot\delta t} \cdot (\pi_u f_{1,3} + \pi_d f_{0,3})$$
$$= 0.9934 \cdot \bigl[0.4481 \cdot 23.73 + (1 - 0.4481) \cdot 42.48\bigr] \approx 33.85.$$

However, this should be checked against the intrinsic value:

$$f_{0,2} = \max\{f^c_{0,2}, K - S_{0,2}\} = \max\{33.85, 70 - 35.68\}$$
$$= \max\{33.85, 34.32\} = 34.32,$$

from which we see that early exercise occurs at node $(0,2)$. As a consequence, the resulting option price is a bit larger than the price of the corresponding European-style put.

13.4 A continuous-time model: The Black–Scholes–Merton pricing formula

The binomial model allows us to gain essential insights into option pricing and hedging and is a useful numerical tool. However, it does not provide us with an analytical pricing formula that would be of great practical use, not only in terms of efficiency. In risk management, we need to assess the sensitivity of an asset price with respect to the relevant risk factors, and an analytical formula would be of great value from this viewpoint. In this section, we apply the machinery of stochastic calculus and derive the celebrated and controversial Black–Scholes–

13.4 A continuous-time model: The Black–Scholes–Merton pricing formula

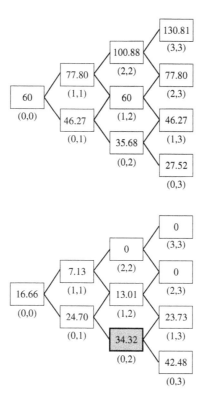

FIGURE 13.8 Pricing an American-style put by a binomial lattice.

Merton (BSM) pricing formula.[12] The model is based on the assumption that the underlying asset price follows a geometric Brownian motion,

$$dS_t = \mu S_t\, dt + \sigma S_t\, dW_t.$$

We may consider the BSM model as the continuous-time limit of the binomial model when the time step δt goes to zero or, going the other way around, we may consider the binomial model as a discretization of the continuous model. Curiously enough, it is by now standard to start textbook expositions with the more intuitive binomial model, but historically this was developed after the BSM model.

---

[12] The formula was published by Fisher Black and Myron Scholes in 1973. A similar research line had been pursued by Robert Merton. Scholes and Merton were awarded the Nobel prize in Economics in 1997. By that time, unfortunately, Fisher Black was deceased.

13.4.1 THE DELTA-HEDGING VIEW

Here, we apply the hedging-based approach of Section 13.3.1, taking the viewpoint of the writer of a vanilla, European-style call option, written on a stock share that does not pay any dividend. Let us denote by $f(S_t, t)$ the fair option price at time t, when the underlying asset price is S_t. Note that we are taking for granted that the option price depends on these two variables only, which makes sense for a path-independent vanilla option. In the case of Asian options, where the payoff depends on average prices, things are not that simple.

Using Itô's lemma, we may write a stochastic differential equation for $f(\cdot, \cdot)$:

$$df = \frac{\partial f}{\partial t} dt + \frac{\partial f}{\partial S_t} dS_t + \frac{1}{2}\sigma^2 S_t^2 \frac{\partial^2 f}{\partial S_t^2} dt. \qquad (13.15)$$

Just as in the binomial case, what we know is the option value at maturity,

$$f(S_T, T) = \max\{S_T - K, 0\},$$

and what we would like to know is $f(S_t, t)$, the fair option price at time $t < T$, in particular, the current fair price at time $t = 0$. Just like we did in the binomial case, we consider hedging the writer's risk by taking a position in Δ stock shares, so that the value of the hedged portfolio at time t is

$$\Pi(S_t, t) = -f(S_t, t) + \Delta \cdot S_t.$$

Unlike previous applications of Itô's lemma, we do *not* know the function $f(\cdot, \cdot)$. Hence, Eq. (13.15) looks like an ugly partial differential equation involving a random term, and it does not suggest an immediate way to find the option price. However, from a formal viewpoint, this equation would look a little bit nicer, without the random term dS_t. From a financial viewpoint, we can hedge risk away, by eliminating the dependence of Π_t on random variations in S_t. This may be accomplished by choosing

$$\Delta = \frac{\partial f}{\partial S_t},$$

which is the continuous-time counterpart of Eq. (13.8). To see this, let us differentiate the portfolio value Π and take advantage of our choice of Δ:

$$\begin{aligned} d\Pi &= -df + \Delta \, dS_t \\ &= \left(-\frac{\partial f}{\partial S_t} + \Delta\right) dS_t - \left(\frac{\partial f}{\partial t} + \frac{1}{2}\sigma^2 S_t^2 \frac{\partial^2 f}{\partial S_t^2}\right) dt \\ &= -\left(\frac{\partial f}{\partial t} + \frac{1}{2}\sigma^2 S_t^2 \frac{\partial^2 f}{\partial S_t^2}\right) dt \end{aligned} \qquad (13.16)$$

Thanks to the choice of Δ, the term multiplying the random increment dS_t in the second line of Eq. (13.16) vanishes, so that the portfolio is riskless. Then, by no-arbitrage arguments, it must earn the risk-free interest rate r:

$$d\Pi = r\Pi \, dt. \qquad (13.17)$$

13.4 A continuous-time model: The Black–Scholes–Merton pricing formula

Eliminating $d\Pi$ between Eqs. (13.16) and (13.17), we obtain

$$-\left(\frac{\partial f}{\partial t} + \frac{1}{2}\sigma^2 S_t^2 \frac{\partial^2 f}{\partial S_t^2}\right) dt = r\left(-f + S_t \frac{\partial f}{\partial S_t}\right) dt,$$

which can be simplified by eliminating dt and rearranged as

$$\frac{\partial f}{\partial t} + rS_t \frac{\partial f}{\partial S_t} + \frac{1}{2}\sigma^2 S_t^2 \frac{\partial^2 f}{\partial S_t^2} = rf. \qquad (13.18)$$

Now, we have a deterministic partial differential equation (PDE for short) describing an option value $f(S_t, t)$. This is the **Black–Scholes–Merton equation** (BSM equation for short), which must be solved, subject to suitable boundary conditions. Before doing so, the following key observations are in order:

- The choice of Δ as the derivative of the option value with respect to the price of the underlying asset is consistent with the findings in the binomial case.
- This eliminates the dependence with respect to the true drift μ, which is related to the expected return of the underlying asset under the real probability measure; indeed, μ does not occur in Eq. (13.18). This is also consistent with the binomial model, where the expected value of the future asset price, based on the objective probabilities, does not play any role.

What does not quite look consistent with the binomial model is that, in that case, we obtain the price as an expected value, whereas here we have to solve a PDE. As we shall show later, in Section 13.4.2, there is a theorem bridging the gap between the solution of a certain class of PDEs and conditional expectations related to stochastic processes. We will use this theorem in Section 13.4.2.1, in order to prove the BSM formula in the risk-neutral expectation setting. However, the proof was originally obtained by taking advantage of the analogy between the BSM equation and a fundamental equation of mathematical physics, the heat equation, which is the prototypical equation for diffusion processes. Since this kind of proof is a bit involved, we shall pursue a different strategy in Section 13.4.2.1. Nevertheless, we provide some intuition in Section 13.4.1.1.

Like any differential equation, we need some additional condition to pinpoint a specific solution. In fact, the BSM equation is fairly generic and, for instance, it does not discriminate between a call and a put option. The domain on which we have to solve the equation is the unbounded strip depicted in Fig. 13.9, for $t \in [0, T]$ and $S_t \in [0, +\infty)$. Note that the strip is bounded in time but unbounded in price since, in principle, there is no upper bound to the price that a stock share may attain. In the case of a vanilla call, we have to solve the equation with a *terminal* condition related to the payoff:

$$C_T^e = \max\{S_T - K, 0\}.$$

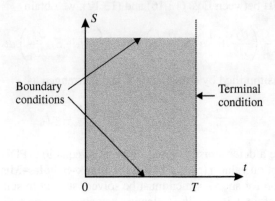

FIGURE 13.9 The domain for the solution of the BSM PDE.

In fact, the BSM equation has to be solved backward in time, just like we do with binomial lattices. We also have the following boundary conditions:

$$\lim_{S_t \to 0} C_t^e = 0, \qquad \lim_{S_t \to +\infty} C_t^e = S_t - Ke^{-r\cdot(T-t)} = S_t - \text{PV}(K).$$

These are consistent with the bounds on call option prices that we have discussed in Section 13.2.1. Solving the PDE leads to the celebrated **Black–Scholes–Merton formula** (BSM formula for short), which gives the price C_t^e of a vanilla European-style call option:

$$C_t^e = S_t \Phi(d_1) - Ke^{-r(T-t)} \Phi(d_2), \qquad (13.19)$$

where

$$d_1 = \frac{\log(S_t/K) + (r + \sigma^2/2)(T-t)}{\sigma\sqrt{T-t}},$$

$$d_2 = \frac{\log(S_t/K) + (r - \sigma^2/2)(T-t)}{\sigma\sqrt{T-t}} = d_1 - \sigma\sqrt{T-t},$$

and $\Phi(x)$ is the cumulative distribution function (CDF) for the standard normal distribution:

$$\Phi(x) = \frac{1}{\sqrt{2\pi}} \int_{-\infty}^{x} e^{-z^2/2}\, dz.$$

As usual in this book, we denote the natural logarithm by log, but ln is also often used. We should note that Eq. (13.19) is written in a general form, referring to time t. Hence, $T - t$ is to be understood as time-to-maturity. When we price the option at time $t = 0$, time-to-maturity boils down to T, as in the following example.

13.4 A continuous-time model: The Black–Scholes–Merton pricing formula

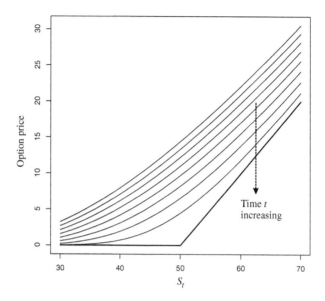

FIGURE 13.10 Call option prices as a function of time t and underlying asset price S_t. When maturity is approached, the smooth price curve converges to the kinky option payoff.

Example 13.6 A numerical example

Let us consider the same setting as Example 13.4: $S_0 = K = 50$, $r = 0.1$, $\sigma = 0.4$, and time-to-maturity is $T = 5/12$ (here we take $t = 0$). The calculation is as follows:

$$d_1 = \frac{\log(50/50) + (.10 + 0.4^2/2) \times 5/12}{0.4 \times \sqrt{5/12}} = 0.2905,$$

$d_2 = 0.2905 - 0.4\sqrt{5/12} = 0.0323,$

$\Phi(0.2905) = 0.6143,$

$\Phi(0.0323) = 0.5129,$

$C_0 = 50 \times 0.6143 - 50 \times e^{-0.10 \times 5/12} \times 0.5129 = 6.1165.$

This exact result may be compared with the binomial approximation, which gave a price 6.36, which proves to be not too bad. It is important to realize that "exact" should be taken as relative to the BSM model.

If we apply the BSM formula to a call option for different prices S_t and different time instants t, we obtain the plots depicted in Fig. 13.10. Each curve corresponds to a different time-to-maturity and shows the fair option value as a function of S_t. When t is approaching maturity, the plots converge to the option payoff. We observe that the price, for a given S_t, is a decreasing function of time. These plots are consistent with the bounds and the basic intuition that we developed in Section 13.2.1. We also notice that the price is a continuous, differentiable function, which converges to a kinky one. This is a property due to the diffusion character of the BSM equation.

The same kind of reasoning can be applied to price a put option, but we may spare ourselves the effort by resorting to put–call parity. By plugging the BSM formula for the call option price into Eq. (13.4) we find, after some rearrangement, a similar formula for the put option price:

$$P_t^e = Ke^{-r(T-t)}\Phi(-d_2) - S_t\Phi(-d_1), \qquad (13.20)$$

where d_1 and d_2 are the same expressions occurring in the call option price.

13.4.1.1 The analogy with the heat equation

The BSM PDE belongs to a wide class of quite common equations of mathematical physics, related to diffusion processes. In the PDE parlance, it is a linear, second-order, and **parabolic** equation. The prototypical equation in this class is the nondimensional heat equation,

$$\frac{\partial u}{\partial t} = \frac{\partial^2 u}{\partial x^2}, \qquad (13.21)$$

which describes the evolution of temperature $u(x,t)$ in a one-dimensional body, like a bar, at position x and time t. This form of the equation is referred to as nondimensional (or dimensionless), in the sense that we are not including physical constants related to the specific material of the bar. Parabolic equations like the BSM equation, by suitable changes of coordinates, can be recast in the form of Eq. (13.21). Indeed, this is how the BSM formula was derived originally, but since it involves plenty of technicalities, we will not pursue this approach.

Nevertheless, it is quite useful to get an intuitive feeling for the link between the heat equation and the BSM equation. As we have seen in Fig. 13.10, even though the terminal condition involves a nondifferentiable function, the option value looks like a smooth function of S_t before maturity. This is due to the fact that parabolic equations are related to diffusion processes, which involve some form of smoothing. On the contrary, hyperbolic equations are related to wave propagation, where singularities and shocks (think of an earthquake) may be propagated in space. The diffusion property is due to the deep link between Brownian motion, which is a diffusion process, and the heat equation.[13] As we shall see, this smoothing feature is fundamental for hedging option risk.[14]

---

[13] This link was discussed in a celebrated paper written by Albert Einstein and published in 1905 on the Annalen der Physik.

[14] Option risk may be managed by dynamic delta-hedging strategies, but if the option delta swings too much, dynamic hedging turns out to be problematic, as we shall see in Section 13.8.3.

13.4 A continuous-time model: The Black–Scholes–Merton pricing formula

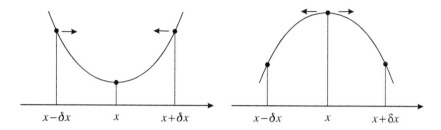

FIGURE 13.11 Intuitive interpretation of the heat equation.

The heat equation may be derived by physical arguments, but we may settle for an intuitive interpretation, clarifying its diffusion feature. Let us consider the two temperature profiles depicted in Fig. 13.11. The first one is convex, and intuition suggests that heat will flow from the neighboring high-temperature points $x - \delta x$ and $x + \delta x$ toward the low-temperature point x. Therefore, temperature will increase at point x. Hence, a positive second order partial derivative with respect to space corresponds to a positive partial derivative with respect to time. The contrary happens with a concave profile, like in the second case, where heat should flow from the high-temperature point x toward its low-temperature neighbors. Thus, the heat equation is related with a diffusion process that, in some sense, works like an "averaging" process. As we shall see in Section 13.4.2, this view is reinforced by the link between a certain class of parabolic equations and stochastic differential equations driven by the Wiener process (the prototypical diffusion process), which allows us to solve the PDE by taking an expectation.

Now, let us consider a bar of length L, mapped on the interval $x \in [0, L]$. The equation must be solved on the unbounded strip depicted in Fig. 13.12. We have an initial condition $u(x, 0) = f(x)$, $x \in [0, L]$, for a given function $f(\cdot)$, and boundary conditions at the endpoints of the bar, $u(0, t) = g_0(t)$ and $u(L, t) = g_L(t)$, $t \in [0, +\infty)$. We may notice that, in this case, time is unbounded and space is bounded, whereas, in the BSM equation, roles are swapped, and time is bounded and price is unbounded. Furthermore, time is going forward in the heat equation, whereas we price options backward in time. Let us compare Fig. 13.10 against Fig. 13.13, which shows the solution of the heat equation when the boundary conditions prescribe zero temperature at points $x = 0$ and $x = L$ (the body of the bar is perfectly insulated and there is no loss of heat along the bar), and the initial condition is a triangular, kinky temperature profile. Here, the bar length is $L = 1$, and we show the solution at different consecutive time instants in Figs. 13.13(a)–(d). Figure 13.13(a) gives the initial condition at $t = 0$. Then, the temperature profile is a smooth function that decays to zero on the whole bar, as heat is lost at the endpoints. Modulo a change of coordinate, we observe a definite similarity with the price of a call option, where we converge toward a kinky terminal condition, given by the option payoff (see Fig. 13.10).

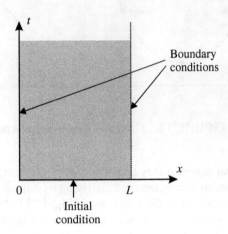

FIGURE 13.12 The domain for the solution of the heat PDE.

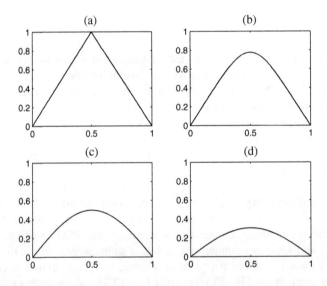

FIGURE 13.13 Solution of the heat equation at different time instants: Plot (a) gives the initial condition.

13.4.1.2 Option pricing and delta-hedging

In Section 13.4.1, we have motivated the BSM equation by considering a dynamic hedging strategy to manage the risk stemming from a short position in an option. The value of the hedged portfolio at time t depends on the value of the option and the price of the underlying asset:

$$\Pi_t = -f_t + \Delta_t S_t.$$

In Example 13.3, we have seen that a static hedging strategy may hedge risk in a single-step model. Here, we need a *dynamic* hedging strategy, where Δ_t is adjusted dynamically. The choice $\Delta_t = \frac{\partial f_t}{\partial S_t}$ makes Π_t riskless, and the option value is just the cost of setting up the hedge. The strategy is called delta-hedging, due to the role of the option delta, which may be evaluated by analytical formulas, when available, or approximated numerically. In theory, the hedging ratio Δ_t should be adjusted dynamically in continuous-time, which is not practically feasible because of transaction costs. A discrete-time approximation results in residual risk due to hedging errors. Furthermore, as we shall see in Section 13.8.3, the strategy breaks down when the option delta swings a lot, resulting in a very nervous hedge.

Our reasoning is somewhat heuristic, but very instructive for a financial engineer. From a mathematical viewpoint, a more rigorous analysis should rely on the concept of self-financing portfolios. The self-financing condition is practically relevant since, after using the option premium (and, maybe, some additional borrowed cash) to initialize the hedging strategy, we may dynamically adjust the hedge, but we should not use additional funds. Mathematically, this leads to the application of stochastic calculus concepts, namely, stochastic integrals and martingales, to find the option price. We have preferred to use a less rigorous, but more financially motivated approach, which is easier to extend when we consider more realistic incomplete market models, transaction costs, and model risk.

13.4.2 THE RISK-NEUTRAL VIEW: FEYNMAN–KAČ REPRESENTATION THEOREM

In this section, we describe a version of the Feynman–Kač stochastic representation theorem, which expresses the solution of a parabolic PDE, like the BSM equation, as a conditional expectation related with the state of a stochastic process. This allows us to bridge the gap between the PDE framework and risk-neutral pricing.

THEOREM 13.2 Feynman–Kač representation theorem. *Consider the partial differential equation*

$$\frac{\partial f}{\partial t} + \mu(x,t) \cdot \frac{\partial f}{\partial x} + \frac{1}{2}\sigma^2(x,t) \cdot \frac{\partial^2 f}{\partial x^2} = rf,$$

and let $f(x,t)$ be a solution satisfying the terminal condition

$$f(x,T) = H(x).$$

Then, under technical conditions, $f(x,t)$ can be represented as a conditional expectation,

$$f(x,t) = \mathrm{E}_{x,t}\big[H(X_T)\big],$$

where X_t is a stochastic process satisfying the differential equation

$$dX_\tau = \mu(X_\tau, \tau)\, d\tau + \sigma(X_\tau, \tau)\, dW_\tau,$$

with initial condition $X_t = x$.

The notation $\mathrm{E}_{x,t}[\,\cdot\,]$ represents the *conditional* expectation, given that, at time t, the value of the stochastic process is $X_t = x$. Clearly, in a financial setting, the function $H(\cdot)$ corresponds to the value of a derivative at maturity, but the theorem was developed with applications to physics in mind. From a physical viewpoint, the theorem is a consequence of the connection between Brownian diffusions and parabolic PDEs. From a mathematical viewpoint, the proof relies on Itô's lemma and stochastic integration (see, e.g., [1]).

Example 13.7 Applying Feynman–Kač theorem

Let us consider the PDE
$$\frac{\partial f}{\partial t} + \frac{1}{2}\sigma^2 \frac{\partial^2 f}{\partial x^2} = 0,$$
with terminal condition
$$f(x, T) = x^2.$$

In order to apply the representation theorem, we observe that $r = 0$, $\sigma(x,t) = \sigma$, and $\mu(x,t) = 0$, so that the underlying stochastic process boils down to a martingale described by
$$dX(\tau) = \sigma\, dW(\tau).$$

Therefore, conditional on $X(t) = x$, integration over the time interval $[t, T]$ yields
$$\int_t^T dX(\tau) = \sigma \int_t^T dW(\tau) \quad \Rightarrow \quad X(T) \stackrel{d}{=} x + \sigma\sqrt{T-t}\cdot\epsilon,$$
where $\epsilon \sim \mathsf{N}(0,1)$. Therefore,
$$\begin{aligned} f(X,t) &= \mathrm{E}\!\left[X^2(T)\,\big|\, X(t) = x\right] \\ &= \mathrm{Var}\!\left[X(T)\,\big|\, X(t) = x\right] + \mathrm{E}^2\!\left[X(T)\,\big|\, X(t) = x\right] \\ &= \sigma^2 \cdot (T-t) + x^2. \end{aligned}$$

It is easy to check that this function satisfies the PDE and the terminal condition.

The application of the representation theorem to the BSM equation, for an option with payoff function $H(\cdot)$, immediately yields
$$f(S_0, 0) = e^{-rT} \cdot \mathrm{E}_{\mathbb{Q}_n}\left[H(S_T)\right],$$

13.4 A continuous-time model: The Black–Scholes–Merton pricing formula

which is consistent with Eq. (13.6). We point out once more that the expectation is taken under a risk-neutral measure, and we see that this change of measure amounts to replacing the drift μ by the risk-free rate r in the stochastic differential equation for S_t,

$$dS_t = rS_t\, dt + \sigma S_t\, dW_t.$$

By recalling the result of Eq. (11.33), we conclude that, under the risk-neutral measure, the price S_t is lognormal and may be written as

$$S_t = S_0 \cdot \exp\left\{\left(r - \frac{\sigma^2}{2}\right)t + \sigma\sqrt{t}\,\epsilon\right\},$$

where $\epsilon \sim \mathsf{N}(0,1)$. Then, proving the BSM formula is a fairly easy, though a bit tedious exercise in integration.

13.4.2.1 A proof of the BSM formula

As a first step, let us prove a theorem concerning a generic lognormal variable X with parameters μ and σ^2, which may be expressed as

$$X = e^{\mu + \sigma\epsilon},$$

where $\epsilon \sim \mathsf{N}(0,1)$.

THEOREM 13.3 *If X is a lognormal random variable with parameters (μ, σ^2), then*

$$\mathrm{E}\big[\max\{X - K, 0\}\big] = \mathrm{E}[X] \cdot \Phi(d_1) - K \cdot \Phi(d_2),$$

where $\Phi(\cdot)$ is the CDF of the standard normal and

$$\mathrm{E}[X] = e^{\mu + \sigma^2/2},$$
$$d_1 = \frac{\log\big(\mathrm{E}[X]/K\big) + \sigma^2/2}{\sigma},$$
$$d_2 = \frac{\log\big(\mathrm{E}[X]/K\big) - \sigma^2/2}{\sigma}.$$

PROOF Using the representation of X as a function of a standard normal, we have

$$\mathrm{E}[\max\{X - K, 0\}] = \int_{-\infty}^{+\infty} \max\big[e^{\mu + \sigma z} - K, 0\big]\,\phi(z)\,dz,$$

where $\phi(z)$ is the PDF of a standard normal. We also note that $X = K$ when

$$e^{\mu + \sigma\epsilon} = K \quad\Rightarrow\quad \epsilon = \frac{\log K - \mu}{\sigma} \doteq q.$$

Hence, we may get rid of the max in the integrand function by setting q as the lower extreme of the integration interval. The desired expected value may be written as a difference of two integrals:

$$\int_q^{+\infty} \left[e^{\mu+\sigma z} - K\right] \frac{1}{\sqrt{2\pi}} e^{-z^2/2} \, dz$$
$$= \int_q^{+\infty} e^{\mu+\sigma z} \frac{1}{\sqrt{2\pi}} e^{-z^2/2} \, dz - K \int_q^{+\infty} \frac{1}{\sqrt{2\pi}} e^{-z^2/2} \, dz. \quad (13.22)$$

In order to find the first integral, we rewrite the product of the two exponentials by completing the square,

$$\mu + \sigma z - \frac{z^2}{2} = \mu - \frac{-\sigma^2 + \sigma^2 - 2\sigma z + z^2}{2} = \mu + \frac{\sigma^2}{2} - \frac{1}{2}(z-\sigma)^2.$$

Hence,

$$\int_q^{+\infty} e^{\mu+\sigma z} \frac{1}{\sqrt{2\pi}} e^{-z^2/2} \, dz = e^{\mu+\sigma^2/2} \frac{1}{\sqrt{2\pi}} \int_q^{+\infty} \phi(z-\sigma) \, dz$$
$$= \mathrm{E}[X] \cdot \frac{1}{\sqrt{2\pi}} \int_{q-\sigma}^{+\infty} \phi(y) \, dy, \quad (13.23)$$

where we use the variable substitution $y = z - \sigma$, so that Eq. (13.23) involves the probability that a standard normal variable ϵ is larger than $q - \sigma$. Hence, we obtain

$$\mathrm{E}[X] \cdot \mathrm{P}\{\epsilon \geq q - \sigma\} = \mathrm{E}[X] \cdot \mathrm{P}\left\{\epsilon \geq \frac{\log K - \mu - \sigma^2}{\sigma}\right\}$$
$$= \mathrm{E}[X] \cdot \mathrm{P}\left\{\epsilon \geq \frac{\log K - (\mu + \sigma^2/2) - \sigma^2/2}{\sigma}\right\}$$
$$= \mathrm{E}[X] \cdot \mathrm{P}\left\{\epsilon \geq -\frac{\log(\mathrm{E}[X]/K) + \sigma^2/2}{\sigma}\right\}$$
$$= \mathrm{E}[X] \cdot \mathrm{P}\left\{\epsilon \leq \frac{\log(\mathrm{E}[X]/K) + \sigma^2/2}{\sigma}\right\}$$
$$= \mathrm{E}[X] \cdot \Phi(d_1),$$

where we use symmetry of the standard normal distribution as usual, in order to write the expression in terms of its CDF $\Phi(\cdot)$ By a similar token, we may rewrite the second integral in Eq. (13.22) as

$$K \cdot \mathrm{P}\{\epsilon \geq q\} = K \cdot \mathrm{P}\left\{\epsilon \geq \frac{\log K - \mu}{\sigma}\right\}$$
$$= K \cdot \mathrm{P}\left\{\epsilon \geq \frac{\log K - (\mu + \sigma^2/2) + \sigma^2/2}{\sigma}\right\}$$
$$= K \cdot \mathrm{P}\left\{\epsilon \geq -\frac{\log(\mathrm{E}[X]/K) - \sigma^2/2}{\sigma}\right\}$$
$$= K \cdot \Phi(d_2).$$

13.4 A continuous-time model: The Black–Scholes–Merton pricing formula

By taking the difference of the two integrals, the result follows. ∎

In the BSM case, by risk-neutral expectation, the call option price at time $t = 0$ (so that T is time-to-maturity) can be written as

$$e^{-rT} \cdot \mathrm{E}_{\mathbb{Q}_n}\big[\max\{S_T - K, 0\}\big],$$

which involves a lognormal variable S_T with parameters $(r - \sigma^2/2)T$ and $\sigma^2 T$, rather than just μ and σ, so that $\mathrm{E}_{\mathbb{Q}_n}[S_T] = S_0 e^{rT}$. In order to prove the BSM formula, we have just to plug these parameters into the formula of Theorem 13.3:

$$e^{-rT} \cdot \big\{S_0 e^{rT} \Phi(d_1) - K \Phi(d_2)\big\} = S_0 \Phi(d_1) - K e^{-rT} \Phi(d_2),$$

where, in this case,

$$d_1 = \frac{\log(S_0 e^{rT}/K) + \sigma^2 T/2}{\sigma \sqrt{T}} = \frac{\log(S_0/K) + (r + \sigma^2/2)T}{\sigma \sqrt{T}}$$

and

$$d_2 = \frac{\log(S_0 e^{rT}/K) - \sigma^2 T/2}{\sigma \sqrt{T}} = \frac{\log(S_0/K) + (r - \sigma^2/2)T}{\sigma \sqrt{T}} = d_1 - \sigma \sqrt{T}.$$

13.4.3 INTERPRETING THE FACTORS IN THE BSM FORMULA

Let us write the BSM formula of Eq. (13.19) for a call at time $t = 0$:

$$C_0^e = S_0 \Phi(d_1) - K e^{-rT} \Phi(d_2), \qquad (13.24)$$

where we set $t = 0$ also in the expressions of d_1 and d_2. The option price depends on the current stock price, multiplied by a factor $\Phi(d_1)$, and the discounted strike, i.e., the value of a zero with face value K, multiplied by a factor $\Phi(d_2)$. It is natural to wonder whether these two factors have a specific meaning. In both the binomial and the BSM model, when we take the hedging-based view, we build a portfolio with Δ stock shares, where Δ is the increment ratio or the derivative of the option value with respect to the underlying asset price. Hence, it is not quite surprising that we may prove the following result:

$$\frac{\partial C_0^e}{\partial S_0} = \Phi(d_1). \qquad (13.25)$$

At first sight, this may look like an immediate consequence of Eq. (13.24), where S_0 occurs multiplied by $\Phi(d_1)$. However, this way of reasoning is wrong, as both d_1 and d_2 depend on S_0. Nevertheless, by taking derivatives correctly (which is left as a boring, yet useful exercise), we do find the result of Eq. (13.25), which actually applies to a generic time instant t. In Section 13.5 we will see that the option delta is the basic sensitivity measure of the option price.

The second term, too, has an interpretation, as it is the probability that the call option is in-the-money, under the risk-neutral measure:

$$\Phi(d_2) = P_{\mathbb{Q}_n}\{S_T \geq K\}. \qquad (13.26)$$

To see this, we may express S_T as a function of a standard normal variable:[15]

$$\begin{aligned}
P_{\mathbb{Q}_n}\{S_T \geq K\} &= P_{\mathbb{Q}_n}\left\{S_0 \cdot \exp\left[\left(r - \frac{\sigma^2}{2}\right)T + \sigma\sqrt{T}\epsilon\right] \geq K\right\} \\
&= P_{\mathbb{Q}_n}\left\{\left(r - \frac{\sigma^2}{2}\right)T + \sigma\sqrt{T}\epsilon \geq \log\left(\frac{K}{S_0}\right)\right\} \\
&= P_{\mathbb{Q}_n}\left\{\epsilon \geq -\frac{\log(S_0/K) + (r - \sigma^2/2)T}{\sigma\sqrt{T}}\right\} \\
&= P_{\mathbb{Q}_n}\{\epsilon \geq -d_2\} \\
&= P_{\mathbb{Q}_n}\{\epsilon \leq d_2\} \equiv \Phi(d_2),
\end{aligned}$$

where we use symmetry of the standard normal in the last equality.

Remark. We note that, since we have used the risk-neutral drift r, we have obtained the probability that the call option is in-the-money in the risk-neutral world. If we use the actual drift μ, we obtain the corresponding objective probability. It is important to realize that we should use a risk-adjusted measure for pricing purposes, but the objective measure should be used when generating scenarios for risk measurement purposes. This difference is illustrated in the following examples.

Example 13.8 Probability of exercise

> The current price of a stock share that pays no dividend is €50. The price follows a GBM with drift 12% and volatility 35%; the continuously compounded risk-free rate is 5%. Consider a call and a put options, both European-style, with strike €55, maturing in nine months. Which option is more likely to be exercised?
>
> The probability of exercising the call is $P\{S_T \geq K\}$, which is given by $\Phi(d_2)$, *provided that we use the true drift.* Hence, we should compute d_2 as
>
> $$d_2 = \frac{\log\left(\frac{S_0}{K}\right) + \left(\mu - \frac{\sigma^2}{2}\right)T}{\sigma\sqrt{T}}$$

---

[15] The careful reader will notice that this calculation has already been used in the proof of Theorem 13.3.

$$= \frac{\log\left(\frac{50}{55}\right) + \left(0.12 - \frac{0.35}{2}\right) \cdot \frac{9}{12}}{\sigma\sqrt{\frac{9}{12}}}$$

$$= -0.1691.$$

Since $d_2 < 0$, there is no need for further calculations, as this implies $\Phi(d_2) < 0.5$. Hence, the put is more likely to be exercised.

Example 13.9 Pricing a binary option

An immediate consequence of Eq. (13.26), is that (modulo a discount factor) it gives, for free, the BSM price of a digital (or binary) option, i.e., an option paying \$1 if $S_T \geq K$. Using the indicator function

$$\mathbf{1}_{\{S_T \geq K\}} \doteq \begin{cases} 1, & \text{if } S_T \geq K, \\ 0, & \text{otherwise,} \end{cases}$$

we find

$$e^{-rT} \mathrm{E}_{\mathbb{Q}_n}[\mathbf{1}_{\{S_T \geq K\}}] = e^{-rT} \mathrm{P}_{\mathbb{Q}_n}\{S_T \geq K\}$$
$$= e^{-rT} \Phi(d_2). \qquad (13.27)$$

It is important to realize that, in this case, we must calculate d_2 using the risk-neutral drift r, as we are *pricing* the option.

Also note that the binary call features a *discontinuous* payoff at maturity, but thanks to the parabolic nature of the BSM equation, we find a continuous price as a function of S_t. As one may expect, when approaching maturity, the derivative of the option price with respect to S_t will get steeper and steeper, as shown in Fig. 13.18. As we discuss in Section 13.8.3, this could make hedging difficult.

13.5 Option price sensitivities: The Greeks

The BSM formula shows that the price of a vanilla European-style option depends on five factors:

1. Current price S_t of the underlying asset
2. Volatility σ

3. Time-to-maturity $T - t$
4. Risk-free rate r
5. Strike price K

In this section, we deal exclusively with European-style options, since American-style options require numerical methods to evaluate their sensitivities. A fundamental advantage of the BSM formula, apart from its computational appeal, is that we may evaluate the sensitivity (first-order derivative) of the option price with respect to each of these factors. These sensitivities are collectively known as the **option Greeks**. We have seen, in Section 2.2.1, how such sensitivities may be used for risk management and, in Chapter 6, we have investigated the use, as well as the limitations, of duration and convexity for managing interest rate risk in fixed-income portfolios. In this section, we do the same for option price sensitivities with respect to underlying asset price, volatility, and time. Time, strictly speaking, is not a random risk factor, but how an option price changes over time is quite relevant. We do not consider sensitivity with respect to the constant parameter K, and we also disregard the sensitivity with respect to the risk-free rate.[16]

13.5.1 DELTA AND GAMMA

The option **delta** is defined as the first-order sensitivity of the option price with respect to the current price of the underlying asset. For a vanilla call at time t,[17]

$$\Delta_C = \frac{\partial C_t^e}{\partial S_t} = \Phi(d_1). \qquad (13.28)$$

The delta of a vanilla call option is given by the CDF of the standard normal distribution, which is a probability. Hence, the call delta is in the interval $[0, 1]$. This makes sense, as it should be the number of stock shares that the option writer should be long for each call option. The more the option is in-the-money, the closer this number is to 1. The call option Δ_C is plotted in Fig. 13.14, for an option with strike price $K = 50$. By differentiating the put–call parity relationship with respect to S_t, we immediately find the corresponding delta for the put option:

$$\Delta_P = \Delta_C - 1 = \Phi(d_1) - 1. \qquad (13.29)$$

As shown in Fig. 13.14, the put option delta is identical to the call delta, modulo a vertical shift. The put delta must be in the interval $[-1, 0]$ and corresponds to a short hedge in the underlying asset. When the put is in-the-money, delta gets

---

[16] The sensitivity of the option price with respect to r is called rho. The sensitivity with respect to K may look irrelevant, but it plays a role in extracting the risk-neutral density from observed option prices.

[17] We have introduced the formula for $t = 0$, but this is actually irrelevant. The BSM formula depends on the time-to-maturity $\tau = T - t$, rather than time t. Whenever we consider $t = 0$, maturity and time-to-maturity are just the same. A case in which we must be careful, as we shall see, is the option theta, which gives the sensitivity of the option price with respect to t.

13.5 Option price sensitivities: The Greeks

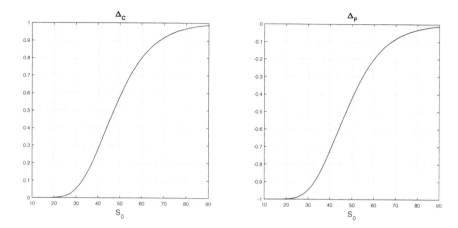

FIGURE 13.14 Delta for call and put options with strike $K = 50$.

closer to -1, since the option writer is subject to the risk of being forced to buy a stock share at a strike price that is quite larger than the spot price. A short position in the stock compensates the resulting loss.

Delta is to options what duration is to bonds, more or less. The main difference is that the former is an absolute sensitivity, whereas the latter is a relative one. The role of convexity is played by the option **gamma**, which is a second-order sensitivity:

$$\Gamma = \frac{\partial^2 C_t^e}{\partial S_t^2} = \frac{\phi(d_1)}{S_t \sigma \sqrt{T-t}}. \tag{13.30}$$

Since call and put deltas differ by a constant, gamma is the same for both kinds of options. It is important to notice that the option gamma involves the PDF $\phi(\cdot)$ of a standard normal distribution, rather than the CDF $\Phi(\cdot)$. Hence, gamma is always non-negative, which is consistent with the increasing monotonicity of delta. This also shows that both call and put prices are convex functions of S_t. In Fig. 13.15, we plot the option gamma for different times-to-maturity. We notice that the plot gets more and more peaked when the option is approaching its maturity. This makes sense since, for instance, the option delta at maturity is 0 for an out-of-the-money call and 1 for an in-the-money call, and it is undefined for $S_T = K$. This has a practical implication for delta-hedging close to maturity, since when delta gets more "nervous," delta-hedging may become difficult and expensive for at-the-money options. In practice, option books may be hedged as a whole, rather than single options, easing the problem.

Example 13.10 Making an option book delta-neutral

An investment bank has written 10,000 call options (strike 40, maturing in three months), and 5000 put options (strike 33, maturing

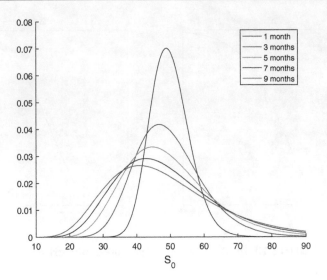

FIGURE 13.15 Gamma for call and put options.

in seven months), written on a stock share whose current price is $S_0 = 36$. The risk-free rate is 3%, with continuous compounding, and the stock price volatility is 36%. How many stock shares should they hold in order to make the portfolio delta-neutral?

The value of the hedged portfolio depends on the prices of the two options and the underlying stock share:

$$V_t = -10{,}000 \times C_t^e - 5000 \times P_t^e + \phi S_t,$$

where the negative signs reflect the short positions in the two options, and ϕ is the number of stock shares, which may be positive or negative, depending on the yet unknown net exposure. The delta of the portfolio depends on the deltas of the two options:

$$\Delta \doteq \frac{\partial V_t}{\partial S_t} = -10{,}000 \times \frac{\partial C_t^e}{\partial S_t} - 5000 \times \frac{\partial P_t^e}{\partial S_t} + \phi \times \frac{\partial S_t}{\partial S_t}$$
$$= -10{,}000 \times \Delta_C - 5000 \times \Delta_P + \phi.$$

We find the deltas of the two options by using Eqs. (13.28) and (13.29):

$$\Delta_C = 0.3250, \qquad \Delta_P = -0.6750.$$

Hence, the delta of the hedged portfolio is

$$\Delta = -10{,}000 \times 0.3250 + 5000 \times 0.6750 + \phi = 124.4998 + \phi.$$

13.5 Option price sensitivities: The Greeks

> Note that the risk exposures from writing the two options tend to cancel each other, but the exposure due to the put options prevails. By setting $\Delta = 0$, we find
> $$\phi \approx -125,$$
> i.e., they should hold a short position in 125 stock shares.

It is important to consider Example 13.10 within the more general framework of first-order immunization, which was introduced in Section 2.2.3.3. We may use option delta to build a delta-neutral option portfolio, just like we use bond duration to make a fixed-income portfolio insensitive to interest rate risk, and the stock beta to obtain a market-neutral equity portfolio.

The portfolio delta gives a first-order approximation of the change δV in the value of an option portfolio, as a function of the change δS in the underlying asset price. The quality of the approximation may be improved by using gamma, which is a second-order sensitivity:

$$\delta V \approx \Delta \cdot \delta S + \tfrac{1}{2}\Gamma \cdot (\delta S)^2. \tag{13.31}$$

We note that the gamma of stock shares is zero, so that in order to change the gamma of an option portfolio, we need to use other options. For instance, we may change the exposure from writing an exotic option, by taking a position in vanilla options written on the same underlying asset.

Option Greeks may be used for both risk management and risk measurement, as the following example suggests.

Example 13.11 Approximating V@R of an option portfolio

> The approximation of Eq. (13.31) is sometimes suggested as a possible way to approximate V@R of option portfolios. For instance, with the data of Example 13.10, we may apply the following first-order approximation:
> $$\delta V \approx \Delta \cdot \delta S = \Delta \cdot S_0 \cdot \tilde{r},$$
> where we link the change in the portfolio value, δV, with the random return on the stock share, $\tilde{r} = \delta S / S_0$. Say that we need the 99% daily V@R. If we assume that the stock return is normally distributed, we find
> $$\text{V@R}_{0.99,1} = z_{0.99} \cdot \sqrt{\text{Var}(\delta V)} \approx z_{0.99} \cdot \Delta \cdot S_0 \cdot \sigma_d,$$

> where σ_d is the *daily* volatility of the stock share price. This may be estimated by scaling the annual volatility as follows:
>
> $$\sigma_d = \frac{0.40}{\sqrt{252}} = 0.0252,$$
>
> where we use the square-root rule introduced in Example 2.1. The square-root rule is also the reason why we do not consider the stock share daily drift. Note that, to scale annual volatility down to daily volatility, we use 252, which is roughly the number of trading days in one year. Hence,
>
> $$\text{V@R}_{0.99,1} \approx 2.3263 \cdot 124.4998 \cdot 36 \cdot 0.0252 = 262.7274.$$

The approximation of Example 13.11 should be taken with some skepticism. It might be improved by including gamma, but in this case we lose normality, which is in any case a rather questionable assumption. Furthermore, risk measures mainly deal with extreme events, where both normality and the use of approximations valid for small perturbations are unlikely to yield sensible results. The estimation of risk measures for option portfolios is a difficult endeavor, requiring heavy use of numerical methods.

13.5.2 THETA

Another relevant Greek is the option **theta**, which is related to the passage of time:

$$\Theta_C = \frac{\partial C_t^e}{\partial t} = -\frac{S_t \phi(d_1) \sigma}{2\sqrt{T-t}} - rKe^{-r(T-t)} \cdot \Phi(d2), \tag{13.32}$$

$$\Theta_P = \frac{\partial P_t^e}{\partial t} = -\frac{S_t \phi(d_1) \sigma}{2\sqrt{T-t}} + rKe^{-r(T-t)} \cdot \Phi(-d2). \tag{13.33}$$

We see that different formulas are required for vanilla call and put options, but there is a subtle point that should be stressed. We have often been rather liberal in our writing of BSM formulas, which have been written in some cases with reference to time $t = 0$, and in other cases for a generic time t. In the first case, T is both the maturity and the time-to-maturity. In the more general case, we should write $\tau = T - t$ as time-to-maturity. If we want to find the sensitivity with respect to t, we have to use the more general formulas. It is also possible to measure the sensitivity with respect to time-to-maturity τ, which is readily obtained by changing the sign in the above formulas.

Given our choice, if theta is negative, it means that the option is losing value when time is moving *forward*. Typically, option thetas are negative, which

13.5 Option price sensitivities: The Greeks

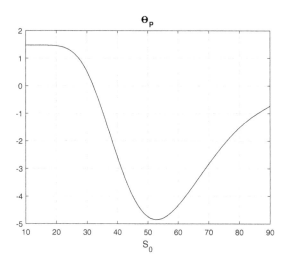

FIGURE 13.16 Theta for a put option. Time-to-maturity is six months and the strike price is 40.

shows that option values are subject to a decay over time. However, this not always the case. If we consider Fig. 13.1, we observe that the asymptote corresponding to the lower bound $S_t - \text{PV}(K)$ on the call option price will move downward in time. Indeed, the call option price will be reduced over time, all other factors being equal, which is confirmed by the plots in Fig. 13.10. A quick look at Eq. (13.32) proves that Θ_C cannot be positive. However, the picture is different for a put option. In Fig. 13.2, we observe that the lower bound $\text{PV}(K) - S_t$ will *increase* over time. Hence, we may find a positive theta for in-the-money put options. This is confirmed by the plot of Fig. 13.16, where theta is positive for small prices of the underlying asset.

13.5.3 RELATIONSHIP BETWEEN DELTA, GAMMA, AND THETA

Imagine that we build a delta-neutral option portfolio, consisting of vanilla options on the same underlying asset, and that the portfolio is long gamma. This means that the portfolio, whose value we denote by Π_t, has zero delta and positive gamma. What happens when the price of the underlying asset changes over time? We have seen something similar when dealing with interest rate risk in Chapter 6. A bond with a large convexity looks quite appealing, since it will gain a lot when rates drop, and it will not lose much when rates rise. If we set the duration of a fixed-income portfolio to zero, and keep convexity positive, the portfolio value will increase for any (small) change in the interest rate. By a similar token, if the delta of an option portfolio is zero and gamma is positive, the value Π_t is a (locally) convex function with respect to S_t. Hence, it seems that we should always have a positive profit, no matter which (small) change we observe in S_t.

Clearly, this is too good to be true, and there must be a fly somewhere in the ointment. We may find the fly by rewriting the BSM equation (13.18) for Π_t in terms of the Greeks:

$$\Theta + r\Pi_t \Delta + \frac{1}{2}\sigma^2 \Pi_t^2 \Gamma = r\Pi_t.$$

This shows that the Greeks are related, and if we choose a delta-neutral portfolio, this PDE becomes

$$\Theta + \frac{1}{2}\sigma^2 \Pi_t^2 \Gamma = r\Pi_t.$$

Now, imagine that we increase gamma, which is nice. The problem is that to maintain the equation, there must be an increase in Π_t, which means that the portfolio is expensive, or a decrease in Θ. If theta is negative, this means that a portfolio with large gamma will lose value very quickly. Whatever the case, we see that gamma has a cost.

This is also relevant from the viewpoint of an option writer, who faces a negative gamma (in the common parlance, we say that she is short gamma). This seems like a very uncomfortable place to be, even if the writer adopts a delta-hedging strategy. However, the decay in the option value, captured by a negative theta, may help writers. If the option book is losing value from the viewpoint of the option holders, this is good news to the option writer. Moreover, option writers rely on more careful strategies than naive delta-hedging.

13.5.4 VEGA

Last, but not least, the option **vega** measures the sensitivity of the option price with respect to volatility. For both vanilla call and put options, vega is

$$\mathcal{V} = \frac{\partial C_t^e}{\partial \sigma} = \frac{\partial P_t^e}{\partial \sigma} = \phi(d_1) S_t \sqrt{T-t}, \qquad (13.34)$$

where again $\phi(\cdot)$ is the PDF of a standard normal, not to be confused with the CDF $\Phi(\cdot)$.

Vega is important because, despite the assumption of constant volatility in the GBM model, volatility is stochastic in real life. We may resort to stochastic volatility models, or we may use the BSM model while keeping an eye on the volatility as a risk factor. The role of volatility is so important that we shall further elaborate on it in Section 13.6. From Eq. (13.34), we immediately observe that vega cannot be negative. Hence, an increase in volatility will result in an increase in the prices of both call and put options. The intuitive explanation is that an increase of volatility makes extreme events, i.e., the observation of very low or very high stock prices, more likely. Given the asymmetric nature of option payoffs, this increases the potential profit associated with in-the-money scenarios, whereas the payoff in the out-of-the-money scenarios is not affected. This may be used in volatility trading strategies, as illustrated later in Section 13.8.2. If we expect an increase in volatility, we may invest in call and put options. Note that this is a *nondirectional* trade, as we are not betting on the market going either up or down. It is important to realize that the behavior of exotic options, with respect to volatility, may be more complex than the behavior of vanilla options.

13.6 The role of volatility

Volatility has a significant impact on option prices. The BSM formula relies on a simple GBM model, where volatility is assumed constant, but every professional is well aware of its role as a risk factor. In fact, volatility trading strategies are used to take long or short positions with respect to this elusive factor.[18] Over the years, volatility derivatives have been introduced to transfer volatility risk between speculators and hedgers. Clearly, volatility is not a tradable asset, and pricing volatility derivatives may be a tricky business, as is the task of *measuring* a non-observable parameter like volatility, in such a way that market players may agree on it. Common sense would suggest the use of simple statistics applied to time series data. On the contrary, as we show below, volatility may be used in a rather surprising way. Furthermore, while volatility has a straightforward impact on vanilla options, its effect on certain exotic options may be less obvious.

13.6.1 THE IMPLIED VOLATILITY SURFACE

The BSM formula relies on five input arguments. Two of them, strike price and time-to-maturity, are easy to observe. The risk-free rate is a bit trickier, as we should find the continuous yield of a risk-free asset, with a maturity corresponding to the option maturity. Estimating yields based on bond prices is not as simple as it may sound, but there are sensible ways for carrying out this task. Apparently, the current stock price S_t is easily observed, too, but we should consider the fact that it is changing all the time. Possibly, the stock price may change as a consequence of our very hedging actions. However, by far, the most elusive parameter is volatility. Apparently, we may use statistical analysis of a time series of stock prices to measure its volatility. However, by doing so, we may estimate the **historical volatility**. Rather than this backward-looking approach, we need a forward-looking view on *future* volatility.

If we look at the BSM formula, we see volatility as an input, and the option price as an output. Actually, we may see things the other way around. Since vanilla options are widely traded and liquid securities, we may find the volatility implied by the observed option prices. This is an example of an *inverse* problem, and it is an example of a more general approach taken in model calibration.[19] In order to find this **implied volatility**, we invert the BSM pricing formula by

---

[18]See Section 13.8.2.
[19]Model calibration is a fundamental step when pricing in incomplete markets. See Section 14.4.

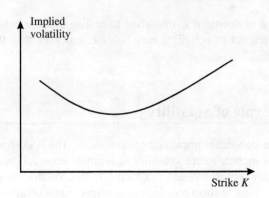

FIGURE 13.17 Qualitative shape of a volatility smile.

solving the following equation with respect to σ:

$$h(\sigma) \doteq \widehat{C}_{\text{BSM}}(\sigma; S_0, K, r, T) = C_{\text{obs}},$$

where the estimate $\widehat{C}_{\text{BSM}}$, based on the BSM model, is matched against the observed price C_{obs}. This may be easily accomplished by numerical methods.[20]

The implied volatility may be regarded as a market-consensus volatility forecast. For instance, it has been suggested to use implied volatility in V@R calculations, rather than a historical estimate. Actually, some doubts have been cast on the actual predictive power of implied volatility. Nevertheless, there is no doubt that implied volatility may be a market signal. As a matter of fact, at the peak of the subprime crisis, implied volatility was something like 80%.

Implied volatility may also be used to check the consistency among option prices. According to the BSM model, volatility is a property of the underlying asset, and it should be constant across any option, irrespective of strike prices and times-to-maturity. Actually, if we plot the implied volatility of a vanilla option with respect to the strike price K, we might obtain a nonlinear plot like the one depicted in Fig. 13.17. The precise shape may depend on the underlying asset, but we observe what has been nicknamed a **volatility smile** (or smirk). One possible explanation of this anomaly is that the BSM model relies on a lognormality assumption, which is inconsistent with empirically observed fatter tails. Since extreme events are more likely than what the GBM model would suggest, there is more risk in writing options with small or large strikes than the model accounts for. This is translated to larger option prices, which, in turn, imply a larger volatility. Out-of-the-money options may be relatively riskier to option writers than what a low BSM price would suggest, and the same happens for long-maturity options. Plotting implied volatility as a function of both K and $\tau = T - t$ results in a **volatility surface**.. Volatility surfaces may be used to compensate for the shortcomings of the BSM model, without resorting to more

---

[20] A safe way for doing so is relying on the bisection method to solve nonlinear equations. See, e.g., [4].

complex models. The surface is fitted against quoted prices of exchange-traded and liquid options, and then it is used to price OTC derivatives, or to check the price consistency among other options. We see that the machinery of option pricing is actually a way to interpolate/extrapolate internally consistent prices.

A further use of implied volatility is for quoting purposes. In the case of bonds, the price may not carry much information, as we may observe an array of quite different bond prices, depending on coupon rates and maturities. Despite of all of its shortcomings, yield-to-maturity gives a more concrete feeling for the driving factors of bond prices. By comparing yields for different bonds and observing how they change over time, a trader may get a better feeling for how markets are behaving. By the same token, an option trader may observe implied volatilities, in order to compare options and get a feeling, even though the trader may be quite skeptical about the validity of the plain BSM model. In fact, some options are quoted in terms of the volatility, which would give the actual market price when plugged into a BSM-like formula.

13.6.2 THE IMPACT OF VOLATILITY ON BARRIER OPTIONS

We have seen that the vega of vanilla options is always positive, which implies that the prices of vanilla call and put options are increasing functions of volatility. We may observe a less obvious behavior for exotic options. Let us consider barrier options, for which some analytical pricing formulas are available. As an example, let us consider a down-and-out put with strike price K, expiring in T time units, with a barrier set to S_b. The following formula has been proven, where S_0, r, and σ have the usual meaning:

$$P = Ke^{-rT} \cdot \left\{ \Phi(d_4) - \Phi(d_2) - a\left[\Phi(d_7) - \Phi(d_5)\right] \right\} \\ - S_0 \cdot \left\{ \Phi(d_3) - \Phi(d_1) - b\left[\Phi(d_8) - \Phi(d_6)\right] \right\},$$

where

$$a = \left(\frac{S_b}{S_0}\right)^{-1+2r/\sigma^2}, \qquad b = \left(\frac{S_b}{S_0}\right)^{1+2r/\sigma^2},$$

$$d_1 = \frac{\log(S_0/K) + (r + \sigma^2/2)T}{\sigma\sqrt{T}}, \qquad d_2 = \frac{\log(S_0/K) + (r - \sigma^2/2)T}{\sigma\sqrt{T}},$$

$$d_3 = \frac{\log(S_0/S_b) + (r + \sigma^2/2)T}{\sigma\sqrt{T}}, \qquad d_4 = \frac{\log(S_0/S_b) + (r - \sigma^2/2)T}{\sigma\sqrt{T}},$$

$$d_5 = \frac{\log(S_0/S_b) - (r - \sigma^2/2)T}{\sigma\sqrt{T}}, \qquad d_6 = \frac{\log(S_0/S_b) - (r + \sigma^2/2)T}{\sigma\sqrt{T}},$$

$$d_7 = \frac{\log(S_0 K/S_b^2) - (r - \sigma^2/2)T}{\sigma\sqrt{T}}, \qquad d_8 = \frac{\log(S_0 K/S_b^2) - (r + \sigma^2/2)T}{\sigma\sqrt{T}}.$$

This formula relies on the BSM model and is valid under the rather impractical assumption of continuous barrier monitoring. Its proof takes advantage of cer-

Table 13.1 The interaction of barrier level and volatility in a down-and-out put option.

	$S_b = 40$	$S_b = 35$	$S_b = 30$	$S_b = 1$	Vanilla
$\sigma = 0.4$	0.5735	1.9964	3.5610	4.5841	4.5841
$\sigma = 0.3$	0.9629	2.3928	3.1656	3.3235	3.3235

tain properties of Brownian motion,[21] but both the mathematical underpinnings and the formula are not important for our purposes. What is important is that we may use it to get a glimpse of the impact of volatility on the option price. To this aim, let us compare a vanilla and a down-and-out put with similar features: Both options are at-the-money ($S_0 = K = 50$), they mature in five months ($T = 5/12$), and the risk-free rate is 5%. We consider different barrier levels, $S_b \in \{40, 35, 30, 1\}$, and two values of volatility, $\sigma \in \{0.4, 0.3\}$. The resulting option prices are reported in Table 13.1.

If we read the table row by row from the right, we notice that the price of the vanilla and barrier options are the same, when the barrier level is very low, so that crossing the barrier is very unlikely. If we consider larger and larger barrier levels, the barrier option gets cheaper and cheaper, which makes intuitive sense, as an option knock-out is more likely. However, a different pattern emerges if we look at the table column by column. For the vanilla option, a decrease in volatility implies a decrease in price, as expected. This also applies to the exotic option when the barrier level is low, suggesting a positive vega. However, when the barrier level is large, vega seems negative: For $S_b = 40$ and $S_b = 35$, decreasing volatility increases price.

The explanation is that, for a vanilla put, a lower volatility implies a lower price, as there are less chances to observe a very low stock price at maturity. For the barrier option, this effect does contribute to the price, but there is another effect, which drives the option price into the opposite direction: A lower volatility *may* imply a higher price, since it makes crossing the barrier less likely. This second effect is negligible for low barrier levels, but when the barrier level is not too far from the current stock price, it is the dominating one.

Thus, vega may be both negative and positive, and the volatility effect is not monotonic for barrier options. In Section 13.8.3, we will observe that barrier options may also feature a rather weird behavior in terms of delta.

13.7 Options on assets providing income

We have seen, in Section 12.1.2, that when the underlying asset of a forward contract provides income, the spot–forward parity relationship is affected ac-

---
[21] See, e.g., [6].

cordingly. This applies, e.g., to stock shares paying dividends, stock indexes with a dividend yield, and foreign currencies. The same consideration applies to options, which are (usually) not dividend-protected. The case of a lump dividend payment applies to stock options. When the stock goes ex-dividend, no-arbitrage models predict a jump in the stock price. This may be taken into account by numerical methods, including binomial lattices. We will not consider this case here, but only the idealized case of a continuous dividend yield, which is relevant to index options. The case of a currency option is quite similar. An interesting case arises when we consider derivatives written on other derivatives, specifically, options written on futures.

13.7.1 INDEX OPTIONS

In Section 12.1.2, we have discussed spot–forward parity for an asset providing income at a continuous yield rate q. Under the usual stylized assumptions, the parity relationship is

$$F_0 = S_0 \cdot e^{(r-q)T}, \qquad (13.35)$$

which may be interpreted in two equivalent ways, related to hedging a short position in the contract:

- We do not need to buy one unit of the underlying asset for each short position, but only e^{-qT}, which grows to 1 unit, if we reinvest income in the underlying asset itself. This is equivalent to assuming that the initial price of the asset is $S_0 \cdot e^{-qT}$.
- We collect the income from the underlying asset over the life of contract, which lowers the cost-of-carry. Hence, we incur carrying cost at rate $r-q$, rather than r.

In order to price an option using the BSM framework, we need to write a model for the underlying asset price under a risk-neutral measure. We may translate the above intuition as a change in the drift coefficient,

$$dS_t = (r-q)S_t \, dt + \sigma S_t \, dW_t. \qquad (13.36)$$

This is also reflected in the corresponding BSM equation:

$$\frac{\partial f}{\partial t} + (r-q)S_t \frac{\partial f}{\partial S_t} + \frac{1}{2}\sigma^2 S_t^2 \frac{\partial^2 f}{\partial S_t^2} = rf.$$

By going through the BSM drill again, we obtain a simple modification of the BSM formula:

$$C_0^e = S_0 e^{-qT} \Phi(d_1) - K e^{-rT} \Phi(d_2), \qquad (13.37)$$

where

$$d_1 = \frac{\log(S_0/K) + (r - q + \sigma^2/2)T}{\sigma\sqrt{T}},$$

$$d_2 = \frac{\log(S_0/K) + (r - q - \sigma^2/2)T}{\sigma\sqrt{T}} = d_1 - \sigma\sqrt{T}.$$

Put–call parity becomes

$$P_t^e + S_t e^{-qT} = C_t^e + Ke^{-r\cdot(T-t)}, \qquad (13.38)$$

which allows us to immediately find the put option price:

$$P_0^e = Ke^{-rT}\Phi(-d_2) - S_0 e^{-qT}\Phi(-d_1). \qquad (13.39)$$

Index options may be used for hedging purposes, just like index futures. As described in Section 12.3.3, the hedge may be adjusted to account for the portfolio beta. Indeed, portfolio insurance strategies have been devised to hedge risk of equity portfolios, based on the creation of synthetic puts based on dynamic trading. In Section 13.8.1, we illustrate the mechanics, as well as the pitfalls, of portfolio insurance.

An interesting observation is that, by using spot–forward parity, we may rewrite the BSM formula in terms of the forward price $F_0 \equiv F(0,T)$:

$$C_0^e = e^{-rT}[F_0\Phi(d_1) - K\Phi(d_2)], \qquad (13.40)$$

where

$$d_1 = \frac{\log(F_0/K) + \sigma^2 T/2}{\sigma\sqrt{T}},$$

$$d_2 = \frac{\log(F_0/K) - \sigma^2 T/2}{\sigma\sqrt{T}}.$$

It is easy to see that, by plugging Eq. (13.35) into Eq. (13.40), we find Eq. (13.37). If spot–forward parity applies, this is just a way to rewrite BSM and adds nothing new. However, if we consider options on non-traded assets, the picture becomes much less obvious. The usual hedging arguments break down, if we cannot build a hedge based on the underlying asset. However, even if the asset itself is not traded, but forward contracts are available, by using risk-neutral arguments and changes of measure[22] we may justify the use of Eq. (13.40), which is known as **Black's formula**.

13.7.2 CURRENCY OPTIONS

A foreign currency may be interpreted as an asset providing income at the foreign risk-free rate r_f, and the forward price (in the domestic currency) is

$$F_0 = S_0 \cdot e^{(r_d - r_f)T},$$

where we denote the domestic risk-free rate by r_d for the sake of clarity. To price a currency option we have just to use formulas (13.37) and (13.39) for index options, by replacing the income rate q with r_f.

---

[22] See Section 14.3.4.

13.7 Options on assets providing income

A class of popular options related with foreign currencies are **quanto options**, a name stemming from *quantity adjusting options*. Imagine that we are a European investor, whose reference currency is EUR, who wants to invest on the Japanese stock market, possibly by using options on the Nikkei index. These options provide a payoff in Japanese yen (JPY), and the contract is settled in cash by multiplying the index by a given factor. Therefore, we are exposed to a twofold risk: (a) the risk of the Japanese stock market, and (b) the exchange risk, as the underlying asset is denominated in JPY. With a quanto option, the payoff is converted to EUR by a fixed exchange rate:

$$E_0 \cdot \max\{S_T - K, 0\},$$

where S_T and K are the values of the index at maturity and the strike, respectively, denominated in JPY, and E_0 is a fixed exchange rate converting the payoff in EUR. Quanto options are also called **cross-currency** options..

13.7.3 FUTURES OPTIONS

A call futures option maturing at T gives the holder the right to enter into a *long* futures contract at a specified futures price K; the corresponding put option gives the right to enter into a *short* futures position. We should not confuse the maturities of the two contracts: T is the maturity of the option, whereas the underlying futures contract matures at a later time $T_F > T$. Futures options are popular and sometimes even preferred to spot options on the underlying asset, because of the liquidity of futures. We should first clarify the mechanics of futures options. Then, we get to a key point: In the risk-neutral world, we may interpret a futures contract as an asset providing us with a risk-free income. Hence, the risk-neutral dynamics of the futures price has no drift and is a martingale.

13.7.4 THE MECHANICS OF FUTURES OPTIONS

From a mathematical viewpoint, the payoff of the call and put futures options may be written as
$$\max\{F_T - K, 0\},$$
and
$$\max\{K - F_T, 0\},$$
respectively. If the option is American-style, we should replace the futures price F_T at maturity by the current futures price F_t, when the option is exercised. This is true mathematically, but we should understand how this is accomplished exactly:

- If the holder exercises a call futures option, she will receive a long position in the underlying futures contract, plus a cash amount corresponding to the strike price minus the last settlement price.

- If the holder exercises a put futures option, she will receive a short position, as well as a cash amount corresponding to the last settlement price minus the strike price.

Thus, we see that the actual transaction is a bit more involved than the case of a spot option (i.e., an option on the underlying asset). Let us illustrate the mechanism with an example.

Example 13.12 Exercising a call futures option

Let us assume that at time t (now), the futures price of an underlying asset is

$$F_t = \$151.$$

The strike of the futures option is $K = \$140$, and F_{last} is the last settlement futures price that was observed at the end of the previous trading day. Let assume that it was

$$F_{\text{last}} = \$150.$$

This means that, if the call option is exercised, the holder will receive a cash amount

$$F_{\text{last}} - K = 150 - 140 = \$10,$$

plus a long position in the underlying contract. If the holder immediately closes out her long position in the futures, she will earn

$$F_t - F_{\text{last}} = 151 - 150 = \$1.$$

Hence, the total profit is

$$(F_t - F_{\text{last}}) + (F_{\text{last}} - K) = F_t - K = \$11,$$

corresponding to the familiar call option payoff on a spot asset.

13.7.5 A BINOMIAL VIEW OF FUTURES OPTIONS

To acquire the fundamental intuition about pricing a futures option, we may start with the simple single-step binomial model, where we assume that the futures price now is F_0, and it will be either $F_0 u$ or $F_0 d$ in the future. We may use the same setup as the spot option, with one major difference: Taking a position Δ in the underlying futures, in order to hedge the short position in the futures option, does not require any upfront payment. Hence, let us assume that we hold a short position in a futures option and we hedge with an amount Δ of the

13.7 Options on assets providing income

futures contract. The value of the hedged portfolio now is just

$$\Pi_0 = -f_0,$$

where f_0 is the unknown value of the option. In the up state, the portfolio value is

$$\Pi_u = \Delta \cdot (F_0 u - F_0) - f_u,$$

since, in that state, we receive a cash amount related to the difference in futures prices $F_0 u - F_0$. Note that, strictly speaking, this is true only if the time elapsed is one day, otherwise we are ignoring the effect of daily marking-to-market. By the same token, in the down state, the portfolio value is

$$\Pi_d = \Delta \cdot (F_0 d - F_0) - f_d.$$

The portfolio is riskless if $\Pi_u = \Pi_d$, i.e., if

$$\Delta = \frac{f_u - f_d}{F_0 u - F_0 d}, \qquad (13.41)$$

which is formally the same expression that we find for a spot option. However, when we write the value Π_0, by risk-neutral discounting, we find

$$-f_0 = e^{-rT}[\Delta \cdot (F_0 u - F_0) - f_u]. \qquad (13.42)$$

We have to plug the expression of Δ given by Eq. (13.41) into Eq. (13.42) and rearrange:

$$\begin{aligned} f_0 &= e^{-rT} \cdot \left[f_u - \frac{f_u - f_d}{F_0 u - F_0 d} \cdot (F_0 u - F_0) \right] \\ &= e^{-rT} \cdot \left[f_u \cdot \frac{(F_0 u - F_0 d) - (F_0 u - F_0)}{F_0 u - F_0 d} + f_d \cdot \frac{F_0 u - F_0}{F_0 u - F_0 d} \right] \\ &= e^{-rT} \cdot \left[f_u \cdot \frac{1-d}{u-d} + f_d \cdot \frac{u-1}{u-d} \right] \\ &= e^{-rT} \cdot [\pi_u f_u + \pi_d f_d] \\ &= e^{-rT} \cdot \mathrm{E}_{\mathbb{Q}_n}[f_T], \end{aligned}$$

with risk-neutral probabilities

$$\pi_u = \frac{1-d}{u-d}, \quad \pi_d = 1 - \pi_u = \frac{u-1}{u-d}. \qquad (13.43)$$

A comparison with the corresponding risk-neutral probability π_u for a spot stock option suggests that a term e^{rT} has been replaced by 1. The same term, in an option on an asset providing a continuous income at rate q, would be $e^{(r-q)T}$, where $r - q$ is the risk-neutral drift coefficient in Eq. (13.36). Hence, we may interpret the risk-neutral probability of Eq. (13.43) as

$$\pi_u = \frac{e^{(r-q)T} - d}{u - d},$$

where the income rate q from marking-to-market is actually given by r, under the risk-neutral measure. Therefore, $e^{(r-q)T} = e^{(r-r)T} = 1$, and the risk-neutral drift coefficient is $r - r = 0$.

13.7.6 A RISK-NEUTRAL VIEW OF FUTURES OPTIONS

To extend the intuition from the single-step binomial model to a continuous-time, risk-neutral framework, let us consider a short time step δt and the payoff from a long futures position held from time $t = 0$ to time $t = \delta t$,

$$F_{\delta t} - F_0.$$

The discounted expectation, under the risk-neutral measure, is

$$e^{-r \cdot \delta t} \cdot \mathrm{E}_{\mathbb{Q}_n}[F_{\delta t} - F_0].$$

However, the initial investment is zero, since there is no upfront payment, and so we must have

$$e^{-r \cdot \delta t} \cdot \mathrm{E}_{\mathbb{Q}_n}[F_{\delta t} - F_0] = 0 \quad \Rightarrow \quad \mathrm{E}_{\mathbb{Q}_n}[F_{\delta t}] = F_0.$$

Hence, the futures price process under the risk-neutral measure is a martingale. If we consider a GBM model, the drift must be zero,

$$dF_t = \sigma F_t \, dW_t,$$

which is reflected in the related form of the pricing PDE for the option price $f(F_t, t)$,

$$\frac{\partial f}{\partial t} + \frac{1}{2}\sigma^2 F_t^2 \frac{\partial^2 f}{\partial F_t^2} = rf.$$

By solving the PDE, or by finding the risk-neutral expectation, we obtain the price of a call option on a futures contract:

$$C_0^e = e^{-rT}[F_0 \Phi(d_1) - K \Phi(d_2)], \tag{13.44}$$

where

$$d_1 = \frac{\log(F_0/K) + \sigma^2 T/2}{\sigma \sqrt{T}},$$

$$d_2 = \frac{\log(F_0/K) - \sigma^2 T/2}{\sigma \sqrt{T}}.$$

This is a familiar formula by now, related to the Black's model. In this case, however, F_0 is the *futures*, rather than the forward price.

Remark. We see that the futures price is a martingale under the traditional risk-neutral measure. The case of a forward is different, as it requires the introduction of a *forward risk-neutral* measure, as shown in Section 14.3.3.

13.8 Portfolio strategies based on options

In this section, we outline some option portfolio management strategies, in order to give a feel for the potential applications of the theory that we have developed, as well as some of its pitfalls.

13.8.1 PORTFOLIO INSURANCE AND THE BLACK MONDAY OF 1987

Put options can be used to protect the value of an equity portfolio. Index options may be used to this aim, and the portfolio beta may help in setting up the hedge, if the portfolio is well diversified, but not quite in line with the index. However, we may hold a portfolio for which we feel that index options are inappropriate, because they can only protect against systematic risk and we may not trust the beta estimate, or we may not find options matching the desired maturity. Portfolio insurance is a strategy to create a *synthetic* protective put by dynamic trading.

Here we give a rough-cut description of a simple portfolio insurance strategy. We know from put–call parity that

$$P_0 + S_0 = C_0 + Ke^{-rT},$$

where S_0 is the current value of the underlying asset, which in our case is an equity portfolio, and P_0 and C_0 are the hypothetical values of put and call options, with strike K and maturity T, written on the portfolio. Note that $P_0 + S_0$ is the value of a protective put strategy, with strike K, which cannot actually be implemented, if the option we need is not traded. Plugging the BSM price of the call option and rearranging, the protective put is equivalent to

$$S_0 \Phi(d_1) - Ke^{-rT}\Phi(d_2) + Ke^{-rT} = S_0 \Phi(d_1) + Ke^{-rT}\left[1 - \Phi(d_2)\right].$$

The last expression suggests that our portfolio should consist of $\Phi(d_1) < 1$ "portfolio shares" and an amount invested in a risk-free asset. This implies that $1 - \Phi(d_1)$ portfolio shares should be sold and invested in the risk-free asset. We also know that $\Phi(d_1)$ is the delta of a call option and, when the value of the underlying asset drops, delta is reduced. Thus, more portfolio shares should be sold. On the contrary, when the portfolio value goes up, delta will, too, and shares will be purchased back.

The idea looks very simple, but it is actually simplistic for a few reasons:

- We have assumed that we implement a protective put with value $P_0 + S_0$, which actually implies that additional cash (for an amount P_0) should be added. Who is going to sustain that cost? A more sensible framework should start with a portfolio with overall value S_0, and part of that value should be allocated to the put.

- Actually, selling and buying back the portfolio may be expensive. Hence, futures contracts may be used to that purpose to create synthetic short positions. Again, index futures may achieve limited effectiveness in case of peculiar portfolios.

Example 13.13 Who was to blame on October 1987?

October 19, 1987, is an infamous day in financial history, better known as "the Black Monday of 1987." On that single day, the Dow Jones

> Industrial Average (DJIA) dropped by 22.61%, and other indexes did not fare much better. The analysis of the causes of this global market crash is controversial, but the use of automated trading to implement portfolio insurance strategies has been often blamed as one of the potential culprits.
>
> Whether this is true or not, it is clear that the strategy that we have outlined may suffer from a fatal flaw: We take for granted that uncertainty and risk are exogenous and not affected by our trading activities. In other words, we disregard the market impact of our trades. However, on the one hand, markets can lose liquidity and depth, and on the other one, even if each individual trader is no big fish enough to affect prices in the pond, risk may be endogenously generated if many traders pursue similar strategies. In portfolio insurance, we sell assets when prices go down, which in turn may trigger further selling orders if market impact is significant. This may create a vicious feedback cycle.
>
> Whether automated trading was the only reason behind the Black Monday is debatable, but one thing is sure: Some traders who held true put options, not synthetic ones, made a fortune on that day.

13.8.2 VOLATILITY TRADING

The fact that vega is positive for vanilla options suggests that options may be used in order to take long or short positions with respect to volatility.

Suppose that an at-the-money put option on a stock share matures in 60 days,[23] and that its strike price is $90. The risk-free rate is 4%, and the put option sells for $4.4955. Then, implied volatility is 33%. However, let us assume that this level of implied volatility is too low, in our view, and we believe that it will increase to 35%. If our view will prove correct, the put price will increase to $4.785, with a corresponding profit of $0.29 for each put option (assuming that the BSM formula applies). Note that we assume an instantaneous change in volatility, i.e., we neglect the effect of theta.

Taking a position in a call or a put option is more or less the same, since vega is identical for both options (assuming that the maturity and the strike are the same, too). However, changes in the underlying asset price have quite different impacts. If we buy the put, we might lose money if the stock price increases, possibly overwhelming the impact of volatility. Let us assume that we

---

[23] Here we replicate the numerical results for an example that we are borrowing from [2]. They are obtained by the BSM formula, but time-to-maturity should be taken as 60/365 years, rather than two months.

13.8 Portfolio strategies based on options

Table 13.2 Hedging a volatility bet.

Stock price	89	90	91
Put price ($\sigma = 35\%$)	5.2536	4.7846	4.3467
Profit on each put	0.7581	0.2891	-0.1488
Value of 1000 puts	5253.6	4784.6	4346.7
Value of 453 shares	40,317	40,770	41,223
Total	45,570.6	45,554.6	45,569.7
Profit	305.1	289.1	304.2

are confident in our view on volatility, but we have no clue about the direction of the underlying asset price. To hedge this risk, we may make the portfolio delta-neutral, by purchasing a few stock shares. The portfolio value, if we hold N shares and one put option, is

$$NS_0 + P_0,$$

and its delta is

$$N + \Delta_P.$$

Hence, we should hold $N = -\Delta_P$ stock shares for each put option to obtain a delta-neutral portfolio. If we assume that our assessment of volatility is correct,[24] i.e., we evaluate delta with $\sigma = 0.35$, we find $\Delta_P = -0.4533$. Note that the actual position in the stock share is long, as we must gain from the hedge, when the put price suffers from an increase in the stock share price.

Assume that we buy 1000 put options. The initial value of the hedged portfolio is

$$1000 \times 4.4955 + 453 \times 90 = \$45{,}265.5.$$

Let us check the effectiveness of the hedge by considering three scenarios:

1. The stock price decreases to $89.
2. The stock price does not move.
3. The stock price increases to $91.

The results are summarized in Table 13.2. For *small* changes of the stock price, the hedged position performs a little better than anticipated, due to the positive gamma of the option.[25] This looks very nice, but what about large movements in the price of the underlying asset? Furthermore, we have ignored the time decay in the value of options. The strategy should work fast, lest we suffer from theta, i.e., the decay in the option value over time.

---

[24] A more sophisticated approach would also consider the cross-sensitivity, i.e., the derivative of vega with respect to changes in the stock price. This sensitivity is called **vanna**; see Problem 13.15.
[25] We have observed a similar convexity effect for bonds. See Section 6.5.

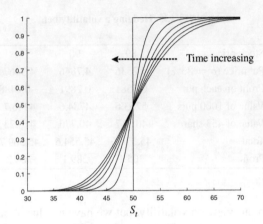

FIGURE 13.18 The evolution of a binary option price over time. The smooth price function converges to the discontinuous payoff when maturity is approached.

13.8.3 DYNAMIC VS. STATIC HEDGING

We have derived the BSM equation in a rather informal way, which has the advantage of emphasizing the relationship between fair option price and hedging cost, as well as giving a recipe for dynamic delta-hedging. In principle, we should keep a delta-neutral portfolio to hedge the option. As we pointed out in Section 13.4.1.2, continuous-time hedging is out of the question, because of transaction costs. Delta-hedging has been the subject of further criticism. Apart from limitations of delta estimates based on the BSM model, it has been argued that other approaches are needed to achieve an approximate, but robust hedging, rather than perfect hedging for small risks. From a practical viewpoint, we should consider that an option book may be hedged, rather than a single contract, with corresponding opportunities for saving.

Anyway, even if we remain within the safe BSM world, delta-hedging may prove difficult to achieve. A clue is given in Fig. 13.15, where we see how gamma gets more and more peaked as maturity is approached. This implies that delta may change abruptly for at-the-money options, since the option value gets closer and closer to the nondifferentiable payoff function. Things are even uglier in the case of a discontinuous payoff. Equation (13.27) gives the price of a binary option, which is plotted in Fig. 13.18 as a function of the underlying asset price in the range $[30, 70]$, when the strike price is set to $K = 50$, for different times-to-maturity. When time-to-maturity goes to zero, the price function gets steeper and steeper, even though it preserves its continuity properties due to the diffusion nature of the pricing PDE. The problem is that delta-hedging in such a situation may get very difficult, if there are huge swings in the hedge, corresponding to modest changes in the underlying asset price.

When we consider exotic options, things may get ugly, too. Let us consider an example of barrier option, the up-and-out call option. When solving the

13.8 Portfolio strategies based on options

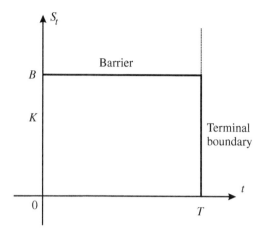

FIGURE 13.19 Boundaries for an up-and-out call option.

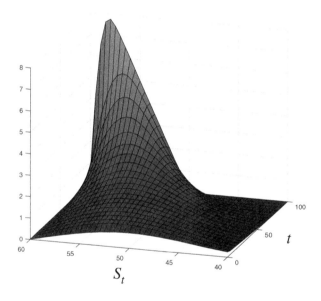

FIGURE 13.20 Surface plot of an up-and-out call option price.

pricing PDE, the boundary conditions illustrated in Fig. 13.19 should be set to reflect the option specification. The terminal boundary is just the usual call option payoff, but along the barrier, the knock-out option value is zero. The resulting option value is displayed in Fig. 13.20. The surface plot gives the value of the option for a range of underlying asset prices (from 40 to 60) and time instants (from $t = 0$ to $t = 100$ days, which is the option maturity). The strike is $K = 50$, the barrier is $S_b = 60$, and we use an analytical formula available for

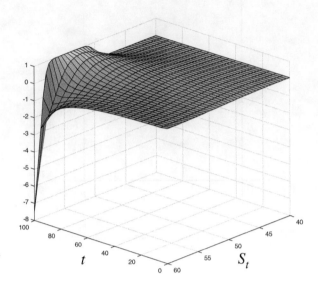

FIGURE 13.21 Surface plot of an up-and-out call option delta.

continuous-time monitoring in the BSM world. Far from the barrier, we notice the usual call option payoff at maturity. Close to the barrier, the option value drops to zero. We also notice a different pattern of theta. The option price tends to increase over time, as there are less and less chances to cross the barrier. However, the most striking effect is on the option delta, which is displayed in Fig. 13.21. At maturity, far from the barrier, we notice the usual pattern of a call delta, which is related to a normal CDF: Delta is zero below the strike and is one above the strike. However, we notice a striking a singularity close to the barrier. There, delta is negative and quite large in absolute value. This suggests that delta-hedging the up-and-out call option is a nightmare, if we are close to the barrier and close to maturity.

A further complicating factor is **model risk**. The simplest modeling error is related to the use of a wrong value for a parameter, like volatility. A thornier issue is the use of a wrong model altogether. When pricing a barrier option, we might want to consider a stochastic volatility model, possibly the Heston model. However, the more sophistication we introduce in the model, the more uncertain parameters we have to estimate.

All of these difficulties have suggested the opportunity to replace dynamic delta-hedging with alternative approaches. One such approach is **static hedging**, where a portfolio based on vanilla options is chosen once and for all in order to hedge an exotic option. In practice, we may pursue semi-static hedging strategies; furthermore, if certain events happen, like a knock-out of a barrier option, this will affect the hedge. Static hedging may be accomplished in a variety of ways, including the solution of an optimization model. Model risk may be tackled by robust optimization techniques,[26] whose application to finance is an active research area.

13.9 Option pricing by numerical methods

In this chapter, we have relied on analytical formulas, for the sake of convenience. Most options, especially when early exercise features are included, must be priced numerically. All numerical methods rely on some form of discretization, which may be deterministic or stochastic. There is a huge variety of numerical methods for financial engineering. Sometimes, a considerable amount of adaptation is required to devise an efficient and robust procedure for a specific contract. All of this is definitely outside the scope of this book, but it is useful to link the three general classes of pricing methods with the concepts that we are already familiar with, so that the reader may appreciate the tradeoffs among the alternatives.

Lattice methods. The simple binomial lattice approach that we have considered in Section 13.3 can be extended to multiple dimensions, and trinomial lattices offer additional flexibility. The advantages of lattice methods are the computational efficiency, since they are based on a careful deterministic discretization, and the possibility of dealing with early exercise. However, they are not well suited to high-dimensional options, and they also require some adaptation to cope with path dependency. Furthermore, they do not yield Greeks naturally.

Finite difference methods. An alternative class of methods relies on the discretization of the pricing PDE. Finite difference methods are based on a grid discretization of the domain on which the PDE must be solved, and the discretization of partial derivatives by increment ratios. When the starting PDE is linear, as the BSM equation is, we either have to solve a sequence of systems of linear equations, or we apply a simple set of explicit formulas, similar to those we obtain from lattice methods. The first case occurs with implicit schemes; the second case occurs with explicit schemes. Implicit schemes may be more time consuming, but are more accurate and more robust in terms of numerical stability. In both cases, we find the option price by a backward process starting from the option payoff. Finite difference methods are rather efficient in general since, just like lattice methods, they rely on a deterministic discretization. Using finite differences, we find some Greeks, like delta and gamma, for free. These methods are able to cope with early exercise, but are limited to low-dimensional problems. Furthermore, they can be applied to path-dependent options, provided that the correct PDE is devised. For weakly path-dependent, barrier options, the standard

---

[26] See Section 15.9.

BSM equation is valid, whereas a different PDE applies to strongly path-dependent, Asian options. Sometimes, careful analysis is needed to find suitable boundary conditions. More sophisticated approaches, like finite-elements methods, are also applied.

Monte Carlo methods. Unlike the previous approaches, Monte Carlo methods rely on random sampling. Starting from a dynamic model based on stochastic differential equations, a discrete-time sample path generation procedure is selected,[27] random paths are sampled, and the sample mean of the resulting payoff is used to estimate the option price. This sounds quite simple and, in fact, Monte Carlo methods are quite flexible and powerful. They are actually the only viable solution method for high-dimensional problems, especially when complex path dependencies must be accounted for. The disadvantage is the possibly large sample size needed to obtain reliable estimates. Furthermore, the adaptation to American-style options is difficult and computationally intensive,[28] and additional work is needed to find reliable estimates of the option price sensitivities. Nevertheless, some methods to reduce variance of the estimators are available, and a significant advantage of Monte Carlo methods is that they may be easily parallelized, taking advantage of the availability of multiple processors.

Problems

13.1 A stock price is currently $40. It is known that, in one month, the price will be either $42 or $38. The annual risk-free interest rate is 8%, with continuous compounding. What is the value of a one-month European call option with a strike price of $39? How many stock shares should a writer hold in order to hedge risk? Repeat in the case of a put option, and check the working of the hedge in detail.

13.2 The following figure shows the payoff of a common strategy called *spread*.

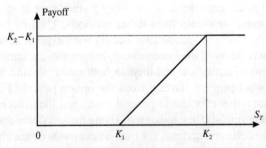

---

[27] See Section 11.6.
[28] This requires numerical methods to solve stochastic dynamic programming problems. See Section 15.7.

Explain how you could create the strategy by using either vanilla calls or vanilla puts. What is the difference between the two implementations in terms of cash flow timing?

13.3 Consider the payoff depicted in the following figure, depending on S_T, the price of an underlying stock share at time T.

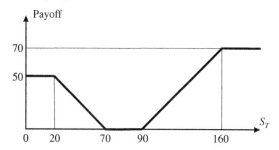

Find a trading strategy that replicates this payoff. You may use European-style call and put options on the stock, the stock itself, and a risk-free zero coupon with maturity T.

13.4 Consider a forward contract on a stock, with given delivery price and delivery date. How can we synthesize the payoff of a long position in a forward, by using vanilla options?

13.5 Consider a call and a put option written on an underlying asset with current price $60. The options have strike $55 and mature in nine months. The risk-free rate (with continuous compounding) is 5%. If you think that it is necessary, you may assume any drift and volatility you like for the underlying asset price. The call price is $12, the put price is $4. Is there an arbitrage opportunity? If so, devise a strategy to take advantage of it and check that it will work in every scenario.

13.6 Extend the put–call parity relationship of Eq. (13.4) to the case in which the underlying stock share pays dividends, with present value D, between now and the maturity of the two options.

13.7 Let us consider a European-style call option on an underlying stock share whose price dynamics is represented by a three-step binomial lattice (Note: The lattice has 4 nodes in the last time layer, corresponding to maturity). Time-to-maturity is one year, and the continuously compounded (annual) risk-free rate is 3%. The current underlying asset price is $30 and the strike is $30. We do not consider dividends and, at each time step, the stock share either gains 15%, with probability 0.6, or loses 10%, with probability 0.4.

- Price the option.
- Consider the sample path (up, up, down), and imagine that we are at the beginning of the last time period (after two steps, when time-to-maturity is four months). How many stock shares should be bought or sold (at this time instant) by the option writer to hedge risk?

13.8 Consider an American-style put option written on a stock share that does not pay dividends. The continuously compounded risk-free rate is 3%. The option matures in nine months and its strike is $60. The current stock price is $50 and its annualized drift and volatility are 10% and 50%, respectively. Price the option using a time step of three months. If we had to price a European-style Asian option, would the lattice calibration or anything else change?

13.9 The price of a stock share (no dividend, current price is €50) follows a GBM with drift 12% and volatility 35%; the continuously compounded risk-free rate is 5%. Consider a (European-style) lookback call with floating strike, whose payoff is $(S_{max} - S_{min})$. Find the option price, by approximating the underlying stochastic process by a three-step binomial tree (the option matures in nine months, and each time step is three months).

13.10 The price of a non-dividend-paying stock share is currently 55 and its volatility is 31%, while the drift is 10%. The risk-free rate, continuously compounded, is 6%. We want to find the price of an as-you-like-it option, with strike 55 and maturity T_2 of six months. In such an option, the holder may chose at a given time T_1, before maturity $(T_1 < T_2)$, whether the option is a call or a put. In other words, at time T_1 the holder may choose whether the uncertain payoff at time T_2 will be given by a call or a put payoff. Both payoffs have the same strike, maturity, and underlying asset, of course.

- For which price of the underlying asset you are indifferent between choosing the put or the call option at T_1?
- Estimate the value of the option at time $t = 0$, using a two-step binomial lattice, assuming that the choice has to be made after three months, i.e., half-way to maturity, which is six months (Clearly, this is a very rough estimate!).

13.11 Compute the delta of a European-style call option with strike 40, maturing in four months, written on an underlying asset whose current price is 37. The stock price follows a GBM with drift and volatility coefficients 13% and 30%, respectively. The annual risk-free rate is 6%, with continuous compounding.

13.12 Compute the delta of a European-style put option with strike price 35, maturing in five months, written on a stock share (no dividends) whose current price is 47, with drift and volatility 12% and 45%, respectively. The annual risk-free return is 3% (all rates are continuously compounded). If an investor holds 100 such puts and wishes a portfolio that is not sensitive to variations of the underlying asset (on the short term), how many stock shares should she hold?

13.13 You hold a portfolio of (vanilla, European-style) options written on the same stock share, whose price follows a geometric Brownian motion with drift 9% and volatility 25%. At present, the stock price is $30, and the risk-free rate, with continuous compounding, is 3%. The portfolio consists of:

- A short position in 1000 put options, strike $27, maturing in three months
- A long position in 500 call options, strike $30, maturing in four months
- A short position in 1500 call options, strike $28, maturing in two months

How many stock shares do you need to make the portfolio delta-neutral? Can you also make the portfolio gamma-neutral by using stock shares? If so, explain how. Otherwise, how should you change the position in the last call to make the portfolio gamma neutral?

13.14 You hold a long position in a call and a short position in a put, written on the same non-dividend-paying stock share, but having different strikes and maturities, as reported in the table below (note that the number of puts held is negative):

Option	Call	Put
Maturity	six months	two months
Strike (€)	55	45
Number	1000	−1500

The underlying asset price follows a GBM with drift coefficient 13%. The current volatility is 35%, while the continuously compounded risk-free rate is 4% for all maturities; the current underlying asset price is €50. Is your position long or short with respect to volatility? Note: When you are long a risk factor, it means that you gain if the factor increases; you are short the risk factor, if you gain when its value drops. We consider only small variations.

13.15 In risk management, we sometimes need second-order sensitivities, like gamma. Vanna is defined as

$$\frac{\partial^2 V}{\partial S \, \partial \sigma},$$

where V is the option value. This may be regarded as the derivative of delta with respect to volatility, or the derivative of vega with respect to the current price of the underlying asset. Let us assume that the BSM model applies.

- Is vanna different for call and put options?
- Find a formula for the option vanna.

13.16 Consider a European-style option on a non-dividend-paying stock share, whose price follows a geometric Brownian motion with drift 5% and volatility 35% (per year); the continuously compounded risk-free rate is 3%; the option matures in 4 months, and the current underlying asset price is $50. The payoff (in USD) is given by the following contingency table depending on the terminal price of the underlying asset:

Condition	Payoff
$S_T < 50$	0
$50 \leq S_T < 60$	5
$60 \leq S_T < 70$	10
$70 \leq S_T$	1

Find the option price.

13.17 Let us consider a European-style put option on a non-dividend-paying asset whose price follows a GBM with drift 10% and volatility 40% (annualized). The risk-free rate is 5% with continuous compounding. The option matures in six months, the current underlying asset price is €40, and the strike is €50. Find the probability that the payoff is between €10 and €20.

13.18 Consider a European-style derivative, depending on S_T, the price of a non-dividend-paying stock share at time T, characterized by the payoff depicted in the figure below.

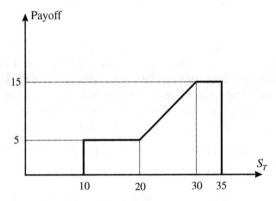

Find a simple expression for the option price.

13.19 An investment bank offers a derivative whose payoff at maturity T is given by S_T^2, where S_T is the price of the underlying asset, a non-dividend-paying stock share with a price following a GBM.

- Find the price of the derivative at time 0.
- Check the correctness of your result by verifying that it satisfies the BSM partial differential equation.

13.20 Using the Feynman–Kač representation theorem, solve the PDE

$$\frac{\partial V}{\partial t} + ax\frac{\partial V}{\partial x} + \frac{1}{2}b^2 x^2 \frac{\partial^2 V}{\partial x^2} = 0,$$

for an unknown function $V(x,t)$ with terminal condition

$$V(x,T) = \log(x^4) + k,$$

where a, b, and k are known constants. Note that we have no right-hand side in the PDE (like rV), so that the formula can be simplified.

13.21 Prove that the price of a European-style vanilla call option is a convex function of its strike, without any assumption on the model (e.g., you should not assume that the BSM model applies). *Hint:* Consider the butterfly-spread in Example 2.11.

13.22 Assume that an index value is 100 at present, and that an at-the-money put option on the index is available and used to hedge a portfolio with a beta of 1.5. The put option matures in two months, the index volatility is 25%, and the risk-free rate is 5% (with continuous compounding). For the sake of simplicity, let us assume that, in pricing and settling the option, the index is multiplied by 1 (hence, we may identify the index value with the value of the underlying asset, which is $100). We have invested $100,000 in the portfolio.

- Build a delta-neutral portfolio, without adding cash. This means that a part of the current wealth is used to buy the put options. How many options are bought? What are the dollar values of the two positions?
- Check the effectiveness of the hedge and explain the results in the following scenarios:
  - Instantaneous return on the index of $\pm 5\%$
  - Return on the index of $\pm 5\%$, after one month

Hint: You may assume that you initially hold 1000 units of the index, and that the portfolio return is only affected by systematic risk (i.e., there is no contribution from specific risk). Neglect dividend yield.

Further reading

- We have just scratched the surface of option pricing. A standard and quite extensive coverage of options can be found in [11]. You may also see [17] or [18].
- We have followed a rather liberal and intuitive approach to the application of stochastic calculus to financial engineering. This is arguably the best way to get going and build useful intuition, but more advanced applications, in terms of both contract types and underlying dynamic models, require a more solid foundation. See, e.g., [1] or [6].
- Apart from a theoretical understanding of options, it is very useful to read about real-life stories. For instance, more details about the Black Monday of 1987 can be found in [3].
- Static hedging is described in [7], for the case of barrier options. An application based on robust optimization is proposed in [12]. Robust optimization, which we outline in Chapter 15, is used to address parametric

uncertainty. See also [14] for a discussion of static hedging and model risk. A comprehensive reference on model risk is [13].

- An introductory reference on numerical methods is [4], where standard approaches are covered. See also [8]. More specific references about Monte Carlo methods are [5] and [9], whereas [16] deals with numerical methods based on the solution of PDEs.

- After the 2008 credit crunch, new concepts have been introduced to adjust the valuation of derivatives, taking a range of factors into account. They are collectively known as xVA and include credit, debit, and funding valuation adjustments (CVA, DVA, and FVA, respectively). See, e.g., [10].

Bibliography

1 T. Björk. *Arbitrage Theory in Continuous Time* (2nd ed.). Oxford University Press, Oxford, 2004.

2 Z. Bodie, A. Kane, and A. Marcus. *Investments* (9th ed.). McGraw-Hill, New York, 2010.

3 R. Bookstaber. *A Demon of Our Own Design: Markets, Hedge Funds, and the Perils of Financial Innovation*. Wiley, New York, 2008.

4 P. Brandimarte. *Numerical Methods in Finance and Economics: A MATLAB-Based Introduction* (2nd ed.). Wiley, Hoboken, NJ, 2006.

5 P. Brandimarte. *Handbook in Monte Carlo Simulation: Applications in Financial Engineering, Risk Management, and Economics*. Wiley, Hoboken, NJ, 2014.

6 G. Campolieti and R.N. Makarov. *Financial Mathematics: A Comprehensive Treatment*. CRC Press, Boca Raton, FL, 2014.

7 P. Carr, K. Ellis, and V. Gupta. Static hedging of exotic options. *Journal of Finance*, 53:1165–1190, 1998.

8 L. Clewlow and C. Strickland. *Implementing Derivatives Models*. Wiley, Chichester, 1998.

9 P. Glasserman. *Monte Carlo Methods in Financial Engineering*. Springer, New York, 2004.

10 Jon Gregory. *The xVA Challenge: Counterparty Credit Risk, Funding, Collateral, and Capital* (3rd ed.). Wiley, Chichester, 2015.

11 J.C. Hull. *Options, Futures, and Other Derivatives* (8th ed.). Prentice Hall, Upper Saddle River, NJ, 2011.

12 J.H. Maruhn and E.W. Sachs. Robust static hedging of barrier options in stochastic volatility models. *Mathematical Methods of Operations Research*, 70:405–433, 2009.

13 M. Morini. *Understanding and Managing Model Risk: A Practical Guide*

for Quants, Traders and Validators. Wiley, Chichester, 2011.

14 M. Nalholm and R. Poulsen. Static hedging and model risk for barrier options. *Journal of Futures Markets*, 26:449–463, 2006.

15 S. Shreve. *Stochastic Calculus for Finance (vol. I)*. Springer, New York, 2003.

16 D. Tavella and C. Randall. *Pricing Financial Instruments: The Finite Difference Method*. Wiley, New York, 2000.

17 P. Wilmott. *Derivatives: The Theory and Practice of Financial Engineering*. Wiley, Chichester, West Sussex, England, 1999.

18 P. Wilmott. *Quantitative Finance (vols. I and II)*. Wiley, Chichester, West Sussex, England, 2000.

Chapter Fourteen

Option Pricing: Incomplete Markets

In Chapter 13, we have introduced the essential mathematical machinery required to price options. We have done so under the somewhat self-contradictory assumption of market completeness. The practical implication of market completeness is that we price derivatives under the condition that they are of no use, as they can be perfectly replicated, and that any risk can be hedged away. Needless to say, markets are not complete, as a consequence of the following reasons:

- The derivative may be written on a non-traded asset, which cannot be included in a hedging/replication portfolio. This is the case, e.g., for interest rate derivatives.
- The BSM model relies on geometric Brownian motion (GBM), which features continuous sample paths, a gentle tail behavior with moderate kurtosis, as well as a deterministic volatility. Actually, volatility is an unobservable risk factor, which may be hard to hedge.
- Diffusion models with stochastic volatility may yield heavy-tailed distributions that are more compatible with observed behavior than GBM. A more radical approach is to rule out continuity and introduce jumps. Jumps introduce a non-predictable component that may disrupt replication approaches.
- Transaction costs preclude continuous-time hedging, resulting in hedging errors and residual risk.

From a theoretical viewpoint, market incompleteness implies the existence of multiple probability measures compatible with no-arbitrage. From a practical viewpoint, market incompleteness implies that the option writer cannot fully hedge the whole risk of her business. Hence, risk aversion creeps back into the game, in one form or another.

Lack of market completeness gives rise to different approaches to option pricing and hedging:

- One possible approach relies on the adaptation of the pricing machinery of complete markets. We use the same tools, like PDEs or risk-neutral expectations, but we have to spot a suitable measure by model calibration, i.e., by matching observed prices of liquid securities, and then using the model to price illiquid and OTC derivatives.

- Another approach relies on the solution of an optimization model, in which we want to devise a hedging portfolio minimizing a risk measure. The price of the derivative should be related to the cost of the resulting hedging strategy. There are alternative ways of doing so, depending on the trading strategy that we want to pursue (static or dynamic), and on the risk measure that we select.

Note that these two views coincide in the ideal setting of complete markets, but lead to different frameworks in a more realistic setting. We will pursue the first approach, which is more common and aims at finding a system of internally consistent prices.[1] The machinery of option pricing is, essentially, an interpolation/extrapolation approach based on relative pricing. We do not claim that we find the "right" prices, but only that those prices are consistent, in the sense that they do not allow for arbitrage opportunities.

The limited aim of this chapter is to provide the reader with the essential concepts, which will be illustrated by simple examples. The sheer variety of derivatives and pricing approaches precludes an exhaustive treatment. Furthermore, a more advanced knowledge of stochastic calculus than the very basic one provided in Chapter 11 would be needed. We shall only deal with interest rate derivatives, showing how the approaches devised for complete markets may be generalized. Even so, there is a significant variety of modeling approaches to cope with interest rate derivatives. The difficulty in modeling interest rates is that there are plenty of them, even if we only consider risk-free rates and do not model credit spreads. In principle, we should model the dynamics of a full term structure $r(t, t + \tau)$, as a function of t, for a conceptually infinite set of time to maturities τ. This is considerably more complicated than modeling a single stock price, and it should be done in a credible and arbitrage-free way. Clearly, an infinite-dimensional problem must be discretized and boiled down to a finite-dimensional one.

- An extreme choice is to use a single factor and relate all of the relevant rates to it. Models based on the (instantaneous) short rate have been proposed in this vein. Clearly, this approach is limited in terms of both ability to fit market prices and dynamic realism, as all rates are related to a single risk factor and certain twists in the term structure are precluded.
- An alternative approach is to use a limited set of factors, such as a subset of spot rates or a suitable combination of spot rates (possibly obtained by data reduction algorithms, like principal component analysis). An alternative is to model *forward* rates and obtain spot rates as a consequence.

---

[1] We shall hint at financial optimization models in Chapter 15.

- With respect to short-rate models, a multifactor model trades off model tractability and fitting flexibility. A further limitation of short-rate models is that they rely on a factor that is not really observable. Market models rely on rates, such as LIBOR rates, that may be directly observed on markets.

We will mostly rely on short-rate models, despite the fact that these models may have limited practical value. Short-rate models have a definite pedagogical motivation, beside their historical value, as they provide us with a simple way to understand essential concepts.

In Section 14.1, we pursue a PDE approach, which essentially generalizes the hedging view of Section 13.4.1. We will illustrate how this framework may be applied in Section 14.2, where we price fixed-income assets using single-factor models based on the short rate. Then, in Section 14.3, we offer another viewpoint based on martingale measures, which generalizes the risk-neutral approach of Section 13.4.2. Finally, in Section 14.4, we consider potential pitfalls in model calibration, which is an essential task when dealing with incomplete markets.

14.1 A PDE approach to incomplete markets

In Section 13.4.1, we have derived a pricing PDE on the basis of a hedging argument. The aim was to hedge a short position in a derivative, by taking a position in the underlying asset. Here, we consider a single risk factor that, unlike a stock share, cannot be included in an investment portfolio. To be concrete, we will refer to a continuously compounded risk-free short rate r_t. This may be interpreted as an instantaneous rate applying to a small time slice $[t, t + \delta t]$, for $\delta t \to 0$. It is important not to confuse the short rate r_t with the rate $r(t, T)$ applying over a possibly long time period $[t, T]$. Both of them are annualized, as usual, but if we invest over the time period $[t, T]$ at rate $r(t, T)$, we face no interest rate risk. On the contrary, we are subject to reinvestment risk, if we roll the short rate r_t over the time period $[t, T]$. In fact, one of the things we have to check is which kind of term structure dynamics can be generated by a single risk factor.

We assume that the short rate may be modeled by the following Itô diffusion model[2]:

$$dr_t = m(r_t, t)\, dt + s(r_t, t)\, dW_t, \qquad (14.1)$$

for given drift function $m(r_t, t)$ and volatility function $s(r_t, t)$. This family of models includes GBM, which is not a very sensible model for interest rates, as it does not display mean reversion, which is a commonly observed feature of interest rates. We will provide specific examples in Section 14.2.

---

[2] Much of the treatment in this section follows [11].

Let us denote by $Z_1(r_t, t)$ and $Z_2(r_t, t)$ the prices of two traded assets, whose value depends on the short rate r_t at time t. These assets can be as simple as two zero-coupon bonds, which explains the notation, or more complex interest rate derivatives. Note that a zero may be interpreted as an interest rate derivative with respect to the short rate r_t, and that we do *not* want to price the zero using the classical discount factor depending on a known term structure $r(t,T)$ at time t.

Imagine that we hold a long position in Z_1. We cannot build a hedging portfolio for Z_1 including the underlying risk factor, which is not tradable. Nevertheless, we still can get rid of risk if we include Z_2 in the portfolio, as it is a traded asset, subject to the *same* risk factor as Z_1. Apart from this difference, we essentially use the same mathematical machinery as in the BSM model for equity derivatives. Using Itô's lemma as usual, we may find the stochastic differential equations for prices $Z_i(r_t, t)$, $i = 1, 2$:

$$dZ_i = \left\{ \frac{\partial Z_i}{\partial t} + m(r_t, t) \cdot \frac{\partial Z_i}{\partial r_t} + \frac{1}{2} s^2(r_t, t) \cdot \frac{\partial^2 Z_i}{\partial r_t^2} \right\} dt + s(r_t, t) \cdot \frac{\partial Z_i}{\partial r_t} dW_t. \tag{14.2}$$

In order to hedge risk, we may set up a portfolio involving both Z_1 and Z_2 in the right proportions. Let us build a hedged portfolio as follows:

$$\Pi(r_t, t) = Z_1(r_t, t) - \phi_t \cdot Z_2(r_t, t), \tag{14.3}$$

for a suitable choice of the hedging ratio ϕ_t (which is going to change over time). If we apply Itô's lemma to the portfolio, we find

$$d\Pi = \left\{ \frac{\partial \Pi}{\partial t} + m(r_t, t) \cdot \frac{\partial \Pi}{\partial r_t} + \frac{1}{2} s^2(r_t, t) \cdot \frac{\partial^2 \Pi}{\partial r_t^2} \right\} dt + s(r_t, t) \cdot \frac{\partial \Pi}{\partial r_t} dW_t. \tag{14.4}$$

By differentiating Eq. (14.3), we find

$$\frac{\partial \Pi}{\partial r_t} = \frac{\partial Z_1}{\partial r_t} - \phi_t \cdot \frac{\partial Z_2}{\partial r_t}, \tag{14.5}$$

$$\frac{\partial^2 \Pi}{\partial r_t^2} = \frac{\partial^2 Z_1}{\partial r_t^2} - \phi_t \cdot \frac{\partial^2 Z_2}{\partial r_t^2}. \tag{14.6}$$

From Eq. (14.5), we see that we obtain $\partial \Pi / \partial r_t = 0$, i.e., we make the portfolio locally immune to changes in the risk factor, if we choose

$$\phi_t = \frac{\partial Z_1 / \partial r_t}{\partial Z_2 / \partial r_t}. \tag{14.7}$$

This choice of ϕ_t eliminates both the term multiplying dW_t and the term involving the drift function $m(r_t, t)$ in Eq. (14.4), which boils down to

$$d\Pi = \left\{ \frac{\partial \Pi}{\partial t} + \frac{1}{2} s^2(r_t, t) \cdot \frac{\partial^2 \Pi}{\partial r_t^2} \right\} dt. \tag{14.8}$$

14.1 A PDE approach to incomplete markets

Since this portfolio is risk-free, by no-arbitrage, we can also write

$$d\Pi = r_t \Pi\, dt. \tag{14.9}$$

By linking Eqs. (14.8) and (14.9), taking advantage of Eqs. (14.5) and (14.6), and rearranging, we find

$$\left\{ \frac{\partial Z_1}{\partial t} + \frac{1}{2} s^2(r_t, t) \cdot \frac{\partial^2 Z_1}{\partial r_t^2} - r_t Z_1 \right\} = \phi_t \left\{ \frac{\partial Z_2}{\partial t} + \frac{1}{2} s^2(r_t, t) \cdot \frac{\partial^2 Z_2}{\partial r_t^2} - r_t Z_2 \right\}. \tag{14.10}$$

Substituting ϕ_t and rearranging again yield

$$\frac{\frac{\partial Z_1}{\partial t} + \frac{1}{2} s^2(r_t, t) \frac{\partial^2 Z_1}{\partial r_t^2} - r_t Z_1}{\frac{\partial Z_1}{\partial r_t}} = \frac{\frac{\partial Z_2}{\partial t} + \frac{1}{2} s^2(r_t, t) \frac{\partial^2 Z_2}{\partial r_t^2} - r_t Z_2}{\frac{\partial Z_2}{\partial r_t}}. \tag{14.11}$$

We note that the drift term $m(r_t, t)$ does not play any role, which is consistent with the BSM case. However, here we are relating *two* derivative prices. Equation (14.11) is not a pricing equation that we may immediately use. It only provides us with partial information, which is essentially a consistency condition for pricing different derivatives depending on the same risk factor r_t. To see this, let us observe that the two ratios in Eq. (14.11) are identical and not specific of any derivative written on r_t. They should be associated with the risk factor r_t itself.[3] Therefore, for a generic derivative with value $Z(r_t, t)$, we may define the following function of the short rate and time:

$$m^*(r_t, t) \doteq -\frac{\frac{\partial Z}{\partial t} + \frac{1}{2} s^2(r_t, t) \frac{\partial^2 Z}{\partial r_t^2} - r_t Z}{\frac{\partial Z}{\partial r_t}}, \tag{14.12}$$

where we change the sign just for the sake of convenience, as this allows to write the **pricing equation** in a familiar form:

$$\frac{\partial Z}{\partial t} + m^*(r_t, t) \frac{\partial Z}{\partial r_t} + \frac{1}{2} s^2(r_t, t) \frac{\partial^2 Z}{\partial r_t^2} = r_t Z. \tag{14.13}$$

We notice that this is quite close to the BSM partial differential equation, but it involves an unknown function $m^*(r_t, t)$. By using the Feynman–Kač representation theorem, we can express the derivative price as the expected value of the payoff, where the risk factor follows the risk-adjusted dynamics

$$dr_t = m^*(r_t, t)\, dt + s(r_t, t)\, dW_t. \tag{14.14}$$

Once again, we notice that there is a change in drift, which reflects a change in probability measure. However, due to market incompleteness, we do not

---
[3] We have used the same kind of reasoning when dealing with APT in Example 10.2.

know a priori which measure we should use, as uniqueness of the risk-neutral measure requires completeness. The difficulty is that the drift is related to a **market price of risk**, as we shall see later. Unlike the BSM case, risk does play a role here, since we cannot hedge using the underlying factor directly. On the contrary, in the complete market case of Chapter 13, we find a unique risk-neutral measure with a well-defined drift. The way out of the dilemma requires calibrating the model against the quoted prices of liquid and exchange-traded derivatives. In the case of interest rates, we might use the quoted prices of bonds or other liquid assets to formulate a **model calibration** problem and solve the corresponding optimization model to estimate the function $m^*(r_t, t)$. Having pinned down the right risk-adjusted measure, we may price other derivatives by using analytical formulas, when available, or by using numerical methods to solve the PDE, like finite differences, or by estimating the price in expectation form by Monte Carlo methods.

Model calibration is an example of an inverse problem and may be tackled by different approaches, as we discuss in Section 14.4. A basic choice is between parametric and non-parametric approaches. The function $m^*(r_t, t)$ in its full generality is an infinite-dimensional object, and we may fit an arbitrary function, in principle. Unfortunately, this may require plenty of data and result in some overfitting. In terms of robustness, a possibly better approach is to parameterize the function by choosing a functional form depending on a limited set of parameters. A natural choice, from this viewpoint, is to use a model that makes financial sense, like the short-rate models that we describe in Section 14.2. Before doing so, it is useful to understand how the pricing PDE may be tackled in an artificially simple case.

14.1.1 PRICING A ZERO-COUPON BOND IN A DRIFTLESS WORLD

To illustrate the overall approach based on the pricing PDE (14.13), let us price a simple zero-coupon bond, assuming a driftless short-rate process

$$dr_t = \sigma \, dW_t,$$

whose simplicity will lead to an analytical solution. Note that we consider this as the risk-adjusted model, and for now we neglect model calibration issues. This model, per se, is clearly *not* a realistic one, as it yields a normally distributed rate, which may easily get negative.[4]

We consider a zero-coupon bond with unit face value, whose price will be denoted as $Z(r_t, t; T)$ to make the dependence on the maturity T explicit. As usual in derivative pricing, we need a terminal condition, which in this case

---

[4]We may consider this model as a subcase of the Ho–Lee model, which assumes the differential equation $dr_t = \theta_t \, dt + \sigma \, dW_t$, for some given function θ_t that may be calibrated against market data.

14.1 A PDE approach to incomplete markets

refers to the face value of the bond, which is repaid at maturity:

$$Z(r_T, T; T) = 1, \quad \forall r_T. \tag{14.15}$$

Since we assume zero drift, the pricing equation (14.13) becomes

$$\frac{\partial Z}{\partial t} + \frac{1}{2}\sigma^2 \frac{\partial^2 Z}{\partial r_t^2} = r_t Z.$$

In order to get a clue about the solution of a differential equation, reasoning by analogy may be useful. We know that, in a deterministic setting with a constant rate r, we should find the same formula of elementary financial mathematics in terms of a discount factor:

$$Z(r, t; T) = e^{-r \cdot (T-t)}.$$

Therefore, we may guess a solution of a similar form, in terms of unknown quantities, and see where this *ansatz* may lead us. In this case, a sensible guess, based on analogy with the deterministic price, is

$$Z(r_t, t; T) = e^{A(t;T) - B(t;T) \cdot r_t},$$

for a pair of unknown functions $A(t; T)$ and $B(t; T)$. This functional form makes financial sense, and it separates the dependence on time from the dependence on the state variable r_t. In fact, this is an example of a general PDE solution strategy based on **separation of variables**. Furthermore, exponential functions are particularly well-behaved in terms of their derivatives, so there should be hope![5]

The idea is to plug the selected functional form into the pricing PDE and hope for the best. Given the terminal condition (14.15) on the bond price, we require

$$B(T; T) = 0, \qquad A(T; T) = 0.$$

Then, we calculate the partial derivatives that occur in the pricing PDE:

$$\frac{\partial Z}{\partial t} = \left[A'(t; T) - B'(t; T) r_t\right] \cdot Z,$$
$$\frac{\partial Z}{\partial r_t} = -B(t; T) \cdot Z,$$
$$\frac{\partial^2 Z}{\partial r_t^2} = B^2(t; T) \cdot Z.$$

By plugging these partial derivatives into the pricing PDE and simplifying a bit, we find a condition that does not involve the bond price:

$$\left[A'(t; T) - B'(t; T) \cdot r_r\right] + \frac{1}{2}\sigma^2 \cdot B^2(t; T) = r_t.$$

---

[5]In fact, this form of solution applies to a whole family of short-rate models yielding a so-called *affine term structure*.

This equation may be rearranged in order to factor the state variable r_t:

$$r_t \cdot [1 + B'(t;T)] = A'(t;T) + \frac{1}{2}\sigma^2 \cdot B^2(t;T). \tag{14.16}$$

A fundamental observation is that this condition must hold for whatever value of the state variable r_t. For this to be the case, both factors in Eq. (14.16) must be zero, which yields two ordinary differential equations:

$$B'(t;T) = -1, \tag{14.17}$$

$$A'(t;T) = -\frac{1}{2}\sigma^2 \cdot B^2(t;T). \tag{14.18}$$

This is the essence of separation of variables: We start from a PDE and boil it down to ordinary differential equations (ODEs). In this very easy case, the two ODEs are simple and may be solved in sequence. In fact, Eq. (14.17) is easily solved:

$$B(t;T) = -t + K_1,$$

for some integration constant K_1, which is found from the terminal condition:

$$B(T;T) = -T + K_1 = 0 \quad \Rightarrow \quad B(t;T) = T - t.$$

Then, we plug this solution into Eq. (14.18) and find the following equation for $A(t;T)$:

$$A'(t;T) = -\frac{1}{2}\sigma^2(T-t)^2,$$

which yields

$$A(t;T) = \frac{1}{6}\sigma^2(T-t)^3 + K_2,$$

for another integration constant K_2, which is found again by enforcing the terminal condition. Here we have $K_2 = 0$, so that

$$A(t;T) = \frac{1}{6}\sigma^2(T-t)^3.$$

By putting everything together, we end up with the bond price

$$Z(r_t, t; T) = \exp\left\{\frac{1}{6}\sigma^2(T-t)^3 - (T-t)r_t\right\}. \tag{14.19}$$

This formula looks a bit weird, when compared with the classical bond price formula, so it is worthwhile to examine it in some detail and check whether some basic intuition is preserved. To this aim, it is useful to derive the differential equation of the bond price using Itô's lemma. From Eq. (14.19), we find the required partial derivatives, as usual:

$$\frac{\partial Z}{\partial t} = \left\{-\frac{1}{2}\sigma^2(T-t)^2 + r_t\right\} \cdot Z, \tag{14.20}$$

$$\frac{\partial Z}{\partial r_t} = -(T-t) \cdot Z, \tag{14.21}$$

$$\frac{\partial^2 Z}{\partial r_t^2} = (T-t)^2 \cdot Z. \tag{14.22}$$

14.1 A PDE approach to incomplete markets

As a first reality check, we observe that Eqs. (14.21) and (14.22) are consistent with the familiar bond duration and convexity for a zero. A less intuitive fact is that the bond's theta, given in Eq. (14.20), could be negative for a large volatility σ of the interest rate. We would expect a positive theta, as a zero price should converge to the face value from below, unless interest rates are negative. Indeed, in the driftless model that we are considering, the probability of a negative rate is related to volatility.

If we plug these derivatives into Itô's lemma, we find

$$dZ = Z \cdot \left\{ \left[-\frac{1}{2}\sigma^2(T-t)^2 + r_t \right] + \frac{1}{2}\sigma^2(T-t)^2 \right\} dt - Z \cdot \sigma \cdot (T-t) \, dW_t$$
$$= Z \cdot r_t \, dt - Z \cdot \sigma \cdot (T-t) \, dW_t.$$

If we consider a small time slice δt, we may rewrite the differential equation in approximate form as

$$\frac{\delta Z}{Z} \approx r_t \, \delta t - \sigma(T-t)\sqrt{\delta t}\, \epsilon_{t+\delta t},$$

where the increment of the Wiener process is written, as usual, in the form $W_{t+\delta t} - W_t = \sqrt{\delta t}\, \epsilon_{t+\delta t}$, where $\epsilon_{t+\delta t}$ is standard normal. This allows us to interpret the ratio $\delta Z / Z$ on the right-hand side as an approximation of return on a small time step δt and to check qualitative properties:

- The expected return is

$$\mathrm{E}\left[\frac{\delta Z_t}{Z_t}\right] \approx r_t \, \delta t,$$

which states that the bond return is essentially the risk-free rate and does not depend on time-to-maturity $T - t$.

- The volatility of return is

$$\sqrt{\mathrm{Var}\left(\frac{\delta Z_t}{Z_t}\right)} \approx (T-t)\sigma\sqrt{\delta t}$$

and depends on the underlying volatility σ, as well as on time-to-maturity. As we should expect, the volatility function for the bond price goes to zero as we approach maturity.

As we have pointed out, the short rate is actually an unobservable state variable. The quality of an interest rate model can be measured in terms of fit against market prices, but also in terms of realism of the generated term structure. Hence, let us check which kind of term structure we may generate with the simple model we are working with. To this aim, we must consider the continuously compounded interest rate $r(t, t+\tau)$, with time-to-maturity $\tau = T - t$. This may be inferred by inverting the classical bond price formula,

$$Z(r_t, t; t+\tau) = e^{-\tau \cdot r(t, t+\tau)},$$

which, given Eq. (14.19), implies

$$r(t, t+\tau) = -\frac{\log Z(r_t, t; t+\tau)}{\tau} = -\frac{\sigma^2 \tau^2}{6} + r_t.$$

For a small time-to-maturity $\tau \to 0$, we find the short rate, but the term structure is unbounded and decreasing with τ. For any given τ, we see that the increments in the rate are directly related to the increments in the short rate,

$$dr(t, t+\tau) = dr_t,$$

which implies that interest rates for all maturities are perfectly correlated, resulting in parallel shifts. This is no big surprise, since we have used a single risk factor, but the term structure does not look at all realistic. At the very least, a sensible drift term must be introduced into the short-rate model; anyway, there is only so much we can obtain by a single-factor model, and we might resort to a multifactor approach.

14.2 Pricing by short-rate models

While a single-factor model based on the short rate looks not quite realistic, it may provide us with valuable explicit pricing formulas. Analytical formulas help in building intuition, and they also make the task of model calibration much easier. A few models for the short rate have been proposed in the literature, and here we consider two popular ones:

1. The **Vasicek model**, characterized by a stochastic differential equation featuring mean reversion:

$$dr_t = \gamma(\bar{r} - r_t)\,dt + \sigma\,dW_t. \qquad (14.23)$$

    The Vasicek model is based on the Ornstein–Uhlenbeck process.

2. The **Cox–Ingersoll–Ross (CIR) model**, which is quite similar to the Vasicek model, but differs by a slight change in the volatility term:

$$dr_t = \gamma(\bar{r} - r_t)\,dt + \sqrt{\alpha r_t}\,dW_t. \qquad (14.24)$$

    The CIR model is an example of a square-root diffusion.

Both models feature mean reversion toward the long term average $\bar{r}$, with speed γ. The Vasicek model is more tractable, but it leads to normally distributed short rates, which may be negative. The inclusion of a square root term in Eq. (14.24), for a suitable choice of the parameters, prevents negative interest rates. The square-root diffusion is related to a more complicated distribution, the noncentral chi-square.

14.2.1 THE VASICEK SHORT-RATE MODEL

Given the process described by Eq. (14.23), the first step is to figure out how short rates are distributed according to this model. To try and solve this differential equation, we may resort to a typical trick of the trade, related to the method of **integrating factors** for deterministic ordinary differential equations. The trick is to apply Itô's lemma to the process

$$f(r_t, t) = r_t e^{\gamma t}.$$

The usual drill requires the calculation of partial derivatives:

$$\frac{\partial f}{\partial t} = \gamma r_t e^{\gamma t}, \qquad \frac{\partial f}{\partial r_t} = e^{\gamma t}, \qquad \frac{\partial^2 f}{\partial r_t^2} = 0.$$

Hence, the differential equation for $f(r_t, t)$ is

$$df = \left[\gamma r_t e^{\gamma t} + \gamma(\bar{r} - r_t)e^{\gamma t}\right] dt + \sigma e^{\gamma t} \, dW_t = \gamma \bar{r} e^{\gamma t} \, dt + \sigma e^{\gamma t} \, dW_t.$$

This equation does not look too bad and can be integrated over the interval $(0, t)$:

$$\int_0^t d(r_\tau e^{\gamma \tau}) = r_t e^{\gamma t} - r_0 = \gamma \bar{r} \int_0^t e^{\gamma \tau} \, d\tau + \sigma \int_0^t e^{\gamma \tau} \, dW_\tau. \qquad (14.25)$$

Since

$$\gamma \int_0^t e^{\gamma \tau} \, d\tau = e^{\gamma t} - 1,$$

Eq. (14.25) can be rewritten as

$$r_t = r_0 e^{-\gamma t} + \bar{r}(1 - e^{-\gamma t}) + \sigma \int_0^t e^{\gamma(\tau - t)} \, dW_\tau.$$

This expression includes a stochastic integral with respect to the Wiener process, which is normally distributed and has zero expected value. Thus, we may immediately conclude that the short rate r_t, under the Vasicek model, is normally distributed with expected value[6]:

$$\mathrm{E}[r_t] = r_0 e^{-\gamma t} + \bar{r}(1 - e^{-\gamma t}). \qquad (14.26)$$

Finding variance is a bit trickier and requires the application of Theorem 11.2, known as Itô's isometry:

$$\mathrm{E}\left[\left(\int_0^t X_\tau \, dW_\tau\right)^2\right] = \mathrm{E}\left[\int_0^t X_\tau^2 \, d\tau\right],$$

---

[6] By the way, we can find this expectation by setting $\sigma = 0$ and solving the resulting deterministic equation.

where X_t is a stochastic process adapted to the standard Wiener process W_t. The result allows us to find the variance of the short rate:

$$\text{Var}(r_t) \equiv \text{E}\left[(r_t - \text{E}[r_t])^2\right] = \text{E}\left[\left(\sigma \int_0^t e^{\gamma(\tau-t)}\, dW_\tau\right)^2\right]$$

$$= \text{E}\left[\sigma^2 \int_0^t e^{2\gamma(\tau-t)}\, d\tau\right] = \frac{\sigma^2}{2\gamma}\left(1 - e^{-2\gamma t}\right). \qquad (14.27)$$

Again, it is important to get an intuitive feeling for Eqs. (14.26) and (14.27):

- At time $t = 0$, the short rate is r_0 and its variance is zero, as it should be, since it is a given initial state.
- When $t \to +\infty$, the short rate tends to the long-term average value $\bar{r}$; the larger the speed of mean reversion γ, the faster the convergence.
- Long-term variance is given by $\sigma^2/(2\gamma)$. As expected, it is influenced by the volatility σ. However, γ plays a role, too, and a strong mean reversion tends to kill volatility.

14.2.1.1 Pricing a zero under the Vasicek model

Since a coupon-bearing bond can be priced as a portfolio of zero-coupon bonds, the first interest rate derivative that we should price is a zero-coupon bond. We insist once again that the parameters of the Vasicek model are *not* the parameters of the real world, which could be estimated by analyzing a time series of short rates, but those in the risk-neutral world. Let assume that we have calibrated a Vasicek model. This means that we have parameters γ^* and $\bar{r}^*$ in the pricing measure, replacing the original γ and $\bar{r}$, and σ, which is not affected by the change of measure.[7] Let $Z(r_t, t; T)$ be the price of a zero maturing at T. The bond pricing equation is

$$\frac{\partial Z}{\partial t} + \gamma^*(\bar{r}^* - r_t) \cdot \frac{\partial Z}{\partial r_t} + \frac{1}{2}\sigma^2 \cdot \frac{\partial^2 Z}{\partial r_t^2} = r_t Z,$$

with terminal condition $Z(r_t, T; T) = 1$.

A common strategy to solve pricing equations is the following, which we have introduced in Section 14.1.1:

1. We guess the *structure* of the solution, using insights and experience. In our case, we may again use analogy with naive bond pricing and guess a solution of the form

$$Z(r_t, t; T) = e^{A(t;T) - B(t;T) \cdot r_t},$$

for unknown functions $A(t; T)$ and $B(t; T)$.

---

[7]This claim should be backed by a formal statement, based on the Girsanov theorem, but this is beyond the scope of this book. We rely on the intuition from the BSM case, where we end up replacing the drift coefficient in the GBM, but not the volatility.

14.2 Pricing by short-rate models

2. Note that our guess implies a form of separation of variables. Then, we may plug our guess and derive ordinary differential equations (ODEs) for the unknown functions.

To begin with, we find the partial derivatives of the bond price:

$$\frac{\partial Z}{\partial t} = \left[A'(t;T) - B'(t;T) \cdot r_t\right] e^{A(t;T) - B(t;T) \cdot r_t}$$
$$= \left[A'(t;T) - B'(t;T) \cdot r_t\right] \cdot Z,$$

$$\frac{\partial Z}{\partial r_t} = -B(t;T) \cdot e^{A(t;T) - B(t;T) \cdot r_t} = -B(t;T) \cdot Z,$$
$$\frac{\partial^2 Z}{\partial r_t^2} = B^2(t;T) \cdot e^{A(t;T) - B(t;T) \cdot r_t} = B^2(t;T) \cdot Z.$$

Plugging these partial derivatives into the pricing equation yields the following equation:

$$\left[A'(t;T) - B'(t;T) \cdot r_t\right] \cdot Z - \gamma^*(\bar{r}^* - r_t) \cdot B(t;T) \cdot Z + \frac{1}{2}\sigma^2 B^2(t;T) \cdot Z = r_t Z,$$

which may be simplified, by eliminating Z, and rearranged by collecting terms in r_t:

$$\left[1 + B'(t;T) - \gamma^* B(t;T)\right] \cdot r_t = A'(t;T) - \gamma^* \bar{r}^* B(t;T) + \frac{1}{2}\sigma^2 B^2(t;T). \tag{14.28}$$

Since Eq. (14.28) must hold for any value of r_t, we may rewrite it as two ODEs:

$$B'(t;T) = \gamma^* B(t;T) - 1, \tag{14.29}$$

$$A'(t;T) = \gamma^* \bar{r}^* B(t;T) - \frac{1}{2}\sigma^2 B(t;T)^2, \tag{14.30}$$

with terminal conditions $A(T;T) = 0$ and $B(T;T) = 0$.

Equation (14.29) is a first-order, nonhomogeneous ODE, which may be solved as we show in the technical note below. This yields

$$B(t;T) = \frac{1 - e^{-\gamma^*(T-t)}}{\gamma^*}.$$

Then, we plug this solution into Eq. (14.30) and integrate, which yields

$$A(t;T) = \left[B(t;T) - (T-t)\right]\left(\bar{r}^* - \frac{\sigma^2}{2\gamma^{*2}}\right) - \frac{\sigma^2 B(t,T)^2}{4\gamma^*}.$$

Thus, to summarize what we have achieved, the price at time t of a zero-coupon bond maturing in T, with face value $1, under the Vasicek short-rate model, can be expressed as follows:

$$Z(r_t, t; T) = e^{A(t;T) - B(t;T) r_t},$$

where

$$B(t;T) = \frac{1}{\gamma}\left[1 - e^{-\gamma(T-t)}\right], \tag{14.31}$$

$$A(t;T) = \left[B(t;T) - (T-t)\right]\left(\bar{r} - \frac{\sigma^2}{2\gamma^2}\right) - \frac{\sigma^2 B(t;T)^2}{4\gamma}. \tag{14.32}$$

Given the linearity of pricing, we may use this formula to price a coupon-bearing bond as well. Thus, we might calibrate the Vasicek model against observed bond prices.

Technical note: Solving first-order nonhomogeneous ODEs

For the sake of completeness, here we give a clue about the approach to solve an equation like Eq. (14.29). Consider the ODE $y' + ay = f(x)$, for an unknown function $y(x)$, with initial condition $y(0) = y_0$. Due to the term $f(x)$, this is a nonhomogeneous equation, which may be solved by summing the general solution of the homogeneous part and one specific solution of the nonhomogeneous equation. The homogeneous part $y' + ay = 0$ is easily solved, and its general solution is

$$y = Ce^{-ax},$$

where the constant C depends on the initial condition. To find a specific solution, we may use the method of undetermined coefficients, whose details depend on the form of $f(x)$. For instance, when $f(x)$ is a polynomial, we may try a polynomial of the same order, with unknown coefficients that are found by plugging the polynomial into the differential equation and solving the resulting algebraic equations. This yields a specific solution, which is added to the general one.

In our specific case, we have $B' = \gamma^* B - 1$, along with a *terminal* condition. The solution to the homogeneous equation is

$$B(t;T) = Ce^{\gamma^* t}.$$

Since the additional term is a constant, we may try a specific solution like $B(t;T) = k$, for an unknown constant k. Plugging this into the equation, we easily find

$$0 = \gamma^* k - 1 \quad\Rightarrow\quad k = \frac{1}{\gamma^*}.$$

The general solution is

$$B(t;T) = Ce^{\gamma^* t} + \frac{1}{\gamma^*},$$

and the terminal condition $B(T;T) = 0$ yields

$$Ce^{\gamma^* T} + \frac{1}{\gamma^*} = 0 \quad\Rightarrow\quad C = -\frac{e^{-\gamma^* T}}{\gamma^*}.$$

Thus, the desired solution is

$$B(t; T) = -\frac{e^{-\gamma^* T}}{\gamma^*} e^{\gamma^* t} + \frac{1}{\gamma^*} = \frac{1 - e^{-\gamma^*(T-t)}}{\gamma^*}.$$

14.2.1.2 Pricing a bond option under the Vasicek model

The Vasicek model can be criticized, as it relies on a single factor and features normal, possibly negative interest rates. However, it is simple enough to yield analytical pricing formulas for some elementary derivatives. The next logical step is to price a relatively simple option, namely, a call option on a zero-coupon bond. We have to realize that there are two maturities involved:

- The option maturity, T_O, i.e., the time at which the option can be exercised, by purchasing a zero-coupon bond at the strike price K.
- The bond maturity, T_B. Clearly, for the option to make sense, we must have $T_O < T_B$.

Therefore, the option payoff at T_O is

$$\max\{Z(r_{T_O}, T_O; T_B) - K, 0\}. \tag{14.33}$$

It can be shown that the price at time $t = 0$ of a European-style call option maturing at time T_O on a zero-coupon bond maturing in T_B is, under the Vasicek model,

$$Z(0, r_0; T_B) \cdot \Phi(d_1) - KZ(0, r_0; T_O) \cdot \Phi(d_2), \tag{14.34}$$

where r_0 is the value of the short rate at $t = 0$, $\Phi(\cdot)$ is the familiar CDF of the standard normal distribution, and

$$d_1 = \frac{1}{\mathcal{S}(T_O)} \log\left[\frac{Z(0, r_0; T_B)}{KZ(0, r_0; T_O)}\right] + \frac{\mathcal{S}(T_O)}{2}, \tag{14.35}$$

$$d_2 = d_1 - \mathcal{S}(T_O), \tag{14.36}$$

$$\mathcal{S}(T_O) = B(T_O; T_B) \sqrt{\frac{\sigma^2}{2\gamma}\left(1 - e^{-2\gamma T_O}\right)}. \tag{14.37}$$

To interpret this formula, it is quite useful to note its deep similarity with the Black–Scholes–Merton price of a vanilla call option on a stock, given in Eq. (13.19), and note the following:

- In Eq. (14.34), $Z(0, r_0; T_B)$ should be interpreted as the price of the underlying asset, i.e., the bond maturing in T_B, whereas $Z(0, r_0; T_O)$ should be interpreted as a discount factor from the option maturity T_O to time $t = 0$.
- The terms d_1 and d_2 in Eqs. (14.35) and (14.36) look much like the similar terms in the BSM formula.
- The term $\mathcal{S}(T_O)$, where $B(T_O; T_B)$ is just the function given in Eq. (14.31), plays the role of a volatility.

It is also useful to recall that short rates under the Vasicek model, which relies on an Ornstein–Uhlenbeck process, are normally distributed, and that the price of a zero-coupon bond is an exponential of these rates. Hence, the bond price is lognormally distributed, just like stock prices under the BSM model, and this is the essential reason behind the observed similarity with the BSM formula.

14.2.2 THE COX–INGERSOLL–ROSS SHORT-RATE MODEL

The CIR model relies on a square-root diffusion process, which is not as easy to deal with as the Ornstein–Uhlenbeck process underlying the Vasicek model. Because of this additional difficulty, in this section, unlike the case of the Vasicek model, we will state results without any proof. Apart from technicalities, the essential concepts are not really different. By studying the transition density of the square-root diffusion process, a link with the noncentral chi-square distribution can be established. We recall that the chi-square distribution with n degrees of freedom arises as a sum of n independent squared standard normals,

$$\sum_{i=1}^{n} Z_i^2 \sim \chi_n^2.$$

If we sum nonstandard squared normals, we get the noncentral chi-squared distribution. More precisely, let us consider a random variable

$$X = \sum_{i=1}^{n} (Z_i + a_i)^2.$$

The corresponding variable is labeled noncentral chi-square with n degrees of freedom and noncentrality parameter $\lambda = \sum_{i=1}^{n} a_i^2$. It is possible to generalize this distribution to a noninteger ν, resulting in a $\chi_\nu'^2(\lambda)$ variable. The following formulas for the relevant moments can be proved:

$$\mathrm{E}[\chi_\nu'^2(\lambda)] = \nu + \lambda, \qquad \mathrm{Var}(\chi_\nu'^2(\lambda)) = 2(\nu + 2\lambda). \qquad (14.38)$$

THEOREM 14.1 (CIR short-rate model) *The transition law from r_0 to r_t for the CIR model can be expressed as*

$$r_t = \frac{\alpha(1 - e^{-\gamma t})}{4\gamma} \cdot \chi_\nu'^2(\lambda), \qquad (14.39)$$

where the degrees of freedom are $\nu = 4\bar{r}\gamma/\alpha$ and the noncentrality parameter is

$$\lambda = \frac{4\gamma e^{-\gamma t}}{\alpha(1 - e^{-\gamma t})} \cdot r_0.$$

By applying this theorem to the transition from r_0 to r_t and using Eq. (14.38), we find the conditional expectation and variance of r_t, conditional on r_0. For the expected value, we have

$$\begin{aligned} \mathrm{E}[r_t|r_0] &= \frac{\alpha(1-e^{-\gamma t})}{4\gamma}(\nu+\lambda) \\ &= \frac{\alpha(1-e^{-\gamma t})}{4\gamma}\left[\frac{4\bar{r}\gamma}{\alpha}+\frac{4\gamma e^{-\gamma t}}{\alpha(1-e^{-\gamma t})}r_0\right] \\ &= r_0 e^{-\gamma t}+\bar{r}(1-e^{-\gamma t}). \end{aligned} \qquad (14.40)$$

We notice that the expected value has the same intuitive form as in the Vasicek model. Variance is a bit trickier:

$$\begin{aligned} \mathrm{Var}(r_t|r_0) &= \left[\frac{\alpha(1-e^{-\gamma t})}{4\gamma}\right]^2 \cdot 2(\nu+2\lambda) \\ &= \left[\frac{\alpha(1-e^{-\gamma t})}{4\gamma}\right]^2 \cdot 2\left[\frac{4\bar{r}\gamma}{\alpha}+2\frac{4\gamma e^{-\gamma t}}{\alpha(1-e^{-\gamma t})}r_0\right] \\ &= \frac{r_0\alpha}{\gamma}\left(e^{-\gamma t}-e^{-2\gamma t}\right)+\frac{\bar{r}\alpha}{2\gamma}\left(1-e^{-\gamma t}\right)^2. \end{aligned} \qquad (14.41)$$

14.2.2.1 Pricing a zero under the CIR model

Since the short rate is no longer normal under the square-root diffusion process, everything turns out to be more complicated. Nevertheless, a zero-coupon bond can be priced analytically under the CIR model, even though the formulas are more complicated than those for the Vasicek model:

$$Z(r_t, t; T) = e^{A(t;T) - B(t;T)r_t},$$

where

$$B(t;T) = \frac{2(e^{\psi(T-t)}-1)}{(\gamma+\psi)(e^{\psi(T-t)}-1)+2\psi},$$

$$A(t;T) = \frac{2\bar{r}\gamma}{\alpha}\log\left[\frac{2\psi e^{(\psi+\gamma)\frac{T-t}{2}}}{(\gamma+\psi)(e^{\psi(T-t)}-1)+2\psi}\right],$$

$$\psi = \sqrt{\gamma^2+2\alpha}.$$

The above formulas show how the CIR model belongs, just like the Vasicek model, to the general family of short-rate models yielding an affine term structure.

14.3 A martingale approach to incomplete markets

In Chapter 13, we have seen that a vanilla option on equity may be priced by solving a PDE or by resorting to a risk-neutral expectation. Conceptually, by the Feynman–Kač representation theorem, the two approaches are equivalent. When resorting to numerical methods, both approaches have their merits, advantages, and disadvantages, as we have hinted at in Section 13.9. So far, in this chapter, we have taken a PDE view. Now, let us consider how we can extend the risk-neutral pricing approach to incomplete markets. We have already seen that, in PDE terms, we have to replace the drift function with a new one, somehow reflecting risk. We know that this is related to a change of probability measure, which in the complete market case, under the BSM model, means replacing the drift coefficient by the risk-free rate. The BSM equation was obtained under the assumption of a *constant* risk-free rate, which we have to relax when dealing with interest rate derivatives. Using risk-neutral pricing within the complete market setting, the price at time t of a path-independent, European-style option maturing in T can be written as

$$V(S_t, t) = e^{-r(T-t)} \mathrm{E}_{\mathbb{Q}_n}[f(S_T) \,|\, S_t], \qquad (14.42)$$

where $f(S_T)$ is the payoff at maturity, depending on the stock price S_T, $\mathbb{Q}_n$ is the pricing measure, and the expectation is conditional on the price S_t of the underlying asset at time t.

In the incomplete market setting, possibly with stochastic interest rates, additional complications arise. As we already know, the risk-neutral measure $\mathbb{Q}_n$ should be replaced by some risk-adjusted measure $\mathbb{Q}$, which is not unique in an incomplete market and must be calibrated against market data, in order to reflect risk factors that cannot be fully hedged away. Furthermore, since we are dealing with stochastic interest rates, unlike Eq. (14.42), we cannot take the discount factor outside the expectation. If we use a single-factor short-rate model, the risk-adjusted pricing approach leads here to a formula like

$$V(r_t, t) = \mathrm{E}_{\mathbb{Q}}\left[\exp\left(-\int_t^T r_\tau \, d\tau \right) \cdot f(r_T) \,\bigg|\, r_t \right]. \qquad (14.43)$$

If we consider a zero-coupon bond, then $f(r_T) = 1$, and we find that the zero price is

$$Z(t; T) = \mathrm{E}_{\mathbb{Q}}\left[\exp\left(-\int_t^T r_\tau \, d\tau \right) \,\bigg|\, r_t \right]. \qquad (14.44)$$

This formula is a natural generalization of the simple exponential discount factor, $e^{-r \cdot (T-t)}$, which we would have with a constant risk-free rate r. Note how this depends on the whole sample path of the short rate up to maturity, even if the payoff is path independent. In the more general case of a derivative, the price in Eq. (14.43) is the expectation of a *product* of random variables, which is not quite convenient. It would be nice to factor the expectation of the product into the product of expectations, but the two random variables are certainly *not* independent when dealing with an interest rate derivative.

Pricing may be considerably simplified by taking advantage of the following observation. Let us consider again the complete market case, with a constant risk-free rate r. Under the risk-neutral measure $\mathbb{Q}_n$, if we assume a GBM

14.3 A martingale approach to incomplete markets

model, the asset price dynamics is given by the stochastic differential equation

$$dS_t = rS_t\,dt + \sigma S_t\,dW_t$$

Its solution is

$$S_t = S_0 \exp\left\{\left(r - \frac{\sigma^2}{2}\right)\cdot t + \sigma\sqrt{t}\cdot\epsilon\right\}, \tag{14.45}$$

where $\epsilon \sim \mathsf{N}(0,1)$. The bank account equation leads to

$$B_t = B_0 \exp(rt). \tag{14.46}$$

Now, let us consider a process given by the ratio of S_t and B_t and use Eqs. (14.45) and (14.46):

$$M_t \doteq \frac{S_t}{B_t} = \frac{S_0}{B_0}\exp\left\{-\frac{\sigma^2}{2}\cdot t + \sigma\sqrt{t}\cdot\epsilon\right\}. \tag{14.47}$$

Note that process M_t gives the price of asset S_t using a different **numeraire**. We are used to express prices in monetary units, but we may express prices using any unit we wish. We might divide asset prices by the unit price of a banana, and express stock prices in bananas. Here we do so using the bank account as a numeraire and, in order to avoid a certain degree of arbitrariness, it is natural to choose the initial condition $B_0 = 1$. To find the stochastic differential equation of M_t, we might use a generalized version of Itô' lemma, to cope with a ratio of stochastic processes. However, by comparing Eqs. (14.45) and (14.47), it is easy to see that the process M_t has no drift. Hence, it is the solution of the differential equation

$$dM_t = \sigma M_t\,dW_t.$$

This process is driftless and is a **martingale**.[8] Indeed, we may also use properties of the lognormal distribution to see that Eq. (14.47) implies $\mathsf{E}[M_t\mid M_0] = M_0$. More generally, it turns out that, under the standard risk-neutral measure, the ratio of any price and the bank account is a martingale. In fact, we may rewrite Eq. (14.42) at time $t=0$ as

$$\frac{V(S_0,0)}{B_0} = \mathsf{E}_{\mathbb{Q}_n}\left[\frac{V(S_T,T)}{B_T}\right],$$

where $B_0 = 1$ and $B_T = e^{rT}$.

In an incomplete setting, we may consider different measures and different numeraires. In particular, it would be very helpful to find a numeraire process g_t, and a corresponding probability measure, such that the ratio f_t/g_t is a martingale. Indeed, if are able to do so, the martingale property implies

$$\frac{f_0}{g_0} = \mathsf{E}_g\left[\frac{f_T}{g_T}\right],$$

---

[8] See Section 11.1.5.

where the notation $E_g[\cdot]$ emphasizes the change of measure in the expectation. This immediately provides us with a pricing formula,

$$f_0 = g_0 \, E_g\left[\frac{f_T}{g_T}\right].$$

In concrete terms, we must choose a numeraire with price process g_t, in such a way that computing this expectation is reasonably easy. However, doing this in a rigorous way requires a mathematical machinery to change measure, i.e., tools like Radon–Nikodym derivatives and Girsanov theorem. In order to avoid too many technicalities, we follow [4] and pursue a heuristic, but financially motivated approach in the following section.[9]

14.3.1 AN INFORMAL APPROACH TO MARTINGALE EQUIVALENT MEASURES

We consider two derivatives depending on r_t (say, the short rate, but this is not necessary), with price processes given by

$$\frac{df_t}{f_t} = m_f(r_t, t) \, dt + s_f(r_t, t) \, dW_t, \qquad (14.48)$$

$$\frac{dg_t}{g_t} = m_g(r_t, t) \, dt + s_g(r_t, t) \, dW_t, \qquad (14.49)$$

where the standard Wiener process W_t drives both equations, and it is also the driving risk factor of r_t. Hence, we have a common risk factor, shared by all of the processes involved. There is a slight change of notation with respect to Eq. (14.2), since we use drift functions m_f and m_g, as well as volatility functions s_f and s_g, without referring to explicit partial derivatives. We do so in order to focus on the dynamics of the prices of the two derivative assets, leaving the dynamics of the underlying risk factor aside. Furthermore, Eqs. (14.48) and (14.49) are related to *returns* on the two derivatives rather than absolute changes (we consider relative increments like df/f, rather than increments df). This will help us in building financial intuition. Apart from the difference in notation, the core of the arguments does not change: Since there is a single underlying risk factor W_t, we may build a risk-free portfolio by following the usual drill. Using a hedge ratio ϕ_t, the value of the hedged portfolio is

$$\Pi_t = f_t - \phi_t g_t.$$

The hedged portfolio process is manipulated as follows (easing notation a bit):

$$d\Pi_t = df_t - \phi_t \, dg_t$$
$$= \left(f_t m_f - \phi_t g_t m_g\right) dt + \left(f_t s_f - \phi_t g_t s_g\right) dW_t. \qquad (14.50)$$

---
[9] A more careful treatment along this line can also be found in [11, Chapter 21], which is only based on the link between PDEs and expectations of certain stochastic processes.

14.3 A martingale approach to incomplete markets

If we choose
$$\phi_t = \frac{f_t s_f}{g_t s_g},$$
we make the portfolio riskless and, by no-arbitrage, we obtain
$$d\Pi_t = r_t \Pi_t \, dt = r_t (f_t - \phi_t g_t) \, dt. \tag{14.51}$$

By putting Eqs. (14.50) and (14.51) together and simplifying, we find
$$f_t m_f - \phi_t g_t m_g = r_t (f_t - \phi_t g_t),$$
which may be rewritten as
$$f_t(m_f - r_t) = \phi_t g_t (m_g - r_t) = \frac{f_t s_f}{g_t s_g} g_t (m_g - r_t),$$
which finally implies
$$\frac{m_f(r_t, t) - r_t}{s_f(r_t, t)} = \frac{m_g(r_t, t) - r_t}{s_g(r_t, t)} \doteq \lambda(r_t, t). \tag{14.52}$$

As before, we find that there is a ratio that does not depend on the specific derivative, but only on the underlying risk factor. We denote this ratio by $\lambda(r_t, t)$. Now, since the drift function $m(r_t, t)$ is related to expected return and the function $s(r_t, t)$ to volatility, this ratio looks much like a Sharpe ratio, and it may interpreted as a **market price of risk**, which depends on the underlying risk factor and the market risk aversion, rather than on the specific derivative. Hence, for a generic derivative depending on r_t, we may write
$$m(r_t, t) = r_t + \lambda(r_t, t) \cdot s(r_t, t),$$
which should be interpreted as follows: The required (instantaneous) expected return $m(r_t, t)$ is given by the risk-free short rate r_t, plus a risk premium depending on volatility $s(r_t, t)$ and market price of risk $\lambda(r_t, t)$.[10]

Now let us put intuition to good use. What is going to happen in a risk-neutral world? As we have argued, there is no risk premium in such a world. Hence, $\lambda = 0$, and the drift to be used for pricing purposes is just the risk-free rate r_t. This is what happens under the standard risk-neutral measure, but in an incomplete market setting, there are multiple measures compatible with no-arbitrage. These measures correspond to different choices of λ. By choosing the market price of risk, we consider alternative worlds where the prices of derivatives f_t and g_t are internally consistent. We may find a remarkably convenient choice of λ as follows. Let us rewrite the stochastic differential equations (14.48) and (14.49) for the derivative prices, by making λ explicit:
$$df_t = \left[r_t + \lambda s_f(r_t, t)\right] \cdot f_t \, dt + s_f(r_t, t) \cdot f_t \, dW_t,$$
$$dg_t = \left[r_t + \lambda s_g(r_t, t)\right] \cdot g_t \, dt + s_g(r_t, t) \cdot g_t \, dW_t.$$

Now, we may state the key result.

---
[10] Please note again the similarity with APT in Example 10.2.

THEOREM 14.2 *If we choose $\lambda(r_t, t) = s_g(r_t, t)$, then the ratio f_t/g_t is a martingale.*

PROOF With such a choice, the two price processes are given by

$$df_t = (r_t + s_g s_f) \cdot f_t\, dt + s_f f_t\, dW_t,$$
$$dg_t = (r_t + s_g^2) \cdot g_t\, dt + s_g g_t\, dW_t,$$

where we streamline notation a bit. In order to find the equation of the ratio process f_t/g_t, let us switch to log-prices using Itô' lemma:

$$d \log f_t = \left(r_t + s_g s_f - \frac{s_f^2}{2} \right) dt + s_f\, dW_t,$$
$$d \log g_t = \left(r_t + \frac{s_g^2}{2} \right) dt + s_g\, dW_t.$$

By taking the difference of logs, we find

$$d \log \left(\frac{f_t}{g_t} \right) = d(\log f_t - \log g_t) = \left(s_g s_f - \frac{s_f^2}{2} - \frac{s_g^2}{2} \right) dt + (s_f - s_g)\, dW_t$$
$$= -\frac{(s_f - s_g)^2}{2} dt + (s_f - s_g)\, dW_t.$$

Finally, by switching back to prices we find:

$$d \left(\frac{f_t}{g_t} \right) = (s_f - s_g) \cdot \frac{f_t}{g_t}\, dW_t,$$

showing that the ratio process f_t/g_t is a martingale under this probability measure, called the **equivalent martingale measure**. ∎

The equivalent martingale measure generalizes the standard risk-neutral measure and accounts for the overall market risk aversion. In fact, an equivalent name is *risk-adjusted* measure. What we should do is choose a numeraire asset, on the basis of convenience, so that the calculation of the expected value is relatively simple. This is not necessarily trivial, as have to work with an alternative probability measure, which may not be easy to characterize without more sophisticated tools from stochastic calculus. However, we may show the potential of the approach by considering two simple, but quite relevant cases.

14.3.2 CHOICE OF NUMERAIRE: THE BANK ACCOUNT

If we choose a bank account as the numeraire, earning the risk-free short rate r_t, we have

$$dg_t = r_t g_t\, dt,$$

and the market price of risk is $\lambda = 0$. As we said, this corresponds to the standard risk-neutral world, and

$$f_0 = g_0 \cdot \mathrm{E}_{\mathbb{Q}_n}\left[\frac{f_T}{g_T}\right], \qquad \text{where } g_0 = 1, \ g_T = \exp\left\{\int_0^T r_t\, dt\right\}.$$

Therefore,

$$f_0 = \mathrm{E}_{\mathbb{Q}_n}\left[e^{-\int_0^T r_t\, dt} f_T\right]. \tag{14.53}$$

If r_t is constant and we assume GBM, we find a BSM-like formula. However, when r_t is stochastic, the expectation involves the product of random variables. Pricing might be simplified if we choose a more appropriate numeraire.

14.3.3 CHOICE OF NUMERAIRE: THE ZERO-COUPON BOND

The problem with the bank account is that its initial price is given (we choose $g_0 = 1$ for the sake of convenience), but its future value is random due to reinvestment risk. If we instead choose, as the numeraire, a risk-free zero-coupon bond with unit face value, maturing at time T, we have

$$g_0 = Z(0,T), \qquad g_T = 1.$$

Note that we are choosing a bond whose maturity is the same as that of the derivative we want to price. The terminal value of the numeraire is just $g_T = 1$, the face value of the zero. The current price of the zero is given by the expectation in Eq. (14.44), but it may also be directly observed on the market. Hence, the pricing formula becomes

$$f_0 = Z(0,T) \cdot \mathrm{E}_{\mathbb{Q}_T}[f_T], \tag{14.54}$$

which is much better-looking than Eq. (14.53), as the stochastic nature of the interest rate is encapsulated in the price of the zero. However, the downside is that we should use a possibly non-obvious equivalent martingale measure, where expectation is denoted by $\mathrm{E}_{\mathbb{Q}_T}[\cdot]$.

To get a clue on how to change measure and calculate the expectation, let us consider a forward contract on an asset with price process θ_t, for delivery at time T. The asset may be a commodity, rather than an investment asset. We know that the forward price $F(0,T)$ is chosen in such a way that the value of the contract now is zero, and that the payoff is $\theta_T - F(0,T)$ for the long position. Under the pricing measure, we have

$$f_0 = Z(0,T) \cdot \mathrm{E}_{\mathbb{Q}_T}[\theta_T - F(0,T)] = 0 \quad \Rightarrow \quad F(0,T) = \mathrm{E}_{\mathbb{Q}_T}[\theta_T],$$

i.e., under this measure the forward price is just the expected value of θ_T. This is why this martingale measure is call **forward risk-neutral**. Given the spot–forward convergence condition $\theta_T = F(T,T)$, the above result implies

$$F(0,T) = \mathrm{E}_{\mathbb{Q}_T}[F(T,T)],$$

which suggests that forward prices are martingales under the corresponding forward risk-neutral measure.[11] This result may be proved rigorously using tools from stochastic calculus. This choice of numeraire provides us with a justification for a pricing formula, the Black's model, which was proposed as an extension of the BSM formula.

14.3.4 PRICING OPTIONS WITH STOCHASTIC INTEREST RATES: BLACK'S MODEL

As an application of martingale concepts, we may derive a theoretical justification for the **Black's model**, originally published in 1976 as an extension of the BSM pricing model (published in 1973). The model allowed to price options under stochastic interest rates, and as such it was used by practitioners. However, a sound justification was obtained later, by using a forward risk-neutral measure, which corresponds to selecting a zero as the numeraire.

Let us consider a European-style call option, written on an asset with spot price process $S(t)$ and current forward price $F(0, T)$. Using Eq. (14.54), as well as the spot–forward convergence condition $S(T) = F(T, T)$, we may write the current price of the option as

$$C(0) = Z(0, T) \cdot \mathrm{E}_{\mathbb{Q}_T}\big[\max\{S(T) - K, 0\}\big]$$
$$= Z(0, T) \cdot \mathrm{E}_{\mathbb{Q}_T}\big[\max\{F(T, T) - K, 0\}\big].$$

Now we need to express the expectation under measure $\mathbb{Q}_T$, which requires some distributional assumption. In order to extend the BSM model, we assume that $F(T, T)$ is lognormal under the forward risk-neutral measure. Since forward prices are martingales under this measure, we have

$$\mathrm{E}_{\mathbb{Q}_T}[S(T)] = \mathrm{E}_{\mathbb{Q}_T}[F(T, T)] = F(0, T).$$

Furthermore, we assume that the standard deviation (volatility) of the *logarithm* of $F(T, T)$ is $\sigma_F \cdot \sqrt{T}$. Then, we may use Theorem 13.3 and write

$$C(0) = Z(0, T) \cdot \big[F(0, T) \cdot \Phi(d_1) - K \cdot \Phi(d_2)\big], \qquad (14.55)$$

where $\Phi(\cdot)$ is the usual CDF of a standard normal and

$$d_1 = \frac{\log\left[\frac{F(0,T)}{K}\right] + \frac{1}{2}\sigma_F^2 \cdot T}{\sigma_F \cdot \sqrt{T}},$$

$$d_2 = \frac{\log\left[\frac{F(0,T)}{K}\right] - \frac{1}{2}\sigma_F^2 \cdot T}{\sigma_F \cdot \sqrt{T}}.$$

As a reality check, the reader may verify that, if we use the simple spot–forward parity and plug $F(0, T) = S(0) \cdot e^{rT}$, we recover the familiar BSM formula. In

---

[11] Futures prices are martingales under the usual risk-neutral measure. See Section 13.7.6.

the Black's model, we use the forward price of the underlying asset, rather than its spot price. This accounts for the effect of stochastic interest rates, as well as the fact that the underlying asset may be a commodity that is not an investment asset. No drift is involved, because of the martingale property, and the critical input to the model is the volatility σ_F, which is not readily available. A further complication is that the formula requires a forward price, whereas only futures prices may be available (or more reliable due to liquidity). Sometimes, we may assume that the futures price is a sensible proxy for the forward price. In other cases, correction factors have been devised.

14.3.5 EXTENSIONS

We have seen that, within an incomplete market setting, we should choose the pricing measure by a calibration procedure. Apparently, if we choose a numeraire, we also immediately choose a martingale measure. However, we still need critical inputs, like the volatility σ_F in the Black's model of Eq. (14.55). In fact, one way traders used this model was to quote σ_F as an implied volatility, which may be a more useful information for traders than the price itself. When dealing with more complex derivatives, we need a range of volatilities for different maturities, which must be properly calibrated against liquid securities.

The concepts behind Black's model have been extended to price more complicated interest rate derivatives, like caps, floors, and swaptions. Standard market models have been developed, like the LIBOR market model, where one tries to rely on inputs that may be obtained by market quotes, rather than by complicate calibration procedures. An alternative is to extend single-factor models based on the short rate to multifactor models. Another possibility is to model the dynamics of the forward rates, from which the spot rates may be obtained. There is a multitude of available models, and no single model seems to be the best one for every conceivable application. Please see the end-of-chapter references for a more specific treatment.

14.4 Issues in model calibration

In principle, model calibration may be achieved by solving a nonlinear least-squares problem. We select a model depending on a vector β of parameters, and a set $\mathcal{K}$ of traded and sufficiently liquid securities with observed price P_k^o, $k \in \mathcal{K}$. The model should estimate prices $\widehat{P}_k(\beta)$ as close as possible to the observed ones, which we try to achieve by solving the optimization problem

$$\min_{\beta} \sum_{k \in \mathcal{K}} \left[\widehat{P}_k(\beta) - P_k^o \right]^2. \qquad (14.56)$$

If necessary, parameters may be constrained to be within a feasible set. Given the possible nonlinearity of model prices with respect to parameters, this is not necessarily a (seemingly) trivial linear regression problem solved by least-squares methods. We may apply methods from nonlinear programming, as we discuss in Section 15.10. Unfortunately, model calibration often results in a nonconvex optimization problem, which may require expensive global optimization methods, which are outlined in Section 16.2. This may be problematic if model calibration must be repeatedly updated and CPU time becomes an issue. Luckily, the availability of cheap and more and more powerful hardware, as well as the progress in optimization software, may ease computational difficulties.

In this section, however, we want to point out some difficulties that go well beyond the computational side of the coin. These difficulties also apply to the familiar least-squares case. Hence, we first review the traditional linear regression problem, pointing out some difficult tradeoffs that must be addressed. Then, we point out some additional issues that arise when a financial model is calibrated for pricing purposes.

14.4.1 BIAS–VARIANCE TRADEOFF AND REGULARIZED LEAST-SQUARES

Estimating a linear model by least-squares looks like a simple business. We assume that data are generated by an underlying statistical model like

$$Y = \beta_0 + \beta_1 X_1 + \beta_2 X_2 + \cdots + \beta_q X_q + \epsilon, \tag{14.57}$$

where a target (regressed) variable Y is related to features (regressors) X_j, $j = 1, \ldots, q$, and an error term ϵ. If we introduce an additional fictitious variable $X_0 \equiv 1$, we may rewrite the model (14.57) in compact vector form as

$$Y = \beta^{\mathrm{T}} \mathbf{X} + \epsilon.$$

The error is assumed to be a random variable independent from $\mathbf{X}$, with $\mathrm{E}[\epsilon] = 0$ and $\mathrm{Var}[\epsilon] = \sigma_\epsilon^2$. We also assume that different realizations of the error term are mutually independent. Hence, conditional on $\mathbf{X} = \mathbf{x}_0$, we would have

$$\mathrm{E}[Y \mid \mathbf{X} = \mathbf{x}_0] = \beta^{\mathrm{T}} \mathbf{x}_0,$$

and

$$\mathrm{Var}[Y \mid \mathbf{X} = \mathbf{x}_0] = \sigma_\epsilon^2.$$

The model is estimated on the basis of a sample of n joint observations,

$$(Y_i, X_{i1}, X_{i2}, \ldots, X_{iq}), \quad i = 1, \ldots, n,$$

indexed by i, which are collected into vector $\mathbf{Y}$ and matrix $\mathcal{X}$:

$$\mathbf{Y} = \begin{bmatrix} Y_1 \\ Y_2 \\ Y_3 \\ \vdots \\ Y_n \end{bmatrix}, \quad \mathcal{X} = \begin{bmatrix} 1 & X_{11} & X_{12} & \cdots & X_{1q} \\ 1 & X_{21} & X_{22} & \cdots & X_{2q} \\ 1 & X_{31} & X_{32} & \cdots & X_{3q} \\ \vdots & \vdots & \vdots & & \vdots \\ 1 & X_{n1} & X_{n2} & \cdots & X_{nq} \end{bmatrix},$$

14.4 Issues in model calibration

where the data matrix $\mathcal{X} \in \mathbb{R}^{n,q+1}$ collects observed values of the regressors and includes a leading column of ones. The data set used to estimate (or fit) a model is called **training** or **learning** sample. Sometimes, a subset of data is kept apart for *out-of-sample* model validation purposes. This set is called **test sample**.

In order to estimate the parameters β, we may apply ordinary least-squares, on the basis of the following regression equations:

$$Y_i = b_0 + b_1 X_{i1} + b_2 X_{i2} + \cdots + b_1 X_{iq} + e_i, \qquad i = 1, \ldots, n, \qquad (14.58)$$

where e_i is the residual for the ith observation and b_j is the estimate of parameter β_j, $j = 0, \ldots, q$. We note that the residuals are the observable counterpart of errors, and that the approach is justified if we take for granted the above assumptions about errors. If we collect residuals into vector $\mathbf{e} \in \mathbb{R}^n$ and coefficients into vector $\mathbf{b}$, the regression equations may be rewritten in the following convenient matrix form:

$$\mathbf{Y} = \mathcal{X}\mathbf{b} + \mathbf{e}. \qquad (14.59)$$

Then, we minimize the squared Euclidean norm[12] of vector $\mathbf{e}$,

$$\min_{\mathbf{b}} \sum_{i=1}^{n} e_i^2 = \|\mathbf{e}\|_2^2 = \mathbf{e}^\mathsf{T}\mathbf{e}. \qquad (14.60)$$

This is easily accomplished by ordinary least-squares (OLS), which yield

$$\mathbf{b} = \left(\mathcal{X}^\mathsf{T}\mathcal{X}\right)^{-1}\mathcal{X}^\mathsf{T}\mathbf{Y}. \qquad (14.61)$$

Here, we are taking for granted that the matrix $\mathcal{X}^\mathsf{T}\mathcal{X}$ is invertible, which is not the case if features are not linearly independent.

Then, in standard statistical theory, some questions are addressed, including the following ones:

- Under which conditions can we assume that estimators are unbiased, i.e., $\mathrm{E}[\mathbf{b}] = \beta$?
- What about the uncertainty in the estimates, i.e., the standard error of the least-squares estimators?
- How can we check the assumptions about the unobservable errors by analyzing observed residuals?
- How can we check if the model fits data well?

One standard way to answer the last question is by checking the coefficient of determination $R^2 \in [0, 1]$. This coefficient measures the ratio of explained variance against total variance, and it may be used as a measure of fit. Unfortunately, relying on how well data are fitted *in sample* may be quite misleading.

---

[12] We recall that the Euclidean norm L_2 of a vector $\mathbf{x} \in \mathbb{R}^n$ is $\|\mathbf{x}\|_2 \doteq \sqrt{\sum_{i=1}^{n} x_i^2}$. We may also use the L_1 norm $\|\mathbf{x}\|_1 \doteq \sum_{i=1}^{n} |x_i|$, based on absolute values. This may improve robustness, but we lose the advantage of an analytical solution.

There are some basic flaws in this simple picture. To begin with, as evident from Eqs. (14.57) and (14.58), we take for granted that the functional form of the estimated model is the same as the underlying data generating process. Clearly, this is hardly the case. The "true" data generating model may be assumed to be something like

$$Y = f(\mathbf{X}) + \epsilon, \tag{14.62}$$

where Y is again the target variable, $\mathbf{X}$ is a vector of random variables (features), and we assume that errors are independent from the features. However, we do not really know the *form* of function $f(\cdot)$, which is nonlinear in general. The seemingly natural answer is to estimate a model based on a functional form $\widehat{f}(\cdot)$ extending the simple linear model of Eq. (14.58). Indeed, we may apply nonlinear transformations to the original variables, and we may also account for potential interactions among them by using products of variables as features. Using a richer functional form cannot yield a worse R^2, as the more parameters we use, the better the fit against sampled data. We may even push the idea to the extreme and fit a *nonparametric* model, like local regression models or regression trees. But is this a sound approach? The answer is negative, in general, as long as this may result in overfitting and poor *out-of-sample* performance. A large R^2 may give a false sense of security, and it does not help in **model selection**, i.e., the choice of the functional form of the model we want to estimate.

To properly frame the problem, we should realize that, in choosing a statistical model, there are some basic tradeoffs that have to be addressed:

- Bias vs. variance
- Flexibility/fitting vs. interpretability

The last point is relevant to nonparametric models, which may fit data very well, but lack the interpretability and intuitive appeal of a parsimonious linear model. To understand what the **bias–variance tradeoff** is about, let us assume that data are generated by the unknown data generating process of Eq. (14.62), and that the usual assumptions about the error term apply. Hence, conditional on $\mathbf{X} = \mathbf{x}_0$, we have

$$\mathrm{E}[Y \mid \mathbf{X} = \mathbf{x}_0] = f(\mathbf{x}_0), \qquad \mathrm{Var}[Y \mid \mathbf{X} = \mathbf{x}_0] = \sigma_\epsilon^2.$$

In practice, given a random sample of joint observations of $(\mathbf{X}, Y)$, we estimate a model $\widehat{f}(\mathbf{X})$, for a prespecified functional form $\widehat{f}(\cdot)$. This yields a forecast, conditional on $\mathbf{X} = \mathbf{x}_0$, which we denote by

$$\widehat{Y}_{\mid \mathbf{X} = \mathbf{x}_0} = \widehat{f}(\mathbf{x}_0).$$

We often suppress conditioning information to ease notational burden, but it is important to understand its meaning. In fact, we should keep in mind that the forecast is also conditional on other information. The estimated model depends on the random sample of joint observations that we use in learning, and this

14.4 Issues in model calibration

in turn depends on multiple realizations of the error term. Hence, we should regard the forecast $\widehat{Y}$ as a random variable depending on the learning sample. The actual realization Y that we want to forecast, given that $\mathbf{X} = \mathbf{x}_0$, is also a random variable, which is independent from $\widehat{Y}$, since the future realization of the error that affects Y is independent from the past realizations that affect the estimated model. It is important to understand that, in the following, we must regard *both* the model and the forecast as random. Conceptually, we should imagine a random experiment consisting of: (a) multiple estimations of $\widehat{f}(\cdot)$, based on different random learning samples, and (b) forecasting with the resulting models for the same setting $\mathbf{X} = \mathbf{x}_0$. Therefore, to be (overly) precise, we could say that:

- $\widehat{Y}_{|\mathbf{X}=\mathbf{x}_0}$ is a random variable,
- and $\widehat{Y}_{|\mathbf{X}=\mathbf{x}_0,\text{learning sample}}$ is a number.

To assess the quality of a forecast, we may consider the expected value of the squared deviation $\mathrm{E}\big[(Y - \widehat{Y})^2\big]$, also known as MSE (**mean squared error**). Note that, in order to compare alternative models with different levels of complexity, MSE should be evaluated *out of sample*. The MSE may be rewritten as follows, where everything is conditional on $\mathbf{X} = \mathbf{x}_0$:

$$\mathrm{E}\big[(Y - \widehat{Y})^2\big] = \mathrm{E}\big[(f(\mathbf{x}_0) + \epsilon - \widehat{f}(\mathbf{x}_0))^2\big]$$
$$= \mathrm{E}\big[(f(\mathbf{x}_0) - \widehat{f}(\mathbf{x}_0))^2\big] + 2\mathrm{E}[\epsilon] \cdot \mathrm{E}\big[f(\mathbf{x}_0) - \widehat{f}(\mathbf{x}_0)\big] + \mathrm{E}[\epsilon^2]$$
$$= \mathrm{E}^2\big[f(\mathbf{x}_0) - \widehat{f}(\mathbf{x}_0)\big] + \mathrm{Var}\big[f(\mathbf{x}_0) - \widehat{f}(\mathbf{x}_0)\big] + \mathrm{Var}[\epsilon],$$

where we use the independence of the future error ϵ from the errors that had an impact on the estimate of the model, as well as the assumption $\mathrm{E}[\epsilon] = 0$. Since $f(\mathbf{x}_0)$ is a constant, even though an unknown one, its variance is zero, and we may express MSE as follows:

$$\mathrm{MSE} = \mathrm{E}^2\big[f(\mathbf{x}_0) - \widehat{f}(\mathbf{x}_0)\big] + \mathrm{Var}[\widehat{f}(\mathbf{x}_0)] + \sigma_\epsilon^2. \qquad (14.63)$$

We see that MSE, which is what really matters, may be decomposed into three terms:

1. The **bias term**, $\mathrm{E}^2\big[f(\mathbf{x}_0) - \widehat{f}(\mathbf{x}_0)\big]$, which is related to the difference between the functional form $f(\cdot)$ of the true data generating process and the functional form $\widehat{f}(\cdot)$ that we estimate.
2. The **variance term**, $\mathrm{Var}[\widehat{f}(\mathbf{x}_0)]$, which is related to the uncertainty of the model we estimate.
3. The **irreducible variance**, σ_ϵ^2, which is related to the uncertainty in the error term ϵ.

While the irreducible variance is what it is, we may try to reduce the bias by using a richer set of variables (features) or a rich nonparametric model. However, this comes with a price, as it can increase variance. Variance is a consequence of the model complexity, since the more parameters we fit, the more estimation uncertainty we introduce. Furthermore, we also risk an overfitting of the model to the data in the learning sample.

A consequence of the above discussion is that we may give up a bit of unbiasedness in order to reduce variance. This may be achieved by assuming simpler models, but also by resorting to a **regularization** approach, which is somehow related to shrinkage estimators, which we have introduced in Example 9.1. The idea is to penalize large regression coefficients to avoid overfitting. We mention two popular regularized regression methods, differing in the kind of penalization:

- In **ridge regression**, the classical objective function based on squared residuals is augmented as follows:

$$\sum_{i=1}^{n} \left(Y_i - \beta_0 - \sum_{j=1}^{q} \beta_j X_{ij} \right)^2 + \lambda \sum_{j=1}^{q} \beta_j^2.$$

- In **lasso regression**, we use absolute values, rather than squared coefficients:

$$\sum_{i=1}^{n} \left(Y_i - \beta_0 - \sum_{j=1}^{q} \beta_j X_{ij} \right)^2 + \lambda \sum_{j=1}^{q} |\beta_j|.$$

In both cases, λ is a penalty coefficient, driving coefficients toward zero (in the limit). Needless to say, this increases bias, but it possibly improves MSE because of a reduction in variance. Note that we do not penalize β_0, but only $\beta_1, \ldots, \beta_q$, as including β_0 would make the model sensitive to where the origin of the data is placed. The standard least-squares problem can be written as

$$\min_{\boldsymbol{\beta}} \|\mathbf{Y} - \mathcal{X}\boldsymbol{\beta}\|_2^2,$$

where we minimize the Euclidean norm of the residual vector. In regularized regression, we augment this objective by a norm of vector $\boldsymbol{\beta}_{-0}$, i.e., the vector of coefficients $\boldsymbol{\beta}$, with the first component β_0 omitted. The ridge regression problem is

$$\min_{\boldsymbol{\beta}} \|\mathbf{Y} - \mathcal{X}\boldsymbol{\beta}\|_2^2 + \lambda \|\boldsymbol{\beta}_{-0}\|_2^2,$$

whereas the lasso regression problem uses a different norm in the penalty term

$$\min_{\boldsymbol{\beta}} \|\mathbf{Y} - \mathcal{X}\boldsymbol{\beta}\|_2^2 + \lambda \|\boldsymbol{\beta}_{-0}\|_1.$$

Additional light can be shed on regularized regression by considering its link with worst-case robust optimization.[13] Financial model calibration is often regarded as an **ill-conditioned problem**, because quite different estimated models may yield a similar performance in terms of fit. In general, if we allow for

---
[13] See Section 15.9.

several, possibly correlated, features, the variance of the estimates will translate in an ill-conditioned problem. This means that a little perturbation in the input data may result in drastically different output, i.e., a quite different estimated model. Given the possible noise affecting market data, this is an obvious reason for concern. To tackle the issue within the linear regression context, let us assume that the matrix $\mathcal{X} \in \mathbb{R}^{n \times q}$, collecting the observations of features, is affected by a perturbation matrix $\boldsymbol{\Delta}$ with the same dimensions. Hence, we should consider the matrix $\mathcal{X}$ as the *nominal* value of data, which are uncertain and expressed as

$$\mathcal{X} + \boldsymbol{\Delta}.$$

Depending on $\boldsymbol{\Delta}$, we will find different parameter estimates $\widehat{\boldsymbol{\beta}}$. We would like to find a robust estimate, which yields a fairly good fit for a suitable range of perturbations. Hence, let us define an **uncertainty set** $\mathcal{U}$ for $\boldsymbol{\Delta}$. Now, the problem is how we may define an uncertainty set for matrices.

We discuss uncertainty sets for vector data at length in Chapter 15. For a vector, the task is fairly easy. Given a vector $\mathbf{a}_0 \in \mathbb{R}^n$ of nominal data, we may consider a neighborhood of points close to $\mathbf{a}_0$, which may be specified as

$$\mathcal{U} = \big\{ \mathbf{a} \in \mathbb{R}^n : \mathbf{a} = \mathbf{a}_0 + \boldsymbol{\delta},\ \|\boldsymbol{\delta}\| \leq \alpha \big\},$$

for a suitable vector norm and a given value of α, which constrains the norm of the perturbation $\boldsymbol{\delta} \in \mathbb{R}^n$ and is related to the amount of uncertainty. If we deal with data matrices, we should resort to matrix norms. We may safely omit the related technicalities and understand that the resulting uncertainty set defines a neighborhood of the nominal data matrix $\mathcal{X}$. In a robust setting, optimization is carried out in a worst-case sense, resulting in the following min–max problem:

$$\min_{\boldsymbol{\beta}} \max_{\boldsymbol{\Delta} \in \mathcal{U}} \|\mathbf{Y} - (\mathcal{X} + \boldsymbol{\Delta})\boldsymbol{\beta})\|. \tag{14.64}$$

It can be shown that, depending on the choice of the involved norms, this robust optimization model is equivalent to ridge or lasso regression.[14]

14.4.2 FINANCIAL MODEL CALIBRATION

From a technical viewpoint, calibrating a pricing model may be regarded as solving a regression model on steroids, as we may have to cope with nasty nonlinearities and possibly nonconvexity. These additional difficulties exacerbate the bias–variance issues and the possible ill-conditioning of statistical learning. However, there is a financial side to it, too, which is best illustrated by a seemingly simple example.[15]

---

[14] See [12] for details.

[15] The example is worked out in full detail, on the basis of actual market data, in [11, Chapter 15], to which we refer the interested reader.

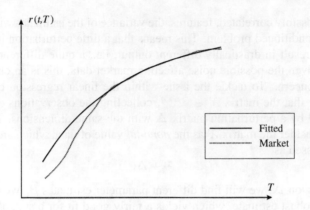

FIGURE 14.1 Hypothetical qualitative shape of a term structure (continuous line) calibrated from a short-rate model against market data (dotted line).

Example 14.1 Calibrating a short-rate model

To appreciate the subtle issues that we may face in financial model calibration, imagine the task of calibrating a short-rate model, like Vasicek or CIR. A simple approach could be to collect a set of prices of zero-coupon bonds with a range of maturities, for which we have analytical pricing formulas, and solve the nonlinear least-squares problem (14.56). This requires finding a set of liquid zeros with comparable risk (actually, they should be risk-free).

As a check, we should verify the fit between the observed term structure of interest rates, which is implicit in the bond prices, and the one generated by the model. Imagine that we obtain the term structure depicted in Fig. 14.1. As one should expect, the fit is not perfect, since we have used a single-factor model. However, we do note a pattern, as the lack of fit is not uniform across maturities. It seems that the long term rates are matched well enough, whereas the shorter term rates leave much to be desired. Is there an explanation for this pattern?

We may find an explanation in terms of bond duration. The zeros with long maturities have a large duration, and their prices are quite sensitive to changes in yield. Hence, the calibration should really match the long term rates well. On the contrary, the zeros with smaller duration are less sensitive, and a mismatch in term of underlying interest rates does not affect the quality of the calibration, which is carried out in term of *prices*, not rates.

The example is trivial with respect to what is carried out in practice, but it does raise an important point. The quality of the calibration depends on its use.

14.4 Issues in model calibration

We may fit a range of prices, which are nonlinear functions of underlying rates, fairly well, but what if we use the model to price other assets? The mismatch we observe on the short end of Fig. 14.1 may not be so critical if we price bonds, but what if we price a derivative depending on the rates themselves? There must be a consistency between the assets that we use in calibrating the model and how the model is used. Careless calibration introduces **model risk**. We must take due care in selecting the assets that we use in calibration and in expressing the objective function that we optimize. Weighted least-squares may be used to this purpose, as well as some form of regularization.

As we have seen, regularization may be used to ease overfitting and ill-conditioning issues. To see how overfitting may arise in financial model calibration, let us compare two approaches to deal with stochastic volatility.

Stochastic volatility models. We have already mentioned the Heston stochastic volatility model,

$$dS_t = \mu S_t \, dt + \sigma_t S_t \, dW_{1t},$$
$$dV_t = \alpha \left[\bar{V} - V_t \right] dt + \xi \sqrt{V_t} \, dW_{2t},$$

where $V_t = \sigma_t^2$ is the variance process and $\bar{V}$ is its long-term average. This is a parametric extension of the GBM model, allowing for a stochastic volatility, which is related to a mean-reverting square root diffusion. Fitting the model means selecting a value for a handful of parameters against market data.

Local volatility models. A local volatility model, may be considered as a non-parametric approach to deal with stochastic volatility. The model is again an extension of GBM,

$$dS_t = \mu S_t \, dt + \sigma(S_t, t) \cdot S_t \, dW_t,$$

but now we are representing volatility as a function $\sigma(S_t, t)$ depending *explicitly* on time.

Local volatility models have theoretical support as they are related to the identification of a risk-adjusted probability measure from prices. Clearly, they offer excellent calibration flexibility, but one should wonder about the potential overfitting. After all, what is the financial motivation of volatility as a deterministic function of time? Furthermore, as argued in [8], there is a considerable danger in fitting marginals of a multivariate distribution with respect to time, neglecting the transition probabilities over time itself. If we are pricing a European-style derivative, what really matters is what happens at maturity. But what if we use a model calibrated against European-style derivatives (possibly, because that is the case where we have analytical pricing formulas) to price American-style derivatives?

Getting back to the mathematical side of financial model calibration, the problem may be ill-posed and yield unstable solutions over time. This may be the result of multiple local optima with similar fitting quality or flat regions, where different sets of parameters may give similar performance measures for the assets in the learning sample (but possibly quite different prices for other derivatives). Regularization by an additional penalty term may be used to make the optimization model convex and improve problem conditioning. In this case, the choice of the regularization term is a tad more difficult than with a plain linear regression by least-squares, and it is beyond the scope of this book. When nonparametric models are calibrated, regularization may involve the distance between probability measures, which may be expressed by entropy (see, e.g., [2]). As a further application of regularization, we mention its use in procedures that improve the consistency of model calibration over time.

Further reading

- Derivative pricing in incomplete markets is discussed in advanced chapters of [4], which has been the basis of our treatment in Section 14.3. The case of interest rate derivatives is dealt with, e.g., in [11], which has been the basis of our treatment in Section 14.1.
- An extensive treatment of interest rate models can be found in [6]. Interest rate derivatives are also treated in depth by [9]. For the use of market models, you may refer to [10].
- You may also find a concise treatment of interest rate models and interest rate derivatives in [7].
- Our discussion on the bias–variance tradeoff is shaped after [5], where you may also find a treatment of regularized regression and nonparametric models.
- In this introductory book, we do not deal with models featuring jumps. For this class of models, see [1]. In Chapter 10 of this reference, you may also find a discussion of pricing and hedging in incomplete markets, whereas Chapter 13 discusses calibration, including its regularization.
- We illustrate a few global optimization methods for nonconvex optimization in Section 16.2. See, e.g., [3] for an application to model calibration.

Bibliography

1 R. Cont and P. Tankov. *Financial Modeling with Jump Processes*. Chapman & Hall/CRC Press, Boca Raton, FL, 2004.

2 R. Cont and P. Tankov. Non-parametric calibration of jump-diffusion option pricing models. *Journal of Computational Finance*, 7(3):1–49, 2004.

3 S.B. Hamida and R. Cont. Recovering volatility from option prices by evolutionary optimization. *Journal of Computational Finance*, 8(4):43–76, 2005.

4 J.C. Hull. *Options, Futures, and Other Derivatives* (8th ed.). Prentice Hall, Upper Saddle River, NJ, 2011.

5 G. James, D. Witten, T. Hastie, and R. Tibshirani. *An Introduction to Statistical Learning: With Applications in R*. Springer, New York, 2013.

6 J. James and N. Webber. *Interest Rate Modelling*. Wiley, Chichester, 2000.

7 D. McInerney and T. Zastawniak. *Stochastic Interest Rates*. Cambridge University Press, Cambridge, 2015.

8 M. Morini. *Understanding and Managing Model Risk: A Practical Guide for Quants, Traders and Validators*. Wiley, Chichester, 2011.

9 R. Rebonato. *Interest-Rate Option Models* (2nd ed.). Wiley, Chichester, 2000.

10 R. Rebonato. *Modern Pricing of Interest-Rate Derivatives: The LIBOR Market Model and Beyond*. Princeton University Press, Princeton, NJ, 2002.

11 P. Veronesi. *Fixed Income Securities: Valuation, Risk, and Risk Management*. Wiley, Hoboken, NJ, 2010.

12 H. Xu, C. Caramanis, and S. Mannor. Robust regression and lasso. *IEEE Transactions on Information Theory*, 56:3561–3574, 2010.

Part Five

Advanced optimization models

Chapter Fifteen

Optimization Model Building

In this chapter, we describe a range of optimization models aimed at financial applications. Emphasis will be almost exclusively on model *building*, rather than model *solving*, which is deferred to the next chapter. To this aim, it is important to have a broad and clear picture of the types of model formulations that may be solved by available software tools. Intuition suggests that a large problem is a harder nut to crack than a small one and that a linear problem is easier than a nonlinear one. Actually, this is not necessarily true, and we shall learn that the main problem feature, drawing the line between relatively easy and difficult problems, is convexity.

We begin, in Section 15.1, with a classification of optimization problems, revolving around the concepts of convex sets and convex functions. The important class of linear programming (LP) models is the topic of Section 15.2. We have already appreciated the role of LP models in the mathematics of arbitrage in Section 2.4. Here, we see how it may be an essential tool in solving possibly large-scale problems. The next level of the hierarchy is convex quadratic programming (QP), which is dealt with in Section 15.3. Convex QP models are the foundation of the mean–variance portfolio theory illustrated in Chapter 8. In Section 15.4, we take a detour into the realm of hard, nonconvex problems. While LPs and convex QPs can be solved quite efficiently, here we add an integrality restriction on a subset of variables. The resulting modeling framework of integer programming is extremely flexible and powerful, but it may result in quite challenging problems from a computational viewpoint. When dealing with integer programming models, it may be essential to formulate the model in such a way that the available solutions methods may work efficiently. Because of this reason, in that section, we get a glimpse of the interaction between proper model formulation and the performances of solution methods, without going into much details about the latter ones. We get back to convex models in Section 15.5, where we describe the possibly less familiar family of conic optimization models. The class of conic optimization problems is a generalization of LP, and it includes second-order cone programming (SOCP) and semidefinite programming (SDP) models, which may be used, among other things, to tackle optimization problems under uncertainty. Conic optimization models were not practical tools in the past, but they are now, thanks to the development of new interior-point solvers, extending those originally introduced as LP solution methods.

In the next part of this chapter, we tackle optimization modeling under uncertainty. This is a quite challenging endeavor, and there are multiple ways to tackle it. In this case, too, the modeling approach cannot be considered as independent from the solution method we intend to apply. In Section 15.6, we consider stochastic programming models, which are related with the stochastic dynamic programming framework of Section 15.7. The two approaches have comparative advantages and disadvantages, but may be fruitfully integrated. Whatever choice we make in dealing with stochastic optimization, we may face a severe computational challenge. To reduce the computational burden, a possible modeling and solution alternative is outlined in section 15.8, based on the optimization of simple decision rules. This idea can be exploited for both stochastic and robust optimization (described later), and it is halfway between a modeling strategy and a solution strategy. Whatever solution approach we pursue, stochastic optimization models assume a probabilistic characterization of uncertainty. Sometimes, this kind of knowledge is difficult to build on the basis of statistical tools, resulting in some ambiguity in the uncertainty itself. An alternative framework is worst-case robust optimization, which is the subject of Section 15.9 and relies heavily on conic optimization.

Finally, in Section 15.10 we provide some examples of generic, not necessarily convex, nonlinear programming models.

15.1 Classification of optimization models

A fairly general statement of an optimization problem is

$$\min_{\mathbf{x} \in S} f(\mathbf{x}), \tag{15.1}$$

which highlights the following three building blocks:

- A vector of **decision variables** $\mathbf{x} \in \mathbb{R}^n$, representing a solution of the problem in mathematical form
- A **feasible set** $S \subseteq \mathbb{R}^n$, also called **feasible region**, to which $\mathbf{x}$ must belong
- An **objective function** $f(\cdot)$, mapping each feasible solution $\mathbf{x} \in S$ to a performance measure $f(\mathbf{x}) \in \mathbb{R}$, which we are interested in optimizing

This rather dry and abstract statement encompasses an incredible variety of decision problems. In this book, we consider only finite-dimensional problems, where decisions are represented by an n-dimensional tuple of real numbers, i.e., an element of a finite-dimensional space. In the academic literature, infinite-dimensional optimal control problems are also considered, where we must choose functions of continuous time, like $u(t)$, $t \in [0, T]$. We shall not consider such problems here. However, in the context of robust optimization, we must tackle **semi-infinite problems**, where we have to cope with an infinite

15.1 Classification of optimization models

number of constraints. Some authors refer to optimization in finite-dimensional spaces as **mathematical programming**.

Note that there is no loss of generality in considering minimization problems, as a maximization problem can be easily converted into a minimization one:

$$\max f(\mathbf{x}) \quad \Rightarrow \quad -\min\left[-f(\mathbf{x})\right].$$

When $S \equiv \mathbb{R}^n$, we speak of an **unconstrained** optimization problem; otherwise, we have a **constrained** optimization problem.

The feasible region S, in concrete terms, is typically described by a set of equality and inequality constraints, as in the following mathematical program:

$$\begin{align}
\min \quad & f(\mathbf{x}) \tag{15.2}\\
\text{s.t.} \quad & h_i(\mathbf{x}) = 0, \quad i \in E,\\
& g_i(\mathbf{x}) \le 0, \quad i \in I,\\
& \mathbf{x} \in \mathcal{X} \subset \mathbb{R}^n.
\end{align}$$

The condition $\mathbf{x} \in \mathcal{X}$ may include additional restrictions, such as the integrality of a subset of decision variables. The most common form of this restriction concerns binary decision variables, i.e., $x_j \in \{0, 1\}$; this trick of the trade is useful to model logical decisions like "we do it" vs. "we do not." Simple lower and upper bounds, describing **box constraints** like $l_j \le x_j \le u_j$, are usually considered apart from general inequalities, for the sake of computational efficiency. So, we might also have $\mathcal{X} = \{\mathbf{x} \in \mathbb{R}^n \mid \mathbf{l} \le \mathbf{x} \le \mathbf{u}\}$, where vectors $\mathbf{l}$ and $\mathbf{u}$ collect the lower and upper bounds on decision variables, respectively.

An optimization problem can be very simple or on the verge of intractability, depending on some essential features of its building blocks. Such features are often more important than the sheer size of the problem. This will be clearer later, when we consider concrete modeling examples and illustrate the astonishing variety of problems that we may address. For now, it suffices to say that an essential classification framework involves the following dimensions:

- Convex vs. nonconvex problems
- Linear vs. nonlinear problems
- Deterministic vs. stochastic problems

By far, the most important property of an optimization problem is convexity. Convexity is a property of sets, in particular of the feasible set S, which can be extended to a property of functions, as is relevant to the objective function.

DEFINITION 15.1 (Convex sets) *A set $S \subseteq \mathbb{R}^n$ is said to be a* convex set *if*

$$\mathbf{x}, \mathbf{y} \in S \quad \Rightarrow \quad \lambda \mathbf{x} + (1-\lambda)\mathbf{y} \in S, \quad \forall \lambda \in [0, 1].$$

Convexity can be grasped intuitively by observing that points of the form $\lambda \mathbf{x} + (1-\lambda)\mathbf{y}$, where $0 \le \lambda \le 1$, are simply the points on the straight-line segment joining $\mathbf{x}$ and $\mathbf{y}$. So, a set S is convex if the line segment joining any pair of

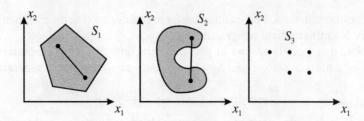

FIGURE 15.1 An illustration of set convexity.

points $\mathbf{x}, \mathbf{y} \in S$ is also contained in S. This is illustrated in Fig. 15.1: S_1 is convex, but S_2 is not. S_3 is a discrete set and it is not convex; this fact has important consequences for discrete optimization problems. It is easy to see that the intersection of convex sets is a convex set. For instance, a polyhedron is an intersection of half-planes, and it is a convex set. This does not necessarily apply to set union. For instance, a possible restriction on a decision variable is

$$x_j \in \{0\} \cup [l_j, u_j],$$

where l_j and u_j are both strictly positive. This restriction specifies a nonconvex set, and it should not be confused with the box constraint $l_j \leq x_j \leq u_j$, which corresponds to a convex set. An interval is a convex set, but if we add the singleton $\{0\}$, we lose convexity, since the segment joining the origin and a point in the interval $[l_j, u_j]$ includes points outside the set. A variable subject to such a restriction is called **semicontinuous**. As a concrete example, imagine that x_j is the weight of asset j in a portfolio. What we want to express is *not* that we must invest at least a fraction l_j of our wealth in that asset. The point $x_j = 0$ is feasible, i.e., we may not invest in asset j, but if we do, there are a lower bound l_j and an upper bound u_j on its weight. We face an example of a semicontinuous variable in everyday life, when a minimum purchase quantity is required by the seller of a good.

Set convexity can be extended to function convexity as follows.

DEFINITION 15.2 (Convex functions) *A function* $f: \mathbb{R}^n \to \mathbb{R}$, *defined over a convex set* $S \subseteq \mathbb{R}^n$, *is a convex function on* S *if, for any* $\mathbf{y}$ *and* $\mathbf{z}$ *in* S, *and for any* $\lambda \in [0, 1]$, *we have*

$$f(\lambda \mathbf{y} + (1 - \lambda)\mathbf{z}) \leq \lambda f(\mathbf{y}) + (1 - \lambda)f(\mathbf{z}). \tag{15.3}$$

The definition can be interpreted by looking at Fig. 15.2. If we join any two points on the function graph with a line segment, all of the segment lies above the function graph. In other words, a function is convex if its epigraph, i.e., the region above the function graph, is a convex set. If the condition (15.3) is satisfied with strict inequality for all $\mathbf{y} \neq \mathbf{z}$, the function is **strictly convex**. We illustrate examples of convex and nonconvex functions in Fig. 15.3. The function in Fig. 15.3(a) is convex, whereas the function in Fig. 15.3(b) is not. The function in Fig. 15.3(c) is a *polyhedral* convex function, and it is kinky. This

15.1 Classification of optimization models

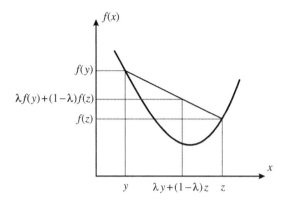

FIGURE 15.2 An illustration of function convexity.

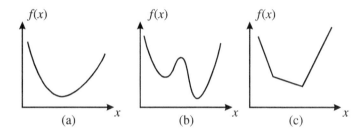

FIGURE 15.3 Convex and nonconvex functions.

example shows that a convex function need not be differentiable everywhere. Note that, in the second case, we have a local minimum that is not a global one. Indeed, convexity is so relevant in minimization problems because it rules out local minima. When solving a minimization problem, it is nice to have a convex objective function. When the objective has to be maximized, it is nice to deal with a concave function.

DEFINITION 15.3 (Concave function) *A function $f(\cdot)$ is concave if the function $-f(\cdot)$ is convex.*

Thus, a concave function is just a convex function turned upside down. We have observed that a convex function features a convex epigraph; hence, function convexity relies on set convexity. A further link between convex sets and convex functions is that the set $S = \{\mathbf{x} \in \mathbb{R}^n \mid g(\mathbf{x}) \leq 0\}$ is convex if g is a convex function.[1]

A related property of a convex function is that its **sublevel sets** $L_f(t)$, defined as
$$L_f(t) \doteq \{\mathbf{x} \in S : f(\mathbf{x}) \leq t\},$$

---

[1] See Problem 15.2.

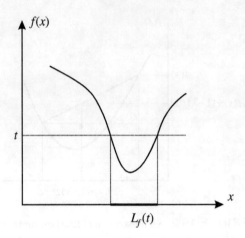

FIGURE 15.4 A quasiconvex function and a sublevel set.

are convex (S is the domain on which f is defined and convex). As it turns out, this property is shared by a larger family of functions, which we call **quasiconvex**. A convex function is quasiconvex, but the converse is not true, as we see from Fig. 15.4. The function is not really convex, but it features convex sublevel sets and no local optima. By flipping a quasiconvex function upside down, we create a **quasiconcave function**.[2]

In general, equality constraints like

$$h(\mathbf{x}) = 0$$

do not describe convex sets. We may understand why, by rewriting the equality as two inequalities:

$$h(\mathbf{x}) \leq 0,$$
$$-h(\mathbf{x}) \leq 0.$$

We know that an inequality constraint like $g(\mathbf{x}) \leq 0$ defines a convex set if g is convex. However, there is no way in which $h(\cdot)$ and $-h(\cdot)$ can be both convex, unless the involved function is **affine**:

$$h(\mathbf{x}) = \mathbf{a}^\mathsf{T}\mathbf{x} - b.$$

It is pretty intuitive that convexity of the objective function makes an unconstrained problem relatively easy, since local minima are ruled out. But what about convexity of the feasible set of a constrained problem? Actually, convexity must be ensured in both the objective function and the feasible set.

DEFINITION 15.4 (Convex optimization problem) *The minimization problem $\min_{\mathbf{x} \in S} f(\mathbf{x})$ is convex if both S and $f(\cdot)$ are convex. The maximization problem $\max_{\mathbf{x} \in S} f(\mathbf{x})$ is convex if S is convex and f is concave.*

---

[2] See the Sharpe ratio plotted in Fig. 8.8.

15.1 Classification of optimization models

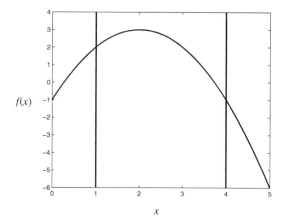

FIGURE 15.5 A concave problem may feature local optima, but they lie on the boundary of the convex feasible set.

Note that we also speak of a convex problem when we are maximizing a concave objective. We speak of a **concave problem** when we *minimize* a concave objective on a convex set. Concave problems are not as nice as convex problems, as the following example illustrates.

Example 15.1 A concave problem

Consider the following one-dimensional problem:

$$\min \quad -(x-2)^2 + 3$$
$$\text{s.t.} \quad 1 \leq x \leq 4$$

This is a concave problem, since the leading term in the quadratic objective is negative, so that the second-order derivative is negative everywhere. In Fig. 15.5, we show the objective function and the feasible set. The stationarity point $x = 2$ is of no use to us, since it is a maximizer. We see that local minimizers are located at the boundary of the feasible set. A local minimizer lies at the left boundary, $x = 1$, and the global minimizer is located at the right boundary, $x = 4$.

It turns out that the property illustrated in Example 15.1, i.e., we can find an optimal solution on the boundary of the (convex) feasible set, applies to concave problems in general. This kind of structure may be exploited in global optimization procedures.

In the following, we will consider the following classes of optimization problems:

- **Linear programming** (LP), where the objective function and the constraints are represented by linear (affine) functions. The class of LP problems is convex and can be solved by quite robust and efficient procedures implementing the simplex method or interior-point methods.

- **Quadratic programming** (QP), where constraints are linear but the objective function involves a quadratic form. QP is convex if the quadratic form is convex, which is the case when the involved matrix is positive semidefinite. In finance, this is mostly the case, as the quadratic form we consider involves a covariance matrix. The covariance is also symmetric. We shall denote the set of symmetric positive semidefinite $n \times n$ matrices by $\mathbb{S}_+^n$. If constraints involve quadratic forms, we have a **quadratically constrained QP** (QCQP) problem, which is convex if the quadratic forms define a convex set. Convex QP and QCQP problems may be solved by classical nonlinear programming methods or by more recent interior-point methods.

- **Mixed-integer LP/QP** (MILP, MIQP) problems are obtained when we require that a subset of decision variables take integer values. There are quite sophisticated software packages based on branch-and-bound or branch-and-cut methods to solve such problems. However, they are not convex and may not be solvable within a reasonable amount of time, when their structure is difficult or their size is too large.

- **Second-order cone programming** (SOCP) and **semidefinite programming** (SDP) problems are a relatively recent class of convex optimization problems that can be solved by interior-point methods. As we shall see, they involve feasible sets defined by cones. Semidefinite programming deals with the rather peculiar cone of positive semidefinite matrices. At present, large-scale SDPs are not easily solvable, but several real-life problems may be tackled with a reasonable computational effort.

- Generic **nonlinear programming** (NLP) problems are obtained when the model formulation involves generic nonlinear functions. They are non-convex in general, and most commercial solvers only find local optima. For some specific structures, global optimization procedures guaranteeing the global optimality of the solution may be devised. In general, we may adopt heuristic global optimization approaches, which do not guarantee true optimality.

All of the above classes may be formulated as deterministic problems, provided that there is no uncertainty in the problem data, or uncertain problems. In the latter case, we may formulate a stochastic or a robust optimization problem, depending on the way we characterize uncertainty.

15.2 Linear programming

We deal with a linear programming (LP) model, when the objective function and all of the constraints are expressed by linear (affine) functions. An LP model can be formulated as

$$\min \sum_{j=1}^{n} c_j x_j$$

$$\text{s.t.} \sum_{j=1}^{n} d_{ij} x_j \leq e_i, \quad i \in I$$

$$\sum_{j=1}^{n} h_{ij} x_j = q_i, \quad i \in E.$$

An LP problem may involve a number of equalities and inequalities, but it is easy to see that the former ones can be transformed into the latter ones, and vice versa. The equation

$$\sum_{j=1}^{n} h_{ij} x_j = q_i$$

can be rewritten as a pair of inequalities:

$$\sum_{j=1}^{n} h_{ij} x_j \leq q_i, \qquad \sum_{j=1}^{n} h_{ij} x_j \geq q_i.$$

By introducing slack variables $s_i \geq 0$, we can also get rid of inequalities:

$$\sum_{j=1}^{n} d_{ij} x_j \leq e_i \quad \Rightarrow \quad \sum_{j=1}^{n} d_{ij} x_j + s_i = e_i.$$

Note that we trade an inequality for another one, $s_i \geq 0$, which is however a simple non-negativity bound. In fact, we usually have non-negative decision variables, but there are exceptions, as is the case when we allow short-selling. If necessary, a generic unrestricted variable x_j can be expressed as a difference of unrestricted variables, corresponding to the positive and the negative part of the variable:

$$x_j = x_j^+ - x_j^-; \qquad x_j^+, x_j^- \geq 0.$$

Per se, this is not quite correct, as we should also enforce a complementarity condition, stating that at least one of the two parts is zero, $x_j^+ \cdot x_j^- = 0$; however, this would introduce a nonlinearity. Usually, these variables are involved in the objective function, and the complementarity condition will be enforced by optimality. To see why, if both variables have a cost, expressing $x_j = 3$ by $x_j^+ = 10$ and $x_j^- = 7$ is more expensive than using $x_j^+ = 3$ and $x_j^- = 0$.

We may cast LPs into a more compact matrix-based form:

$$\min \quad \mathbf{c}^T\mathbf{x} \tag{15.4}$$
$$\text{s.t.} \quad \mathbf{A}\mathbf{x} = \mathbf{b},$$
$$\mathbf{x} \geq \mathbf{0},$$

where $\mathbf{x} \in \mathbb{R}^n$, $\mathbf{c} \in \mathbb{R}^n$, $\mathbf{b} \in \mathbb{R}^m$, and $\mathbf{A} \in \mathbb{R}^{m \times n}$. Clearly, such a problem makes sense if $m < n$, i.e., if the involved system of linear equations is underdetermined. The model of Eq. (15.4) is an LP in **standard form**, involving only equality constraints and non-negative decision variables. Another way to state the problem is the **canonical form**, involving only inequalities:

$$\min \quad \mathbf{c}^T\mathbf{x} \tag{15.5}$$
$$\text{s.t.} \quad \mathbf{A}\mathbf{x} \geq \mathbf{b},$$
$$\mathbf{x} \geq \mathbf{0}.$$

The standard and canonical forms are equivalent, as we may transform either form into the other one. The standard form is preferred from an algorithmic viewpoint, since we may take advantage of familiar concepts from linear algebra, as we shall see. Essentially, the constraints state that we must express vector $\mathbf{b}$ as a linear combination of the columns of matrix $\mathbf{A}$, with non-negative coefficients. The canonical form allows us to visualize the feasible region as an intersection of halfplanes, i.e., a polyhedron.

A remarkable fact is that an LP is both a convex and concave programming problem. This implies that any local optimum is also a global one, and that we may just bother with points on the boundary of the feasible region. More precisely, we have an optimal solution corresponding to a vertex of the feasible region, which is polyhedral.

Example 15.2 Illustrating the geometry of LP

> Figure 15.6 shows a possible feasible set for an LP where we maximize an objective function with respect to two decision variables, x_1 and x_2. A polyhedron like this may be generated by three linear inequalities and two sign restrictions on the decisions. We show the level curves of the objective, which are parallel lines in an LP problem. The optimal solution corresponds to the extreme point C. A change in the coefficients of the objective function would imply a rotation of the level curves, so that the optimal solution would shift to B or D. Note that when the level curves are parallel to a face of the polyhedron, we have multiple optima (actually, a whole edge of them).

The geometry of LP is exploited in the celebrated simplex algorithm, which is a clever way to explore vertices of the feasible set and is available in plenty

15.2 Linear programming

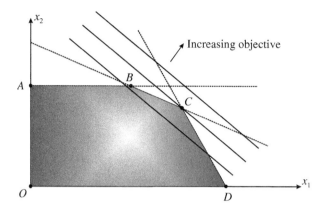

FIGURE 15.6 Geometry of linear programming.

of commercial packages. There are also alternative strategies based on interior-point methods. LP solution methods are outlined in Section 16.3.

We have seen how LP may be used to characterize no-arbitrage conditions in a simple one-period, discrete-state market model in Section 2.3. In the next section, we describe a possible model for financial planning.

15.2.1 CASH FLOW MATCHING

In Section 6.3.1, we have considered a simple LP model, which we recall here for the sake of convenience, to tackle a simplified asset–liability management problem. The aim is make optimal use of assets to meet a stream of deterministic liabilities L_t at times $t = 1, \ldots, T$. A portfolio exactly matching the liabilities would be easy to build, if we had a rich set of zeros but, in practice, we have to use coupon-bearing bonds. Given a set of n bonds, each with current price P_i ($i = 1, \ldots, n$) and paying a cash flow F_{it} at time t, we may consider the following cash flow matching model:

$$\min \sum_{i=1}^{n} P_i x_i$$
$$\text{s.t.} \sum_{i=1}^{n} F_{it} x_i \geq L_t, \quad \forall t$$
$$x_i \geq 0.$$

The decision variable x_i represents the amount of bond i purchased.

As we have observed in Section 6.3.1, this model is quite oversimplified and based on an inflexible approach. When there is a cash surplus in a time period, this is not used in any way. In order to reduce the cost of the portfolio, we might introduce the possibility of reinvesting surplus at time t, as well as financing shortfall. Let us introduce cash surplus v_t^+ at time t, which can be reinvested at the risk-free rate r_{ft}, and cash shortfall v_t^-, which can be financed at the risk free rate $r_{ft} + \delta$. Here, r_{ft} is not an annual rate, but it refers to the time bucket size selected to build the model. We use the spread δ to take bid–ask spreads into account, as the rate at which we may lend is not the rate at which we may borrow money. Given an initial budget b_0, the objective could be to maximize the terminal cash surplus v_T^+:

$$\max \quad v_T^+$$

$$\text{s.t.} \quad b_0 - \sum_{i=1}^{n} P_i x_i = v_0^+ - v_0^-, \tag{15.6}$$

$$\sum_{i=1}^{n} F_{it} x_i + (1 + r_{f,t-1}) v_{t-1}^+ + v_t^-$$
$$= L_t + v_t^+ + (1 + r_{f,t-1} + \delta) v_{t-1}^-, \quad t = 1, \ldots, T \tag{15.7}$$

$$v_T^- = 0; \quad v_t^+, v_t^-, x_i \geq 0.$$

Constraint (15.6) states that, if the cost of the portfolio we buy exceeds the budget b_0, we have to finance an initial shortfall v_0^-; otherwise, we have a surplus v_0^+. Constraint (15.7) may be interpreted as a cash flow balance for each successive time period. We have cash inflows due to coupons and face values of matured bonds, surplus invested in the previous period, and borrowed money. We have cash outflows due to liabilities, surplus invested now, and the repayment of debt contracted in the previous time period. Note that we rule out any outstanding debt at the last time bucket T, since we require $v_T^- = 0$.

This model is a slight improvement with respect to the trivial cash flow matching, but it is hardly satisfying. To begin with, it is deterministic and does not account for stochastic interest rates. The main limitation is that we cannot either buy or sell bonds at arbitrary time periods. We may introduce this possibility, but then we should account for the uncertainty in future bond prices. A demanding, but flexible approach to cope with uncertain bond prices is to build a stochastic programming model; see Section 15.6.3.2.

15.3 Quadratic programming

Linear programming (LP) is an important case of convex optimization. The next level in the hierarchy is a quadratic programming (QP) problem like

$$\min \quad \tfrac{1}{2} \mathbf{x}^T \mathbf{Q} \mathbf{x} + \mathbf{f}^T \mathbf{x}$$
$$\text{s.t.} \quad \mathbf{A}\mathbf{x} \leq \mathbf{b},$$
$$\mathbf{C}\mathbf{x} = \mathbf{d},$$
$$\mathbf{x} \geq \mathbf{0}.$$

15.3 Quadratic programming

This is a convex optimization problem, provided that $\mathbf{Q} \in \mathbb{S}_+^n$, where $\mathbb{S}_+^n$ is the set of symmetric positive semidefinite $n \times n$ matrices. There is no loss in generality from considering symmetric matrices, since any quadratic form may be rewritten in terms of a symmetric matrix $\mathbf{Q}$. We have already met QP problems when dealing with mean–variance portfolio optimization. We may trace the efficient mean–variance frontier by solving a sequence of QPs, which is a computationally easy task. Actually, given state-of-the-art solvers, we can say that solving a QP is no harder than solving an LP. The real issue, as far as mean–variance optimization is concerned, is related to statistical estimation, as we have seen in Sections 9.1 and 10.3. In the next section, we consider a related problem, Sharpe ratio maximization.

15.3.1 MAXIMIZING THE SHARPE RATIO

We have tackled the maximization of the Sharpe ratio in Section 8.4, where we have seen that it may be trivially accomplished by solving a system of linear equations, assuming an unconstrained portfolio. However, additional constraints should be considered in practice, which preclude such a simple solution. Furthermore, as we have observed in Section 8.4.1, the approach we followed was built on shaky foundations, since the Sharpe ratio is not really a concave function.[3] Thus, the first-order optimality conditions need not be sufficient.

Here, we show that maximizing the Sharpe ratio can be accomplished by solving a QP, too. This is not quite a trivial fact, as the problem may be formulated as

$$\max \quad \frac{\boldsymbol{\mu}^\mathsf{T}\mathbf{w} - r_f}{\sqrt{\mathbf{w}^\mathsf{T}\boldsymbol{\Sigma}\mathbf{w}}} = \frac{(\boldsymbol{\mu} - r_f\mathbf{i})^\mathsf{T}\mathbf{w}}{\sqrt{\mathbf{w}^\mathsf{T}\boldsymbol{\Sigma}\mathbf{w}}} \tag{15.8}$$

$$\text{s.t.} \quad \mathbf{i}^\mathsf{T}\mathbf{w} = 1, \tag{15.9}$$

$$\mathbf{A}\mathbf{w} = \mathbf{b}, \tag{15.10}$$

$$\mathbf{D}\mathbf{w} \geq \mathbf{c}, \tag{15.11}$$

where $\mathbf{w}$ is the vector of portfolio weights, $\mathbf{i} = [1, 1, \ldots, 1]^\mathsf{T}$ is a vector with n elements set to 1, and we also consider additional equality and inequality constraints on portfolio composition, possibly including no short-selling restrictions or bound on exposures to industrial or geographical sectors.

Under sensible assumptions, we may recast the maximization of Sharpe ratio as a quadratic program. The logical steps of the transformation are as follows:

- The objective function in Eq. (15.8) looks nasty, but we might try improving it a bit by flipping the objective upside down and minimizing

$$\frac{\sqrt{\mathbf{w}^\mathsf{T}\boldsymbol{\Sigma}\mathbf{w}}}{(\boldsymbol{\mu} - r_f\mathbf{i})^\mathsf{T}\mathbf{w}}. \tag{15.12}$$

---

[3] We have very informally argued that the objective function is quasiconcave. See [33] for a full and rigorous treatment of these issues.

This is not quite correct, unless we can guarantee that $(\boldsymbol{\mu} - r_f\mathbf{i})^T\mathbf{w} > 0$, i.e., that the risk premium for all portfolios is strictly positive. In fact, it is *not* true that this condition applies to any portfolio. Nevertheless, the *optimal* risky portfolio must be such that its expected return exceeds the risk-free return. Otherwise, on the mean–risk plane, we would have a tangency portfolio below the risk-free asset, and a capital allocation line with a negative slope. If so, in the optimal portfolio, wealth would clearly be completely allocated to the risk-free asset. Hence, we assume that there is a feasible portfolio such that its expected return is strictly larger than the risk-free rate and restrict our attention to portfolios with positive risk premium.

- Under the previous assumption, we may even square the objective in Eq. (15.12) to get rid of the square root, since the objective function is guaranteed to be non-negative. By introducing an auxiliary variable,

$$t \doteq \frac{1}{(\boldsymbol{\mu} - r_f\mathbf{i})^T\mathbf{w}} > 0, \tag{15.13}$$

we may further improve the look of the objective function by recasting it as

$$\min \quad t^2 \cdot \mathbf{w}^T \boldsymbol{\Sigma} \mathbf{w}.$$

Then, by the change of variables

$$\mathbf{x} = t\mathbf{w}, \tag{15.14}$$

we can write the objective as a quadratic form

$$\min \quad \mathbf{x}^T \boldsymbol{\Sigma} \mathbf{x}.$$

The new variables $\mathbf{x}$ can be interpreted as pseudoweights, whose sum is not necessarily 1.

- The change of variables of Eq. (15.14), implies

$$\mathbf{w} = \frac{\mathbf{x}}{t},$$

which is legitimate under the condition $t > 0$ of Eq. (15.13). Under the same assumption, we may transform constraints (15.9) (15.10), and (15.11) as follows:

$$\mathbf{i}^T\mathbf{x}/t = 1 \quad \Rightarrow \quad \mathbf{i}^T\mathbf{x} = t,$$
$$\mathbf{A}\mathbf{x}/t = \mathbf{b} \quad \Rightarrow \quad \mathbf{A}\mathbf{x} = t\mathbf{b},$$
$$\mathbf{D}\mathbf{x}/t \geq \mathbf{c} \quad \Rightarrow \quad \mathbf{D}\mathbf{x} \geq t\mathbf{c},$$

which are linear constraints.

- Finally, the definition of Eq. (15.13) may be rewritten as a linear constraint, too:

$$(\boldsymbol{\mu} - r_f\mathbf{i})^T \cdot \mathbf{w}t = 1 \quad \Rightarrow \quad (\boldsymbol{\mu} - r_f\mathbf{i})^T \cdot \mathbf{x} = 1.$$

15.3 Quadratic programming

Putting all of this together, we conclude that the Sharpe ratio may be maximized by solving the following QP:

$$\min \quad \mathbf{x}^\mathsf{T} \boldsymbol{\Sigma} \mathbf{x} \tag{15.15}$$
$$\text{s.t.} \quad \mathbf{i}^\mathsf{T} \mathbf{x} = t,$$
$$(\boldsymbol{\mu} - r_f \mathbf{i})^\mathsf{T} \cdot \mathbf{x} = 1,$$
$$\mathbf{A}\mathbf{x} = t\mathbf{b},$$
$$\mathbf{D}\mathbf{x} \geq t\mathbf{c},$$
$$t \geq 0. \tag{15.16}$$

The inequality constraint of Eq. (15.16) replaces the strict condition $t > 0$. This is done to make the feasible set closed, as required by any optimization software. If we solve problem (15.15) and find a solution $\mathbf{x}^*$, $t^* > 0$, then the desired portfolio is obtained by setting

$$\mathbf{w}^* = \frac{\mathbf{x}^*}{t^*}.$$

Under the assumption of a positive sloped CAL (capital allocation line), we may rule out $t^* < 0$, and the case $t^* = 0$ would also correspond to a pathology, a portfolio with infinite risk premium, not to be expected in the real world.

15.3.2 QUADRATICALLY CONSTRAINED QUADRATIC PROGRAMMING

In mean–variance optimization, rather than minimizing variance subject to a lower bound on expected return, we may swap the objectives and maximize expected return subject to an upper bound on variance. Hence, we end up with a quadratic constraint like

$$\mathbf{w}^\mathsf{T} \boldsymbol{\Sigma} \mathbf{w} \leq \beta,$$

which is convex, since $\boldsymbol{\Sigma} \in \mathbb{S}_+$ (we use $\mathbb{S}_+$ to denote a symmetric, positive semidefinite matrix without reference to its size). More generally, we may consider a quadratically constrained quadratic programming model like

$$\min \quad \tfrac{1}{2} \mathbf{x}^\mathsf{T} \mathbf{Q} \mathbf{x} + \mathbf{f}^\mathsf{T} \mathbf{x}$$
$$\text{s.t.} \quad \tfrac{1}{2} \mathbf{x}^\mathsf{T} \mathbf{H}_i \mathbf{x} + \mathbf{h}_i^\mathsf{T} \mathbf{x} \leq d_i, \quad i = 1, \ldots, m$$
$$\mathbf{A}\mathbf{x} = \mathbf{b},$$
$$\mathbf{x} \geq \mathbf{0},$$

which is convex, provided that all $\mathbf{H}_i \in \mathbb{S}_+$ as well. Efficient solvers, based on interior-point methods, are available for QCQPs.

A word of caution is needed when dealing with a constraint like

$$\| \mathbf{A}_i \mathbf{x} + \mathbf{b}_i \|_2 \leq \mathbf{c}_i^\mathsf{T} \mathbf{x} + d_i, \tag{15.17}$$

where $\|\cdot\|_2$ denotes Euclidean norm. For $\mathbf{y} \in \mathbb{R}^n$,

$$\|\mathbf{y}\|_2 = \sqrt{\mathbf{y}^T\mathbf{y}} = \sqrt{\sum_{j=1}^{n} y_j^2}.$$

A simple case of Eq. (15.17) is an upper bound on standard deviation,

$$\sqrt{\mathbf{w}^T\boldsymbol{\Sigma}\mathbf{w}} \leq \gamma,$$

which could be squared to yield a bound on variance. Unfortunately, this transformation is unwise in general, since it may transform a convex problem into a nonconvex one. We will investigate this issue in Section 15.5, where we show how to deal with a constraint like (15.17) within the framework of conic optimization.

15.4 Integer programming

LPs and QPs are convex problems that we may solve with robust commercial software. Things get much less pleasant, when we have to restrict some variables to take only integer values. For instance, we obtain a mixed-integer LP (MILP) problem when a subset of variables, collected into vector $\mathbf{y}$, is restricted to assume non-negative integer values:

$$P(S) \quad \min \quad \mathbf{c}^T\mathbf{x} + \mathbf{d}^T\mathbf{y} \qquad (15.18)$$
$$\text{s.t.} \quad \mathbf{Ax} + \mathbf{Ey} \leq \mathbf{b},$$
$$\mathbf{x} \in \mathbb{R}_+^{n_1}, \quad \mathbf{y} \in \mathbb{Z}_+^{n_2}.$$

Here n_1 and n_2 are the numbers of continuous and integer variables respectively, and $\mathbb{Z}_+$ is the set of non-negative integers. The notation $P(S)$ emphasizes the feasible set S, which fails to be convex. A mixed-integer QP (MIQP) problem is obtained similarly, by restricting some variables of a QP model to integer values. Let us consider a couple of typical examples. As we shall see, the most common restriction is $y_j \in \{0, 1\}$, which may be used to enforce qualitative properties on a portfolio.

Example 15.3 Modeling fixed charges

> Let us consider an activity, whose level is measured by a continuous decision variable $x \geq 0$. A concrete example would be the amount of an asset that we purchase. The cost of the activity may include both a fixed charge f and a variable component c. By a fixed charge, we mean a cost that is incurred whenever $x > 0$, in which case the cost

15.4 Integer programming

does not depend on the specific level x of the activity. Note that this is not the same as a fixed (sunk) cost that is always paid for whatever level x, including $x = 0$. The cost function is

$$C(x) = \begin{cases} 0, & \text{if } x = 0, \\ f + cx, & \text{if } x > 0. \end{cases}$$

This function is discontinuous at the origin, and it is clearly not convex.

We may express the cost within an LP framework by introducing an auxiliary binary decision variable δ, related to x:

$$\delta = \begin{cases} 0, & \text{if } x = 0, \\ 1, & \text{if } x > 0. \end{cases}$$

If we introduce a suitably large constant M, playing the role of an upper bound on x, we may relate the continuous variable x and the binary variable δ by a linear inequality:

$$x \leq M\delta.$$

To see how this works, observe that if $\delta = 0$, then x is forced down to zero, whereas the bound $x \leq M$ is enforced when $\delta = 1$. This bound is irrelevant if M is large enough. Then, cost is expressed by the linear function

$$cx + f\delta.$$

It may be argued that this does not forbid the nonsensical case in which we pay the fixed charge f, but leave $x = 0$. However, it is easy to see that such a solution will never be optimal. This kind of modeling trick is called **big-M constraint**, as we require a suitably large M (when no sensible bound can be devised). From a computational viewpoint, the smaller the big-M, the better, as we shall see later.

Example 15.4 Semicontinuous variables

Let us consider how we may represent a semicontinuous variable within an LP framework. As we have mentioned, a common requirement on the level of an activity is that, if it is undertaken, its level should be in the interval $[m, M]$. Note that this is *not* equivalent to

> requiring that $m \leq x \leq M$. Rather, we want something like
>
> $$x \in \{0\} \cup [m, M],$$
>
> which is a non-convex set (recall that the union of convex sets need not be convex). Using the same big-M trick as above, we may introduce a binary variable δ and write
>
> $$x \geq m\delta, \qquad x_i \leq M\delta.$$
>
> Semicontinuous variables may be used when the amount of an asset in a portfolio must be above a minimum threshold, if the asset is included in the portfolio.

15.4.1 A MIQP MODEL TO MINIMIZE TEV UNDER A CARDINALITY CONSTRAINT

Sometimes, our task is not really to be smart and build a portfolio beating a benchmark, but rather to *track* a benchmark at low cost. This is the case, e.g., when we manage a passive fund tracking an index.[4] In other cases, we should replicate a target portfolio by reducing the number of assets involved, so that trading and management costs are minimized.

So, let us consider a universe of n assets, indexed by $i = 1, \ldots, n$. We are given a target or benchmark portfolio with weights $w_i^b \geq 0$, which must be tracked with a limited number, at most $C_{\max}$, of assets. Thus, we need a tracking portfolio, with weights $w_i, i, \ldots, n$, minimizing some distance measure between the target and tracking portfolios, subject to a cardinality constraint. A trivial distance metric can be defined by a L_1 norm:

$$\sum_{i=1}^{n} |w_i - w_i^b|. \tag{15.19}$$

By proper model formulation, this metric yields a MILP model. We have just to introduce auxiliary variables w_i^+ and w_i^-, both non-negative, representing positive and negative deviations from the benchmark, allowing us to express the absolute value as a piecewise linear function:

$$w_i - w_i^b = w_i^+ - w_i^-, \quad i = 1, \ldots, n.$$

---

[4] As a relevant example, an exchange-traded fund (ETF) is a passive fund related to a benchmark that should be tracked as closely as possible. The fund is cheap because it is traded as a security, rather than distributed through a possibly expensive retail network.

15.4 Integer programming

Then, we may transform the objective function of Eq. (15.19) into

$$\sum_{i=1}^{n} \left(w_i^+ + w_i^- \right).$$

However, this distance metric does not take any relationship between asset returns into account. For instance, let us consider two assets, and assume that in the benchmark we have

$$w_1^b = 0.2, \quad w_2^b = 0.1,$$

whereas in the tracking portfolio we have

$$w_1 = 0.3, \quad w_2 = 0.$$

Then, these two assets contribute the following amount to the distance:

$$|0.3 - 0.2| + |0 - 0.1| = 0.2.$$

However, imagine that the two asset returns are perfectly correlated. Then, the true distance would be zero!

We must account for correlations, and an alternative distance metric is the **tracking error variance** (TEV), defined as

$$\text{Var}(\tilde{r}_p - \tilde{r}_b),$$

where $\tilde{r}_p$ and $\tilde{r}_b$ are the random returns of the tracking and benchmark portfolios, respectively. Clearly, variance of the tracking error would be zero for a perfect tracking. TEV can be expressed as

$$\sum_{i=1}^{n} \sum_{j=1}^{n} \left(w_i - w_i^b \right) \sigma_{ij} \left(w_j - w_j^b \right),$$

where σ_{ij} is the covariance between the returns of assets i and j. This objective is a convex quadratic form.

In order to express the cardinality constraint, we need to introduce a set of binary variables δ_i, one for each asset, modeling the inclusion of asset i in the tracking portfolio:

$$\delta_i = \begin{cases} 1, & \text{if asset } i \text{ is included in the tracking portfolio,} \\ 0, & \text{otherwise.} \end{cases}$$

Binary variables δ_i and continuous variables w_i should be linked by the linear inequality

$$w_i \leq M \delta_i,$$

where M is a suitably large constant. Clearly, if we rule out short-selling, $M = 1$ would do. As we show later, performance of solution methods depends on the strength of the model formulation, which suggests using a smaller upper bound:

$$w_i \leq \overline{w}_i \delta_i.$$

Straightforward minimization of TEV, subject to a cardinality constraint, yields the following model:

$$\min \sum_{i=1}^{n}\sum_{j=1}^{n}(w_i - w_i^b)\sigma_{ij}(w_j - w_j^b) \qquad (15.20)$$

$$\text{s.t.} \quad \sum_{i=1}^{n} w_i = 1,$$

$$\sum_{i=1}^{n} \delta_i \leq C_{\max},$$

$$w_i \leq \overline{w}_i \delta_i, \qquad \forall i$$

$$w_i \geq 0, \qquad \delta_i \in \{0, 1\}, \qquad \forall i.$$

This model is an MIQP problem. Note that we assume that short-selling is forbidden; see Problem 15.4 for an extension of the model.

15.4.2 GOOD MILP MODEL BUILDING: THE ROLE OF TIGHT MODEL FORMULATIONS

As we have pointed out, the feasible set of an MILP problem is nonconvex, which spells trouble in general. Furthermore, we may expect that there is no easy-to-check optimality condition associated with integer programming models. Traditional optimality conditions involve derivatives of the objective function, or some augmented version of it, with respect to decision variables; however, the idea itself does not make any sense, when variables are restricted to integer values. Indeed, even if we are handed the optimal solution, we cannot easily check that it is truly optimal. To solve the problem, we have to rely on some form of enumeration of the feasible solutions. This is implemented in commercial software based on branch-and-bound methods. These methods are described in Chapter 16. Usually, we do not need to know too many details about solution methods, but MILP models are an exception. We have to realize what is needed to build a model that is solvable. The key is the "tightness" of the model.

Example 15.5 **Rounding noninteger solutions may not work**

The get a concrete feeling for the difficulties in integer programming, let us consider a pure integer LP borrowed from [35]:

$$\max \quad x_1 + x_2$$

15.4 Integer programming

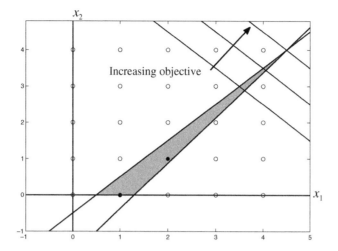

FIGURE 15.7 An integer LP featuring a large integrality gap.

$$\text{s.t.} \quad 10x_1 - 8x_2 \leq 13,$$
$$2x_1 - 2x_2 \geq 1,$$
$$x_1, x_2 \in \mathbb{Z}_+.$$

We might think that a simple way to get a good solution, even though not necessarily an optimal one, is to ignore the integrality constraints and solve the corresponding continuous LP. This need not yield an integer solution, but by judicious rounding we should be done. If we relax the integrality requirement, i.e., we just require $x_1, x_2 \geq 0$, we find the optimal solution

$$x_1^* = 4.5, \qquad x_2^* = 4,$$

with an optimal objective value 8.5. This looks a fairly good starting point, as one variable is already integer, and we may easily round the first one up or down to see what we get. Unfortunately, the solution $(4, 4)$ is not feasible with respect to the second constraint, and the solution $(5, 4)$ is not feasible with respect to the first one. Therefore, both trivially rounded solutions are not even feasible, and we are lost in our quest for optimality. In fact, the integer optimal solution is $x_1^* = 2, x_2^* = 1$, with optimal value 3. We get a clear picture from Fig. 15.7, where we plot the feasible integer set and the polyhedron corresponding to the continuous LP. The true feasible set consists of

> two points, and there is a large gap between the integer feasible set
> and the continuous LP polyhedron. This implies a significant gap
> between the objective function value of the optimal integer solution
> and the solution of the corresponding continuous LP.

Example 15.5 may look somewhat pathological, but we have to realize that in a large-dimensional problem, possibly involving a mix of binary and continuous variables, the geometry is expected to be weird.

We describe the basic ideas of branch-and-bound methods in Section 16.5, but, for now, it is sufficient to realize that their performance depends on the quality of the bounds on the optimal value, which we find by relaxing the feasible set to a polyhedron. In Example 15.5, we find an *upper* bound, with value 9, on the optimal value of the objective function, which is 3, by relaxing the integrality requirement, since we are maximizing it. Unfortunately, it is a rather weak and loose bound, in this specific case. In a minimization problem, we find a lower bound on the optimal value of the objective. In both cases, we find an optimistic estimate by solving the following **continuous (LP) relaxation** of problem $P(S)$ in Eq. (15.18):

$$P(\overline{S}) \qquad \min \quad \mathbf{c}^T\mathbf{x} + \mathbf{d}^T\mathbf{y} \qquad (15.21)$$
$$\text{s.t.} \quad \mathbf{Ax} + \mathbf{Ey} \leq \mathbf{b},$$
$$\begin{bmatrix} \mathbf{x} \\ \mathbf{y} \end{bmatrix} \in \mathbb{R}_+^{n_1+n_2}.$$

This LP problem features a relaxed feasible set $\overline{S}$, which includes the feasible set S of the MILP problem (15.18). Since $S \subseteq \overline{S}$, by solving the relaxed problem $P(\overline{S})$, we find a lower bound on the optimal value of the objective for $P(S)$.

In Fig. 15.7, we observe a huge gap between $\overline{S}$ and S. In Fig. 15.8, we depict a typical example for a pure integer LP. The feasible set consists of the bullets, which are contained by the outermost polyhedron $\overline{S}$. If we were able to find the smallest convex polyhedron containing S, corresponding to the shaded polyhedron, enclosed by the dotted line in Fig. 15.8, we would be done: The LP relaxation would provide us with an integer solution, as all of the vertices of the inside polyhedron have integer coordinates. Unfortunately, finding the inside polyhedron, which is called the **convex hull**[5] of S, is quite hard. If there is a large gap between the two sets, as in Example 15.5, bounds turns out to be weak and the search process is computationally expensive. It is also difficult

---

[5] The convex hull of a set $S \subset \mathbb{R}^n$ is the smallest convex set including S. In principle, it is the intersection of all convex sets including S. In slightly more concrete terms, this may be found by taking convex combinations of the points in S. If S is a discrete and finite set, its convex hull is a bounded polyhedron.

15.4 Integer programming

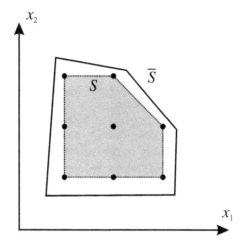

FIGURE 15.8 Integrality gap in a pure integer LP.

to find good approximations by rounding a nearly integer solution and carrying out a limited local search around it. However, we may at least find a model formulation bridging the gap as far as possible. There are two approaches to improve the tightness of a model formulation:

- We may formulate the model in a more careful way, by choosing alternative decision variables and rewriting constraints.
- We may resort to the automatic generation of *cuts*, i.e., additional constraints that cut some portion of the outer polyhedron without eliminating integer solutions. The deeper the cut, the closer we get to the convex hull. Cut generation is a common feature of state-of-the-art software packages.

Example 15.6 Disaggregated vs. aggregated constraints

Consider the four binary variables $x_0, x_1, x_2, x_3 \in \{0, 1\}$ and the constraints
$$3x_0 \leq x_1 + x_2 + x_3, \tag{15.22}$$
and
$$x_0 \leq x_1, \quad x_0 \leq x_2, \quad x_0 \leq x_3. \tag{15.23}$$
The two feasible sets of Eqs. (15.22) and (15.23) may look different but they are, in fact, the same. If we interpret the variables as the decision to start an activity, we are expressing the fact that we *may* start activity 0 only if all activities 1, 2, and 3 are started. If even one of the preconditions is not met, and the corresponding variable is set to 0, then the variable x_0 must be set to 0 as well.

> Since the two formulations are equivalent, common sense would suggest that having to do with one constraint is better than dealing with three of them. This intuition is wrong, and many MILP packages automatically transform the aggregate constraint (15.22) into the three disaggregated constraints (15.23). To see why, observe that the aggregated constraint is just the sum of the three constraints. Generally, when we add inequality constraints, we relax the feasible set. For instance, a point $\mathbf{x}^\circ$ satisfying the individual inequalities
>
> $$g_1(\mathbf{x}^\circ) \leq 0, \quad g_2(\mathbf{x}^\circ) \leq 0$$
>
> will certainly satisfy the aggregate inequality
>
> $$g_1(\mathbf{x}^\circ) + g_2(\mathbf{x}^\circ) \leq 0,$$
>
> but the converse is not true (the sum of negative numbers is negative, but we may get a negative number by summing a small positive one and a large negative one).
>
> In the present case, the integrality restriction has the effect that we do not really relax the constraints by aggregating them, in terms of integer feasible points; however, this will happen in the continuous LP relaxation. The result is that the integrality gap will increase.

The kind of reformulation of Example 15.6 can be performed manually, but state-of-the-art software may do this automatically. Leaving the task to the software may be a good choice, as only the relevant constraints will be added, without generating too many of them. In fact, in large-scale cases, the benefit of improving the bound may be overwhelmed by the number of additional constraints, which may also get into the way of rounding heuristics.

Example 15.7 Making the big-M smaller

> In TEV minimization, we have introduced the big-M constraint
>
> $$x \leq M\delta,$$
>
> linking the continuous variable $x \geq 0$ to the binary variable $\delta \in \{0, 1\}$. Here, M should be a suitably large constant, such that the constraint is practically ineffective when $\delta = 1$. Choosing a huge M is no harm, in principle, but it is a bad choice computationally. To see why, observe the geometry in Fig. 15.9. The feasible set consists of the origin (corresponding to $\delta = 0$), and a segment of a vertical line (corresponding to $\delta = 1$). However, the continuous relaxation

15.4 Integer programming

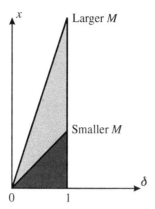

FIGURE 15.9 The effect of the big-M on the tightness of the continuous LP relaxation.

corresponds to a shaded triangle, below the line with slope M. The smaller the big-M, the tighter the LP bound.

Example 15.8 Cover cuts for a knapsack constraint

Let us consider the pure binary problem:

$$\begin{aligned}
\max \quad & 10x_1 + 7x_2 + 25x_3 + 24x_4 \\
\text{s.t.} \quad & 2x_1 + 1x_2 + 6x_3 + 5x_4 \leq 7, \\
& x_j \in \{0, 1\}.
\end{aligned}$$

This kind of model is known as knapsack problem. We have a set of items, with given weight and value. We want to find a subset of items with maximum value, without exceeding the weight capacity of the knapsack. The problem can be considered as a simple version of a capital budgeting problem, where investments are indivisible and we must select the best subset, subject to a single resource budget. If we relax the integrality condition to $x_j \in [0, 1]$ and solve the corresponding LP, we obtain

$$x_1 = 1, \quad x_2 = 1, \quad x_3 = 0, \quad x_4 = 0.8,$$

with objective value 36.2. This objective value is an *upper bound* on the optimal value. The solution is easy to interpret in the light of a possible greedy heuristic. We would like to choose items with a large value, but this must be traded off against the resource consumption.

Thus, we may compute a priority p_j for each item, by taking ratios of value and weight:

$$p_1 = \frac{10}{2} = 5, \quad p_2 = \frac{7}{1} = 7, \quad p_3 = \frac{25}{6} = 4.167, \quad p_4 = \frac{24}{5} = 4.8.$$

So, we may select the highest-priority items 2 and 1, but then only 80% of item 4 can be included. The actual solution we find includes items 2 and 1, providing a rather low value, 17, which is very far from the upper bound 36.2 and leaves a lot of budget unused. This rule does not guarantee optimality. In fact, it is easy to see that we are better off by selecting items 1 and 4, as this choice uses all of the budget and has value 34. Since the upper bound is 36.2, and all of the problem data are integer, we are sure that there is no way to find a solution with profit larger than 36. However, we cannot rule out the existence of a solution with a value larger than 34, yet, and we wonder whether the solution $\{1, 4\}$ is really optimal.

To generate some additional cuts, let us observe that items 1 and 3 cannot be both selected, as their total weight is 8, larger than the budget. The same applies to items 1, 2, and 4. Hence we may add the cuts

$$x_3 + x_4 \leq 1,$$
$$x_1 + x_2 + x_4 \leq 2.$$

They are obviously redundant in the discrete domain, but they are not in the continuous relaxation. This kind of cut is called a cover cut. By adding these two cover cuts, the LP relaxation yields a stronger bound, $34.66667 < 36.2$. The bound does show that the solution $\{1, 4\}$ is optimal.

Cover and other kinds of cuts are automatically generated by state-of-the-art commercial software for integer programming. This may improve performance considerably, and it also helps the application of clever heuristics that may generate high-quality integer solutions from almost integer ones. As we shall see in Section 16.5, this is another key ingredient in branch-and-bound methods.

15.5 Conic optimization

The classes of convex optimization models we have considered so far (LP, QP) are standard topics in optimization and operations research courses, and are

15.5 Conic optimization

also standard material for quantitative finance courses, even though their use in financial practice has raised some controversy. Robust and powerful solution software have been available for a while, and the difficulties in their application are mostly of a statistical nature. MILP problems are not convex, but even though they may be harder to solve to optimality, finding very good solutions is now feasible with a reasonable computational effort. As far as nonlinear programming is concerned, most algorithms available in the past were actually *local* optimizers. More recently, solvers originally developed for LP have been generalized to a wider class of convex problems, the family of conic optimization models. This family includes second-order cone programming (**SOCP**) models and semidefinite programming (**SDP**) models, which may play a relevant role in financial optimization under uncertainty and are the subject of a remarkable amount of research. Rather unsurprisingly, conic optimization relies on the concept of a cone, which we define in full generality as follows.

DEFINITION 15.5 (Cones in $\mathbb{R}^n$) *A set $C \subseteq \mathbb{R}^n$ is called a cone if, for every $\mathbf{x} \in C$ and $\lambda \geq 0$, we have $\lambda \mathbf{x} \in C$.*

We discuss cones in more detail, along with some examples, in Section 15.5.1, but we may immediately notice the following:

- The whole space $\mathbb{R}^n$ is a cone.
- The non-negative orthant $\mathbb{R}^n_+$ is a cone.
- The singleton $\{\mathbf{0}\}$, consisting of the origin of $\mathbb{R}^n$, is a cone.

As we shall see, we may consider cones in a more general setting, as subsets of a linear space.

It is also useful to get an immediate feeling for the relevance of cones, by taking a more abstract view of an LP model in standard form:

$$\min \quad \mathbf{c}^\mathsf{T}\mathbf{x} \qquad (15.24)$$
$$\text{s.t.} \quad \mathbf{A}\mathbf{x} = \mathbf{b}, \qquad (15.25)$$
$$\mathbf{x} \geq \mathbf{0}. \qquad (15.26)$$

The following observations allow us to cast LP within a wider class of convex problems:

- The objective function (15.24) involves the inner product of two vectors $\mathbf{c}, \mathbf{x} \in \mathbb{R}^n$. To generalize the concept, we may consider a finite-dimensional linear space V, equipped with an inner product $\langle \cdot, \cdot \rangle_V$ that maps pairs of elements in V into the real line $\mathbb{R}$. Hence, the minimization of the objective function may be written as

$$\min \quad \langle \mathbf{c}, \mathbf{x} \rangle_V.$$

- The equality constraint (15.25) involves a matrix $\mathbf{A} \in \mathbb{R}^{m \times n}$, mapping a vector in $\mathbb{R}^n$ into a vector in $\mathbb{R}^m$. If we identify $\mathbb{R}^n$ and $\mathbb{R}^m$ with linear spaces V and W, respectively, we may generalize $\mathbf{A}$ to a linear operator

$A: V \longrightarrow W$. The equality constraint may be thought as

$$A(\mathbf{x}) - \mathbf{b} = \mathbf{0}_W,$$

where $\mathbf{0}_W$ is the zero element of W. Since, as we have seen, the singleton $\{\mathbf{0}\}$ is a cone, we may rewrite the equality constraint in a more general and abstract form as

$$A(\mathbf{x}) - \mathbf{b} \in L,$$

where $L \equiv \{\mathbf{0}_W\} \subseteq W$ is a cone. We notice that this also applies to an LP model in canonical form, involving the inequality constraint $\mathbf{Ax} - \mathbf{b} \geq \mathbf{0}$, which may be written as

$$A(\mathbf{x}) - \mathbf{b} \in \mathbb{R}^n_+.$$

The non-negative orthant is a cone, too, playing the role of L in this case.

- Finally, the non-negativity condition (15.26) may be written in abstract form as

$$\mathbf{x} \in K,$$

where the cone K is just the non-negative orthant $\mathbb{R}^n_+$. If variables are unrestricted in sign, we take K as the whole space $\mathbb{R}^n$, which is a cone, too.

Putting all of this together, we define an abstract conic optimization problem as follows.

DEFINITION 15.6 (Conic optimization problem) *Let V and W be (finite-dimensional) linear spaces equipped with an inner product. Let $K \subseteq V$ and $L \subseteq W$ be closed convex cones. A conic optimization problem may be expressed as the following mathematical program:*

$$\begin{align}
\min \quad & \langle \mathbf{c}, \mathbf{x} \rangle_V \\
\text{s.t.} \quad & A(\mathbf{x}) - \mathbf{b} \in L, \\
& \mathbf{x} \in K,
\end{align}$$

where $\mathbf{c} \in V$ and $\mathbf{b} \in W$, and $A(\cdot)$ is linear mapping from V to W.

In this definition, cones are required to be convex. We consider convex cones in the next section.

15.5.1 CONVEX CONES

Definition 15.5 of a cone includes what we may naturally associate with the intuitive idea of a cone, and some more. As we have already pointed out, the whole space $\mathbb{R}^n$, the non-negative orthant $\mathbb{R}^n_+$, and the singleton $\{\mathbf{0}\}$ are cones. The cone in Fig. 15.10(a) corresponds to the natural view of a cone. Given any vector $\mathbf{x}$ in the shaded region, if we move along the positive direction given by

15.5 Conic optimization

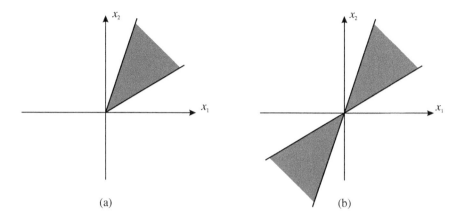

FIGURE 15.10 Two cones in $\mathbb{R}^2$.

$\mathbf{x}$, i.e., we consider $\lambda \mathbf{x}$, for $\lambda \geq 0$, we stay in the cone. However, the same applies to the cone in Fig. 15.10(b), which is also a cone, according to the definition, even if we would intuitively think of it as two joined cones. Figure 15.10 suggests that cones can be:

- **Pointed** or not: The cone of Fig. 15.10(b) is *not* pointed, as it includes a whole line, whereas a pointed cone does not include any line.
- **Convex** or not: The cone of Fig. 15.10(b) is not a convex set.
- **Closed** or open: A cone may include its boundary or not.
- **Polyhedral** or not.

The last feature deserves a few more comments. A polyhedral cone has a sort of pyramidal shape and may be visualized in a couple of ways:

1. As the intersection of half-spaces associated with hyperplanes passing through the origin.
2. As the conic hull of a finite set of vectors, $\mathbf{x}^{(1)}, \ldots, \mathbf{x}^{(m)}$, which are called the generators of the cone. The conic hull of a set is the set of points that can be generated as conic combinations of points in the set, i.e., linear combinations
$$\sum_{j=1}^{m} \lambda_j \mathbf{x}^{(j)},$$
with non-negative coefficients $\lambda_j \geq 0$.

A relevant example of nonpolyhedral cone is the Lorentz (or ice-cream) cone:
$$C = \left\{ (\mathbf{x}, t) \in \mathbb{R}^{n+1} : \|\mathbf{x}\|_2 \leq t \right\}.$$

This cone, displayed in Fig. 15.11, is really what we associate with the idea of a cone, and we may think of this cone as generated by an infinite number of generators. For each value of $t \geq 0$, we have a horizontal slice of the cone,

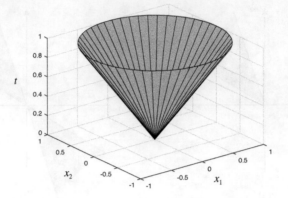

FIGURE 15.11 The second-order cone.

which is a circle of radius t. This is also called **second-order cone** and, as we shall see, is a particular case of a norm cone.

DEFINITION 15.7 (Norm cone) *Given a norm $\|\cdot\|$ on $\mathbb{R}^n$, we may define the norm cone*

$$K = \{(\mathbf{x}, t) \in \mathbb{R}^{n+1} : \|\mathbf{x}\| \leq t\}.$$

To see that the definition makes sense, we easily check that if $(\mathbf{x}, t) \in K$, then $\lambda(\mathbf{x}, t)$ is in the cone as well, since

$$\|\mathbf{x}\| \leq t \quad \Rightarrow \quad \|\lambda \mathbf{x}\| = \lambda \|\mathbf{x}\| \leq \lambda t,$$

where we use the condition $\lambda \geq 0$ and the properties of a norm.[6]

In optimization, we need to work with proper cones that are pointed, convex, and closed. The following example shows a less familiar and intuitive cone, which is not a subset of $\mathbb{R}^n$.

Example 15.9 The convex cone $\mathbb{S}_+^n$

> Let us consider the set $\mathbb{S}_+^n$ of **symmetric positive semidefinite matrices** $n \times n$. It is easy to see that this is a convex cone. Indeed, if $\mathbf{Q}_1, \mathbf{Q}_2 \in \mathbb{S}_+^n$, i.e., they satisfy the condition $\mathbf{x}^T \mathbf{Q}_i \mathbf{x} \geq 0$, for any $\mathbf{x}$ and $i = 1, 2$, then $\lambda_1 \mathbf{Q}_1 + \lambda_2 \mathbf{Q}_2 \in \mathbb{S}_+^n$, for any $\lambda_1, \lambda_2 \geq 0$. In particular, this holds when the non-negative weights add up to 1, which implies convexity. To streamline notation, we often write $\mathbf{Q} \succeq \mathbf{0}$ rather than $\mathbf{Q} \in \mathbb{S}_+^n$.

---

[6]A norm $\|\cdot\|$ is a mapping from a linear space into $\mathbb{R}_+$, satisfying the following properties: (a) $\|\mathbf{x}\| = 0$ implies $\mathbf{x} = \mathbf{0}$; (b) $\|\lambda \mathbf{x}\| = |\lambda| \cdot \|\mathbf{x}\|$, for any real number λ; (c) $\|\mathbf{x} + \mathbf{y}\| \leq \|\mathbf{x}\| + \|\mathbf{y}\|$. The last property is known as *triangular inequality*.

15.5 Conic optimization

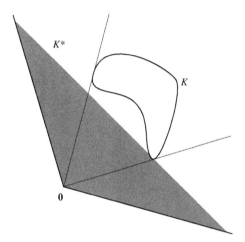

FIGURE 15.12 A generic set $K \subset \mathbb{R}^n$ and its dual cone K^*. Here, 0 is the origin of $\mathbb{R}^n$.

By using the concept of a cone, we may generalize inequalities as follows. Given two $n \times n$ matrices **A** and **B**, we may write the **generalized inequality**

$$\mathbf{A} \succeq \mathbf{B}$$

which should be read as

$$\mathbf{A} - \mathbf{B} \in \mathbb{S}^n_+.$$

More generally, given a cone K within a linear space V and two elements **X** and **Y** of V, we may write

$$\mathbf{A} \succeq_K \mathbf{B},$$

which means

$$\mathbf{A} - \mathbf{B} \in K.$$

15.5.1.1 Dual norms and dual cones

A key concept in optimization is duality, which we discuss later, in Section 16.1.4. In model building, we do not need a full understanding of duality, but there is a concept that proves useful, especially when applying conic optimization models to robust optimization.

DEFINITION 15.8 (Dual cone) *Given a subset $K \subset \mathbb{R}^n$, its* dual cone *is the set*

$$K^* = \{\mathbf{y} \in \mathbb{R}^n : \mathbf{y}^T\mathbf{x} \geq 0, \ \forall \mathbf{x} \in K\}$$

In plain English, the dual cone of set K is the set of vectors forming an acute angle (i.e., the inner product is non-negative) with all vectors in K. The idea is illustrated in Fig. 15.12. Note that the set K need not be a cone. If K is a cone and $K^* = K$, then we say that K is a **self-dual cone**.

Example 15.10 Trivial dual cones

It is easy to see that $(\mathbb{R}^n)^* = \{\mathbf{0}\}$ and $(\{\mathbf{0}\})^* = \mathbb{R}^n$. The non-negative orthant $\mathbb{R}^n_+$ is self-dual, i.e., $(\mathbb{R}^n_+)^* = \mathbb{R}^n_+$. To see this, let us first consider $\mathbf{y} \in \mathbb{R}^n_+$. If $\mathbf{y} \geq \mathbf{0}$, then we have $\mathbf{y}^T\mathbf{x} \geq 0$ for all $\mathbf{x} \in \mathbb{R}^n_+$, so $\mathbb{R}^n_+ \subseteq (\mathbb{R}^n_+)^*$. On the other hand, if we have a component $y_i < 0$, then $\mathbf{y}^T\mathbf{e}_i < 0$, where $\mathbf{e}_i$ is the ith unit vector. This implies that, if $\mathbf{y} \notin \mathbb{R}^n_+$, then $\mathbf{y} \notin (\mathbb{R}^n_+)^*$, so $(\mathbb{R}^n_+)^* \subseteq \mathbb{R}^n_+$, and the result follows.

Finding the dual cone may not be easy in general, but in the case of dual cones of norm cones, there is a general strategy based on dual norms.

DEFINITION 15.9 (Dual norm) *Given a dual norm $\|\cdot\|$, the corresponding dual norm $\|\cdot\|_*$ is defined as*

$$\|\mathbf{u}\|_* \doteq \max\{\mathbf{u}^T\mathbf{x} : \|\mathbf{x}\| \leq 1\}.$$

Example 15.11 The dual norms of L_1 and L_∞

Let us consider the L_∞ norm

$$\|\mathbf{x}\|_\infty \doteq \max_{i=1,\ldots,n} |x_i|.$$

To find its dual norm, we have to solve the problem

$$\max\{\mathbf{u}^T\mathbf{x} : \|\mathbf{x}\|_\infty \leq 1\}.$$

The objective function, in a more explicit form, is

$$\sum_{j=1}^n u_j x_j,$$

to be maximized under the condition that

$$|x_j| \leq 1, \qquad j = 1,\ldots,n.$$

Given the linearity of the objective and the independence of the constraints, the maximizer is clearly

$$x_j^* = \pm 1, \qquad j = 1,\ldots,n,$$

where the sign is the same as the sign of the corresponding component u_j, so that their product is positive. Thus, the optimal value is

$$\sum_{i=1}^n |u_i| = \|\mathbf{u}\|_1,$$

> showing that the dual of the L_∞ norm is the L_1 norm. By a similar token, the dual of the L_1 norm is the L_∞ norm (see Problem 15.3).

Example 15.12 *The Euclidean norm is self-dual*

> To find the dual of the L_2 norm we have to solve the problem
>
> $$\max\{\mathbf{u}^\top\mathbf{x} : \|\mathbf{x}\|_2 \leq 1\},$$
>
> which requires to find a vector $\mathbf{x}$ on the unit circle, with maximal projection on $\mathbf{u}$. Clearly, the optimal solution $\mathbf{x}^*$ must be a unit norm vector, parallel to $\mathbf{u}$. Hence,
>
> $$\mathbf{x}^* = \frac{\mathbf{u}}{\|\mathbf{u}\|_2} \;\Rightarrow\; \mathbf{u}^\top \cdot \frac{\mathbf{u}}{\|\mathbf{u}\|_2} = \frac{\|\mathbf{u}\|_2^2}{\|\mathbf{u}\|_2} = \|\mathbf{u}\|_2.$$
>
> Thus, we conclude that the Euclidean norm L_2 is self-dual.

Given a norm cone, and the dual of the involved norm, the following result shows how to find the dual cone.

THEOREM 15.10 (Dual of a norm cone) *Let $\|\cdot\|$ be a norm on $\mathbb{R}^n$, with dual norm $\|\cdot\|_*$. Given the norm cone $K \subseteq \mathbb{R}^{n+1}$ defined by norm $\|\cdot\|$, its dual cone is*

$$K^* = \{(\mathbf{u}, v) \in \mathbb{R}^{n+1} : \|\mathbf{u}\|_* \leq v\}.$$

Why are dual cones relevant? On the one hand, they are relevant in duality theory for conic optimization problems, which in turn provides us with the necessary theoretical background for efficient methods to solve conic programs. On the other hand, they allow us to write an infinite set of constraints in a compact and convenient way (see Example 15.13). Indeed, Definition 15.8 states that a vector is in a dual cone if it satisfies an infinite set of inequalities.

15.5.1.2 Matrix inner product and the dual cone of $\mathbb{S}_+^n$

If we want to apply the concepts that we have just introduced to the cone of positive semidefinite matrices, we need to define an inner product and a norm for matrices. This is essential in semidefinite programming but, since we do not consider SDP in much detail in this book, we just give an intuitive glimpse of the relevant concepts.

One way to define an inner product between two $m \times n$ matrices is the **Frobenius product**,

$$\langle \mathbf{A}, \mathbf{B} \rangle_F \doteq \sum_{i,j=1}^{n} a_{ij} b_{ij} = \operatorname{tr}(\mathbf{A}^\mathsf{T} \mathbf{B}), \tag{15.27}$$

where the trace of a square matrix $\mathbf{C} \in \mathbb{R}^{n \times n}$ is defined as[7]

$$\operatorname{tr}(\mathbf{C}) \doteq \sum_{i=1}^{n} c_{ii},$$

i.e., the sum of elements on the diagonal. The transposition in Eq. (15.27) is needed to make the matrices conform, and the Frobenius product is commutative, since we get the same result by taking $\operatorname{tr}(\mathbf{B}^\mathsf{T} \mathbf{A})$. In fact, this inner product may be regarded as the standard inner product for vectors, applied to "vectorized" matrices, obtained by stacking matrix columns.

If we consider square and symmetric matrices in the set $\mathbb{S}^n$, the above inner products may be slightly simplified, as we do not need transposition:

$$\langle \mathbf{A}, \mathbf{B} \rangle_F = \operatorname{tr}(\mathbf{A}\mathbf{B}) = \sum_{i,j=1}^{n} a_{ij} b_{ij}.$$

The notation

$$\mathbf{A} \bullet \mathbf{B}$$

is often used to denote inner product between matrices when stating SDP models.

We have observed that the set $\mathbb{S}^n_+$ is a cone. A notable fact is that this cone is self-dual, i.e., $(\mathbb{S}^n_+)^* = \mathbb{S}^n_+$. This is a consequence of the following fact, which we state without proof:

$$\operatorname{tr}(\mathbf{A}\mathbf{B}) \geq 0, \; \forall \mathbf{A} \succeq \mathbf{0} \iff \mathbf{B} \succeq \mathbf{0}.$$

The essential message is that some formal properties and concepts that apply to vectors may be extended to matrices, which paves the way to solve SDPs.

15.5.2 SECOND-ORDER CONE PROGRAMMING

Second-order cone programming (SOCP) problems are a generalization of LP problems, whereby we require that the vector of decision variables is constrained by a second-order cone, modulo an affine transformation. As we have seen, the Lorentz cone defined by the inequality

$$\|\mathbf{x}\|_2 \leq t$$

---

[7] Among the properties of matrix trace, we only recall that $\operatorname{tr}(\mathbf{A}\mathbf{B}) = \operatorname{tr}(\mathbf{B}\mathbf{A})$, assuming that $\mathbf{A}$ and $\mathbf{B}$ are conformable matrices, so that the products are well defined.

15.5 Conic optimization

is a convex cone in $\mathbb{R}^{n+1}$. Since convexity is preserved by an affine mapping,[8] we may generalize the above inequality by considering the following affine mapping from $\mathbb{R}^n$ to $\mathbb{R}^{k+1}$:

$$\mathbf{x} \longrightarrow \begin{bmatrix} \mathbf{Ax} + \mathbf{b} \\ \mathbf{c}^\mathsf{T}\mathbf{x} + d \end{bmatrix},$$

where $\mathbf{A} \in \mathbb{R}^{k \times n}$, $\mathbf{b} \in \mathbb{R}^k$, $\mathbf{c} \in \mathbb{R}^n$, and $d \in \mathbb{R}$. Then, we require that the mapped point lies in the second-order (Lorentz) cone in $\mathbb{R}^{k+1}$:

$$\|\mathbf{Ax} + \mathbf{b}\|_2 \leq \mathbf{c}^\mathsf{T}\mathbf{x} + d.$$

This kind of second-order cone constraint defines a convex set. It is important to realize that we should *not* square the constraint in order to get a QCQP problem, as the resulting quadratic form need not be convex. Since the intersection of convex sets is convex, we may enforce multiple second-order cone constraints. A general statement of an SOCP model is

$$\begin{aligned} \min \quad & \mathbf{f}^\mathsf{T}\mathbf{x} \\ \text{s.t.} \quad & \|\mathbf{A}_i\mathbf{x} + \mathbf{b}_i\|_2 \leq \mathbf{c}_i^\mathsf{T}\mathbf{x} + d_i, \quad i = 1, \ldots, n \\ & \mathbf{Fx} = \mathbf{g}, \end{aligned} \qquad (15.28)$$

where we see that we may also include linear constraints. A non-negativity constraint $\mathbf{x} \geq \mathbf{0}$, by a proper setting of parameters, may be expressed as a second-order constraint like (15.28), showing that SOCP is a generalization of LP. SOCP problems may be solved by interior-point methods, which we will not consider in detail in this book. Our aim is just to illustrate their use in optimization modeling.

Example 15.13 Robust constraints in LPs

Consider an uncertain constraint $\mathbf{a}^\mathsf{T}\mathbf{x} \leq b$, where

$$\mathbf{a} = \mathbf{a}_0 + \mathbf{u}, \quad \|\mathbf{u}\|_\infty \leq 1.$$

In other words, we have a linear inequality where the data vector $\mathbf{a}$ is uncertain, but rather than giving a stochastic characterization, we assume that it lies within an uncertainty set. The nominal value is $\mathbf{a}_0$, and the uncertainty set is a simple box around the nominal value. The feasible set is the collection of points $\mathbf{x}$, such that the inequality is satisfied for every $\mathbf{a}$ in the uncertainty set. This kind of constraint, as we shall see later, is common in robust optimization, and it can be generalized to data matrices; see Section 14.4.1 for its relationship with regularized regression.

---

[8] See [7].

The difficulty is that we have to cope with an *infinite* collection of inequalities:

$$a_0^T x + u^T x \leq b, \quad \forall u : \|u\|_\infty \leq 1. \tag{15.29}$$

Apparently, we have to solve a difficult semi-infinite optimization problem, but we may find a nice reformulation based on the dual norm of L_∞. Since constraint (15.29) must hold in the worst case, we may rewrite it as follows:

$$a_0^T x + \left(\max_{\|u\|_\infty \leq 1} u^T x \right) \leq b.$$

However, the maximization within the parentheses is just the definition of the dual norm of L_∞, which is L_1. Thus, the uncertain constraint is equivalent to the LP constraint

$$a_0^T x + \|x\|_1 \leq b.$$

We see that, in the case of box uncertainty, an uncertain linear constraint may be rewritten as a linear constraint. What if we assume an ellipsoidal uncertainty set for a? As a clue, let us consider a simple circle with unit radius centered on the nominal data a_0, i.e.,

$$a = a_0 + u, \quad \|u\|_2 \leq 1.$$

This uncertainty set is related to the norm L_2, which is self-dual. Hence, the uncertain constraint may be expressed as

$$a_0^T x + \|x\|_2 \leq b,$$

which is a second-order cone constraint.

Example 15.14 A chance constraint in LP

Let us consider a random linear constraint $a^T x \leq b$, where $a \sim N(\mu, \Sigma)$, and say that we require $P\{a^T x \leq b\} \geq \eta$. This kind of constraint is called *chance constraint*, and it enforces a probabilistic guarantee, with reliability η, that an uncertain constraint is satisfied. For a given vector x, we have $a^T x \sim N(\nu, \sigma^2)$, where $\nu = \mu^T x$ and $\sigma^2 = x^T \Sigma x$. Using the standard normal CDF $\Phi(z)$, we may rewrite

15.5 Conic optimization

> the constraint as
>
> $$P\left\{\frac{\mathbf{a}^T\mathbf{x} - \nu}{\sigma} \leq \frac{b - \nu}{\sigma}\right\} = \Phi\left(\frac{b - \nu}{\sigma}\right) \geq \eta$$
>
> $$\iff \frac{b - \nu}{\sigma} \geq \Phi^{-1}(\eta).$$
>
> Assuming $\eta > 0.5$, as it is sensible to do, so that $\Phi^{-1}(\eta) > 0$, this is a second-order cone constraint:
>
> $$\boldsymbol{\mu}^T\mathbf{x} + \Phi^{-1}(\eta) \|\boldsymbol{\Sigma}^{1/2}\mathbf{x}\|_2 \leq b,$$
>
> where $\boldsymbol{\Sigma}^{1/2}$ is the square root of the covariance matrix, i.e., a symmetric matrix such that $\boldsymbol{\Sigma}^{1/2}\boldsymbol{\Sigma}^{1/2} = \boldsymbol{\Sigma}$. Hence, an LP with *disjoint* (individual) chance constraints of this kind, under a normality assumption, is a convex SOCP.

We shall further discuss chance constraints in Section 15.6.1, and more general uncertainty sets in Section 15.9.

15.5.3 SEMIDEFINITE PROGRAMMING

In semidefinite programming (SDP) we deal with symmetric semidefinite positive matrices $\mathbf{X}$ in the convex cone $\mathbb{S}_+^n$. There are two ways of stating an SDP model.

1. In the **inequality form**, we rely on a linear matrix inequality (LMI):

$$\min \ \mathbf{c}^T\mathbf{x} \quad (15.30)$$
$$\text{s.t.} \ x_1\mathbf{F}_1 + \cdots + x_n\mathbf{F}_n + \mathbf{G} \preceq \mathbf{0},$$
$$\mathbf{A}\mathbf{x} = \mathbf{b},$$

   where $\mathbf{F}_1, \ldots, \mathbf{F}_n, \mathbf{G} \in \mathbb{S}_k$ are symmetric matrices in $\mathbb{R}^{k \times k}$, and $\mathbf{x} \in \mathbb{R}^n$ is a vector with elements $x_1, \ldots, x_n$. In this form, we do not observe a matrix variable explicitly, but the scalar decision variables must generate a matrix that is negative semidefinite (of course we may reverse the matrix inequality to require positive semidefiniteness). Note that $\mathbf{A} \preceq \mathbf{0}$ means $-\mathbf{A} \succeq \mathbf{0}$, i.e., $-\mathbf{A} \in \mathbb{S}_+^n$.

2. Alternatively, we rely on the matrix inner product and formulate the model in **standard form**:

$$\max \ \mathbf{C} \bullet \mathbf{X} \quad (15.31)$$

$$\text{s.t.} \quad \mathbf{A}_1 \bullet \mathbf{X} = b_1,$$
$$\vdots$$
$$\mathbf{A}_m \bullet \mathbf{X} = b_m,$$
$$\mathbf{X} \succeq \mathbf{0}.$$

It turns out that the forms (15.30) and (15.31) are equivalent, since they are related by conic duality. SDPs arise, among other things, in certain robust optimization problems with ellipsoidal uncertainty. The mathematics involved is a bit out of the scope of this book, so we will not consider these cases. Nevertheless, we can show a simple example, where we may appreciate the power of SDP model formulations. The motivation behind the following toy example is that covariance and correlation matrices may be subject to estimation uncertainty, resulting in a distributional ambiguity, e.g., in portfolio optimization. Anyway, they are required to stay positive semidefinite.

Example 15.15 **The most negative correlation coefficient**

Consider a correlation matrix for asset returns. Correlation coefficients may be positive or negative, but common sense suggests that cycles of negative correlation coefficients may be problematic. For instance, say that random returns $\tilde{r}_1$ and $\tilde{r}_2$ are negatively correlated, and $\tilde{r}_2$ and $\tilde{r}_3$ are, too. Can $\tilde{r}_1$ and $\tilde{r}_3$ be negatively correlated as well? Maybe yes, but only up to a point. Let us consider a fictitious correlation matrix involving a single correlation coefficient $-\rho$:

$$\mathbf{R} = \begin{bmatrix} 1 & -\rho & -\rho & \cdots & -\rho \\ -\rho & 1 & -\rho & \cdots & -\rho \\ \vdots & \vdots & \vdots & \ddots & \vdots \\ -\rho & -\rho & -\rho & \cdots & 1 \end{bmatrix}.$$

In this correlation matrix, all of the elements on the diagonal are equal to 1, since they correspond to correlations of a variable with itself, and we assume that the correlation is always the same for any pair of variables (which is not quite realistic) What is the largest value of ρ, such that $\mathbf{R}$ stays in $\mathbb{S}_+^n$? This problem may be stated as an SDP:

$$\max \quad \rho \qquad (15.32)$$
$$\text{s.t.} \quad \rho \mathbf{F} - \mathbf{I} \preceq \mathbf{0}, \qquad (15.33)$$

where $\mathbf{I}$ is the identity matrix and

$$\mathbf{F} = \begin{bmatrix} 0 & 1 & 1 & \cdots & 1 \\ 1 & 0 & 1 & \cdots & 1 \\ \vdots & \vdots & \vdots & \ddots & \\ 1 & 1 & 1 & \cdots & 0 \end{bmatrix} = \mathbf{i}\mathbf{i}^\mathsf{T} - \mathbf{I}.$$

> Here, we use **i** to denote a vector with all elements set to 1 as usual. When $n = 3$, we find $\rho = \frac{1}{2}$, and when $n = 4$, we find $\rho = \frac{1}{4}$. For these values of ρ, the matrix **R** is singular.

15.6 Stochastic optimization

Uncertainty is a feature that may not be ignored in most real-life decision problems, but in finance it is really crucial. If we consider mean–variance optimization, uncertainty does not pose severe difficulties, at least, if we disregard statistical estimation issues. We only consider the first two moments of the random asset returns, which are collected into the vector of expected returns and the variance–covariance matrix, boiling down the problem to a simple convex QP model. However, we may wish to consider alternative risk measures, like conditional value-at-risk (CV@R), or a dynamic model of uncertainty, or a better model of uncertainty that cannot be summarized by the mean–variance inputs. Hence, we may need the ability to introduce uncertainty in a more explicit and direct way into an optimization model.

As a starting point, let us consider how we might introduce uncertainty into an LP model written in standard form,

$$\begin{aligned} \min \quad & \mathbf{c}^\mathsf{T}\mathbf{x} \\ \text{s.t.} \quad & \mathbf{A}\mathbf{x} = \mathbf{b}, \\ & \mathbf{x} \geq \mathbf{0}. \end{aligned}$$

If we assume that the problem data are stochastic, we could represent them by random variables and consider the following stochastic model:[9]

$$\begin{aligned} \min \quad & \mathbf{c}(\omega)^\mathsf{T}\mathbf{x} \\ \text{s.t.} \quad & \mathbf{A}(\omega)\mathbf{x} = \mathbf{b}(\omega), \quad \text{a.s.} \\ & \mathbf{x} \geq \mathbf{0}. \end{aligned}$$

Unfortunately, stated as such, this model is not even posed in a sensible way. To begin with, we cannot minimize a random objective. By selecting **x**, we do not choose a numerical value of the objective function, but only its probability distribution. Anyway, we could rank probability distributions by considering the expected value of the linear objective, a nonlinear expected utility, or a

---

[9] The notation "a.s." stands for *almost surely*, which means that the constraints are satisfied with probability 1, i.e., with the possible exception of events with zero probability measure. If we model uncertainty using a discrete set of scenarios, so that the sample space Ω is finite, we may write $\forall \omega \in \Omega$. With continuous random variables, some technical issues may arise, which we will ignore.

mean–risk objective, as we have discussed in Chapter 7. What we did not really consider so far is the *feasibility* of the solution. Can we find any **x** such that the equality constraints are satisfied for *every* realization of the random data? Clearly, this is not feasible in practice, and the constraints must be relaxed in some way. The case of inequality constraints is a bit easier. Requiring

$$\mathbf{A}(\omega)\mathbf{x} \geq \mathbf{b}(\omega), \quad \text{a.s.}$$

might lead to a feasible solution. However, this *fat* solution could well be overly expensive. Therefore, we must find some more clever way to formulate constraints under uncertainty. There are two standard approaches to deal with feasibility in stochastic optimization:

- To allow for a limited probability of violating constraints
- To introduce *recourse* actions, i.e., decisions that are made after the realization of uncertainty, in order to restore feasibility

The first approach leads to chance-constrained models. The second one leads to stochastic programming with recourse.

15.6.1 CHANCE-CONSTRAINED LP MODELS

One possibility to build a sensible stochastic optimization model is to settle for a solution with a probabilistic guarantee of feasibility.

- We may require that the whole set of constraints is satisfied with a suitably large probability, i.e.,

$$\mathrm{P}\left\{\mathbf{A}(\omega)\mathbf{x} \geq \mathbf{b}(\omega)\right\} \geq 1 - \alpha,$$

for a suitably small α. In this case, we have a **joint chance constraint**.

- We may require that each individual constraint is satisfied with sufficient reliability, i.e.,

$$\mathrm{P}\left\{\mathbf{a}_i^\mathsf{T}(\omega)\mathbf{x} \geq b_i(\omega)\right\} \geq 1 - \alpha_i, \quad i = 1, \ldots, m.$$

In this case, we have a set of **individual chance constraints**.

As a rule, individual chance-constraints are easier to deal with. This leads to chance-constrained models, which do have a sensible interpretation in terms of solution reliability. For instance, in portfolio optimization, we may set a target return, along with a small chance of not achieving it. In an ALM problem, we might allow for the possibility of not covering the liabilities in every scenario. The idea sounds natural and quite appealing, but it raises some serious technical and nontechnical issues.

We have seen in Example 15.14 that some chance constraints may be easily dealt with by rewriting them as a convex constraint. Unfortunately, this is not the rule, as the chance-constrained approach may lead to nonconvex optimization problems. The essential reason for this nonconvexity is that the union of convex sets need not be convex. However, it is sometimes possible to find convex approximations of chance-constrained problems, e.g., by resorting to robust optimization models. Apart from these technical issues, chance-constrained models have definite limitations from the financial viewpoint:

- They do not consider the flow of information and the sequence of decision stages. In real life, we plan things ahead of the realization of random scenarios, but we allow for the possibility of adjusting our decisions. This dynamic decision process is not considered in chance-constrained models.
- They can lead to quite risky solutions. This is a consideration related to critical remarks on value-at-risk as a risk measure. An unpleasing event may be quite unlikely, but can we ignore its consequences? We are not considering at all what disaster may really occur, even though its probability is very small. If we miss liabilities, we would like to know by how much. This issue is clearly related to the reasons why value-at-risk might be complemented by conditional value-at-risk.
- Last but not least, even if we consider a very small α to make the solution robust, can we really trust our ability to estimate very small probabilities under distributional ambiguity?

Because of these difficulties, we do not treat chance-constrained optimization in detail.

15.6.2 TWO-STAGE STOCHASTIC LINEAR PROGRAMMING WITH RECOURSE

In optimization models with recourse, we take advantage of the dynamic flow of information and distinguish decision stages. In Fig. 15.13, we depict the simplest case, where a two-stage decision process is followed and uncertainty is modeled by a scenario fan. At the first stage, corresponding to the root of the tree, we take here-and-now, immediate decisions. At the second stage, on the leaves of the tree, recourse actions are taken, after observing the realized random variables. We may formalize a two-stage LP model with recourse as follows:

- The **first-stage decision** $\mathbf{x} \geq \mathbf{0}$ must satisfy immediate constraints $\mathbf{A}\mathbf{x} = \mathbf{b}$ and incur an immediate (first-stage) cost $\mathbf{c}^\mathsf{T}\mathbf{x}$.
- At the second stage, a random scenario ω occurs, associated with random data. Given this information, a **second-stage decision** (recourse action) $\mathbf{y}(\omega) \geq \mathbf{0}$ is made.
- The second-stage decisions are related to the first-stage decision by inter-stage constraints, like $\mathbf{W}\mathbf{y}(\omega) + \mathbf{T}(\omega)\mathbf{x} = \mathbf{h}(\omega)$.
- The second stage decision incurs a cost $\mathbf{q}(\omega)^\mathsf{T}\mathbf{y}(\omega)$, which is random from the viewpoint of the root node.

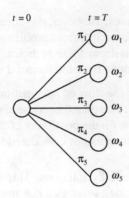

FIGURE 15.13 A scenario fan for two-stage stochastic programming.

- The overall objective is to minimize the sum of the first-stage cost and the expected value of second-stage cost.

Thus, we obtain the following optimization model:

$$\min \quad c^T x + E_\omega \left[q(\omega)^T y(\omega) \right]$$
$$\text{s.t.} \quad Ax = b,$$
$$\quad Wy(\omega) + T(\omega)x = h(\omega), \quad \text{a.s.}$$
$$\quad x, y(\omega) \geq 0.$$

When the **recourse matrix** W is deterministic, as above, we have a *fixed recourse* problem. The more general case with a random recourse matrix $W(\omega)$ may present additional difficulties, which we will ignore in this book.

The model seems to make sense, but it is not quite clear how it could be solved. In particular, the formulation involves an expectation, which is a multidimensional integral, when the underlying data are described by a continuous multivariate distribution. As the reader may imagine, a simple computational approach is to discretize uncertainty by a finite set of scenarios. Before doing so, in order to shed more light on the nature of stochastic programming with recourse, we may recast the model in different way. Let us introduce a recourse function $\mathcal{Q}(x)$ and rewrite the model as the following deterministic equivalent:

$$\min \quad c^T x + \mathcal{Q}(x)$$
$$\text{s.t.} \quad Ax = b,$$
$$\quad x \geq 0,$$

where we define the **recourse function**

$$\mathcal{Q}(x) \doteq E_{\boldsymbol{\xi}} \left[Q(x, \boldsymbol{\xi}(\omega)) \right],$$

and

$$Q(x, \boldsymbol{\xi}(\omega)) \doteq \min_y \left\{ q(\boldsymbol{\xi}(\omega))^T y \mid Wy = h(\boldsymbol{\xi}(\omega)) - T(\boldsymbol{\xi}(\omega))x, \ y \geq 0 \right\}.$$

15.6 Stochastic optimization

Here, we make the presence of random variables (risk factors) $\boldsymbol{\xi}(\omega)$ explicit. This formulation shows that stochastic linear programming is, in general, a nonlinear programming problem. In fact the recourse function $\mathcal{Q}(\mathbf{x})$ is nonlinear, in general, and looks like a "hopeless function:"

- It is an expectation with respect to the joint distribution of $\boldsymbol{\xi}(\omega)$; hence, it is a multidimensional integral, if random variables are continuous.
- It is a multidimensional integral of a function that we do not really know, as it is implicitly defined by an optimization problem.

Luckily, in many cases of practical interest, we can prove interesting properties of the recourse function, most notably convexity. In some cases, $\mathcal{Q}(\mathbf{x})$ is differentiable; in other cases it is polyhedral. This does not imply that it is easy to evaluate the recourse function, but we may resort to statistical sampling (scenario generation) and take advantage of both convexity and problem structure. Another advantage of this view is that it makes the transition to multistage stochastic programming and stochastic dynamic programming more natural. We make a decision $\mathbf{x}$ here and now, but we cannot only focus on its immediate cost: The future cost-to-go must be accounted for, which is the purpose of the recourse function.

For computational purposes, we may represent uncertainty by a discrete probability distribution, resulting in a scenario tree (fan), where ω_s is the outcome corresponding to scenario s.[10] Each scenario $s \in \mathcal{S}$ is associated with a probability π_s and a set of scenario-dependent random data, as shown in the following LP:

$$\begin{aligned}
\min \quad & \mathbf{c}^\mathsf{T}\mathbf{x} + \sum_{s \in \mathcal{S}} \pi_s \mathbf{q}_s^\mathsf{T} \mathbf{y}_s \\
\text{s.t.} \quad & \mathbf{A}\mathbf{x} = \mathbf{b}, \\
& \mathbf{W}\mathbf{y}_s + \mathbf{T}_s\mathbf{x} = \mathbf{h}_s, \quad \forall s \in \mathcal{S} \\
& \mathbf{x}, \mathbf{y}_s \geq \mathbf{0}.
\end{aligned}$$

This is a plain LP, even though a possibly large-scale one. Note that first-stage decisions $\mathbf{x}$ do not depend on the scenarios, whereas second-stage decisions $\mathbf{y}_s$ are subscripted by $s \in \mathcal{S}$. Since $\mathcal{S}$ is a finite set, we can do without the "almost surely" technicality and enforce the interstage constraints for all of the scenarios.

15.6.2.1 An example: CV@R optimization

We have introduced conditional value-at-risk (CV@R) in Section 7.4.3.3, as the expected loss conditional on the loss being larger than V@R$_{1-\alpha}$, i.e., value-at-risk at confidence level $1 - \alpha$. We leave the holding horizon implicit. Note that we use α to denote the small probability of the "bad" tail, associated with large

---

[10] We have outlined basic scenario generation concepts in Section 11.6.

losses. Thus, if we consider V@R at level 95%, $\alpha = 0.05$. We have mentioned that CV@R is a convex (and coherent) risk measure. We might consider a mean–risk portfolio optimization model where CV@R replaces variance (or, rather, volatility) as a risk measure. Given the role of convexity in optimization, one might expect that the resulting model might not be that bad, but there is no obvious way to express it. The following result[11] provides us with a surprisingly simple answer.

THEOREM 15.11 (Minimization of CV@R) *Let $L(\mathbf{x}, \mathbf{Y})$ be a loss or cost function, depending on a vector of decision variables $\mathbf{x}$ and a vector of random variables $\mathbf{Y}$, with joint density $f_{\mathbf{Y}}(\mathbf{y})$. Let us define the following function:*

$$H_{1-\alpha}(\mathbf{x}, \zeta) = \zeta + \frac{1}{\alpha} \int \left[L(\mathbf{x}, \mathbf{y}) - \zeta\right]^+ f_{\mathbf{Y}}(\mathbf{y}) \, d\mathbf{y}, \qquad (15.34)$$

where $[z]^+ \doteq \max\{z, 0\}$ and $\zeta \in \mathbb{R}$ is an auxiliary variable.

1. *For a given choice of $\mathbf{x}$, the corresponding CV@R is obtained by minimizing $H_{1-\alpha}(\mathbf{x}, \zeta)$ with respect to ζ. The resulting value of the auxiliary variable ζ is the corresponding V@R$_{1-\alpha}$.*
2. *Minimization of CV@R at confidence level $1 - \alpha$ is accomplished by the minimization of $H_{1-\alpha}(\mathbf{x}, \zeta)$ with respect to both of its arguments.*

We shall not prove the theorem, but we may get a clue about it in the case of discrete scenarios, under a simplifying assumption.[12] To be more specific, in our setting, the decision variables $\mathbf{x}$ may correspond to the allocation of an initial wealth W_0 to a set of n assets, indexed by $i = 1, \ldots, n$. If we denote the initial asset prices by P_{i0}, and ignore transaction costs, we have

$$\sum_{i=1}^n P_{i0} x_i = W_0.$$

The role of the random variables $\mathbf{Y}$ is played by the asset prices at the end of the holding period T. So, let P_{iT}^s denote the future price of asset i in scenario $s \in \mathcal{S}$, associated with probability π^s. Then, the terminal wealth in scenario s is

$$W_T(\mathbf{x}; s) = \sum_{i=1}^n P_{iT}^s x_i,$$

and loss in scenario s is

$$L(\mathbf{x}; s) = W_0 - W_T(\mathbf{x}; s).$$

When terminal wealth is larger than the initial wealth, we have a profit, i.e., a negative loss. Letting $\zeta \equiv \text{V@R}_{1-\alpha}(\mathbf{x})$, we find

$$\text{CV@R}_{1-\alpha}(\mathbf{x}) = \mathrm{E}\bigl[L(\mathbf{x}; \cdot) \,|\, L(\mathbf{x}; \cdot) > \zeta\bigr], \qquad (15.35)$$

---

[11] See [30, 31] for a proof and a more careful treatment.
[12] Here we follow the treatment by [37, pp. 125–127].

15.6 Stochastic optimization

where, with some abuse of notation, we let $L(\mathbf{x};\cdot)$ denote the random loss (rather than its realization in a specific scenario $s \in \mathcal{S}$). Also note that we emphasize the dependence of the risk measure on the portfolio choice $\mathbf{x}$.

Remark. When considering discrete distributions, a technical difficulty arises, related to "splitting atoms." In general, if we sum the probabilities of scenarios in a subset of $\mathcal{S}$, we may not be able to match probabilities $1-\alpha$ and α exactly. Furthermore, in the same discrete setting, using the strict inequality $>$ rather than $\geq$ makes a difference, and we are led to slightly different definitions of risk measures.[13] We will disregard these difficulties and take for granted that we may obtain $1-\alpha$ and α exactly by summing probabilities. This is true in the case that we generate a fairly large number scenarios by Monte Carlo sampling, so that the probabilities are uniform, and we consider probability levels like 95% or 99%, without too many decimals.

Given our assumption, for every portfolio choice $\mathbf{x}$, we may consider the subset $\mathcal{B}(\mathbf{x}) \subset \mathcal{S}$ of bad scenarios where loss exceeds ζ:

$$\mathcal{B}(\mathbf{x}) \doteq \{s \in \mathcal{S} \mid L(\mathbf{x};s) > \zeta\},$$

where, by the definition of $\text{V@R}_{1-\alpha}$,

$$\sum_{s \in \mathcal{B}(\mathbf{x})} \pi^s = \alpha.$$

We can rewrite the conditional expectation of Eq. (15.35) as

$$\text{CV@R}_{1-\alpha}(\mathbf{x}) = \frac{\sum_{s \in \mathcal{B}(\mathbf{x})} \pi^s L(\mathbf{x};s)}{\sum_{s \in \mathcal{B}(\mathbf{x})} \pi^s} = \frac{1}{\alpha} \sum_{s \in \mathcal{B}(\mathbf{x})} \pi^s L(\mathbf{x};s). \qquad (15.36)$$

Let us introduce an auxiliary variable

$$z^s \doteq \max\{0, L(\mathbf{x};s) - \zeta\}, \qquad s \in \mathcal{S}, \qquad (15.37)$$

representing the excess loss with respect to $\text{V@R}_{1-\alpha}$, which is zero if loss does not exceed the threshold ζ. Then, we may write

$$\sum_{s \in \mathcal{S}} \pi^s z^s = \sum_{s \notin \mathcal{B}(\mathbf{x})} \pi^s z^s + \sum_{s \in \mathcal{B}(\mathbf{x})} \pi^s z^s$$

$$= 0 + \sum_{s \in \mathcal{B}(\mathbf{x})} \pi^s \cdot \big[L(\mathbf{x};s) - \zeta\big]$$

$$= \sum_{s \in \mathcal{B}(\mathbf{x})} \pi^s \cdot L(\mathbf{x};s) - \zeta \cdot \sum_{s \in \mathcal{B}(\mathbf{x})} \pi^s$$

$$= \sum_{s \in \mathcal{B}(\mathbf{x})} \pi^s \cdot L(\mathbf{x};s) - \zeta \cdot \alpha. \qquad (15.38)$$

---

[13] See, e.g., [12].

Finally, we may rearrange Eq. (15.38) as

$$\zeta + \frac{1}{\alpha}\sum_{s\in\mathcal{S}}\pi^s z^s = \frac{1}{\alpha}\sum_{s\in\mathcal{B}(\mathbf{x})}\pi^s \cdot L(\mathbf{x};s),$$

where the last expression, by Eq. (15.36), is just $\text{CV@R}_{1-\alpha}(\mathbf{x})$. This may be regarded as a discretized version of Eq. (15.34), and it is justified by Theorem 15.11.

Now, let us put all of this together and formulate a computationally viable portfolio optimization model. Given the discretized price scenarios, we define the expected future price of asset i as

$$\overline{P}_{iT} \doteq \sum_{s\in\mathcal{S}}\pi^s P_{iT}^s,$$

and write the following mean–risk model:

$$\min \quad \zeta + \frac{1}{\alpha}\sum_{s\in\mathcal{S}}\pi^s z^s \tag{15.39}$$

$$\text{s.t.} \quad z^s \geq W_0 - \sum_{i=1}^{n} P_{iT}^s x_i - \zeta, \qquad \forall s \in \mathcal{S} \tag{15.40}$$

$$z^s \geq 0, \qquad \forall s \in \mathcal{S} \tag{15.41}$$

$$\sum_{i=1}^{n}\overline{P}_{iT}x_i \geq W_T^{\min},$$

$$\sum_{i=1}^{n}P_{i0}x_i = W_0,$$

$$\mathbf{x}\in\mathcal{X},$$

where $W_T^{\min}$ is a minimum target expected wealth, $\mathcal{X}$ denotes a feasible set of portfolios accounting for additional constraints on portfolio composition, and we use Eqs. (15.40) and (15.41) to linearize the definition of excess loss z^s in Eq. (15.37).

So, we see that the minimization of CV@R boils down to the solution of a linear programming model. This important result should be tempered by the difficulty in getting a quantile-based estimate right, when using a limited number of scenarios. See, e.g., [14] for some critical remarks on the coherence of risk measures and their estimates. We may also notice that this is only formally a two-stage optimization model. In fact, the second-stage variables z^s are not true decisions, but only "accounting" variables that we need to express the objective function. In the following, we consider multistage problems where actual scenario-dependent decisions are to be made.

15.6.3 MULTISTAGE STOCHASTIC LINEAR PROGRAMMING WITH RECOURSE

Multistage stochastic programming formulations arise naturally as a generalization of two-stage models. Conceptually, we just have to nest recourse functions corresponding to decision stages, and a rather imprecise statement of a multistage stochastic LP model would look as follows:[14]

$$\min_{\substack{\mathbf{A}_{00}\mathbf{x}_0 = \mathbf{b}_0 \\ \mathbf{x}_0 \geq \mathbf{0}}} \mathbf{c}_0^T\mathbf{x}_0 + \mathrm{E}\left[\min_{\substack{\mathbf{A}_{10}\mathbf{x}_0 + \mathbf{A}_{11}\mathbf{x}_1 = \mathbf{b}_1 \\ \mathbf{x}_1 \geq \mathbf{0}}} \mathbf{c}_1^T\mathbf{x}_1 \right. \quad (15.42)$$

$$\left. + \mathrm{E}\left[\cdots + \mathrm{E}\left[\min_{\substack{\mathbf{A}_{H,H-1}\mathbf{x}_{H-1} + \mathbf{A}_{HH}\mathbf{x}_H = \mathbf{b}_H \\ \mathbf{x}_H \geq \mathbf{0}}} \mathbf{c}_H^T\mathbf{x}_H \right]\right]\right].$$

To understand this model, we may go through the sequence of decision stages:

- At the beginning of the first time period, at time $t = 0$, we select the decision vector $\mathbf{x}_0$; this decision has a deterministic immediate cost $\mathbf{c}_0^T\mathbf{x}_0$ and must satisfy the constraint

$$\mathbf{A}_{00}\mathbf{x}_0 = \mathbf{b}_0.$$

- At the beginning of the second time period, at time $t = 1$, we observe random data $(\mathbf{A}_{10}, \mathbf{A}_{11}, \mathbf{c}_1, \mathbf{b}_1)$; then, on the basis of this information, we make decision $\mathbf{x}_1$; this second decision has an immediate cost $\mathbf{c}_1^T\mathbf{x}_1$ and must satisfy the constraint

$$\mathbf{A}_{10}\mathbf{x}_0 + \mathbf{A}_{11}\mathbf{x}_1 = \mathbf{b}_1.$$

Note that these data are not known at time $t = 0$, but only at time $t = 1$; the new decision depends on the realization of these random variables and is also affected by the previous decision.

- We repeat the above scheme for all of the time periods up to $H - 1$, where H is our planning horizon.

- Finally, at the beginning of the last time period H, we observe random data $(\mathbf{A}_{H,H-1}, \mathbf{A}_{HH}, \mathbf{c}_H, \mathbf{b}_H)$; then, on the basis of this information, we make decision $\mathbf{x}_H$, which has an immediate cost $\mathbf{c}_H^T\mathbf{x}_H$ and must satisfy the constraint

$$\mathbf{A}_{H,H-1}\mathbf{x}_{H-1} + \mathbf{A}_{HH}\mathbf{x}_H = \mathbf{b}_H.$$

---

[14] See, e.g., [32] for a more detailed discussion.

The statement of the model is, as we said, rather loose, since we are not really clarifying the dependence of the random matrices and vectors (**A**, **b**, and **c**) on the underlying stochastic process $\boldsymbol{\xi}_t$, $t = 1, \ldots, H$, followed by the risk factors. A critical modeling choice concerns the mutual dependence of this sequence of random variables:

- In the easiest case, they are mutually independent.
- In the Markovian case, $\boldsymbol{\xi}_t$ depends only on $\boldsymbol{\xi}_{t-1}$.
- In the most general case, a possibly complicated path dependency should be accounted for.

By a similar token, in this formulation, we observe that a decision $\mathbf{x}_t$ depends directly, though interstage constraints, only on the previous decision $\mathbf{x}_{t-1}$. In general, decisions may depend on all of the past history, leading to a slightly more complicated model. However, we may often introduce additional state variables, in such a way that the above formulation applies. The multistage stochastic programming framework, in principle, can deal with arbitrary path dependencies, whereas stochastic dynamic programming (to be discussed in Section 15.7) deals only with Markovian dependence structures. This is due to the fact that, in stochastic programming with recourse, uncertainty is represented by a scenario tree, which can be generated by any path-dependent process we like. However, we should stress that, in practice, multistage scenario generation is a delicate and complicated task, even more so for financial applications, as scenarios must be arbitrage-free. Furthermore, the relevant output of a stochastic programming model is the vector of immediate decisions $\mathbf{x}_0$. The remaining decision variables could be regarded as contingency plans, but the actual realizations of risk factors need not match what we include in the scenario tree. Hence, it is more likely that the model will be solved again and again, according to a rolling-horizon logic. On the contrary, dynamic programming provides us with a way to generate new decisions dynamically in a state-dependent manner, as we shall see.

Last but not least, in model (15.42) we do not clarify an important point. From the perspective of time period $t = 0$, the decisions $\mathbf{x}_1, \ldots, \mathbf{x}_H$ are random variables, as they will be adapted to the realization of the underlying stochastic process. However, the only information that we may use in making each decision consists of the observed history so far. Decisions cannot be anticipative, a point that may be formalized by using measurability concepts that we have outlined in Supplement S11.1. However, rather than dealing with abstract formulations, it is best to consider a simple example.

15.6.3.1 A toy example: Asset–liability management

In order to illustrate model building in multistage stochastic programming, let us consider a simple ALM example borrowed from [6]:

- We are given an initial wealth, say, $W_0 = 55,000$, which we may invest in a set $\mathcal{I}$ of broad asset classes, like stocks and bonds.

15.6 Stochastic optimization

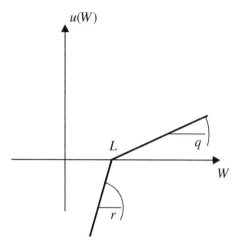

FIGURE 15.14 A piecewise linear utility function.

- Our aim is to generate enough wealth to meet a single and deterministic liability, $L = 80,000$, in $T = 3$ years.
- The portfolio is rebalanced at the beginning of each year. This means that we will allocate wealth to assets at times $t = 0, 1, 2$. At time $T = 3$, we liquidate the portfolio and (hopefully) meet the liability. We assume that there are no transaction costs.
- Asset returns are uncertain, and we assume that they are a sequence of independent and identically distributed random variables. In good years, the return is 25% for stocks and 14% for bonds; in bad years, the return is 6% for stocks and 12% for bonds. Good and bad years are equally likely. Clearly, these data are not realistic, but an important feature should be appreciated. No asset dominates the other one and these scenarios are arbitrage-free.
- Our aim is to meet the liability and keep some surplus if possible; however, we are risk-averse and really do not want to end up with any shortfall.

One way to represent risk aversion is by a concave utility function, as we have seen in Chapter 7. To keep it simple and avoid nonlinearities, let us adopt the piecewise linear utility function depicted in Fig. 15.14. We specify the utility $u(W)$ of the random terminal wealth. The utility is zero, when the terminal wealth W matches the liability L exactly. The slope $r = 4$ penalizes shortfall and is larger than the reward $q = 1$ for surplus. As to uncertainty, the above assumptions correspond to the (familiar) scenario tree depicted again in Fig. 15.15.

- Each node n_k corresponds to an event, and a scenario consists of an event sequence, i.e., a sequence of asset returns. We have eight scenarios in

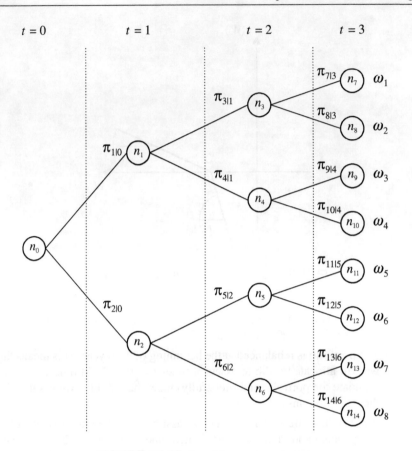

FIGURE 15.15 A multistage scenario tree.

the example. For instance, scenario ω_2 consists of the node sequence (n_0, n_1, n_3, n_8). Let us denote the set of nodes by $\mathcal{N}$. In our example,

$$\mathcal{N} = \{n_0, n_1, n_2, \ldots, n_{14}\}.$$

However, we do not make rebalancing decisions at every node. We choose the initial allocation at root node n_0. Then, we rebalance the portfolio at a subset $\mathcal{T} \subset \mathcal{N}$ of intermediate nodes, which in our case is

$$\mathcal{T} = \{n_1, \ldots, n_6\}.$$

Finally, there is a set $\mathcal{S} \subset \mathcal{N}$ of leaf (terminal) nodes at time $t = 3$,

$$\mathcal{S} = \{n_7, \ldots, n_{14}\}.$$

Each node in $\mathcal{S}$ corresponds to a scenario, where we just compare our final wealth to the liability and assess the utility value.

15.6 Stochastic optimization

- For each node, we have a set of branches, labeled by a conditional probability of occurrence, $P\{n_k \mid n_i\}$, where $n_i = a(n_k)$ is the immediate parent of node n_k. Each node $n \in \mathcal{N}$, with the exception of the root node n_0, has a unique direct parent node, denoted by $a(n)$; for instance, $a(n_3) = n_1$.
- The probability of each scenario depends on the conditional probability of each node on its path. In the example, each branch at each node is equiprobable, i.e., the conditional probability is $1/2$. Therefore, each scenario in our tree has probability $1/8$.

Each node corresponds to a state of the stochastic process, where we have to make decisions. Let us introduce the following decision variables:

- Since we have no transaction costs, we may consider decision variables x_{in}, the monetary amount invested in asset $i \in \mathcal{I}$ at each trading node $n \in \{n_0\} \cup \mathcal{T}$. This kind of decision variable is not appropriate in the presence of transaction costs, as we shall see later. The initial allocation x_{i,n_0} is the here-and-now decision, i.e., the initial asset allocation at time $t = 0$.
- For each terminal (leaf) node $s \in \mathcal{S}$, we define surplus and shortfall variables w_+^s and w_-^s.

Finally, let p^s be the probability of reaching the terminal node $s \in \mathcal{S}$, and $R_{i,n}$ be the multiplicative gain (1 plus return) for asset i during the period that *leads to* node $n \in \mathcal{T} \cup \mathcal{S}$. Then, our model can be stated as follows:

$$\max \sum_{s \in \mathcal{S}} p^s (q w_+^s - r w_-^s) \tag{15.43}$$

$$\text{s.t.} \sum_{i \in \mathcal{I}} x_{i,n_0} = W_0, \tag{15.44}$$

$$\sum_{i \in \mathcal{I}} R_{i,n} x_{i,a(n)} = \sum_{i \in \mathcal{I}} x_{in}, \quad \forall n \in \mathcal{T} \tag{15.45}$$

$$\sum_{i \in \mathcal{I}} R_{is} x_{i,a(s)} = L + w_+^s - w_-^s, \quad \forall s \in \mathcal{S} \tag{15.46}$$

$$x_{in}, w_+^s, w_-^s \geq 0.$$

The objective function of Eq. (15.43) is the expected utility, depending on surplus and shortfall at terminal nodes. The initial wealth W_0 is allocated at node n_0 among the available assets, as expressed by Eq. (15.44). The constraints in Eq. (15.45) have a similar meaning, but now the available wealth depends on the allocation at the *parent* node $a(n)$, and the return is observed *after* that allocation, in the successor node n. This way of expressing constraints ensures nonanticipativity of the decisions quite naturally. We make the asset allocation decision before observing asset returns. Finally, in Eq. (15.46), we check terminal wealth at each scenario against the target liability and measure surplus and

Table 15.1 Solution of the toy ALM example (amounts are measures in thousands).

Node	Stocks	Bonds
n_0	41.4793	13.5207
n_1	65.0946	2.16814
n_2	36.7432	22.368
n_3	83.8399	0
n_4	0	71.4286
n_5	0	71.4286
n_6	64	0

shortfall. Here, we are using a typical trick of the trade to measure positive and negative deviations from a target.

The solution of this model, for the data in the toy example, yields the decisions listed in Table 15.1. The actual decision to be implemented here and now would correspond to the initial asset allocation at node n_0, but it is worth looking at the solution structure to check the sensibility of the model. A striking fact is that, while the initial portfolio is diversified, the portfolios at the last rebalancing nodes are not, which may sound weird. Actually, this is a consequence of two features of this toy model:

- We are approximating a nonlinear utility function by a piecewise linear function, and this may imply "local" risk neutrality, so that we only care about expected return. For instance, node n_3 corresponds to the beginning of the last year after two consecutive good years. There, we already have more money than necessary and, given our scenario data, we cannot lose money. Thus, we only see the second linear segment of the utility function, and we end up investing everything in stocks, for which expected return is larger. To overcome this issue, we could use either a nonlinear programming model or a more accurate representation of utility by more linear pieces.

- The scenario tree has a very low branching factor, and this does not represent uncertainty accurately. Increasing the branching factor may ease the problem, but it clearly increases computational burden and requires a careful check of arbitrage opportunities.

15.6.3.2 A multistage financial planning model with proportional transaction costs

We may generalize the asset–liability management model of Section 15.6.3.1 as follows:

15.6 Stochastic optimization

- We consider a stream of stochastic liabilities. Let L^n be the liability to be met at each node $n \in \mathcal{N}$, where $\mathcal{N}$ is again the set of nodes in the tree.
- For the sake of simplicity, we do not consider the possibility of receiving new cash deposits along the way, as it would be the case, e.g., for a pension fund receiving new contributions. The only way to raise cash is by selling assets.
- We consider a simple form of transaction cost. Whenever we buy or sell assets, we incur proportional (linear) transaction costs; the transaction cost is a percentage c of the traded value, for both buying and selling any asset.[15]

As we did in the model of Eq. (15.43), we index assets by $i \in \mathcal{I}$. The root node is n_0, the set of terminal nodes is $\mathcal{S}$, and we rebalance the portfolio at trading nodes in the set

$$\mathcal{T} = \mathcal{N} \setminus (\{n_0\} \cup \mathcal{S}),$$

where $\setminus$ denotes set difference. We want to maximize the expected utility of terminal wealth at the leaves of the tree.

The main change in the model is due to the introduction of transaction costs. Now we cannot just represent the amount of money allocated to each asset. We have to explicitly account for asset holdings, i.e., number of bonds or stock shares. We must also specify how many units of each asset we buy or sell. Thus, we introduce the following decision variables:

- z_i^n, the amount of asset i purchased at node n
- y_i^n, the amount of asset i sold at node n
- x_i^n, the amount of asset i held at node n, after rebalancing
- W^s, the wealth at terminal node $s \in \mathcal{S}$

Variables z_i^n, y_i^n, and x_i^n are only defined at nodes $n \in \mathcal{N} \setminus \mathcal{S}$, as we assume that no rebalancing occurs at terminal nodes, where the portfolio is liquidated, we pay the liability L^s, and we measure the terminal wealth W^s. Let $u(W)$ be the nonlinear utility of wealth W. The introduction of transaction costs has two more implications. We need to consider the initial holdings $\overline{h}_i^{n_0}$ for each asset $i \in \mathcal{I}$ at the root node, and we must represent asset prices, rather than just gains or returns. Let P_i^n be the price for asset i at node n.

On the basis of this notation, we may write the following model:

$$\max \quad \sum_{s \in \mathcal{S}} \pi^s u(W^s) \tag{15.47}$$

$$\text{s.t.} \quad x_i^{n_0} = \overline{h}_i^{n_0} + z_i^{n_0} - y_i^{n_0}, \qquad \forall i \in \mathcal{I} \tag{15.48}$$

---

[15] It is easy to generalize the model by introducing proportional transaction costs depending on the liquidity of each single asset, as well as on the sign of the trade.

$$x_i^n = x_i^{a(n)} + z_i^n - y_i^n, \qquad \forall i \in \mathcal{I}, \forall n \in \mathcal{T} \qquad (15.49)$$

$$(1-c)\sum_{i \in \mathcal{I}} P_i^n y_i^n - (1+c)\sum_{i \in \mathcal{I}} P_i^n z_i^n = L^n, \quad \forall n \in \mathcal{N} \setminus \mathcal{S} \qquad (15.50)$$

$$W^s = (1-c)\sum_{i \in \mathcal{I}} P_i^s x_i^{a(s)} - L^s, \qquad \forall s \in \mathcal{S} \qquad (15.51)$$

$$x_i^n, z_i^n, y_i^n, W^s \geq 0. \qquad (15.52)$$

The objective (15.47) is the expected utility of terminal wealth. Equation (15.48) expresses the initial asset balance, taking the current holdings into account; the asset balance at intermediate trading dates, i.e., at all nodes with the exception of root and leaves, is taken into account by Eq. (15.49). Equation (15.50) ensures that enough cash is generated by selling assets in order to meet the liabilities; we may also reinvest the proceeds of what we sell in new asset holdings. Note how proportional transaction costs are expressed for selling and purchasing. Effectively, we sell at a price that is lower than the price at which we buy. Equation (15.51) is used to evaluate terminal wealth at leaf nodes, after portfolio liquidation and payment of the last liability. In practice, we would repeatedly solve the model on a rolling-horizon basis, so the exact expression of the objective function is a bit debatable. The role of terminal utility is just to ensure that we are left in a good position at the end of the planning horizon, in order to avoid nasty end-of-horizon effects.

15.6.4 SCENARIO GENERATION AND STABILITY IN STOCHASTIC PROGRAMMING

The key ingredient in stochastic programming models is a scenario tree, and the quality of the solution depends critically on how well uncertainty is captured. This topic is related with sample path generation, which was discussed in Section 11.6. In principle, we might just sample a dynamic model of risk factors using Monte Carlo simulation. However, there are some nasty complications with respect to the easy case in which Monte Carlo sampling is used to price a European-style option:

- The sample size must be kept limited, because we are going to associate decision variables with states, and the computational burden of solving an optimization problem, rather than taking a sample mean, may be remarkable.
- We need to generate non-anticipative decisions, which requires a tree structure prone to an exponential increase of complexity. If we branch 100 successor nodes out of each node in the tree, we have one million scenarios after three steps.
- When pricing a European-style option, we do not simulate any decision process. An optimization solver, however, will take advantage of any inconsistency in the scenario tree (most notably, arbitrage opportunities)

to reap a reward that cannot be actually obtained in the real world, leading to portfolios that might perform very poorly out of sample.

Plenty of research effort has been dedicated to scenario generation, which we can only hint at. Whatever scenario generation strategy we select, it is important to check the stability of the resulting solution, as we discuss in Section 15.6.4.2.

15.6.4.1 Deterministic scenario generation

There are two basic classes of scenario generation approaches.

Stochastic scenario generation. Stochastic scenario generation is based on random Monte Carlo sampling, which relies on the brute force of a large sample size. This is something that we may afford in a two-stage problem, not so much in a multistage setting. However, performance may be sometimes improved by applying variance reduction strategies, such as importance sampling.

Deterministic scenario generation. Rather than using brute force, we may try to come up with a clever choice of scenarios, using an array of methods including Gaussian quadrature, low-discrepancy sequences (like Sobol sequences), or optimized scenario generation.

All of these strategies may be interpreted within the general framework of numerical integration, also called numerical quadrature. To see why, let us consider the expected value of a function $f(\mathbf{x}, \boldsymbol{\xi})$ depending on a vector $\mathbf{x}$ of decision variables and a vector of random variables[16] $\boldsymbol{\xi}$ with joint density $h(\boldsymbol{\xi})$. By the integration with respect to $\boldsymbol{\xi}$, we define a function $F(\mathbf{x}) \doteq \mathrm{E}_{\boldsymbol{\xi}}[f(\mathbf{x}, \boldsymbol{\xi})]$, which may be approximated by a random sample of size S as follows:

$$F(\mathbf{x}) = \int f(\mathbf{x}, \boldsymbol{\xi}) h(\boldsymbol{\xi}) \, d\boldsymbol{\xi} \approx \frac{1}{S} \sum_{s=1}^{S} f(\mathbf{x}, \boldsymbol{\xi}_s),$$

where $\boldsymbol{\xi}_s$ is an observation in the sample. However, we may use a clever (possibly deterministic) sample of points $\boldsymbol{\xi}_s$, associated with weights π_s, and use the approximation

$$F(\mathbf{x}) \approx \sum_{s=1}^{S'} \pi_s f(\mathbf{x}, \boldsymbol{\xi}_s),$$

where $S' < S$. In the context of stochastic optimization, the pairs $(\pi_s, \boldsymbol{\xi}_s)$ may be understood in terms of probabilities and values, i.e., as a discrete distribution approximating the original continuous distribution. Gaussian quadrature formulas provide us with ways to select the discretization, and low-discrepancy sequences are used in quasi-Monte Carlo simulation, where the sample is generated in order to guarantee a "good" coverage of the integration domain.

---

[16] We allow some abuse of notation and do not distinguish the random variable $\widetilde{\boldsymbol{\xi}}$ from its realization $\boldsymbol{\xi}$.

Space does not allow to cover all of these topics, but we have already appreciated the role of moment matching in the calibration of a binomial lattice.[17] The idea can be applied to scenario generation as well, as shown in the following example.

Example 15.16 Scenario generation by moment matching

Consider a random variable with a multivariate normal distribution, $\boldsymbol{\xi} \sim N(\boldsymbol{\mu}, \boldsymbol{\Sigma})$. The expected values and covariances of the discrete approximation should match those of the continuous distribution. Furthermore, since we are dealing with a normal distribution, we know that skewness and kurtosis for each individual variable should be 0 and 3, respectively.

Let us denote by ξ_i^s the realization of the component ξ_i of $\boldsymbol{\xi}$, $i = 1, \ldots, n$, in scenario s, $s = 1, \ldots, S$. Natural requirements are:

$$\frac{1}{S} \sum_{s=1}^{S} \xi_i^s \approx \mu_i, \qquad \forall i$$

$$\frac{1}{S} \sum_{s=1}^{S} (\xi_i^s - \mu_i)(\xi_j^s - \mu_j) \approx \sigma_{ij}, \qquad \forall i,j$$

$$\frac{1}{S} \sum_{s=1}^{S} \frac{(\xi_i^s - \mu_i)^3}{\sigma_i^3} \approx 0, \qquad \forall i$$

$$\frac{1}{S} \sum_{s=1}^{S} \frac{(\xi_i^s - \mu_i)^4}{\sigma_i^4} \approx 3, \qquad \forall i.$$

Note that, in order to simplify the model, we assume uniform probabilities π_s for each scenario. Approximate moment matching is obtained by minimizing the following squared error, with respect to the set of realized values ξ_i^s:

$$w_1 \sum_{i=1}^{n} \left[\frac{1}{S} \sum_{s=1}^{S} \xi_i^s - \mu_i \right]^2$$

$$+ w_2 \sum_{i=1}^{n} \sum_{j=1}^{n} \left[\frac{1}{S} \sum_{s=1}^{S} (\xi_i^s - \mu_i)(\xi_j^s - \mu_j) - \sigma_{ij} \right]^2$$

$$+ w_3 \sum_{i=1}^{n} \left[\frac{1}{S} \sum_{s=1}^{S} \left(\frac{\xi_i^s - \mu_i}{\sigma_i} \right)^3 \right]^2 + w_4 \sum_{i=1}^{n} \left[\frac{1}{S} \sum_{s=1}^{S} \left(\frac{\xi_i^s - \mu_i}{\sigma_i} \right)^4 - 3 \right]^2.$$

---

[17] See Section 13.3.2.

15.6 Stochastic optimization

> The objective function includes four weights w_k, $k = 1, \ldots, 4$, which may be used to fine tune performance. It should be mentioned that the resulting scenario optimization problem need not be convex. However, if we manage to find any solution with a low value of the "error" objective function, this is arguably a satisfactory solution, even though it is not necessarily the globally optimal one. The idea can be generalized to "property matching," since we can match other features that are not necessarily related to moments; see, e.g., [19].

Moment (or property) matching has been criticized, since it is possible to build counterexamples, i.e., pairs of distributions that share the first few moments, but are actually quite different. An alternative idea is to rely on metrics fully capturing the distance between probability distributions. Thus, given a scenario tree topology, we should find the assignment of values and probabilities minimizing some distance with respect to the "true" distribution. Alternatively, given a large scenario tree, we could try to reduce it to a more manageable size by optimal scenario reduction. See, e.g., [17] for more details. We should mention that a counterargument to proponents of the optimal approximation approach is that we are not really interested in the distance between the ideal distribution and the scenario tree. What really matters is the quality of the solution. For instance, if we are considering a plain mean–variance portfolio optimization problem, it can be argued that matching the first two moments is all we need. See, e.g., [22, Chapter 4] for a discussion of these issues.

We also remark a couple of possibly useful guidelines in shaping the structure of the scenario tree:

1. If the model just aims at a robust first-stage decision, it may be advisable to use a rich branching factor at the first stage, and a limited branching factor at later stages (provided that the tree is arbitrage-free).

2. The number of stages can also be limited by using non-uniform time steps. For instance, the first two time steps could correspond to one month, and the later ones to one year.

3. A clever way to limit the number of stages is to augment the objective function with a term measuring the quality of the terminal state. By doing so, we may avoid myopic decisions and reduce the size of the tree by limiting the number of stages.

4. Sometimes, we may reduce the effective dimensionality of the problem by applying data reduction techniques, such as principal component analysis. However, we must be able to sample the few principal components accurately and generate corresponding scenarios for all the random variables involved in the model.

15.6.4.2 In- and out-of-sample stability

Whatever scenario generation strategy we employ, it is important to check the stability of the resulting solution. There are quite sophisticated studies on the stability of stochastic optimization, relying on formal metric spaces concepts. Here, we only consider a down-to-earth approach following the discussion in [22].

Let us denote the exact stochastic optimization problem, in a succinct way, by

$$\min_{\mathbf{x}} \mathrm{E}^{\mathbb{P}}[f(\mathbf{x}, \boldsymbol{\xi})],$$

which involves the expectation of the objective function $f(\cdot, \cdot)$ under the true measure $\mathbb{P}$, i.e., the "exact" distribution of the vector $\boldsymbol{\xi}$ of risk factors. The actual problem we solve is based on an approximate scenario tree $\mathcal{T}$, and can be denoted by

$$\min_{\mathbf{x}} \widehat{f}(\mathbf{x}; \mathcal{T}),$$

where the notation $\widehat{f}$ emphasizes that we are just estimating the true expected value of the objective, given a generated scenario tree $\mathcal{T}$ replacing the true measure $\mathbb{P}$. Each scenario tree induces an optimal solution of the approximate problem. Thus, if we sample trees $\mathcal{T}_i$ and $\mathcal{T}_j$, we obtain solutions $\mathbf{x}_i^*$ and $\mathbf{x}_j^*$, respectively, as well as corresponding values of the objective function. If the tree generation mechanism is reliable, we should observe a certain stability in the solution output. We may consider stability in the solution itself, but it is easier, and perhaps more relevant, to check stability in the value of the objective function. When the objective function is relatively flat, we may find rather different solutions with comparable performance. After all, if we want to minimize cost or maximize profit, the value of the objective is what matters most.[18]

Two concepts of stability should be considered. **In-sample stability** means that, if we sample two scenario trees, the values of the solutions that we obtain should not be too different:

$$\widehat{f}(\mathbf{x}_i^*; \mathcal{T}_i) \approx \widehat{f}(\mathbf{x}_j^*; \mathcal{T}_j). \tag{15.53}$$

This definition does not apply directly, if the scenario tree is generated deterministically; in that case, we might compare trees with slightly different branching structures, in order to check whether the tree structure that we are using is rich enough to ensure solution stability. This concept of stability is called in-sample, as we evaluate a solution using the *same* tree that we have used to find the solution itself. But since the tree is only a limited representation of uncertainty, we should wonder what happens when we apply the solution in the real world, where a different scenario may unfold. This leads to **out-of-sample stability**, where we compare the objective value that we obtain from the optimization model to the actual expected performance of the solution.

$$\mathrm{E}^{\mathbb{P}}[f(\mathbf{x}_i^*, \boldsymbol{\xi})] \approx \mathrm{E}^{\mathbb{P}}[f(\mathbf{x}_j^*, \boldsymbol{\xi})]. \tag{15.54}$$

---

[18] A different view may be taken when calibrating a pricing model.

If the trees are reliable, we should not notice a significant difference in performance. This kind of check is not too hard in two-stage models, as we should just plug the first-stage solution into several second-stage problems, each one corresponding to an observation of the risk factors $\boldsymbol{\xi}$. Typically, solving a large number of second-stage problems is not too demanding, especially if they are just linear programs; as a rule, it takes more time to solve one large stochastic LP than a large number of deterministic LPs.

Unfortunately, this is not so easy in a multistage setting, where realistic performance evaluation should be carried out by a costly rolling horizon simulation. To be more precise, we should consider two alternatives:

- In the **sliding** or **rolling horizon** approach, we solve a multistage problem with H stages, apply the first-stage solution, observe an out-of-sample realization of the risk factors for the first time period, and solve again a multistage problem with H stages starting from the second time period of the original model.
- In the **shrinking horizon** approach, we solve a multistage problem with H stages, apply the first-stage solution, observe an out-of-sample realization of the risk factors, but then we solve a multistage problem with just $H - 1$ stages. At each time step, we reduce the number of stages.

The second approach makes sense when we really have a well-defined horizon for performance evaluation, and it is computationally cheaper than the first one. Anyway, while a rolling horizon is technically feasible, it is quite expensive computationally. A possible (cheaper) alternative is to generate solutions $\mathbf{x}_i^*$ and $\mathbf{x}_j^*$ from trees $\mathcal{T}_i$ and $\mathcal{T}_j$, respectively, and then to check the values of the (approximate) objective functions by swapping the trees. In other words, we may check whether

$$\widehat{f}(\mathbf{x}_i^*; \mathcal{T}_j) \approx \widehat{f}(\mathbf{x}_j^*; \mathcal{T}_i).$$

The need for out-of-sample evaluation shows a possible trouble with the stochastic programming approach. Stochastic programming provides us with a first-stage, here-and-now solution, but there is no obvious way to adapt decisions at later stages when uncertainty unfolds. All of the decisions are generated explicitly and are associated with nodes in the scenario tree. On the contrary, alternative approaches, like dynamic programming and decision rules (to be discussed next), provide us with decisions in feedback form, which makes them very suitable for out-of-sample simulation and evaluation of policies.

15.7 Stochastic dynamic programming

Multistage stochastic programming with recourse is just one possible approach to cope with dynamic decision-making under uncertainty, and we may summarize its features as follows:

1. On the one hand, stochastic programming is a very flexible framework, since we may consider (in principle) any kind of intertemporal dependence among risk factors.
2. On the other hand, the approach yields a first-stage solution which is meant to be implemented here-and-now, and a set of future decisions associated with each node on the scenario tree.
3. However, when we move forward in time and risk factors are realized, we will be in some state that was not included in the scenario tree, possibly with no clue about what we should do. We have a decision for each state of the world in the scenario tree, and nothing else.

It may be argued that, in practice, the last point is not a real problem, as we shall solve a new multistage problem with new information, and that the role of decisions at later stages is just to avoid making a myopic immediate decision. However, as we mentioned, this makes performance evaluation difficult, not to mention the need for a possibly huge scenario tree. Thus, it would be nice to have a decision policy in a different form, making the adaptation of decisions to realized states easier. Ideally, we would like to have a function mapping every possible future state into the optimal decision. Clearly, it is impossible to find such a mapping in explicit form, with the exception of problems with a rather simple structure, especially for problems with a multidimensional and continuous state space. Nevertheless, we may look for suitable approximations. In this section, we consider one such approach, stochastic dynamic programming. In Section 15.8, we shall consider parameterized decision rules.

Dynamic programming (**DP**) is, arguably, the most powerful optimization *principle* available. In fact, DP is not a specific algorithm, but rather a principle that should be adapted to the specific problem at hand. In theory, DP can be used to deal with:

- Deterministic and stochastic problems
- Discrete- and continuous-time models
- Finite and infinite time horizons
- Continuous and discrete state variables
- Continuous and discrete decision variables

The price we pay for this generality is that DP is often very difficult, if not impossible, to apply exactly, and we may have to resort to some approximate version in order to break the so-called curses of dimensionality.

15.7.1 THE DYNAMIC PROGRAMMING PRINCIPLE

We consider here only stochastic DP[19] in discrete-time, over a finite horizon. States and decision variables are assumed to be continuous variables, but a

---

[19]We refrain from using the acronym SDP, when referring to stochastic dynamic programming, in order to avoid confusion with semidefinite programming problems.

15.7 Stochastic dynamic programming

discrete state component may be easily dealt with.[20] The starting point of a stochastic DP approach is a dynamic model based on a state transition function, like those we have introduced in Chapter 11:

$$\mathbf{S}_{t+1} = \mathbf{g}_t(\mathbf{S}_t, \mathbf{x}_t, \boldsymbol{\epsilon}_{t+1}), \tag{15.55}$$

where $\mathbf{S}_t$ is the random state at time t, $\mathbf{x}_t$ is the decision made after observing the state, and $\boldsymbol{\epsilon}_{t+1}$ is a random disturbance occurring *after* we have made our decision. We use capital $\mathbf{S}_t$ to point out that we are dealing with a partially controlled stochastic process with random states.[21] States may be influenced by our decisions, but they may be a purely exogenous process, e.g., a scenario of stock prices. It is important to notice that we are assuming a Markovian dependence structure, as the next state only depends on the current state, not on the whole past history; we also assume that the stochastic process of disturbances $\boldsymbol{\epsilon}_t$ consists of a sequence of i.i.d. random variables.[22] We have to make decisions at times $t = 0, 1, \ldots, T$, over a finite time horizon. When we make decision $\mathbf{x}_t$ in state $\mathbf{S}_t$, we incur an immediate cost or reward $f_t(\mathbf{S}_t, \mathbf{x}_t)$. In finite-horizon problems, it is also possible to assign a value or cost $F_{T+1}(\mathbf{S}_{T+1})$ to the terminal state.

In a stochastic multistage decision problem, we cannot build the sequence of optimal decisions $\mathbf{x}_t^*$ in advance, as they should adapt to information that is made progressively available over time. In stochastic DP, we look for an *optimal policy in feedback form*, i.e., a mapping $\pi(\cdot)$ from state to decisions:

$$\mathbf{x}_t = \pi(\mathbf{S}_t). \tag{15.56}$$

The mapping must be admissible, in the sense that we may have to comply with constraints on decisions and state variables. Let Π be the set of admissible policies. We want to solve the problem

$$\max_{\pi \in \Pi} \ \mathrm{E}\left[\sum_{t=0}^{T} \beta^t f_t(\mathbf{S}_t, \mathbf{x}_t) + \beta^{T+1} F_{T+1}(\mathbf{S}_{T+1})\right], \tag{15.57}$$

where we allow for a discount factor $\beta \in (0, 1]$.

Problem (15.57) is intractable, in general, but we may take advantage of its Markovian structure to decompose it with respect to time, obtaining a sequence of single-period subproblems. We refrain from a rigorous treatment, complete with proofs, but the intuition behind stochastic DP is fairly easy to grasp. Let us define the **value function** $V_t(\mathbf{S}_t)$, for time $t = 0, 1, \ldots, T$, as the optimal value obtained by applying an optimal policy starting from state $\mathbf{S}_t$ at time t. Then, the following **Bellman's equation** allows us to find the sequence of value

---

[20] See the model in Section 2.1.2 for an example that lends itself naturally to stochastic DP.
[21] As we have mentioned before, we could also use capital $\mathbf{X}_t$, since the decision process is a stochastic process, too, in a multistage model, where decisions depend on realized states.
[22] Actually, this assumption may be partially relaxed.

functions:

$$V_t(\mathbf{S}_t) = \max_{\mathbf{x}_t \in \mathcal{X}(\mathbf{S}_t)} \left\{ f_t(\mathbf{S}_t, \mathbf{x}_t) + \beta \cdot \mathrm{E}_t\left[V_{t+1}(\mathbf{S}_{t+1})\big|\mathbf{S}_t, \mathbf{x}_t\right] \right\}. \qquad (15.58)$$

Eq. (15.58) is a recursive functional equation that should be interpreted as follows:

- We are in state $\mathbf{S}_t$ at time t, and we must choose $\mathbf{x}_t$ within the set $\mathcal{X}(\mathbf{S}_t)$ of actions that are feasible in this state and at this time.
- We do so in order to maximize an objective function which includes not only the immediate reward $f_t(\mathbf{S}_t, \mathbf{x}_t)$, but also the discounted expected value of the reward that we shall collect in the future.
- The future reward depends on what the next state $\mathbf{S}_{t+1}$ shall be, conditional on the current state and decision. Indeed, the expectation is conditional, at time t, on $\mathbf{S}_t$ and $\mathbf{x}_t$. The next state will be given by the transition function of Eq. (15.55), so we might also replace the conditional expectation by

$$\mathrm{E}\left[V_{t+1}\Big(\mathbf{g}_t(\mathbf{S}_t, \mathbf{x}_t, \boldsymbol{\epsilon}_{t+1})\Big)\right].$$

- The value $V_{t+1}(\mathbf{S}_{t+1})$ of the next state is defined under the assumption that we shall apply an optimal policy from time $t+1$ onward. Given the Markovian structure of the system, the value function only depends on the initial state, not on the past sample path that leads us to that state.
- By optimizing the sum of the immediate reward and the discounted expected value function, we find an optimal decision and obtain the value function $V_t(\mathbf{S}_t)$ of the current state.

Equation (15.58) allows us to decompose a multistage problem into a sequence of single-stage problems, provided that we know the sequence of value functions for each time instant and each state. This is the key **Bellman's principle** underlying DP, and it should be carefully justified, but the intuition is rather simple. Under Markovian dynamics, and assuming an additive objective function, lending itself to decomposition with respect to time, we may argue (rather informally) as follows. Let us consider the optimal policy π_t^*, from time t onward, and the optimal policy π_{t+1}^*, from time $t+1$ onward. If we consider the restriction of π_t^* from time $t+1$ onward, can it be different from π_{t+1}^*? The answer is "no," otherwise we might replace this restriction with π_{t+1}^* and improve π_t^*, contradicting the assumption that π_t^* is optimal. In a sense, we may build an optimal policy by a recursive assembling process of optimal subpolicies.

Assuming knowledge of the value functions, we have an optimal policy in feedback form, which is applied as follows:

- At time $t = 0$, starting from the initial state $\mathbf{s}_0$ (we use the lowercase $\mathbf{s}$, since this state is known at time $t = 0$), we find the optimal decision $\mathbf{x}_0^*$ by solving the single-stage problem

$$\max_{\mathbf{x}_0 \in \mathcal{X}(\mathbf{s}_0)} \left\{ f_0(\mathbf{s}_0, \mathbf{x}_0) + \beta \cdot \mathrm{E}\left[V_1\Big(\mathbf{g}_0(\mathbf{s}_0, \mathbf{x}_0, \boldsymbol{\epsilon}_1)\Big)\right] \right\}.$$

We earn an immediate reward $f_0(\mathbf{s}_0, \mathbf{x}_0^*)$.

- After the application of $\mathbf{x}_0^*$, we will be at time $t = 1$ in the observed state $\mathbf{s}_1$ and solve

$$\max_{\mathbf{x}_1 \in \mathcal{X}(\mathbf{s}_1)} \left\{ f_1(\mathbf{s}_1, \mathbf{x}_1) + \beta \cdot \mathrm{E}\left[V_2\Big(\mathbf{g}_1(\mathbf{s}_1, \mathbf{x}_1, \boldsymbol{\epsilon}_2)\Big)\right] \right\}.$$

We earn an immediate reward $f_1(\mathbf{s}_1, \mathbf{x}_1^*)$.

- The process is repeated until we are in state $\mathbf{s}_T$ at time $t = T$, where we solve the last problem,

$$\max_{\mathbf{x}_T \in \mathcal{X}(\mathbf{s}_T)} \left\{ f_T(\mathbf{s}_T, \mathbf{x}_T) + \beta \cdot \mathrm{E}\left[V_{T+1}\Big(\mathbf{g}_T(\mathbf{s}_T, \mathbf{x}_T, \boldsymbol{\epsilon}_{T+1})\Big)\right] \right\}.$$

We earn an immediate reward $f_T(\mathbf{s}_T, \mathbf{x}_T^*)$.

- Finally, we end up in state $\mathbf{s}_{T+1}$, where we collect the terminal state reward $F_{T+1}(\mathbf{s}_{T+1})$, if specified.

If necessary, the performance of this policy may be evaluated by Monte Carlo simulation: We just have to generate sample paths of states, apply the policy step by step, and cumulate the single-step discounted rewards. Monte Carlo simulation is very useful to check the performance of approximate policies or to assess the robustness of a policy against misspecified dynamics of the risk factors (model risk). This is a significant advantage with respect to multistage stochastic programming with recourse. However, stochastic DP relies on Markovian assumptions that are not needed for stochastic programming. It is also natural to draw an analogy between the value function of stochastic DP and the recourse function of stochastic programming. However, when we use a scenario tree and solve a discretized stochastic program, we do not really need to find the whole recourse function.[23] Stochastic DP, on the contrary, requires finding the value functions in some form. Depending on the problem at hand, this may be an advantage or a disadvantage.

15.7.2 SOLVING BELLMAN'S EQUATION: THE THREE CURSES OF DIMENSIONALITY

Clearly, the difficulty of stochastic DP is finding the sequence of value functions. This may be done by unfolding the recursive equation *backward* from the last time period. If a terminal state value $F_{T+1}(\cdot)$ is given, then we start from the terminal condition

$$V_{T+1}(\mathbf{s}) = F_{T+1}(\mathbf{s}), \quad \forall \mathbf{s}.$$

Otherwise, we just set $V_{T+1}(\mathbf{s}) \equiv 0$. To be precise, the condition applies to every possible terminal state $\mathbf{s}$. Then, in order to find the function $V_T(\cdot)$, we

---

[23] Some specific solution methods for stochastic programming, like L-shaped decomposition, only require a *local* approximation of the recourse function near the optimal solution.

should solve

$$V_T(\mathbf{s}) = \max_{\mathbf{x} \in \mathcal{X}_T(\mathbf{s})} \left\{ f_T(\mathbf{s}, \mathbf{x}) + \beta \cdot \mathrm{E}\left[V_{T+1}\Big(\mathbf{g}_T(\mathbf{s}, \mathbf{x}, \boldsymbol{\epsilon}_{T+1})\Big)\right] \right\},$$

for all possible values of the state variable $\mathbf{s}$ at time $t = T$. Given $V_T(\cdot)$, we step backwards and find $V_{T-1}(\cdot)$ by solving

$$V_{T-1}(\mathbf{s}) = \max_{\mathbf{x} \in \mathcal{X}_{T-1}(\mathbf{s})} \left\{ f_T(\mathbf{s}, \mathbf{x}) + \beta \cdot \mathrm{E}\left[V_T\Big(\mathbf{g}_{T-1}(\mathbf{s}, \mathbf{x}, \boldsymbol{\epsilon}_T)\Big)\right] \right\},$$

for all possible values of the state variable $\mathbf{s}$ at time $t = T - 1$. We proceed recursively and finally find the value function V_1,

$$V_1(\mathbf{s}) = \max_{\mathbf{x} \in \mathcal{X}_1(\mathbf{s})} \left\{ f_1(\mathbf{s}, \mathbf{x}) + \beta \cdot \mathrm{E}\left[V_2\Big(\mathbf{g}_1(\mathbf{s}, \mathbf{x}, \boldsymbol{\epsilon}_2)\Big)\right] \right\},$$

for all possible states $\mathbf{s}$ at time $t = 1$.

The above recursion does not look too complicated, conceptually, but it is not generally feasible for a continuous and multidimensional state space. This is commonly referred to as the **curse of dimensionality**, but actually there are three such curses:

1. We should find the value function for a huge number of states. Even if we try to discretize a continuous state space by a grid, we are in trouble when the state space is not low-dimensional.
2. When the disturbance is a multidimensional random variable, the expectation in the recursive equations is hard to compute.
3. The single-step optimization problem itself may have to cope with many decision variables and possibly difficult constraints.

When the problem is low-dimensional, classical tools from numerical analysis may be able to crack the nut. In general, approximate dynamic programming strategies need to be applied. This is beyond the scope of this book, but we may illustrate stochastic DP by a familiar option pricing problem.

15.7.3 APPLICATION TO PRICING OPTIONS WITH EARLY EXERCISE FEATURES

We have considered a binomial model to price an American-style option in Section 13.3.4. The approach can be best interpreted (and generalized) within a stochastic DP framework. We recall that the price of an American-style option is given by

$$\max_{\tau} \mathrm{E}_{\mathbb{Q}_n}\left[e^{-r\tau} f(S_\tau)\right],$$

where the function $f(\cdot)$ is the option payoff, the expectation is taken under the risk-neutral measure $\mathbb{Q}_n$, and τ is a stopping time. Formally, a stopping time

15.7 Stochastic dynamic programming

is a random variable meeting some measurability properties.[24] In practice, it corresponds to a strategy by which we decide, at each time instant, if we should exercise the option or not. In the case of a Bermudan-style option, the set of early exercise opportunities is finite, which simplifies the problem. In the case of an American-style option, we discretize time and consider sample paths

$$(S_0, S_1, \ldots, S_t, \ldots, S_T),$$

where, with some abuse of notation, we interpret t as a discrete time index, leaving the discretization step δt implicit. The price S_t is the state variable, which is purely exogenous and not influenced by our exercise decisions. The transition function, if we assume a GBM model, under the risk-neutral measure, is

$$S_{t+1} = S_t \cdot \exp\left\{\left(r - \frac{\sigma^2}{2}\right)\delta t + \sigma\sqrt{\delta t} \cdot \epsilon_{t+1}\right\}.$$

The control decision is quite simple: Either we exercise or we do not.[25]

The option value $V_t(S_t)$ corresponds to the value function of stochastic DP. If we denote by $h_t(S_t)$ the intrinsic value of the option at time t, the dynamic programming recursion for the value function $V_t(S_t)$ is

$$V_t(S_t) = \max\left\{h_t(S_t),\, \mathrm{E}_{\mathbb{Q}_n}\!\left[e^{-r\,\delta t} \cdot V_{t+1}(S_{t+1})\big| S_t\right]\right\}. \tag{15.59}$$

The maximization problem is trivial, as we have to choose between two alternatives, the intrinsic value and the continuation value, which is the discounted expectation of the value function at time $t+1$. However, we may still suffer from the other two curses of dimensionality, which are related in this case. On the one hand, we may have a vector $\mathbf{S}_t$ of state variables, possibly the prices of the underlying assets of a rainbow option, rather than the scalar S_t, so that the value function is a multidimensional object. On the other hand, these prices will be related with a multidimensional driving stochastic process, rather than a scalar process ϵ_t, which complicates the expectation.

The binomial lattice recursion that we have introduced in Section 13.3.4,

$$f_{i,j} = \max\{K - S_{ij},\, e^{-r\cdot\delta t}(\pi_u f_{i+1,j+1} + \pi_d f_{i,j+1})\},$$

is a simple approximation of Eq. (15.59) and can be applied to low-dimensional problems. When the number of early exercise opportunities is limited, we may also use a scenario tree. Otherwise, we have to resort to a finite-dimensional approximation of the value function. In this case, a common strategy is to choose

---

[24] Formally, a stopping time is a measurable random variable with respect to the filtration generated by the stochastic process of stock prices. In plain English, each stopping time is associated with a non-anticipative exercise policy, which relies on the sample path observed so far, but cannot look into the future.

[25] To be precise, we should introduce a state variable stating if we have exercised or not along the sample path. Since we consider options that may be exercised at most once, we can dispense with that. However, there are some energy derivatives that may be exercised multiple times.

a set of L basis functions $\psi_k(\mathbf{S}_t)$, $k = 1, \ldots, L$, and to approximate the value function at each time instant t as follows:

$$V_t(\mathbf{S}_t) \approx \sum_{k=1}^{L} \alpha_{kt} \psi_k(\mathbf{S}_t).$$

The regression coefficients α_{kt}, for time t and basis function k, may be learned by Monte Carlo simulation of sample paths and least-squares regression.[26]

Strictly speaking, approximate DP yields a low-biased estimate of the fair option price, since we find only an approximate optimal policy for a maximization problem. The low-biased estimator can be complemented by a high-biased one, in order to define a proper confidence interval.

15.8 Decision rules for multistage SLPs

In a multistage decision problem under uncertainty, the decisions $\mathbf{x}_t$ at time $t > 0$ (i.e., with the exception of first-stage decision at time $t = 0$) are random variables, possibly depending on the whole history of risk factors up to time t, which we may denote by $\boldsymbol{\xi}_{[t]}$, as well as the sequence of past decisions $\mathbf{x}_{[t-1]}$. We may make this functional dependence explicit by using a notation like

$$\mathbf{x}_t\left(\mathbf{x}_{[t-1]}, \boldsymbol{\xi}_{[t]}\right).$$

Finding the exact function providing us with the optimal $\mathbf{x}_t$ is too demanding in practice, and some simplification is in order.

- In stochastic DP, we simplify the functional dependence by assuming a Markovian structure embodied in state variables, and the optimal decisions are implicitly defined, in state feedback form, by the sequence of value functions. The approach is very powerful, but plagued by the curses of dimensionality.
- In stochastic programming, the state space is discretized, and the above functional dependence boils down to the direct association of decision variables with the nodes in the scenario tree [see, e.g., the model of Eq. (15.47)]. The approach is quite flexible, but the scenario tree is subject to an exponential explosion, and it is very expensive to assess performance by out-of-sample simulations.

Both approaches have advantages and disadvantages, and they may be considered as two extremes, leaving room for something in between.

One possible intermediate approach is obtained by assuming a simplified functional dependence of decision variables with respect to the realization of risk factors. The simplest functional form is a linear affine function, which is

---

[26] See [16, Chapter 8] or [8, Chapter 10].

15.8 Decision rules for multistage SLPs

fully specified by a limited set of parameters. Then, given a set of sample paths, we may optimize the parameters of the decision rules. On the one hand, decision rules do not provide us with the full power of stochastic programming with recourse or dynamic programming, since the parameters of the policy function are given here-and-now and a simple functional form is adopted. On the other hand, the approach is intrinsically non-anticipative and does not suffer from the need to sample huge scenario trees.

The idea is best illustrated by a simple example, so let us outline a decision-rule based optimization approach proposed in [11, Chapter 14]), to which we refer for more details. We consider a multistage portfolio optimization with no transaction costs, much like the problem of Section 15.6.3.1. However, here we deal with a pure asset management problem, where we just optimize a given function of terminal wealth. To cast the problem within our framework, let us specify the following:

State variables. We introduce state variables S_{it}, representing the monetary amount allocated to asset $i = 1, \ldots, n$ at the beginning of time period $t = 0, 1, 2, \ldots, T$. The (given) initial wealth is

$$W_0 = \sum_{i=1}^{n} S_{i0},$$

and the (random) terminal wealth is

$$W_T = \sum_{i=1}^{n} S_{iT}.$$

Since we are not considering transaction costs, we might just consider *wealth* as the only state variable. However, let us allow for some more generality and use individual asset allocations as state variables. State variables at each time instant may be collected into vector $\mathbf{S}_t \in \mathbb{R}^n$. The careful reader should object that these are *decision*, rather than state variables. Indeed, they are the decision variables in the model of Eq. (15.43). However, as we shall see in Eq. (15.61), we are following a different framework, where they are a consequence of the adopted decision rules, whose parameters are the true decision/control variables.

Control variables. The control variables x_{it}, collected into vector $\mathbf{x}_t$, are the monetary amounts of asset i that we buy or sell at the beginning of time period t. When we buy an asset, we have a positive value of x_{it}, and negative values correspond to asset sales. Since we are not considering transaction costs, we do not need to introduce separate control variables for what we buy and what we sell, unlike the model of Section 15.6.3.2. We will have to make (nonanticipative) portfolio rebalancing decisions at times $t = 0, 1, \ldots, T-1$, and we will observe terminal wealth at time $t = T$.

Disturbances. The role of disturbances (risk factors) is played by $\tilde{r}_{it}$, the random return of asset $i = 1, \ldots, n$ in period $t = 1, \ldots, T$ (i.e., the period from time instant $t-1$ to time instant t). We may collect the multiplicative gains $g_{it} = 1 + r_{it}$ into vector

$$\mathbf{g}_t \doteq \begin{bmatrix} 1+r_{1t} \\ 1+r_{2t} \\ \vdots \\ 1+r_{nt} \end{bmatrix} \equiv \begin{bmatrix} g_{1t} \\ g_{2t} \\ \vdots \\ g_{nt} \end{bmatrix} \in \mathbb{R}^n,$$

as well in the *diagonal matrix*

$$\mathbf{G}_t \doteq \begin{bmatrix} g_{1t} & 0 & 0 & \cdots & 0 \\ 0 & g_{2t} & 0 & \cdots & 0 \\ 0 & 0 & g_{3t} & \cdots & 0 \\ \vdots & \vdots & \vdots & \ddots & \vdots \\ 0 & 0 & 0 & \cdots & g_{nt} \end{bmatrix} \in \mathbb{R}^{n \times n}.$$

The role of the diagonal matrix will be clear in the following.

We may enforce a set of constraints on state and control variables. If we rule out short-selling, we must have $\mathbf{S}_t \geq \mathbf{0}$. We may easily introduce any additional linear constraint on portfolio composition. We also assume that the portfolio is self-financing, which requires the constraint

$$\sum_{i=1}^{n} x_{it} = 0, \qquad t = 0, 1, \ldots, T-1. \tag{15.60}$$

Since we assume frictionless markets, the state transition equation may be written as

$$\mathbf{S}_{t+1} = \mathbf{G}_{t+1} \cdot (\mathbf{S}_t + \mathbf{x}_t), \qquad t = 0, 1, \ldots, T-1. \tag{15.61}$$

Now, in order to apply numerical optimization, we need a random sample of returns or gains. Hence, we assume that a scenario generation mechanism yields a set of m sample paths, indexed by k and collected into vectors and matrices

$$\mathbf{g}_t^k, \ \mathbf{G}_t^k, \qquad k = 1, \ldots, m; \ t = 1, \ldots, T.$$

Let us also introduce the vector $\bar{\mathbf{g}}_t$ of (unconditional) expected gains at time t. It is important to point out that these scenarios are *not* tree structured. They are just independent and identically distributed sample paths. We do not need a scenario tree because of the mechanism by which the control decisions are generated, which does not allow anticipative decisions.

The simplest structure of decision rules is an **affine policy**:

$$\mathbf{x}_t = \bar{\mathbf{x}}_t + \mathbf{\Theta}_t \cdot (\mathbf{g}_t - \bar{\mathbf{g}}_t), \qquad t = 1, \ldots, T-1, \tag{15.62}$$

15.8 Decision rules for multistage SLPs

where $\bar{\mathbf{x}}_t \in \mathbb{R}^n$ and $\boldsymbol{\Theta}_t \in \mathbb{R}^{n \times n}$ are the policy parameters. For $t = 0$, we also introduce $\bar{\mathbf{x}}_0$, which is the initial allocation, made before observing the first realization of the random gains. The key idea is that we have a nominal value $\bar{\mathbf{g}}_t$ of the risk factors, to which the decision vector $\bar{\mathbf{x}}_t$ should correspond. If there are deviations $\mathbf{g}_t - \bar{\mathbf{g}}_t$ from what is expected, a control action will be taken, depending on the deviations and on the policy parameters in matrix $\boldsymbol{\Theta}_t$. The optimization of the decision rules calls for the optimization of $\bar{\mathbf{x}}_t$ and $\boldsymbol{\Theta}_t$ for $t = 1, \ldots, T - 1$, as well as $\bar{\mathbf{x}}_0$. We remark that these parameters are not scenario-dependent, so that nonanticipativity is built into the model itself. Using the sampled scenarios, the state transition equations may be rewritten as

$$\mathbf{S}_1^k = \mathbf{G}_1^k \cdot \left[\mathbf{S}_0 + \bar{\mathbf{x}}_0\right], \qquad k = 1, \ldots, m$$
$$\mathbf{S}_{t+1}^k = \mathbf{G}_{t+1}^k \cdot \left[\mathbf{S}_t^k + \bar{\mathbf{x}}_t + \boldsymbol{\Theta}_t \cdot \left(\mathbf{g}_t^k - \bar{\mathbf{g}}_t\right)\right],$$
$$k = 1, \ldots, m;\ t = 1, \ldots, T - 1.$$

Given a set of scenarios and a set of policy parameters, we will find the terminal wealth for each scenario,

$$W_T^k = \sum_{i=1}^n S_T^k, \qquad k = 1, \ldots, m,$$

which we may use in defining a suitable objective function. We might set a target end-of-horizon wealth W^* and penalize shortfall with respect to this target. We may also use piecewise linear functions, like we did in Section 15.6.3.1, which will result in a LP problem. To illustrate a variation, let us introduce a quadratic penalty for shortfall:

$$\frac{1}{m}\sum_{k=1}^m \left[\max\{0, W^* - W_T^k\}\right]^2.$$

This is a sort of lower partial second-order moment, which may be optimized by introducing auxiliary shortfall variables Z^k and solving the following QP problem:

$$\min\ \frac{1}{m}\sum_{k=1}^m (Z^k)^2$$

$$\text{s.t.}\ Z^k \geq W^* - W_T^k, \qquad k = 1, \ldots, m$$

$$\sum_{i=1}^n S_{i0} = W_0$$

$$\sum_{i=1}^n S_{iT}^k = W_T^k, \qquad k = 1, \ldots, m$$

$$\mathbf{S}_1^k = \mathbf{G}_1^k \cdot \left[\mathbf{S}_0 + \bar{\mathbf{x}}_0\right], \qquad k = 1, \ldots, m$$

$$\mathbf{S}_{t+1}^k = \mathbf{G}_{t+1}^k \cdot \left[\mathbf{S}_t^k + \bar{\mathbf{x}}_t + \boldsymbol{\Theta}_t \cdot \left(\mathbf{g}_t^k - \bar{\mathbf{g}}_t\right)\right],$$

$$k = 1, \ldots, m;\ t = 1, \ldots, T - 1$$

$$\mathbf{i}^T \bar{\mathbf{x}}_t = 0, \qquad t = 0, \ldots, T-1 \qquad (15.63)$$

$$\mathbf{i}^T \boldsymbol{\Theta}_t = \mathbf{0}, \qquad t = 1, \ldots, T-1 \qquad (15.64)$$

$$Z^k, S_{it}^k \geq 0.$$

The model should be rather self-explanatory, and the only constraints that we should clarify are Eqs. (15.63) and (15.64), which result from the application of the self-financing constraint of Eq. (15.60) to the affine policy of Eq. (15.62). Here, as before, $\mathbf{i} \in \mathbb{R}^n$ is a vector with all components set to one.

The clear downside of affine decision rules is some degree of suboptimality, but we obtain a moderate size LP or a QP, depending on the objective function, with a considerable reduction of computational effort. Furthermore, once we have optimized the policy parameters, out-of-sample performance evaluation by simulation is easily accomplished. More complex rule structures have been proposed in the literature.

15.9 Worst-case robust models

There are different concepts of robust optimization, as illustrated in [13]. Here, we only consider robust optimization in the *worst-case* sense, as proposed in [2]. The idea is to deal with uncertain parameters $\boldsymbol{\xi}$ that are not associated with a stochastic characterization of uncertainty, but with an **uncertainty set** Ξ. The only thing we know is that value of the uncertain parameters will belong to the uncertainty set. The emphasis is on constraint robustness in an **uncertain optimization problem** like

$$\min_{\mathbf{x}} \quad f(\mathbf{x}; \boldsymbol{\xi})$$
$$\text{s.t.} \quad \mathbf{g}(\mathbf{x}; \boldsymbol{\xi}) \leq \mathbf{0},$$

which should be considered as a *collection* of problems, indexed by $\boldsymbol{\xi} \in \Xi$.

In order to assign a concrete meaning to the above uncertain problem, we consider the following worst-case optimization model:

$$\min_{\mathbf{x}} \quad \left\{ \max_{\boldsymbol{\xi} \in \Xi} f(\mathbf{x}; \boldsymbol{\xi}) \right\}$$
$$\text{s.t.} \quad \mathbf{g}(\mathbf{x}; \boldsymbol{\xi}) \leq \mathbf{0}, \qquad \forall \boldsymbol{\xi} \in \Xi.$$

This problem is called the **robust counterpart** of the uncertain optimization problem. We observe that there are two sides of the coin:

- Robustness in terms of optimality, i.e., we strive for a solution that will perform reasonably well for every value of the uncertain parameters.
- Robustness in terms of feasibility, i.e., our choice of $\mathbf{x}$ should be feasible for every value of the uncertain parameters.

It is customary to transform the problem in such a way that uncertainty affects only the constraints, by introducing an auxiliary variable z and solving the

15.9 Worst-case robust models

equivalent problem

$$\begin{aligned}\min \quad & z \\ \text{s.t.} \quad & f(\mathbf{x}; \boldsymbol{\xi}) \leq z, && \forall \boldsymbol{\xi} \in \Xi \\ & \mathbf{g}(\mathbf{x}; \boldsymbol{\xi}) \leq \mathbf{0}, && \forall \boldsymbol{\xi} \in \Xi,\end{aligned}$$

To be more specific, we may consider the following robust counterpart of an uncertain LP:

$$\begin{aligned}\min \quad & \mathbf{c}^\mathsf{T}\mathbf{x} \\ \text{s.t.} \quad & \mathbf{A}(\boldsymbol{\xi})\mathbf{x} \leq \mathbf{b}(\boldsymbol{\xi}), && \forall \boldsymbol{\xi} \in \Xi,\end{aligned}$$

where we assume a certain objective without loss of generality. The difficulty in solving the robust LP depends on the kind of uncertainty set that we consider. We have seen a couple of simple uncertainty sets in Example 15.13. Typical uncertainty sets are as follows:

- **Finite set of scenarios**: $\Xi = \{\boldsymbol{\xi}_1, \boldsymbol{\xi}_2, \ldots, \boldsymbol{\xi}_k\}$. Scenarios may be obtained by observations of past outcomes, in which case we often talk about **data-driven optimization**.
- **Interval (box) uncertainty**: $\Xi = \{\mathbf{l} \leq \boldsymbol{\xi} \leq \mathbf{u}\}$. In this case, for each uncertain parameter ξ_k, we give a lower bound l_k and an upper bound u_k, collected into vectors $\mathbf{l}$ and $\mathbf{u}$, respectively. We can characterize box uncertainty by using the L_∞ norm as follows:

$$\Xi = \{\boldsymbol{\xi} \mid \boldsymbol{\xi} = \boldsymbol{\xi}_0 + \mathbf{M}\mathbf{u}, \|\mathbf{u}\|_\infty \leq 1\},$$

where $\boldsymbol{\xi}_0$ is the vector of nominal values, and $\mathbf{M}$ is a *diagonal* matrix. The matrix is diagonal, since, in this case, we do not consider any relationship (e.g., correlation) among parameters.
- **Polyhedral uncertainty**, where Ξ is a (bounded) polyhedron. One way of specifying the polyhedron is by taking the convex hull of a finite set of points, $\Xi = \text{conv}(\boldsymbol{\xi}_1, \mathbf{x}_2, \ldots, \mathbf{x}_k)$. This may generalize data-driven optimization by allowing combinations of past observations. We often use the term **polytopic uncertainty**.
- In box uncertainty, we use the L_∞ norm. If we switch to the Euclidean norm L_2, we may describe **ellipsoidal uncertainty** sets,

$$\Xi = \{\boldsymbol{\xi} \mid \boldsymbol{\xi} = \boldsymbol{\xi}_0 + \mathbf{M}\mathbf{u}, \|\mathbf{u}\|_2 \leq 1\}.$$

Note that we obtain an ellipse by an affine mapping of the unit ball $\|\mathbf{u}\|_2 \leq 1$. The matrix $\mathbf{M}$ need not be diagonal, which allows for correlations among the parameters. This form of uncertainty set may remind us of confidence regions in multivariate statistics.

When we use a finite set of scenarios, the robust counterpart has the same basic form of the corresponding uncertain optimization problem. We just replicate the constraints a few times. In the other cases, the uncertainty sets consists of an infinite number of points, and the robust counterpart is a **semi-infinite programming** problem, with a finite number of variables, but an infinite number of constraints. Clearly, we may always resort to some form of sampling to make the semi-infinite problem manageable, but there are lucky cases in which the robust counterpart turns out to be a tractable convex optimization problem like an LP, SOCP, or SDP problem. In the next sections, we will illustrate a couple of examples, without laying down a general theory: We show how a robust LP problem may be transformed into another LP, at the cost of introducing some additional variables, or an SOCP problem.

Here, we are considering robust optimization without recourse decisions: We make a choice of **x**, which is not adapted in any way after the discovery of the actual values of the uncertain data. This is fine if we are interested in the robustness of the solution, but not in the flexibility of adaptation. We may apply the ideas of stochastic programming with recourse by considering **adjustable robust optimization**. Unfortunately, multistage adjustable robust optimization problems are intractable, in general, and we have to resort to decision rules in the vein of Section 15.8. In fact, decision rules were originally introduced in this context and were later proposed for stochastic programming. A fundamental result, which we claim without proof,[27] is that, if we rule out adjustability, we may evaluate feasibility *constraint-wise*. This means that we may consider robustness with respect to each individual constraint, which is a significant simplification. This is a natural consequence of the worst-case approach. For instance, in robust LP we consider individual uncertain constraints like

$$\{\mathbf{a}_i^T \mathbf{x} \leq b_i\}_{[\mathbf{a}_i; b_i] \in \mathcal{U}_i},$$

which is a collection of constraints for an uncertainty set $\mathcal{U}_i$. Each uncertainty set $\mathcal{U}_i$ consists of a collection of vectors $\mathbf{a}_i$ and scalars b_i, which are stacked into the column vector[28] $[\mathbf{a}_i; b_i]$. It is customary to express the uncertain data in terms of affine combinations of uncertain underlying factors $\boldsymbol{\xi} \in \Xi$,

$$\mathcal{U}_i = \left\{ [\mathbf{a}_i; b_i] = [\mathbf{a}_i^0; b_i^0] + \sum_{k=1}^{L} \xi_k [\mathbf{a}_i^k; b_i^k] : \boldsymbol{\xi} \in \Xi \subset \mathbb{R}^L \right\}, \quad (15.65)$$

where $[\mathbf{a}_i^0; b_i^0]$ is the vector of nominal data, and we consider L risk factors ξ_k, $k = 1, \ldots, L$, collected into vector $\boldsymbol{\xi} \in \Xi$.

As a further consideration, a possible objection to the framework of worst-case robust optimization is that it may lead to overly conservative solutions, especially in finance. Actually, this depends on the choice of the uncertainty set, which should be done wisely, in order to trade off risk and reward. As it turns out, stochastic and robust optimization are not mutually exclusive. For

---

[27] See [2] for a discussion, as well as for adjustable robust optimization models.
[28] This kind of notation, where the semicolon means vertical stacking of column vectors, is not typical in mathematics, but it comes in handy and is reminiscent of MATLAB syntax.

15.9 Worst-case robust models

instance, given a probabilistic chance constraint, we may find an uncertainty set that guarantees satisfaction of the uncertain constraint with the required probability. The advantage is that chance-constrained stochastic programming may lead to difficult nonconvex problems, whereas robust optimization may provide us with a convex approximation of the original problem.

15.9.1 UNCERTAIN LPS: POLYHEDRAL UNCERTAINTY

Let us consider the robust counterpart

$$\begin{aligned} \min \quad & \mathbf{c}^\mathsf{T}\mathbf{x} \\ \text{s.t.} \quad & \max_{\mathbf{a}_i \in \mathcal{U}_i} \mathbf{a}_i^\mathsf{T}\mathbf{x} \leq b_i, \quad i = 1,\ldots,m \end{aligned} \quad (15.66)$$

where $\mathbf{c} \in \mathbb{R}^n$, $\mathbf{x} \in \mathbb{R}^n$, and

$$\mathcal{U}_i = \{\mathbf{a}_i \mid \mathbf{C}_i \mathbf{a}_i \leq \mathbf{d}_i\}$$

is a nonempty polyhedron, where $\mathbf{C}_i \in \mathbb{R}^{m_i,n}$ and $\mathbf{d}_i \in \mathbb{R}^{m_i}$. A given vector $\mathbf{x}$ is feasible, if the constraint (15.66) is satisfied for every $\mathbf{a}_i$ in the polyhedron. Note that we consider here only uncertainty in the vectors $\mathbf{a}_i$, while b_i is assumed given (for the sake of simplicity). The uncertainty set for each individual constraint is a given polyhedron, and we do not need to resort to the more general formulation of Eq. (15.65).

The key to deal with this problem is LP duality, which was outlined in Supplement S2.2 and will be more thoroughly discussed in Sections 16.1.4 and 16.3.2. A feasible choice of $\mathbf{x}$ must guarantee satisfaction of the constraint (15.66) for every $\mathbf{a}_i \in \mathcal{U}_i$. Thus, given a choice of $\mathbf{x}$, the worst-case $\mathbf{a}_i$ is found by solving a maximization problem, which may be transformed into the dual minimization problem as follows:

$$\begin{aligned} \max \quad & \mathbf{x}^\mathsf{T}\mathbf{a}_i \\ \text{s.t.} \quad & \mathbf{C}_i \mathbf{a}_i \leq \mathbf{d}_i \end{aligned} \quad \Rightarrow \quad \begin{aligned} \min \quad & \mathbf{d}_i^\mathsf{T}\mathbf{z}_i \\ \text{s.t.} \quad & \mathbf{C}_i^\mathsf{T}\mathbf{z}_i = \mathbf{x}, \\ & \mathbf{z}_i \geq \mathbf{0}. \end{aligned}$$

Please note that, in the primal maximization problem, $\mathbf{x}$ is given and the decision variables are the components of vector $\mathbf{a}_i$; the decision variables in the dual minimization problem are collected into vector $\mathbf{z}_i \in \mathbb{R}^{m_i}$. The advantage of the minimization form of the dual is that a given $\mathbf{x}$ is feasible, with respect to the ith constraint, if we can find any $\mathbf{z}_i$ satisfying the constraints in the dual and such that its objective function is not larger than the right-hand side b_i. The reason is that, in this duality relationship, any dual feasible solution $\mathbf{z}_i$ gives an upper bound on the optimal value of the primal maximization problem. Formally, by strong duality we have

$$\mathbf{x}^\mathsf{T}\mathbf{a}_i^* = \mathbf{d}_i^\mathsf{T}\mathbf{z}_i^* \leq \mathbf{d}_i^\mathsf{T}\mathbf{z}_i,$$

where $\mathbf{a}_i^*$ and $\mathbf{z}_i^*$ are optimal solutions of the primal and dual problems, respectively, and $\mathbf{z}_i$ is any dual feasible solution. Therefore, if we find a dual feasible solution, such that the corresponding value $\mathbf{d}_i^T \mathbf{z}_i$ does not exceed b_i, we are sure that the maximum value of the primal will not exceed b_i, too. Hence, constraint (15.66) is satisfied for any $\mathbf{a}_i \in \mathcal{U}_i$, for a given choice of $\mathbf{x}$, if and only if there exists a $\mathbf{z}_i$ such that

$$\mathbf{d}_i^T \mathbf{z}_i \leq b_i, \quad \mathbf{C}_i^T \mathbf{z}_i = \mathbf{x}, \quad \mathbf{z}_i \geq \mathbf{0}.$$

Therefore, we may formulate the robust LP as

$$\begin{aligned}
\min \quad & \mathbf{c}^T \mathbf{x} \\
\text{s.t.} \quad & \mathbf{d}_i^T \mathbf{z}_i \leq b_i, \quad i = 1, \ldots, m \\
& \mathbf{C}_i^T \mathbf{z}_i = \mathbf{x}, \quad i = 1, \ldots, m \\
& \mathbf{z}_i \geq \mathbf{0}, \quad i = 1, \ldots, m
\end{aligned}$$

with variables $\mathbf{x}$ and $\mathbf{z}_i$. The good news is that, with polyhedral uncertainty, the robust counterpart of an uncertain LP is just another LP, with a moderate increase in size due to the introduction of the auxiliary variables $\mathbf{z}_i$. We obtained a similar result in the simpler setting of Example 15.13.

15.9.2 UNCERTAIN LPS: ELLIPSOIDAL UNCERTAINTY

Here, we introduce another straightforward generalization of Example 15.13, where we considered uncertainty sets given by "unit balls" with respect to L_1 and L_2 norms. We consider an ellipsoidal uncertainty set obtained by an affine mapping of a Euclidean ball,

$$\Xi = \left\{ \boldsymbol{\xi} \in \mathbb{R}^L : \|\boldsymbol{\xi}\|_2 \leq \Omega \right\},$$

with radius Ω. We recall that an affine mapping of a ball is an ellipsoid. Applying the constraint structure of Eq. (15.65), we find the following collection of constraints:

$$[\mathbf{a}^0]^T \mathbf{x} + \sum_{k=1}^{L} \xi_k [\mathbf{a}^k]^T \mathbf{x} \leq b^0 + \sum_{k=1}^{L} \xi_k b^k, \quad \forall \boldsymbol{\xi} : \|\boldsymbol{\xi}\|_2 \leq \Omega.$$

All of these constraints are satisfied, for a given $\mathbf{x}$, if

$$\max_{\|\boldsymbol{\xi}\|_2 \leq \Omega} \left\{ \sum_{k=1}^{L} \xi_k \cdot \left[[\mathbf{a}^k]^T \mathbf{x} - b^k \right] \right\} \leq b^0 - [\mathbf{a}^0]^T \mathbf{x}.$$

Here we do not need to resort to LP duality, as we may just recall that the L_2 norm is self-dual. In fact, it is easy to see that the solution of a problem of the form

$$\begin{aligned}
\max \quad & \boldsymbol{\beta}^T \mathbf{y} \\
\text{s.t.} \quad & \|\mathbf{y}\|_2 \leq \Omega
\end{aligned}$$

is
$$\mathbf{y}^* = \Omega \, \frac{\beta}{\|\beta\|_2},$$
with value $\Omega \cdot \|\beta\|_2$. By plugging the value of the maximum into the constraint, we find the equivalent form

$$\Omega \cdot \sqrt{\sum_{k=1}^{L} \left([\mathbf{a}^k]^\mathsf{T}\mathbf{x} - b^k\right)^2} \leq b^0 - [\mathbf{a}^0]^\mathsf{T}\mathbf{x},$$

which can be recast as the SOCP constraint

$$[\mathbf{a}^0]^\mathsf{T}\mathbf{x} + \Omega \sqrt{\sum_{k=1}^{L} \left([\mathbf{a}^k]^\mathsf{T}\mathbf{x} - b^k\right)^2} \leq b^0.$$

Hence, an uncertain LP with ellipsoidal uncertainty can be reformulated as an SOCP.

15.10 Nonlinear programming models in finance

Some financial problems do not fit within the class of nice convex problems (like LP, QP, and SOCP) for which extremely efficient algorithms are available. Then, we have to resort to generic nonlinear programming (NLP), where we have both relatively easy and quite complex problems. Such problems may arise in the following domains:

Parameter estimation. We are all quite familiar with linear regression by least-squares, whose solution is so easy that this is not even considered as an optimization problem. If we consider regularized regression, like ridge or lasso regression,[29] the picture does not change really, as we stay within the class of easy convex problems. The matter is different if we have to resort to estimation by maximum likelihood, which may still lead to easy problems (when we maximize a concave likelihood function), but may get more challenging when estimating certain time series models. It is interesting to note that, by framing the estimation problem within numerical optimization, we may also enforce constraints on the parameters.

Model calibration. We have discussed calibration of pricing models in Section 14.4, where we have seen that it may lead to nonconvex, nonlinear least-squares problems. The matter may get even more involved if we take a nonparametric approach to model calibration, e.g., when we deal with pricing measures directly. Then, an infinite-dimensional problem must be somehow boiled down to a finite-dimensional one.

Portfolio optimization. Mean–variance problems may be formulated as convex QP models, but if we want to maximize a generic utility function, we

---
[29] See Section 14.4.1.

face an NLP problem. Since typical utility functions are concave, we are still within the domain of convex optimization problems. However, we may face nonconvex problems when we use optimization to learn certain portfolio management decision rules.

In the next section, we show how a mean–variance problem may result in a tough nonconvex problem, if we adopt a simple portfolio management rule within a multiperiod setting. For another example of NLP in model calibration, see Section 3.5.2, where bond prices are used to fit a term structure of interest rates.

15.10.1 FIXED-MIX ASSET ALLOCATION

In Chapter 8, we have considered static, single-period asset allocation models, whereas in Section 15.6.3.2, we have considered a full-fledged multistage model, allowing for dynamic portfolio adjustments. An intermediate strategy can be devised by specifying a fixed-mix, i.e., by keeping portfolio weights w_i constant over time. Note that this does not imply that the monetary amount allocated to each asset is constant. Wealth will change along the way, but the fractions allocated to each asset are kept constant. This results in a contrarian, sell-high/buy-low strategy, since when rebalancing the portfolio, we will sell assets that overperformed and buy assets that have underperformed over the last time period.

Let us also assume that we only care about mean and variance of terminal wealth. Unlike the static case, expressing variance of the terminal wealth is not easily accomplished. Hence, we resort to a set of S return scenarios. A fixed-mix policy is intrinsically nonanticipative, so that we do not need to generate a scenario tree. The approach may be considered as an instance of the more general strategy of learning decision rules. Let $G_{it}^s = 1 + r_{it}^s$ denote the multiplicative gain for asset i, in scenario s, during time period $t = 1, \ldots, T$. Each scenario is associated with a probability π_s, $s = 1 \ldots, S$. The model we describe here is due to [26], to which we refer for further information and computational experiments.

Let W_0 be the initial wealth and denote the decision variables, the selected portfolio weights, by w_i. Then, wealth at the end of time period 1 in scenario s will be

$$W_1^s = W_0 \cdot \sum_{i=1}^{n} G_{i1}^s w_i.$$

Note that wealth is scenario-dependent, but the asset allocation is not. In general, when we consider two consecutive time periods, we have

$$W_t^s = W_{t-1}^s \cdot \sum_{i=1}^{n} G_{it}^s w_i, \qquad \forall t, s.$$

Unfolding the recursion, we see that wealth at the end of the planning horizon is

$$W_T^s = W_0 \cdot \prod_{t=1}^{T}\left(\sum_{i=1}^{n} G_{it}^s w_i\right), \quad \forall s.$$

Within a mean–variance framework, we build a risk-adjusted objective function depending on terminal wealth. Given a parameter λ related to risk aversion, the objective function is

$$\max \; \mathrm{E}[W_T] - \lambda \cdot \mathrm{Var}(W_T).$$

To express the objective function, we recall that $\mathrm{Var}(X) = \mathrm{E}[X^2] - \mathrm{E}^2[X]$ and write the model as

$$\max \; W_0 \cdot \sum_{s=1}^{S} \pi_s \cdot \left[\prod_{t=1}^{T}\left(\sum_{i=1}^{n} G_{it}^s w_i\right)\right]$$
$$+ \lambda W_0^2 \cdot \left\{\left[\sum_{s=1}^{S} \pi_s \cdot \left[\prod_{t=1}^{T}\left(\sum_{i=1}^{n} G_{it}^s w_i\right)\right]\right]^2 \right.$$
$$\left. - \sum_{s=1}^{S} \pi_s \cdot \left[\prod_{t=1}^{T}\left(\sum_{i=1}^{n} G_{it}^s w_i\right)\right]^2\right\}$$

$$\text{s.t.} \quad \sum_{i=1}^{n} w_i = 1,$$
$$w_i \geq 0.$$

The objective function is a polynomial function of portfolio weights, which is a difficult nonconcave function to maximize. However, while the objective function is a bit messy, the constraints are quite simple. Different global optimization methods may be applied to solve this problem. See Section 16.2.

Problems

15.1 Prove that the intersection of convex sets is a convex set.

15.2 Prove that the set $S = \{\mathbf{x} \in \mathbb{R}^n \mid g(\mathbf{x}) \leq 0\}$ is convex, if $g(\cdot)$ is a convex function.

15.3 Prove that the dual of the L_1 norm is the L_∞ norm.

15.4 We have considered a basic TEV (tracking error variance) minimization model in Section 15.4.1, where we are given the weights of a benchmark (target) portfolio that must be tracked, the covariance matrix between assets returns (assumed known), and the maximum cardinality of the tracking portfolio (number of assets included). Short-selling was not allowed and we did not consider transaction costs. Extend the model as follows:

- Short-selling is allowed.
- We hold a current portfolio, with given weights, which should be rebalanced in order to improve tracking. However, we want to trade off tracking against turnover, i.e., changes with respect to the current portfolio. We do not want to include explicit transaction costs, but we do not want to change the portfolio too much. Thus, in the objective function, we want to include the L_1 norm distance between the new and the current portfolio, suitably penalized.
- We do not really trust our estimate of the covariance matrix, and we consider a finite uncertainty set consisting of m covariance matrices (thus, we consider a simple form of distributional ambiguity). The model should be robust and minimize the worst-case performance over this set of alternative matrices.

Formulate the model in such a way that it is solvable by a commercial software tool implementing a branch-and-bound approach for mixed-integer problems with linear or quadratic objective and constraints.

15.5 Consider a Bermudan-style Asian option, written on a non-dividend-paying stock, whose price, under the risk-neutral measure, follows the usual GBM process. The payoff is based on the arithmetic average of the prices at M time instants, $t_i = i \cdot T/M$, $i = 1, \ldots, M$. At maturity $t_M \equiv T$, the payoff is

$$\max\left\{\frac{1}{M}\sum_{i=1}^{M} S(t_i) - K, 0\right\}.$$

The option can only be exercised at the above time instants t_i, after observing the current underlying asset price, based on which the average is updated. At time t_j, the payoff is related to the average cumulated so far, i.e., the intrinsic value is

$$\max\left\{\frac{1}{j}\sum_{i=1}^{j} S(t_i) - K, 0\right\}.$$

- Define the relevant state variable(s) and write the transition function for each time step.
- Define the relevant control variable(s).
- Write a dynamic programming recursive equation to define the value function at a generic time instant when the option can be exercised.

Note: You may consider pricing at time $t = 0$, just when the option is written.

15.6 Consider the consumption–saving problem of Section 2.1.2. How would you formulate the problem using multistage stochastic programming with recourse or stochastic dynamic programming?

15.7 In the classical mean–variance portfolio optimization model, we use variance/standard deviation as a risk measure, but an alternative is mean absolute deviation (MAD), which is defined as $E[|X - \mu_X|]$, for a random variable

X with expected value μ_X. In other words, we get rid of the sign of deviations by taking their absolute value, rather than by squaring them. This may be more robust to outliers and leads to an LP, rather than a QP model.

We need an optimization model to minimize MAD in terms of monetary wealth (not return), with a constraint on a target expected wealth at the end of the holding horizon, while keeping transaction costs under control. We consider a single holding horizon from $t = 0$ to $t = T$, so that the model is static, rather than dynamic. We have the following information, for a set of N stock shares in which we may invest:

- Current holding of each stock (number of stock shares we hold now, before rebalancing the portfolio at time $t = 0$)
- Current price of each stock share at time $t = 0$
- A set of M equally likely price scenarios for stock share prices at the end $t = T$ of the holding horizon
- Expected wealth that we wish to achieve at time $t = T$ (a minimum desired target)

In the objective function, we consider only MAD (the selected risk measure), but we also have a budget (upper bound) constraint on transaction costs, in order to limit the cost of trading. Let us assume that we may reduce transaction costs by trading with a broker that accepts to buy/sell only multiples of a basic lot of shares. For instance, if the basic lot is 1000, the broker will buy/sell 1000, 2000, 3000, ..., stock shares. For each lot (of 1000 shares in the example), there is a fixed transaction cost. We can also buy or sell an arbitrary number of stock shares using a standard trading platform, where we incur a proportional cost for each trade (buy/sell), given by a percentage of the amount traded for each stock share (e.g, 0.5% on the total value of the shares we buy/sell). We are free to use any mix of broker/platform we want, as the only limit is on the total transaction cost. The model must be in MILP form.

Further reading

- While there are plenty of excellent books on optimization methods, the coverage of model building is much less extensive. An excellent source of tricks of the trade in optimization modeling is [35]. A simple introduction to optimization models, in a general setting, can also be found in [9]. See also [11], which includes a chapter on financial applications.
- More specific references on model building for financial optimization are [13], [25], and [37].
- The many facets of linear programming are covered in [34]. See, e.g., [1] for a classical account on nonlinear programming, and [7] for a review of recent convex optimization approaches, ranging from LP to SOCP and SDP. A nice treatment of semidefinite programming can also be found in [15].

- A comprehensive reference on integer programming is [36]. An example of application to finance can be found in [5].
- The classical references on stochastic programming are [6] and [21].
- We have assumed that a stochastic program is solved *after* a scenario tree has been generated. There are, however, alternative strategies based on interleaving optimization and sampling, until a convergence criterion is met. Two approaches in this vein are stochastic decomposition [18] and stochastic dual dynamic programming [28].
- An extensive coverage of dynamic programming can be found in [3, 4]. Standard numerical methods for dynamic programming are discussed in [20] and [27], whereas [29] covers both model building and innovative solution approaches. See also [10, Chapter 10] for some R code to solve simple problems by standard numerical stochastic DP.
- Option pricing by dynamic programming is the foundation of [24], an early reference on Monte Carlo pricing of American-style derivatives.
- For robust optimization, see [2]. The intractability of multistage robust optimization models lead to the development of approaches based on decision rules, which are applied to stochastic optimization in [23].

Bibliography

1 M.S. Bazaraa, H.D. Sherali, and C.M. Shetty. *Nonlinear Programming. Theory and Algorithms* (3rd ed.). Wiley, Hoboken, NJ, 2006.

2 A. Ben-Tal, L. El Ghaoui, and A. Nemirovski. *Robust Optimization*. Princeton University Press, Princeton, NJ, 2009.

3 D.P. Bertsekas. *Dynamic Programming and Optimal Control Vol. 1* (3rd ed.). Athena Scientific, Belmont, MA, 2005.

4 D.P. Bertsekas. *Dynamic Programming and Optimal Control Vol. 2* (4th ed.). Athena Scientific, Belmont, MA, 2012.

5 D. Bertsimas, C. Darnell, and R. Stoucy. Portfolio construction through mixed-integer programming at Grantham, Mayo, Van Otterloo and Company. *Interfaces*, 29:49–66, 1999.

6 J.R. Birge and F. Louveaux. *Introduction to Stochastic Programming* (2nd ed.). Springer, New York, 2011.

7 S. Boyd and L. Vandenberghe. *Convex Optimization*. Cambridge University Press, New York, 2004. The book pdf can be downloaded from http://www.stanford.edu/~boyd/cvxbook/.

8 P. Brandimarte. *Numerical Methods in Finance and Economics: A MATLAB-Based Introduction* (2nd ed.). Wiley, Hoboken, NJ, 2006.

9 P. Brandimarte. *Quantitative Methods: An Introduction for Business Management*. Wiley, Hoboken, NJ, 2011.

10 P. Brandimarte. *Handbook in Monte Carlo Simulation: Applications in Financial Engineering, Risk Management, and Economics*. Wiley, Hoboken, NJ, 2014.

11 G. Calafiore and L. El Gahoui. *Optimization Models*. Cambridge University Press, Cambridge, 2014.

12 M. Capiński and E. Kopp. *Portfolio Theory and Risk Management*. Cambridge University Press, Cambridge, 2014.

13 G. Cornuejols and R. Tütüncü. *Optimization Methods in Finance*. Cambridge University Press, New York, 2007.

14 F.J. Fabozzi and R. Tunaru. On risk management problems related to a coherence property. *Quantitative Finance*, 6:75–81, 2006.

15 B. Gärtner and J. Matousek. *Approximation Algorithms and Semidefinite Optimization*. Springer, Heidelberg, 2012.

16 P. Glasserman. *Monte Carlo Methods in Financial Engineering*. Springer, New York, 2004.

17 H. Heitsch and W. Roemisch. Scenario reduction algorithms in stochastic programming. *Computational Optimization and Applications*, 24:187–206, 2003.

18 J.L. Higle and S. Sen. *Stochastic Decomposition*. Kluwer Academic Publishers, Dordrecht, 1996.

19 K. Hoyland and S.W. Wallace. Generating scenario trees for multistage decision problems. *Management Science*, 47:296–307, 2001.

20 K.L. Judd. *Numerical Methods in Economics*. MIT Press, Cambridge, MA, 1998.

21 P. Kall and S.W. Wallace. *Stochastic Programming*. Wiley, Chichester, 1994. The book pdf can be downloaded from http://stoprog.org/index.html?resources.html.

22 A.J. King and S.W. Wallace. *Modeling with Stochastic Programming*. Springer, Berlin, 2012.

23 D. Kuhn, W. Wiesemann, and A. Georghiu. Primal and dual linear decision rules in stochastic and robust optimization. *Mathematical Programming*, 130:177–209, 2011.

24 F.A. Longstaff and E.S. Schwartz. Valuing American options by simulation: A simple least-squares approach. *Review of Financial Studies*, 14:113–147, 2001.

25 R. Mansini, W. Ogryczak, and M.G. Speranza. *Linear and Mixed-Integer Programming for Portfolio Optimization*. Springer, Heidelberg, 2015.

26 C.D. Maranas, I.P. Androulakis, C.A. Floudas, A.J. Berger, and J.M. Mulvey. Solving long-term financial planning problems via global optimization. *Journal of Economic Dynamics and Control*, 21:1405–1425, 1997.

27 M.J. Miranda and P.L. Fackler. *Applied Computational Economics and Fi-

nance. MIT Press, Cambridge, MA, 2002.

28. M.V.F. Pereira and L.M.V.G. Pinto. Multi-stage stochastic optimization applied to energy planning. *Mathematical Programming*, 52:359–375, 1991.

29. W.B. Powell. *Approximate Dynamic Programming: Solving the Curses of Dimensionality* (2nd ed.). Wiley, Hoboken, NJ, 2011.

30. R.T. Rockafellar and S. Uryasev. Optimization of conditional value-at-risk. *The Journal of Risk*, 2:21–41, 2000.

31. R.T. Rockafellar and S. Uryasev. Conditional value-at-risk for general loss distributions. *Journal of Banking and Finance*, 26:1443–1471, 2002.

32. A. Ruszczyński and A. Shapiro. Stochastic programming models. In A. Ruszczyński and A. Shapiro, editors, *Stochastic Programming*. Elsevier, Amsterdam, 2003.

33. S.V. Stoyanov, S.T. Rachev, and F.J. Fabozzi. Optimal financial portfolios. *Applied Mathematical Finance*, 14:401–436, 2007.

34. R.J. Vanderbei. *Linear Programming: Foundations and Extensions* (3rd ed.). Springer, Heidelberg, 2010.

35. H.P. Williams. *Model Building in Mathematical Programming* (5th ed.). Wiley, Chichester, 2013.

36. L.A. Wolsey. *Integer Programming*. Wiley, New York, 1998.

37. S.A. Zenios. *Practical Financial Optimization: Decision Making for Financial Engineers*. Blackwell Publishing, Oxford, 2007.

Chapter Sixteen

Optimization Model Solving

In this final chapter, we provide the interested reader with some background on the most common methods to solve an optimization model. Given the limited space, we can only hope to scratch the surface of this huge body of knowledge, and priority has been given to the essential concepts enabling the user of commercial optimization software to select a specific method among those offered.

In section 16.1, we outline the basic concepts of classical local approaches to nonlinear programming. By "local," we mean that these methods aim at finding a locally optimal solution, which may depend on an initial solution provided by the user and is not guaranteed to be a globally optimal solution. We first deal with gradient-based as well as derivative-free methods for unconstrained optimization. Then, we outline ideas for dealing with constrained problems, such as penalty functions and Lagrangian methods. A most important topic that we discuss is duality theory. In Section 16.2, we describe a few simple approaches for global optimization of nonconvex functions of continuous variables, namely genetic algorithms and particle swarm optimization. Section 16.3 deals with the important topic of linear programming. We describe both the classical simplex algorithm and interior-point methods, and we also show how the general framework of duality theory may be applied to this specific case. In Section 16.4, we illustrate how concepts from linear programming can be extended to deal with the more general case of conic programming. Finally, we describe the basic ideas of branch-and-bound methods for mixed-integer LP models in Section 16.5.

We shall never discuss convergence of any algorithm, as this is a more theoretical topic. Moreover, some of the methods that we describe have a more general applicability than we illustrate. For instance, we describe global optimization methods, like genetic algorithms and particle swarm optimization, in the context of continuous optimization, but they may also be applied to discrete optimization problems. Going the other way around, we present branch-and-bound methods for integer programming, but they may also be used for continuous global optimization. The concepts that we outline are all available in quite popular optimization software, both commercial and academic, for which we provide web links in Section 16.6. Sometimes, building blocks must be assembled to deal with a nonstandard problem. For instance, difficult stochastic programming problems and stochastic dynamic programming require adaptation of general ideas, but this is beyond the scope of this book. Nevertheless, a grasp of general concepts should enable the reader to tackle the more advanced literature.

16.1 Local methods for nonlinear programming

Nonlinear programming models are, in general, nonconvex. However, most nonlinear programming theory and computational methods aim at finding local optima. Despite these limitations, we do need concepts like penalty functions, Lagrange multipliers, and duality theory to lay down the foundations of state-of-the-art interior-point methods for convex optimization. Furthermore, a local solver may be used as the building block of multistart methods for global optimization.

16.1.1 UNCONSTRAINED NONLINEAR PROGRAMMING

Let us consider the unconstrained problem $\min_{\mathbf{x} \in \mathbb{R}^n} f(\mathbf{x})$. In principle, if the objective function is differentiable, the following first-order optimality condition is necessary for local optimality:

$$\nabla f(\mathbf{x}) = \mathbf{0}. \tag{16.1}$$

This stationarity condition yields, in general, a system of n nonlinear equations in n unknown variables, which we might solve by numerical methods. Actually, numerical analysis is sometimes counterintuitive. Given the potential difficulty in solving systems of nonlinear equations, we do not tackle optimization problems in this way. In fact, we may even go the other way around. For instance, to solve the system of nonlinear equations

$$h_i(\mathbf{x}) = 0, \qquad i = 1, \ldots, n,$$

a possible strategy is to solve the optimization problem

$$\min \quad f(\mathbf{x}) = \sum_{i=1}^{n} h_i^2(\mathbf{x}).$$

If we find an optimal solution such that $f(\mathbf{x}^*) = 0$, then we have a solution of the original system of nonlinear equations.[1]

Nevertheless, the stationarity condition of Eq. (16.1) is useful, as it provides us with a condition that we may test to check optimality of a given solution. Classical methods for unconstrained optimization are iterative in nature

---

[1] For instance, the `fsolve` function of the MATLAB Optimization Toolbox adopts this approach.

16.1 Local methods for nonlinear programming

and generate a sequence of solutions $\mathbf{x}^{(k)}$, $k = 1, 2, 3, \ldots$, starting from a user-supplied initial point $\mathbf{x}^{(0)}$. If the gradient is available, it is quite useful in generating this sequence, as it gives the direction of steepest ascent for a maximization problem, and a direction $-\nabla f(\mathbf{x})$ of steepest descent for a minimization problem. However, the gradient may not be available because the function is discontinuous, or not given in analytical form, as is the case with stochastic optimization, or when we can only estimate $f(\cdot)$ by possibly expensive simulation runs. In such a case, we may resort to derivative-free methods. Some methods in this class work with a set of solutions, which evolves over iterations, rather than with a single solution at a time.

16.1.1.1 Gradient-based methods

We want to generate a sequence of points $\mathbf{x}^{(k)}$ leading to a local minimizer of $f(\cdot)$. Given an initial point $\mathbf{x}^{(0)}$, we would like to find a search direction $\boldsymbol{\delta}$, along which the function is decreasing, i.e.,

$$f(\mathbf{x}^{(0)} + \alpha \boldsymbol{\delta}) < f(\mathbf{x}^{(0)}),$$

for a suitable stepsize $\alpha > 0$. In a minimization problem, since the gradient gives the direction of steepest ascent, we may change its sign and adopt the **steepest descent** method:

$$\mathbf{x}^{(k+1)} = \mathbf{x}^{(k)} - \alpha^{(k)} \cdot \frac{\nabla f(\mathbf{x}^{(k)})}{\|\nabla f(\mathbf{x}^{(k)})\|}.$$

Here, we divide the gradient by its norm to define a unit search direction, but it is not essential. In order to find the step-size $\alpha^{(k)}$, we may use one among the several available one-dimensional line search strategies. We could stop when the norm of the gradient is sufficiently close to zero. This condition must be assessed with care, as it depends on the unit of measurement of $f(\cdot)$: In principle, we might just multiply the objective function by a small positive number, which does not change the optimal solution, and satisfy the condition immediately!

Clearly, steepest descent is not expected to work with nondifferentiable functions or nonconvex problems. Unfortunately there are additional issues: Despite its intuitive appeal, the gradient method may suffer from difficulties in convergence and zig-zagging behavior, even when it is supposed to work in theory. The root of the evil is that it relies on a local first-order approximation (linearization) of the objective function. Among the possible remedies, we mention:

Newton method. In this approach, we rely on a second-order local model of the objective, in order to find a displacement $\boldsymbol{\delta}$ to be applied to the current point $\mathbf{x}^{(k)}$:

$$f(\mathbf{x}^{(k)} + \boldsymbol{\delta}) \approx f(\mathbf{x}^{(k)}) + [\nabla f(\mathbf{x}^{(k)})]^\mathsf{T} \boldsymbol{\delta} + \frac{1}{2} \boldsymbol{\delta}^\mathsf{T} \mathbf{H}(x^{(k)}) \boldsymbol{\delta},$$

where **H** is the Hessian matrix, collecting the second-order derivatives

$$H_{ij} = \frac{\partial^2 f}{\partial x_i \partial x_j}.$$

Since (for well-behaved functions) $H_{ij} = H_{ji}$, the Hessian matrix is symmetric. If **H** is also positive definite, we find a minimizer for the convex quadratic approximation by solving the system of linear equations

$$\mathbf{H}(\mathbf{x}^{(k)}) \boldsymbol{\delta} = -\nabla f(\mathbf{x}^{(k)}),$$

and then we set $\mathbf{x}^{(k+1)} = \mathbf{x}^{(k)} + \boldsymbol{\delta}$. In quasi-Newton methods, finite differences are used to approximate derivatives, without placing the burden of finding them (and writing the required code) on the user.

Trust region methods. Another approach is to restrict the step α taken along the direction provided by the gradient. The rationale is that the first-order approximation is valid and can be trusted only in a neighborhood of the current iterate $\mathbf{x}^{(k)}$. To find the displacement $\boldsymbol{\delta}$, we could consider the restricted minimization subproblem:

$$\min_{\boldsymbol{\delta}} \quad f(\mathbf{x}^{(k)}) + \left[\nabla f(\mathbf{x}^{(k)})\right]^\mathsf{T} \boldsymbol{\delta}$$
$$\text{s.t.} \quad \|\boldsymbol{\delta}\| \leq h^{(k)}.$$

The trust region is delimited by the parameter $h^{(k)}$, which controls the step length and should be adjusted dynamically. We may compare the predicted improvement in the objective function (according to the linearized function) with the actual improvement. A large difference suggests that the approximation is not reliable and that the step length should be reduced. Otherwise, the step length can be increased.

16.1.1.2 Derivative-free methods

Derivative-free methods can be used when the objective function might be discontinuous, or when it is only possible to evaluate it by simulation. This class of methods includes several variants of pattern search, as well as some population-based methods that we describe later, within the framework of global optimization. Here we just give a clue about the **simplex search** algorithm, also known as Nelder–Mead method.[2] This algorithm was originally proposed to optimize industrial processes, for which a reliable mathematical was not available, and only experimental performance measures could be taken. Rather than working with a single point, it uses a simplex of $n+1$ points in $\mathbb{R}^n$. A simplex in $\mathbb{R}^n$ is the convex hull of a set of $n+1$ affinely independent points $\mathbf{x}_1, \ldots, \mathbf{x}_{n+1}$.[3] The

---

[2]This method should not be confused with the simplex algorithm for linear programming.

[3]Affine independence of $n+1$ vectors in $\mathbb{R}^n$ means that the vectors $(\mathbf{x}_2-\mathbf{x}_1), \ldots, (\mathbf{x}_{n+1}-\mathbf{x}_1)$ are linearly independent. For $n = 2$, this means that the three points $\mathbf{x}_1$, $\mathbf{x}_2$, and $\mathbf{x}_3$ do not lie on the same line and form a triangle. For $n = 3$, this means that the four points do not lie on the same plane.

16.1 Local methods for nonlinear programming

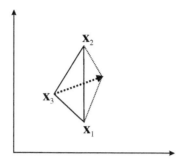

FIGURE 16.1 Reflection of the worst value point in the Nelder–Mead simplex search procedure.

convex hull of a set of points is just the set of points that may be obtained by taking convex combinations (linear combinations whose weights are non-negative and add up to 1) of the elements of the set. For instance, in two dimensions, a simplex is a triangle, whereas in three dimensions, it is a tetrahedron.

The rationale behind the method is illustrated in Fig. 16.1, for a minimization problem in $\mathbb{R}^2$. In this case, the simplex search generates a sequence of sets consisting of three points, in which the worst one is discarded and replaced by a new one. For instance, let us assume that $\mathbf{x}_3$ is associated with the worst objective value; then, it seems reasonable to move away from $\mathbf{x}_3$, by reflecting it through the center of the face formed by the other two points, as shown in Fig. 16.1. Then, a new simplex is obtained and the process is repeated. The generation of the new point is easily accomplished algebraically. If $\mathbf{x}_{n+1}$ is the worst point, we compute the centroid of the other n points as

$$\mathbf{c} = \frac{1}{n} \sum_{i=1}^{n} \mathbf{x}_i,$$

and we try a new point of the form

$$\mathbf{x}_r = \mathbf{x}_{n+1} + \alpha(\mathbf{c} - \mathbf{x}_{n+1}).$$

Clearly, the key issue is finding the right reflection coefficient $\alpha > 0$. If $\mathbf{x}_r$ turns out to be even worse than $\mathbf{x}_{n+1}$, we may argue that the step was too long and the simplex should be contracted. If $\mathbf{x}_r$ turns out to be the new best point, we have found a good search direction and the simplex might be expanded.

16.1.2 PENALTY FUNCTION METHODS

Armed with methods for unconstrained optimization, the next logical step is dealing with constraints. It would be nice to find a way to apply whatever unconstrained optimization method we like to the constrained case, too. Penalty function methods are based on the idea of transforming a constrained optimization problem into an unconstrained one by penalizing constraint violations.

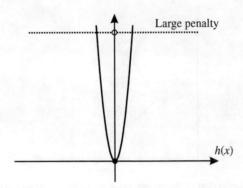

FIGURE 16.2 Quadratic approximation of the ideal penalty function.

Penalty functions may be a clever idea from a modeling viewpoint, as quite often treating a constraint as a soft requirement, rather than a hard one, is a sensible thing to do. They are quite useful when using derivative-free methods, and they also play a key role in modern interior-point methods for convex optimization.

Let us start with a problem featuring equality constraints only:

$$\min \quad f(\mathbf{x})$$
$$\text{s.t.} \quad h_i(\mathbf{x}) = 0, \qquad i \in E.$$

Ideally, we would like to introduce a penalty function that goes to infinity for points $\mathbf{x}$ where any $h_i(\mathbf{x}) \neq 0$, and is zero otherwise. This is illustrated as a dashed line in Fig. 16.2. Clearly, such a discontinuous function is numerically impossible to deal with, and must be approximated by a smoother one. One possibility is to resort to a quadratic approximation, which yields the unconstrained problem

$$\min \Psi(\mathbf{x}, \sigma) = f(\mathbf{x}) + \sigma \sum_{i \in E} h_i^2(\mathbf{x}).$$

If σ is large enough, the optimization algorithm will, in some sense, first drive the solution toward the feasible region by minimizing the penalty term; then it will try to minimize the objective $f(\cdot)$. Actually, convergence difficulties will arise if we try solving the unconstrained problem with a large value of the penalty coefficient σ. So, it is advisable to solve a sequence of unconstrained problems: We choose an increasing sequence of penalty coefficients σ_k, where $\lim_{k \to +\infty} \sigma_k = +\infty$, which yields a sequence of relaxed unconstrained problems with optimal solutions $\mathbf{x}_k^*$; the solution $\mathbf{x}_k^*$ is used to initialize the problem for σ_{k+1}.[4]

---

[4]This is an instance of a more general strategy called homotopy continuation, where a numerically hard problem is approximated by a sequence of easier ones.

16.1 Local methods for nonlinear programming

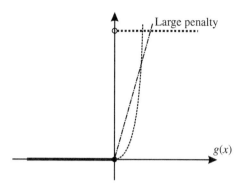

FIGURE 16.3 Ideal penalty function for the inequality constraint $g(\mathbf{x}) \leq 0$, and two possible approximations.

In the case of inequality constraints,

$$\min \quad f(\mathbf{x})$$
$$\text{s.t.} \quad g_i(\mathbf{x}) \leq 0, \quad i \in I,$$

we must only penalize positive values of the constraint functions g_i, as shown in Fig. 16.3. Again, we may find a continuous approximation of the ideal penalty

$$P(\mathbf{x}) = \begin{cases} +\infty, & \text{if } g_i(\mathbf{x}) > 0, \\ 0, & \text{if } g_i(\mathbf{x}) \leq 0. \end{cases}$$

Using the notation $y^+ = \max\{y, 0\}$, we may use an augmented objective function like

$$f(\mathbf{x}) + \sigma \sum_{i \in I} \left[g_i^+(\mathbf{x}) \right]^2,$$

or

$$f(\mathbf{x}) + \sigma \sum_{i \in I} g_i^+(\mathbf{x}). \tag{16.2}$$

The two approximated penalty functions are also shown in Fig. 16.3. The first penalty has the advantage of being a smooth function, whereas the second one introduces a non-differentiability (even though it is still a continuous function). However, the kinky function has the advantage of requiring smaller values of σ, possibly easing numerical difficulties, and may be tackled by non-smooth optimization techniques.[5]

All of the penalty functions that we have considered so far are called **exterior penalty** functions, as the feasible set is approached from outside for increasing values of the penalty coefficient σ. If the optimal solution is on the

---

[5]For equality constraints, we could also take the absolute value $|h_i(\mathbf{x})|$. In fact, one of the motivations for research work that was carried out on non-differentiable optimization was the need to cope with kinky penalty functions.

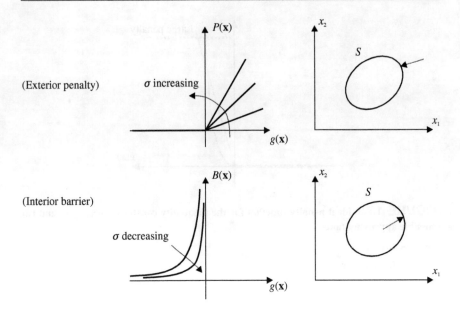

FIGURE 16.4 Exterior vs. interior penalty functions.

boundary of the feasible set (which is usually the case, since some inequality constraints are typically active[6] at the optimal solution), a feasible solution is obtained only in the limit. This may be quite fine when constraints are soft and express desirable features, rather than a hard requirement. However, in other cases, we would like to be able to stop the algorithm whenever we want and still come up with a strictly feasible solution. To overcome this difficulty, an **interior penalty** approach can be pursued. The difference between interior and exterior penalties is illustrated in Fig. 16.4. Both functions approximate the ideal penalty, but the exterior penalty is the kinky function of Eq. (16.2), which should be used with a large penalty coefficient σ, whereas the interior penalty is a **barrier function**, which should be used with a coefficient σ going to zero. The barrier function goes to infinity when **x** tends to the boundary of the feasible region from the inside. Clearly, interior penalties have the additional requirement that we must be able start the algorithm with a feasible point, and we must avoid long steps driving the search outside the feasible set. One possible barrier function is

$$B(\mathbf{x}) = -\sum_{i \in I} \frac{1}{g_i(\mathbf{x})}.$$

An alternative is the logarithmic barrier function

$$B(\mathbf{x}) = -\sum_{i \in I} \log\bigl(-g_i(\mathbf{x})\bigr).$$

---

[6] An inequality constraint $g(\mathbf{x}) \leq 0$ is said to be active at point $\mathbf{x}^*$, if $g(\mathbf{x}^*) = 0$.

Then, an unconstrained problem,

$$\min f(\mathbf{x}) + \sigma B(\mathbf{x}),$$

is solved for decreasing values of σ, until the term $\sigma B(\mathbf{x})$ is small enough. Logarithmic barriers are widely used in state-of-the art interior-point methods for conic optimization.

16.1.3 LAGRANGE MULTIPLIERS AND CONSTRAINT QUALIFICATION CONDITIONS

Penalty function methods are a quite useful numerical tool, but for a sound theory of constrained optimization we must introduce Lagrangian methods. As before, let us consider first the equality constrained case:

$$\min \; f(\mathbf{x}) \qquad (16.3)$$
$$\text{s.t.} \; h_i(\mathbf{x}) = 0, \qquad i = 1, \ldots, m.$$

The classical Lagrange method requires associating **Lagrange multipliers** λ_i with each constraint, and building the **Lagrangian function**

$$\mathcal{L}(\mathbf{x}, \boldsymbol{\lambda}) \doteq f(\mathbf{x}) + \sum_{i=1}^{m} \lambda_i h_i(\mathbf{x}),$$

where $\boldsymbol{\lambda} \in \mathbb{R}^m$ collects the multipliers. Then a *necessary* condition for *local* optimality of a feasible point $\mathbf{x}^*$ is that there exist multipliers $\lambda_i^*, i = 1, \ldots, m$, such that the Lagrangian function is stationary,

$$\nabla f(\mathbf{x}^*) + \sum_{i=1}^{m} \lambda_i^* \nabla h_i(\mathbf{x}^*) = \mathbf{0}. \qquad (16.4)$$

In other words, we need $n + m$ numbers $\mathbf{x}^*$ and $\boldsymbol{\lambda}^*$ satisfying the m equality constraints of problem (16.3) and the n stationarity condition (16.4) for the Lagrangian function. We will not prove the result, but it is more relevant to observe its weakness, since it gives a necessary (not sufficient) condition for local (not global) optimality, assuming differentiability and some additional regularity condition on the constraints (whose role we will appreciate in Example 16.2). Local optimality may yield global optimality in the convex case, but care should be taken to avoid pathologies. The next example is free from such difficulties, and it is useful to understand the rationale behind Eq. (16.4) intuitively.

Example 16.1 Quadratic programming

Consider the quadratic programming problem

$$\min \; x_1^2 + x_2^2 \qquad (16.5)$$
$$\text{s.t.} \; x_1 + x_2 = 4. \qquad (16.6)$$

The quadratic form of the objective is convex, and we may use condition (16.4) to find the global optimum. We associate a multiplier λ with the linear constraint and form the Lagrangian function,

$$\mathcal{L}(x_1, x_2, \lambda) = x_1^2 + x_2^2 + \lambda(x_1 + x_2 - 4).$$

The stationarity conditions,

$$\frac{\partial \mathcal{L}}{\partial x_1} = 2x_1 + \lambda = 0,$$

$$\frac{\partial \mathcal{L}}{\partial x_2} = 2x_2 + \lambda = 0,$$

$$\frac{\partial \mathcal{L}}{\partial \lambda} = x_1 + x_2 - 4 = 0,$$

are just a system of linear equations, whose solution yields $x_1^* = x_2^* = 2$ and $\lambda^* = -4$. We may notice that the equality constraint can also be written as $4 - x_1 - x_2 = 0$; if we do so, we have only a change in the sign of the multiplier, which is inconsequential.

We may get an intuitive feeling for the Lagrange conditions by taking a look at Fig. 16.5, where we can see the level curves of the objective function (16.5), a set of concentric circles, and the feasible region corresponding to Eq. (16.6), a line. From a geometric perspective, the problem calls for finding the closest point to the origin on the line $x_1 + x_2 = 4$. We note that the optimizer is where this line is tangent to the level curve associated with the lowest value of the objective. From an analytical viewpoint, the gradient of the objective function $f(\mathbf{x}) = x_1^2 + x_2^2$ is

$$\nabla f(x_1, x_2) = \begin{bmatrix} \dfrac{\partial f}{\partial x_1} \\ \dfrac{\partial f}{\partial x_2} \end{bmatrix} = \begin{bmatrix} 2x_1 \\ 2x_2 \end{bmatrix}.$$

This gradient, changed in sign, gives a vector pointing toward the origin, which is the steepest-descent direction for the objective. At point $\mathbf{x}^* = (2, 2)$, the gradient is $[4, \ 4]^\mathsf{T}$. The gradient of the constraint $h(\mathbf{x}) = x_1 + x_2 - 4$ is

$$\nabla h(x_1, x_2) = \begin{bmatrix} \dfrac{\partial h}{\partial x_1} \\ \dfrac{\partial h}{\partial x_2} \end{bmatrix} = \begin{bmatrix} 1 \\ 1 \end{bmatrix}.$$

Note that this vector is orthogonal to the feasible region and is parallel to the gradient of the objective at the optimizer. In Fig. 16.5, we show

16.1 Local methods for nonlinear programming

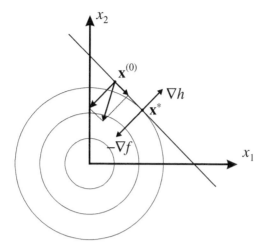

FIGURE 16.5 Geometric interpretation of the Lagrange optimality conditions for a simple quadratic programming problem.

> a non-optimal point $\mathbf{x}^{(0)}$, where we cannot move along the steepest descent direction. However, this direction may be decomposed into two vectors, one orthogonal and one parallel to the constraint. The orthogonal direction would lead outside the feasible set, but we may move along the direction parallel to the constraint and improve the objective. At the optimal point $\mathbf{x}^*$, we cannot do the same, as the only available component is orthogonal to the feasible set.

Example 16.1 provides us with an intuitive interpretation of Eq. (16.4). At the optimal solution, the gradient of the objective should be expressed as a linear combination of the gradients of the constraints. However, nothing guarantees that the latter allow us to do so, as they may fail to be a basis. The following counterexample shows that we should also qualify the constraint in some way.

Example 16.2 The role of constraint qualification conditions

Consider the problem

$$\min \quad x_1 + x_2$$
$$\text{s.t.} \quad h_1(\mathbf{x}) = x_2 - x_1^3 = 0,$$
$$h_2(\mathbf{x}) = x_2 = 0,$$

> and build the Lagrangian function
>
> $$\mathcal{L}(x_1, x_2, \lambda_1, \lambda_2) = x_1 + x_2 + \lambda_1(x_2 - x_1^3) + \lambda_2 x_2.$$
>
> The stationarity conditions yield the following system of equations:
>
> $$\frac{\partial \mathcal{L}}{\partial x_1} = 1 - 3\lambda_1 x_1^2 = 0, \qquad \frac{\partial \mathcal{L}}{\partial x_2} = 1 + \lambda_1 + \lambda_2 = 0,$$
> $$\frac{\partial \mathcal{L}}{\partial \lambda_1} = x_2 - x_1^3 = 0, \qquad \frac{\partial \mathcal{L}}{\partial \lambda_2} = x_2 = 0.$$
>
> However, this system has no solution.

In Example 16.2, the feasible set boils down to the origin $(0,0)$, which is the (trivial) optimal solution. Unfortunately, both gradients of the two constraints at the origin are parallel to vector $[0, 1]^\mathsf{T}$, whereas the gradient of the objective is $[1, 1]^\mathsf{T}$. Thus, the gradients of the constraints are not a basis able to express the gradient of $f(\cdot)$ at the optimal solution. In fact, a proper statement of the Lagrange theorem requires additional regularity conditions on the constraints, which are known as **constraint qualification** conditions. There are alternative constraint qualification conditions that may be required, and a full listing is definitely beyond the scope of this book,[7] but we mention the following ones:

- The gradients of the constraint functions $h_i(\cdot)$ are linearly independent at $\mathbf{x}^*$ (which is where we fail in Example 16.2).
- The constraints are linear.
- The feasible set includes an interior point. This condition is known as **Slater condition** and is quite popular, since it may be easy to check (even though it may be stronger than necessary). Actually, it applies only to problems with inequality constraints, and a different statement is needed for equality constraints.

Indeed, the technicalities involved in constrained optimization may get quite involved, but for our purposes, it is much more important to get a deeper understanding of the economic meaning of Lagrange multipliers. To this aim, let us consider the perturbed problem

$$\min \quad f(\mathbf{x})$$
$$\text{s.t.} \quad h_i(\mathbf{x}) = \epsilon_i, \qquad i = 1, \ldots, m.$$

The value of the optimal solution, $f^* = f(\mathbf{x}^*)$, will change as a function of the perturbations ϵ_i. Under suitable differentiability conditions, the following result

---

[7] See, e.g., [1].

16.1 Local methods for nonlinear programming

may be proved:
$$\frac{\partial f^*}{\partial \epsilon_i} = -\lambda_i. \tag{16.7}$$

Thus, apart from the sign, which is not too relevant in the case of equality constraints, Lagrange multipliers play the role of sensitivities of the optimal value with respect to changes in the right-hand side of the constraints. As we show below, this interpretation is fundamental in the case of inequality constraints.

Let us now turn to the general constrained problem (P_{EI}) including inequality constraints:

$$\begin{aligned}
\min \quad & f(\mathbf{x}) \\
\text{s.t.} \quad & h_i(\mathbf{x}) = 0, \quad i \in E \\
& g_i(\mathbf{x}) \leq 0, \quad i \in I.
\end{aligned}$$

Note that we arrange inequalities as "less-than" constraints, which is the usual convention and helps the economic interpretation, as inequalities are often related to budget and resource availability constraints. Then, as in the previous case, we introduce Lagrange multipliers and build the Lagrangian function

$$\mathcal{L}(\mathbf{x}, \boldsymbol{\lambda}, \boldsymbol{\mu}) = f(\mathbf{x}) + \sum_{i \in E} \lambda_i h_i(\mathbf{x}) + \sum_{i \in I} \mu_i g_i(\mathbf{x}). \tag{16.8}$$

Subject to the aforementioned constraint qualification and differentiability conditions, a necessary condition for the local optimality of $\mathbf{x}^*$ is that there exist numbers λ_i^* ($i \in E$) and $\mu_i^* \geq 0$ ($i \in I$) such that

$$\begin{aligned}
\nabla f(\mathbf{x}^*) + \sum_{i \in E} \lambda_i^* \nabla h_i(\mathbf{x}^*) + \sum_{i \in I} \mu_i^* \nabla g_i(\mathbf{x}^*) &= \mathbf{0}, \\
\mu_i^* g_i(\mathbf{x}^*) &= 0, \quad \forall i \in I.
\end{aligned} \tag{16.9}$$

These conditions are known as **Karush–Kuhn–Tucker** (KKT) conditions and are similar to those in the classical Lagrange theorem, with two differences:

1. Multipliers μ_i associated with inequality constraints are restricted in sign, $\mu_i \geq 0$.
2. There is an additional condition, Eq. (16.9), known as **complementary slackness**.

To understand these conditions intuitively, let us interpret inequalities as budget constraints with a right-hand side ϵ_i:

$$g_i(\mathbf{x}) \leq \epsilon_i. \tag{16.10}$$

At an optimal solution $\mathbf{x}^*$, the inequality constraint (16.10) can be in two states:

1. We say that an inequality constraint is **active** at a solution $\mathbf{x}^*$ if it is binding, i.e., $g_i(\mathbf{x}^*) = \epsilon_i$.

2. We say that an inequality constraint is **inactive** if it is nonbinding, i.e., $g_i(\mathbf{x}^*) < \epsilon_i$.

Now, let us consider what may happen, if we perturb an active inequality constraint by increasing its right-hand side ϵ_i. Since the constraint is active, it basically behaves as an equality constraint, to which Eq. (16.7) applies. What we are doing, economically, is increasing the budget, which means that we are relaxing the feasible set. Since we are dealing with a minimization problem, the objective function may not change or it may improve, but it certainly cannot get worse, since the optimal solution of the original problem is still feasible in the perturbed one. Hence,

$$\frac{\partial f^*}{\partial \epsilon_i} = -\mu_i \leq 0,$$

which implies non-negativity of multipliers associated with inequalities. This fact may be interpreted economically, if we regard the objective function as a cost, measured in monetary units. Then, the budget constraint is measured in resource units, and each multiplier can be interpreted as a resource price (e.g., euro per unit of a resource). Therefore, Eq. (16.7) suggests that the multiplier gives the maximum price that which we should be willing to pay for an additional unit of resource (to be precise, in a nonlinear problem, this is only true for a small increment). This is why, in economics, Lagrange multipliers are known as **shadow prices**. It is also clear that the shadow price μ_i should be non-negative.

If the constraint (16.10) is not active, we are not using the whole budget, so there is no point in increasing it. If we perturb the constraint by a small amount, the solution is not going to change, so that the derivative of f^* with respect to ϵ_i is zero. By the same token, we are not willing to buy any additional amount of resource, which we are not going to use anyway, so that the shadow price is zero. Therefore, we are also able to interpret the complementary slackness condition of Eq. (16.9):

- If a resource is not fully used, i.e., $g_i(\mathbf{x}^*) < 0$, its shadow price μ_i^* must be zero. Indeed, there is no point in increasing the availability of a resource that is not fully used.

- If the shadow price μ_i^* is strictly positive, then the resource budget constraint must be active, i.e., $g_i(\mathbf{x}^*) = 0$.

Thus, the complementary slackness condition (16.9) rules out the case of a strictly positive multiplier μ_i^* associated with an inactive inequality constraint, for which $g_i(\mathbf{x}^*) < 0$.

From an algorithmic viewpoint, the KKT conditions are not solved directly, but they are the conceptual basis for computationally viable methods. There are methods integrating multipliers with penalty functions, known as augmented Lagrangians. From our viewpoint, the most interesting development related to Lagrange multipliers is duality theory.

16.1.4 DUALITY THEORY

Duality is a fundamental concept in optimization, both from a theoretical and an algorithmic viewpoint.[8] We cannot offer a sound theoretical development, but we may grasp the basics intuitively, with the aim of understanding the role of duality in linear programming and in the development of primal-dual interior-point methods.

As a first step, let us fix the nomenclature, by considering the inequality-constrained[9] problem

$$(P) \quad \begin{array}{ll} \min & f(\mathbf{x}) \\ \text{s.t.} & g_i(\mathbf{x}) \leq 0, \quad i \in I \\ & \mathbf{x} \in S \subseteq \mathbb{R}^n. \end{array} \quad (16.11)$$

Problem (P) is what we want to solve and is called the **primal problem**. The actual decision variables $\mathbf{x}$ are called **primal variables**. The set S is any subset of $\mathbb{R}^n$, possibly a *discrete* one. Here, we neither assume differentiability nor convexity of the objective function. The results we get are therefore extremely general. A good way to motivate what we are going to do is to assume that the constraints (16.11) are "complicating" constraints, in the sense that the minimization of $f(\cdot)$ subject to $\mathbf{x} \in S$ would be easy. Thus, we would like to relax these nasty constraints, which we do by *dualizing* them and building the Lagrangian function:

$$\mathcal{L}(\mathbf{x}, \boldsymbol{\mu}) = f(\mathbf{x}) + \sum_{i \in I} \mu_i g_i(\mathbf{x}) = f(\mathbf{x}) + \boldsymbol{\mu}^\mathsf{T} \mathbf{g}(\mathbf{x}).$$

Within this framework, the multipliers $\boldsymbol{\mu}$ are called **dual variables**, and the minimization of the Lagrangian function with respect to the primal variables $\mathbf{x} \in S$, for a given setting of $\boldsymbol{\mu}$, is called the **relaxed problem**. The solution of the relaxed problem defines a function $w(\boldsymbol{\mu})$, called the **dual function**:

$$w(\boldsymbol{\mu}) \doteq \min_{\mathbf{x} \in S} \mathcal{L}(\mathbf{x}, \boldsymbol{\mu}). \quad (16.12)$$

It is easy to see that the dual function, for any $\boldsymbol{\mu} \geq \mathbf{0}$, provides us with a lower bound on the value of the primal optimal solution.

THEOREM 16.1 (Weak duality) *For any $\boldsymbol{\mu} \geq \mathbf{0}$, the dual function is a lower bound for the optimum $f(\mathbf{x}^*)$ of the primal problem (P), i.e.,*

$$w(\boldsymbol{\mu}) \leq f(\mathbf{x}^*), \qquad \forall \boldsymbol{\mu} \geq \mathbf{0}.$$

---

[8] To be precise, we deal here only with Lagrangian duality. There are other forms of duality, like conjugate duality.

[9] The case of equality constraints is treated similarly. The only difference is that multipliers are unrestricted in sign.

PROOF Let us adopt the notation $\nu(P)$ to denote the optimal value of the objective function for an optimization problem P.

Under the hypothesis $\boldsymbol{\mu} \geq \mathbf{0}$, it is easy to see that

$$\nu(P) \geq \nu \left(\begin{array}{ll} \min & f(\mathbf{x}) \\ \text{s.t.} & \mathbf{x} \in S \\ & \boldsymbol{\mu}^T \mathbf{g}(\mathbf{x}) \leq 0 \end{array} \right) \qquad (16.13)$$

$$\geq \nu \left(\begin{array}{ll} \min & f(\mathbf{x}) + \boldsymbol{\mu}^T \mathbf{g}(\mathbf{x}) \\ \text{s.t.} & \mathbf{x} \in S \\ & \boldsymbol{\mu}^T \mathbf{g}(\mathbf{x}) \leq 0 \end{array} \right) \qquad (16.14)$$

$$\geq \nu \left(\begin{array}{ll} \min & f(\mathbf{x}) + \boldsymbol{\mu}^T \mathbf{g}(\mathbf{x}) \\ \text{s.t.} & \mathbf{x} \in S \end{array} \right). \qquad (16.15)$$

The inequality (16.13) is justified by the fact that, whenever we aggregate constraints, we enlarge the feasible set.[10] The inequality (16.14) is due to the fact that we add a non-positive term to the objective function, and we obtain the final inequality (16.15) by dropping a constraint and relaxing the feasible set once more. ∎

A lower bound on the optimal value of the primal problem may be useful in a number of ways. For instance, given a feasible solution of the primal, possibly obtained by a cheap heuristic approach, we may estimate how far it is from the optimum. To find a tight and informative lower bound, it is natural to look for the largest lower bound by solving the **dual problem**:

$$(D) \qquad \max_{\boldsymbol{\mu} \geq \mathbf{0}} w(\boldsymbol{\mu}) = \max_{\boldsymbol{\mu} \geq \mathbf{0}} \left\{ \min_{\mathbf{x} \in S} \mathcal{L}(\mathbf{x}, \boldsymbol{\mu}) \right\}. \qquad (16.16)$$

The efficient maximization of the dual function should be carried out by algorithms that may depend on the specific problem. However, the dual function enjoys a nice property[11]:

THEOREM 16.2 *The dual function $w(\boldsymbol{\mu})$ is a concave function.*

We have obtained a very general, but weak relationship, since weak duality only yields a lower bound. So, it is quite natural to wonder whether there are lucky cases in which a stronger relationship,

$$\nu(D) = w(\boldsymbol{\mu}^*) = f(\mathbf{x}^*) = \nu(P),$$

holds. Indeed, under suitable conditions, this stronger property, known as **strong duality**, does apply. Common wisdom says that this is the case with primal convex problems, which is not quite correct. There are convex problems for which

---

[10] Recall the discussion in Example 15.6.
[11] See Problem 16.1.

16.2 Global methods for nonlinear programming

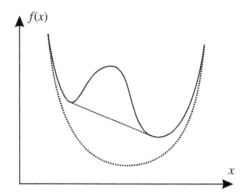

FIGURE 16.6 A convex lower bounding function.

a *duality gap* is observed, i.e., the maximum of the dual is strictly less than the minimum of the primal. We should require some additional condition, like the Slater constraint qualification, for strong duality to hold. It can be shown that strong duality applies to linear programming problems and, in Section 16.3.2, we shall see how we may take advantage of it.

16.2 Global methods for nonlinear programming

When we deal with tough nonconvex problems, like some model calibration tasks or the fixed-mix portfolio optimization problem of Section 15.10.1, the local minimizers provided by classical methods may not be satisfactory. In order to tackle a global optimization problem, we may pursue one of the following alternatives:

- If an optimality guarantee is essential, and we are lucky, the problem may feature some specific structure that is exploited by ad hoc methods. One such example is the minimization of a difference of convex functions, and the minimization of a concave function over a convex set is another one (where we know that an optimal solution can be found on the boundary of the feasible set). Otherwise, we have to resort to computationally demanding approaches, like branch-and-bound methods. We only describe these strategies for MILP models later, but the idea is fairly simple. The domain is partitioned into subregions and a convex lower bounding function is looked for, as shown in Fig. 16.6. Ideally, we would like to find the convexified function displayed in the picture, but this is not usually feasible, and a weaker underestimator is obtained. This is used to determine which subregions may be discarded, when the lower bound has a higher cost than a feasible solution we already know.

- A simpler alternative is to apply a local optimization strategy with multiple starting points. For each possible starting point, we will end up at a locally optimal solution. Thus, we may think of the domain as partitioned in attraction basins. Multiple start methods may be rather effective, but they do not ensure optimality.
- We may also resort to stochastic search strategies, trying to escape from local optimizers. We mention tabu search, simulated annealing, genetic algorithms, and particle swarm optimization. Some methods in this class have a physical motivation, like simulated annealing, whereas others have a biological motivation, like genetic algorithms and particle swarm optimization. A significant advantage of this kind of algorithms is their flexibility, which explains why they are collectively known as **metaheuristics**, where the name suggests that they are generic principles that require adaptation. Metaheuristics may also be used for discrete optimization problems.

In the next two sections, we outline two metaheuristics, namely, genetic algorithms and particle swarm optimization. There are several variants, differing in the number of parameters to be set and the corresponding complexity of fine tuning, though some self-adaptive methods are available. Sometimes, metaheuristics require a considerable amount of computational effort to find near-optimal solutions. Furthermore, we have to select a suitable representation of the solution, in order to apply metaheuristics, adding another layer of customization. As the reader can imagine, it is easy to get lost because of the sheer number of proposals and combinations. The key is to experiment with some of these variants and find the best combination for the specific problem at hand.

16.2.1 GENETIC ALGORITHMS

Unlike other solution methods, genetic algorithms work with a set of solutions at each iteration, rather than a single point. The idea is based on the survival-of-the-fittest mechanism of biological evolution, whereby a population of individuals evolves by a combination of the following elements:

- **Selection** according to a fitness criterion, which in an optimization problem is related to the objective function.
- **Mutation**, whereby some features of an individual are randomly modified.
- **Crossover**, whereby two individuals are selected for reproduction and generate offsprings with features resulting from a mix of their parents' characteristics.

There is a considerable freedom in the way this idea can be translated into a concrete algorithm. A critical issue is how to balance exploration and exploitation. On the one hand, biasing the selection of the new population toward high quality individuals may lead to premature termination; on the other hand, unless a certain degree of elitism is used to keep good individuals in the population, the algorithm might wander around and skip the opportunity of exploring promising subregions of the solution space.

A key ingredient is the encoding of a feasible solution as a string of features, corresponding to genes of a chromosome. For continuous optimization in $\mathbb{R}^n$, the choice of representing each solution by the corresponding vector $\mathbf{x}$ is fairly natural.[12] Let us consider the simplest form of crossover: Given two individuals $\mathbf{x}$ and $\mathbf{y}$ in the current pool, a "breakpoint" position $k \in \{1, 2, \ldots, n\}$ is randomly selected and two offsprings are generated as follows:

$$\left\{ \begin{array}{l} x_1, x_2, \ldots, x_k, x_{k+1}, \ldots, x_n \\ y_1, y_2, \ldots, y_k, y_{k+1}, \ldots, y_n \end{array} \right\} \Rightarrow \left\{ \begin{array}{l} x_1, x_2, \ldots, x_k, y_{k+1}, \ldots, y_n \\ y_1, y_2, \ldots, y_k, x_{k+1}, \ldots, x_n \end{array} \right\}.$$

Variations on the theme are possible; for instance, a double crossover may be exploited, in which two breakpoints are selected for the crossover. The two new solutions are clearly feasible for an unconstrained optimization problem in $\mathbb{R}^n$. When constraints are involved, this guarantee is lost in general. We might enforce hard constraints by eliminating noncomplying individuals, but a usually better alternative is relaxation by a suitable penalty function.

16.2.2 PARTICLE SWARM OPTIMIZATION

Particle swarm optimization (**PSO**) methods are another class of stochastic search algorithms, based on a population of m particles exploring the space of solutions, which we assume is a subset of $\mathbb{R}^n$. The *position* of each particle j in the swarm is a vector

$$\mathbf{x}_j(t) = \Big[x_{1j}(t),\, x_{2j}(t),\, \ldots,\, x_{nj}(t)\Big]^\mathsf{T} \in \mathbb{R}^n,$$

corresponding to a solution of the optimization problem, which changes in discrete time ($t = 1, 2, 3, \ldots$) according to three factors:

1. **Inertia**. Each particle is associated with a velocity vector $\mathbf{v}_j(t)$, which tends to be maintained.
2. **Cognitive factor**. Each particle tends to move toward its personal best $\mathbf{p}_j^*$, i.e., the best point that it has visited so far.
3. **Social factor**. Each particle tends to move toward the global best $\mathbf{g}^*$ of the swarm, i.e., the best point that has been visited so far by the whole set of particles.

---

[12] The choice is not that obvious for many problems in combinatorial optimization, e.g., when solutions correspond to permutations of objects.

The idea is to mimic the interactions of members of a swarm looking for food, and a classical statement of the algorithm is the following:

$$\mathbf{v}_j(t+1) = \mathbf{v}_j(t) + c_1 r_{1j}(t) \cdot \left[\mathbf{p}_j^*(t) - \mathbf{x}_j(t)\right]$$
$$+ c_2 r_{2j}(t) \cdot \left[\mathbf{g}^*(t) - \mathbf{x}_j(t)\right], \qquad (16.17)$$
$$\mathbf{x}_j(t+1) = \mathbf{x}_j(t) + \mathbf{v}_j(t+1). \qquad (16.18)$$

Equation (16.17) governs the evolution of the velocity of each particle $j = 1, \ldots, m$. The new velocity $\mathbf{v}_j(t+1)$ depends on:

- The previous velocity $\mathbf{v}_j(t)$, i.e., the inertia factor.
- The difference $\mathbf{p}_j^*(t) - \mathbf{x}_j(t)$ between the personal best and the current position, i.e., the cognitive factor, scaled by a coefficient c_1 and multiplied by a random variable $r_{1j}(t)$; a typical choice is a uniform distribution $\mathsf{U}(0,1)$.
- The difference $\mathbf{g}^*(t) - \mathbf{x}_j(t)$ between the global best and the current position, i.e., the social factor, scaled by a coefficient c_2 and multiplied by a random variable $r_{2j}(t)$; in this case, too, a typical choice is a uniform distribution $\mathsf{U}(0,1)$.

Equation (16.18) simply changes each component of the current position according to the new velocity. At each iteration, the personal and global bests are updated when necessary, and other adjustments are used in order to keep velocities within a given range, as well as the positions, if bounds on the variables are specified.

PSO can be regarded as a Monte Carlo search approach, and several variants have been proposed:

- In small world methods, the global best position for neighboring particles is used, rather than the whole swarm.
- In quantum PSO, different rules are used to evolve particle positions. A version of the algorithm generates a new particle position as

$$\mathbf{x}_j(t+1) = \mathbf{p}_j(t) \pm \beta \cdot \left|\mathbf{p}_j(t) - \mathbf{x}_j(t)\right| \cdot \log(1/U),$$

where U is a uniformly distributed random variable, β is a coefficient to be chosen, and $\mathbf{p}_j(t)$ is a random combination of the personal and the global best. The $\pm$ is resolved by the flip of a fair coin, i.e., $+$ and $-$ have both 50% probability.
- Some algorithms use Lévy flights, i.e., the shocks to current positions are generated by using heavy-tailed distributions, which is contrasted against the normal distribution underlying geometric Brownian motion and standard random walks.
- In the firefly algorithm, the quality of each solution corresponds to the light intensity of a firefly attracting other particles.

16.3 Linear programming

There are a few alternative methods to solve LP problems, but the following two are the most popular ones:

1. The **simplex method** is the "original" method, developed by George Dantzig in 1947. The feasible set for an LP problem is a polyhedron, and we know that, if the problem is feasible and the optimal value is bounded, a vertex will be an optimal solution (see Fig. 15.6). The simplex method implements a clever strategy to explore vertices until a locally optimal solution is found, which is also globally optimal, given problem convexity. The number of vertices is finite but potentially huge, so one should wonder about the efficiency of the approach. Indeed, a potential pitfall of the simplex method is that its computational complexity is, in the worst-case, exponential in the size of the problem. We cannot discuss computational complexity in any detail here, but it sufficient to say that it relates the size of a problem with the number of operations carried out by a solution algorithm. Algorithms featuring polynomial complexity are typically preferred to those with exponential complexity. However, the number of vertices actually explored is, in practice, usually quite limited. Furthermore, the performance of the method has been impressively improved over the years, thanks to developments in numerical linear algebra.

2. Given the exponential complexity of the simplex algorithm, a natural question was whether a polynomial algorithm could be developed. The quest for a polynomial algorithm for LP problems ended with a success in 1984, many years after the development of the simplex method, with Karmarkar's projective algorithm. However, despite nice theoretical properties, this is not an industrial strength approach. Truly competitive methods were developed later, based on logarithmic barriers and primal–dual algorithms. These methods are called **interior-point** methods, since they follow a path within the interior of the feasible set. Later, interior-point methods for LP were extended to the more general class of conic optimization problems.

For several large-scale problems, interior-point methods do outperform the simplex method, but this is not guaranteed to happen, as exponential complexity refers to the worst case, rather than to the average case. Apart from computational efficiency, a solution from simplex method tends to be qualitatively different, as it is more *sparse*, which means that more variables are set to zero. When there are alternative optima, located on a face of the feasible polyhedron, the simplex method yields an extreme solution corresponding to a vertex, whereas an interior-point method will yield a solution corresponding to the center of this face. Since a more extreme solution may be sometimes preferred, software packages offer crossover strategies to find an optimal vertex, in case of alternative optima, given an optimal solution found by an interior-point method. Furthermore, simplex has better warm-start capability, which is very useful for integer programming, as we shall see later. They rely on LP duality, which is also fundamental in primal-dual interior-point methods and is explored in Section 16.3.2.

16.3.1 THE SIMPLEX METHOD

The simplex method deals with LP problems in standard form,

$$\min \quad \mathbf{c}^T\mathbf{x}$$
$$\text{s.t.} \quad \mathbf{A}\mathbf{x} = \mathbf{b},$$
$$\mathbf{x} \geq \mathbf{0},$$

where $\mathbf{A} \in \mathbb{R}^{m,n}$ and $m < n$. Any LP model may be transformed into the standard form, so there is no loss of generality. As we have pointed out before, we may rely on geometric intuition to solve the problem:

- An LP problem is a convex problem; hence, if we find a locally optimal solution, we have also found a global optimizer.
- An LP problem is a concave problem, too; hence, we know that we may restrict the search for the optimal solution to the boundary of the feasible set, which is a polyhedron. Actually, it can be shown that there is an optimal solution corresponding to a *vertex*, or extreme point, of the polyhedron.
- So, we need a clever way to explore extreme points of the feasible set. Geometrically, we may imagine moving from a vertex to a neighboring one, trying to improve the objective function. When there is no neighboring vertex improving the objective, we have found a local optimizer, which is also global.

This geometric intuition can be translated into algebraic terms as follows. Let us consider the matrix $\mathbf{A}$ as the collection of its column vectors $\mathbf{a}_j \in \mathbb{R}^m$, for $j = 1, \ldots, n$:

$$\mathbf{A} = \begin{bmatrix} \vdots & \vdots & & \vdots \\ \mathbf{a}_1 & \mathbf{a}_2 & \cdots & \mathbf{a}_n \\ \vdots & \vdots & & \vdots \end{bmatrix}.$$

The system of linear equations $\mathbf{A}\mathbf{x} = \mathbf{b}$ states that we are expressing the right-hand side vector $\mathbf{b}$ as a linear combination of columns of $\mathbf{A}$,

$$\sum_{j=1}^{n} \mathbf{a}_j x_j = \mathbf{b}.$$

Since we also require $\mathbf{x} \geq \mathbf{0}$, this not just a linear combination, but it is a *conic* combination. There are n columns in matrix $\mathbf{A}$, but a subset $\mathcal{B}$ of m columns suffices to express $\mathbf{b}$ as

$$\sum_{j \in \mathcal{B}} \mathbf{a}_j x_j = \mathbf{b}.$$

16.3 Linear programming

To be precise, we should make sure that the subset of columns $\mathcal{B}$ is a basis, i.e., that the columns are linearly independent; let us cut a few corners and assume that this is the case.

Given a basis $\mathcal{B}$, we have a subset of **basic variables**, $\{x_j \mid j \in \mathcal{B}\}$, which may take a nonzero value, and a set of **nonbasic variables**, $\{x_j \mid j \notin \mathcal{B}\}$, which are forced to zero. A solution corresponding to a basis is called a **basic solution**. If, in addition, the basic variables are non-negative, as it is required, we have a **basic feasible solution**. The simplex algorithm relies on a fundamental result, which we state loosely and without any proof:

> *Basic feasible solutions correspond to extreme points of the polyhedral feasible set of the LP problem.*

Thus, we may explore vertices by enumerating the finite set of possible bases in a clever way. We may do that, because it can also be shown that we move from a vertex to an adjacent vertex by swapping a basic variable for a nonbasic one, i.e., by changing a single element of the basis (provided that the resulting basis is also feasible).

Let us formalize the intuition in an algorithmically workable way. If we have a feasible basis, we can partition the vector $\mathbf{x}$ into two subvectors: the subvector $\mathbf{x}_B \in \mathbb{R}^m$ of the basic variables and the subvector $\mathbf{x}_N \in \mathbb{R}^{n-m}$ of the nonbasic variables. Using a suitable permutation of variables and corresponding columns, we may rewrite the system of linear equations

$$\mathbf{A}\mathbf{x} = \mathbf{b}$$

as

$$\begin{bmatrix} \mathbf{A}_B \mathbf{A}_N \end{bmatrix} \begin{bmatrix} \mathbf{x}_B \\ \mathbf{x}_N \end{bmatrix} = \mathbf{A}_B \mathbf{x}_B + \mathbf{A}_N \mathbf{x}_N = \mathbf{b}, \qquad (16.19)$$

where $\mathbf{A}_B \in \mathbb{R}^{m,m}$ is nonsingular and $\mathbf{A}_N \in \mathbb{R}^{m,n-m}$.

Solving an LP amounts to finding a way to express $\mathbf{b}$ as a least-cost linear combination of at most m columns of $\mathbf{A}$, with non-negative coefficients. Assume that we have a basic feasible solution $\mathbf{x}$; we will consider later how to obtain an initial basic feasible solution. If $\mathbf{x}$ is basic feasible, it may be written as

$$\mathbf{x} = \begin{bmatrix} \mathbf{x}_B \\ \mathbf{x}_N \end{bmatrix} = \begin{bmatrix} \widehat{\mathbf{b}} \\ \mathbf{0} \end{bmatrix},$$

where

$$\widehat{\mathbf{b}} = \mathbf{A}_B^{-1} \mathbf{b} \geq \mathbf{0}.$$

Then, the value of the objective function corresponding to $\mathbf{x}$ is

$$\widehat{f} = \begin{bmatrix} \mathbf{c}_B^\mathsf{T} & \mathbf{c}_N^\mathsf{T} \end{bmatrix} \begin{bmatrix} \widehat{\mathbf{b}} \\ \mathbf{0} \end{bmatrix} = \mathbf{c}_B^\mathsf{T} \widehat{\mathbf{b}}. \qquad (16.20)$$

Now, we should check whether there are adjacent vertices improving this value. Adjacent vertices may be obtained by swapping a column in the basis with a column outside the basis. This means that one nonbasic variable is brought into the basis, and one basic variable leaves the basis.

To assess the potential benefit of introducing a nonbasic variable into the basis, we should express the objective function in terms of nonbasic variables. To this aim, we rewrite the objective function in Eq. (16.20), making its dependence on nonbasic variables explicit. Using Eq. (16.19), we may express the basic variables as

$$\mathbf{x}_B = \mathbf{A}_B^{-1}(\mathbf{b} - \mathbf{A}_N \mathbf{x}_N) = \widehat{\mathbf{b}} - \mathbf{A}_B^{-1} \mathbf{A}_N \mathbf{x}_N. \tag{16.21}$$

Then, we rewrite the objective function in terms of nonbasic variables only:

$$\begin{aligned} \mathbf{c}^\mathsf{T} \mathbf{x} &= \mathbf{c}_B^\mathsf{T} \mathbf{x}_B + \mathbf{c}_N^\mathsf{T} \mathbf{x}_N \\ &= \mathbf{c}_B^\mathsf{T} \left(\widehat{\mathbf{b}} - \mathbf{A}_B^{-1} \mathbf{A}_N \mathbf{x}_N \right) + \mathbf{c}_N^\mathsf{T} \mathbf{x}_N \\ &= \mathbf{c}_B^\mathsf{T} \widehat{\mathbf{b}} + \left(\mathbf{c}_N^\mathsf{T} - \mathbf{c}_B^\mathsf{T} \mathbf{A}_B^{-1} \mathbf{A}_N \right) \mathbf{x}_N \\ &= \widehat{f} + \widehat{\mathbf{c}}_N^\mathsf{T} \mathbf{x}_N, \end{aligned}$$

where

$$\widehat{\mathbf{c}}_N^\mathsf{T} \doteq \mathbf{c}_N^\mathsf{T} - \mathbf{c}_B^\mathsf{T} \mathbf{A}_B^{-1} \mathbf{A}_N. \tag{16.22}$$

The quantities in vector $\widehat{\mathbf{c}}_N$ are called **reduced costs**, as they measure the marginal variation of the objective function, when the value of a nonbasic variable is changed, while preserving overall feasibility of the solution. If $\widehat{\mathbf{c}}_N \geq \mathbf{0}$, then bringing any nonbasic variable into the basis at some positive value cannot reduce the overall cost. Hence, it is not possible to improve the current objective function and the current basis is optimal. If, on the contrary, there exists a $q \in N$ such that $\widehat{c}_q < 0$, then, it is possible to improve the objective function by bringing x_q into the basis. When we bring x_q into the basis, with a positive value, two cases may occur:

1. The increase of x_q has the effect of decreasing some currently basic and positive variable. Since we must preserve non-negativity of the solution, the limit value of x_q in the new basis is determined by the first basic value which takes a zero value, leaving the basis. Now we have a new feasible basis and the whole process is repeated.

2. The increase of x_q does not create any feasibility problem. However, this means that we may bring x_q into the basis, improving the objective without any bound. If this happens, the problem has an unbounded solution (which means that a profit goes to $+\infty$ or a cost goes to $-\infty$). For this to happen, the feasible set must be an unbounded polyhedron. This is usually a sign of a modeling error, but in finance, it may also be the consequence of an arbitrage opportunity.

A simple strategy is to choose q such that

$$\widehat{c}_q = \min_{j \in N} \widehat{c}_j. \tag{16.23}$$

In practice, this approach is not efficient, as it ignores the actual value that x_q will take in the new basis, and it may lead to cycles in the algorithm. Commercial software packages avoid this difficulty and implement the simplex method in an efficient and numerically accurate way.

As a final touch, an initial basis is needed in order to start the iterations. One possibility is to rely on a penalty function approach. We introduce a set of auxiliary *artificial variables* **z** into the constraints:

$$\mathbf{Ax} + \mathbf{z} = \mathbf{b}, \tag{16.24}$$
$$\mathbf{x}, \mathbf{z} \geq \mathbf{0}.$$

The artificial variables can be regarded as residuals, in the same vein as residuals in linear regression, which we should bring to zero. Assume also that the equations have been rearranged in such a way that $\mathbf{b} \geq \mathbf{0}$. Clearly, a basic feasible solution of the system (16.24), where $\mathbf{z} = \mathbf{0}$, is also a basic feasible solution for the original system $\mathbf{Ax} = \mathbf{b}$. In order to find such a solution, we can introduce an auxiliary function,

$$\psi = \sum_{i=1}^{m} z_i, \tag{16.25}$$

which is minimized by using the simplex method itself. Finding an initial basic feasible solution for this artificial problem is trivial: $\mathbf{z} = \mathbf{b} \geq \mathbf{0}$. If the optimal value of (16.25) is $\psi^* = 0$, then we have found a starting point for the original problem; otherwise, the original problem is infeasible. The minimization of the auxiliary function is called **phase I** of the simplex algorithm. The actual optimization, starting from a basic feasible solution, is called **phase II** of the simplex algorithm.

16.3.2 DUALITY IN LINEAR PROGRAMMING

We have appreciated the role of LP duality in Section 2.4, where we have discussed the mathematics of arbitrage. LP duality can be derived from first principles of convex analysis,[13] but it may be preferable to cast it within the more general nonlinear framework and apply the concepts of Section 16.1.4.

Let us consider an LP problem (P_1) in the following *canonical* form:

$$(P_1) \quad \min \quad \mathbf{c}^\mathsf{T}\mathbf{x}$$
$$\text{s.t.} \quad \mathbf{Ax} \geq \mathbf{b}.$$

If we dualize the inequality constraints, we obtain the dual problem

$$\max_{\boldsymbol{\mu} \geq \mathbf{0}} \left\{ \min_{\mathbf{x}} \left[\mathbf{c}^\mathsf{T}\mathbf{x} + \boldsymbol{\mu}^\mathsf{T}(\mathbf{b} - \mathbf{Ax}) \right] \right\} = \max_{\boldsymbol{\mu} \geq \mathbf{0}} \left\{ \boldsymbol{\mu}^\mathsf{T}\mathbf{b} + \min_{\mathbf{x}} \left(\mathbf{c}^\mathsf{T} - \boldsymbol{\mu}^\mathsf{T}\mathbf{A} \right) \mathbf{x} \right\}.$$

---

[13]The key concepts are separation theorems and Farkas' lemma.

However, **x** is unrestricted in sign, so the inner minimization problem with respect to **x** will take only two possible values, $-\infty$ or 0; the latter case occurs when all of the coefficients multiplying the primal variables are zero. Since we want to maximize the dual function, we should choose the dual variables μ such that

$$\mathbf{c}^T - \mu^T \mathbf{A} = \mathbf{0}.$$

Then, the dual problem (D_1) turns out to be

$$(D_1) \quad \max \quad \mathbf{b}^T \mu$$
$$\text{s.t.} \quad \mathbf{A}^T \mu = \mathbf{c},$$
$$\mu \geq \mathbf{0}.$$

We observe that the dual problem is still an LP problem, resulting from:

1. The exchange of **b** with **c**
2. The transposition of **A**
3. A change in the sense of the objective

Using the same reasoning, we may build the dual of an arbitrary LP problem.[14] Note, in particular, that dual variables associated with equality constraints will be unrestricted in sign. We may also observe that the dual of the dual is the primal. Hence, by flipping problems (P_1) and (D_1) a bit, we see that the primal problem

$$(P_2) \quad \min \quad \mathbf{c}^T \mathbf{x}$$
$$\text{s.t.} \quad \mathbf{A}\mathbf{x} = \mathbf{b},$$
$$\mathbf{x} \geq \mathbf{0}$$

corresponds to the dual

$$(D_2) \quad \min \quad \mathbf{b}^T \mu$$
$$\text{s.t.} \quad \mathbf{A}^T \mu \leq \mathbf{c}.$$

In the LP case, strong duality applies and, since there is no duality gap, if both primal and dual problems have a finite optimum, we have

$$\mathbf{c}^T \mathbf{x}^* = \mathbf{b}^T \mu^*. \tag{16.26}$$

However, other cases are possible, assuming a primal minimization problem:

- The primal is unbounded below, and the dual is infeasible
- The dual is unbounded above, and the primal is infeasible
- Both problems are infeasible

---

[14] See Section S2.2.

16.3 Linear programming

In fact, in Section 2.4, we relied on these properties to find conditions ensuring absence of arbitrage opportunities. The trick was based on setting up an LP problem with a trivial objective function identically zero, finding conditions under which it was feasible or not, and drawing a conclusion about the dual.

Another interesting observation is that the optimality condition for the primal LP, i.e., non-negativity of the reduced costs, boils down to dual feasibility. To see this, observe that, if the primal and the dual are both feasible, we may rewrite Eq. (16.26) using Eq. (16.20):

$$\mathbf{c}_B^\mathsf{T} \mathbf{A}_B^{-1} \mathbf{b} = (\boldsymbol{\mu}^*)^\mathsf{T} \mathbf{b},$$

which implies (for the optimal basis)

$$(\boldsymbol{\mu}^*)^\mathsf{T} = \mathbf{c}_B^\mathsf{T} \mathbf{A}_B^{-1}.$$

Hence, Eq. (16.22) becomes (for nonbasic variables)

$$\widehat{\mathbf{c}}_N^\mathsf{T} = \mathbf{c}_N^\mathsf{T} - (\boldsymbol{\mu}^*)^\mathsf{T} \mathbf{A}_N.$$

Furthermore, the reduced costs for basic variables are zero, so, from the non-negativity condition of reduced costs, we obtain the condition

$$\mathbf{c}^\mathsf{T} - \boldsymbol{\mu}^\mathsf{T} \mathbf{A} \geq \mathbf{0},$$

which is just feasibility in problem (D_2).

Apart from theoretical properties, LP duality has a deep computational value.

- Rather than solving the primal problem, we may apply the simplex algorithm to the dual. By doing so, we obtain the **dual simplex algorithm**. In fact, this is the default option in several software packages, as this is often faster than the primal simplex.

- Another interesting observation is that, if we solve an LP problem and add a constraint that is violated by the optimal solution, we lose primal feasibility. Hence, if we want to solve the problem with the additional constraint by primal simplex, we should start again from Phase I. However, it turns out that the previous optimal solution is still dual feasible and, if we apply dual simplex, we may warm-start directly with phase II. As we shall see later, this is what we do in branch-and-bound methods, in order to cut a noninteger optimal solution of the LP relaxation of a MILP model.

- Primal and dual algorithms may be merged into **primal-dual** algorithms. It can be shown that if we have a primal feasible solution $\mathbf{x}^*$ and a dual feasible solution $\boldsymbol{\mu}^*$, which also satisfy the complementary slackness condition [see Eq. (16.9)], then they are optimal for the primal and the dual LP, respectively. This is used in the primal-dual interior-point algorithm of the next section.

16.3.3 INTERIOR-POINT METHODS: PRIMAL-DUAL BARRIER METHOD FOR LP

Interior-point methods, unlike the simplex method, move *inside* the feasible region and approach the optimal solution on its boundary by decreasing a barrier function. They mix different ideas based on Lagrangian methods, duality theory, and penalty functions. Actually, there are plenty of variations of the theme, which are very well treated in [6]. Here, we describe a primal-dual version, following the treatment in [23].

Let us consider a primal problem and its transformation to standard form:

$$\begin{array}{ll} \max & \mathbf{c}^T\mathbf{x} \\ \text{s.t.} & \mathbf{Ax} \leq \mathbf{b}, \\ & \mathbf{x} \geq \mathbf{0} \end{array} \quad \Rightarrow \quad \begin{array}{ll} \max & \mathbf{c}^T\mathbf{x} \\ \text{s.t.} & \mathbf{Ax} + \mathbf{w} = \mathbf{b}, \\ & \mathbf{x}, \mathbf{w} \geq \mathbf{0}. \end{array}$$

Here, we introduce slack variables $\mathbf{w} \geq \mathbf{0}$ in order to characterize the feasible set in terms of a system of linear equations, which we are good at dealing with numerically, and a non-negativity constraint stating that the solution must lie in the positive orthant (which is a cone). By a similar token, we consider the dual and its transformation to standard form:

$$\begin{array}{ll} \min & \mathbf{b}^T\mathbf{y} \\ \text{s.t.} & \mathbf{A}^T\mathbf{y} \geq \mathbf{c}, \\ & \mathbf{y} \geq \mathbf{0} \end{array} \quad \Rightarrow \quad \begin{array}{ll} \min & \mathbf{b}^T\mathbf{y} \\ \text{s.t.} & \mathbf{A}^T\mathbf{y} - \mathbf{z} = \mathbf{c}, \\ & \mathbf{y}, \mathbf{z} \geq \mathbf{0}. \end{array}$$

The vector $\mathbf{z}$ collects the slack variables. We also note that the dual variables $\mathbf{y}$ correspond to Lagrange multipliers of the constraints of the primal, and vice versa. Multipliers are restricted in sign, as the original problem includes inequality constraints.

Let us consider the primal problem in standard form. We can get rid of the non-negativity constraints by using an interior penalty function based on a logarithmic barrier:

$$\begin{array}{ll} \max & \mathbf{c}^T\mathbf{x} + \sigma \sum_j \log x_j + \sigma \sum_i \log w_i \\ \text{s.t.} & \mathbf{Ax} + \mathbf{w} = \mathbf{b}. \end{array}$$

Then, equality constraints can be dualized by Lagrange multipliers $\mathbf{y}$, yielding the Lagrangian function

$$\mathcal{L}(\mathbf{x}, \mathbf{w}, \mathbf{y}) = \mathbf{c}^T\mathbf{x} + \sigma \sum_j \log x_j + \sigma \sum_i \log w_i + \mathbf{y}^T(\mathbf{b} - \mathbf{Ax} - \mathbf{w}).$$

Now, we can apply the first-order stationarity conditions on the Lagrangian:

$$\frac{\partial \mathcal{L}}{\partial x_j} = c_j + \sigma \frac{1}{x_j} - \sum_i y_i a_{ij} = 0, \quad \forall j$$

16.3 Linear programming

$$\frac{\partial \mathcal{L}}{\partial w_i} = \sigma \frac{1}{w_i} - y_i = 0, \qquad \forall i$$

$$\frac{\partial \mathcal{L}}{\partial y_i} = b_i - \sum_j a_{ij} x_j - w_i = 0, \qquad \forall i.$$

These optimality conditions may be rewritten in a compact matrix form,

$$\mathbf{A}^T \mathbf{y} - \sigma \mathbf{X}^{-1} \mathbf{i} = \mathbf{c}$$
$$\mathbf{y} = \sigma \mathbf{W}^{-1} \mathbf{i}$$
$$\mathbf{A}\mathbf{x} + \mathbf{w} = \mathbf{b},$$

where we introduce a diagonal matrix $\mathbf{X}$ and use the familiar vector $\mathbf{i}$:

$$\mathbf{X} = \begin{bmatrix} x_1 & & & \\ & x_2 & & \\ & & \ddots & \\ & & & x_n \end{bmatrix}, \qquad \mathbf{i} = \begin{bmatrix} 1 \\ 1 \\ \vdots \\ 1 \end{bmatrix}$$

To make the meaning of the optimality conditions clearer, let us introduce the auxiliary vector

$$\mathbf{z} = \sigma \mathbf{X}^{-1} \mathbf{i},$$

and let us rearrange the conditions as follows:

$$\mathbf{A}\mathbf{x} + \mathbf{w} = \mathbf{b}, \tag{16.27}$$
$$\mathbf{A}^T \mathbf{y} - \mathbf{z} = \mathbf{c}, \tag{16.28}$$
$$\mathbf{XZi} = \sigma \mathbf{i}, \tag{16.29}$$
$$\mathbf{YWi} = \sigma \mathbf{i}. \tag{16.30}$$

These equations have a nice interpretation in terms of:

1. Primal feasibility, Eq. (16.27)
2. Dual feasibility, Eq. (16.28)
3. Complementary slackness (in the limit, for a penalty coefficient $\sigma \to 0$) for the dual, Eq. (16.29), and the primal, Eq. (16.30)

For $\sigma > 0$, we have a set of nonlinear equations,

$$\mathbf{F}(\boldsymbol{\xi}) = \mathbf{0},$$

where

$$\boldsymbol{\xi} = \begin{bmatrix} \mathbf{x} \\ \mathbf{y} \\ \mathbf{w} \\ \mathbf{z} \end{bmatrix},$$

which may be tackled by numerical iterative methods, like Newton method for nonlinear equations. In principle, by solving this system of nonlinear equations for different values of σ, we generate a path $(\mathbf{x}_\sigma, \mathbf{y}_\sigma, \mathbf{w}_\sigma, \mathbf{z}_\sigma)$. This path is called the **central path** and, for $\sigma \to 0$, it leads to the optimal solution of the original LP. There are different approaches to manage the interplay between the Newton's steps and the adjustment of the penalty parameter σ.

Interior-point methods feature a polynomial complexity, which is an advantage over the older simplex method. However, they are less suited to warm start after adding a constraint, which is relevant, as we shall see, for branch-and-bound methods.

16.4 Conic duality and interior-point methods

Both LP duality and interior-point algorithms can be generalized to deal with conic optimization problems. In this section, we give a glimpse of how this is accomplished, in order to illustrate the beauty and elegance of computational optimization. We refrain from giving algorithmic details, which are better left to the specialized literature.

We have already pointed out, in Section 15.5, that we may frame an LP model in standard form within the more general framework of conic programming. Let us express a primal conic programming problem as follows:

$$(P) \quad \min \quad \langle \mathbf{c}, \mathbf{x} \rangle \tag{16.31}$$
$$\text{s.t.} \quad \mathbf{b} - A(\mathbf{x}) \in L,$$
$$\mathbf{x} \in K,$$

where $\mathbf{c} \in V$, $\mathbf{b} \in W$, V and W are finite-dimensional linear spaces equipped with an inner product, $K \subseteq V$ and $L \subseteq W$ are closed convex cones, and $A : V \longrightarrow W$ is a linear operator. This abstract framework includes LP as well as SOCP and SDP problems, which have wide applicability, including some stochastic and robust optimization problems, as we have shown in Chapter 15.

In the next section, we show how LP duality may be generalized to find the dual of problem (16.31). Then, we show how logarithmic barrier functions may be devised for the second-order cone and the cone of positive semidefinite matrices, paving the way to interior-point methods for SOCP and SDP.

16.4.1 CONIC DUALITY

To build the dual of the primal problem (P) defined in Eq. (16.31), we need a couple of ingredients:

Dual cones. We have introduced dual cones in Section 15.5.1.1. We may find the dual cones L^* and K^*, corresponding to L and K, respectively. In

16.4 Conic duality and interior-point methods

particular, we recall the following pairs of dual cones:

$$[\mathbb{R}^n]^* = \{\mathbf{0}\}, \quad [\mathbb{R}^n_+]^* = \mathbb{R}^n_+, \quad [\mathbb{S}^n_+]^* = \mathbb{S}^n_+,$$

i.e., the dual of the whole space $\mathbb{R}^n$ is the origin, and vice versa, and the positive orthant and the cone of symmetric, positive semidefinite matrices are self-dual.

Adjoint operator. Given a linear operator $A : V \longrightarrow W$, we define its adjoint operator A^T as a linear operator $A^\mathsf{T} : W \longrightarrow V$ such that

$$\langle \mathbf{y}, A(\mathbf{x}) \rangle_W = \langle A^\mathsf{T}(\mathbf{y}), \mathbf{x} \rangle_V,$$

for every $\mathbf{x} \in V$ and $\mathbf{y} \in W$. Note that we use the inner products in spaces W and V, which need not be the same.

The idea of an adjoint operator may sound quite mysterious, but we may regard it as a generalization of the matrix transposition that is involved in LP duality. We should interpret matrices as linear mappings between the familiar vector spaces $\mathbb{R}^m$ and $\mathbb{R}^n$, in which case the adjoint boils down to the familiar transpose of a matrix. To see this, consider a matrix $\mathbf{A} \in \mathbb{R}^{m \times n}$ as a linear operator mapping a vector $\mathbf{x} \in \mathbb{R}^n$ into the vector $\mathbf{y} = \mathbf{A}\mathbf{x} \in \mathbb{R}^m$, and the standard inner products in $\mathbb{R}^m$ and $\mathbb{R}^n$,

$$\langle \mathbf{x}_a, \mathbf{x}_b \rangle_{\mathbb{R}^n} \doteq \mathbf{x}_a^\mathsf{T} \mathbf{x}_b = \sum_{i=1}^n x_{ai} x_{bi},$$

$$\langle \mathbf{y}_a, \mathbf{y}_b \rangle_{\mathbb{R}^m} \doteq \mathbf{y}_a^\mathsf{T} \mathbf{y}_b = \sum_{j=1}^m y_{aj} y_{bj}.$$

Then, it is easy to see that the adjoint operator of $\mathbf{A}$ is the transpose $\mathbf{A}^\mathsf{T}$:

$$\langle \mathbf{y}, \mathbf{A}\mathbf{x} \rangle_{\mathbb{R}^m} = \mathbf{y}^\mathsf{T} \mathbf{A}\mathbf{x} = \mathbf{x}^\mathsf{T} \mathbf{A}^\mathsf{T} \mathbf{y} = \langle \mathbf{x}, \mathbf{A}^\mathsf{T} \mathbf{y} \rangle_{\mathbb{R}^n}.$$

Given all of the above, we can define a conic dual problem, and the following conic duality theorem can be proven.

THEOREM 16.3 (Conic duality) *Consider the primal conic optimization problem* (P) *of Eq. (16.31) and its conic dual,*

$$
\begin{aligned}
(D) \quad & \max \quad \langle \mathbf{b}, \mathbf{y} \rangle \\
& \text{s.t.} \quad A^\mathsf{T}(\mathbf{y}) - \mathbf{c} \in K^*, \\
& \quad\quad\;\; \mathbf{y} \in L^*.
\end{aligned}
$$

If the primal problem (P) *is feasible, has a finite value* γ, *and has an interior (Slater) point* $\widehat{\mathbf{x}}$, *then the dual problem* (D) *is also feasible and has the same value* γ.

The Slater constraint qualification condition is necessary, in general, to rule out some pathological cases resulting in duality gaps, but it may be a stronger condition than necessary. For instance, it is not required in LP duality, but it is essential in SDP duality.

Now, let us see how the conic duality theorem generalizes LP duality:

- If the primal is a minimization problem, the dual is a maximization problem.
- The matrix **A** is transposed and there is a swap between the cost vector **c** and the right-hand side vector **b**.
- If we have inequality constraints in the primal ($L = \mathbb{R}^n_+$), the dual variables **y** are restricted in sign ($L^* = \mathbb{R}^n_+$).
- If we have equality constraints in the primal ($L = \{\mathbf{0}\}$), the dual variables **y** are unrestricted in sign ($L^* = \mathbb{R}^n$).
- If the primal variables **x** are restricted in sign ($K = \mathbb{R}^n_+$), we have inequality constraints in the dual ($L^* = \mathbb{R}^n_+$).
- If the primal variables **x** are not restricted in sign ($K = \mathbb{R}^n$), we have equality constraints in the dual ($L^* = \{\mathbf{0}\}$).

As an illustration, let us see how conic duality may be applied to SDP problems. Consider the primal SDP problem in standard form:

$$\begin{aligned}
\max \quad & \mathbf{C} \bullet \mathbf{X} \\
\text{s.t.} \quad & \mathbf{A}_1 \bullet \mathbf{X} = b_1, \\
& \vdots \\
& \mathbf{A}_m \bullet \mathbf{X} = b_m, \\
& \mathbf{X} \succeq \mathbf{0},
\end{aligned} \tag{16.32}$$

where we interpret the matrix inner product • as in Section 15.5.1.2.

Strong duality applies to SDP, since the cone $\mathbb{S}^n_+$ of semidefinite positive matrices has an interior, the open cone $\mathbb{S}^n_{++}$ of positive definite matrices. To be precise, there must be a symmetric positive definite matrix $\widehat{\mathbf{X}}$ such that the equality constraints hold, $\mathbf{A}(\widehat{\mathbf{X}}) = \mathbf{b}$, where $\mathbf{b} = [b_1, \ldots, b_m]^\mathsf{T}$, and $\mathbf{A}(\cdot)$ collects the matrices $\mathbf{A}_i$ and maps $\mathbb{S}^n_+$ to $\mathbb{R}^m$.

We also have to find the adjoint operator corresponding to the linear operator implicit in the equality constraints of the primal, which may be written as the following map from $\mathbb{S}^n_+$ to $\mathbb{R}^m$:

$$A(\mathbf{X}) = \begin{bmatrix} \mathbf{A}_1 \bullet \mathbf{X} \\ \mathbf{A}_2 \bullet \mathbf{X} \\ \vdots \\ \mathbf{A}_m \bullet \mathbf{X} \end{bmatrix} = \begin{bmatrix} b_1 \\ b_2 \\ \vdots \\ b_m \end{bmatrix} \in \mathbb{R}^m.$$

In order to find the required adjoint operator, we may start from the familiar inner product in $\mathbb{R}^m$. So, let us introduce a vector $\mathbf{y} \in \mathbb{R}^m$ and proceed as follows:

$$\langle \mathbf{y}, A(\mathbf{X}) \rangle_{\mathbb{R}^m} = \langle \mathbf{y}, \mathbf{b} \rangle_{\mathbb{R}^m} = \sum_{i=1}^m y_i b_i = \sum_{i=1}^m y_i (\mathbf{A}_i \bullet \mathbf{X})$$

$$= \left(\sum_{i=1}^m y_i \mathbf{A}_i \right) \bullet \mathbf{X} = \Big\langle \sum_{i=1}^m y_i \mathbf{A}_i, \mathbf{X} \Big\rangle_{\mathbb{S}_+^n},$$

where we have used linearity of the inner product. We observe that the adjoint operator $\mathbf{A}^\mathsf{T} : \mathbb{R}^m \to \mathbb{S}_+^n$ is the matrix $\sum_{i=1}^m y_i \mathbf{A}_i$. Thus, the dual of problem (16.32) is:

$$\begin{aligned} \min \quad & \mathbf{b}^\mathsf{T} \mathbf{y} \\ \text{s.t.} \quad & \sum_{i=1}^m y_i \mathbf{A}_i - \mathbf{C} \succeq 0, \quad \mathbf{y} \in \mathbb{R}^m, \end{aligned}$$

which is an SDP written in linear matrix inequality form. Thus, we see that the two SDP forms introduced in Section 15.5.3 are actually equivalent.

16.4.2 INTERIOR-POINT METHODS FOR SOCP AND SDP

If we look at statements of SOCP and SDP problems, they are basically LP problems with complicating conic constraints. In SOCP, we have

$$\| \mathbf{x} \|_2 \leq y, \tag{16.33}$$

where $\mathbf{x} \in \mathbb{R}^n$, and in SDP we have

$$\mathbf{X} \succeq \mathbf{0}, \tag{16.34}$$

where $\mathbf{X} \in \mathbb{S}^n$. To devise primal-dual interior-point methods, a barrier function is needed to relax these constraints. It turns out that logarithmic barriers are quite convenient.

For the second-order cone, the following barrier function can be used:

$$B(\mathbf{x}, y) = \log \left(y^2 - \sum_{i=1}^n x_i^2 \right).$$

For SDP, we observe that the determinant of a positive definite matrix is strictly positive, and it goes to zero for a singular matrix, which is on the boundary of the cone $\mathbb{S}_+^n$. Hence, just like the case of non-negativity constraints in LP, we might consider the logarithmic barrier

$$\log\big(\det(\mathbf{X}) \big).$$

At first sight, this is a weird and pretty bad choice, as computing the determinant of a large matrix may be cumbersome. However, we do not really need to evaluate the determinant. What we actually need, is to apply the Newton method for nonlinear equations, in order to follow the central path, just like in the LP case. To this aim, we need derivatives of the logarithmic barrier function. We recall that, for a non-negative scalar variable $x \in \mathbb{R}_+$, we have

$$\frac{d \log x}{dx} = \frac{1}{x}.$$

It turns out that a surprisingly similar formula applies to the log-barrier for positive semidefinite matrices:

$$\frac{d \log(\det(\mathbf{X}))}{d\mathbf{X}} = \mathbf{X}^{-1},$$

i.e., we need the inverse of the matrix $\mathbf{X}$. This is not so difficult to prove, but it requires the definition of a matrix derivative, which we wish to skip. The essential message is that the same machinery that we use in interior-point methods for LP can be extended to SOCP and SDP models.

16.5 Branch-and-bound methods for integer programming

We have already pointed out, in Section 15.4, that the MILP model

$$P(S) \quad \min \quad \mathbf{c}^\mathsf{T}\mathbf{x} + \mathbf{d}^\mathsf{T}\mathbf{y} \quad (16.35)$$
$$\text{s.t.} \quad \mathbf{Ax} + \mathbf{Ey} \leq \mathbf{b},$$
$$\mathbf{x} \in \mathbb{R}_+^{n_1}, \quad \mathbf{y} \in \mathbb{Z}_+^{n_2}.$$

is not associated with any easy-to-check optimality condition. A possible solution approach relies on some form of enumeration of the feasible solutions, by successive partitions of the feasible region S, which may be explored by a tree search procedure. The idea is more general and can be applied to continuous global optimization as well. To be more specific, let us consider a partition of the feasible set S, i.e., a collection of subsets $S_k \subset S$, $k = 1, \ldots, m$, such that:

1. Subsets are mutually disjoint,

$$S_i \cap S_j = \emptyset, \quad i \neq j.$$

2. Subsets are collectively exhaustive,

$$\bigcup_{k=1}^{m} S_k = S.$$

We associate a subproblem $P(S_k)$ with each subset of S. In practice, the partition is usually created by **branching** on an integer decision variable. For a

16.5 Branch-and-bound methods for integer programming

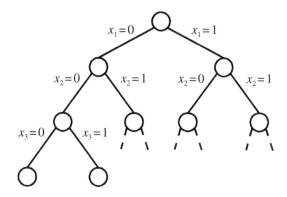

FIGURE 16.7 Search tree for a pure binary LP model.

binary decision variable $x_j \in \{0,1\}$, we may split the feasible set S into two subsets, where we set $x_j = 0$ and $x_j = 1$, respectively. In the case of a pure binary LP model like the knapsack problem of Example 15.8, where all variables are either 0 or 1, it is easy to see that, by successive partitions, we may explore the whole feasible set by a search tree like the one in Fig. 16.7. To branch on a general integer variable $x_j \in \mathbb{Z}_+ = \{0, 1, 2, \ldots\}$, we may choose a positive integer number q and generate two subsets by adding the constraints $x_j \leq q$ and $x_j \geq q + 1$, respectively. Note that, by adding these constraints, we are cutting away a strip of the feasible set, which does not contain any feasible integer solution. In an MILP problem, there is no need to branch on continuous variables, as they do not create any difficulty.[15] By branching, we generate a tree of subproblems. We need not generate exactly two child subproblems per node, but this is the most common branching choice.

A search tree generated in this way, barring some pathological cases with an unbounded feasible set, allows us to explore the whole feasible set S systematically, but it has exponential complexity. For an LP problem involving n binary decision variables, we may create up to 2^n subproblems, even though some of them will turn out to be infeasible. To avoid excessive branching, we adopt a **bounding** strategy to eliminate subproblems. Let us assume that we know a feasible (not necessarily optimal) solution $\mathbf{x}^\circ$. In a minimization problem

$$\min_{\mathbf{x} \in S} f(\mathbf{x}),$$

the objective value $f(\mathbf{x}^\circ)$ is clearly an **upper bound** on the optimal value f^*, as an optimal solution $\mathbf{x}^*$ cannot be worse than the feasible solution $\mathbf{x}^\circ$. Let us also assume that, for each subproblem $P(S_k)$, we can find a **lower bound** $\beta(S_k)$ such that

$$\beta(S_k) \leq \min_{\mathbf{x} \in S_k} f(\mathbf{x}).$$

---

[15] See, e.g., [13] for branching strategies in continuous global optimization.

Any subset with a bound that is *not* smaller than the upper bound is not worth exploring, as we have a guarantee that we will not improve the solution $\mathbf{x}^\circ$ by solving the subproblem corresponding to that subset. Formally, if

$$\beta(S_k) \geq f(\mathbf{x}^\circ),$$

then the subproblem $P(S_k)$ may be eliminated from further consideration. This corresponds to **pruning** a branch of the search tree, with a corresponding computational saving. For a maximization problem, the roles of lower and upper bounds are swapped.

One way to find lower bounds is to relax the feasible set, since

$$S \subseteq T \quad \Rightarrow \quad \min_{\mathbf{x} \in S} f(\mathbf{x}) \geq \min_{\mathbf{x} \in T} f(\mathbf{x}).$$

For an MILP model like problem (16.35), a lower bound is easily found by taking its **continuous (LP) relaxation**:

$$P(\overline{S}) \quad \min \quad \mathbf{c}^T\mathbf{x} + \mathbf{d}^T\mathbf{y}$$
$$\text{s.t.} \quad \mathbf{A}\mathbf{x} + \mathbf{E}\mathbf{y} \leq \mathbf{b},$$
$$\begin{bmatrix} \mathbf{x} \\ \mathbf{y} \end{bmatrix} \in \mathbb{R}_+^{n_1+n_2},$$

characterized by a relaxed feasible set $\overline{S}$, which is easily solved by the simplex or the primal–dual barrier methods. Commercial branch-and-bound software for MILP models is based on such LP relaxations. The relaxed problem at the root of the search tree will (usually) yield a noninteger solution, i.e., a solution where the integrality requirement is violated for some integer variable (otherwise, we already have an optimal solution and stop immediately). A variable that does not meet the integrality requirement is selected and branched on, in order to generate two child subproblems, and the process is repeated recursively. When we branch, we add constraints to each subproblem and generate a finer and finer partition, consisting of smaller and smaller subsets. This has two effects:

1. Lower bounds tend to increase when we proceed down the search tree.
2. Sooner or later (ruling out pathologies) an integer solution will be found, providing us with an **incumbent** solution and an upper bound.

The incumbent solution is the current best integer solution, not necessarily an optimal one. When solving a subproblem on the search tree, sometimes we find a better incumbent, which helps pruning the tree; in other cases, we eliminate a subproblem, because it is infeasible or cannot improve the incumbent. Otherwise, we choose a branching variable and add the child nodes to the tree of open problems.

Here is an outline of a possible branch-and-bound strategy.

1. *Initialization.* The list of open subproblems is initialized to $P(S)$; the value of the incumbent solution ν° is set to $+\infty$. At each step, the incumbent solution is the best integer solution found so far.

2. *Selecting a candidate subproblem.* If the list of open subproblems is empty, stop: The incumbent solution $\mathbf{x}^\circ$, if any has been found, is reported as the optimal solution $\mathbf{x}^*$, along with its value; if $\nu^\circ = +\infty$, the original problem was infeasible. Otherwise, select a subproblem $P(S_k)$ from the list.

3. *Bounding.* Compute a lower bound $\beta(S_k)$ on $\nu[P(S_k)]$ by solving a relaxed problem $P(\overline{S}_k)$. Let $\overline{\mathbf{x}}_k$ be the optimal solution of the relaxed subproblem.

4. *Prune by optimality.* If $\overline{\mathbf{x}}_k$ is feasible, prune subproblem $P(S_k)$. Furthermore, if $f(\overline{\mathbf{x}}_k) < \nu^\circ$, update the incumbent solution $\mathbf{x}^\circ$ and its value ν°. Go to Step 2.

5. *Prune by infeasibility.* If the relaxed subproblem $P(\overline{S}_k)$ is infeasible, eliminate $P(S_k)$ from further consideration. Go to Step 2.

6. *Prune by bound.* If $\beta(S_k) \geq \nu^\circ$, eliminate subproblem $P(S_k)$ and go to Step 2.

7. *Branching.* Replace $P(S_k)$ in the list of open subproblems with a list of child subproblems $P(S_{k1}), P(S_{k2}), \ldots, P(S_{kq})$, obtained by partitioning S_k; go to Step 2.

This is just the statement of a principle, and we have to pay attention to many more issues to come up with a working algorithms. Clearly, efficiency is obtained if (in the case of minimization):

1. We obtain tight upper bounds, possibly by invoking high quality heuristics trying to find a good integer solution from a noninteger (but close to feasible) solution.

2. We obtain tight lower bounds by strong model formulations and cut generation, as we discussed in Section 15.4.2.

Furthermore, the exact way in which we select a branching variable and select the next subproblem to be solved may have a significant impact on performance. We may strive for feasibility, in which case we explore subproblems where the number of noninteger variables is small, or optimality, in which case we give priority to promising lower bounds. Another issue is how we solve the LP models at each node of the tree. This is a choice under user control, and a possible strategy is to use an interior-point method for the root problem, then cross over to a basic feasible solution, and use dual simplex at each node to take advantage of warm start.

16.5.1 A MATHEURISTIC APPROACH: FIX-AND-RELAX

Despite the efficiency of branch-and-bound implementations, it may be the case that we cannot find a provably optimal solution within a reasonable amount of time. We may resort to problem-specific heuristic approaches, or possibly adapt one of the metaheuristic principles that we have described for continuous nonconvex optimization. An alternative, however, is to use MILP modeling and solution methods to devise high-quality heuristics, without giving up the power of mathematical programming. Such approaches are generically labeled as **matheuristics**.

The simplest idea is to relax the pruning condition in the exact branch-and-bound method. As we have pointed out, we may prune a branch of the search tree, if LB $\geq$ UB, when dealing with a minimization problem. This condition may be relaxed to

$$\text{LB} \geq \text{UB}(1-\epsilon),$$

where $\epsilon \in (0,1)$ is a suboptimality tolerance. By doing so, we look for solutions improving the incumbent by a minimal relative amount ϵ. For instance, if we have a feasible solution with UB $= 100$ and set $\epsilon = 0.05$, we only explore subproblems (subtrees) that could deliver a solution with cost at most 95. If we stop the search process with the incumbent solution of cost 100, we cannot be sure that we have not missed a solution with a cost of, say, 97. However, we know that the percentage deviation from the optimal value is at most ϵ.

An alternative approach is the fix-and-relax heuristic, originally proposed in [10]. Consider an MILP where variables x_j in set V are required to be binary, and the remaining variables are continuous. Then, partition the set V into subsets V_i, $i = 1, \ldots, k$, and solve a sequence of k problems, where problem $r = 1, \ldots, k$ is subject to the following requirements:

$$\begin{aligned}
x_j &= \widehat{x}_j, \quad \forall j \in V_i, \ i = 1, \ldots, r-1 \quad (\text{for } r > 1) \\
x_j &\in \{0,1\}, \quad \forall j \in V_r \\
x_j &\in [0,1], \quad \forall j \in V_i, \ i = r+1, \ldots, k \quad (\text{for } r < k)
\end{aligned}$$

In practice, we are performing a restricted branch-and-bound search with respect to binary variables in subset V_r, where the other variables are either frozen to a given value $\widehat{x}_j$ or relaxed to the interval $[0,1]$. The way in which variables are partitioned is problem-dependent. We may devise a natural hierarchy of decision variables, as is the case when they refer to successive time periods.

16.6 Optimization software

There is an abundance of both commercial and free optimization software, and there is no one-fits-all choice. Here, we do not aim at an exhaustive overview, which would get out-of-date very soon. We just want to give some clues about the different tools that we may look for, with a short list that is admittedly *very* biased by this author's personal taste and experience, and as such should be taken with care.

Optimization software come in many guises and building blocks, including:

- The solver itself, i.e., the numerical libraries that solve a specific problem instance

16.6 Optimization software

- An interface through an imperative programming language
- An interface through a non-imperative algebraic language
- Additional interfaces, e.g., modules offering connectivity to an external database, or a graphical user interface (GUI) to help the user

As it will be clear from the following, all of these elements are mixed and there is no sharp boundary.

16.6.1 SOLVERS

There are several solvers matching the different kinds of optimization problems that we have considered. Actually, we have libraries of solvers, and there is no best option, as every bundle has strengths and weaknesses.

- For LP and MILP problems, state-of-the-art packages feature Gurobi[16] and CPLEX.[17] They include simplex, interior-point, and branch-and-bound methods, as well as some support for conic problems like QP, QCQP, and SOCP. They are not able to cope with generic nonlinear programming problems.
- Nonlinear programming libraries, like Minos and CONOPT, are usually also able to deal with LP, but not integer programming.
- SDP solvers are available, but they are not yet industrial strength libraries, but rather the result of academic research efforts. The same applies to nonlinear integer programming and global optimization.
- Numerical computing environments, like MATLAB, offer optimization libraries. The MATLAB suite includes an Optimization Toolbox, which is able to deal with most problems we have described, using a mix of gradient-based and derivative-free algorithms, as well as simplex and interior-point methods. The more recent Global Optimization Toolbox includes multiple start, pattern search, simulated annealing, genetic algorithms, and PSO. As a general rule, the libraries within such generic numerical computing environments are not competitive with the more specific products, but they are a reasonable choice for problems up to moderate size. Furthermore, the availability of a surrounding environment makes interfacing with the library rather easy and convenient.
- Free libraries are also available, such as plenty of R packages. Needless to say, they are usually not recommended for large-scale applications. Especially for large-scale MILP problems, the difference in performance between free and state-of-the-art commercial code can be impressive. However, there are problems for which free software is quite adequate.

---

[16] See http://www.gurobi.com/
[17] See https://www-01.ibm.com/software/commerce/optimization/cplex-optimizer/index.html

16.6.2 INTERFACING THROUGH IMPERATIVE PROGRAMMING LANGUAGES

Old numerical libraries came in the form of subroutines that could be called by generic scientific programming languages like FORTRAN. These languages are classified as imperative, since a program consists of a sequence of instructions to be executed in a precise order.

Using these old-style interfaces was awkward, time-consuming, and error-prone. To understand why, imagine specifying the matrix **A** of a large-scale LP model. We have to associate variables with columns and constraints with rows, and figure out in which position a certain coefficient should be written. Furthermore, the matrix should be stored as a *sparse* matrix, since most elements are zero. Writing down the required code in Fortran or C quickly becomes a nightmare, debugging is difficult, and model maintenance when additional requirements pop up is overly painful. The interface between solvers like Gurobi or CPLEX with environments like MATLAB and R is just a bit better, and it is recommended only for problems with a simple and regular structure.

Luckily, the development of modern object oriented programming has improved things considerably. They provide high-level object classes like decision variables, constraints, expressions, and models, with plenty of convenient attributes and methods. Class libraries are available within the Microsoft .NET environment (in C++, Visual Basic, etc.). Gurobi also has a very nice interface with Python, which makes data manipulation quite convenient, possibly using `pandas`, and allows the user to develop customized solution procedures, possibly based on the solution of a sequence of related subproblems. CPLEX supports this kind of approach, too.

16.6.3 INTERFACING THROUGH NON-IMPERATIVE ALGEBRAIC LANGUAGES

When we state an optimization problem, we *declare* things, like constraints and objective functions, without specifying a solution procedure. An object-oriented interface in, say, Python allows to describe a model in this vein, but a more radical approach is based on purely declarative algebraic languages. In an algebraic language, we define sets, parameters, variables, constraints, as well as the objective function, without any reference to a solution procedure. Even better, we may separate the model *structure* from a specific model *instance* which is obtained by associating numerical values with data.

An excellent example of the approach is AMPL,[18] which offers a very powerful language to manipulate data and create a wide array of optimization models. AMPL is *not* a solver, but a declarative language which may be interfaced with a rich set of specific solvers, including Gurobi, CPLEX, Minos, CONOPT, and many others. AMPL also includes an imperative language to build scripts.

---

[18] See http://ampl.com/

This is useful to control solver options and to write applications in which several models are solved in sequence.

There are some proprietary environments, like IBM OPL, which are specific to a given solver (IBM CPLEX in this case). GAMS[19] is similar to AMPL, ad it can be interfaced with a range of solvers, whereas CVX[20] is a less powerful, but quite interesting product, since it is fully integrated within MATLAB, making the development of full-fledged applications quite convenient. CVX was originally meant for convex optimization, but it has been extended to cope with integer programming as well. CVX may be used with MATLAB Optimization Toolbox solvers, as well as Gurobi, and other solvers for SOCP and SDP problems.

16.6.4 ADDITIONAL INTERFACES

In order to develop an industrial application, we need to be able to instantiate an optimization model with data. Thus, we need a way to connect with popular relational databases or, at the very least, with an Excel spreadsheet. By the same token, the model solution must be saved in a suitable form somewhere. AMPL, for instance, offers database connectivity for reading and writing relational database files. The Gurobi Python interface can also take advantage of data manipulation libraries like `pandas`.

Especially in finance, we have to rely on external databases that are accessed through the web. R and Python offer libraries to read from external websites, like Yahoo's Finance. MATLAB offers a specific Datafeed Toolbox to connect with professional financial data providers including Thomson Reuters's Datastream or Bloomberg. The extracted data can be used to populate an optimization model, among other things.

We may also consider interfaces toward the end user, in the form of a GUI (graphical user interface). MATLAB and Microsoft .NET make the development of graphical user interfaces quite convenient. The MATLAB Optimization Toolbox also offers a simple GUI, but this is meant more as an experimentation tool than an application interface. IBM also offers the CPLEX Development Studio, aimed at making model management and experimentation with models convenient. At the time of writing, other vendors are offering (or developing) similar user interfaces.

Problems

16.1 Prove that the dual function of Eq. (16.12) is concave.

---

[19]See `https://www.gams.com/`
[20]See `http://cvxr.com/`

16.2 Consider the quadratic programming problem

$$\min \quad x_1^2 + x_2^2$$
$$\text{s.t.} \quad x_1 + x_2 \geq 4,$$
$$x_1, x_2 \geq 0.$$

- Solve the problem using KKT conditions.
- Observe that, if we could get rid of the first inequality, we could decompose the problem into two independent minimizations with respect to each variable. Hence, dualize the coupling constraint, find the dual function, and check that the maximum of the dual function is the same as the optimal value of the primal.

16.3 Consider the discrete optimization problem

$$\min \quad \mathbf{c}^T\mathbf{x}$$
$$\text{s.t.} \quad \mathbf{a}^T\mathbf{x} \leq b,$$
$$\mathbf{x} \in S = \left\{\mathbf{x}^1, \mathbf{x}^2, \ldots, \mathbf{x}^m\right\}.$$

The feasible set consists of a finite collection of m points in $\mathbb{R}^n$, and we have one inequality constraint. Dualize the inequality and show that the dual function is a piecewise linear, nondifferentiable concave function. Where is the dual function not differentiable?

Further reading

- An extensive coverage of linear programming is offered by [21].
- A classical account of nonlinear programming methods can be found in [1]. For an exhaustive treatment of modern convex optimization, the standard reference is [6]. For a somewhat easier treatment you may have a look at [7].
- A nice treatment of semidefinite programming can also be found in [12].
- Mathematical approaches to global optimization are covered, e.g., in [13].
- A useful reference for genetic algorithms is [16]. For the original references on particle swarm optimization, see [11] and [18]. Several variants of swarm optimization have been proposed; see, e.g., [24] for the firefly variant and the use of Levy flights, or [20] for the quantum variant.
- An extensive treatment of integer programming can be found in [22].
- For more about optimization model building and solving, with specific reference to financial applications, see [9].
- A concrete example of heuristic solution algorithms applied to a financial problem is provided in [8].

- We did not consider at all any specific algorithm to solve optimization models under uncertainty.
  - For stochastic programming methods, the reader may refer to [5] and [15].
  - Numerical dynamic programming is discussed in [14] and [17], whereas more recent approximate DP is covered by [19]. You may also refer to [3, 4].
  - Robust optimization is dealt with in [2].

Bibliography

1 M.S. Bazaraa, H.D. Sherali, and C.M. Shetty. *Nonlinear Programming. Theory and Algorithms* (3rd ed.). Wiley, Hoboken, NJ, 2006.

2 A. Ben-Tal, L. El Ghaoui, and A. Nemirovski. *Robust Optimization*. Princeton University Press, Princeton, NJ, 2009.

3 D.P. Bertsekas. *Dynamic Programming and Optimal Control Vol. 1* (3rd ed.). Athena Scientific, Belmont, MA, 2005.

4 D.P. Bertsekas. *Dynamic Programming and Optimal Control Vol. 2* (4th ed.). Athena Scientific, Belmont, MA, 2012.

5 J.R. Birge and F. Louveaux. *Introduction to Stochastic Programming* (2nd ed.). Springer, New York, 2011.

6 S. Boyd and L. Vandenberghe. *Convex Optimization*. Cambridge University Press, New York, 2004. The book pdf can be downloaded from http://www.stanford.edu/~boyd/cvxbook/.

7 G. Calafiore and L. El Gahoui. *Optimization Models*. Cambridge University Press, Cambridge, 2014.

8 T.-J. Chang, N. Meade, J.E. Beasley, and Y.M. Sharaiha. Heuristics for cardinality constrained portfolio optimization. *Computers and Operations Research*, 27:1271–1302, 2000.

9 G. Cornuejols and R. Tütüncü. *Optimization Methods in Finance*. Cambridge University Press, New York, 2007.

10 C. Dillenberger, L.F. Escudero, A. Wollensak, and W. Zhang. On practical resource allocation for production planning and scheduling with period overlapping setups. *European Journal of Operational Research*, 75:275–286, 1994.

11 R.C. Eberhart and J. Kennedy. Particle swarm optimization. In *Proceedings of the IEEE International Conference on Neural Networks*, pp. 1942–1948, 1995.

12 B. Gärtner and J. Matousek. *Approximation Algorithms and Semidefinite Optimization*. Springer, Heidelberg, 2012.

13. R. Horst, P.M. Pardalos, and N.V. Thoai. *Introduction to Global Optimization*. Kluwer Academic, Dordrecht, The Netherlands, 1995.
14. K.L. Judd. *Numerical Methods in Economics*. MIT Press, Cambridge, MA, 1998.
15. P. Kall and S.W. Wallace. *Stochastic Programming*. Wiley, Chichester, 1994. The book pdf can be downloaded from http://stoprog.org/index.html?resources.html.
16. Z. Michalewicz. *Genetic Algorithms + Data Structures = Evolution Programs*. Springer, Berlin, 1996.
17. M.J. Miranda and P.L. Fackler. *Applied Computational Economics and Finance*. MIT Press, Cambridge, MA, 2002.
18. R. Poli, J. Kennedy, and T. Blackwell. Particle swarm optimization: An overview. *Swarm Intelligence*, 1:33–57, 2007.
19. W.B. Powell. *Approximate Dynamic Programming: Solving the Curses of Dimensionality* (2nd ed.). Wiley, Hoboken, NJ, 2011.
20. J. Sun, C.-H. Lai, and X.-J. Wu. *Particle Swarm Optimisation: Classical and Quantum Perspectives*. CRC Press, Boca Raton, FL, 2011.
21. R.J. Vanderbei. *Linear Programming: Foundations and Extensions* (3rd ed.). Springer, Heidelberg, 2010.
22. L.A. Wolsey. *Integer Programming*. Wiley, New York, 1998.
23. S.J. Wright. *Primal–Dual Interior-Point Methods*. Society for Industrial and Applied Mathematics, Philadelphia, 1997.
24. X.-S. Yang. Firefly algorithm, Lévy flights and global optimization. In M. Bramer, R. Ellis, and M. Petridis, editors, *Research and Development in Intelligent Systems XXVI*, pp. 209–218. Springer, 2010.

Index

σ-field, 475

ABS, *see* asset-backed security
absolute loss, 88
acceptability functional, 301
accrued interest, 230, 233
active constraint, 711
active portfolio management, 49, 358
actuarially fair, 282
adapted process, 447
adjustable robust optimization, 688
admissible policy, 677
affine
    function, 622
    independence, 702
    transformation, 290
affine policy, 684
affine term structure, 585
algebra (of sets), 430, 470
alpha, 354
ambiguity, 69
    distributional, 281
American-style option, 507
anchoring, 401
annual percentage rate, 150
annuity, 177
APR, *see* annual percentage rate
APT, *see* arbitrage pricing theory
arbitrage
    dynamic, 105
    instantaneous, 105
    opportunity, 50, 52, 125, 155, 468, 722, 725
    static, 105
    strategy, 104
arbitrage pricing theory, 388

arbitrageur, 52
artificial variable, 723
ask price, 51
asset, 13, 15
    exchange-traded vs. over-the-counter., 14
    liquid vs. illiquid, 14
    marketable, 20
    real vs. financial, 14
    tradable, 20
    tradable vs. nontradable, 14
    valuation vs. pricing, 102
asset–liability management, 50, 69, 262, 263, 266, 417, 668
asset-backed security, 20, 240
attainable
    payoff, 128
    portfolio, 327
    set, 327
augmented Lagrangians, 712
average
    arithmetic, 7
    geometric, 7

backwardation, 490
balance sheet, 15
bank
    commercial, 48
    investment, 48
    retail, 48
bank discount, 234
banker's acceptance, 239
barrier function, 706, 726
barrier option, 555
basic
    solution, 721
    variable, 721
basis point, 160, 238

743

price value, 257
basis risk, 494
Bayes' theorem, 406
Bayesian estimation
    normal distribution, 408
Bayesian statistics, 404
behavioral factor, 365
Bellman
    equation, 677
    principle, 678
bequest (utility from), 79
Bermudan-style option, 507
beta
    distribution, 407
    function, 407
beta-neutral portfolio, 367, 370
bias–variance tradeoff, 356, 606
bid price, 51
bid–ask spread, 5, 29, 51, 108, 146, 234
bid–offer spread, see bid–ask spread, 51
big-M constraint, 633
binary option, 545
binomial distribution, 407, 527
binomial model, 112, 519
bisection method, 554
Black Monday (of 1987), 36, 563
Black's formula, 558
Black's model, 562, 602
Black–Litterman estimator, 385
Black–Scholes–Merton
    equation, 533
    formula, 531, 534
bond, 24, 109, 584
    callable, 44, 240
    clean price, 233
    convertible, 43, 237
    convexity, 99, 587
    coupon bearing, 25
    coupon rate, 25
    dirty price, 233
    duration, 587, 610
    face value, 25
    floating-rate, 26, 188, 221, 232
    duration, 254
    futures, 239, 265
    high yield, 233
    indenture, 26, 232
    inflation-indexed, 163, 232
    junk, 233
    maturity, 25
    nominal value, 25
    off-the-run, 22
    on-the-run, 21
    option, 593
    par value, 25
    price, 165
    rating, 233
    structured, 44
    theta, 587
    trading at discount, 167
    trading at par, 167
    trading at premium, 167
    zero-coupon, 25, 83
book value, 16, 19
book-to-market ratio, 16, 365
bootstrapping (interest rates), 93, 169
box constraint, 619
box uncertainty, 687
branch-and-bound method, 624, 715
broker, 51
BSM, see Black–Scholes–Merton
bundling, 45, 119
butterfly spread, 110
buying on margin, 55

càdlàg function, 426
CAL, see capital allocation line
calibration, 128
call option, 24
canonical form, 723
capital allocation, 320
capital allocation line, 320, 323, 336, 376, 630, 631
capital asset pricing model, 191, 337, 375

capital budgeting, 641
capital gain, 22
capital market, 4, 232
capital market line, 376
capital structure, 23
caplet, 238, 518
CAPM, *see* capital asset pricing model
CARA, *see* constant absolute risk aversion
cardinality constraint, 132, 329, 634
cash flow
    matching, 262
    unbundling, 45
cash-and-carry, 483
CDF, *see* cumulative probability function
CDO, *see* collateralized debt obligation
CDS, *see* credit default swap
central path, 728
certainty effect, 402
certainty equivalent, 290
certificate of deposit, 239
chain rule, 258, 447, 452
chance constraint, 284, 652
    individual, 656
    joint, 284, 656
chance-constrained optimization, 656
change of measure, 541
cheapest-to-deliver bond, 239, 265
chi-square
    distribution
        noncentral, 594
    noncentral, 588, 594
CIR, *see* Cox–Ingersoll–Ross
clearinghouse, 35
CML, *see* capital market line
CMS, *see* swap, constant maturity
coefficient of variation, 285
collateral, 26, 36, 240
collateralized debt obligation, 241

common risk factor, 353
complementary slackness, 711, 725
complete market, 115
compounding, 150
    daily, 152
    quarterly, 150
    semiannual, 150
computational complexity, 719
concave
    optimization problem, 623
    function, 289, 340, 621, 714
    problem, 720
conditional
    distribution, 423
    expectation, 432, 540
    probability, 433
conditional value-at-risk, 304, 308, *see also* value-at-risk, conditional, 659
cone, 130, 643
    closed, 645
confidence interval, 404, 465
confidence level, 87, 404
conic
    combination, 645, 720
    duality, 654
    hull, 645
    optimization, 643, 728
    optimization problem, 644
consol bond, 180
constant absolute risk aversion, 294
constant relative risk aversion, 294
constraint qualification, 710
consumption–saving problem, 69, 77
contango, 489
contingent claim, 118
continuation
    region, 516
    value, 515
continuous compounding, 151
continuous relaxation, 638, 734

continuous-time stochastic process, 420
contrarian strategy, 55, 692
control variable, 76
convenience yield, 490
conversion factor, 239
convex
    combination, 289
    function, 252, 460, 620
        polyhedral, 620
        strictly, 620
    hull, 638, 702
    optimization problem, 622
    problem, 720
    set, 132, 619
convexity, 184, 301, 619
    bond, 84, 266
    correction, 219
    dollar, 267
    of a bond, 185
cornering, 239
corporate finance, 23
correlation, 96
    instantaneous, 461
correlation risk, 19, 241, 242
cost-of-carry, 557
counterparty risk, 35
counting process, 426
coupon stripping, 45
cover cut, 642
coverage, 404
covered position, 520
Cox–Ingersoll–Ross model, 461, 588, 610
credit default swap, 233, 241
credit rating, 418, 424
credit risk, 146
cross rate, 29
cross-currency option, 559
cross-hedging, 95, 495
cross-sectional data, 421
crossover strategy, 719
CRR lattice calibration, 524
CRRA, *see* constant relative risk aversion

cumulative probability function, 421
currency
    base, 28
    quoted, 28
currency risk, 171
curse of dimensionality, 680
CV@R, *see* conditional value-at-risk

DARA, *see* decreasing absolute risk aversion
data-driven optimization, 687
day count, 209, 231
dealer, 51
debt restructuring, 26
debtor, 241
decision
    adapted to a filtration, 476
    nonanticipative, 476
decision problem
    multistage, 277
    static, 277
decision variable, 70, 76, 618
    semicontinuous, 633
decreasing absolute risk aversion, 294
decreasing relative risk aversion, 294
default, 307
default correlation, 241
default risk, 26, 46, 172, 233
delta, 84
delta-hedging, 532, 538, 547, 566
delta-neutral portfolio, 548, 565
deposit- vs. non-deposit taking intermediary, 48
derivative, 29
diagonal model, 353
difference equation, 418
differential (of a stochastic process), 441
differential equation, 418
digital option, 545
dimensionality, 353

INDEX **747**

discount curve, 169
discount factor, 154
    forward, 199
    subjective, 79
discounted price process, 118
discounting, 154
discrete-event stochastic process, 420, 426
discrete-time stochastic process, 420
discretization error, 445, 464
distribution
    beta, 407
    binomial, 407
    cumulative function, 534
    geometric, 436
distributional ambiguity, 694
disturbance, 76
diversification, 301
dividend, 22
    policy, 23
divisor, 61
DJIA, *see* Dow Jones Industrial Average
dollar duration, 255, 265
dollar-neutral portfolio, 368, 370
dominant strategy, 121
Dow Jones Industrial Average, 60
down-and-out put, 555
DP, *see* dynamic programming
drift, 458, 459
    coefficient, 445
    function, 457
DRRA, *see* decreasing relative risk aversion, *see* increasing relative risk aversion
dual
    cone, 647, 728
    function, 713
    norm, 648
    problem, 134, 714
    simplex algorithm, 725
    variable, 713, 726
duality, 133, 647
    gap, 715, 724, 730
    linear programming, 124, 133, 689, 723
    theory, 713
dualization (of a constraint), 713
duration, 84, 182, 248
    as investment horizon, 258
    analytical formula, 186
    Macauley, 183
    modified, 183
dynamic problem, 68
dynamic programming, 676

EAR, *see* effective annual rate
earnings per share, 17
effective annual rate, 151
efficient
    frontier, 74, 129, 130, 319, 325, 328, 332
    solution, 129, 130
efficient market hypothesis, 103, 382, 399, 423
ellipsoidal uncertainty, 687
elliptical distribution, 343
EMH, *see* efficient market hypothesis
epigraph, 620
epoch, 3
EPS, *see* earnings per share
equity, 15, 263
equity tranche, 241
equivalent martingale measure, 438, 600
ETF, *see* exchange-traded fund, *see* exchange-traded fund
Euclidean norm, 605, 632, 649
Euler discretization scheme, 464
EURIBOR, 60, 208
eurodollar, 217
    futures, 216, 265
European-style option, 507
event, 469
    family of, 470
    field of, 470
ex-dividend date, 22
excess return, 321, 353
exchange-traded fund, 50, 634

execution uncertainty, 116
exercise region, 516
expectation hypothesis, 199
expected shortfall, 300
expected utility, 401
exponential
    distribution, 436
    random variable, 425

factor analysis, 398
factor model, 80, 353
    linear, 82
fat tail, 85
feasible
    region, 618
    set, 70, 618
Feynman–Kač theorem, 539, 583
field, 430, 470
    generated by a partition, 472
filtration, 432, 447, 475
    decision adapted to, 476
financial ratio, 16
firefly algorithm, 718
first-order immunization, 549
Fisher's equation, 162
fix-and-relax, 736
fixed charge, 632
fixed leg, 220
fixed-income market, 26
fixed-mix portfolio, 692
floater, 26, 188, 232
    reverse, 232
floating leg, 220
floorlet, 238, 518
foreign exchange
    direct quotation, 28
    indirect quotation, 28
    market, 27
foreign-exchange risk, 26
FOREX, see foreign exchange
forward
    contract, 31, 601
    curve, 197
    discount factor, 212
    price, 31, 482
forward contract, 193

forward rate, 580
    discretely compounded, 197
forward rate agreement, 194, 209
forward risk-neutral measure, 484, 562, 601
FRA, see forward rate agreement
Frobenius product, 650
functional, 288, 299
    risk, 283
    vs. function, 281
fund
    hedge, 50
    mutual, 49
fundamental analysis, 399
fundamental factor, 365
fundamental theorem of calculus, 441
futures
    eurodollar, 217
    option, 559
    price, 35

gain, 6, 296
    additive, 120
    discounted, 121
    vs. profit, 7
gamma, 84
Gaussian process, 423
Gaussian quadrature, 463, 671
GBM, see geometric Brownian motion
generalized inequality, 647
generalized inverse function, 87, 306
generalized Wiener process, 443
genetic algorithm, 716
geometric Brownian motion, 88, 445, 457, 523, 531
geometric random variable, 425
geometric series, 148
Girsanov theorem, 420, 590
Glass–Steagall Act, 49
global optimization, 604
global optimum, 708
growth stock, 366

haircut, 240
Harry Markowitz, 75
heavy tail, 85
hedge tailing, 95
hedger, 51
hedging, 33, 483, 521, 581
    minimum variance, 95
    perfect, 94
    quantity-based, 493, 501
    ratio, 494, 496, 582
    value-based, 501
Heston model, 462, 568, 611
Ho–Lee model, 219
holding period return, 6
homogeneous function, 337
hurdle rate, 191, 380
hyperbola, 345
hyperparameter, 409

IARA, *see* increasing absolute risk aversion
idiosyncratic risk, 326
    factor, 353
ill-conditioned problem, 608
implied volatility, 60, 104
importance sampling, 671
inactive constraint, 712
income statement, 16
incomplete market, 115
increasing absolute risk aversion, 294, 341
increasing relative risk aversion, 294
independent increments, 427, 439, 440, 462
index
    futures, 326, 368
    market-value-weighted, 61
    price-based, 61
    tracking, 634
infinitely divisible distribution, 441
inflation, 161, 365
    rate, 161
inflation risk, 18, 26, 172
information ratio, 361

initial public offering, 47
inner product, 643
    for matrices, 650, 653
innovation (in a stochastic process), 423
inside quote, 54
insurance, 11
integrating factor, 589
integrator function, 448
interest rate, 421
    cap, 238, 518
    compounded, 147
    floor, 238, 518
    forward, 194
    simple, 147
    spot, 193
interest rate risk, 18, 26, 172
interest rate swap, 30, 220, 255, 266, 518
interior-point method, 627, 719, 728
internal rate of return, 192
interval uncertainty, 687
intrinsic value, 529
inverse problem, 553, 584
IPO, *see* initial public offering
IRR, *see* internal rate of return
irreducible variance, 607
Islamic finance, 145
Itô
    isometry, 452, 589
    lemma, 454, 455, 532, 582, 589, 597
    process, 446
    stochastic integral, 449

Jensen's inequality, 289, 460
jump–diffusion process, 427, 440

Karush–Kuhn–Tucker conditions, 711
KKT, *see* Karush–Kuhn–Tucker conditions
knapsack problem, 641
kurtosis, 86, 91, 455, 467
    excess, 438

L'Hôpital's rule, 294
Lagrange multiplier, 330, 707, 726
Lagrangian function, 707
lasso regression, 608, 691
lattice calibration, 523
law of large numbers, 285
law of one price, 108, 113, 120, 517, 519
learning sample, 605
least-squares, 97, 171
    nonlinear, 603
    ordinary, 605
Lebesgue integral, 473
leverage, 16, 55
leveraged portfolio, 324
Lévy flight, 441, 718
Lévy process, 427, 440, 462
LGD, *see* loss given default
liability, 15
LIBOR rate, 60, 202, 208, 225, 238, 581
life insurance, 50
likelihood function, 406
limit order book, 53, 462
limit-buy order, 55
limit-sell order, 55
limited liability, 8, 23, 117
linear algebra, 117
linear combination, 118
linear matrix inequality, 653, 731
linear pricing, 165
linear program, 123
linear programming, 73, 117, 263, 300, 624, 625, 662, 719
    canonical form, 626
    relaxation, 638
    standard form, 133, 626
linear regression, 99, 353
linear scenarios, 463
linear space, 118
linker, 26, 232
liquidity, 14, 20, 116, 168
liquidity preference theory, 200
local regression, 606

local volatility, 611
log-return, *see* logarithmic return
logarithm (natural), 9
logarithmic return, 9
lognormal
    distribution, 10, 445, 458, 459
    variable, 460
long position, 31
Long Term Capital Management, 21, 65, 116
long–short portfolio, 367
longevity risk, 18
longitudinal data, 421
lookback
    call, 510
    option, 434
    put, 510
Lorentz cone, 645
loss given default, 233
lottery, 278
low-discrepancy sequence, 463, 671
LP, *see* linear programming
LP relaxation, 734
LTCM, *see* Long Term Capital Management

Macauley duration, 248
Maclaurin series, 9
macroeconomic factor, 365
maintenance margin, 56
margin, 36, 55
    buying on, 55
    call, 36, 56, 108
    maintenance, 36
    ratio, 56
    trading, 20
marginal distribution, 428
mark to market, 36
market
    capitalization, 24
    complete, 119
    completeness, 128
    friction, 51

incomplete, 119
index, 487
maker, 51, 237
micro-structure, 69
neutral portfolio, 367
portfolio, 376
price of risk, 321, 378, 383, 398, 584, 599
primary, 47, 53
secondary, 47
Markov chain
continuous-time, 425
discrete-time, 79, 424
Markov process, 77, 423, 430, 433
Markov property, 433
martingale, 115, 430, 436, 451, 457, 562, 597, 600
equivalent measure, 115, 128
property, 125
mathematical programming, 619
matheuristic, 736
matrix
positive semidefinite, 624
transposed, 71
maturity
bond, 5
derivative, 30
maximization problem, 619
maximum likelihood, 691
MBS, *see* mortgage-backed security
mean reversion, 461, 588
mean squared error, 607
mean value function, 444, 460
mean–risk
model, 73
plane, 129
mean–variance
efficient frontier, 74
portfolio optimization, 70, 319, 629, 673
mean-preserving spread, 288
memoryless
continuous distribution, 425
discrete distribution, 425

property, 436
mental accounting, 401
mergers and acquisitions, 24
metaheuristics, 716
mezzanine tranche, 241
MILP, *see* mixed-integer linear programming
minimum variance curve, 331
minimum variance hedging, 495
MIQP, *see* mixed-integer quadratic programming
mixed-integer linear programming, 624
mixed-integer quadratic programming, 624, 636
model calibration, 171, 553, 584, 603, 691, 715
model risk, 100, 107, 202, 231, 506, 568, 611, 679
model selection, 606
Modern Portfolio Theory, 75
modern portfolio theory, 279, 319, 341
moment generating function, 455
moment matching, 463, 467, 523, 672
momentum, 365, 399
momentum strategy, 55
money market, 4, 232
monotonicity, 301
Monte Carlo sampling, 444, 463, 671
mortgage-backed security, 45, 240
MPT, *see* modern portfolio theory
MSE, *see* mean squared error
multifactor model, 270
multiobjective optimization, 74
multiperiod vs. multistage, 76
multistage decision model, 277
vs. multiperiod, 277

naked position, 520
NASDAQ, 60
NAV, *see* net asset value

Nelder–Mead method, 702
Nelson–Siegel model, 171
net asset value, 50
net income, 16
net present value, 153, 191, 380
Newton method
    for nonlinear equations, 728
    optimization, 701
NLP, *see* nonlinear programming
no-arbitrage argument, 154
no-arbitrage principle, 102, 165
non-convex set, 634
non-satiation, 289
nonbasic variable, 721
noncentrality parameter, 594
nondirectional trade, 552
nondominated solution, 129, 130
nonhomogeneous equation, 592
nonlinear programming, 604, 624, 691, 700
nonparametric risk model, 93
norm
    cone, 646
    properties, 646
    self-dual, 649
normal backwardation, 490
normal distribution
    Bayesian estimation, 408
    standard, 534
notional, 209
NPV, *see* net present value
numeraire, 115, 118, 129, 161, 597

objective function, 618
obligor, 233
ODE, *see* ordinary differential equation
OLS, *see* least-squares, ordinary
optimal stopping, 511
optimality condition
    first-order, 700
optimization
    constrained problem, 619
    model, 70, 580, 584
    multiobjective, 129
    semi-infinite, 652
    uncertain problem, 686
    unconstrained problem, 619
    vector, 129
option
    American-style, 40
    American-style put call, 529
    Asian, 43, 45, 508
    at-the-money, 507
    average rate, 510
    average strike, 510
    barrier, 508, 566
    Bermudan-style, 45
    binary, 566
    call, 39, 106
    delta, 521, 546, 563
    European-style, 39
    European-style call, 526
    exotic, 41, 508
    gamma, 99, 547, 566
    Greeks, 546
    hedging, 417
    holder, 39
    in-the-money, 507
    intrinsic value, 507
    knock-in, 508
    knock-out, 508
    long position, 39
    lookback, 508
    maturity, 39
    out-of-the-money, 507
    own-and-out put, 508
    put, 39, 94, 106
    quanto, 559
    rainbow, 45, 508
    sensitivity, 546
    short position, 39
    theta, 550, 564
    time value, 507
    up-and-out, 566
    vanilla, 39, 41
    vanna, 565, 573
    vega, 552, 564
    writer, 39
ordinary differential equation, 586

ordinary least-squares, 353
Ornstein–Uhlenbeck process, 461, 588
orthodox statistics, 404
OTC, *see* over the counter
out-of-sample performance, 606
over the counter, 14, 43, 508
overnight rate, 239

panel data, 421
par yield, 222
parity relationship, 509, 517, 518
partial differential equation, 533
particle swarm optimization, 717
partition, 432, 471
passive fund, 634
passive portfolio management, 49, 358
payoff (derivative), 30
PDE, *see* partial differential equation
PDF, *see* probability density function
PE, *see* price to earnings
pecking order, 232
penalty function, 703
    exterior, 705
    interior, 706
pension capital, 148
pension fund, 148
    defined-benefit, 50
    defined-contribution, 50
perpetuity, 179
PMF, *see* probability mass function
pointed cone, 645
Poisson
    process, 421, 425
        compound, 427, 462
        inhomogeneous, 427
        random variable, 426
polyhedral cone, 645
polyhedral uncertainty, 687
polytopic uncertainty, 687
portable alpha, 370
portfolio

fully invested, 71
insurance, 563
weight, 71
positive homogeneity, 301
positive semidefinite matrix, 624
posterior distribution, 406
precision, 410
prepayment risk, 240
price to earnings, 17
pricing
    vs. valuation, 30
pricing equation, 583
pricing functional, 118
    linear, 120
    non-negative, 124
primal problem, 134, 713
primal variable, 713
primal-dual algorithm, 726
prime rate, 146
principal component analysis, 100, 271, 398, 580, 673
prior distribution, 406
private equity, 23
probability
    measure, 469
    space, 468
probability density function, 421
probability mass function, 421
probability measure, 430
probability space, 430, 469
prospect theory, 402
protective put, 42
PSO, *see* particle swarm optimization
purchasing power, 161
put–call parity, 107, 536, 546, 558, 563

QCQP, *see* quadratic program, quadratically constrained
QP, *see* quadratic programming
quadratic program
    quadratically constrained, 74
quadratic programming, 73, 300, 624, 628, 707

quadratic utility, 341
quadratically constrained quadratic
    programming, 624, 631
quantile, 87
quanto, 559
quantum PSO, 718
quasi-Monte Carlo
    sampling, 463
    simulation, 671
quasi-Newton method, 702
quasiconcave function, 341, 622, 629
quasiconvex function, 622

Radon–Nikodym derivative, 420
rainbow option, 681
random field, 163, 422
random variable, 5, 471
    measurable, 473
    notation, 72
random walk, 423, 441
rating agency, 233
real option, 193
recombining lattice, 463
recourse
    action, 657
    function, 658
    matrix, 658
reduced cost, 722
regime-switching model, 435
regression tree, 606
regularization, 608
reinsurance, 11
reinvestment risk, 27, 44, 46, 150, 172, 240, 581, 601
replicating portfolio, 113
repo
    agreement, 57
    market, *see also* repurchase agreement, 269
    rate, 240
repurchase agreement, 57, 240
reset date, 188, 232
return
    continuously compounded, 459
    gross, 6
    holding period, 6
    logarithmic, *see* logarithmic return
    net, 6
    rate of, 6
    total, 6
return attribution, 398
return on assets, 17, 57
return on equity, 17, 49, 57
ridge regression, 608, 691
Riemann integral, 446, 448, 473
Riemann–Stieltjes integral, 448
risk
    aversion, 287, 290, 322, 693
    coefficient of absolute aversion, 292
    coefficient of relative aversion, 293
    coherent measure, 93
    common factor, 80, 82
    counterparty, 100
    credit, 100
    currency, 100
    factor, 98
    functional, 283, 300
    inflation, 100
    interest rate, 100
    linear model, 81
    management, 11, 80
    market, 100
    measure, 80, 300, 580
        coherent, 300
    model, 80, 101
    noise trader, 116
    operational, 102
    political, 101
    pooling, 11, 46
    premium, 11, 290, 321, 332, 437, 630
        relative, 293
    regulatory, 101
    specific factor, 80, 82
    subadditive measure, 93
    symmetric measure, 84
    systemic, 49

transfer, 11
volatility, 100
volume, 101
risk factor, 659
    common, 353
    idiosyncratic, 353
    specific, 353
    systematic, 353
risk nonlinear model, 81
risk-free
    asset, 10, 171, 296
    rate, 10, 171
    return, 10
risk-neutral
    decision maker, 73
    decision-maker, 79
    measure, 114, 438, 519, 541, 561, 562
    option pricing, 380
    probability measure, 127
ROA, *see* return on assets
robust counterpart, 686
robust optimization, 569
ROE, *see* return on equity
rolling horizon, 675

sample covariance, 352
sample mean, 352
sample path, 12, 117, 421
sample path generation, 444
sample space, 430
sampling error, 464
scalarization, 131, 347
scenario, 12, 117, 421, 473
    fan, 6
    generation, 671
    tree, 12, 463
SDP, *see* semidefinite programming
seasoned offering, 53
second-order cone, 646
    programming, 74, 624
securitization, 11, 20, 46, 240
security, 20
security market line, 379
self-dual cone, 647

self-financing portfolio, 539, 684
semi-infinite problem, 618
semi-infinite programming, 688
semicontinuous variable, 620, 633
semidefinite programming, 624, 653
    inequality form, 653
    standard form, 653
senior tranche, 241
sensitivity analysis, 192
separation of variables, 585
separation property, 337, 376
set
    difference, 469
    intersection, 469
    union, 469, 471
shadow price, 712
shareholder, 19
Sharpe ratio, 323, 336, 337, 356, 361, 378, 599, 622, 629
    maximization, 629
short position, 31, 94
short rate, 443
short-selling, 58
short-squeeze, 58
shortfall probability, 300
shrinkage
    estimation, 374, 382
    estimator, 356, 608
shrinking horizon, 675
sigma
    algebra, 471
    field, 471
simplex, 702
    algorithm, 626
    method, 719
    search, 702
single-index model, 353, 409, 498
skew (positive), 280
skewness, 85
slack variable, 625, 726
Slater
    condition, 710, 715
    constraint qualification, 730

sliding horizon, 675
small world PSO, 718
SML, *see* security market line
smooth pasting, 516
SOCP, *see* second-order cone programming
sojourn time, 425, 436
solution
    efficient, 130
    nondominated, 130
solvency, 263
special purpose vehicle, 240
specialist, 51
specific risk factor, 353
speculator, 51
spinoff, 24
spot price, 31
spot rate curve, 163
spot–forward convergence, 31, 194, 484
spot–forward parity, 482, 511, 557
spread option, 508
SPV, *see* special purpose vehicle
square-root diffusion, 461, 588, 594
square-root rule, 89, 550
St. Petersburg paradox, 287
stability
    in sample, 674
    out of sample, 674
stable distribution, 92
stack-and-roll, 495
standard form (in LP), 720
standard Wiener process, 438
state of the world, 117
state variable, 76, 423
static decision problem, 277
static model, 75
static problem, 68
stationarity condition, 700
stationary increments, 427, 439, 440, 462
steepest descent, 701
    direction, 708

stochastic continuity, 440
stochastic differential equation, 443, 582
stochastic dominance
    first-order, 311
    second-order, 313
stochastic dynamic optimization, 511
stochastic dynamic programming, 676
stochastic integral, 442, 447
stochastic process, 12, 117, 163, 419, 420, 473
    predictable, 173
    self-similar, 89
stochastic programming, 263, 300
stochastic volatility, 568, 611
stock price, 473
stock repurchase, 24
stock share
    common, 23
    preferred, 23
    split, 24
stock-index futures, 497
stockholder, 19
stop-buy order, 55
stop-loss order, 54
stopping time, 527, 680
Stratonovich stochastic integral, 449, 453
strike price, 39
stripping, 26, 168
strong duality, 714, 724
subadditivity, 301
sublevel set, 621
subordinated bond, 232
subprime mortgage, 46, 240
swap
    accrediting, 238
    amortizing, 238
    basis, 238
    constant maturity, 238
    curve, 225
    dollar duration, 255
    forward start, 220
    rate, 220, 237

spot start, 220
valuation, 221
swaption, 225, 238
systematic risk, 496
systematic risk factor, 353

T-bill, 26
tail expectation, 308
tail risk, 91
tailing the hedge, 219, 265, 499
tangency portfolio, 336, 376
Taylor expansion, 9, 84, 98, 291, 454
    first-order, 248
    second-order, 266
technical analysis, 399
tenor, 209
term structure, 248
    estimation, 168
    of interest rates, 163, 580
test sample, 605
TEV, *see* tracking error variance
time
    bucket, 3
    instant, 3
    period, 3
time consistency, 301
time value, 4
TIPS, 232
total probability theorem, 406
trace (of a matrix), 650
tracking error variance, 635
tranching, 46, 241
transaction cost, 116, 669
transition density, 433
transition function, 418
transition probability, 424, 433
translation invariance, 301
transposition, 71
treasury
    bond, 26
    note, 26
Treynor–Black model, 358
triangular inequality, 646
trust region method, 702
two-fund separation theorem, 348

unbiased estimator, 356
unbounded solution, 722
unbundling, 119, 168
unbundling cash flows, 26
uncertainty set, 609, 686
unconditional distribution, 423
uniform prior, 407
up-and-out call, 566
utility
    CARA, 294
    CRRA, 294, 297
    DARA, 294
    DRRA, 294
    expected, 288, 670
    function, 79, 322, 341, 665, 669, 691
    IARA, 294
    IRRA, 294
    logarithmic, 294, 296
    ordinal, 290
    power, 294
    quadratic, 295
    Von Neumann–Morgenstern, 288

value function, 677
value process, 120
    discounted, 121
value stock, 366
value-at-risk, 87, 260, 303, 659
    absolute, 88
    conditional, 304
    historical, 93
    relative, 88
vanilla option, 507
variance reduction, 465
Vasicek model, 461, 588, 610
vector norm, 605
vector transposition, 71
vega, 84
VIX, 60
volatility, 84, 260, 404, 435, 458, 459
    coefficient, 445
    function, 457
    historical, 553

implied, 553
smile, 554
stochastic, 462
surface, 554
trading, 552, 564

warrant, 43, 237
Wiener process
correlated, 461
generalized, 443
standard, 438
worst-case optimization, 686

yield, 83, 155
volatility, 260
yield-to-maturity, 145, 175, 179, 250
YTM, *see* yield-to-maturity

zero, *see* bond, zero-coupon
zero curve, 163
zero-coupon bond, 154, 582, 590, 601